SECOND EDITION

MODERN POWER SYSTEM ANALYSIS

SECOND EDITION

MODERN POWER SYSTEM ANALYSIS

TURAN GÖNEN

CRC Press
Taylor & Francis Group
Boca Raton London New York

CRC Press is an imprint of the
Taylor & Francis Group, an **informa** business

MATLAB® is a trademark of The MathWorks, Inc. and is used with permission. The MathWorks does not warrant the accuracy of the text or exercises in this book. This book's use or discussion of MATLAB® software or related products does not constitute endorsement or sponsorship by The MathWorks of a particular pedagogical approach or particular use of the MATLAB® software.

CRC Press
Taylor & Francis Group
6000 Broken Sound Parkway NW, Suite 300
Boca Raton, FL 33487-2742

© 2013 by Taylor & Francis Group, LLC
CRC Press is an imprint of Taylor & Francis Group, an Informa business

No claim to original U.S. Government works

Printed in the United States of America by Edwards Brothers Malloy, Lillington, NC 27546
Version Date: 20130111

International Standard Book Number-13: 978-1-4665-7081-8 (Hardback)

This book contains information obtained from authentic and highly regarded sources. Reasonable efforts have been made to publish reliable data and information, but the author and publisher cannot assume responsibility for the validity of all materials or the consequences of their use. The authors and publishers have attempted to trace the copyright holders of all material reproduced in this publication and apologize to copyright holders if permission to publish in this form has not been obtained. If any copyright material has not been acknowledged please write and let us know so we may rectify in any future reprint.

Except as permitted under U.S. Copyright Law, no part of this book may be reprinted, reproduced, transmitted, or utilized in any form by any electronic, mechanical, or other means, now known or hereafter invented, including photocopying, microfilming, and recording, or in any information storage or retrieval system, without written permission from the publishers.

For permission to photocopy or use material electronically from this work, please access www.copyright.com (http://www.copyright.com/) or contact the Copyright Clearance Center, Inc. (CCC), 222 Rosewood Drive, Danvers, MA 01923, 978-750-8400. CCC is a not-for-profit organization that provides licenses and registration for a variety of users. For organizations that have been granted a photocopy license by the CCC, a separate system of payment has been arranged.

Trademark Notice: Product or corporate names may be trademarks or registered trademarks, and are used only for identification and explanation without intent to infringe.

Visit the Taylor & Francis Web site at
http://www.taylorandfrancis.com

and the CRC Press Web site at
http://www.crcpress.com

To an excellent engineer,
a great teacher, and a dear friend,
Late Dr. Paul M. Anderson
and
to a lady who had never had any education
but yet had the wisdom,
my mother.

Only the simplest tools can have the directions
written on the handle;
the theoretical tools of engineering
require professional preparation for their use.

JOHN H. FIELDER

Man is nothing but what he makes of himself.

JEAN PAUL SARTRE

Contents

Preface xiii
Acknowledgments xv
Author xvii

Chapter 1 General Considerations 1
    1.1 Introduction 1
    1.2 Power System Planning 5
    References 10
    General References 11

Chapter 2 Basic Concepts 13
    2.1 Introduction 13
    2.2 Complex Power in Balanced Transmission Lines 13
    2.3 One-Line Diagram 16
    2.4 Per-Unit System 19
        2.4.1 Single-Phase System 20
        2.4.2 Converting from Per-Unit Values to Physical Values 24
        2.4.3 Change of Base 24
        2.4.4 Three-Phase Systems 25
    2.5 Constant Impedance Representation of Loads 38
    2.6 Three-Winding Transformers 40
    2.7 Autotransformers 41
    2.8 Delta–Wye and Wye–Delta Transformations 43
    2.9 Short-Circuit MVA and Equivalent Impedance 44
        2.9.1 Three-Phase Short-Circuit MVA 45
            2.9.1.1 If Three-Phase Short-Circuit MVA Is Already Known 45
        2.9.2 Single-Phase-to-Ground Short-Circuit MVA 46
            2.9.2.1 If Single-Phase Short-Circuit MVA Is Already Known 46
    References 48
    General References 48

Chapter 3 Steady-State Performance of Transmission Lines 51
    3.1 Introduction 51
    3.2 Conductor Size 51
    3.3 Transmission Line Constants 58
    3.4 Resistance 58
    3.5 Inductance and Inductive Reactance 59
        3.5.1 Single-Phase Overhead Lines 59
        3.5.2 Three-Phase Overhead Lines 60
    3.6 Capacitance and Capacitive Reactance 61
        3.6.1 Single-Phase Overhead Lines 61
        3.6.2 Three-Phase Overhead Lines 64

3.7	Tables of Line Constants	65
3.8	Equivalent Circuits for Transmission Lines	68
3.9	Transmission Lines	68
	3.9.1 Short Transmission Lines (up to 50 mi or 80 km)	68
	3.9.2 Steady-State Power Limit	71
	3.9.3 Percent Voltage Regulation	73
	3.9.4 Representation of Mutual Impedance of Short Lines	79
3.10	Medium-Length Transmission Lines (up to 150 mi or 240 km)	80
3.11	Long Transmission Lines (above 150 mi or 240 km)	90
	3.11.1 Equivalent Circuit of Long Transmission Line	100
	3.11.2 Incident and Reflected Voltages of Long Transmission Line	103
	3.11.3 Surge Impedance Loading of Transmission Line	107
3.12	General Circuit Constants	110
	3.12.1 Determination of A, B, C, and D Constants	111
	3.12.2 Measurement of ABCD Parameters by Test	112
	3.12.3 A, B, C, and D Constants of Transformer	116
	3.12.4 Asymmetrical π and T Networks	117
	3.12.5 Networks Connected in Series	119
	3.12.6 Networks Connected in Parallel	121
	3.12.7 Terminated Transmission Line	123
	3.12.8 Power Relations Using A, B, C, and D Line Constants	127
3.13	EHV Underground Cable Transmission	134
3.14	Gas-Insulated Transmission Lines	142
3.15	Bundled Conductors	147
3.16	Effect of Ground on Capacitance of Three-Phase Lines	151
3.17	Environmental Effects of Overhead Transmission Lines	152
References		153
General References		153

Chapter 4 Disturbance of Normal Operating Conditions and Other Problems ... 159

4.1	Introduction	159
4.2	Fault Analysis and Fault Types	161
4.3	Balanced Three-Phase Faults at No Load	164
4.4	Fault Interruption	168
4.5	Balanced Three-Phase Faults at Full Load	175
4.6	Application of Current-Limiting Reactors	181
4.7	Insulators	185
	4.7.1 Types of Insulators	185
	4.7.2 Testing of Insulators	187
	4.7.3 Voltage Distribution over a String of Suspension Insulators	189
	4.7.4 Insulator Flashover due to Contamination	194
	4.7.5 Insulator Flashover on Overhead High-Voltage DC Lines	196
4.8	Grounding	197
	4.8.1 Electric Shock and Its Effects on Humans	197
	4.8.2 Reduction of Factor C_s	204
	4.8.3 GPR and Ground Resistance	206
	4.8.4 Ground Resistance	207
	4.8.5 Soil Resistivity Measurements	209
	4.8.5.1 Wenner Four-Pin Method	209
	4.8.5.2 Three-Pin or Driven-Ground Rod Method	213

Contents ix

	4.9	Substation Grounding	214
	4.10	Ground Conductor Sizing Factors	218
	4.11	Mesh Voltage Design Calculations	221
	4.12	Step Voltage Design Calculations	223
	4.13	Types of Ground Faults	223
		4.13.1 Line-to-Line-to-Ground Fault	223
		4.13.2 Single-Line-to-Ground Fault	224
	4.14	Ground Potential Rise	224
	4.15	Transmission Line Grounds	233
	4.16	Types of Grounding	235
	References		238
	General References		239

Chapter 5 Symmetrical Components and Sequence Impedances ... 245

	5.1	Introduction	245
	5.2	Symmetrical Components	245
	5.3	Operator **a**	247
	5.4	Resolution of Three-Phase Unbalanced System of Phasors into Its Symmetrical Components	248
	5.5	Power in Symmetrical Components	252
	5.6	Sequence Impedances of Transmission Lines	255
		5.6.1 Sequence Impedances of Untransposed Lines	255
		5.6.2 Sequence Impedances of Transposed Lines	257
		5.6.3 Electromagnetic Unbalances due to Untransposed Lines	260
		5.6.4 Sequence Impedances of Untransposed Line with Overhead Ground Wire	267
	5.7	Sequence Capacitances of Transmission Line	268
		5.7.1 Three-Phase Transmission Line without Overhead Ground Wire	268
		5.7.2 Three-Phase Transmission Line with Overhead Ground Wire	271
	5.8	Sequence Impedances of Synchronous Machines	275
	5.9	Zero-Sequence Networks	280
	5.10	Sequence Impedances of Transformers	281
	References		288
	General References		288

Chapter 6 Analysis of Unbalanced Faults ... 293

	6.1	Introduction	293
	6.2	Shunt Faults	293
		6.2.1 SLG Fault	293
		6.2.2 Line-to-Line Fault	302
		6.2.3 DLG Fault	307
		6.2.4 Symmetrical Three-Phase Faults	312
		6.2.5 Unsymmetrical Three-Phase Faults	317
	6.3	Generalized Fault Diagrams for Shunt Faults	323
	6.4	Series Faults	329
		6.4.1 One Line Open	330
		6.4.2 Two Lines Open	330
	6.5	Determination of Sequence Network Equivalents for Series Faults	332
		6.5.1 Brief Review of Two-Port Theory	332

		6.5.2	Equivalent Zero-Sequence Networks	333
		6.5.3	Equivalent Positive- and Negative-Sequence Networks	334
	6.6	Generalized Fault Diagram for Series Faults		339
	6.7	System Grounding		343
	6.8	Elimination of SLG Fault Current by Using Peterson Coils		349
	6.9	Six-Phase Systems		352
		6.9.1	Application of Symmetrical Components	353
		6.9.2	Transformations	353
		6.9.3	Electromagnetic Unbalance Factors	355
		6.9.4	Transposition on the Six-Phase Lines	357
		6.9.5	Phase Arrangements	358
		6.9.6	Overhead Ground Wires	358
		6.9.7	Double-Circuit Transmission Lines	358
References				361
General References				361

Chapter 7 System Protection 373

	7.1	Introduction		373
	7.2	Basic Definitions and Standard Device Numbers		377
	7.3	Factors Affecting Protective System Design		380
	7.4	Design Criteria for Protective Systems		380
	7.5	Primary and Backup Protection		382
	7.6	Relays		385
	7.7	Sequence Filters		394
	7.8	Instrument Transformers		396
		7.8.1	Current Transformers	397
			7.8.1.1 Method 1. The Formula Method	400
			7.8.1.2 Method 2. The Saturation Curve Method	401
		7.8.2	Voltage Transformers	402
	7.9	R–X Diagram		403
	7.10	Relays as Comparators		409
	7.11	Duality between Phase and Amplitude Comparators		409
	7.12	Complex Planes		410
	7.13	General Equation of Comparators		412
	7.14	Amplitude Comparator		413
	7.15	Phase Comparator		414
	7.16	General Equation of Relays		418
	7.17	Distance Relays		419
		7.17.1	Impedance Relay	422
		7.17.2	Reactance Relay	427
		7.17.3	Admittance (Mho) Relay	429
		7.17.4	Offset Mho (Modified Impedance) Relay	431
		7.17.5	Ohm Relay	433
	7.18	Overcurrent Relays		439
	7.19	Differential Protection		450
	7.20	Pilot Relaying		459
	7.21	Computer Applications in Protective Relaying		462
		7.21.1	Computer Applications in Relay Settings and Coordination	462
		7.21.2	Computer Relaying	462

Contents   xi

References 464
General References..465

Chapter 8 Power Flow Analysis...471
8.1 Introduction 471
8.2 Power Flow Problem..473
8.3 Sign of Real and Reactive Powers 475
8.4 Gauss Iterative Method..476
8.5 Gauss–Seidel Iterative Method..477
8.6 Application of Gauss–Seidel Method: Y_{bus}................................478
8.7 Application of Acceleration Factors 482
8.8 Special Features...482
   8.8.1 LTC Transformers 483
   8.8.2 Phase-Shifting Transformers...483
   8.8.3 Area Power Interchange Control 484
8.9 Application of Gauss–Seidel Method: Z_{bus} 488
8.10 Newton–Raphson Method 489
8.11 Application of Newton–Raphson Method......................................493
   8.11.1 Application of Newton–Raphson Method to Load Flow Equations in Rectangular Coordinates.............................493
   8.11.2 Application of Newton–Raphson Method to Load Flow Equations in Polar Coordinates..504
      8.11.2.1 Method 1. First Type of Formulation of Jacobian Matrix 505
      8.11.2.2 Method 2. Second Type of Formulation of Jacobian Matrix 509
8.12 Decoupled Power Flow Method 510
8.13 Fast Decoupled Power Flow Method..511
8.14 The DC Power Flow Method..513
References 525
General References..527

Appendix A: Impedance Tables for Overhead Lines, Transformers, and Underground Cables 533

Appendix B: Standard Device Numbers Used in Protection Systems 621

Appendix C: Unit Conversions from English System to SI System 623

Appendix D: Unit Conversions from SI System to English System 625

Appendix E: Prefixes 627

Appendix F: Greek Alphabet Used for Symbols 629

Appendix G: Additional Solved Examples of Shunt Faults 631

Appendix H: Additional Solved Examples of Shunt Faults Using MATLAB........655

Appendix I: Glossary for Modern Power System Analysis Terminology 683

Index...705

Preface

The structure of the electric power system is very large and complex. Nevertheless, its main components (or subsystems) can be identified as the generation system, the transmission system, and the distribution system. These three systems are the basis of the electric power industry. Today, there are various textbooks dealing with a broad range of topics in the power system area of electrical engineering. Some of these are considered to be classics.

However, they do not particularly concentrate on the topics specifically dealing with electric power transmission engineering. Therefore, this text is unique in that it is written specifically for the in-depth study of modern power transmission engineering.

This book has evolved from the content of courses given by the author at California State University, Sacramento, the University of Missouri at Columbia, the University of Oklahoma, and Florida International University. It has been written for senior-level undergraduate and beginning-level graduate students, as well as practicing engineers in the electrical power utility industry. It can serve as a text for a two-semester course, or by a judicious selection; the material can also be condensed to suit a single-semester course.

This book has been particularly written for a student or a practicing engineer who may want to teach himself. Basic material has been explained carefully, clearly, and in detail with numerous examples. Each new term is clearly defined when it is first introduced. Special features of the book include ample numerical examples and problems designed to use the information presented in each chapter. A special effort has been made to familiarize the reader with the vocabulary and symbols used by the industry.

The addition of numerous appendices, including impedance tables for overhead lines, transformers, and underground cables makes the text self-sufficient. The book includes topics such as power system planning, basic concepts, transmission line parameters and the steady-state performance of transmission lines, disturbance of the normal operating conditions and other problems, symmetrical components and sequence impedances, in-depth analysis of balanced and unbalanced faults, and an extensive review of transmission system protection. A detailed review of load flow analysis is also included. A complete and separate solution's manual is available for the instructors.

Turan Gönen
Sacramento, California

MATLAB® is a registered trademark of The MathWorks, Inc. For product information, please contact:

The MathWorks, Inc.
3 Apple Hill Drive
Natick, MA 01760-2098 USA
Tel: 508 647 7000
Fax: 508-647-7001
E-mail: info@mathworks.com
Web: www.mathworks.com

Acknowledgments

The author wishes to express his appreciation to his mentor, late Dr. Paul M. Anderson of Power Math Associates.

The author is most grateful to numerous colleagues and his professors, particularly Dr. Dave D. Robb of D. D. Robb and Associates for his encouragement and invaluable suggestions and friendship over the years; to his dear friend, late Dr. Don O. Koval of the University of Alberta; and for their interest, encouragement, and invaluable suggestions, late Dr. Adly Girgis of Clemson University and Dr. Anjan Bose.

A special thank you is extended to C.J. Baldwin, Manager, Advanced Systems Technology, Westinghouse Electric Corporation.

The author is also indebted to numerous undergraduate and graduate students who studied portions of the book at California State University, Sacramento, the University of Missouri at Columbia, and the University of Oklahoma, and made countless contributions and valuable suggestions for improvements. Among them are Alan Escoriza, Mira Konjevod, Saud Alsairari, Joel Ervine of Pacific Gas & Electric Inc., and Tom Lyons of Sacramento Municipal Utility.

Author

Turan Gönen is professor of electrical engineering and director of the Electrical Power Educational Institute at California State University, Sacramento. Previously, Dr. Gönen was professor of electrical engineering and director of the Energy Systems and Resources Program at the University of Missouri–Columbia. Professor Gönen also held teaching positions at the University of Missouri–Rolla, the University of Oklahoma, Iowa State University, Florida International University, and Ankara Technical College. He has taught electrical machines and electric power engineering for over thirty-nine years.

Dr. Gönen also has a strong background in the power industry; for eight years, he worked as a design engineer in numerous companies both in the United States and abroad. He has served as a consultant for the United Nations Industrial Development Organization (UNIDO), Aramco, Black & Veatch Consultant Engineers, and the public utility industry. Professor Gönen has written over 100 technical papers as well as four other books: *Electric Power Distribution System Engineering*, *Electrical Machines with MATLAB*, *Electric Power Transmission System Engineering: Analysis and Design*, and *Engineering Economy for Engineering Managers*.

Turan Gönen is a Life Fellow member of the Institute of Electrical and Electronics Engineers and the Institute of Industrial Engineers. He has served on several Committees and Working Groups of the IEEE Power Engineering Society, and is a member of numerous honor societies, including Sigma Xi, Phi Kappa Phi, Eta Kappa Nu, and Tau Alpha Pi.

Dr. Gönen holds a BS and MS in electrical engineering from Istanbul Technical College (1964 and 1966, respectively), and a PhD in electrical engineering from Iowa State University (1975). Dr. Gönen also received an MS in industrial engineering (1973) and a PhD co-major in industrial engineering (1978) from Iowa State University, and a Master of Business Administration (MBA) degree from the University of Oklahoma (1980). Professor Gönen received the Outstanding Teacher Award twice at the California State University, Sacramento, in 1997 and 2009.

1 General Considerations

He is not only dull himself; he is the cause of dullness in others.

Samuel Johnson

Some cause happiness wherever they go; others, whenever they go.

Oscar Wilde

1.1 INTRODUCTION

Until the beginning of the twentieth century, energy generation was based on combustion of fuel at the point of energy use. However, thanks to Thomas Edison, a new industry was born when the first electric power station, Pearl Street Electric Station in New York City, went into operation in 1882. The electric utility industry developed rapidly, and generating stations have spread across the entire country.

There are several reasons for the rapid increase in demand for electrical energy. First, electrical energy is, in many ways, the most convenient energy form. It can be transported by wire to the point of consumption and then transformed into mechanical work, heat, radiant energy, light, or other forms. Electrical energy cannot be very effectively stored, and this has also contributed to its increasing use.

Generating facilities have to be designed for peak use. In the late 1950s and early 1960s, this peak came in the winter, when nights were longer and more lighting and heating were needed. It was economically sound to heavily promote off-peak use, such as summer use of air conditioners. This promotion was so effective that summer is now the peak time. The rate structure has also contributed to increasing use of electrical energy where rate reductions are offered to attract hulk, that is, industrial, consumers. Considering the energy needs and available fuels that are forecasted for the next century, energy is expected to be increasingly converted to electricity in the near future.

The structure of the electric power or energy* system is very large and complex. Nevertheless, it can be divided into basically five main stages, or components, or subsystems, as shown in Figure 1.1.

Here, the first major component is the energy source or fuel whose energy content is used for conversion. The energy source may be coal, gas, or oil burned in a furnace to heat water and generate steam in a boiler; it may be fissionable material, which in a nuclear reactor will heat water to produce steam; it may be water in a dam; or it may be oil or gas burned in a combustion turbine.

The second component is the energy converter or generation system, which transforms the energy from (in) the energy source into electrical energy. This is usually accomplished by converting the energy in the fuel to heat energy and then using this heat energy to produce steam to drive a steam turbine, which in turn drives a generator to produce electrical energy. There are also some other possible energy conversion methods.

---

* The term *energy* is being increasingly used in the electric power industry to replace the conventional term *power*. Here, they are used interchangeably.

FIGURE 1.1 Structure of an electrical power system.

The third main component is the transmission system, which transmits this bulk electrical energy from the generation system where it is produced to main load centers where it is to be distributed through high-voltage lines.

The fourth main component is the distribution system, which distributes this energy to consumers by means of lower-voltage networks.

Finally, the fifth component is the load or energy sink, which utilizes the energy by converting it to a "useful" or desirable form. The electrical load, or energy sink, may be lights, motors, heaters, or other devices, alone or in combination.

Energy is never used up; it is merely successively converted to different forms. Thus, in accordance with the principle of conservation of energy, all energy use ends up as unrecoverable waste heat. The final heat sink for the earth is radiation to space. It should be noted that the utilization of energy depends on two factors: (1) available resources and (2) technological skill to convert the resources to useful heat and work. Energy resources have always been generally available, and the heating process is ancient. Power devices able to convert energy to useful work have been a recent historical development.

The currently available energy conversion methods can be classified into four different groups. The first group covers the conventional and technologically feasible conversion methods. They produce >99% of today's electrical energy. They convert either hydro energy to electrical energy, or thermal energy from fossil fuels or from nuclear fission energy to mechanical energy via thermal energy and then to electrical energy. The second group includes methods that are technically possible but have low-energy conversion efficiency. They include the internal combustion engine and the gas turbine. The third group contains the methods capable of supplying only very small amounts of energy. Batteries, fuel cells, and photovoltaic solar cells are typical examples of this group. Despite the potential of these methods for bulk power generation, currently their use is limited to special applications, such as for space vehicles. The fourth and last group includes methods that are not technologically feasible but seem to have great potential. This group includes fusion reactors, magneto hydrodynamic generators, and electrogasdynamic generators. Some of these methods would solve transmission problems and open new horizons in very long distance transmission.

The fuel supply system industry represents a huge industry. Energy sources are not readily available to the user. They have to be discovered, processed or mined, and transported to demand centers. A general fuel supply system includes the exploration, extraction, processing, and transportation stages, as shown in Figure 1.2. In each of these stages, there are considerable damages to the environment. The fuel supply industry creates basically the same type of pollution problems as do other industrial processes, with the exception of radioactive wastes, land damages, and oil spillage. The disposal of these wastes creates serious problems that must be considered in association with biological safety. The industry continuously searches for practical and better methods to collect, recycle, or safely dispose pollutants. For some energy source, or fuel, some of the stages shown in Figure 1.2 do not exist. For example, in the case of hydropower, "exploration" is still applicable, whereas combustion and its by-products, of course, do not exist. In determining the availability of a fuel for the production of electrical energy, all by-products and their handling must be seriously considered and evaluated.

Utilization of most forms of energy ultimately produces waste heat, which increases the thermal burden on the biosphere. This is not true for an invariant energy source such as hydropower

General Considerations

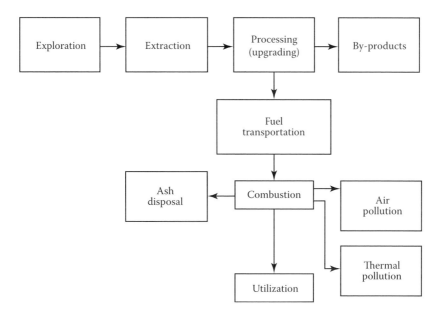

FIGURE 1.2 General fuel supply system.

because it basically circulates through the terrestrial water cycle. Unfortunately, hydropower will never be able to meet a major portion of energy needs. Solar energy could also be an invariant energy resource if either the solar cells or the converters were located on the earth's surface. All other means of energy utilization, other than hydropower and solar conversion, are noninvariant and so would result in the release of waste heat into the environment. This is even true of geothermal energy, which originates in the earth itself.

All the stages shown in Figure 1.2 contribute in various degrees to the total cost of fuel. Of course, they also differ according to the fuel type used. Except where nuclear fuel is employed to generate electricity in power plants located near load centers, the transportation of energy source in one or another form constitutes a significant part of the total cost of electricity. In the case of electrical energy generation, either the fossil fuels (coal, gas, and oil) must be transported from the source to the generating plant or where the electrical energy is generated at the source of fuel, as in the case of mine-mouth plants (coal), wellhead plants (gas), hydropower (water), or plants located near refineries (residual fuel oil), the generated electricity must be transported to load centers by wire. The choice of energy transport is usually one of economics with the environmental impact.

As a simple definition, energy conversion is the changing of the form of energy from one type to another. Currently, all energy conversions are more or less inefficient. In the case of electricity, there are losses at the power plant, in transmission, and at the point of application of the power; in the case of fuels consumed in end uses, the loss comes at the point of use. The waste heat produced at electric power plants, of course, enters the biosphere. Nearly all of the electrical energy that is carried to various points of use degrades to heat.

It is most important, of course, to increase the technical efficiency of the energy conversion devices since the same useful products could thereby be made available with less fuel. In 1920, the average heat rate of electric power plants in the United States was 37,200 Btu/kW h; in 1970, it was increased by about 10,900 Btu/kWh. However, any further increase is not likely, owing to government regulations and environmental concerns.

Today, a large number of electric utilities are not able to generate sufficient cash earnings to even pay for their common-stock dividends, and they finance the payouts from sources such as depreciation, borrowings, or the sale of more stock. Electric utility companies have incurred tens of billions of debt to finance power plant projects so huge and so costly that they threaten to overwhelm the financial capacity of the utilities constructing them. Figure 1.3 shows the total growth in annual electric power system capital in the past.

Figure 1.4 shows the growth in electric utility plants in service in the past. These data represent the privately owned Class A and Class B utilities, which includes 80% of all the electric utilities in

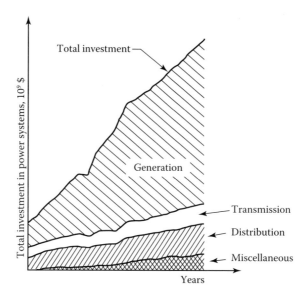

FIGURE 1.3 Growth in total annual electric power system expenditures in the past.

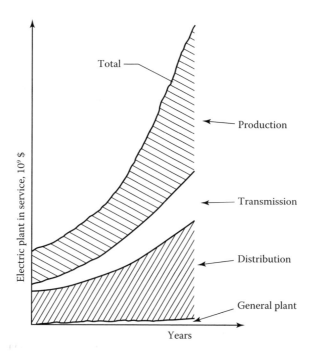

FIGURE 1.4 Growth in electric utility plant in service in the past.

General Considerations

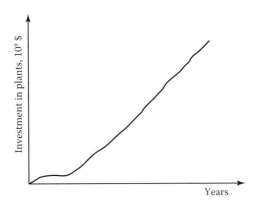

FIGURE 1.5 Investment growth in electric utility plants by investor-owned electric utilities in the past.

the United States. Figure 1.5 shows the actual and projected investment growth in electric utility plants by investor-owned electric utilities in the past. Construction costs have increased tremendously recently. The ever-changing federal regulations, especially for nuclear plants, have forced companies to redesign, at large cost, parts of most facilities to satisfy new rules. Problems such as delays due to such construction problems, blocking actions by environmentalists, inflation-whipped equipment and labor costs, recent recession problems, or the industry's inability to raise money further increase the projected costs.

Today, as reported by an Institute of Electrical and Electronics Engineers committee [5], to cope with these difficulties, utility planning process is changing. These changes affect their planning objectives, planning methods, and even organizational structures in order to adapt to the ever-changing planning environment. Therefore, long-range power system planning has become more of an art than a science owing to the complexity of the problem.

1.2 POWER SYSTEM PLANNING

"The fundamental obligation of the electric utility industry is to provide for an adequate and reliable electric energy supply at a reasonable cost. Adequate electric energy supply *is* essential in assuring a healthy economy, in providing for the health and welfare of the nation's citizens, and for national security" [6]. Therefore, power system planning is vital to assume that the growing demand for electrical energy can be satisfied. In the future, more so than in the past, electric utilities will need fast and economic planning tools to evaluate the consequences of different proposed alternatives and their impact on the rest of the system to provide economic, reliable, and safe electrical energy to consumers. Figure 1.6 shows the factors affecting system planning and its steps. Figure 1.7 shows an organizational chart of the power system planning function in a modern public utility company.

Today, every electric utility company performs both short-term and long-term planning. Here, *short-range planning* is defined as the analytical process that includes an assessment of the near future by evaluating alternative courses of action against desired objectives with the selection of a recommended course of action for the period requiring immediate commitments.

Long-range planning is defined as the analytical process that includes an assessment of the future by evaluating alternative courses of action against desired objectives with the selection of a recommended course of action extending beyond the period requiring immediate commitments.

The necessity for long-range system planning becomes more pressing than in the past because of the very fast changes in technology, fuel availability, and environmental constraints. It enables

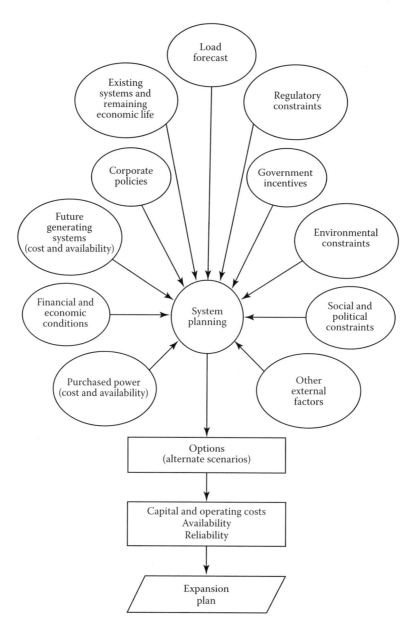

FIGURE 1.6 Factors affecting a typical system planning process.

planners to explore various alternatives in supplying electrical energy and provides them with a guide for making short-range decisions and actions. It must be quantitative as well as qualitative. In general, long-range planning may cover 15 to 20 years into the future. However, the planning horizons are known to be 15 to 30 years for some power plant additions.

The objective of system planning is to optimize the facilities necessary to provide an adequate electrical energy supply at the lowest reasonable cost. In general, system planning activities can be classified as (1) synthesis, that is, the development of initial plans to study the system; (2) analysis, that is, technical evaluation of the system operation in terms of reserve requirements, load flow, and stability, under simulated conditions; and (3) optimization, that is, economic evaluation of

General Considerations

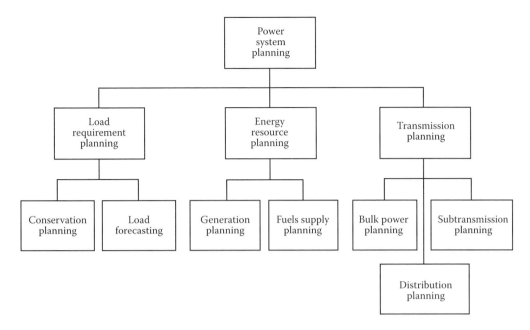

FIGURE 1.7 Modern organizational chart of a power system planning function.

alternatives to determine the minimum-cost alternative. As shown in Figure 1.8, system planning may include the following activities:

1. Load forecasting
2. Generation planning
3. Transmission planning
4. Subtransmission planning
5. Distribution planning
6. Operations planning
7. Fuel supply planning
8. Environmental planning
9. Financial planning
10. Research and development (R&D) planning

The forecasting of load increases and system reaction to these increases is essential to the system planning process. In general, long-range forecasting has a time horizon of between 15 and 20 years, whereas the time horizon of short-range forecasting is somewhere between 1 and 5 years. The basic function of load forecasting is to analyze the raw historical load data to develop models of peak demand for capacity planning, of energy for capacity planning, and of energy for production costing. At the present time, very sophisticated techniques are available to analyze the past trends of load components separately and to determine the effects on each of such factors as weather, general economic conditions, and per capita income.

The end result can be a load projection with its probability distribution. However, the degree of uncertainty planners perceive in their load forecasts has grown noticeably in the past 10 years. The risk of planning based on too low load projections is poor system reliability, whereas the risk of planning based on too high load projections is an uneconomical operation.

Generation planning helps identify the technology, size, and timing of the next generating plants to be added to the power system in order to ensure that adequate generation capacity is

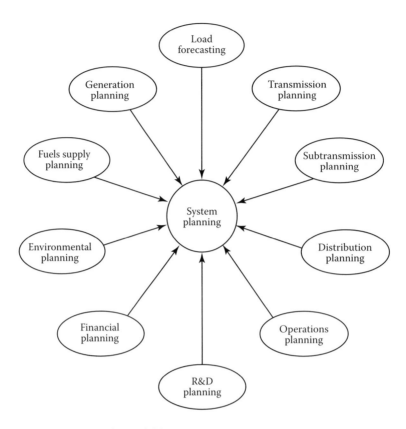

FIGURE 1.8 Power system planning activities.

available to meet future demands for electricity and that the future cost of power generation is economical.

Thus, generation planning activities include (1) generation capacity planning, (2) production costing, and (3) calculating investment and operation and maintenance costs. In generation planning, the planner combines load and generation models, including scheduled maintenance, to determine the capacity needed to meet the system reliability criterion. On the other hand, in production costing, the planner determines the costs of generating energy requirements of the system, including effects of maintenance and forced outages. Finally, the present value of investment and operating and maintenance costs are calculated and consequently predetermined.

The objective is an optimum generation planning that combines the aforementioned functions into one, to develop economically optimum generation expansion patterns year by year over the planning horizon. The variety of synthesis tools that are available to system planners include (1) target plant mix, (2) expert judgment (the planner's past experience), and (3) computer-based mathematical programming models. On the basis of a given generation expansion plan, simulation models are employed to predict (1) power system reliability, (2) capital and production costs, and (3) power plant operation for each year of the planning horizon.

Transmission planning is closely related to generation planning. The objectives of transmission system planning is to develop year-by-year plans for the transmission system on the basis of existing systems, future load and generation scenarios, right-of-way constraints, cost of construction, line capabilities, and reliability criteria.

In general, transmission lines have two primary functions: (1) to transmit electrical energy from the generators to the loads within a single utility and (2) to provide paths for electrical energy to flow between utilities. These latter lines are called "tie lines" and enable the utility companies to

General Considerations

operate as a team to gain benefits that would otherwise not be obtainable. Interconnections, or the installation of transmission circuits across utility boundaries, influence both the generation and transmission planning of each utility involved [8].

When power systems are electrically connected by transmission lines, they must operate at the same frequency, that is, the same number of cycles per second, and the pulse of the alternating current must be coordinated. As a corollary, generator speeds, which determine frequency, must also be coordinated. The various generators are then said to be operating "in parallel" or "in synchronism," and the system is said to be "stable."

A sharp and sudden change in loading at a generator will affect the frequency; however, if the generator is strongly interconnected with other generators, they will normally help absorb the effect of the changed loading so that the change in frequency will be negligible and the system's stability will be unaffected. Therefore, the installation of an interconnection substantially affects generation planning in terms of the amount of generation capacity required, the reserve generation capacity, and the type of generation capacity required for operation.

Also, interconnections may affect generation planning through the installation of apparatus owned jointly by neighboring utilities, and the planning of generating units with greater capacity than would be feasible for a single utility without interconnections. Furthermore, interconnection planning affects transmission planning by requiring bulk power deliveries away from or to interconnection substations, that is, bulk power substations, and often the addition of circuits on a given utility's own network [9].

Subtransmission planning includes planning activities for the major supply of bulk stations, subtransmission lines from the stations to distribution substations, and the high-voltage portion of the distribution substations.

Distribution planning must not only take into consideration substation siting, sizing, number of feeders to be served, voltage levels, and arid type and size of the service area, but also the coordination of overall subtransmission, and even transmission planning efforts, in order to ensure the most reliable and cost-effective system design. Today, many distribution system planners use computer programs such as load flow programs, radial or loop network load programs, short-circuit programs, voltage drop and voltage regulation programs, load forecasting, regulator setting, and capacitor planning. Figure 1.9 shows a functional block diagram of the distribution system planning process [7].

The planning starts at the customer level. The demand, type, load factor, and other customer load characteristics dictate the type of distribution system required. Once the customer loads are determined, secondary lines are then defined that connect to distribution transformers. The distribution transformer loads are then combined to determine the demands on the primary distribution system. The primary distribution system loads are then assigned to substations. The distribution system loads determine the size and location (siting) of the substations, as well as the route and capacity of the associated subtransmission lines.

The objective of both fuel supply and operations planning is to provide generation planning with information necessary for the proper modeling of power system operation. In fuel supply planning, the price and availability of various types of fuels are estimated and information with respect to long-term fuel contracts is gathered. In operations planning, information about heat rates, plant capacity factor restrictions, and maintenance of the existing power plants, as well as energy sales and purchases is provided.

Environmental planning provides information on environmental regulations and responsibilities in order to establish the plant type, plant location, and design requirements and available fuel types. This information is used as an essential input to generation planning to limit the available alternatives in developing system expansion plans.

In financial planning, financial tools such as corporate models are used to develop annual and monthly financial reports and cash flow information on system expansion (from generation planning), including tax and regulatory constraints. It can indirectly control the generation planning

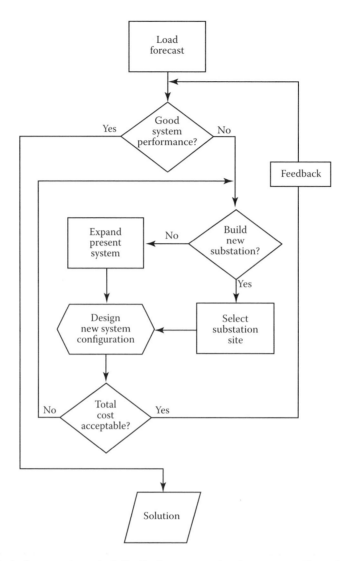

FIGURE 1.9 Block diagram of a typical distribution system planning process. (From Gönen, T., *Electrical Power Distribution System Engineering*, 2nd ed., CRC Press, Boca Raton, FL, 2008.)

process by limiting the construction budget that the company can afford. R&D planning provides information to generation planning in terms of costs, characteristics, and availability of alternative energy sources and future technological developments in generation.

REFERENCES

1. U.S. Department of Energy, *The National Electric Reliability Study: Technical Study Reports*, DOEIEP-0005. USDOE, Office of Emergency Operations, Washington, D.C., 1981.
2. Clair, M. L., Annual statistical report. *Electr. World* 193 (6) 49–80 (1980).
3. Energy Information Administration, *Energy Data Reports—Statistics of Privately-Owned Electric Utilities in the United States*. U.S. Dept. of Energy, Washington, D.C., 1975–1978.
4. Vennard, E., *Management of the Electric Energy Business*. McGraw-Hill, New York, 1979.
5. Institute of Electrical and Electronics Engineers Committee Report, The significance of assumptions implied in long-range electric utility planning studies. *IEEE Trans. Power Appar. Syst.* PAS-99, 1047–1056 (1980).

6. National Electric Reliability Council, *Tenth Annual Review of Overall Reliability and Adequacy of the North American Bulk Power Systems.* NERC, Princeton, NJ, 1980.
7. Gönen, T., *Electrical Power Distribution System Engineering.* McGraw-Hill, New York, 1986.
8. Electric Power Research Institute, *Transmission Line Reference Book: 345 kV and Above.* EPRI, Palo Alto, CA, 1979.
9. Gönen, T., *Electrical Power Distribution System Engineering*, 2nd ed. CRC Press, Boca Raton, FL, 2008.
10. Gönen, T., *Electrical Power Transmission System Engineering: Analysis and Design*, 2nd ed. CRC Press, Boca Raton, FL, 2009.

GENERAL REFERENCES

Edison Electric Institute, *EHV Transmission Line Reference Book.* EEI, New York, 1968.

Electric Power Research Institute, *Transmission Line Reference Book: 115–138 kV Compact Line Design.* EPRI, Palo Alto, CA, 1978.

Electric Power Research Institute, *Transmission Line Reference Book: HVDC to ±600 kV.* EPRI, Palo Alto, CA, 1978.

Fink, D. G., and Beaty, H. W., *Standard Handbook for Electrical Engineers*, 11th ed. McGraw-Hill, New York, 1978.

Gönen, T., and Anderson, P. M., The impact of advanced technology on the future electric energy supply problem. *Proc. IEEE Energy '78 Conf.* 1978, pp. 117–121 (1978).

Gönen, T., and Bekiroglu, H., Some views on inflation and a Phillips curve for the U.S. economy. *Proc. Am. Inst. Decision Sci. Conf.,* 1977, pp. 328–331 (1977).

Gönen, T., Anderson, P. M., and Bowen, D. W., Energy and the future. *Proc. World Hydrogen Energy Cont., 1st,* 1976 Vol. 3(2C), pp. 55–78 (1976).

2 Basic Concepts

2.1 INTRODUCTION

In this chapter, a brief review of fundamental concepts associated with steady-state alternating current circuits, especially with three-phase circuits, is presented. It is hoped that this brief review is sufficient to provide a common base, in terms of notation and references, that is necessary to be able to follow the subsequent chapters.

2.2 COMPLEX POWER IN BALANCED TRANSMISSION LINES

Figure 2.1a shows a per-phase representation (or a one-line diagram) of a short three-phase balanced transmission line connecting buses i and j. Here, the term *bus* defines a specific nodal point of a transmission network. Assume that the bus voltages $\mathbf{V}_i$ and $\mathbf{V}_j$ are given in phase values (i.e., line-to-neutral values) and that the line impedance is $\mathbf{Z} = R + jX$ per phase. Since the transmission line is a short one, the line current $\mathbf{I}$ can be assumed to be approximately the same at any point in the line. However, because of the line losses, the complex powers $\mathbf{S}_{ij}$ and $\mathbf{S}_{ji}$ are not the same. Thus, the complex power per phase* that is being transmitted from bus i to bus j can be expressed as

$$\mathbf{S}_{ij} = P_{ij} + jQ_{ij} = \mathbf{V}_i \mathbf{I}^* \tag{2.1}$$

Similarly, the complex power per phase that is being transmitted from bus j to bus i can be expressed as

$$\mathbf{S}_{ji} = P_{ji} + jQ_{ji} = \mathbf{V}_j(-\mathbf{I})^* \tag{2.2}$$

Since

$$\mathbf{I} = \frac{\mathbf{V}_i - \mathbf{V}_j}{\mathbf{Z}} \tag{2.3}$$

substituting Equation 2.3 into Equations 2.1 and 2.2,

$$\begin{aligned}\mathbf{S}_{ij} &= \mathbf{V}_i \left(\frac{\mathbf{V}_i^* - \mathbf{V}_j^*}{\mathbf{Z}^*} \right) \\ &= \frac{|\mathbf{V}_i|^2 - |\mathbf{V}_i||\mathbf{V}_j| \angle \theta_i - \theta_j}{R - jX}\end{aligned} \tag{2.4}$$

---

* For an excellent treatment of the subject, see Elgerd [1].

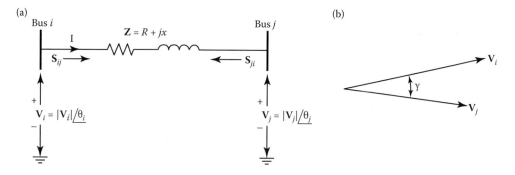

FIGURE 2.1 Per-phase representation of short transmission line.

and

$$S_{ji} = V_j \left(\frac{V_j^* - V_i^*}{Z^*} \right) \quad (2.5)$$

$$= \frac{|V_j|^2 - |V_j||V_i|\angle \theta_j - \theta_i}{R - jX}$$

However, as shown in Figure 2.1b, if the power angle (i.e., the phase angle between the two bus voltages) is defined as

$$\gamma = \theta_i - \theta_j \quad (2.6)$$

then the real and the reactive power per-phase values can be expressed, respectively, as

$$P_{ij} = \frac{1}{R^2 + X^2} \left(R|V_i|^2 - R|V_i||V_j|\cos\gamma + X|V_i||V_j|\sin\gamma \right) \quad (2.7)$$

and

$$Q_{ij} = \frac{1}{R^2 + X^2} \left(X|V_i|^2 - X|V_i||V_j|\cos\gamma + R|V_i||V_j|\sin\gamma \right) \quad (2.8)$$

Similarly,

$$P_{ji} = \frac{1}{R^2 + X^2} \left(R|V_j|^2 - R|V_i||V_j|\cos\gamma + X|V_i||V_j|\sin\gamma \right) \quad (2.9)$$

and

$$Q_{ji} = \frac{1}{R^2 + X^2} \left(X|V_j|^2 - X|V_i||V_j|\cos\gamma + R|V_i||V_j|\sin\gamma \right) \quad (2.10)$$

The three-phase real and reactive power can directly be found from Equations 2.7 through 2.10 if the phase values are replaced by the line values.

Basic Concepts

In general, the reactance of a transmission line is much greater than its resistance. Therefore, the line impedance value can be approximated as

$$\mathbf{Z} = jX \tag{2.11}$$

by setting $R = 0$. Therefore, Equations 2.7 through 2.10 can be expressed as

$$P_{ij} = \left(\frac{|\mathbf{V}_i||\mathbf{V}_j|}{X}\right)\sin\gamma \tag{2.12}$$

$$Q_{ij} = \frac{1}{X}\left(|\mathbf{V}_i|^2 - |\mathbf{V}_i||\mathbf{V}_j|\cos\gamma\right) \tag{2.13}$$

and

$$P_{ji} = -\left(\frac{|\mathbf{V}_i||\mathbf{V}_j|}{X}\right)\sin\gamma = -P_{ij} \tag{2.14}$$

$$Q_j = \frac{1}{X}\left(|\mathbf{V}_j|^2 - |\mathbf{V}_i||\mathbf{V}_j|\cos\gamma\right) \tag{2.15}$$

EXAMPLE 2.1

Assume that the impedance of a transmission line connecting buses 1 and 2 is $100\angle 60°\ \Omega$, and that the bus voltages are $73{,}034.8\angle 30°$ and $66{,}395.3\angle 20°$ V per phase, respectively. Determine the following:

(a) Complex power per phase that is being transmitted from bus 1 to bus 2
(b) Active power per phase that is being transmitted
(c) Reactive power per phase that is being transmitted

Solution

(a)

$$S_{12} = \mathbf{V}_1\left(\frac{\mathbf{V}_1^* - \mathbf{V}_2^*}{\mathbf{Z}^*}\right)$$

$$= (73{,}034.8\angle 30°)\left(\frac{73{,}034.8\angle -30° - 66{,}395.3\angle -20°}{100\angle -60°}\right)$$

$$= 10{,}104{,}280.766.7\angle 3.56°$$

$$= 10{,}085{,}280.6 + j627{,}236.51\,\text{VA}$$

(b) Therefore,

$$P_{12} = 10,085,280.6 \text{ W}$$

(c)

$$Q_{12} = 627,236.51 \text{ vars}$$

2.3 ONE-LINE DIAGRAM

In general, electrical power systems are represented by a one-line diagram, as shown in Figure 2.2a. The one-line diagram is also referred to as the single-line diagram. Figure 2.2b shows the three-phase equivalent impedance diagram of the system given in Figure 2.2a. However, the need for the three-phase equivalent impedance diagram is almost nil in usual situations. This is because a balanced three-phase system can always be represented by an equivalent impedance diagram per phase, as shown in Figure 2.2c. Furthermore, the per-phase equivalent impedance can also be simplified by neglecting the neutral line and representing the system components by standard symbols rather than by their equivalent circuits. The result is, of course, the one-line diagram shown in Figure 2.2a.

Table 2.1 provides some of the symbols that are used in one-line diagrams. Additional standard symbols can be found in Institute of Electrical and Electronics Engineers Standard 315-1971 [2]. At times, as a need arises, the one-line diagram may also show peripheral apparatus such as instrument transformers (i.e., current and voltage transformers), protective relays, and lighting arrestors.

Thus, the details shown on a one-line diagram depend on its purpose. For example, the one-line diagrams that will be used in load flow studies do not show circuit breakers or relays, contrary to those that will be used in stability studies. Furthermore, those that will be used in unsymmetrical fault studies may even show the positive-, negative-, and zero-sequence networks separately.

Note that the buses (i.e., the nodal points of the transmission network) that are shown in Figure 2.2a have been identified by their bus numbers. Also note that the neutral of generator 1 has been "solidly grounded," that is, the neutral point has been directly connected to the earth, whereas the neutral of generator 2 has been "grounded through impedance" using a resistor. Sometimes, it is grounded using an inductance coil. In either case, they are used to limit the current flow to ground under fault conditions.

Usually, the neutrals of the transformers used in transmission lines are solidly grounded. In general, proper generator grounding for generators is facilitated by burying a ground electrode system made of grids of buried horizontal wires. As the number of meshes in the grid is increased, its conductance becomes greater. Sometimes, a metal plate is buried instead of a mesh grid (especially in European applications).

Transmission lines with overhead ground wires have a ground connection at each supporting structure to which the ground wire is connected. In some circumstances, a "counterpoise," that is, a bare conductor, is buried under a transmission line to decrease the ground resistance. The best-known example is the one that has been installed for the transmission line crossing the Mohave Desert. The counterpoise is buried alongside the line and connected directly to the towers and the overhead ground wires.

Note that the equivalent circuit of the transmission line shown in Figure 2.2c has been represented by a nominal π*. The line impedance, in terms of the resistance and the series reactance of a single conductor for the length of the line, has been lumped. The line-to-neutral capacitance (or shunt capacitive reactance) for the length of the line has been calculated, and half of this value

---

* Read Chapter 3 for further information.

Basic Concepts

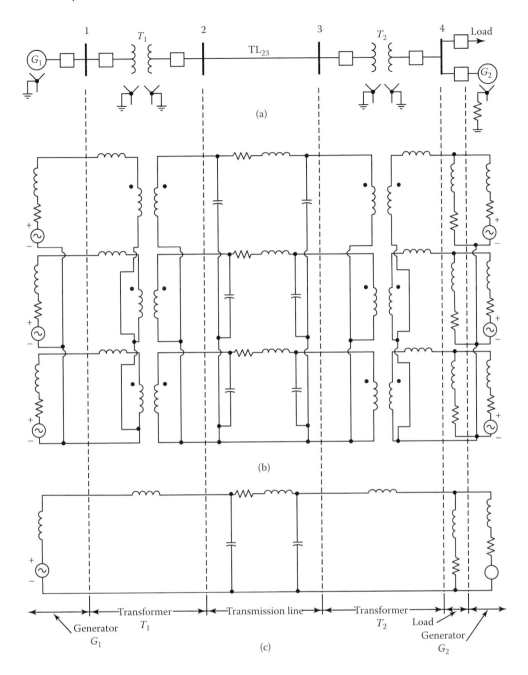

FIGURE 2.2 Power system representations: (a) one-line diagram; (b) three-phase equivalent impedance diagram; (c) equivalent impedance diagram per phase.

has been put at each end of the line. The transformers have been represented by their equivalent reactances, neglecting their magnetizing currents and consequently their shunt admittances. Also neglected, are the resistance values of the transformers and generators, because their inductive reactance values are much greater than their resistance values.

Also, not shown in Figure 2.2c, is the ground resistor. This is because of no current flowing in the neutral under balanced conditions. The impedance diagram shown in Figure 2.2c is also referred to as the positive-sequence network or diagram. The reason is that the phase order of the balanced

TABLE 2.1
Symbols Used in One-Line Diagrams

Symbol	Usage	Symbol	Usage
○	Rotating machine	—▭—	Circuit breaker
—┼—	Bus	—⌒—	Circuit breaker (air)
(two-winding transformer symbol)	Two-winding transformer	—/—	Disconnect
(three-winding transformer symbol)	Three-winding transformer	⌇ or ▯	Fuse
△	Delta connection (3φ, three wire)	—∫—	Fused disconnect
Y	Wye connection (3φ, neutral ungrounded)	(lightning arrester symbol)	Lightning arrester
Y with ground	Wye connection (3φ, neutral grounded)	—⌒⌒—	Current transformer (CT)
———	Transmission line	—)⌇(or)—	Potential transformer (VT)
—▸	Static load	—┴— with ground	Capacitor

Basic Concepts

voltages at any point in the system is the same as the phase order of the generated voltage, and they are positive. The per-phase impedance diagrams may represent a system given either in ohms or in per unit.

At times, as need arises, the one-line diagram may also show peripheral apparatus such as instrument transformers (i.e., current and voltage transformers), protective relays, and lighting arrestors. Therefore, the details shown on a one-line diagram depend on its purpose. For example, the one-line diagrams that will be used in load flow studies do not show circuit breakers or relays, contrary to those that will be used in stability studies. Furthermore, those that will be used in unsymmetrical fault studies may even show the positive-, negative-, and zero-sequence networks separately.

2.4 PER-UNIT SYSTEM

Because of various advantages involved, it is customary in power system analysis calculations to use impedances, currents, voltages, and powers in per-unit values (which are scaled or normalized values) rather than in physical values of ohms, amperes, kilovolts, and megavolt-amperes (MVA; or megavars, or megawatts). A per-unit system is a means of expressing quantities for ease in comparing them. The per-unit value of any quantity is defined as the ratio of the quantity to an "arbitrarily" chosen base (i.e., reference) value having the same dimensions. Therefore, the per-unit value of any quantity can be defined as

$$\text{Quantity in per unit} = \frac{\text{physical quantity}}{\text{base value of quantity}} \qquad (2.16)$$

where *physical quantity* refers to the given value in ohms, amperes, volts, etc. The *base value* is also called unit value since in the per-unit system it has a value of 1, or unity. Therefore, a base current is also referred to as a unit current. Since both the physical quantity and base quantity have the same dimensions, the resulting per-unit value expressed as a decimal has no dimension and therefore is simply indicated by a subscript pu. The base quantity is indicated by a subscript B. The symbol for per unit is pu, or 0/1. The percent system is obtained by multiplying the per-unit value by 100. Therefore,

$$\text{Quantity in percent} = \frac{\text{physical quantity}}{\text{base value of quantity}} \times 100 \qquad (2.17)$$

However, the percent system is somewhat more difficult to work with and more subject to possible error since it must always be remembered that the quantities have been multiplied by 100. Therefore, the factor 100 has to be continually inserted or removed for reasons that may not be obvious at the time. For example, 40% reactance times 100% current is equal to 4000% voltage, which, of course, must be corrected to 40% voltage. Thus, the per-unit system is preferred in power system calculations. The advantages of using the per-unit include the following:

1. Network analysis is greatly simplified since all impedances of a given equivalent circuit can directly be added together regardless of the system voltages.
2. It eliminates the $\sqrt{3}$ multiplications and divisions that are required when balanced three-phase systems are represented by per-phase systems. Therefore, the factors $\sqrt{3}$ and 3 associated with delta and wye quantities in a balanced three-phase system are directly taken into account by the base quantities.
3. Usually, the impedance of an electrical apparatus is given in percent or per unit by its manufacturer on the basis of its nameplate ratings (e.g., its rated volt-amperes and rated voltage).

4. Differences in operating characteristics of many electrical apparatus can be estimated by a comparison of their constants expressed in per units.
5. Average machine constants can easily be obtained since the parameters of similar equipment tend to fall in a relatively narrow range and therefore arc comparable when expressed as per units according to rated capacity.
6. The use of per-unit quantities is more convenient in calculations involving digital computers.

2.4.1 Single-Phase System

In the event that any two of the four base quantities (i.e., base voltage, base current, base volt-amperes, and base impedance) are "arbitrarily" specified, the other two can be determined immediately. Here, the term arbitrarily is slightly misleading since in practice the base values are selected so as to force the results to fall into specified ranges. For example, the base voltage is selected such that the system voltage is normally close to unity. Similarly, the base volt-ampere is usually selected as the kilovolt-ampere (kVA) or MVA rating of one of the machines or transformers in the system, or a convenient round number such as 1, 10, 100, or 1000 MVA, depending on the system size. As aforementioned, on determining the base volt-amperes and base voltages, the other base values are fixed. For example, the current base can be determined as

$$I_B = \frac{S_B}{V_B} = \frac{\mathrm{VA}_B}{V_B} \tag{2.18}$$

where

I_B = current base in amperes
$S_B = \mathrm{VA}_B$ = selected volt-ampere base in volt-amperes
V_B = selected voltage base in volts

Note that,

$$S_B = \mathrm{VA}_B = P_B = Q_B = V_B I_B \tag{2.19}$$

Similarly, the impedance base* can be determined as

$$Z_B = \frac{V_B}{I_B} \tag{2.20}$$

where

$$Z_B = X_B = R_B \tag{2.21}$$

Similarly,

$$Y_B = B_B = G_B = \frac{I_B}{V_B} \tag{2.22}$$

---

* Defined as that impedance across which there is a voltage drop that is equal to the base voltage if the current through it is equal to the base current.

Basic Concepts

Note that by substituting Equation 2.18 into Equation 2.20, the impedance base can be expressed as

$$Z_B = \frac{V_B}{VA_B/V_B} = \frac{V_B^2}{VA_B} \tag{2.23}$$

or

$$Z_B = \frac{(kV_B)^2}{MVA_B} \tag{2.24}$$

where
kV_B = voltage base in kilovolts
MVA_B = volt-ampere base in MVA

The per-unit value of any quantity can be found by the *normalization process*, that is, by dividing the physical quantity by the base quantity of the same dimension. For example, the per-unit impedance can be expressed as

$$Z_{pu} = \frac{Z_{physical}}{Z_B} \tag{2.25}$$

or

$$Z_{pu} = \frac{Z_{physical}}{V_B^2/(kVA_B \times 1000)} \tag{2.26}$$

or

$$Z_{pu} = \frac{(Z_{physical})(kVA_B)(1000)}{V_B^2} \tag{2.27}$$

or

$$Z_{pu} = \frac{(Z_{physical})(kVA_B)}{(kV_B)^2(1000)} \tag{2.28}$$

or

$$Z_{pu} = \frac{(Z_{physical})}{(kV_B)^2/MVA_B} \tag{2.29}$$

or

$$Z_{pu} = \frac{(Z_{physical})(MVA_B)}{(kV_B)^2} \tag{2.30}$$

Similarly, the others can be expressed as

$$I_{pu} = \frac{I_{physical}}{I_B} \tag{2.31}$$

or

$$V_{pu} = \frac{V_{physical}}{V_B} \tag{2.32}$$

or

$$kV_{pu} = \frac{kV_{physical}}{kV_B} \tag{2.33}$$

or

$$VA_{pu} = \frac{VA_{physical}}{VA_B} \tag{2.34}$$

or

$$kVA_{pu} = \frac{kVA_{physical}}{kVA_B} \tag{2.35}$$

or

$$MVA_{pu} = \frac{MVA_{physical}}{MVA_B} \tag{2.36}$$

Note that, the base quantity is always a real number, whereas the physical quantity can be a complex number. For example, if the actual impedance quantity is given as $Z\angle\theta$ Ω, it can be expressed in the per-unit system as

$$Z_{pu} = \frac{Z\angle\theta}{Z_B} = Z_{pu}\angle\theta \tag{2.37}$$

that is, it is the magnitude expressed in per-unit terms. Alternatively, if the impedance has been given in rectangular form as

$$\mathbf{Z} = R + jX \tag{2.38}$$

then

$$\mathbf{Z}_{pu} = R_{pu} + jX_{pu} \tag{2.39}$$

Basic Concepts

where

$$R_{pu} = \frac{R_{physical}}{Z_B} \tag{2.40}$$

and

$$X_{pu} = \frac{X_{physical}}{Z_B} \tag{2.41}$$

Similarly, if the complex power has been given as

$$\mathbf{S} = P + jQ \tag{2.42}$$

then

$$\mathbf{S}_{pu} = P_{pu} + jQ_{pu} \tag{2.43}$$

where

$$P_{pu} = \frac{P_{physical}}{S_B} \tag{2.44}$$

and

$$Q_{pu} = \frac{Q_{physical}}{S_B} \tag{2.45}$$

If the actual voltage and current values are given as

$$\mathbf{V} = V\angle\theta_V \tag{2.46}$$

and

$$\mathbf{I} = I\angle\theta_I \tag{2.47}$$

then the complex power can be expressed as

$$\mathbf{S} = \mathbf{V}\mathbf{I}^* \tag{2.48}$$

or

$$S\angle\theta = (V\angle\theta_V)(I\angle-\theta_I) \tag{2.49}$$

Therefore, dividing through by S_B,

$$\frac{S\angle\phi}{S_B} = \frac{(V\angle\theta_V)(I\angle-\theta_I)}{S_B} \tag{2.50}$$

However,

$$S_B = V_B I_B \tag{2.51}$$

Thus,

$$\frac{S\angle\theta}{S_B} = \frac{(V\angle\theta_V)(I\angle-\theta_I)}{V_B I_B} \tag{2.52}$$

or

$$S_{pu}\angle\theta = (V_{pu}\angle\theta_V)(I_{pu}\angle-\theta_I) \tag{2.53}$$

or

$$S_{pu} = V_{pu} I_{pu}^* \tag{2.54}$$

2.4.2 Converting from Per-Unit Values to Physical Values

The physical values (or system values) and per-unit values are related by the following relations:

$$\mathbf{I} = \mathbf{I}_{pu} \times I_B \tag{2.55}$$

$$\mathbf{V} = \mathbf{V}_{pu} \times V_B \tag{2.56}$$

$$\mathbf{Z} = \mathbf{Z}_{pu} \times Z_B \tag{2.57}$$

$$R = R_{pu} \times Z_B \tag{2.58}$$

$$X = X_{pu} \times Z_B \tag{2.59}$$

$$VA = VA_{pu} \times VA_B \tag{2.60}$$

$$P = P_{pu} \times VA_B \tag{2.61}$$

$$Q = Q_{pu} \times VA_B \tag{2.62}$$

2.4.3 Change of Base

In general, the per-unit impedance of a power apparatus is given on the basis of its own volt-ampere and voltage ratings and consequently on the basis of its own impedance base. When such an apparatus is used in a system that has its own bases, it becomes necessary to refer all the given per-unit

Basic Concepts

values to the system base values. Assume that the per-unit impedance of the apparatus is given on the basis of its nameplate ratings as

$$Z_{pu(given)} = (Z_{physical}) \frac{MVA_{B(given)}}{[kV_{B(given)}]^2} \tag{2.63}$$

and that it is necessary to refer the very same physical impedance to a new set of voltage and volt-ampere bases such that

$$Z_{pu(new)} = (Z_{physical}) \frac{MVA_{B(new)}}{[kV_{B(new)}]^2} \tag{2.64}$$

By dividing Equation 2.63 by Equation 2.64 side by side,

$$Z_{pu(new)} = Z_{pu(given)} \left[\frac{MVA_{B(new)}}{MVA_{B(given)}} \right] \left[\frac{kV_{B(given)}}{kV_{B(new)}} \right]^2 \tag{2.65}$$

In certain situations, it is more convenient to use subscripts 1 and 2 instead of the subscripts "given" and "new," respectively. Then, Equation 2.65 can be expressed as

$$Z_{pu(2)} = Z_{pu(1)} \left[\frac{MVA_{B(2)}}{MVA_{B(1)}} \right] \left[\frac{kV_{B(1)}}{kV_{B(2)}} \right]^2 \tag{2.66}$$

In the event that the kilovolt bases are the same but the MVA bases are different, from Equation 2.65,

$$Z_{pu(new)} = Z_{pu(given)} \frac{MVA_{B(new)}}{MVA_{B(given)}} \tag{2.67}$$

Similarly, if the MVA bases are the same but the kilovolt bases are different, from Equation 2.65,

$$Z_{pu(new)} = Z_{pu(given)} \left[\frac{kV_{B(given)}}{kV_{B(new)}} \right]^2 \tag{2.68}$$

Equations 2.65 through 2.68 must only be used to convert the given per-unit impedance from the base to another but not for referring the physical value of an impedance from one side of the transformer to another [3].

2.4.4 THREE-PHASE SYSTEMS

The three-phase problems involving balanced systems can be solved on a per-phase basis. In that case, the equations that are developed for single-phase systems can be used for three-phase systems as long as per-phase values are used consistently. Therefore,

$$I_B = \frac{S_{B(1\phi)}}{V_{B(L-N)}} \tag{2.69}$$

or

$$I_B = \frac{VA_{B(1\phi)}}{V_{B(L-N)}} \tag{2.70}$$

and

$$Z_B = \frac{V_{B(L-N)}}{I_B} \tag{2.71}$$

or

$$Z_B = \frac{[kV_{B(L-N)}]^2 (1000)}{kVA_{B(1\phi)}} \tag{2.72}$$

or

$$Z_B = \frac{[kV_{B(L-N)}]^2}{MVA_{B(1\phi)}} \tag{2.73}$$

where the subscripts 1ϕ and L–N denote per phase and line to neutral, respectively. Note that, for a *balanced system*,

$$V_{B(L-N)} = \frac{V_{B(L-L)}}{\sqrt{3}} \tag{2.74}$$

and

$$S_{B(1\phi)} = \frac{S_{B(3\phi)}}{3} \tag{2.75}$$

However, it has been customary in three-phase system analysis to use line-to-line voltage and three-phase volt-amperes as the base values. Therefore,

$$I_B = \frac{S_{B(3\phi)}}{\sqrt{3} V_{B(L-L)}} \tag{2.76}$$

or

$$I_B = \frac{kVA_{B(3\phi)}}{\sqrt{3} kV_{B(L-L)}} \tag{2.77}$$

Basic Concepts

and

$$Z_B = \frac{V_{B(L-L)}}{\sqrt{3}I_B} \tag{2.78}$$

$$Z_B = \frac{[kV_{B(L-L)}]^2(1000)}{kVA_{B(3\phi)}} \tag{2.79}$$

or

$$Z_B = \frac{[kV_{B(L-L)}]^2}{MVA_{B(3\phi)}} \tag{2.80}$$

where the subscripts 3φ and L–L denote per three phase and line, respectively. Furthermore, base admittance can be expressed as

$$Y_B = \frac{1}{Z_B} \tag{2.81}$$

or

$$Y_B = \frac{MVA_{B(3\phi)}}{[kV_{B(L-L)}]^2} \tag{2.82}$$

where

$$Y_B = B_B = G_B \tag{2.83}$$

The data for transmission lines are usually given in terms of the line resistance R in ohms per mile at a given temperature, the line inductive reactance X_L in ohms per mile at 60 Hz, and the line shunt capacitive reactance X_c in megohms per mile at 60 Hz. Therefore, the line impedance and shunt susceptance in per units for 1 mi of line can be expressed as*

$$\mathbf{Z}_{pu} = (\mathbf{Z}, \Omega/mi)\frac{MVA_{B(3\phi)}}{[kV_{B(L-L)}]^2} \quad pu \tag{2.84}$$

where

$$\mathbf{Z} = R + jX_L = Z\angle\theta \;\Omega/mi$$

and

$$B_{pu} = \frac{[kV_{B(L-L)}]^2 \times 10^{-6}}{[MVA_{B(3\phi)}][X_c, M\Omega/mi]} \tag{2.85}$$

---

* For further information, see Anderson [4].

In the event that the admittance for a transmission line is given in microsiemens per mile, the per-unit admittance can be expressed as

$$Y_{pu} = \frac{[kV_{B(L-L)}]^2 (Y, \mu S)}{[MVA_{B(3\phi)}] \times 10^6} \tag{2.86}$$

Similarly, if it is given as reciprocal admittance in megohms per mile, the per-unit admittance can be found as

$$Y_{pu} = \frac{[kV_{B(L-L)}]^2 \times 10^{-6}}{[MVA_{B(3\phi)}][Z, M\Omega/mi]} \tag{2.87}$$

Figure 2.3 shows conventional three-phase transformer connections and associated relations between the high-voltage and low-voltage side voltages and currents. The given relations are correct for a three-phase transformer as well as for a three-phase bank of single-phase transformers. Note that, in the figure, n is the turns ratio, that is,

$$n = \frac{N_1}{N_2} = \frac{V_1}{V_2} = \frac{I_2}{I_1} \tag{2.88}$$

where the subscripts 1 and 2 are used for the primary and secondary sides. Therefore, an impedance Z_2 in the secondary circuit can be referred to the primary circuit provided that

$$Z_1 = n^2 Z_2 \tag{2.89}$$

Thus, it can be observed from Figure 2.3 that in an ideal transformer, voltages are transformed in the direct ratio of turns, currents in the inverse ratio, and impedances in the direct ratio squared; and power and volt-amperes are, of course, unchanged. Note that, a balanced delta-connected circuit of Z_Δ Ω/phase is equivalent to a balanced wye-connected circuit of Z_Y Ω/phase as long as

$$Z_Y = \frac{1}{3} Z_\Delta \tag{2.90}$$

The per-unit impedance of a transformer remains the same without taking into account whether it is converted from physical impedance values that are found by referring to the high-voltage side or low-voltage side of the transformer. This can be accomplished by choosing separate appropriate bases for each side of the transformer (whether or not the transformer is connected in wye–wye, delta–delta, delta–wye, or wye–delta since the transformation of voltages is the same as that made by wye–wye transformers as long as the same line-to-line voltage ratings are used).* In other words, the designated per-unit impedance values of transformers are based on the coil ratings.

Since the ratings of coils cannot alter by a simple change in connection (e.g., from wye–wye to delta–wye), the per-unit impedance remains the same regardless of the three-phase connection. The line-to-line voltage for the transformer will differ. Because of the method of choosing the base in various sections of the three-phase system, the per-unit impedances calculated in various sections can be put together on one impedance diagram without paying any attention to whether the transformers are connected in wye–wye or delta–wye.

---

* This subject has been explained in greater depth in an excellent review by Stevenson [3].

Basic Concepts

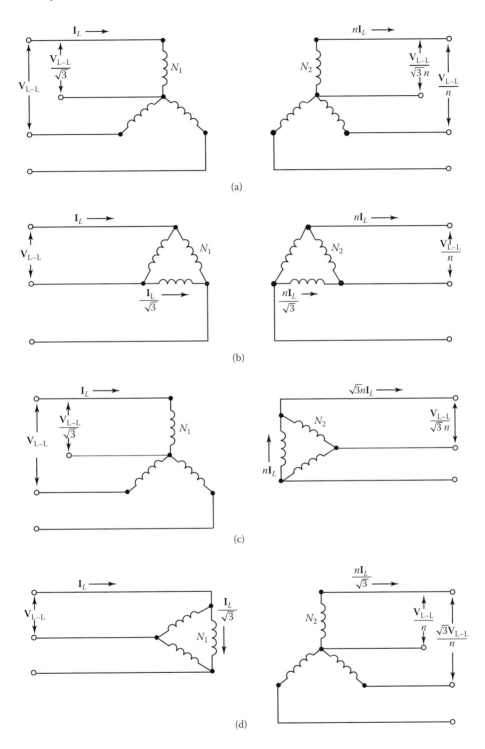

FIGURE 2.3 Conventional three-phase transformer connections: (a) wye–wye connection; (b) delta–delta connection; (c) wye–delta connection; (d) delta–wye connection.

EXAMPLE 2.2

A three-phase transformer has a nameplate rating of 30 MVA, 230Y/69Y kV with a leakage reactance of 10% and the transformer connection is wye–wye. Select a base of 30 MVA and 230 kV on the high-voltage side and determine the following:

(a) Reactance of transformer in per units
(b) High-voltage side base impedance
(c) Low-voltage-side base impedance
(d) Transformer reactance referred to the high-voltage side in ohms
(e) Transformer reactance referred to the low-voltage side in ohms

Solution

(a) The reactance of the transformer in per units is 10/100, or 0.10 pu. Note that, it is the same whether it is referred to the high-voltage or the low-voltage side.

(b) The high-voltage side base impedance is

$$Z_{B(HV)} = \frac{[kV_{B(HV)}]^2}{MVA_{B(3\phi)}}$$

$$= \frac{230^2}{30} = 1763.3333 \ \Omega$$

(c) The low-voltage side base impedance is

$$Z_{B(LV)} = \frac{[kV_{B(LV)}]^2}{MVA_{B(3\phi)}}$$

$$= \frac{69^2}{30} = 158.7 \ \Omega$$

(d) The reactance referred to the high-voltage side is

$$X_{\Omega(HV)} = X_{pu} \times X_{B(HV)}$$

$$= (0.10)(1763.3333) = 176.3333 \ \Omega$$

(e) The reactance referred to the low-voltage side is

$$X_{\Omega(LV)} = X_{pu} \times X_{B(LV)}$$

$$= (0.10)(158.7 \ \Omega) = 15.87 \ \Omega$$

or, from Equation 2.89,

$$X_{\Omega(LV)} = \frac{X_{\Omega(HV)}}{n^2}$$

$$= \frac{176.3333 \ \Omega}{\left(\frac{230/\sqrt{3}}{69/\sqrt{3}}\right)^2} = \frac{176.3333 \ \Omega}{(3.3333)^2} \cong 15.87 \ \Omega$$

where n is defined as the turns ratio of the windings.

Basic Concepts

EXAMPLE 2.3

A three-phase transformer has a nameplate rating of 30 MVA, and the voltage ratings of 230Y kV/69Δ kV with a leakage reactance of 10% and the transformer connection is wye–delta. Select a base of 30 MVA and 230 kV on the high-voltage side and determine the following:

(a) Turns ratio of windings
(b) Transformer reactance referred to the low-voltage side in ohms
(c) Transformer reactance referred to the low-voltage side in per units

Solution

(a) The turns ratio of the windings is

$$n \triangleq \frac{V_{HV(\phi)}}{V_{LV(\phi)}} = \frac{230/\sqrt{3}}{69} = 1.9245$$

(b) Since the high-voltage side impedance base is

$$Z_{B(HV)} = \frac{[kV_{B(HV)}]^2}{MVA_{B(3\phi)}}$$

$$= \frac{[230\,kV]^2}{30} = 1763.3333\,\Omega$$

and

$$X_{\Omega(HV)} = X_{pu} \times X_{B(HV)}$$

$$= (0.10)(1763.3333\,\Omega) = 176.3333\,\Omega$$

Thus, the transformer reactance referred to the delta-connected low-voltage side is

$$X_{\Omega(LV)} = \frac{X_{\Omega(HV)}}{n^2}$$

$$= \frac{176.3333}{(1.9245)^2} = 47.61\,\Omega = X_\Delta$$

(c) From Equation 2.90, the reactance of the equivalent wye connection is

$$Z_Y = \frac{Z_\Delta}{3}$$

$$= \frac{47.61\,\Omega}{3} = 15.87\,\Omega = X'_{\Omega(LV)}$$

where $X'_{\Omega(LV)}$ = reactance per phase at a low voltage of the equivalent wye.
Similarly,

$$Z_{B(LV)} = \frac{[kV_{B(LV)}]^2}{MVA_{B(3\phi)}}$$

$$= \frac{69^2}{30} = 158.7 \, \Omega$$

Thus,

$$X_{pu} = \frac{X'_{\Omega(LV)}}{Z_{B(LV)}}$$

$$= \frac{15.87 \, \Omega}{158.7 \, \Omega} = 0.10 \, pu$$

Alternatively, if the line-to-line voltages are used,

$$X_{(LV)} = \frac{X_{\Omega(HV)}}{n^2}$$

$$= \frac{176.3333 \, \Omega}{\left(\dfrac{230}{69}\right)^2} = 15.87 \, \Omega$$

and thus,

$$X_{pu} = \frac{X'_{\Omega(LV)}}{Z_{B(LV)}}$$

$$= \frac{15.87 \, \Omega}{158.7 \, \Omega} = 0.10 \, pu$$

as before.

EXAMPLE 2.4

Consider the previous example that a three-phase transformer has a nameplate rating of 30 MVA, and voltage ratings of 230Y kV/69Δ kV with a leakage reactance of 10%. Now, assume that this transformer connection is not delta–wye but wye–wye. Select a base of 30 MVA and 230 kV on the high-voltage side for the wye–delta transformer and solve it by converting to its equivalent wye–wye connection first and determine the following:

(a) Turns ratio of windings
(b) Transformer reactance referred to the low-voltage side in ohms
(c) Transformer reactance referred to the low-voltage side in per units

Solution

First converting the delta low-voltage to its corresponding wye low-voltage as

$$\sqrt{3}(69 \, kV) = 119.5115 \, kV$$

Basic Concepts

(a) The turns ratio of the windings is

$$n \triangleq \frac{V_{HV(\phi)}}{V_{LV(\phi)}} = \frac{230/\sqrt{3}}{119.5115/\sqrt{3}} = \frac{230}{119.5115} = 1.9245$$

(b) Since the high-voltage side impedance base is

$$Z_{B(HV)} = \frac{[kV_{B(HV)}]^2}{MVA_{B(3\phi)}}$$

$$= \frac{[230\,kV]^2}{30} = 1763.3333\,\Omega$$

and

$$X_{\Omega(HV)} = X_{pu} \times X_{B(HV)}$$

$$= (0.10)(1763.3333\,\Omega) = 176.3333\,\Omega$$

Thus, the transformer reactance referred to the delta-connected low-voltage side is

$$X_{\Omega(LV)} = \frac{X_{\Omega(HV)}}{n^2}$$

$$= \frac{176.3333}{(1.9245)^2} = 47.61\,\Omega = X_\Delta$$

(c) From Equation 2.90, the reactance of the equivalent wye connection is

$$Z_Y = \frac{Z}{3}$$

$$= \frac{47.61\,\Omega}{3} = 15.87\,\Omega = X'_{\Omega(LV)}$$

where $X'_{\Omega(LV)}$ = reactance per phase at low-voltage of equivalent wye.
Similarly,

$$Z'_{B(LV)} = \frac{[kV_{B(LV)}]^2}{MVA_{B(3\phi)}}$$

$$= \frac{(119.5115)^2}{30} = 476.1\,\Omega$$

Thus,

$$X_{pu} = \frac{X_{\Omega(LV)}}{Z'_{B(LV)}}$$

$$= \frac{47.61\,\Omega}{476.1\,\Omega} = 0.10\,pu$$

EXAMPLE 2.5

Resolve Example 2.3 but violate the definition of turns ratio. Use it as the ratio of the line-to-line voltage of the wye-connected primary voltage to the line-to-line voltage of the delta-connected secondary voltage. Since the transformer is rated as 230Y kV/69Δ kV, solve it without converting to its equivalent wye–wye connection first and determine the following:

(a) Turns ratio of windings
(b) Transformer reactance referred to the low-voltage side in ohms
(c) Transformer reactance referred to the low-voltage side in per units

Solution

(a) The turns ratio of the windings is

$$n \triangleq \frac{V_{HV(L-L)}}{V_{LV(L-V)}} = \frac{230}{69} = 3.3333$$

Here, of course, the definition of the turns ratio has been violated.

(b) Since the high-voltage side impedance base is

$$Z_{B(HV)} = \frac{[kV_{B(HV)}]^2}{MVA_{B(3\phi)}}$$

$$= \frac{[230\,kV]^2}{30} = 1763.3333\,\Omega$$

and

$$X_{\Omega(HV)} = X_{pu} \times X_{B(HV)}$$

$$= (0.10)(1763.3333\,\Omega) = 176.3333\,\Omega$$

Thus, the transformer reactance referred to the delta-connected low-voltage side is

$$X_{\Omega(LV)} = \frac{X_{\Omega(HV)}}{n^2}$$

$$= \frac{176.3333}{(3.3333)^2} = 15.8703\,\Omega = X_\Delta$$

(c) From Equation 2.90, the reactance of the delta connection is

$$Z_{B(LV)} = \frac{[kV_{B(LV)}]^2}{MVA_{B(3\phi)}}$$

$$= \frac{(69)^2}{30} = 158.7\,\Omega$$

Basic Concepts

Thus,

$$X_{pu} = \frac{X_{\Omega(LV)}}{Z_{B(LV)}}$$

$$= \frac{15.8703\,\Omega}{158.7\,\Omega} = 0.10\,\text{pu}$$

This method is obviously a shortcut, but one should apply it carefully.

EXAMPLE 2.6

Figure 2.4 shows a one-line diagram of a three-phase system. Assume that the line length between the two transformers is negligible and the three-phase generator is rated 4160 kVA, 2.4 kV, and 1000 A and that it supplies a purely inductive load of $I_{pu} = 2.08\angle{-90°}$ pu. The three-phase transformer T_1 is rated 6000 kVA, 2.4Y–24Y kV, with leakage reactance of 0.04 pu. Transformer T_2 is made up of three single-phase transformers and is rated 4000 kVA, 24Y–12Y kV, with leakage reactance of 0.04 pu. Determine the following for all three circuits, 2.4-, 24-, and 12-kV circuits:

(a) Base kVA values
(b) Base line-to-line kilovolt values
(c) Base impedance values
(d) Base current values
(e) Physical current values (neglect magnetizing currents in transformers and charging currents in lines)
(f) Per-unit current values
(g) New transformer reactances based on their new bases
(h) Per-unit voltage values at buses 1, 2, and 4
(i) Per-unit apparent power values at buses 1, 2, and 4
(j) Summarize results in a table

Solution

(a) The kVA base for all three circuits is arbitrarily selected as 2080 kVA.
(b) The base voltage for the 2.4-kV circuit is arbitrarily selected as 2.5 kV. Since the turns ratios for transformers T_1 and T_2 are

$$\frac{N_1}{N_2} = 10 \quad \text{or} \quad \frac{N_2}{N_1} = 0.10$$

and

$$\frac{N_1'}{N_2'} = 2$$

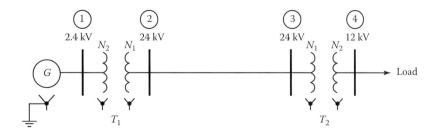

FIGURE 2.4 One-line diagram for Example 2.4.

the base voltages for the 24- and 12-kV circuits are determined to be 25 and 12.5 kV, respectively.

(c) The base impedance values can be found as

$$Z_B = \frac{[kV_{B(L-L)}]^2(1000)}{kVA_{B(3\phi)}}$$

$$= \frac{[2.5\,kV]^2 1000}{2080\,kVA} = 3.005\,\Omega$$

and

$$Z_B = \frac{[25\,kV]^2 1000}{2080\,kVA} = 300.5\,\Omega$$

and

$$Z_B = \frac{[12.5\,kV]^2 1000}{2080\,kVA} = 75.1\,\Omega$$

(d) The base current values can be determined as

$$I_B = \frac{kVA_{B(3\phi)}}{\sqrt{3}kV_{B(L-L)}}$$

$$= \frac{2080\,kVA}{\sqrt{3}(2.5\,kV)} = 480\,A$$

and

$$I_B = \frac{2080\,kVA}{\sqrt{3}(25\,kV)} = 48\,A$$

and

$$I_B = \frac{2080\,kVA}{\sqrt{3}(12.5\,kV)} = 96\,A$$

(e) The physical current values can be found based on the turns ratios as

$$I = 1000\,A$$

$$I = \left(\frac{N_2}{N_1}\right)(1000\,A) = 100\,A$$

$$I = \left(\frac{N_1'}{N_2'}\right)(100\,A) = 200\,A$$

(f) The per-unit current value is the same, 2.08 pu, for all three circuits.

(g) The given transformer reactances can be converted on the basis of their new bases using

$$Z_{pu(new)} = Z_{pu(given)}\left[\frac{kVA_{B(new)}}{kVA_{B(given)}}\right]\left[\frac{kV_{B(given)}}{kV_{B(new)}}\right]^2$$

Basic Concepts

Therefore, the new reactances of the two transformers can be found as

$$Z_{pu(T_1)} = j0.04 \left[\frac{2080\,\text{kVA}}{6000\,\text{kVA}} \right] \left[\frac{2.4\,\text{kV}}{2.5\,\text{kV}} \right]^2 = j0.0128\,\text{pu}$$

and

$$Z_{pu(T_2)} = j0.04 \left[\frac{2080\,\text{kVA}}{4000\,\text{kVA}} \right] \left[\frac{12\,\text{kV}}{12.5\,\text{kV}} \right]^2 = j0.0192\,\text{pu}$$

(h) Therefore, the per-unit voltage values at buses 1, 2, and 4 can be calculated as

$$V_1 = \frac{2.4\,\text{kV}\angle 0°}{2.5\,\text{kV}} = 0.96\angle 0°\,\text{pu}$$

$$V_2 = V_1 - I_{pu} Z_{pu(T_1)}$$
$$= 0.96\angle 0° - (2.08\angle -90°)(0.0128\angle 90°) = 0.9334\angle 0°\,\text{pu}$$

$$V_4 = V_2 - I_{pu} Z_{pu(T_2)}$$
$$= 0.9334\angle 0° - (2.08\angle -90°)(0.0192\angle 90°) = 0.8935\angle 0°\,\text{pu}$$

(i) Thus, the per-unit apparent power values at buses 1, 2, and 4 are

$$S_1 = 2.00\,\text{pu}$$

$$S_2 = V_2 I_{pu} = (0.9334)(2.08) = 1.9415\,\text{pu}$$

$$S_4 = V_4 I_{pu} = (0.8935)(2.08) = 1.8585\,\text{pu}$$

(j) The results are summarized in Table 2.2.

TABLE 2.2
Results of Example 2.4

Quantity	2.4-kV Circuit	24-kV Circuit	12-kV Circuit
$kVA_{B(3\phi)}$	2080 kVA	2080 kVA	2080 kVA
$kV_{B(L-L)}$	2.5 kV	25 kV	12.5 kV
Z_B	3.005 Ω	300.5 Ω	75.1 Ω
I_B	480 A	48 A	96 A
$I_{physical}$	1000 A	100 A	200 A
I_{pu}	2.08 pu	2.08 pu	2.08 pu
V_{pu}	0.96 pu	0.9334 pu	0.8935 pu
S_{pu}	2.00 pu	1.9415 pu	1.8585 pu

2.5 CONSTANT IMPEDANCE REPRESENTATION OF LOADS

Usually, the power system loads are represented by their real and reactive powers, as shown in Figure 2.5a. However, it is possible to represent the same load in terms of series or parallel combinations of its equivalent constant-load resistance and reactance values, as shown in Figure 2.5b and 2.5c, respectively [4].

In the event that the load is represented by the series connection, the equivalent constant impedance can be expressed as

$$\mathbf{Z}_s = R_s + jX_s \tag{2.91}$$

where

$$R_s = \frac{|\mathbf{V}|^2 \times P}{P^2 + Q^2} \tag{2.92}$$

$$X_s = \frac{|\mathbf{V}|^2 \times Q}{P^2 + Q^2} \tag{2.93}$$

where
- R_s = load resistance in series connection in ohms
- X_s = load reactance in series connection in ohms
- $\mathbf{Z}_s$ = constant-load impedance in ohms
- V = load voltage in volts
- P = real, or average, load power in watts
- Q = reactive load power in vars

The constant impedance in per units can be expressed as

$$\mathbf{Z}_{pu(s)} = R_{pu(s)} + jX_{pu(s)} \text{ pu} \tag{2.94}$$

where

$$R_{pu(s)} = (P_{\text{physical}}) \frac{S_B \times (V_{pu})^2}{P^2 + Q^2} \text{ pu} \tag{2.95}$$

and

$$X_{pu(s)} = (Q_{\text{physical}}) \frac{S_B \times (V_{pu})^2}{P^2 + Q^2} \text{ pu} \tag{2.96}$$

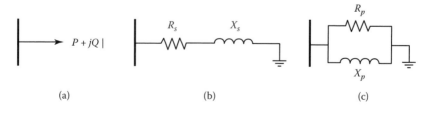

FIGURE 2.5 Load representations as: (a) real and reactive powers; (b) constant impedance in terms of series combination; (c) constant impedance in terms of parallel combination.

Basic Concepts

If the load is represented by the parallel connection, the equivalent constant impedance can be expressed as

$$\mathbf{Z}_p = j \frac{R_p \times X_p}{R_p + X_p} \tag{2.97}$$

where

$$R_p = \frac{V^2}{P}$$

and

$$X_p = \frac{V^2}{Q}$$

where
R_p = load resistance in parallel connection in ohms
X_p = load reactance in parallel connection in ohms
$\mathbf{Z}_p$ = constant-load impedance in ohms

The constant impedance in per units can be expressed as

$$Z_{\text{pu}(p)} = j \frac{R_{\text{pu}(p)} \times X_{\text{pu}(p)}}{R_{\text{pu}(p)} + X_{\text{pu}(p)}} \quad \text{pu} \tag{2.98}$$

where

$$\mathbf{R}_{\text{pu}(p)} = \frac{S_B}{P} \left(\frac{V}{V_B}\right)^2 \quad \text{pu} \tag{2.99}$$

or

$$R_{\text{pu}(p)} = \frac{V_{\text{pu}}^2}{P_{\text{pu}}} \quad \text{pu} \tag{2.100}$$

and

$$X_{\text{pu}(p)} = \frac{S_B}{Q} \left(\frac{V}{V_B}\right)^2 \quad \text{pu} \tag{2.101}$$

or

$$X_{\text{pu}(p)} = \frac{V_{\text{pu}}^2}{Q_{\text{pu}}} \quad \text{pu} \tag{2.102}$$

2.6 THREE-WINDING TRANSFORMERS

Figure 2.6a shows a single-phase three-winding transformer. They are usually used in the bulk power (transmission) substations to reduce the transmission voltage to the subtransmission voltage level. If excitation impedance is neglected, the equivalent circuit of a three-winding transformer can be represented by a wye of impedances, as shown in Figure 2.6b, where the primary, secondary, and tertiary windings are denoted by *P*, *S*, and *T*, respectively.

Note that, the common point 0 is fictitious and is not related to the neutral of the system. The tertiary windings of a three-phase and three-winding transformer bank is usually connected in delta and may be used for (1) providing a path for zero-sequence currents, (2) in-plant power distribution, and (3) application of power-factor-correcting capacitors or reactors. The impedance of any of the branches shown in Figure 2.6b can be determined by considering the short-circuit impedance between pairs of windings with the third open. Therefore,

$$Z_{PS} = Z_P + Z_S \qquad (2.103a)$$

$$Z_{TS} = Z_T + Z_S \qquad (2.103b)$$

$$Z_{PT} = Z_P + Z_T \qquad (2.103c)$$

where

$$Z_P = \frac{1}{2}(Z_{PS} + Z_{PT} - Z_{TS}) \qquad (2.104a)$$

$$Z_S = \frac{1}{2}(Z_{PS} + Z_{TS} - Z_{PT}) \qquad (2.104b)$$

$$Z_T = \frac{1}{2}(Z_{PT} + Z_{TS} - Z_{PS}) \qquad (2.104c)$$

where

Z_{PS} = leakage impedance measured in primary with secondary short-circuited and tertiary open
Z_{PT} = leakage impedance measured in primary with tertiary short-circuited and secondary open
Z_{ST} = leakage impedance measured in secondary with tertiary short-circuited and primary open

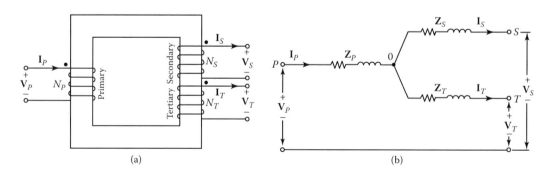

FIGURE 2.6 Single-phase, three-winding transformer: (a) winding diagram; (b) equivalent circuit.

Basic Concepts

Z_P = impedance of primary winding
Z_S = impedance of secondary winding
Z_T = impedance of tertiary winding

In most large transformers, the value of Z_S is very small and can be negative. Contrary to the situation with a two-winding transformer, the kVA ratings of the three windings of a three-winding transformer bank are not usually equal. Therefore, all impedances, as defined above, should be expressed on the same kVA base. For three-winding three-phase transformer banks with delta- or wye-connected windings, the positive- and negative-sequence diagrams are always the same. The corresponding zero-sequence diagrams are shown in Figure 5.10.

2.7 AUTOTRANSFORMERS

Figure 2.7a shows a two-winding transformer. Viewed from the terminals, the same transformation of voltages, currents, and impedances can be obtained with the connection shown in Figure 2.7b. Therefore, in the autotransformer, only one winding is used per phase, the secondary voltage being tapped off the primary winding, as shown in Figure 2.7b. The *common winding* is the winding between the low-voltage terminals, whereas the remainder of the winding, belonging exclusively to the high-voltage circuit, is called the *series winding* and, combined with the common winding, forms the *series-common winding* between the high-voltage terminals.

In a sense, an autotransformer is just a normal two-winding transformer connected in a special way. The only structural difference is that the series winding must have extra insulation. In a *variable autotransformer*, the tap is movable. Autotransformers are increasingly used to interconnect two high-voltage transmission lines operating at different voltages. An autotransformer has two separate sets of ratios, namely, circuit ratios and winding ratios. For circuit ratios, consider the equivalent circuit of an ideal autotransformer (neglecting losses) shown in Figure 2.7b. Viewed from the terminals, the voltage and current ratios can be expressed as

$$a = \frac{V_1}{V_2} \quad (2.105a)$$

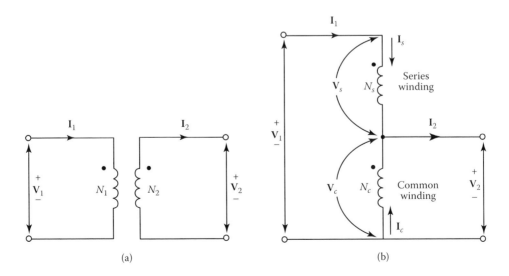

FIGURE 2.7 Schematic diagram of ideal (step-down) transformer connected as: (a) two-winding transformer; (b) autotransformer.

$$a = \frac{N_c + N_s}{N_c} \tag{2.105b}$$

$$= \frac{N_c + N_s}{N_c} \tag{2.105c}$$

and

$$a = \frac{I_2}{I_2} \tag{2.106}$$

From Equation 2.105c, it can be observed that the ratio a is always larger than 1.

For winding ratios, consider the voltages and currents of the series and common windings, as shown in Figure 2.7b. Therefore, the voltage and current ratios can be expressed as

$$\frac{V_s}{V_c} = \frac{N_s}{N_c} \tag{2.107}$$

and

$$\frac{I_c}{I_s} = \frac{I_2 - I_1}{I_1} \tag{2.108a}$$

$$= \frac{I_2}{I_1} - 1 \tag{2.108b}$$

From Equation 2.105c,

$$\frac{N_s}{N_c} = a - 1 \tag{2.109}$$

Therefore, substituting Equation 2.109 into Equation 2.107 yields

$$\frac{V_s}{V_c} = a - 1 \tag{2.110}$$

Similarly, substituting Equations 2.106 and 2.109 into Equation 2.108b simultaneously yields

$$\frac{I_c}{I_s} = a - 1 \tag{2.111}$$

For an ideal autotransformer, the volt-ampere ratings of circuits and windings can be expressed, respectively, as

Basic Concepts

$$S_{circuits} = V_1 I_1 = V_2 I_2 \tag{2.112}$$

and

$$S_{windings} = V_s I_s = V_c I_c \tag{2.113}$$

The advantages of autotransformers are lower leakage reactances, lower losses, smaller exciting currents, and less cost than two-winding transformers when the voltage ratio does not vary too greatly from 1 to 1. For example, if the same core and coils are used as a two-winding transformer and as an autotransformer, the ratio of the capacity as an autotransformer to the capacity as a two-winding transformer can be expressed as

$$\frac{\text{Capacity as autotransformer}}{\text{Capacity as two-winding transformer}} = \frac{V_1 I_1}{V_s I_s} = \frac{V_1 I_1}{(V_1 - V_2) I_1} = \frac{a}{a-1} \tag{2.114}$$

Therefore, maximum advantage is obtained with a relatively small difference between the voltages on the two sides (e.g., 161 kV/138 kV, 500 kV/700 kV, and 500 kV/345 kV). Therefore, a large saving in size, weight, and cost can be achieved over a two-windings per-phase transformer. The disadvantages of an autotransformer are that there is no electrical isolation between the primary and secondary circuits and there is a greater short-circuit current than the one for the two-winding transformer.

Three-phase autotransformer banks generally have wye-connected main windings, the neutral of which is normally connected solidly to the earth. In addition, it is common practice to include a third winding connected in delta, called the tertiary winding.

2.8 DELTA–WYE AND WYE–DELTA TRANSFORMATIONS

The three-terminal circuits encountered so often in networks are the delta and wye* configurations, as shown in Figure 2.8. In some problems, it is necessary to convert delta to wye or vice versa. If the impedances $\mathbf{Z}_{ab}$, $\mathbf{Z}_{bc}$, and $\mathbf{Z}_{ca}$ are connected in delta, the equivalent wye impedances $\mathbf{Z}_a$, $\mathbf{Z}_b$, and $\mathbf{Z}_c$ are

$$\mathbf{Z}_a = \frac{\mathbf{Z}_{ab}\mathbf{Z}_{ca}}{\mathbf{Z}_{ab} + \mathbf{Z}_{bc} + \mathbf{Z}_{ca}} \tag{2.115}$$

$$\mathbf{Z}_b = \frac{\mathbf{Z}_{ab}\mathbf{Z}_{bc}}{\mathbf{Z}_{ab} + \mathbf{Z}_{bc} + \mathbf{Z}_{ca}} \tag{2.116}$$

$$\mathbf{Z}_c = \frac{\mathbf{Z}_{bc}\mathbf{Z}_{ca}}{\mathbf{Z}_{ab} + \mathbf{Z}_{bc} + \mathbf{Z}_{ca}} \tag{2.117}$$

If $\mathbf{Z}_{ab} = \mathbf{Z}_{bc} = \mathbf{Z}_{ca} = \mathbf{Z}$,

$$\mathbf{Z}_a = \mathbf{Z}_b = \mathbf{Z}_c = \frac{\mathbf{Z}}{3} \tag{2.118}$$

---

* In Europe, it is called the *star configuration*.

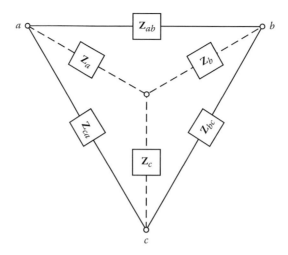

FIGURE 2.8 Delta–wye or wye–delta transformations.

On the other hand, if the impedances $\mathbf{Z}_a$, $\mathbf{Z}_b$, and $\mathbf{Z}_c$ are connected in wye, the equivalent delta impedances $\mathbf{Z}_{ab}$, $\mathbf{Z}_{bc}$, and $\mathbf{Z}_{ca}$ are

$$\mathbf{Z}_{ab} = \mathbf{Z}_a + \mathbf{Z}_b + \frac{\mathbf{Z}_a \mathbf{Z}_b}{\mathbf{Z}_c} \tag{2.119}$$

$$\mathbf{Z}_{bc} = \mathbf{Z}_b + \mathbf{Z}_c + \frac{\mathbf{Z}_b \mathbf{Z}_c}{\mathbf{Z}_a} \tag{2.120}$$

$$\mathbf{Z}_{ca} = \mathbf{Z}_c + \mathbf{Z}_a + \frac{\mathbf{Z}_c \mathbf{Z}_a}{\mathbf{Z}_b} \tag{2.121}$$

If $\mathbf{Z}_a = \mathbf{Z}_b = \mathbf{Z}_c = \mathbf{Z}$, then

$$\mathbf{Z}_{ab} = \mathbf{Z}_{bc} = \mathbf{Z}_{ca} = 3\mathbf{Z} \tag{2.122}$$

2.9 SHORT-CIRCUIT MVA AND EQUIVALENT IMPEDANCE

Often, when a new circuit is to be added to an existing bus in a complex power system, the short-circuit MVA (or kVA) for that bus has to be known or determined. Such short-circuit MVA (or kVA) data provide the equivalent impedance of the existing power system up to that bus. After determination of the short-circuit MVA, one can easily find the associated short-circuit impedance of the system. The short-circuit MVA is found for both three-phase faults and for single line-to-ground faults, separately.

Basic Concepts

2.9.1 Three-Phase Short-Circuit MVA

At a given three-phase bus, three-phase short-circuit MVA can be determined from the following equation:

$$\text{MVA}_{sc(3\phi)} = \frac{\sqrt{3}(kV_{L-L})I_{f(3\phi)}}{1000} \quad (2.123)$$

where
    kV_{L-L} = system line-to-line voltage in kV (rated system voltage)
    $I_{f(3\phi)}$ = total three-phase fault current in amperes

2.9.1.1 If Three-Phase Short-Circuit MVA Is Already Known

Then, the three-phase short-circuit fault current can be determined from

$$I_{f(3\phi)} = \frac{1000\,\text{MVA}_{sc(3\phi)}}{\sqrt{3}kV_{(L-L)}} \quad \text{A} \quad (2.124)$$

then the equivalent impedance can be found as

$$Z_{sc} = \frac{V_{L-N}}{I_{f(3\phi)}} \quad \Omega \quad (2.125a)$$

or

$$Z_{sc} = \frac{1000 kV_{L-L}}{\sqrt{3} I_{f(3\phi)}} \quad \Omega \quad (2.125b)$$

or

$$Z_{sc} = \frac{kV_{L-L}^2}{\text{MVA}_{sc(3\phi)}} \quad \Omega \quad (2.125c)$$

Since base impedance is found from

$$Z_B = \frac{kV_{B(3\phi)}^2}{\text{MVA}_{B(3\phi)}} \quad (2.126)$$

then the per unit impedance can found from

$$Z_{pu} = \frac{Z_{\Omega(sc)}}{Z_B} \quad (2.127a)$$

or

$$Z_{pu} = \frac{MVA_{3\phi} \times Z_{sc}}{kV_{B(L-L)}^2} \tag{2.127b}$$

Thus, the positive-sequence Z to the fault location can be found as

$$Z_{1(pu)} = \frac{MVA_{B(3\phi)}}{MVA_{sc}} \text{ pu} \tag{2.128}$$

In general,

$$Z_1 = Z_2 \tag{2.129}$$

Also, it is assumed that $Z_1 = X_1$ unless the X/R ratio of the system is known so that the angle can be found.

2.9.2 Single-Phase-to-Ground Short-Circuit MVA

At a given single-phase bus, single-phase-to-ground short-circuit MVA can be determined from the following equation:

$$MVA_{f(SLG)SC} = \frac{\sqrt{3}(kV_{L-L})I_{f(SLG)}}{1000} \tag{2.130}$$

where
kV_{L-L} = system line-to-line voltage in kV (rated system voltage)
$I_{f(SLG)}$ = total single-phase fault current in amperes

2.9.2.1 If Single-Phase Short-Circuit MVA Is Already Known

Then the single-phase-to-ground short-circuit fault current can be determined from

$$I_{f(SLG)} = \frac{1000 \, MVA_{f(SLG)SC}}{\sqrt{3}kV_{(L-L)}} \text{ A} \tag{2.131}$$

But,

$$I_{f(SLG)} = I_{a0} + I_{a1} + I_{a2} \tag{2.132}$$

or

$$I_{f(SLG)} = \frac{3V_{L-N}}{Z_0 + Z_1 + Z_2} \tag{2.133}$$

or

$$I_{f(SLG)} = \frac{3V_{L-N}}{Z_G} \tag{2.134}$$

Basic Concepts

where

$$Z_G = Z_0 + Z_1 + Z_2 \tag{2.135}$$

From Equations 2.125c and 2.127,

$$Z_G = \frac{3kV_{L\text{-}L}^2}{\text{MVA}_{f(\text{SLG})\text{SC}}} \quad \Omega \tag{2.136}$$

and

$$Z_G = \frac{3\text{MVA}_B}{\text{MVA}_{f(\text{SLG})\text{SC}}} \quad \text{pu} \tag{2.137}$$

From

$$Z_0 = Z_G - Z_1 - Z_2 \tag{2.138}$$

or in general,

$$X_0 = X_G - X_1 - X_2 \tag{2.139}$$

since the resistance involved is usually very small with respect to the associated reactance value.

EXAMPLE 2.7

A short-circuit (fault) study shows that at bus 15 in a 132-kV system, on a 100 MVA base, short-circuit MVA is 710 MVA and the single-line-to-ground short-circuit MVA is 825 MVA. Determine the following:

(a) The positive and negative reactances of the system
(b) The X_G of the system
(c) The zero-sequence reactance of the system

Solution

(a) The positive- and negative-reactances of the system are

$$X_1 = X_2 = \frac{100 \text{ MVA}}{710 \text{ MVA}} = 0.1408 \text{ pu}$$

(b) The X_G of the system is

$$X_G = \frac{300 \text{ MVA}}{825 \text{ MVA}} = 0.3636 \text{ pu}$$

(c) The zero-sequence of the system is

$$X_0 = 0.3636 - 0.1408 = 0.2228 \text{ pu}$$

Note that, all values above are on a 100-MVA, 132-kV base.

REFERENCES

1. Elgerd, O. I., *Electric Energy Systems Theory: An Introduction*. McGraw-Hill, New York, 1971.
2. Institute of Electrical and Electronics Engineers, *Graphic Symbols for Electrical and Electronics Diagrams*, IEEE Std. 315-1971 [or American National Standards Institute (ANSI) Y32.2-1971]. IEEE, New York, 1971.
3. Stevenson, W. D., *Elements of Power System Analysis*, 4th ed. McGraw-Hill, New York, 1981.
4. Anderson, P. M., *Analysis of Faulted Power Systems*. Iowa State Univ. Press, Ames, IA, 1973.

GENERAL REFERENCES

AIEE Standards Committee Report, *Electr. Eng. (Am. Inst. Electr. Eng.)* 65, 512 (1946).
Clarke, E., *Circuit Analysis of AC Power Systems*, Vol. 1, General Electric Co., Schenectady, New York, 1943.
Eaton, J. R., *Electric Power Transmission Systems*. Prentice-Hall, Englewood Cliffs, NJ, 1972.
Elgerd, O. I., *Basic Electric Power Engineering*. Addison-Wesley, Reading, MA, 1977.
Gönen, T., *Electric Power Distribution System Engineering*, 2nd ed., CRC Press, Boca Raton, FL, 2008.
Gross, C. A., *Power System Analysis*. Wiley, New York, 1979.
Gross, E. T. B., and Gulachenski, E. M., Experience of the New England Electric Company with generator protection by resonant neutral grounding. *IEEE Trans. Power Appar. Syst.* PAS-92 (4) 1186–1194 (1973).
Neuenswander, J. R., *Modern Power Systems*. International Textbook Co., New York, 1971.
Nilsson, J. W., *Introduction to Circuits, Instruments, and Electronics*. Harcourt, Brace & World, New York, 1968.
Skilling, H. H., *Electrical Engineering Circuits*, 2nd ed. Wiley, New York, 1966.
Travis, I., Per unit quantities. *Trans. Am. Inst. Electr. Eng.* 56, 143–151 (1937).
Wagner, C. F., and Evans, R. D., *Symmetrical Components*. McGraw-Hill, New York, 1933.
Weedy, B. M., *Electric Power Systems*, 3rd ed. Wiley, New York, 1979.
Zaborsky, J., and Rittenhouse, J. W., *Electric Power Transmission*. Rensselaer Bookstore, Troy, New York, 1969.

PROBLEMS

1. Assume that the impedance of a line connecting buses 1 and 2 is $50\angle 90°$ Ω and that the bus voltages are $7560\angle 10°$ and $7200\angle 0°$ V per phase, respectively. Determine the following:
   (a) Real power per phase that is being transmitted from bus 1 to bus 2
   (b) Reactive power per phase that is being transmitted from bus 1 to bus 2
   (c) Complex power per phase that is being transmitted
2. Solve Problem 1 assuming that the line impedance is $50\angle 26°$ Ω/phase.
3. Verify the following equations:
   (a) $V_{pu(L-N)} = V_{pu(L-L)}$
   (b) $VA_{pu(1\phi)} = VA_{pu(3\phi)}$
   (c) $Z_{pu(Y)} = Z_{pu(\Delta)}$
4. Verify the following equations:
   (a) Equation 2.24 for a single-phase system
   (b) Equation 2.80 for a three-phase system
5. Show that $Z_{B(\Delta)} = 3Z_{B(Y)}$.
6. Consider two three-phase transmission lines with different voltage levels that are located side by side in close proximity. Assume that the bases of VA_B, $V_{B(1)}$, and $I_{B(1)}$ and the bases of VA_B, $V_{B(2)}$, and $I_{B(2)}$ are designated for the first and second lines, respectively. If the mutual reactance between the lines is X_m Ω, show that this mutual reactance in per unit can be expressed as

$$X_{pu(m)} = (\text{physical } X_m) \frac{MVA_B}{[kV_{B(1)}][kV_{B(2)}]}$$

7. Consider Example 2.3 and assume that the transformer is connected in delta–wye. Use a 25-MVA base and determine the following:
   (a) New line-to-line voltage of low-voltage side
   (b) New low-voltage side base impedance
   (c) Turns ratio of windings
   (d) Transformer reactance referred to the low-voltage side in ohms
   (e) Transformer reactance referred to the low-voltage side in per units
8. Verify the following equations:
   (a) Equation 2.92
   (b) Equation 2.93
   (c) Equation 2.94
   (d) Equation 2.96
9. Verify the following equations:
   (a) Equation 2.100
   (b) Equation 2.102
10. Consider the one-line diagram given in Figure P2.1. Assume that the three-phase transformer T_1 has nameplate ratings of 15,000 kVA, 7.97/13.8Y – 69Δ kV with leakage impedance of $0.01 + j0.08$ pu based on its ratings, and that the three-phase transformer T_2 has nameplate ratings of 1500 kVA, 7.97Δ kV – 277/480Y V with leakage impedance of $0.01 + j0.05$ pu based on its ratings. Assume that the three-phase generator G_1 is rated 10/12.5 MW/MVA, 7.97/13.8Y kV with an impedance of $0 + j1.10$ pu based on its ratings, and that three-phase generator G_2 is rated 4/5 MW/MVA, 7.62/13.2Y kV with an impedance of $0 + j0.90$ pu based on its ratings. The transmission line TL_{23} has a length of 50 mi and is composed of 4/0 ACSR (aluminum conductor steel reinforced) conductors with an equivalent spacing (D_m) of 8 ft and has an impedance of $0.445 + j0.976$ Ω/mi. Its shunt susceptance is given as 5.78 μS/mi. The line connects buses 2 and 3. Bus 3 is assumed to be an infinite bus, that is, the magnitude of its voltage remains constant at given values and its phase position is unchanged regardless of the power and power factor demands that may be put on it. Furthermore, it is assumed to have a constant frequency equal to the nominal frequency of the system studied. Transmission line TL_{14} connects buses 1 and 4. It has a line length of 2 mi and an impedance of $0.80 + j0.80$ Ω/mi.

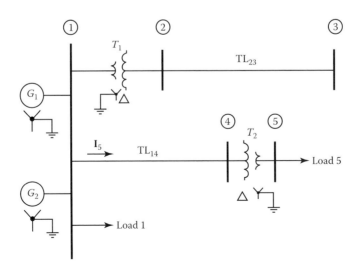

FIGURE P2.1 One-line diagram for Problem 2.10.

TABLE P2.1
Table for Problem 10

Quantity	Nominally 69-kV Circuits	Nominally 13-kV Circuits	Nominally 480-V Circuits
$kVA_{B(3\phi)}$	5000 kVA	5000 kVA	5000 kVA
$kV_{B(L-L)}$	69 kV		
$kV_{B(L-N)}$	39.84 kV		
$I_{B(L)}$			
$I_{B(\phi)}$			
Z_B			
Y_B			

Because of the line length, its shunt susceptance is assumed to be negligible. The load that is connected to bus 1 has a current magnitude $|\mathbf{I}_1|$ of 523 A and a lagging power factor of 0.707. The load that is connected to bus 5 is given as $8000 + j6000$ kVA. Use the arbitrarily selected 5000 kVA as the three-phase kVA base and 39.84/69.00 kV as the line-to-neutral and line-to-line voltage base and determine the following:

(a) Complete Table P2.1 for the indicated values. Note the I_L means line current and I_ϕ means phase currents in delta-connected apparatus.

(b) Draw a single-line positive-sequence network of this simple power system. Use the nominal π circuit to represent the 69-kV line. Show the values of all impedances and susceptances in per units on the chosen bases. Show all loads in per unit $P + jQ$.

11. Assume that a $500 + j200$-kVA load is connected to a load bus that has a voltage of $1.0\angle 0°$ pu. If the power base is 1000 kVA, determine the per-unit R and X of the load:
    (a) When load is represented by parallel connection
    (b) When load is represented by series connection

3 Steady-State Performance of Transmission Lines

3.1 INTRODUCTION

The function of the overhead three-phase electric power transmission line is to transmit bulk power to load centers and large industrial users beyond the primary distribution lines. A given transmission system comprises all land, conversion structures, and equipment at a primary source of supply, including lines, switching, and conversion stations between a generating or receiving point and a load center or wholesale point. It includes all lines and equipment whose main function is to increase, integrate, or tie together power supply sources.

The decision to build a transmission system results from system planning studies to determine how best to meet the system requirements. At this stage, the following factors need to be considered and established:

1. Voltage level
2. Conductor type and size
3. Line regulation and voltage control
4. Corona and losses
5. Proper load flow and system stability
6. System protection
7. Grounding
8. Insulation coordination
9. Mechanical design
    - Sag and stress calculations
    - Conductor composition
    - Conductor spacing
    - Insulator and conductor hardware selection
10. Structural design
    - Structure types
    - Stress calculations

The basic configuration selection depends on many interrelated factors, including esthetic considerations, economics, performance criteria, company policies and practice, line profile, right-of-way restrictions, preferred materials, and construction techniques. Figure 3.1 shows typical compact configurations [1]. Figures 3.2 through 3.5 show typical structures used for extra-high-voltage (EHV) transmission systems [2].

3.2 CONDUCTOR SIZE

Conductor sizes are based on the circular mil. A circular mil (cmil) is the area of a circle that has a diameter of 1 mil. A mil is equal to 1×10^{-3} in. The cross-sectional area of a wire in square inches equals its area in circular mils multiplied by 0.7854×10^{-6}. For the smaller conductors, up

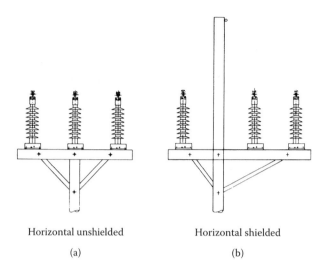

Horizontal unshielded	Horizontal shielded
(a)	(b)

FIGURE 3.1 Typical compact configurations. (From Elgerd, O. I., *Electric Energy Systems Theory: An Introduction*, McGraw-Hill, New York, 1971.)

to 211,600 cmil, the size is usually given by a gage number according to the American Wire Gauge (AWG) standard, formerly known as the Brown and Sharpe Wire Gauge (B&S).

In the AWG standard, gage sizes decrease as the wire increases in size. (The larger the gage size, the smaller the wire.) These numbers start at 40, the smallest, which is assigned to a wire with a diameter of 3.145 mil. The largest size is number 0000, written as 4/0 and read as "four odds." Above 4/0, the size is determined by cross-sectional area in circular mils. In summary

$$1 \text{ linear mil} = 0.001 \text{ in} = 0.0254 \text{ mm}$$

$$1 \text{ circular mil} = \text{area of cirlce 1 linear mil in diameter}$$

$$= \frac{\pi}{4} \text{ mil}^2$$

$$= \frac{\pi}{4} \times 10^{-6} = 0.7854 \times 10^{-6} \text{ in}^2$$

One thousand circular mils is often used as a unit, for example, a size given as 250 kcmil (or MCM) refers to 250,000 cmil.

A given conductor may consist of a single strand, or several strands. If a single strand, it is solid; if of more than one strand, it is stranded. A solid conductor is often called a *wire*, whereas a stranded conductor is called a *cable*. A general formula for the total number of strands in concentrically stranded cables is

$$\text{Number of strands} = 3n^2 - 3n + 1$$

where n is the number of layers, including the single center strand.

In general, distribution conductors larger than 2 AWG are stranded. Insulated conductors for underground distribution or aerial cable lines are classified as cables and usually are stranded. Table 3.1 gives standard conductor sizes.

Conductors may be selected on the basis of Kelvin's law. According to Kelvin's law,* "the most economical area of conductor is that for which the annual cost of the energy wasted is equal to the

---

* Expressed by Sir William Thomson (Lord Kelvin) in 1881.

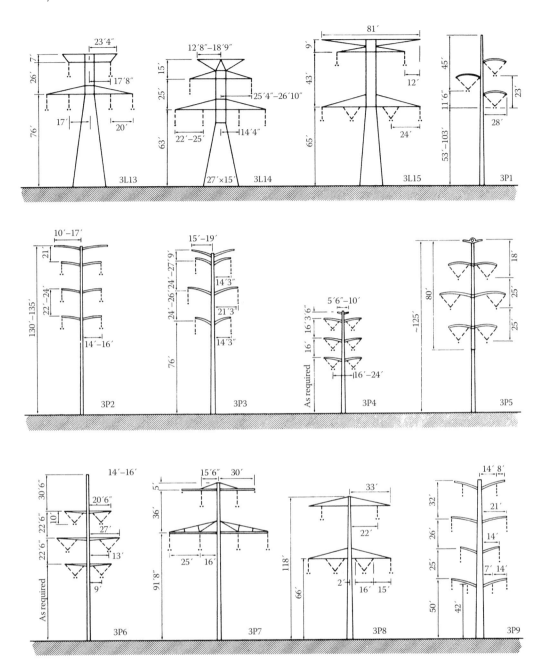

FIGURE 3.2 Typical pole and lattice structures for 345-kV transmission systems. (From Electric Power Research Institute, "Transmission Line Reference Book: 345kv and Above," 2nd ed., EPRI, Palo Alto, California, 1982.)

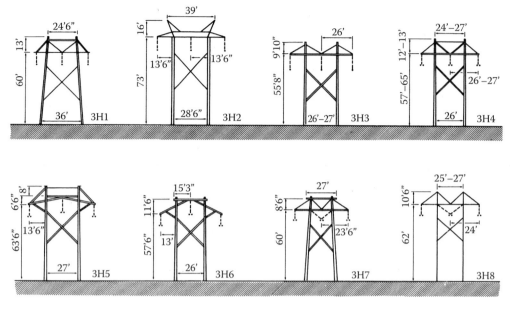

FIGURE 3.3 Typical wood H-frame structures for 345 kV. (From Electric Power Research Institute, "Transmission Line Reference Book: 345kv and Above," 2nd ed., EPRI, Palo Alto, California, 1982.)

interest on that portion of the capital expense which may be considered as proportional to the weight of the conductor" [3]. Therefore,

$$\text{Annual cost} = \frac{3CI^2R}{1000} + \frac{pwa}{1000} \tag{3.1}$$

where
C = cost of energy wasted in dollars per kilowatt-year
I = current per wire
R = resistance per mile per conductor
p = cost per pound conductor
w = weight per mile of all conductors
a = percent annual cost of money

Steady-State Performance of Transmission Lines

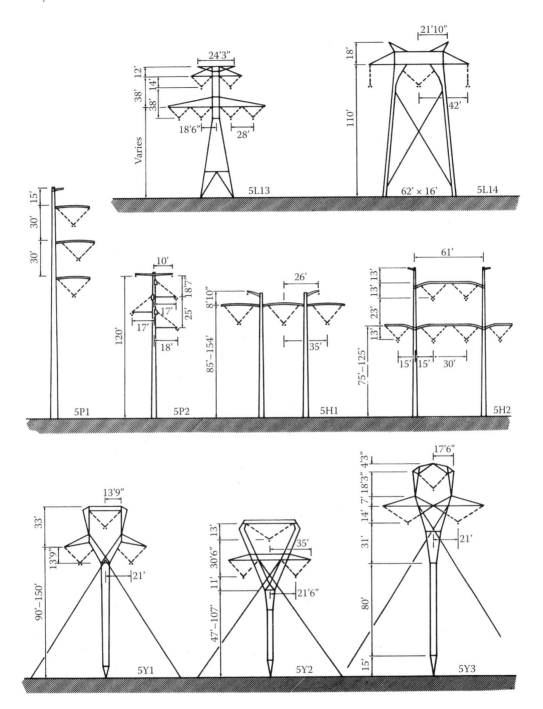

FIGURE 3.4 Typical 500-kV lattice, pole, H-frame, and guyed Y-type structures. (From Electric Power Research Institute, "Transmission Line Reference Book: 345kv and Above," 2nd ed., EPRI, Palo Alto, California, 1982.)

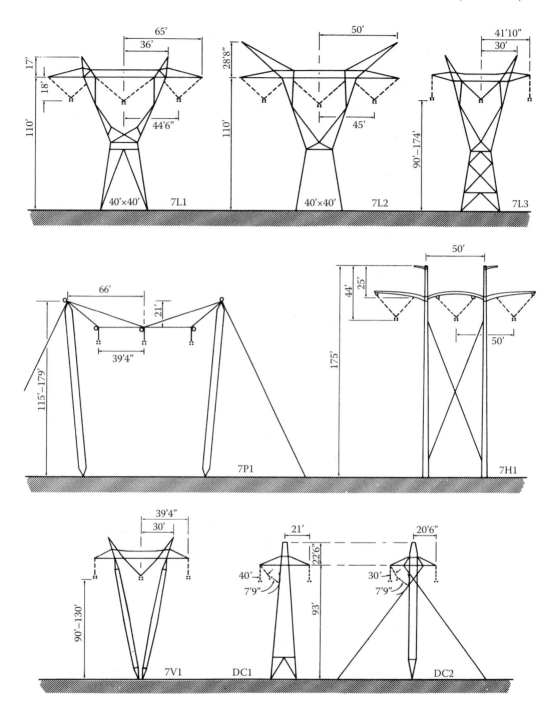

FIGURE 3.5 Typical dc structures (DC1 and DC2) and 735–800 kV ac designs. (From Electric Power Research Institute, "Transmission Line Reference Book: 345kv and Above," 2nd ed., EPRI, Palo Alto, California, 1982.)

TABLE 3.1
Standard Conductor Sizes

Size (AWG or kcmil)	(cmil)	No. of Wires	Solid or Stranded
18	1,620	1	Solid
16	2,580	1	Solid
14	4,110	1	Solid
12	6,530	1	Solid
10	10,380	1	Solid
8	16,510	1	Solid
7	20,820	1	Solid
6	26,250	1	Solid
6	26,250	3	Stranded
5	33,100	3	Stranded
5	33,100	1	Solid
4	41,740	1	Solid
4	41,740	3	Stranded
3	52,630	3	Stranded
3	52,630	7	Stranded
3	52,630	1	Solid
2	66,370	1	Solid
2	66,370	3	Stranded
2	66,370	7	Stranded
1	83,690	3	Stranded
1	83,690	7	Stranded
0 (or 1/0)	105,500	7	Stranded
00 (or 2/0)	133,100	7	Stranded
000 (or 3/0)	167,800	7	Stranded
000 (or 3/0)	167,800	12	Stranded
0000 (or 4/0)	211,600	17	Stranded
0000 (or 4/0)	211,600	12	Stranded
0000 (or 4/0)	211,600	19	Stranded
250	250,000	12	Stranded
250	250,000	19	Stranded
300	300,000	12	Stranded
300	300,000	19	Stranded
350	350,000	12	Stranded
350	350,000	19	Stranded
400	400,000	19	Stranded
450	450,000	19	Stranded
500	500,000	19	Stranded
500	500,000	37	Stranded
600	600,000	37	Stranded
700	700,000	37	Stranded
750	750,000	37	Stranded
800	800,000	37	Stranded
900	900,000	37	Stranded
1,000	1,000,000	37	Stranded

Here, the minimum cost is obtained when

$$\frac{3CI^2R}{1000} = \frac{pwa}{1000}$$

However, in practice, Kelvin's law is seldom used. Instead, the I^2R losses are calculated for the total time horizon.

The conductors used in modern overhead power transmission lines are bare aluminum conductors, which are classified as follows:

AAC: all-aluminum conductor
AAAC: all-aluminum-alloy conductor
ACSR: aluminum conductor steel reinforced
ACAR: aluminum conductor alloy reinforced

3.3 TRANSMISSION LINE CONSTANTS

For the purpose of system analysis, a given transmission line can be represented by its resistance, inductance or inductive reactance, capacitance or capacitive reactance, and leakage resistance (which is usually negligible).

3.4 RESISTANCE

The direct current (dc) resistance of a conductor is

$$R_{dc} = \frac{\rho \times \ell}{A} \quad \Omega$$

where
  ρ = conductor resistivity
  ℓ = conductor length
  A = conductor cross-sectional area

In practice, several different sets of units are used in the calculation of the resistance. For example, in the International System of Units (SI units), ℓ is in meters, A is in square meters, and ρ is in ohm meters. Whereas in power systems in the United States, ρ is in ohm circular mils per foot, ℓ is in feet, and A is in circular mils. The resistance of a conductor at any temperature may be determined by

$$\frac{R_2}{R_1} = \frac{T_0 + t_2}{T_0 + t_1}$$

where
  R_1 = conductor resistance at temperature t_1
  R_2 = conductor resistance at temperature t_2
  t_1, t_2 = conductor temperatures in degrees Celsius
  T_0 = constant varying with conductor material
   = 234.5 for annealed copper
   = 241 for hard-drawn copper
   = 228 for hard-drawn aluminum

The phenomenon by which alternating current (ac) tends to flow in the outer layer of a conductor is called the *skin effect*. The skin effect is a function of conductor size, frequency, and the relative resistance of the conductor material.

Tables given in Appendix A provide the dc and ac resistance values for various conductors. The resistances to be used in the positive- and negative-sequence networks are the ac resistances of the conductors.

3.5 INDUCTANCE AND INDUCTIVE REACTANCE

3.5.1 Single-Phase Overhead Lines

Figure 3.6 shows a single-phase overhead line. Assume that a current flows out in conductor *a* and returns in conductor *b*. These currents cause magnetic field lines that link between the conductors. A change in current causes a change in flux, which in turn results in an induced voltage in the circuit. In an ac circuit, this induced voltage is called the *IX* drop. In going around the loop, if *R* is the resistance of each conductor, the total loss in voltage due to resistance is 2*IR*. Thus, the voltage drop in the single-phase line due to loop impedance at 60 Hz is

$$\text{VD} = 2l\left(R + j0.2794 \log_{10} \frac{D_m}{D_s}\right)I \quad (3.2)$$

where
    VD = voltage drop due to line impedance in volts
    l   = line length in miles
    R   = resistance of each conductor in ohms per mile
    D_m = equivalent or geometric mean distance (GMD) between conductor centers in inches
    D_s = geometric mean radius (GMR) or self-GMD of one conductor in inches,
        = 0.7788*r* for cylindrical conductor
    r   = radius of cylindrical conductor in inches (see Figure 3.6)
    I   = phase current in amperes

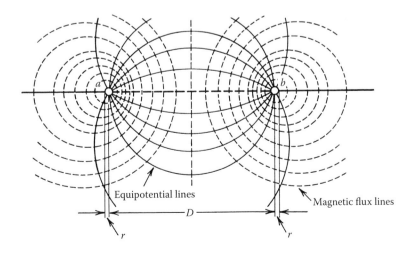

FIGURE 3.6 Magnetic field of single-phase line.

Thus, the inductance of the conductor is expressed as

$$L = 2 \times 10^{-7} \ln \frac{D_m}{D_s} \quad \text{H/m} \tag{3.3}$$

or

$$L = 0.7411 \log_{10} \frac{D_m}{D_s} \quad \text{mH/mi} \tag{3.4}$$

With the inductance known, the inductive reactance* can be found as

$$X_L = 2\pi f L = 2.02 \times 10^{-3} f \ln \frac{D_m}{D_s} \tag{3.5}$$

or

$$X_L = 4.657 \times 10^{-3} f \log_{10} \frac{D_m}{D_s} \tag{3.6}$$

or, at 60 Hz,

$$X_L = 0.2794 \log_{10} \frac{D_m}{D_s} \quad \Omega/\text{mi} \tag{3.7}$$

$$X_L = 0.1213 \ln \frac{D_m}{D_s} \quad \Omega/\text{mi} \tag{3.8}$$

By using the GMR of a conductor, D_s, the calculation of inductance and inductive reactance can be done easily. Tables give the GMR of various conductors readily.

3.5.2 THREE-PHASE OVERHEAD LINES

In general, the spacings D_{ab}, D_{bc}, and D_{ca} between the conductors of three-phase transmission lines are *not* equal. For any given conductor configuration, the average values of inductance and capacitance can be found by representing the system by one with equivalent equilateral spacing. The "equivalent spacing" is calculated as

$$D_{eq} \triangleq D_m = (D_{ab} \times D_{bc} \times D_{ca})^{1/3} \tag{3.9}$$

In practice, the conductors of a transmission line are transposed, as shown in Figure 3.7. The transposition operation, that is, exchanging the conductor positions, is usually carried out at switching stations.

Therefore, the average inductance per phase is

$$L_a = 2 \times 10^{-7} \ln \frac{D_{eq}}{D_s} \quad \text{H/m} \tag{3.10}$$

---

* It is also the same as the positive- and negative-sequence of a line.

Steady-State Performance of Transmission Lines

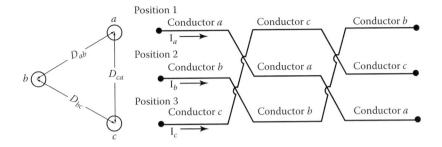

FIGURE 3.7 Complete transposition cycle of three-phase line.

or

$$L_a = 0.7411 \log_{10} \frac{D_{eq}}{D_s} \quad \text{mH/mi} \tag{3.11}$$

and the inductive reactance is

$$X_L = 0.1213 \ln \frac{D_{eq}}{D_s} \quad \Omega/\text{mi} \tag{3.12}$$

or

$$X_L = 0.2794 \log_{10} \frac{D_{eq}}{D_s} \quad \Omega/\text{mi} \tag{3.13}$$

3.6 CAPACITANCE AND CAPACITIVE REACTANCE

3.6.1 SINGLE-PHASE OVERHEAD LINES

Figure 3.8 shows a single-phase line with two identical parallel conductors a and b of radius r separated by a distance D, center to center, and with a potential difference of V_{ab} volts. Let conductors a

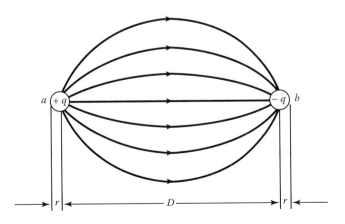

FIGURE 3.8 Capacitance of single-phase line.

and b carry charges of $+q_a$ and $-q_b$ farads per meter, respectively. The capacitance between conductors can be found as

$$C_{ab} = \frac{q_a}{V_{ab}}$$

$$= \frac{2\pi\varepsilon}{\ln\left(\dfrac{D^2}{r_a \times r_b}\right)} \quad \text{F/m} \tag{3.14}$$

If $r_a = r_b = r$,

$$C_{ab} = \frac{2\pi\varepsilon}{2\ln\left(\dfrac{D}{r}\right)} \quad \text{F/m} \tag{3.15}$$

Since

$$\varepsilon = \varepsilon_0 \times \varepsilon_r$$

where

$$\varepsilon_0 = \frac{1}{36\pi \times 10^9} = 8.85 \times 10^{-12} \quad \text{F/m}$$

and

$$\varepsilon_r \cong 1 \text{ for air}$$

Equation 3.15 becomes

$$C_{ab} = \frac{0.0388}{2 \times \log_{10}\left(\dfrac{D}{r}\right)} \quad \mu\text{F/mi} \tag{3.16}$$

or

$$C_{ab} = \frac{0.0894}{2 \times \ln\left(\dfrac{D}{r}\right)} \quad \mu\text{F/mi} \tag{3.17}$$

or

$$C_{ab} = \frac{0.0241}{2 \times \log_{10}\left(\dfrac{D}{r}\right)} \quad \mu\text{F/km} \tag{3.18}$$

Stevenson [3] explains that the capacitance to neutral or capacitance to ground for the two-wire line is twice the line-to-line capacitance or capacitance between conductors, as shown in Figures 3.9 and 3.10. Therefore, the line-to-neutral capacitance is

$$C_n = C_{an} = C_{bn} = \frac{0.0388}{\log_{10}\left(\dfrac{D}{r}\right)} \quad \mu\text{F/mi to neutral} \tag{3.19}$$

This can easily be verified since C_N must equal $2C_{ab}$ so that the capacitance between the conductors can be

$$\begin{aligned}C_{ab} &= \frac{C_n \times C_n}{C_n + C_n} \\ &= \frac{C_n}{2} \\ &= C_{ab}\end{aligned} \tag{3.20}$$

as before.

With the capacitance known, the capacitive reactance between one conductor and neutral can be found as

$$X_c = \frac{1}{2\pi f C_n} \tag{3.21}$$

or, for 60 Hz,

$$X_c = 0.06836 \log_{10} \frac{D}{r} \quad \text{M}\Omega \cdot \text{mi to neutral} \tag{3.22}$$

and the line-to-neutral susceptance is

$$b_c = \omega C_n$$

or

$$b_c = \frac{1}{X_c} \tag{3.23}$$

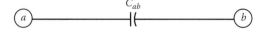

FIGURE 3.9 Line-to-line capacitance.

FIGURE 3.10 Line-to-neutral capacitance.

Or line-to-neutral susceptance as

$$b_c = \frac{14.6272}{\log_{10}\left(\dfrac{D}{r}\right)} \quad \text{m}\Omega/\text{mi to neutral} \tag{3.24}$$

The *charging current of the line* is

$$\mathbf{I}_c = j\omega C_{ab} V_{ab} \text{ A/mi} \tag{3.25}$$

3.6.2 Three-Phase Overhead Lines

Figure 3.11 shows the cross section of a three-phase line with equilateral spacing D. The line-to-neutral capacitance can be found as

$$C_n = \frac{0.0388}{\log_{10}\left(\dfrac{D}{r}\right)} \quad \mu\text{F/mi to neutral} \tag{3.26}$$

which is identical to Equation 3.19.

On the other hand, if the spacings between the conductors of the three-phase line are not equal, the line-to-neutral capacitance is

$$C_n = \frac{0.0388}{\log_{10}\left(\dfrac{D_{eq}}{r}\right)} \quad \mu\text{F/mi to neutral} \tag{3.27}$$

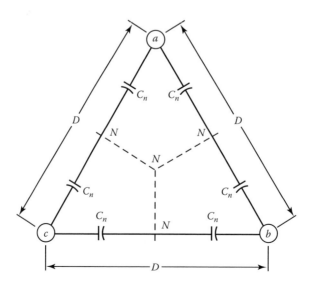

FIGURE 3.11 Three-phase line with equilateral spacing.

where

$$D_{eq} \triangleq D_m = (D_{ab} \times D_{bc} \times D_{ca})^{1/3}$$

The *charging current per phase* is

$$\mathbf{I}_c = j\omega C_n V_{an} \text{ A/mi} \tag{3.28}$$

3.7 TABLES OF LINE CONSTANTS

Tables provide the line constants directly without using equations for calculation. This concept was suggested by W. A. Lewis [6]. According to this concept, Equation 3.8 for inductive reactance at 60 Hz, that is,

$$X_L = 0.1213 \ln \frac{D_m}{D_s} \text{ }\Omega\text{/mi} \tag{3.29a}$$

can be broken down to

$$X_L = 0.1213 \ln \frac{1}{D_s} + 0.1213 \ln D_m \text{ }\Omega\text{/mi} \tag{3.29b}$$

where
  D_s = GMR, which can be found from the tables for a given conductor
  D_m = GMD between conductor centers

Therefore, Equation 3.29b can be rewritten as

$$X_L = x_a + x_d \text{ }\Omega\text{/mi} \tag{3.30}$$

where
  x_a = inductive reactance at 1-ft spacing

$$= 0.1213 \ln \frac{1}{D_s} \text{ }\Omega\text{/mi} \tag{3.31}$$

  x_d = inductive reactance spacing factor

$$= 0.1213 \, D_m \text{ }\Omega\text{/mi} \tag{3.32}$$

For a given frequency, the value of x_a depends only on the GMR, which is a function of the conductor type. However, x_d depends only on the spacing D_m. If the spacing is greater than 1 ft, x_d has a positive value that is added to x_a. On the other hand, if the spacing is less than 1 ft, x_d has a negative value that is subtracted from x_a. Tables given in Appendix A give x_a and x_d directly.

Similarly, Equation 3.22 for shunt capacitive reactance at 60 Hz, that is,

$$x_c = 0.06836 \log_{10} \frac{D_m}{r} \text{ M}\Omega \times \text{mi} \tag{3.33a}$$

can be split into

$$x_c = 0.06836 \log_{10} \frac{1}{r} + 0.06836 \log_{10} D_m \quad \text{M}\Omega \times \text{mi} \quad (3.33\text{b})$$

or

$$x_c = x'_a + x'_d \quad \text{M} \times \text{mi} \quad (3.34)$$

where
  x'_a = capacitive reactance at 1-ft spacing

$$= 0.06836 \log_{10} \frac{1}{r} \quad \text{M}\Omega \times \text{mi} \quad (3.35)$$

x'_d = capacitive reactance spacing factor

$$= 0.06836 \log_{10} D_m \quad \text{M}\Omega \times \text{mi} \quad (3.36)$$

Tables given in Appendix A provide x'_a and x'_d directly. The term x'_d is added or subtracted from x'_a depending on the magnitude of D_m.

EXAMPLE 3.1

A three-phase, 60-Hz, transposed line has conductors that are made up of 4/0, seven-strand copper. At the pole top, the distances between conductors, center-to-center, are given as 6.8, 5.5, and 4 ft. The diameter of the conductor copper used is 0.1739 in. Determine the inductive reactances per mile per phase

(a) By using Equation 3.29a
(b) By using Tables and Equation 3.30

Solution

(a) First calculating the equivalent spacing for the pole top,

$$D_{eq} = D_m = (D_{ab} \times D_{bc} \times D_{ca})^{1/3}$$

$$= (6.8 \times 5.5 \times 4)^{1/3} = 5.3086 \text{ ft}$$

From Table A.1, D_s = 0.01579 ft for the conductor. Hence, its inductive reactance is

$$X_L = 0.1213 \ln \frac{D_{eq}}{D_s}$$

$$= 0.1213 \ln \frac{5.3086 \text{ ft}}{0.01579 \text{ ft}}$$

$$= 0.705688 \ \Omega/\text{mi} \cong 0.7057 \ \Omega/\text{mi}$$

Steady-State Performance of Transmission Lines

(b) From Table A.1, $x_a = 0.503$ Ω/mi and from Table A.8, for $D_{eq} = 5.30086$ ft, by linear interpolation, $x_d = 0.2026$ Ω/mi. Thus, the inductive reactance is

$$X_L = x_a + x_d$$
$$= 0.503 + 0.2026$$
$$= 0.7056 \text{ Ω/mi}$$

EXAMPLE 3.2

Consider the pole-top configuration given in Example 3.1. If the line length is 100 mi, determine the shunt capacitive reactance by using

(a) Equation 3.33a
(b) Using tables in Appendix B

Solution

(a) By using the Equation 3.33a,

$$X_c = 0.06836 \log_{10} \frac{D_m}{r}$$

$$= 0.06836 \log_{10} \frac{5.3086 \text{ ft}}{\left(\frac{0.522}{2 \times 12}\right) \text{ft}}$$

$$= 0.06836 \log_{10} \frac{1}{\left(\frac{0.522}{2 \times 12}\right) \text{ft}} + 0.06836 \log_{10}(5.3086 \text{ ft})$$

$$= 0.113651284 + 0.49559632$$

$$\cong 0.163211 \text{ MΩ} \times \text{mi}$$

(b) From Table A.1, $x'_a = 0.1136$ MΩ × mi and from Table A.9, $x'_d = 0.049543$ MΩ × mi. Hence, from Equation 3.33b

$$x_c = x'_a + x'_d$$
$$= 0.1136 + 0.049543$$
$$= 0.163143 \text{ MΩ} \times \text{mi}$$

(c) The capacitive reactance of the 100-mi-long line is

$$X_c = \frac{x_c}{\ell}$$
$$= \frac{0.163143 \text{ MΩ} \times \text{mi}}{100 \text{ mi}}$$
$$= 1.63143 \times 10^{-3} \text{ MΩ}$$

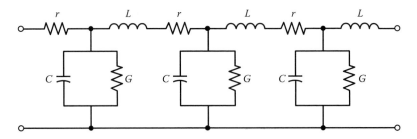

FIGURE 3.12 Distributed constant equivalent circuit of line.

3.8 EQUIVALENT CIRCUITS FOR TRANSMISSION LINES

An overhead line or a cable can be represented as a distributed constant circuit, as shown in Figure 3.12. The resistance, inductance, capacitance, and leakage conductance of a distributed constant circuit are distributed uniformly along the line length. In the figure, L represents the inductance of a line conductor to neutral per unit length, r represents the ac resistance of a line conductor per unit length, C is the capacitance of a line conductor to neutral per unit length, and G is the leakage conductance per unit length.

3.9 TRANSMISSION LINES

A brief review of transmission system modeling is presented in this section. Transmission lines are modeled and classified according to their lengths as

1. Short-transmission lines
2. Medium-length transmission lines
3. Long transmission lines

3.9.1 Short Transmission Lines (up to 50 mi or 80 km)

The short transmission lines are those lines that have lengths up to 50 mi or 80 km. The medium-length transmission lines are those lines that have lengths up to 150 mi or 240 km. Similarly, the long transmission lines are those lines that have lengths above 150 mi or 240 km.

The modeling of a short transmission line is the most simplistic one. Its shunt capacitance is so small that it can be omitted entirely with little loss of accuracy. (Its shunt admittance is neglected since the current is the same throughout the line.) Thus, its capacitance and leakage resistance (or conductance) to the earth are usually neglected, as shown in Figure 3.13. Therefore, the transmission line can be treated as a simple, lumped, and constant impedance, that is,

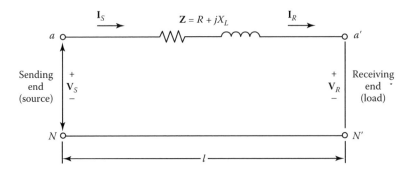

FIGURE 3.13 Equivalent circuit of short transmission line.

Steady-State Performance of Transmission Lines

$$\mathbf{Z} = R + jX_L$$
$$= zl$$
$$= rl + jxl \quad \Omega \tag{3.37}$$

where
- $\mathbf{Z}$ = total series impedance per phase in ohms
- z = series impedance of one conductor in ohms per unit length
- X_L = total inductive reactance of one conductor in ohms
- x = inductive reactance of one conductor in ohms per unit length
- l = length of line

The current entering the line at the sending end of the line is equal to the current leaving at the receiving end. Figures 3.14 and 3.15 show vector (or phasor) diagrams for a short transmission line connected to an inductive load and a capacitive load, respectively. It can be observed from the figures that

$$\mathbf{V}_S = \mathbf{V}_R + \mathbf{I}_R \mathbf{Z} \tag{3.38}$$

$$\mathbf{I}_S = \mathbf{I}_R = \mathbf{I} \tag{3.39}$$

$$\mathbf{V}_R = \mathbf{V}_S - \mathbf{I}_R \mathbf{Z} \tag{3.40}$$

where
- $\mathbf{V}_S$ = sending-end phase (line-to-neutral) voltage
- $\mathbf{V}_R$ = receiving-end phase (line-to-neutral) voltage
- $\mathbf{I}_S$ = sending-end phase current
- $\mathbf{I}_R$ = receiving-end phase current
- $\mathbf{Z}$ = total series impedance per phase

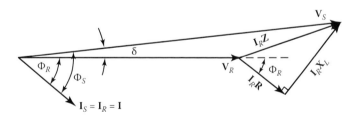

FIGURE 3.14 Phasor diagram of short transmission line to inductive load.

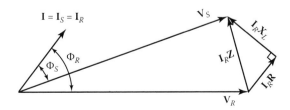

FIGURE 3.15 Phasor diagram of short transmission line connected to capacitive load.

Thus, using $\mathbf{V}_R$ as the reference, Equation 3.38 can be written as

$$\mathbf{V}_S = \mathbf{V}_R + (I_R \cos\theta_R \pm jI_R \sin\theta_R)(R + jX) \tag{3.41}$$

where the plus or minus sign is determined by θ_R, the power factor angle of the receiving end or load. If the power factor is lagging, the minus sign is employed. On the other hand, if it is leading, the plus sign is used.

However, if Equation 3.40 is used, it is convenient to use $\mathbf{V}_S$ as the reference. Therefore,

$$\mathbf{V}_R = \mathbf{V}_S - (I_R \cos\theta_R \pm jI_R \sin\theta_R)(R + jX) \tag{3.42}$$

where θ_S is the sending-end power factor angle, that determines, as before, whether the plus or minus sign will be used. Also, from Figure 3.14, using $\mathbf{V}_R$ as the reference vector,

$$V_S = [(V_R + IR\cos\theta_R + IX\cos\theta_R)^2 + (IX\cos\theta_R \pm IR\sin\theta_R)^2]^{1/2} \tag{3.43}$$

and load angle

$$\delta = \theta_S - \theta_R \tag{3.44}$$

or

$$\delta = \arctan\left(\frac{IX\cos\theta_R \pm IR\sin\theta_R}{V_R + IR\cos\theta_R + IX\sin\theta_R}\right) \tag{3.45}$$

The generalized constants, or **ABCD** parameters, can be determined by inspection of Figure 3.13. Since

$$\begin{bmatrix} \mathbf{V}_S \\ \mathbf{I}_S \end{bmatrix} = \begin{bmatrix} \mathbf{A} & \mathbf{B} \\ \mathbf{C} & \mathbf{D} \end{bmatrix} \begin{bmatrix} \mathbf{V}_R \\ \mathbf{V}_R \end{bmatrix} \tag{3.46}$$

and $\mathbf{AD} - \mathbf{BC} = 1$, where

$$\mathbf{A} = 1 \quad \mathbf{B} = \mathbf{Z} \quad \mathbf{C} = 0 \quad \mathbf{D} = 1 \tag{3.47}$$

then

$$\begin{bmatrix} \mathbf{V}_S \\ \mathbf{I}_S \end{bmatrix} = \begin{bmatrix} 1 & \mathbf{Z} \\ 0 & 1 \end{bmatrix} \begin{bmatrix} \mathbf{V}_R \\ \mathbf{V}_R \end{bmatrix} \tag{3.48}$$

and

$$\begin{bmatrix} \mathbf{V}_R \\ \mathbf{I}_R \end{bmatrix} = \begin{bmatrix} 1 & \mathbf{Z} \\ 0 & 1 \end{bmatrix}^{-1} \begin{bmatrix} \mathbf{V}_S \\ \mathbf{I}_S \end{bmatrix} = \begin{bmatrix} 1 & -\mathbf{Z} \\ 0 & 1 \end{bmatrix} \begin{bmatrix} \mathbf{V}_S \\ \mathbf{I}_S \end{bmatrix}$$

Steady-State Performance of Transmission Lines

The transmission efficiency of the short line can be expressed as

$$\eta = \frac{\text{output}}{\text{input}}$$

$$= \frac{\sqrt{3}V_R I \cos\theta_R}{\sqrt{3}V_S I \cos\theta_S}$$

$$= \frac{V_R \cos\theta_R}{V_S \cos\theta_S} \quad (3.49)$$

Equation 3.49 is applicable whenever the line is single phase. The transmission efficiency can also be expressed as

$$\eta = \frac{\text{output}}{\text{output} + \text{losses}}$$

For a single-phase line,

$$\eta = \frac{V_R I \cos\theta_R}{V_R I \cos\theta_R + 2I^2 R} \quad (3.50)$$

For a three-phase line,

$$\eta = \frac{\sqrt{3}V_R I \cos\theta_R}{\sqrt{3}V_R I \cos\theta_R + 3I^2 R} \quad (3.51)$$

3.9.2 Steady-State Power Limit

Assume that the impedance of a short transmission line is given as $\mathbf{Z} = Z\angle\theta$. Therefore, the real power delivered, at steady state, to the receiving end of the transmission line can be expressed as

$$P_R = \frac{V_S \times V_R}{Z}\cos(\theta - \delta) - \frac{V_R^2}{Z}\cos\theta \quad (3.52)$$

and similarly, the reactive power delivered can be expressed as

$$Q_R = \frac{V_S \times V_R}{Z}\sin(\theta - \delta) - \frac{V_R^2}{Z}\sin\theta \quad (3.53)$$

If V_S and V_R are the line-to-neutral voltages, Equations 3.52 and 3.53 give P_R and Q_R values per phase. Also, if the obtained P_R and Q_R values are multiplied by 3 or the line-to-line values of V_S and V_R are used, the equations give the three-phase real and reactive power delivered to a balanced load at the receiving end of the line.

If, in Equation 3.52, all variables are kept constant with the exception of δ, so that the real power delivered, P_R, is a function of δ only, P_R is maximum when $\delta = \theta$, and the maximum powers* obtainable at the receiving end for a given regulation can be expressed as

$$P_{R,\max} = \frac{V_R^2}{Z^2}\left(\frac{V_S}{V_R}Z - R\right) \tag{3.54}$$

where V_S and V_R are the phase (line-to-neutral) voltages whether the system is single phase or three phase. The equation can also be expressed as

$$P_{R,\max} = \frac{V_S \times V_R}{Z} - \frac{V_R^2 \times \cos\theta}{Z} \tag{3.55}$$

If $V_S = V_R$,

$$P_{R,\max} = \frac{V_R^2}{Z}(1 - \cos\theta) \tag{3.56}$$

or

$$P_{R,\max} = \left(\frac{V_R}{Z}\right)^2 (Z - R) \tag{3.57}$$

and similarly, the corresponding reactive power delivered to the load is given by

$$Q_{R,\max} = -\frac{V_R^2}{Z}\sin\theta \tag{3.58}$$

As can be observed, both Equations 3.57 and 3.58 are independent of V_S voltage. The negative sign in Equation 3.52 points out that the load is a sink of *leading vars*,[†] that is, going to the load or a source of *lagging vars* (i.e., from the load to the supply). The total three-phase power transmitted on the three-phase line is three times the power calculated by using the above equations. If the voltages are given in volts, the power is expressed in watts or vars. Otherwise, if they are given in kilovolts, the power is expressed in megawatts or megavars.

In a similar manner, the real and reactive powers for the sending end of a transmission line can be expressed as

$$P_S = \frac{V_S^2}{Z}\cos\theta - \frac{V_S \times V_R}{Z}\cos(\theta + \delta) \tag{3.59}$$

---

* Also called the steady-state power limit.
† For many decades, the electrical utility industry has declined to recognize two different kinds of reactive power, *leading* and *lagging vars*. Only magnetizing vars are recognized, printed on varmeter scale plates, bought, and sold. Therefore, in the following sections, the leading or lagging vars will be referred to as *magnetizing vars*.

and

$$Q_S = \frac{V_S^2}{Z}\sin\theta - \frac{V_S \times V_R}{Z}\sin(\theta + \delta) \qquad (3.60)$$

If, in Equation 3.59, as before, all variables are kept constant with the exception of δ, so that the real power at the sending end, P_S, is a function of δ only, P_S is a maximum when

$$\theta + \delta = 180°$$

Therefore, the maximum power at the sending end, the maximum input power, can be expressed as

$$P_{S,\max} = \frac{V_S^2}{Z}\cos\theta + \frac{V_S \times V_R}{Z} \qquad (3.61)$$

or

$$P_{S,\max} = \frac{V_S^2 \times R}{Z^2} + \frac{V_S \times V_R}{Z} \qquad (3.62)$$

However, if $V_S = V_R$,

$$P_{S,\max} = \left(\frac{V_S}{Z}\right)^2 (Z + R) \qquad (3.63)$$

and similarly, the corresponding reactive power at the sending end, the maximum input vars, is given by

$$Q_S = \frac{V_S^2}{Z}\sin\theta \qquad (3.64)$$

As can be observed, both Equations 3.63 and 3.64 are independent of V_R voltage, and Equation 3.64 has a positive sign this time.

3.9.3 PERCENT VOLTAGE REGULATION

The voltage regulation of the line is defined by the increase in voltage when full load is removed, that is,

$$\text{Percentage of voltage regulation} = \frac{|\mathbf{V}_S| - |\mathbf{V}_R|}{|\mathbf{V}_R|} \times 100 \qquad (3.65)$$

or

$$\text{Percentage of voltage regulation} = \frac{|\mathbf{V}_{R,\text{NL}}| - |\mathbf{V}_{R,\text{FL}}|}{|\mathbf{V}_{R,\text{FL}}|} \times 100 \qquad (3.66)$$

where
- $|\mathbf{V}_S|$ = magnitude of sending-end phase (line-to-neutral) voltage at no load
- $|\mathbf{V}_R|$ = magnitude of receiving-end phase (line-to-neutral) voltage at full load
- $|\mathbf{V}_{R,\text{NL}}|$ = magnitude of receiving-end voltage at no load
- $|\mathbf{V}_{R,\text{FL}}|$ = magnitude of receiving-end voltage at full load with constant $|\mathbf{V}_S|$

Therefore, if the load is connected at the receiving end of the line,

$$|\mathbf{V}_S| = |\mathbf{V}_{R,\text{NL}}|$$

and

$$|\mathbf{V}_R| = |\mathbf{V}_{R,\text{FL}}|$$

An approximate expression for percentage of voltage regulation is

$$\text{Percentage of voltage regulation} \cong I_R \left(\frac{R \cos \Phi_R \pm X \sin \Phi_R}{V_R} \right) \times 100 \tag{3.67}$$

EXAMPLE 3.3

A three-phase, 60-Hz overhead short transmission line has a line-to-line voltage of 23 kV at the receiving end, a total impedance of 2.48 + j6.57 Ω/phase, and a load of 9 MW with a receiving-end lagging power factor of 0.85.

(a) Calculate the line-to-neutral and line-to-line voltages at the sending end.
(b) Calculate the load angle.

Solution

METHOD 1. USING COMPLEX ALGEBRA:

(a) The line-to-neutral reference voltage is

$$\mathbf{V}_{R(L-N)} = \frac{\mathbf{V}_{R(L-L)}}{\sqrt{3}}$$

$$= \frac{23 \times 10^3 \angle 0°}{\sqrt{3}} = 13{,}294.8 \angle 0° \text{ V}$$

The line current is

$$\mathbf{I} = \frac{9 \times 10^6}{\sqrt{3} \times 23 \times 10^3 \times 0.85} \times (0.85 - j0.527)$$

$$= 265.8(0.85 - j0.527)$$

$$= 225.83 - j140.08 \text{ A}$$

Therefore,

$$\mathbf{IZ} = (225.93 - j140.08)(2.48 + j6.57)$$
$$= (265.8\angle -31.8°)(7.02\angle 69.32°)$$
$$= 1866.8\angle 37.52° \text{ V}$$

Thus, the line-to-neutral voltage at the sending end is

$$\mathbf{V}_{S(L-N)} = \mathbf{V}_{R(L-N)} + \mathbf{IZ}$$
$$= 14{,}803\angle 4.4° \text{ V}$$

The line-to-line voltage at the sending end is

$$\mathbf{V}_{S(L-L)} = \sqrt{3}\mathbf{V}_{S(L-N)}$$
$$= 25{,}640\angle 4.4° + 30° = 25{,}640\angle 34.4° \text{ V}$$

(b) The load angle is 4.4°.

METHOD 2. USING THE CURRENT AS THE REFERENCE PHASOR:

(a)

$$V_R \cos\theta_R + IR = 13{,}279.06 \times 0.85 + 265.8 \times 2.48 = 11{,}946 \text{ V}$$

$$V_R \sin\theta_R + IX = 13{,}294.8 \times 0.527 + 266.1 \times 6.57 = 8744 \text{ V}$$

then

$$V_{S(L-N)} = (11{,}946.39^2 + 8744^2)^{1/2} = 14{,}803 \text{ V/phase}$$

$$V_{S(L-L)} = 25{,}640 \text{ V}$$

(b)

$$\theta_S = \theta_R + \delta = \tan^{-1}\left(\frac{8744}{11{,}946}\right) = 36.2°$$

$$\delta = \theta_S - \theta_R = 36.2 - 31.8 = 4.4°$$

METHOD 3. USING THE RECEIVING-END VOLTAGE AS THE REFERENCE PHASOR:

(a)

$$V_{S(L-N)} = [(V_R + IR\cos\theta_R + IX\sin\theta_R)^2 + (IX\cos\theta_R + IR\sin\theta_R)^2]^{1/2}$$

$$IR\cos\theta_R = 265.8 \times 2.48 \times 0.85 = 560.3$$

$$IR\sin\theta_R = 265.8 \times 2.48 \times 0.527 = 347.4$$

$$IX\cos\theta_R = 265.8 \times 6.57 \times 0.85 = 1484.4$$

$$IX\sin\theta_R = 265.8 \times 6.57 \times 0.527 = 920.3$$

Therefore,

$$V_{S(L-N)} = [(13{,}279 + 560.3 + 920.3)^2 + (1484.4 - 347.4)^2]^{1/2}$$
$$= [14{,}759.7^2 + 1137^2]^{1/2}$$
$$= 14{,}803 \text{ V}$$

$$V_{S(L-L)} = \sqrt{3}\, V_{S(L-L)} = 25{,}640 \text{ V}$$

(b)

$$\delta = \tan^{-1}\left(\frac{1137}{14{,}759.7}\right) = 4.4°$$

METHOD 4. USING POWER RELATIONSHIPS:

The power loss in the line is

$$P_{\text{loss}} = 3I^2 R$$
$$= 3 \times 265.8^2 \times 2.48 \times 10^{-6} = 0.526 \text{ MW}$$

The total input power to the line is

$$P_T = P + P_{\text{loss}}$$
$$= 9 + 0.526 = 9.526 \text{ MW}$$

The var loss in the line is

$$Q_{\text{loss}} = 3I^2 X$$
$$= 3 \times 265.8^2 \times 6.57 \times 10^{-6} = 1.393 \text{ Mvar lagging}$$

The total megavar input to the line is

$$Q_T = \frac{P\sin\theta_R}{\cos\theta_R} + Q_{\text{loss}}$$
$$= \frac{9 \times 0.526}{0.85} + 1.393 = 6.973 \text{ Mvar lagging}$$

Steady-State Performance of Transmission Lines

The total megavolt-ampere input to the line is

$$S_T = \left(P_T^2 + Q_T^2\right)^{1/2}$$
$$= (9.526^2 + 6.973^2)^{1/2} = 11.81 \text{ MVA}$$

(a)

$$V_{S(L-L)} = \frac{S_T}{\sqrt{3}I}$$

$$= \frac{11.81 \times 10^6}{\sqrt{3} \times 265.8} = 25{,}640 \text{ V}$$

$$V_{S(L-N)} = \frac{V_{S(L-L)}}{\sqrt{3}} = 14{,}803 \text{ V}$$

(b)

$$\cos\theta_s = \frac{P_T}{S_T} = \frac{9{,}526}{11.81} = 0.807 \text{ lagging}$$

Therefore,

$$\theta_s = 36.2°$$
$$\delta = 36.2° - 31.8° = 4.4°$$

METHOD 5. TREATING THE THREE-PHASE LINE AS A SINGLE-PHASE LINE AND HAVING V_S AND V_R REPRESENT LINE-TO-LINE VOLTAGES, NOT LINE-TO-NEUTRAL VOLTAGES:

(a) The power delivered is 4.5 MW

$$I_{\text{line}} = \frac{4.5 \times 10^6}{23 \times 10^3 \times 0.85} = 230.18 \text{ A}$$

$$R_{\text{loop}} = 2 \times 2.48 = 4.96 \text{ }\Omega$$

$$X_{\text{loop}} = 2 \times 6.57 = 13.14 \text{ }\Omega$$

$$V_R \cos\theta_R = 23 \times 10^3 \times 0.85 = 19{,}550 \text{ V}$$

$$V_R \sin\theta_R = 23 \times 10^3 \times 0.527 = 12{,}121 \text{ V}$$

$$IR = 230.18 \times 4.96 = 1141.7 \text{ V}$$

$$IX = 230.18 \times 13.14 = 3024.6 \text{ V}$$

Therefore,

$$V_{S(L-L)} = [(V_R \cos \theta_R + IR)^2 + (V_R \sin \theta_R + IX)^2]^{1/2}$$

$$= [(19,550 + 1141.7)^2 + (12,121 + 3024.6)^2]^{1/2}$$

$$= [20,691.7^2 + 15,145.6^2]^{1/2}$$

$$= 25,640 \text{ V}$$

Thus,

$$V_{S(L-N)} = \frac{V_{S(L-L)}}{\sqrt{3}} = 14,803 \text{ V}$$

(b)

$$\theta_s = \tan^{-1} \frac{15,145.6}{20,691.7} = 36.20°$$

and

$$\delta = 36.2° - 31.8° = 4.4°$$

EXAMPLE 3.4

Calculate percentage of voltage regulation for the values given in Example 3.3

(a) Using Equation 3.65
(b) Using Equation 3.66

Solution

(a) Using Equation 3.67,

$$\text{Percentage of voltage regulation} = \frac{|\mathbf{V}_S| - |\mathbf{V}_R|}{|\mathbf{V}_R|} \times 100$$

$$= \frac{14,803 - 13,279.06}{13,279.06} \times 100$$

$$= 11.5$$

(b) Using Equation 3.67,

$$\text{Percentage of voltage regulation} \cong I_R \times \frac{(R \cos \theta_R \pm X \sin \theta_R)}{V_R} \times 100$$

$$= 265.8 \left(\frac{2.48 \times 0.85 + 6.57 \times 0.527}{13,279.06} \right) \times 100$$

$$= 11.1$$

3.9.4 Representation of Mutual Impedance of Short Lines

Figure 3.16a shows a circuit of two lines, x and y, that have self-impedances of $\mathbf{Z}_{xx}$ and $\mathbf{Z}_{yy}$ and a mutual impedance of $\mathbf{Z}_{zy}$. Its equivalent circuit is shown in Figure 3.16b. Sometimes, it may be required to preserve the electrical identity of the two lines, as shown in Figure 3.17. The mutual impedance $\mathbf{Z}_{xy}$ can be in either line and transferred to the other by means of a transformer that has a 1:1 turns ratio. This technique is also applicable for three-phase lines.

EXAMPLE 3.5

Assume that the mutual impedance between two parallel feeders is $0.09 + j0.3$ Ω/mi per phase. The self-impedances of the feeders are $0.604\angle 50.4°$ and $0.567\angle 52.9°$ Ω/mi per phase, respectively. Represent the mutual impedance between the feeders as shown in Figure 3.16b.

Solution

$$\mathbf{Z}_{xy} = 0.09 + j0.3 \text{ Ω}$$

$$\mathbf{Z}_{xx} = 0.604\angle 50.4° = 3.85 + j0.465 \text{ Ω}$$

$$\mathbf{Z}_{yy} = 0.567\angle 52.9° = 0.342 + j0.452 \text{ Ω}$$

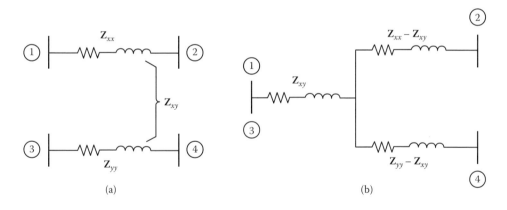

FIGURE 3.16 Representation of mutual impedance between two circuits.

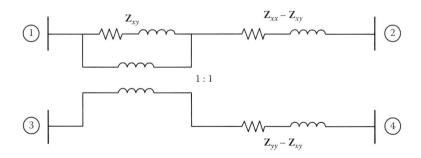

FIGURE 3.17 Representation of mutual impedance between two circuits by means of 1:1 transformer.

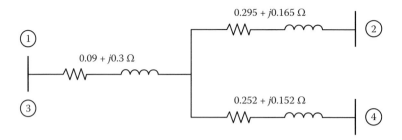

FIGURE 3.18 Resultant equivalent circuit for Example 3.5.

Therefore,

$$\mathbf{Z}_{xx} - \mathbf{Z}_{xy} = 0.295 + j0.165 \ \Omega$$

$$\mathbf{Z}_{yy} - \mathbf{Z}_{xy} = 0.252 + j0.152 \ \Omega$$

Hence, the resulting equivalent circuit is shown in Figure 3.18.

3.10 MEDIUM-LENGTH TRANSMISSION LINES (UP TO 150 MI OR 240 KM)

As the line length and voltage increase, the use of the formulas developed for the short transmission lines give inaccurate results. Thus, the effect of the current leaking through the capacitance must be taken into account for a better approximation. Thus, the shunt admittance is "lumped" at a few points along the line and represented by forming either a T or a π network, as shown in Figures 3.19 and 3.20. In the figures,

$$\mathbf{Z} = zl$$

For the T circuit shown in Figure 3.19,

$$\mathbf{V}_S = \mathbf{I}_S \times \frac{1}{2}\mathbf{Z} + \mathbf{I}_R \times \frac{1}{2}\mathbf{Z} + \mathbf{V}_R$$

$$= \left[\mathbf{I}_R + \left(\mathbf{V}_R + \mathbf{I}_R \times \frac{1}{2}\mathbf{Z}\right)\mathbf{Y}\right]\frac{1}{2}\mathbf{Z} + \mathbf{V}_R + \mathbf{I}_R\frac{1}{2}\mathbf{Z}$$

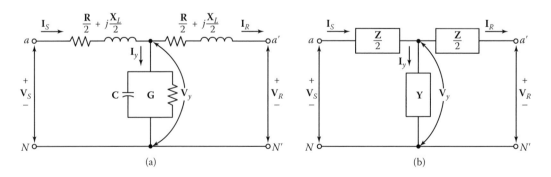

FIGURE 3.19 Nominal T circuit.

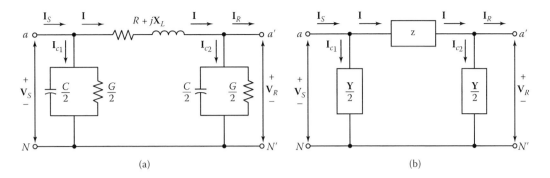

FIGURE 3.20 Nominal π circuit.

or

$$\mathbf{V}_S = \underbrace{\left(1+\frac{1}{2}\mathbf{ZY}\right)}_{A}\mathbf{V}_R + \underbrace{\left(\mathbf{Z}+\frac{1}{4}\mathbf{YZ}^2\right)}_{B}\mathbf{I}_R \qquad (3.68)$$

and

$$\mathbf{I}_S = \mathbf{I}_R + \left(\mathbf{V}_R + \mathbf{I}_R \times \frac{1}{2}\mathbf{Z}\right)\mathbf{Y}$$

or

$$\mathbf{I}_S = \underbrace{\mathbf{Y}}_{C}\times\mathbf{V}_R + \underbrace{\left(1+\frac{1}{2}\mathbf{ZY}\right)}_{D}\mathbf{I}_R \qquad (3.69)$$

Alternatively, neglecting conductance so that

$$\mathbf{I}_C = \mathbf{I}_Y$$

and

$$\mathbf{V}_C = \mathbf{V}_Y$$

yields

$$\mathbf{I}_C = \mathbf{V}_C \times \mathbf{Y}$$

$$\mathbf{V}_C = \mathbf{V}_R + \mathbf{I}_R \times \frac{1}{2}\mathbf{Z}$$

Hence,

$$\mathbf{V}_S = \mathbf{V}_C + \mathbf{I}_S \times \frac{1}{2}\mathbf{Z}$$

$$= \mathbf{V}_R + \mathbf{I}_R \times \frac{1}{2}\mathbf{Z} + \left[\mathbf{V}_R\mathbf{Y} + \mathbf{I}_R\left(1 + \frac{1}{2}\mathbf{YZ}\right)\right]\left(\frac{1}{2}\mathbf{Z}\right)$$

or

$$\mathbf{V}_S = \underbrace{\left(1 + \frac{1}{2}\mathbf{YZ}\right)}_{A}\mathbf{V}_R + \underbrace{\left(\mathbf{Z} + \frac{1}{4}\mathbf{YZ}^2\right)}_{B}\mathbf{I}_R \qquad (3.70)$$

Also,

$$\mathbf{I}_S = \mathbf{I}_R + \mathbf{I}_C$$

$$= \mathbf{I}_R + \mathbf{V}_C \times \mathbf{Y}$$

$$= \mathbf{I}_R + \left(\mathbf{V}_R + \mathbf{I}_R \times \frac{1}{2}\mathbf{Z}\right)\mathbf{Y}$$

Again,

$$\mathbf{I}_S = \underbrace{\mathbf{Y}}_{C} \times \mathbf{V}_R + \underbrace{\left(1 + \frac{1}{2}\mathbf{YZ}\right)}_{D}\mathbf{I}_R \qquad (3.71)$$

Since

$$A = 1 + \frac{1}{2}\mathbf{YZ} \qquad (3.72)$$

$$B = \mathbf{Z} + \frac{1}{4}\mathbf{YZ}^2 \qquad (3.73)$$

$$C = \mathbf{Y} \qquad (3.74)$$

$$D = 1 + \frac{1}{2}\mathbf{YZ} \qquad (3.75)$$

Steady-State Performance of Transmission Lines

for a nominal T circuit, the *general circuit parameter matrix*, or *transfer matrix*, becomes

$$\begin{bmatrix} A & B \\ C & D \end{bmatrix} = \begin{bmatrix} 1 + \frac{1}{2}YZ & Z + \frac{1}{4}YZ^2 \\ Y & 1 + \frac{1}{2}YZ \end{bmatrix}$$

Therefore,

$$\begin{bmatrix} V_S \\ I_S \end{bmatrix} = \begin{bmatrix} 1 + \frac{1}{2}YZ & Z + \frac{1}{4}YZ^2 \\ Y & 1 + \frac{1}{2}YZ \end{bmatrix} \begin{bmatrix} V_R \\ I_R \end{bmatrix} \quad (3.76)$$

and

$$\begin{bmatrix} V_R \\ I_R \end{bmatrix} = \begin{bmatrix} 1 + \frac{1}{2}YZ & Z + \frac{1}{4}YZ^2 \\ Y & 1 + \frac{1}{2}YZ \end{bmatrix} \begin{bmatrix} V_S \\ I_S \end{bmatrix} \quad (3.77)$$

For the π circuit shown in Figure 3.20,

$$V_S = \left(V_R \times \frac{1}{2}Y + I_R \right) Z + V_R$$

or

$$V_S = \underbrace{\left(1 + \frac{1}{2}YZ \right)}_{A} V_R + \underbrace{Z}_{B} \times I_R \quad (3.78)$$

and

$$I_S = \frac{1}{2} Y \times V_S + \frac{1}{2} Y \times V_R + I_R \quad (3.79)$$

By substituting Equation 3.78 into Equation 3.79,

$$I_S = \left[\left(1 + \frac{1}{2}YZ \right) V_R + ZI_R \right] \frac{1}{2}Y + \frac{1}{2}Y \times V_R + I_R$$

or

$$I_S = \underbrace{\left(Y + \frac{1}{4}Y^2Z \right)}_{C} V_R + \underbrace{\left(1 + \frac{1}{2}YZ \right)}_{D} I_R \quad (3.80)$$

Alternatively, *neglecting conductance*,

$$\mathbf{I} = \mathbf{I}_{C2} + \mathbf{I}_R$$

where

$$\mathbf{I}_{C2} = \frac{1}{2}\mathbf{Y} \times \mathbf{V}_R$$

yields

$$\mathbf{I} = \frac{1}{2}\mathbf{Y} \times \mathbf{V}_R + \mathbf{I}_R \tag{3.81}$$

Also,

$$\mathbf{V}_S = \mathbf{V}_R + \mathbf{IZ} \tag{3.82}$$

By substituting Equation 3.81 into Equation 3.82,

$$\mathbf{V}_S = \mathbf{V}_R + \left(\frac{1}{2}\mathbf{Y} \times \mathbf{V}_R + \mathbf{I}_R\right)\mathbf{Z}$$

or

$$\mathbf{V}_S = \underbrace{\left(1 + \frac{1}{2}\mathbf{YZ}\right)}_{\mathbf{A}}\mathbf{V}_R + \underbrace{\mathbf{Z}}_{\mathbf{B}} \times \mathbf{I}_R \tag{3.83}$$

and

$$\mathbf{I}_{C1} = \frac{1}{2}\mathbf{Y} \times \mathbf{V}_S \tag{3.84}$$

By substituting Equation 3.83 into Equation 3.48,

$$\mathbf{I}_{C1} = \frac{1}{2}\mathbf{Y} \times \left(1 + \frac{1}{2}\mathbf{YZ}\right)\mathbf{V}_R + \frac{1}{2}\mathbf{Y} \times \mathbf{ZI}_R \tag{3.85}$$

and since

$$\mathbf{I}_S = \mathbf{I} + \mathbf{I}_{C1} \tag{3.86}$$

by substituting Equation 3.81 into Equation 3.86,

$$\mathbf{I}_S = \frac{1}{2}\mathbf{YV}_R + \mathbf{I}_R + \frac{1}{2}\mathbf{Y}\left(1 + \frac{1}{2}\mathbf{YZ}\right)\mathbf{V}_R + \frac{1}{2}\mathbf{YZI}_R$$

Steady-State Performance of Transmission Lines

or

$$\mathbf{I}_S = \underbrace{\left(\mathbf{Y} + \frac{1}{4}\mathbf{Y}^2\mathbf{Z}\right)}_{C}\mathbf{V}_R + \underbrace{\left(1 + \frac{1}{2}\mathbf{YZ}\right)}_{D}\mathbf{I}_R \qquad (3.87)$$

Since

$$\mathbf{A} = 1 + \frac{1}{2}\mathbf{YZ} \qquad (3.88)$$

$$\mathbf{B} = \mathbf{Z} \qquad (3.89)$$

$$\mathbf{C} = \mathbf{Y} + \frac{1}{4}\mathbf{Y}^2\mathbf{Z} \qquad (3.90)$$

$$\mathbf{D} = 1 + \frac{1}{2}\mathbf{YZ} \qquad (3.91)$$

for a nominal π circuit, the general circuit parameter matrix becomes

$$\begin{bmatrix} \mathbf{A} & \mathbf{B} \\ \mathbf{C} & \mathbf{D} \end{bmatrix} = \begin{bmatrix} 1 + \frac{1}{2}\mathbf{YZ} & \mathbf{Z} \\ \mathbf{Y} + \frac{1}{4}\mathbf{Y}^2\mathbf{Z} & 1 + \frac{1}{2}\mathbf{YZ} \end{bmatrix} \qquad (3.92)$$

Therefore,

$$\begin{bmatrix} \mathbf{V}_S \\ \mathbf{I}_S \end{bmatrix} = \begin{bmatrix} 1 + \frac{1}{2}\mathbf{YZ} & \mathbf{Z} \\ \mathbf{Y} + \frac{1}{4}\mathbf{Y}^2\mathbf{Z} & 1 + \frac{1}{2}\mathbf{YZ} \end{bmatrix} \begin{bmatrix} \mathbf{V}_R \\ \mathbf{I}_R \end{bmatrix} \qquad (3.93)$$

and

$$\begin{bmatrix} \mathbf{V}_R \\ \mathbf{I}_R \end{bmatrix} = \begin{bmatrix} 1 + \frac{1}{2}\mathbf{YZ} & \mathbf{Z} \\ \mathbf{Y} + \frac{1}{4}\mathbf{Y}^2\mathbf{Z} & 1 + \frac{1}{2}\mathbf{YZ} \end{bmatrix}^{-1} \begin{bmatrix} \mathbf{V}_S \\ \mathbf{I}_S \end{bmatrix} \qquad (3.94)$$

As can be proved easily by using a delta–wye transformation, the nominal T and nominal π circuits are not equivalent to each other. This result is to be expected since two different approximations are made to the actual circuit, neither of which is absolutely correct. More accurate results can be obtained by splitting the line into several segments, each given by its nominal T or nominal π circuits and cascading the resulting segments.

Here, the power loss in the line is given as

$$P_{loss} = I^2 R \tag{3.95}$$

which varies approximately as the square of the through-line current. The reactive powers absorbed and supplied by the line are given as

$$Q_L = Q_{absorbed} = I^2 X_L \tag{3.96}$$

and

$$Q_C = Q_{supplied} = V^2 b \tag{3.97}$$

respectively. The Q_L varies approximately as the square of the through line current, whereas the Q_C varies approximately as the square of the mean line voltage. The result is that increasing transmission voltages decrease the reactive power absorbed by the line for heavy loads and increase the reactive power supplied by the line for light loads.

The percentage of voltage regulation for the medium-length transmission lines is given by Stevenson [3] as

$$\text{Percentage of voltage regulation} = \frac{\frac{|\mathbf{V}_{S,LN}|}{|\mathbf{A}|} - |\mathbf{V}_{R,FL}|}{|\mathbf{V}_{R,FL}|} \times 100 \tag{3.98}$$

where

$|\mathbf{V}_{S,LN}|$ = magnitude of sending-end phase (line-to-neutral) voltage
$|\mathbf{V}_{S,FL}|$ = magnitude of receiving-end phase (line-to-neutral) voltage at full load with constant $|\mathbf{V}_S|$
$|\mathbf{A}|$ = magnitude of line constant $\mathbf{A}$

EXAMPLE 3.6

A three-phase 138-kV transmission line is connected to a 49-MW load at a 0.85 lagging power factor. The line constants of the 52-mi-long line are $\mathbf{Z} = 95\angle 78°$ Ω and $\mathbf{Y} = 0.001\angle 90°$ S. Using *nominal T circuit representation*, calculate the

(a) **A, B, C,** and **D** constants of the line
(b) Sending-end voltage
(c) Sending-end current
(d) Sending-end power factor
(e) Efficiency of transmission

Solution

$$V_{R(L-N)} = \frac{138\,\text{kV}}{\sqrt{3}} = 79{,}624.3\,\text{V}$$

Steady-State Performance of Transmission Lines

Using the receiving-end voltage as the reference,

$$\mathbf{V}_{R(L-N)} = 79{,}624.3\angle 0° \text{ V}$$

The receiving-end current is

$$I_R = \frac{49 \times 10^6}{\sqrt{3} \times 138 \times 10^3 \times 0.85}$$

$$= 241.18 \text{ A} \quad \text{or} \quad 241.18\angle -31.80° \text{ A}$$

(a) The **A**, **B**, **C**, and **D** constants for the nominal T circuit representation are

$$\mathbf{A} = 1 + \frac{1}{2}\mathbf{YZ}$$

$$= 1 + \frac{1}{2}(0.001\angle 90°)(95\angle 78°)$$

$$= 0.9535 + j0.0099$$

$$= 0.9536\angle 0.6°$$

$$\mathbf{B} = \mathbf{Z} + \frac{1}{4}\mathbf{YZ}^2$$

$$= 95\angle 78° + \frac{1}{4}(0.001\angle 90°)(95\angle 78°)^2$$

$$= 18.83 + j90.86$$

$$= 92.79\angle 78.3° \text{ }\Omega$$

$$\mathbf{C} = \mathbf{Y} = 0.001\angle 90° \text{ S}$$

$$\mathbf{D} = 1 + \frac{1}{2}\mathbf{YZ} = \mathbf{A}$$

$$= 0.9536\angle 0.6°$$

(b)

$$\begin{bmatrix} \mathbf{V}_{S(L-N)} \\ \mathbf{I}_S \end{bmatrix} = \begin{bmatrix} 0.9536\angle 0.6° & 92.79\angle 78.3° \\ 0.001\angle 90° & 0.9536\angle 0.6° \end{bmatrix} \begin{bmatrix} 79.7674.8\angle 0° \\ 241.46\angle -31.8° \end{bmatrix}$$

The sending-end voltage is

$$\mathbf{V}_{S(L-N)} = 0.9536\angle 0.6° \times 79{,}674.8\angle 0° + 92.79\angle 78.3° \times 241.18\angle -31.8°$$

$$= 91{,}377 + j17{,}028.8 = 92{,}951.2\angle 10.6° \text{ V}$$

or

$$V_{S(L-L)} = 160{,}996.2 \angle 40.6° \text{ V}$$

(c) The sending-end current is

$$I_S = 0.001\angle 90° \times 79{,}674.8\angle 0° + 0.9536\angle 0.6° \times 241.18\angle -31.8°$$

$$= 196.72 - j39.5 = 200.64\angle -11.3° \text{ A}$$

(d) The sending-end power factor is

$$\theta_s = 10.6° + 11.3° = 21.9°$$

$$\cos\Phi_s = 0.928$$

(e) The efficiency of transmission is

$$\eta = \frac{\text{output}}{\text{input}}$$

$$= \frac{\sqrt{3}V_R I_R \cos\Phi_R}{\sqrt{3}V_S I_S \cos\Phi_S} \times 100$$

$$= \frac{138\times 10^3 \times 241.18 \times 0.85}{160{,}996.2 \times 200.64 \times 0.928} \times 100$$

$$= 94.38\%$$

EXAMPLE 3.7

Repeat Example 3.6 using the nominal π circuit representation.

Solution

(a) The **A**, **B**, **C**, and **D** constants for the nominal π circuit representation are

$$A = 1 + \frac{1}{2}YZ$$

$$= 0.9536\angle 0.6°$$

$$B = Z = 95\angle 78° \ \Omega$$

$$C = Y + \frac{1}{4}Y^2 Z$$

$$= 0.001\angle 90° + \frac{1}{4}(0.001\angle 90°)2(95\angle 78°)$$

$$= -4.9379\times 10^{-6} + j9.7677\times 10^{-4} = 0.001\angle 90.3° \text{ S}$$

$$D = 1 + \frac{1}{2}YZ = A$$

$$= 0.9536\angle 0.6°$$

(b)

$$\begin{bmatrix} \mathbf{V}_{S(L-N)} \\ \mathbf{I}_S \end{bmatrix} = \begin{bmatrix} 0.9536\angle 0.6° & 95\angle 78° \\ 0.001\angle 90.3° & 0.9536\angle 0.6° \end{bmatrix} \begin{bmatrix} 79{,}674\angle 0° \\ 241.18\angle -31.8° \end{bmatrix}$$

Therefore,

$$\mathbf{V}_{S(L-N)} = 0.9536\angle 0.6° \times 79{,}674.8\angle 0° + 95\angle 78° \times 241.46\angle -31.8°$$

$$= 91{,}831.7 + j17{,}332.7 = 93{,}453.1\angle 10.7° \text{ V}$$

or

$$\mathbf{V}_{S(L-L)} = 161{,}865.5\angle 40.7° \text{ V}$$

(c) The sending-end current is

$$\mathbf{I}_S = 0.001\angle 90.3° \times 79{,}674.3\angle 0° + 0.9536\angle 0.6° \times 241.18\angle -31.8°$$

$$= 196.31 - j39.47 = 200.24\angle -11.37° \text{ A}$$

(d) The sending-end power factor is

$$\theta_S = 10.7° + 11.37° = 22.07°$$

and

$$\cos\theta_S = 0.927$$

(e) The efficiency of transmission is

$$\eta = \frac{\text{output}}{\text{input}}$$

$$= \frac{\sqrt{3}V_R I_R \cos\theta_R}{\sqrt{3}V_S I_S \cos\theta_S} \times 100$$

$$= \frac{138\times 10^3 \times 241.18 \times 0.85}{161{,}865.5 \times 200.24 \times 0.927} \times 100$$

$$= 94.16\%$$

The discrepancy between these results and the results of Example 4.4 is due to the fact that the nominal T and nominal π circuits of a medium-length line are not equivalent to each other. In fact, neither the nominal T nor the nominal π equivalent circuit exactly represents the actual line because the line is not uniformly distributed. However, it is possible to find the equivalent circuit of a long transmission line and to represent the line accurately.

3.11 LONG TRANSMISSION LINES (ABOVE 150 MI OR 240 KM)

A more accurate analysis of the transmission lines require that the parameters of the lines are not lumped, as before, but are distributed uniformly throughout the length of the line.

Figure 3.21 shows a uniform long line with an incremental section dx at a distance x from the receiving end, its series impedance is $\mathbf{z}dx$, and its shunt admittance is $\mathbf{y}dx$, where $\mathbf{z}$ and $\mathbf{y}$ are the impedance and admittance per unit length, respectively. The voltage drop in the section is

$$d\mathbf{V}_x = (\mathbf{V}_x + d\mathbf{V}_x) - \mathbf{V}_x = d\mathbf{V}_x$$
$$= (\mathbf{I}_x + d\mathbf{I}_x)\mathbf{z}\, dx$$

or

$$d\mathbf{V}_x \cong \mathbf{I}_x \mathbf{z}\, dx \qquad (3.99)$$

Similarly, the incremental charging current is

$$d\mathbf{I}_x = \mathbf{V}_x \mathbf{y}\, dx \qquad (3.100)$$

Therefore,

$$\frac{d\mathbf{V}_x}{dx} = \mathbf{z}\mathbf{I}_x \qquad (3.101)$$

and

$$\frac{d\mathbf{I}_x}{dx} = \mathbf{y}\mathbf{V}_x \qquad (3.102)$$

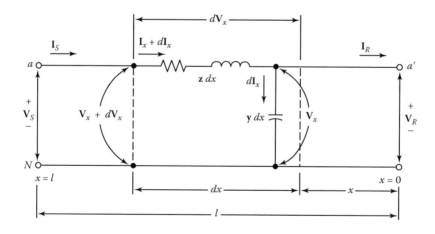

FIGURE 3.21 One phase and neutral connection of three-phase transmission line.

Steady-State Performance of Transmission Lines

Differentiating Equations 3.101 and 3.102 with respect to x,

$$\frac{d^2\mathbf{V}_x}{dx^2} = \mathbf{z}\frac{d\mathbf{I}_x}{dx} \quad (3.103)$$

and

$$\frac{d^2\mathbf{I}_x}{dx^2} = \mathbf{y}\frac{d\mathbf{V}_x}{dx} \quad (3.104)$$

Substituting the values of $d\mathbf{I}_x/dx$ and $d\mathbf{V}_x/dx$ from Equations 3.102 and 3.103 in Equations 3.105 and 3.106, respectively,

$$\frac{d^2\mathbf{V}_x}{dx^2} = \mathbf{yz}\mathbf{V}_x \quad (3.105)$$

and

$$\frac{d^2\mathbf{I}_x}{dx^2} = \mathbf{yz}\mathbf{I}_x \quad (3.106)$$

At $x = 0$, $\mathbf{V}_x = \mathbf{V}_R$ and $\mathbf{I}_x = \mathbf{I}_R$. Therefore, the solution of the ordinary second-order differential Equations 3.105 and 3.106 gives

$$\mathbf{V}(x) = \underbrace{\left(\cosh\sqrt{\mathbf{yz}}x\right)}_{A}\mathbf{V}_R + \underbrace{\left(\sqrt{\frac{\mathbf{z}}{\mathbf{y}}}\sinh\sqrt{\mathbf{yz}}x\right)}_{B}\mathbf{I}_R \quad (3.107)$$

Similarly,

$$\mathbf{I}_{(x)} = \underbrace{\left(\sqrt{\frac{\mathbf{y}}{\mathbf{z}}}\sinh\sqrt{\mathbf{yz}}x\right)}_{C}\mathbf{V}_R + \underbrace{\left(\cosh\sqrt{\mathbf{yz}}x\right)}_{D}\mathbf{I}_R \quad (3.108)$$

Equations 3.107 and 3.108 can be rewritten as

$$\mathbf{V}_{(x)} = (\cosh\gamma x)\mathbf{V}_R + (\mathbf{Z}_c\sinh\gamma x)\mathbf{I}_R \quad (3.109)$$

and

$$\mathbf{I}(x) = (\mathbf{Y}_c\sinh\gamma x)\mathbf{V}_R + (\cosh\gamma x)\mathbf{I}_R \quad (3.110)$$

where
γ = propagation constant per unit length, $= \sqrt{\mathbf{yz}}$
$\mathbf{Z}_c$ = characteristic (or surge or natural) impedance of line per unit length, $= \sqrt{\mathbf{z}/\mathbf{y}}$
$\mathbf{Y}_c$ = characteristic (or surge or natural) admittance of line per unit length, $= \sqrt{\mathbf{y}/\mathbf{z}}$

Further,

$$\gamma = \alpha + j\beta \tag{3.111}$$

where

- α = attenuation constant (measuring decrement in voltage and current per unit length in direction of travel) in nepers per unit length
- β = phase (or phase change) constant in radians per unit length (i.e., change in phase angle between two voltages, or currents, at two points one per unit length apart on infinite line)

When $x = l$, Equations 3.109 and 3.110 become

$$\mathbf{V}_S = (\cosh \gamma l)\mathbf{V}_R + (\mathbf{Z}_c \sinh \gamma l)\mathbf{V}_R \tag{3.112}$$

and

$$\mathbf{I}_S = (\mathbf{Y}_c \sinh \gamma l)\mathbf{V}_R + (\cosh \gamma l)\mathbf{I}_R \tag{3.113}$$

Equations 3.112 and 3.113 can be written in matrix form as

$$\begin{bmatrix} \mathbf{V}_S \\ \mathbf{I}_S \end{bmatrix} = \begin{bmatrix} \cosh \gamma l & \mathbf{Z}_c \sinh \gamma l \\ \mathbf{Y}_c \sinh \gamma l & \cosh \gamma l \end{bmatrix} \begin{bmatrix} \mathbf{V}_R \\ \mathbf{I}_R \end{bmatrix} \tag{3.114}$$

and

$$\begin{bmatrix} \mathbf{V}_R \\ \mathbf{I}_R \end{bmatrix} = \begin{bmatrix} \cosh \gamma l & \mathbf{Z}_c \sinh \gamma l \\ \mathbf{Y}_c \sinh \gamma l & \cosh \gamma l \end{bmatrix} \begin{bmatrix} \mathbf{V}_S \\ \mathbf{I}_S \end{bmatrix} \tag{3.115}$$

or

$$\begin{bmatrix} \mathbf{V}_R \\ \mathbf{I}_R \end{bmatrix} = \begin{bmatrix} \cosh \gamma l & -\mathbf{Z}_c \sinh \gamma l \\ -\mathbf{Y}_c \sinh \gamma l & \cosh \gamma l \end{bmatrix} \begin{bmatrix} \mathbf{V}_S \\ \mathbf{I}_S \end{bmatrix} \tag{3.116}$$

Therefore,

$$\mathbf{V}_R = (\cosh \gamma l)\mathbf{V}_S + (\mathbf{Z}_c \sinh \gamma l)\mathbf{I}_S \tag{3.117}$$

and

$$\mathbf{I}_R = -(\mathbf{Y}_c \sinh \gamma l)\mathbf{V}_S + (\cosh \gamma l)\mathbf{I}_S \tag{3.118}$$

In terms of **ABCD** constants,

$$\begin{bmatrix} \mathbf{V}_S \\ \mathbf{I}_S \end{bmatrix} = \begin{bmatrix} \mathbf{A} & \mathbf{B} \\ \mathbf{C} & \mathbf{D} \end{bmatrix} \begin{bmatrix} \mathbf{V}_R \\ \mathbf{I}_R \end{bmatrix} = \begin{bmatrix} \mathbf{A} & \mathbf{B} \\ \mathbf{C} & \mathbf{A} \end{bmatrix} \begin{bmatrix} \mathbf{V}_R \\ \mathbf{I}_R \end{bmatrix} \tag{3.119}$$

and

$$\begin{bmatrix} \mathbf{V}_R \\ \mathbf{I}_R \end{bmatrix} = \begin{bmatrix} \mathbf{A} & -\mathbf{B} \\ -\mathbf{C} & \mathbf{D} \end{bmatrix} \begin{bmatrix} \mathbf{V}_S \\ \mathbf{I}_S \end{bmatrix} = \begin{bmatrix} \mathbf{A} & -\mathbf{B} \\ -\mathbf{C} & \mathbf{A} \end{bmatrix} \begin{bmatrix} \mathbf{V}_S \\ \mathbf{I}_S \end{bmatrix} \quad (3.120)$$

where

$$\mathbf{A} = \cosh \gamma l = \cosh \sqrt{\mathbf{YZ}} = \cosh \theta \quad (3.121)$$

$$\mathbf{B} = \mathbf{Z}_c \sinh \gamma l = \sqrt{\mathbf{Z}/\mathbf{Y}} \sinh \sqrt{\mathbf{YZ}} = \mathbf{Z}_c \sinh \theta \quad (3.122)$$

$$\mathbf{C} = \mathbf{Y}_c \sinh \gamma l = \sqrt{\mathbf{Y}/\mathbf{Z}} \sinh \sqrt{\mathbf{YZ}} = \mathbf{Y}_c \sinh \theta \quad (3.123)$$

$$\mathbf{D} = \mathbf{A} = \cosh \gamma l = \cosh \sqrt{\mathbf{YZ}} = \cosh \theta \quad (2.124)$$

$$\theta = \sqrt{\mathbf{YZ}} \quad (3.125)$$

$$\sinh \gamma l = \frac{1}{2}(e^{\gamma l} - e^{-\gamma l}) \quad (3.126)$$

$$\cosh \gamma l = \frac{1}{2}(e^{\gamma l} + e^{-\gamma l}) \quad (3.127)$$

Also,

$$\sinh(\alpha + j\beta) = \frac{e^{\alpha}e^{j\beta} - e^{-\alpha}e^{-j\beta}}{2} = \frac{1}{2}\left[e^{\alpha}\angle\beta - e^{-\alpha}\angle-\beta\right]$$

and

$$\cosh(\alpha + j\beta) = \frac{e^{\alpha}e^{j\beta} + e^{-\alpha}e^{-j\beta}}{2} = \frac{1}{2}\left[e^{\alpha}\angle\beta + e^{-\alpha}\angle-\beta\right]$$

Note that, the β in the above equations is the radian, and the radian is the unit found for β by computing the quadrature component of γ. Since 2π radians = 360°, 1 rad is 57.3°. Thus, the β is converted into degrees by multiplying its quantity by 57.3°. For a line length of l,

$$\sinh(\alpha l + j\beta l) = \frac{e^{\alpha l}e^{j\beta l} - e^{-\alpha l}e^{-j\beta l}}{2} = \frac{1}{2}\left[e^{\alpha l}\angle\beta l - e^{-\alpha l}\angle-\beta l\right]$$

and

$$\cosh(\alpha l + j\beta l) = \frac{e^{\alpha l}e^{j\beta l} + e^{-\alpha l}e^{-j\beta l}}{2} = \frac{1}{2}\left[e^{\alpha l}\angle\beta l + e^{-\alpha l}\angle-\beta l\right]$$

Equations 3.112 through 3.125 can be used if tables of complex hyperbolic functions or pocket calculators with complex hyperbolic functions are available.

Alternatively, the following expansions can be used:

$$\sinh \gamma l = \sinh(\alpha l + j\beta l) = \sinh \alpha l \cos \beta l + j \cosh \alpha l \sin \beta l \qquad (3.128)$$

and

$$\cosh \gamma l = \cosh(\alpha l + j\beta l) = \cosh \alpha l \cos \beta l + j \sinh \alpha l \sin \beta l \qquad (3.129)$$

The correct mathematical unit for βl is the radian, and the radian is the unit found for βl by computing the quadrature component of γl.

Furthermore, substituting for γl and $\mathbf{Z}_c$ in terms of $\mathbf{Y}$ and $\mathbf{Z}$, that is, the total line shunt admittance per phase and the total line series impedance per phase, in Equation 3.119 gives

$$\mathbf{V}_S = \left(\cosh\sqrt{\mathbf{YZ}}\right)\mathbf{V}_R + \left(\sqrt{\frac{\mathbf{Z}}{\mathbf{Y}}}\sinh\sqrt{\mathbf{YZ}}\right)\mathbf{I}_R \qquad (3.130)$$

and

$$\mathbf{I}_S = \left(\sqrt{\frac{\mathbf{Y}}{\mathbf{Z}}}\sinh\sqrt{\mathbf{YZ}}\right)\mathbf{V}_R + \left(\cosh\sqrt{\mathbf{YZ}}\right)\mathbf{I}_R \qquad (3.131)$$

or, alternatively,

$$\mathbf{V}_S = \left(\cosh\sqrt{\mathbf{YZ}}\right)\mathbf{V}_R + \left(\frac{\sinh\sqrt{\mathbf{YZ}}}{\sqrt{\mathbf{YZ}}}\right)\mathbf{ZI}_R \qquad (3.132)$$

and

$$\mathbf{I}_S = \left(\frac{\sinh\sqrt{\mathbf{YZ}}}{\sqrt{\mathbf{YZ}}}\right)\mathbf{YV}_R + \left(\cosh\sqrt{\mathbf{YZ}}\right)\mathbf{I}_R \qquad (3.133)$$

The factors in parentheses in Equations 3.130 through 3.133 can readily be found by using Woodruff's charts, which are not included here but can be found in L. F. Woodruff, *Electric Power Transmission* (Wiley, NY, 1952).

The **ABCD** parameters in terms of infinite series can be expressed as

$$\mathbf{A} = 1 + \frac{\mathbf{YZ}}{2} + \frac{\mathbf{Y}^2\mathbf{Z}^2}{24} + \frac{\mathbf{Y}^3\mathbf{Z}^3}{720} + \frac{\mathbf{Y}^4\mathbf{Z}^4}{40,320} + \cdots \qquad (3.134)$$

$$\mathbf{B} = \mathbf{Z}\left(1 + \frac{\mathbf{YZ}}{6} + \frac{\mathbf{Y}^2\mathbf{Z}^2}{120} + \frac{\mathbf{Y}^3\mathbf{Z}^3}{5040} + \frac{\mathbf{Y}^4\mathbf{Z}^4}{362,880} + \cdots\right) \qquad (3.135)$$

Steady-State Performance of Transmission Lines

$$C = Y\left(1 + \frac{YZ}{6} + \frac{Y^2Z^2}{120} + \frac{Y^3Z^3}{5040} + \frac{Y^4Z^4}{362{,}880} + \cdots\right) \quad (3.136)$$

where
  Z = total line series impedance per phase
   $= zl$
   $= (r + jx_L)l \; \Omega$
  Y = total line shunt admittance per phase
   $= yl$
   $= (g + jb)l \; \text{S}$

In practice, usually not more than three terms are necessary in Equations 3.134 through 3.136. Weedy [7] suggests the following approximate values for the **ABCD** constants if the overhead transmission line is <500 km in length:

$$A = 1 + \frac{1}{2}YZ \quad (3.137)$$

$$B = Z\left(1 + \frac{1}{6}YZ\right) \quad (3.138)$$

$$C = Y\left(1 + \frac{1}{6}YZ\right) \quad (3.139)$$

However, the error involved may be too large to be ignored for certain applications.

EXAMPLE 3.8

A single-circuit, 60-Hz, three-phase transmission line is 150 mi long. The line is connected to a load of 50 MVA at a lagging power factor of 0.85 at 138 kV. The line constants are given as $R = 0.1858 \; \Omega/\text{mi}$, $L = 2.60 \; \text{mH/mi}$, and $C = 0.012 \; \mu\text{F/mi}$. Calculate the following:

(a) **A**, **B**, **C**, and **D** constants of the line
(b) Sending-end voltage
(c) Sending-end current
(d) Sending-end power factor
(e) Sending-end power
(f) Power loss in the line
(g) Transmission line efficiency
(h) Percentage of voltage regulation
(i) Sending-end charging current at no load
(j) Value of receiving-end voltage rise at no load if the sending-end voltage is held constant

Solution

$$z = 0.1858 + j2\pi \times 60 \times 2.6 \times 10^{-3}$$
$$= 0.1858 + j0.9802$$
$$= 0.9977 \angle 79.27° \; \Omega/\text{mi}$$

and

$$y = j2\pi \times 60 \times 0.012 \times 10^{-6}$$
$$= 4.5239 \times 10^{-6} \angle 90° \text{ S/mi}$$

The propagation constant of the line is

$$\gamma = \sqrt{yz}$$
$$= [(4.5239 \times 10^{-6} \angle 90°)(0.9977 \angle 79.27°)]^{1/2}$$
$$= [4.5135 \times 10^{-6}]^{1/2} \angle \left(\frac{90° + 79.27°}{2}\right)$$
$$= 0.00214499 \angle 84.63°$$
$$= 0.0002007 + j0.0021346$$

Thus,

$$\gamma l = \alpha l + j\beta l$$
$$= (0.0002007 + j0.0021346)150$$
$$\cong 0.0301 + j0.3202$$

The characteristic impedance of the line is

$$Z_c = \sqrt{\frac{z}{y}} = \left(\frac{0.9977 \angle 79.27°}{4.5239 \times 10^{-6} \angle 90°}\right)^{1/2}$$
$$= \left(\frac{0.9977 \times 10^6}{4.5239}\right)^{1/2} \angle \left(\frac{79.27° - 90°}{2}\right)$$
$$= 469.62 \angle -5.37° \, \Omega$$

The receiving-end line-to-neutral voltage is

$$V_{R(L-N)} = \frac{138 \text{ kV}}{\sqrt{3}} = 79{,}674.34 \text{ V}$$

Using the receiving-end voltage as the reference,

$$V_{R(L-N)} = 79{,}674.34 \angle 0° \text{ V}$$

The receiving-end current is

$$I_R = \frac{50 \times 10^6}{\sqrt{3} \times 138 \times 10^3}$$
$$= 209.18 \text{ A} \quad \text{or} \quad 209.18 \angle -31.8° \text{ A}$$

Steady-State Performance of Transmission Lines

(a) The **A**, **B**, **C**, and **D** constants of the line

$$\mathbf{A} = \cosh \gamma l$$
$$= \cosh(\alpha + j\beta)l$$
$$= \frac{e^{\alpha l}e^{j\beta l} + e^{-\alpha l}e^{-j\beta l}}{2}$$
$$= \frac{e^{\alpha l}\angle \beta l + e^{-\alpha l}\angle -\beta l}{2}$$

Therefore,

$$\mathbf{A} = \frac{e^{0.0301}e^{j0.3202} + e^{-0.0301}e^{-j0.3202}}{2}$$
$$= \frac{e^{0.0301}\angle 18.35° + e^{-0.0301}\angle -18.35°}{2}$$
$$= \frac{1.0306\angle 18.35° + 0.9703\angle -18.35°}{2}$$
$$= 0.9496 + j0.0095 = 0.9497\angle 0.57°$$

Note that, $e^{j0.3202}$ needs to be converted to degrees. Since 2π rad = 360°, 1 rad is 57.3°. Hence,

$$(0.3202 \text{ rad})(57.3°/\text{rad}) = 18.35°$$

and

$$\mathbf{B} = \mathbf{Z}_C \sinh \gamma l = \mathbf{Z}_C \sinh(\alpha + j\beta)l$$
$$= \mathbf{Z}_C \left[\frac{e^{\alpha l}e^{j\beta l} - e^{-\alpha l}e^{-j\beta l}}{2}\right]$$
$$= \mathbf{Z}_C \left[\frac{e^{\alpha l}\angle \beta l - e^{-\alpha l}\angle -\beta l}{2}\right]$$
$$= (469.62\angle -5.37°)\left[\frac{e^{0.0301}e^{j0.3202} - e^{-0.0301}e^{-j0.3202}}{2}\right]$$
$$= 469.62\angle -5.37°\left[\frac{1.0306\angle 18.35° - 0.9703\angle -18.35°}{2}\right]$$
$$= 469.62\angle -5.37°\left(\frac{0.0572 + j0.63}{2}\right)$$
$$= 469.62\angle -5.37°\left(\frac{0.6326\angle 84.81°}{2}\right)$$
$$= 469.62\angle -5.37°(0.3163\angle 84.81°)$$
$$= 148.54\angle 79.44° \ \Omega$$

and

$$C = Y_c \sinh \gamma l = \frac{1}{Z_c} \sinh \gamma l$$

$$= \frac{1}{469.62 \angle -5.37°} \times \frac{0.63259}{2} \angle 84.81°$$

$$= 0.00067 \angle 90.18° \text{ S}$$

and

$$D = A = \cosh \gamma l = 0.9497 \angle 0.57°$$

(b)

$$\begin{bmatrix} V_{S(L-N)} \\ I_S \end{bmatrix} = \begin{bmatrix} A & B \\ C & D \end{bmatrix} \begin{bmatrix} V_{R(L-N)} \\ I_R \end{bmatrix}$$

$$= \begin{bmatrix} 0.9497 \angle 0.57° & 148.54 \angle 79.44° \\ 0.00067 \angle 90.18° & 0.9497 \angle 0.57° \end{bmatrix} \begin{bmatrix} 79{,}674.34 \angle 0° \\ 209.18 \angle -31.8° \end{bmatrix}$$

Thus, the sending-end voltage is

$$V_{S(L-N)} = (0.9497 \angle 0.57°)(79{,}674.34 \angle 0°) + (148.54 \angle 79.44°)(209.18 \angle -31.8°)$$

$$= 99{,}466.41 \angle 13.79° \text{ V}$$

and

$$V_{S(L-L)} = \sqrt{3} V_{S(L-N)}$$

$$= 172{,}280.87 \angle 13.79° + 30°$$

$$= 172{,}280.87 \angle 43.79° \text{ V}$$

Note that, an additional 30° is added to the angle since a line-to-line voltage is 30° ahead of its line-to-neutral voltage.

(c) The sending-end current is

$$I_S = (0.00067 \angle 90.18°)(79{,}674.34 \angle 0°) + (0.9497 \angle 0.57°)(209.18 \angle -31.8°)$$

$$= 180.88 \angle -16.3° \text{ A}$$

(d) The sending-end power factor is

$$\theta_S = 13.79° + 16.3°$$

$$= 30.09°$$

$$\cos \theta_S = 0.9648$$

(e) The sending-end power is

$$P_S = \sqrt{3}V_{S(L-L)}I_S \cos\theta_S$$
$$= \sqrt{3} \times 172,280.87 \times 180.88 \times 0.9948$$
$$\cong 45,652.79 \text{ kW}$$

(f) The receiving-end power is

$$P_R = \sqrt{3}V_{R(L-L)}I_R \cos\theta_R$$
$$= \sqrt{3} \times 138 \times 10^3 \times 209.18 \times 0.85$$
$$= 42,499 \text{ kW}$$

Therefore, the power loss in the line is

$$P_L = P_S - P_R = 3153.79 \text{ kW}$$

(g) The transmission line efficiency is

$$\eta = \frac{P_R}{P_S} \times 100$$
$$= \frac{42,499}{45,652.79} \times 100$$
$$= 93.1\%$$

(h) The percentage of voltage regulation is

$$\text{Percentage of voltage regulation} = \frac{99,470.05 - 79,674.34}{79,674.34} \times 100$$
$$= 24.8\%$$

(i) The sending-end charging current at no load is

$$I_c = \frac{1}{2}YV_{S(L-N)}$$
$$= \frac{1}{2}(678.585 \times 10^{-6})(99,466.41)$$
$$= 33.75 \text{ A}$$

where

$$Y = y \times l$$
$$= (4.5239 \times 10^{-6} \text{ S/mi})(150 \text{ mi})$$
$$= 678.585 \times 10^{-6} \text{ S}$$

(j) The receiving-end voltage rise at no load is

$$V_{R(L-N)} = V_{S(L-N)} - I_c Z$$
$$= 99{,}466.41\angle 13.79° - (33.75\angle 103.79°)(149.66\angle 79.27°)$$
$$= 104{,}433.09\angle 13.27°\ V$$

Therefore, the line-to-line voltage at the receiving end is

$$V_{R(L-L)} = \sqrt{3}\, V_{R(L-N)} = 180{,}883.42\angle 13.27° + 30°$$
$$= 180{,}883.42\angle 43.27°\ V$$

Note that, in a well-designed transmission line, the voltage regulation and the line efficiency should be not greater than about 5%.

3.11.1 Equivalent Circuit of Long Transmission Line

Using the values of the **ABCD** parameters obtained for a transmission line, it is possible to develop an exact π or an exact T, as shown in Figure 3.22. For the equivalent π circuit,

$$Z_\pi = B = Z_c \sinh\theta \tag{3.140}$$

$$= Z_c \sinh \gamma l \tag{3.141}$$

$$= Z\left(\frac{\sinh\sqrt{YZ}}{\sqrt{YZ}}\right) \tag{3.142}$$

and

$$\frac{Y_\pi}{2} = \frac{A-1}{B} = \frac{\cosh\theta - 1}{Z_c \sinh\theta} \tag{3.143}$$

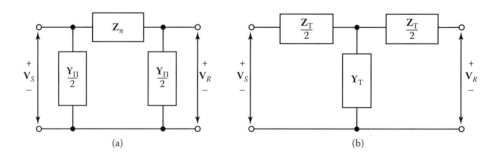

FIGURE 3.22 Equivalent π and T circuits for a long transmission line.

or

$$\mathbf{Y}_\pi = \frac{2\tanh\left(\frac{\gamma l}{2}\right)}{\mathbf{Z}_c} \tag{3.144}$$

or

$$\frac{\mathbf{Y}_\pi}{2} = \frac{\mathbf{Y}}{2}\frac{2\tanh\left(\frac{\sqrt{\mathbf{YZ}}}{2}\right)}{\frac{\sqrt{\mathbf{YZ}}}{2}} \tag{3.145}$$

For the equivalent T circuit,

$$\frac{\mathbf{Z}_T}{2} = \frac{\mathbf{A}-1}{\mathbf{C}} = \frac{\cosh\theta - 1}{\mathbf{Y}_c \sinh\theta} \tag{3.146}$$

or

$$\mathbf{Z}_T = 2\mathbf{Z}_c \tanh\frac{\gamma l}{2} \tag{3.147}$$

or

$$\frac{\mathbf{Z}_T}{2} = \frac{\mathbf{Z}}{2}\left(\frac{\tanh\frac{\sqrt{\mathbf{YZ}}}{2}}{\frac{\sqrt{\mathbf{YZ}}}{2}}\right) \tag{3.148}$$

and

$$\mathbf{Y}_T = \mathbf{C} = \mathbf{Y}_c \sinh\theta \tag{3.149}$$

or

$$\mathbf{Y}_T = \frac{\sinh\gamma l}{\mathbf{Z}_c} \tag{3.150}$$

or

$$\mathbf{Y}_T = \mathbf{Y}\frac{\sinh\sqrt{\mathbf{YZ}}}{\sqrt{\mathbf{YZ}}} \tag{3.151}$$

EXAMPLE 3.9

Find the equivalent π and the equivalent T circuits for the line described in Example 3.8 and compare them with the nominal π and the nominal T circuits.

Solution

Figures 3.23 and 3.24 show the equivalent π and the nominal π circuits, respectively.
For the equivalent π circuit,

$$Z_\pi = B = 148.54 \angle 79.44° \ \Omega$$

$$\begin{aligned}
\frac{Y_\pi}{2} &= \frac{A-1}{B} \\
&= \frac{0.9497\angle 0.57° - 1}{148.54\angle 79.44°} \\
&= 0.000345\angle 89.93° \text{ S}
\end{aligned}$$

For the nominal π circuit,

$$Z = 150 \times 0.9977 \angle 79.27°$$
$$= 149.655 \angle 79.27° \ \Omega$$

$$\frac{Y}{2} = \frac{150(4.5239 \times 10^{-6} \angle 90°)}{2}$$
$$= 0.000339 \angle 90° \text{ S}$$

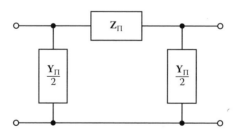

FIGURE 3.23 Equivalent π circuit.

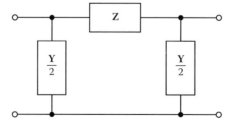

FIGURE 3.24 Nominal π circuit.

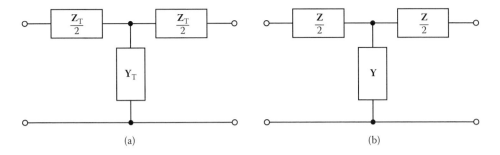

FIGURE 3.25 T circuits: (a) equivalent T; (b) nominal T.

Figure 3.25a and b shows the equivalent T and nominal T circuits, respectively. For the equivalent T circuit,

$$\frac{Z_T}{2} = \frac{A-1}{C}$$

$$= \frac{0.9497\angle 0.57° - 1}{0.00067\angle 90.18°}$$

$$= 76.46\angle 79.19° \ \Omega$$

$$Y_T = C = 0.00067\angle 90.18° \text{ S}$$

For the nominal T circuit,

$$\frac{Z}{2} = \frac{149.655\angle 79.27°}{2}$$

$$= 74.83\angle 79.27° \ \Omega$$

$$Y = 0.000678\angle 90° \text{ S}$$

As can be observed from the results, the difference between the values for the equivalent and nominal circuits is very small for a 150-mi-long transmission line.

3.11.2 Incident and Reflected Voltages of Long Transmission Line

Previously, the propagation constant has been given as

$$\gamma = \alpha + j\beta \text{ per-unit length} \tag{3.152}$$

and also

$$\cosh \gamma l = \frac{e^{\gamma l} + e^{-\gamma l}}{2} \tag{3.153}$$

$$\sinh \gamma l = \frac{e^{\gamma l} - e^{-\gamma l}}{2} \tag{3.154}$$

The sending-end voltage and current have been expressed as

$$\mathbf{V}_S = (\cosh \gamma l)\mathbf{V}_R + (\mathbf{Z}_c \sinh \gamma l)\mathbf{I}_R \quad (3.112)$$

and

$$\mathbf{I}_S = (\mathbf{Y}_c \sinh \gamma l)\mathbf{V}_R + (\cosh \gamma l)\mathbf{I}_R \quad (3.113)$$

By substituting Equations 3.152 through 3.154 in Equations 3.112 and 3.113,

$$\mathbf{V}_S = \frac{1}{2}(\mathbf{V}_R + \mathbf{I}_R \mathbf{Z}_c)e^{\alpha l}e^{j\beta l} + \frac{1}{2}(\mathbf{V}_R - \mathbf{I}_R \mathbf{Z}_c)e^{-\alpha l}e^{-j\beta l} \quad (3.155)$$

and

$$\mathbf{I}_S = \frac{1}{2}(\mathbf{V}_R \mathbf{Y}_c + \mathbf{I}_R)e^{\alpha l}e^{j\beta l} - \frac{1}{2}(\mathbf{V}_R \mathbf{Y}_c - \mathbf{I}_R)e^{-\alpha l}e^{-j\beta l} \quad (3.156)$$

In Equation 3.155, the first and the second terms are called the *incident voltage* and the *reflected voltage*, respectively. They act like *traveling waves* as a function of the line length l. The incident voltage increases in magnitude and phase as the l distance from the receiving end increases, and decreases in magnitude and phase as the distance from the sending end toward the receiving end decreases. Whereas the reflected voltage decreases in magnitude and phase as the l distance from the receiving end toward the sending end increases.

Thus, for any given line length l, the voltage is the sum of the corresponding incident and reflected voltages. Here, the term $e^{\alpha l}$ changes as a function of l, whereas $e^{j\beta l}$ always has a magnitude of 1 and causes a phase shift of β radians per unit length of line.

In Equation 3.155, when the two terms are 180° out of phase, a cancellation will occur. This happens when there is no load on the line, that is, when

$$\mathbf{I}_R = 0 \text{ and } \alpha = 0$$

and when $\beta x = \dfrac{\pi}{2}$ radians, or one-quarter wavelengths.

The wavelength λ is defined as the distance l along a line between two points to develop a phase shift of 2π radians, or 360°, for the incident and reflected waves. If β is the phase shift in radians per mile, the wavelength in miles is

$$\lambda = \frac{2\pi}{\beta} \quad (3.157)$$

Since the propagation velocity is

$$\upsilon = \lambda f \text{ mi/s} \quad (3.158)$$

and is approximately equal to the speed of light, that is, 186,000 mil, at a frequency of 60 Hz, the wavelength is

$$\lambda = \frac{186{,}000 \text{ mi/s}}{60 \text{ Hz}} = 3100 \text{ mi}$$

Steady-State Performance of Transmission Lines

On the other hand, at a frequency of 50 Hz, the wavelength is approximately 6000 km. If a finite line is terminated by its characteristic impedance $\mathbf{Z}_c$, that impedance could be imagined replaced by an infinite line. In this case, there is no reflected wave of either voltage or current since

$$\mathbf{V}_R = \mathbf{I}_R \mathbf{Z}_c$$

in Equations 3.155 and 3.156, and the line is called an *infinite* (or *flat*) line.

Stevenson [3] gives the typical values of $\mathbf{Z}_c$ as 400 ft for a single-circuit line and 200 Ω for two circuits in parallel. The phase angle of $\mathbf{Z}_c$ is usually between 0 and −15° [4].

EXAMPLE 3.10

Using the data given in Example 3.8, determine the following:

(a) Attenuation constant and phase change constant per mile of the line
(b) Wavelength and velocity of propagation
(c) Incident and reflected voltages at the receiving end of the line
(d) Line voltage at the receiving end of the line
(e) Incident and reflected voltages at the sending end of the line
(f) Line voltage at the sending end

Solution

(a) Since the propagation constant of the line is

$$\gamma = \sqrt{\mathbf{yz}}$$
$$= 0.0002 + j0.0021$$

then, the attenuation constant is 0.0002 Np/mi, and the phase change constant is 0.0021 rad/mi.

(b) The wavelength of propagation is

$$\lambda = \frac{2\pi}{\beta}$$
$$= \frac{2\pi}{0.0021}$$
$$= 2991.99 \text{ mi}$$

and the velocity of propagation is

$$\upsilon = \lambda f = 2991.99 \times 60 = 179{,}519.58 \text{ mi/s}$$

(c) From Equation 3.155,

$$\mathbf{V}_S = \frac{1}{2}(\mathbf{V}_R + \mathbf{I}_R \mathbf{Z}_c)e^{\alpha l}e^{j\beta l} + \frac{1}{2}(\mathbf{V}_R - \mathbf{I}_R \mathbf{Z}_c)e^{-\alpha l}e^{-j\beta l}$$

Since, at the receiving end, $I = 0$,

$$\mathbf{V}_S = \frac{1}{2}(\mathbf{V}_R + \mathbf{I}_R\mathbf{Z}_c) + \frac{1}{2}(\mathbf{V}_R - \mathbf{I}_R\mathbf{Z}_c)$$

Therefore, the incident and reflected voltage at the receiving end are

$$\mathbf{V}_{R(\text{incident})} = \frac{1}{2}(\mathbf{V}_R + \mathbf{I}_R\mathbf{Z}_c)$$

$$= \frac{1}{2}[79{,}674.34 \angle 0° + (209.18 \angle -31.8°)(469.62 \angle -5.37°)]$$

$$= 84{,}367.77 \angle -20.59°\ \text{V}$$

and

$$\mathbf{V}_{R(\text{reflected})} = \frac{1}{2}(\mathbf{V}_R - \mathbf{I}_R\mathbf{Z}_c)$$

$$= \frac{1}{2}[79{,}674.34 \angle 0° - (209.18 \angle -31.8°)(469.62 \angle -5.37°)]$$

$$= 29{,}684.15 \angle 88.65°\ \text{V}$$

(d) The line-to-neutral voltage at the receiving end is

$$\mathbf{V}_{R(L-N)} = \mathbf{V}_{R(\text{incident})} + \mathbf{V}_{R(\text{reflected})}$$

$$= 79{,}674 \angle 0°\ \text{V}$$

Therefore, the line voltage at the receiving end is

$$V_{R(L-L)} = \sqrt{3}\, V_{R(L-N)}$$

$$= 138{,}000\ \text{V}$$

(e) At the sending end,

$$\mathbf{V}_{S(\text{incident})} = \frac{1}{2}(\mathbf{V}_R + \mathbf{I}_R\mathbf{Z}_c)e^{\alpha l}e^{j\beta l}$$

$$= (84{,}367.77 \angle -20.59°)e^{0.0301} \angle 18.35°$$

$$= 86{,}946 \angle -2.24°\ \text{V}$$

and

$$\mathbf{V}_{S(\text{reflected})} = \frac{1}{2}(\mathbf{V}_R - \mathbf{I}_R\mathbf{Z}_c)e^{-\alpha l}e^{-j\beta l}$$

$$= (29{,}684.15 \angle 88.65°)e^{-0.0301} \angle -18.35° = 28{,}802.5 \angle 70.3°\ \text{V}$$

Steady-State Performance of Transmission Lines

(f) The line-to-neutral voltage at the sending end is

$$\mathbf{V}_{S(L-N)} = \mathbf{V}_{S(\text{incident})} + \mathbf{V}_{S(\text{reflected})}$$
$$= 86{,}946\angle -2.24° + 28{,}802.5\angle 70.3° = 99{,}458.1\angle 13.8° \text{ V}$$

Therefore, the line voltage at the sending end is

$$V_{S(L-L)} = \sqrt{3}\, V_{S(L-N)} = 172{,}266.5 \text{ V}$$

3.11.3 Surge Impedance Loading of Transmission Line

In power systems, if the line is *lossless*,* the characteristic impedance Z_c of a line is sometimes called *surge impedance*. Therefore, for a loss-free line,

$$R = 0$$

and

$$\mathbf{Z}_L = jX_L$$

Thus,

$$Z_c = \sqrt{\frac{X_L}{Y_c}} \cong \sqrt{\frac{L}{C}}\ \Omega \tag{3.159}$$

and its series resistance and shunt conductance are zero. It is a function of the line inductance and capacitance as shown and is independent of the line length.

The surge impedance loading (SIL) (or the *natural loading*) of a transmission line is defined as the power delivered by the line to a purely resistive load equal to its surge impedance. Therefore,

$$\text{SIL} = \frac{|k\mathbf{V}_{R(L-L)}|^2}{Z_c^*}\ \text{MW} \tag{3.160}$$

or

$$\text{SIL} \cong \frac{|k\mathbf{V}_{R(L-L)}|^2}{\sqrt{\frac{L}{C}}}\ \text{MW} \tag{3.161}$$

or

$$\text{SIL} = \sqrt{3}\,|\mathbf{V}_{R(L-L)}|\,|\mathbf{I}_L|\ \text{W} \tag{3.162}$$

---

* When dealing with high frequencies or with surges due to lightning, losses are often ignored [3].

where

$$|\mathbf{I}_L| = \frac{|\mathbf{V}_{R(L-L)}|}{\sqrt{3} \times \sqrt{\frac{L}{C}}} \text{ A} \qquad (3.163)$$

where
SIL = surge impedance loading in megawatts or watts
$|k\mathbf{V}_{R(L-L)}|$ = magnitude of line-to-line receiving-end voltage in kilovolts
$|\mathbf{V}_{R(L-L)}|$ = magnitude of line-to-line receiving-end voltage in volts
Z_c = surge impedance in ohms $\cong \sqrt{L/C}$
$\mathbf{I}_L$ = line current at SIL in amperes

In practice, the allowable loading of a transmission line may be given as a fraction of its SIL. Thus, SIL is used as a means of comparing the load-carrying capabilities of lines.

However, the SIL in itself is not a measure of the maximum power that can be delivered over a line. For the maximum delivered power, the line length, the impedance of sending- and receiving-end apparatus, and all of the other factors affecting stability must be considered.

Since the characteristic impedance of underground cables is very low, the SIL (or natural load) is far larger than the rated load of the cable. Therefore, a given cable acts as a source of lagging vars.

The best way of increasing the SIL of a line is to increase its voltage level, since, as it can be seen from Equation 3.160, the SIL increases with its square. However, increasing voltage level is expensive. Therefore, instead, the surge impedance of the line is reduced. This can be accomplished by adding capacitors or induction coils. There are four possible ways of changing the line capacitance or inductance, as shown in Figures 3.26 and 3.27.

For a lossless line, the characteristic impedance and the propagation constant can be expressed as

$$Z_c = \sqrt{\frac{L}{C}} \qquad (3.164)$$

and

$$\gamma = \sqrt{LC} \qquad (3.165)$$

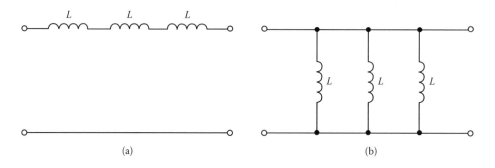

FIGURE 3.26 Transmission line compensation by adding lump inductances in (a) series or (b) parallel (i.e., shunt).

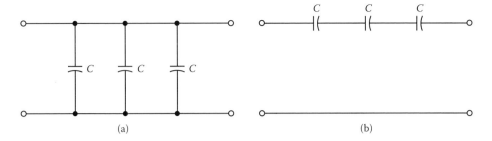

FIGURE 3.27 Transmission line compensation by adding capacitances in (a) parallel (i.e., shunt) or (b) series.

Therefore, the addition of lumped inductances in series will increase the line inductance, and thus, the characteristic impedance and the propagation constant will be increased, which is not desirable.

The addition of lumped inductances in parallel will decrease the line capacitance. Therefore, the propagation constant will be decreased, but the characteristic impedance will be increased, which again is not desirable.

The addition of capacitances in parallel will increase the line capacitance. Hence, the characteristic impedance will be decreased, but the propagation constant will be increased, which negatively affects the system stability. However, for short lines, this method can be used effectively.

Finally, the addition of capacitances in series will decrease the line inductance. Therefore, the characteristic impedance and the propagation constant will be reduced, which is desirable. Thus, the series capacitor compensation of transmission lines is used to improve stability limits and voltage regulation, to provide a desired load division, and to maximize the load-carrying capability of the system. However, having the full line current going through the capacitors connected in series causes harmful overvoltages on the capacitors during short circuits. Therefore, they introduce special problems for line protective relaying.* Under fault conditions, they introduce an impedance discontinuity (negative inductance) and subharmonic currents, and when the capacitor protective gap operates, they impress high-frequency currents and voltages on the system. All of these factors result in incorrect operation of the conventional relaying schemes. The series capacitance compensation of distribution lines has been attempted from time to time for many years. However, it is not widely used.

EXAMPLE 3.11

Determine the SIL of the transmission line given in Example 2.8.

---

* The application of series compensation on the new EHV lines has occasionally caused a problem known as *subsynchronous resonance*. It can be briefly defined as an oscillation due to the interaction between a series capacitor compensated transmission system in electrical resonance and a turbine generator mechanical system in torsional mechanical resonance. As a result of the interaction, a negative resistance is introduced into the electric circuit by the turbine generator. If the effective resistance magnitude is sufficiently large to make the net resistance of the circuit negative, oscillations can increase until mechanical failures take place in terms of flexing or even breaking of the shaft. The event occurs when the electrical subsynchronous resonance frequency is equal or close to 60 Hz minus the frequency of one of the natural torsional modes of the turbine generator. The most well-known subsynchronous resonance problem took place at Mojave Generating Station [8–11].

Solution

The approximate value of the surge impedance of the line is

$$Z_c \cong \sqrt{\frac{L}{C}}$$

$$= \left(\frac{2.6 \times 10^{-3}}{0.012 \times 10^{-6}}\right)^{1/2}$$

$$= 465.5 \, \Omega$$

Therefore,

$$\text{SIL} \cong \frac{|k\mathbf{V}_{R(L-L)}|^2}{\sqrt{\frac{L}{C}}}$$

$$= \frac{|138|^2}{469.5}$$

$$= 0.913 \, \text{MW}$$

which is an approximate value of the SIL of the line. The exact value of the SIL of the line can be determined as

$$\text{SIL} = \frac{|k\mathbf{V}_{R(L-L)}|^2}{Z_c}$$

$$= \frac{|138|^2}{469.62}$$

$$= 40.552 \, \text{MW}$$

3.12 GENERAL CIRCUIT CONSTANTS

Figure 3.28 shows a general two-port, four-terminal network consisting of passive impedances connected in some fashion. From general network theory,

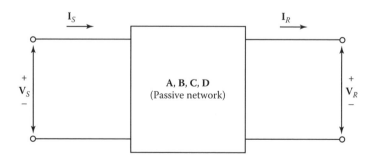

FIGURE 3.28 General two-port, four-terminal network.

Steady-State Performance of Transmission Lines

$$\mathbf{V}_S = \mathbf{A}\mathbf{V}_R + \mathbf{B}\mathbf{I}_R \tag{3.166}$$

and

$$\mathbf{I}_S = \mathbf{C}\mathbf{V}_R + \mathbf{D}\mathbf{I}_R \tag{3.167}$$

Also,

$$\mathbf{V}_R = \mathbf{D}\mathbf{V}_S - \mathbf{B}\mathbf{I}_S \tag{3.168}$$

and

$$\mathbf{I}_R = -\mathbf{C}\mathbf{V}_S + \mathbf{A}\mathbf{I}_S \tag{3.169}$$

It is always true that the determinant of Equations 3.166 and 3.167 or Equations 3.168 and 3.169 is always unity, that is,

$$\mathbf{AD} - \mathbf{BC} = 1 \tag{3.170}$$

In the above equations, **A**, **B**, **C**, and **D** are constants for a given network and are called general circuit constants. Their values depend on the parameters of the circuit concerned and the particular representation chosen. In general, they are complex numbers. For a network that has the symmetry of the uniform transmission line,

$$\mathbf{A} = \mathbf{D} \tag{3.171}$$

3.12.1 Determination of A, B, C, and D Constants

The **A**, **B**, **C**, and **D** constants can be calculated directly by network reduction. For example, when $\mathbf{I}_R = 0$, from Equation 3.167,

$$\mathbf{A} = \frac{\mathbf{V}_S}{\mathbf{V}_R} \tag{3.172}$$

and from Equation 3.167,

$$\mathbf{C} = \frac{\mathbf{I}_S}{\mathbf{V}_R} \tag{3.173}$$

Therefore, the **A** constant is the ratio of the sending- and receiving-end voltages, whereas the **C** constant is the ratio of sending-end current to receiving-end voltage when the receiving end is open-circuited. When $\mathbf{V}_R = 0$, from Equation 3.166,

$$\mathbf{B} = \frac{\mathbf{V}_S}{\mathbf{I}_R} \tag{3.174}$$

When $\mathbf{V}_R = 0$, from Equation 3.167,

$$\mathbf{D} = \frac{\mathbf{I}_S}{\mathbf{I}_R} \tag{3.175}$$

Thus, the **B** constant is the ratio of the sending-end voltage to the receiving-end current when the receiving end is short-circuited, whereas the **D** constant is the ratio of the sending-end and receiving-end currents when the receiving end is short-circuited.

Alternatively, the **A**, **B**, **C**, and **D** generalized circuit constants can be calculated indirectly from a knowledge of the system impedance parameters as shown in the previous sections. Table 3.2 gives general circuit constants for different network types. Table 3.3 gives network conversion formulas to convert a given parameter set into another one.

As can be observed in Equations 3.166 and 3.167, the dimensions of the **A** and **D** constants are numeric. The dimension of the **B** constant is impedance in ohms, whereas the dimension of the **C** constant is admittance in siemens.

3.12.2 Measurement of ABCD Parameters by Test

Since the transmission line is symmetrical, the measurement of the open circuit and short circuit is enough to determine the constants.

Short-circuit test: Short-circuiting one end of the line, the short-circuit impedance is measured at the other end of the line as

$$\mathbf{Z}_{sc} = \frac{\mathbf{B}}{\mathbf{A}} \tag{3.176}$$

since the short-circuited end voltage is zero, and using the reciprocity theorem for a symmetrical network,

$$\mathbf{A} = \mathbf{D} \tag{3.177}$$

then from Equation 3.176,

$$\mathbf{B} = \mathbf{A}\mathbf{Z}_{sc} \tag{3.178}$$

Open-circuit test: Open-circuiting one end of the line, the open-circuit impedance $\mathbf{Z}_{oc}$ is measured at the other end is zero, then

$$\mathbf{Z}_{oc} = \frac{\mathbf{A}}{\mathbf{C}} \tag{3.179}$$

since the current at the open-circuited end is zero, then

$$\mathbf{C} = \frac{\mathbf{A}}{\mathbf{Z}_{oc}} \tag{3.180}$$

For a passive network the determinant is

$$\Delta = \begin{vmatrix} \mathbf{A} & \mathbf{B} \\ \mathbf{C} & \mathbf{D} \end{vmatrix} = \mathbf{AD} - \mathbf{BC} = 1 \tag{3.181}$$

By substituting Equation 3.177 into Equation 3.181,

$$\mathbf{A}^2 - \mathbf{BC} = 1 \tag{3.182}$$

TABLE 3.2
General Circuit Constants for Different Network Types

Network number	Type of network	Equations for general circuit constants in terms of constants of component networks			
		$A =$	$B =$	$C =$	$D =$
1	Series impedance	1	Z	0	1
2	Shunt admittance	1	0	Y	1
3	Transformer	$1 + \dfrac{Z_T Y_T}{2}$	$Z_T\left(1 + \dfrac{Z_T Y_T}{4}\right)$	Y_T	$1 + \dfrac{Z_T Y_T}{2}$
4	Transmission line	$\mathrm{Cosh}\sqrt{ZY}$ $= \left(1 + \dfrac{ZY}{2} + \dfrac{Z^2Y^2}{24} + \cdots\right)$	$\sqrt{Z/Y}\,\mathrm{Sinh}\sqrt{ZY}$ $= Z\left(1 + \dfrac{ZY}{6} + \dfrac{Z^2Y^2}{120} + \cdots\right)$	$\sqrt{Y/Z}\,\mathrm{Sinh}\sqrt{ZY}$ $= Y\left(1 + \dfrac{ZY}{6} + \dfrac{Z^2Y^2}{120} + \cdots\right)$	Same as A
5	General network	A	B	C	D
6	General network and transformer impedance at receiving end	A_1	$B_1 + A_1 Z_{TR}$	C_1	$D_1 + C_1 Z_{TR}$
7	General network and transformer impedance at sending end	$A_1 + C_1 Z_{TS}$	$B_1 + D_1 Z_{TS}$	C_1	D_1
8	General network and transformer impedance at both ends–referred to high voltage	$A_1 + C_1 Z_{TS}$	$B_1 + A_1 Z_{TR} +$ $D_1 Z_{TS} + C_1 Z_{TR} Z_{TS}$	C_1	$D_1 + C_1 Z_{TR}$

(continued)

TABLE 3.2 (Continued)
General Circuit Constants for Different Network Types

	Network Type	Diagram	A	B	C	D
9	General network and transformer impedance at both ends—transformers having different ratios T_R and T_S referred to low voltage	E_S—T_S—$A_1B_1C_1D_1$—T_R—E_R (Z_{TS}, Z_{TR}, E_{RN})	$\frac{T_R}{T_S}(A_1+C_1Z_{TS})$	$\frac{1}{T_RT_S}(B_1+A_1Z_{TR}+D_1Z_{TS}+C_1Z_{TR}Z_{TS})$	$C_1T_RT_S$	$\frac{T_S}{T_R}(D_1+C_1Z_{TR})$
10	General network and shunt admittance at receiving end	E_{SN}—$A_1B_1C_1D_1$—Y_R—E_{RN}	$A_1+B_1Y_R$	B_1	$C_1+D_1Y_R$	D_1
11	General network and shunt admittance at sending end	E_{SN}—Y_S—$A_1B_1C_1D_1$—E_{RN}	A_1	B_1	$C_1+A_1Y_S$	$D_1+B_1Y_S$
12	General network and shunt admittance at both ends	E_{SN}—Y_S—$A_1B_1C_1D_1$—Y_R—E_{RN}	$A_1+B_1Y_R$	B_1	$C_1+A_1Y_S+D_1Y_R+B_1Y_RY_S$	$D_1+B_1Y_S$
13	Two general networks in series	E_S—$A_2B_2C_2D_2$—$A_1B_1C_1D_1$—E_R	$A_1A_2+C_1B_2$	$B_1A_2+D_1B_2$	$A_1C_2+C_1D_2$	$B_1C_2+D_1D_2$
14	Two general networks in series with intermediate impedance	E_S—$A_2B_2C_2D_2$—Z—$A_1B_1C_1D_1$—E_R	$A_1A_2+C_1B_2+C_1A_2Z$	$B_1A_2+D_1B_2+D_1A_2Z$	$A_1C_2+C_1D_2+C_1C_2Z$	$B_1C_2+D_1D_2+D_1C_2Z$
15	Two general networks in series with intermediate shunt admittance	E_{SN}—$A_2B_2C_2D_2$—Y—$A_1B_1C_1D_1$—E_{RN}	$A_1A_2+C_1B_2+A_1B_2Y$	$B_1A_2+D_1B_2+B_1B_2Y$	$A_1C_2+C_1D_2+A_1D_2Y$	$B_1C_2+D_1D_2+B_1D_2Y$
16	Three general networks in series	E_S—$A_3B_3C_3D_3$—$A_2B_2C_2D_2$—$A_1B_1C_1D_1$—E_R	$A_3(A_1A_2+C_1B_2)+B_3(A_1C_2+C_1D_2)$	$A_3(B_1A_2+D_1B_2)+B_3(B_1C_2+D_1D_2)$	$C_3(A_1A_2+C_1B_2)+D_3(A_1C_2+C_1D_2)$	$C_3(B_1A_2+D_1B_2)+D_3(B_1C_2+D_1D_2)$
17	Two general networks in parallel	E_S—$A_1B_1C_1D_1$ || $A_2B_2C_2D_2$—E_R	$\frac{A_1B_2+B_1A_2}{B_1+B_2}$	$\frac{B_1B_2}{B_1+B_2}$	$C_1+C_2+\frac{(A_1-A_2)(D_2-D_1)}{B_1+B_2}$	$\frac{B_1D_2+D_1B_2}{B_1+B_2}$

Note. The exciting current of the receiving end transformers should be added vectorially to the load current, and the exciting current of the sending end transformers should be added vectorially to the sending end current.

General equations: $E_S = E_RA + I_RB$; $E_R = E_SD - I_RB$; $I_S = I_RD + E_RC$; $I_R = I_SA - E_SC$. As a check in the numerical calculation of the A, B, C, and D constants note that in all cases $AD - BC = 1$.

Source: Wagner, C. F., and Evans, R. D., *Symmetrical Components*, Copyright McGraw-Hill Co., 1933. Used with permission of McGraw-Hill Co.

TABLE 3.3
Network Conversion Formulas

	To convert from					To	
	ABCD	Admittance	Impedance	Equivalent π	Equivalent T		
$A =$	ABCD constants	$\dfrac{Y_{11}}{Y_{12}}$	$-\dfrac{Z_{22}}{Z_{12}}$	$1 + ZY_R$	$1 + Z_S Y$		$P_1 + jQ_1 = \hat{A}E_1\hat{I}_1 - \dfrac{1}{B}E_1^2$
$B =$	$E_2 = AE_1 + BI_1$	$\dfrac{1}{Y_{12}}$	$-\dfrac{Z_{11}Z_{22} - Z_{12}^2}{Z_{12}}$	Z	$Z_R + Z_S + YZ_R Z_S$	ABCD	$P_2 + jQ_2 = \dfrac{D}{B}E_2^2 - \dfrac{1}{B}\hat{E}_1\hat{E}_2$
$C =$	$I_2 = CE_1 + DI_1$	$\dfrac{Y_{11}Y_{22} - Y_{12}^2}{Y_{12}}$	$-\dfrac{1}{Z_{12}}$	$Y_R + Y_S + ZY_R Y_S$	Y		
$D =$	$E_1 = DE_2 - BI_2$ $I_1 = -CE_2 + AI_2$	$\dfrac{Y_{22}}{Y_{12}}$	$-\dfrac{Z_{11}}{Z_{12}}$	$1 + ZY_S$	$1 + Z_R Y$		
$Y_{11} =$	$\dfrac{A}{B}$	Admittance constants	$\dfrac{Z_{22}}{Z_{11}Z_{22} - Z_{12}^2}$	$Y_R + \dfrac{1}{Z}$	$\dfrac{1 + Z_S Y}{Z_R + Z_S + YZ_R Z_S}$		$= \hat{Y}_{22}E_1^2 - \hat{Y}_{12}\hat{E}_1\hat{E}_2$
$Y_{12} =$	$\dfrac{1}{B}$		$-\dfrac{Z_{12}}{Z_{11}Z_{22} - Z_{12}^2}$	$\dfrac{1}{Z}$	$\dfrac{1}{Z_R + Z_S + YZ_R Z_S}$	Admittance	$= \hat{Y}_{22}\hat{E}_2^2 - \hat{Y}_{12}\hat{E}_1\hat{E}_2$
$Y_{22} =$	$\dfrac{D}{B}$	$I_1 = Y_{22}E_1 - Y_{12}E_2$ $I_2 = Y_{22}E_2 - Y_{12}E_1$	$\dfrac{Z_{11}}{Z_{11}Z_{22} - Z_{12}^2}$	$Y_S + \dfrac{1}{Z}$	$\dfrac{1 + YZ_R}{Z_R + Z_S + YZ_R Z_S}$		
$Z_{11} =$	$\dfrac{D}{C}$	$\dfrac{Y_{22}}{Y_{11}Y_{22} - Y_{12}^2}$	Impedance constants	$\dfrac{1 + ZY_S}{Y_R + Y_S + ZY_R Y_S}$	$Z_R + \dfrac{1}{Y}$		$= Z_{22}\hat{I}_2^2 - Z_{12}\hat{I}_1\hat{I}_2$
$Z_{12} =$	$-\dfrac{1}{C}$	$-\dfrac{Y_{12}}{Y_{11}Y_{22} - Y_{12}^2}$		$\dfrac{1}{Y_R + Y_S + ZY_R Y_S}$	$\dfrac{1}{Y}$	Impedance	$= Z_{11}\hat{I}_1^2 - Z_{12}\hat{I}_1\hat{I}_2$
$Z_{22} =$	$\dfrac{A}{C}$	$\dfrac{Y_{11}}{Y_{11}Y_{22} - Y_{12}^2}$	$E_1 = Z_{11}I_1 - Z_{12}I_2$ $E_2 = Z_{22}I_2 - Z_{12}I_1$	$\dfrac{1 + ZY_S}{Y_R + Y_S + ZY_R Y_S}$	$Z_S + \dfrac{1}{Y}$		
$Y_R =$	$\dfrac{A - 1}{B}$	$Y_{11} - Y_{12}$	$\dfrac{Z_{22} + Z_{12}}{Z_{11}Z_{22} - Z_{12}^2}$	Equivalent π	$\dfrac{YZ_S}{Z_R + Z_S + YZ_R Z_S}$		
$Z =$	B	$\dfrac{1}{Y_{12}}$	$-\dfrac{Z_{12}}{Z_{11}Z_{22} - Z_{12}^2}$		$Z_R + Z_S + YZ_R Z_S$	Equivalent π	
$Y_S =$	$\dfrac{D - 1}{B}$	$Y_{22} - Y_{12}$	$\dfrac{Z_{11} + Z_{12}}{Z_{11}Z_{22} - Z_{12}^2}$		$\dfrac{YZ_R}{Z_R + Z_S + YZ_R Z_S}$		
$Z_R =$	$\dfrac{D - 1}{C}$	$\dfrac{Y_{22} - Y_{12}}{Y_{11}Y_{22} - Y_{12}^2}$	$Z_{11} + Z_{12}$	$\dfrac{ZY_S}{Y_R + Y_S + ZY_R Y_S}$	Equivalent T		
$Y =$	C	C	$-\dfrac{1}{Z_{12}}$	$\dfrac{1}{Y_R + Y_S + ZY_R Y_S}$		Equivalent T	
$Z_S =$	$\dfrac{A - 1}{C}$	$\dfrac{Y_{11} - Y_{12}}{Y_{11}Y_{22} - Y_{12}^2}$	$Z_{22} + Z_{12}$	$\dfrac{ZY_R}{Y_R + Y_S + ZY_R Y_S}$			

Note 1. P_1 and P_2 are positive in all cases for power flowing into the network from the point considered.
Note 2. P and Q of same sign indicates lagging power factor; that is $P + jQ = EI$.

Source: Wagner, C. F., and Evans, R. D., *Symmetrical Components*, Copyright McGraw-Hill Co., 1933. Used with permission of McGraw-Hill Co.

Thus, by substituting Equation 3.178 and 3.180 into Equation 3.182,

$$\mathbf{A}^2 - \mathbf{A}\mathbf{Z}_{sc}\left(\dfrac{\mathbf{A}}{\mathbf{Z}_{sc}}\right) = 1 \tag{3.183}$$

or

$$\left(\dfrac{\mathbf{Z}_{oc} - \mathbf{Z}_{sc}}{\mathbf{Z}_{oc}}\right)\mathbf{A}^2 = 1 \tag{3.184}$$

Hence,

$$\mathbf{A} = \sqrt{\dfrac{\mathbf{Z}_{oc}}{\mathbf{Z}_{oc} - \mathbf{Z}_{sc}}} \tag{3.185}$$

Substituting Equation 3.185 into Equation 3.178,

$$\mathbf{B} = \mathbf{Z}_{sc}\sqrt{\dfrac{\mathbf{Z}_{oc}}{\mathbf{Z}_{oc} - \mathbf{Z}_{sc}}} \tag{3.186}$$

and substituting Equation 3.185 into Equation 3.180,

$$C = \frac{1}{\sqrt{Z_{oc}(Z_{oc} - Z_{sc})}} \qquad (3.187)$$

3.12.3 A, B, C, and D Constants of Transformer

Figure 3.29 shows the equivalent circuit of a transformer at no load. Neglecting its series impedance,

$$\begin{bmatrix} \mathbf{V}_S \\ \mathbf{I}_S \end{bmatrix} = \begin{bmatrix} \mathbf{A} & \mathbf{B} \\ \mathbf{C} & \mathbf{D} \end{bmatrix} \begin{bmatrix} \mathbf{V}_R \\ \mathbf{I}_R \end{bmatrix} \qquad (3.188)$$

where the transfer matrix is

$$\begin{bmatrix} \mathbf{A} & \mathbf{B} \\ \mathbf{C} & \mathbf{D} \end{bmatrix} = \begin{bmatrix} 1 & 0 \\ \mathbf{Y}_T & 1 \end{bmatrix} \qquad (3.189)$$

since

$$\mathbf{V}_S = \mathbf{V}_R \qquad (3.190)$$

and

$$\mathbf{I}_S = \mathbf{Y}_T \mathbf{V}_R + \mathbf{I}_R \qquad (3.191)$$

and where $\mathbf{Y}_T$ is the magnetizing admittance of the transformer.

Figure 3.30 shows the equivalent circuit of a transformer at full load that has a transfer matrix of

$$\begin{bmatrix} \mathbf{A} & \mathbf{B} \\ \mathbf{C} & \mathbf{D} \end{bmatrix} = \begin{bmatrix} 1 + \dfrac{\mathbf{Z}_T \mathbf{Y}_T}{2} & \mathbf{Z}_T \left(1 + \dfrac{\mathbf{Z}_T \mathbf{Y}_T}{4}\right) \\ \mathbf{Y}_T & 1 + \dfrac{\mathbf{Z}_T \mathbf{Y}_T}{2} \end{bmatrix} \qquad (3.192)$$

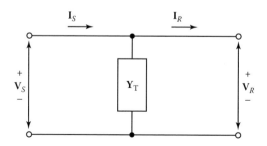

FIGURE 3.29 Transformer equivalent circuit at no load.

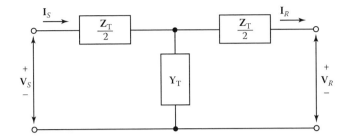

FIGURE 3.30 Transformer equivalent circuit at full load.

since

$$\mathbf{V}_S = \left(1 + \frac{\mathbf{Z}_T \mathbf{Y}_T}{2}\right)\mathbf{V}_R + \mathbf{Z}_T\left(1 + \frac{\mathbf{Z}_T \mathbf{Y}_T}{4}\right)\mathbf{I}_R \qquad (3.193)$$

and

$$\mathbf{I}_S = (\mathbf{Y}_T)\mathbf{V}_R + \left(1 + \frac{\mathbf{Z}_T \mathbf{Y}_T}{2}\right)\mathbf{I}_R \qquad (3.194)$$

where $\mathbf{Z}_T$ is the total equivalent series impedance of the transformer.

3.12.4 Asymmetrical π and T Networks

Figure 3.31 shows an asymmetrical π network that can be thought of as a series (or *cascade*, or *tandem*) connection of a shunt admittance, a series impedance, and a shunt admittance.

The equivalent transfer matrix can be found by multiplying together the transfer matrices of individual components. Thus,

$$\begin{bmatrix} \mathbf{A} & \mathbf{B} \\ \mathbf{C} & \mathbf{D} \end{bmatrix} = \begin{bmatrix} 1 & 0 \\ \mathbf{Y}_1 & 1 \end{bmatrix}\begin{bmatrix} 1 & \mathbf{Z} \\ 0 & 1 \end{bmatrix}\begin{bmatrix} 1 & 0 \\ \mathbf{Y}_2 & 1 \end{bmatrix}$$

$$= \begin{bmatrix} 1 + \mathbf{Z}\mathbf{Y}_2 & \mathbf{Z} \\ \mathbf{Y}_1 + \mathbf{Y}_2 + \mathbf{Z}\mathbf{Y}_1\mathbf{Y}_2 & 1 + \mathbf{Z}\mathbf{Y}_1 \end{bmatrix} \qquad (3.195)$$

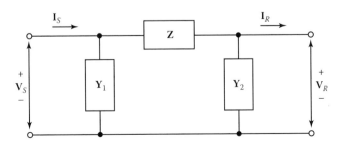

FIGURE 3.31 Asymmetrical π network.

When the π network is symmetrical,

$$Y_1 = Y_2 = \frac{Y}{2}$$

and the transfer matrix becomes

$$\begin{bmatrix} A & B \\ C & D \end{bmatrix} = \begin{bmatrix} 1 + \dfrac{ZY}{2} & Z \\ Y + \dfrac{ZY^2}{4} & 1 + \dfrac{ZY}{2} \end{bmatrix} \qquad (3.196)$$

which is the same as Equation 3.56 for a nominal π circuit of a medium-length transmission line.

Figure 3.32 shows an asymmetrical T network that can be thought of as a cascade connection of a series impedance, a shunt admittance, and a series impedance. Again, the equivalent transfer matrix can be found by multiplying together the transfer matrices of individual components. Thus,

$$\begin{bmatrix} A & B \\ C & D \end{bmatrix} = \begin{bmatrix} 1 & Z_1 \\ 0 & 1 \end{bmatrix} \begin{bmatrix} 1 & 0 \\ Y & 1 \end{bmatrix} \begin{bmatrix} 1 & Z_2 \\ 0 & 1 \end{bmatrix}$$

$$= \begin{bmatrix} 1 + Z_1 Y & Z_1 + Z_2 + Z_1 Z_2 Y \\ Y & 1 + Z_2 Y \end{bmatrix} \qquad (3.197)$$

When the T network is symmetrical,

$$Z_1 = Z_2 = \frac{Z}{2}$$

and the transfer matrix becomes

$$\begin{bmatrix} A & B \\ C & D \end{bmatrix} = \begin{bmatrix} 1 + \dfrac{ZY}{2} & Z + \dfrac{Z^2 Y}{4} \\ Y & 1 + \dfrac{ZY}{2} \end{bmatrix} \qquad (3.198)$$

which is the same as the equation for a nominal T circuit of a medium-length transmission line.

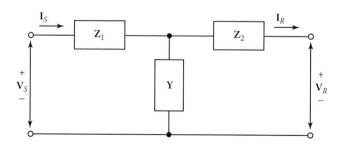

FIGURE 3.32 Asymmetrical T network.

3.12.5 Networks Connected in Series

Two four-terminal transmission networks may be connected in series, as shown in Figure 3.33, to form a new four-terminal transmission network. For the first four-terminal network,

$$\begin{bmatrix} \mathbf{V}_S \\ \mathbf{I}_S \end{bmatrix} = \begin{bmatrix} \mathbf{A}_1 & \mathbf{B}_1 \\ \mathbf{C}_1 & \mathbf{D}_1 \end{bmatrix} \begin{bmatrix} \mathbf{V} \\ \mathbf{I} \end{bmatrix} \qquad (3.199)$$

and for the second four-terminal network,

$$\begin{bmatrix} \mathbf{V} \\ \mathbf{I} \end{bmatrix} = \begin{bmatrix} \mathbf{A}_2 & \mathbf{B}_2 \\ \mathbf{C}_2 & \mathbf{D}_2 \end{bmatrix} \begin{bmatrix} \mathbf{V}_R \\ \mathbf{I}_R \end{bmatrix} \qquad (3.200)$$

By substituting Equation 3.200 into Equation 3.199,

$$\begin{bmatrix} \mathbf{V}_S \\ \mathbf{I}_S \end{bmatrix} = \begin{bmatrix} \mathbf{A}_1 & \mathbf{B}_1 \\ \mathbf{C}_1 & \mathbf{D}_1 \end{bmatrix} \begin{bmatrix} \mathbf{A}_2 & \mathbf{B}_2 \\ \mathbf{C}_2 & \mathbf{D}_2 \end{bmatrix} \begin{bmatrix} \mathbf{V}_R \\ \mathbf{I}_R \end{bmatrix}$$

$$= \begin{bmatrix} \mathbf{A}_1\mathbf{A}_2 + \mathbf{B}_1\mathbf{C}_2 & \mathbf{A}_1\mathbf{B}_2 + \mathbf{B}_1\mathbf{D}_2 \\ \mathbf{C}_1\mathbf{A}_2 + \mathbf{D}_1\mathbf{C}_2 & \mathbf{C}_1\mathbf{B}_2 + \mathbf{D}_1\mathbf{D}_2 \end{bmatrix} \begin{bmatrix} \mathbf{V}_R \\ \mathbf{I}_R \end{bmatrix} \qquad (3.201)$$

Thus, the equivalent **A**, **B**, **C**, and **D** constants for two networks connected in series are

$$\mathbf{A}_{eq} = \mathbf{A}_1\mathbf{A}_2 + \mathbf{B}_1\mathbf{C}_2 \qquad (3.202)$$

$$\mathbf{B}_{eq} = \mathbf{A}_1\mathbf{B}_2 + \mathbf{B}_1\mathbf{D}_2 \qquad (3.203)$$

$$\mathbf{C}_{eq} = \mathbf{C}_3\mathbf{A}_2 + \mathbf{D}_1\mathbf{C}_2 \qquad (3.204)$$

$$\mathbf{D}_{eq} = \mathbf{C}_1\mathbf{B}_2 + \mathbf{D}_1\mathbf{D}_2 \qquad (3.205)$$

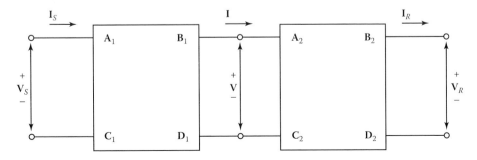

FIGURE 3.33 Transmission networks in series.

EXAMPLE 3.12

Figure 3.34 shows two networks connected in cascade. Determine the equivalent **A**, **B**, **C**, and **D** constants.

Solution

For network 1,

$$\begin{bmatrix} A_1 & B_1 \\ C_1 & D_1 \end{bmatrix} = \begin{bmatrix} 1 & 10\angle 30° \\ 0 & 1 \end{bmatrix}$$

For network 2,

$$Y_2 = \frac{1}{Z_2} = \frac{1}{40\angle -45°} = 0.025\angle 45° \text{ S}$$

Then

$$\begin{bmatrix} A_2 & B_2 \\ C_2 & D_2 \end{bmatrix} = \begin{bmatrix} 1 & 0 \\ 0.025\angle 45° & 1 \end{bmatrix}$$

Therefore,

$$\begin{bmatrix} A_{eq} & B_{eq} \\ C_{eq} & D_{eq} \end{bmatrix} = \begin{bmatrix} 1 & 10\angle 30° \\ 0 & 1 \end{bmatrix} \begin{bmatrix} 1 & 0 \\ 0.025\angle 45° & 1 \end{bmatrix}$$

$$= \begin{bmatrix} 1.09\angle 12.8° & 10\angle 30° \\ 0.025\angle 45° & 1 \end{bmatrix}$$

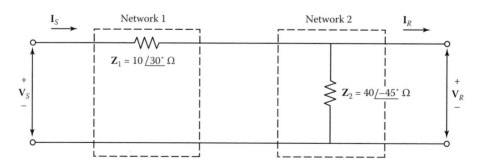

FIGURE 3.34 Network configurations for Example 3.12.

3.12.6 Networks Connected in Parallel

Two four-terminal transmission networks may be connected in parallel, as shown in Figure 3.35, to form a new four-terminal transmission network.

Since

$$\mathbf{V}_S = \mathbf{V}_{S1} + \mathbf{V}_{S2} \tag{3.206}$$

$$\mathbf{V}_R = \mathbf{V}_{R1} + \mathbf{V}_{R2} \tag{3.207}$$

and

$$\mathbf{I}_S = \mathbf{I}_{S1} + \mathbf{I}_{S2} \tag{3.208}$$

$$\mathbf{I}_R = \mathbf{I}_{R1} + \mathbf{I}_{R2} \tag{3.209}$$

for the equivalent four-terminal network,

$$\begin{bmatrix} \mathbf{V}_S \\ \mathbf{I}_S \end{bmatrix} = \begin{bmatrix} \dfrac{\mathbf{A}_1\mathbf{B}_2 + \mathbf{A}_2\mathbf{B}_1}{\mathbf{B}_1 + \mathbf{B}_2} & \dfrac{\mathbf{B}_1\mathbf{B}_2}{\mathbf{B}_1 + \mathbf{B}_2} \\ \mathbf{C}_2 + \mathbf{C}_2 + \dfrac{(\mathbf{A}_1 - \mathbf{A}_2)(\mathbf{D}_2 - \mathbf{D}_1)}{\mathbf{B}_1 + \mathbf{B}_2} & \dfrac{\mathbf{D}_1\mathbf{B}_2 + \mathbf{D}_2\mathbf{B}_1}{\mathbf{B}_1 + \mathbf{B}_2} \end{bmatrix} \begin{bmatrix} \mathbf{V}_R \\ \mathbf{I}_R \end{bmatrix} \tag{3.210}$$

where the equivalent **A**, **B**, **C**, and **D** constants are

$$\mathbf{A}_{eq} = \dfrac{\mathbf{A}_1\mathbf{B}_2 + \mathbf{A}_2\mathbf{B}_1}{\mathbf{B}_1 + \mathbf{B}_2} \tag{3.211}$$

$$\mathbf{B}_{eq} = \dfrac{\mathbf{B}_1\mathbf{B}_2}{\mathbf{B}_1 + \mathbf{B}_2} \tag{3.212}$$

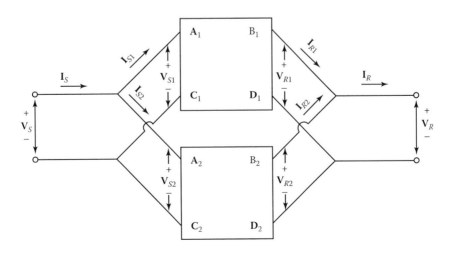

FIGURE 3.35 Transmission networks in parallel.

$$C_{eq} = C_2 + C_2 + \frac{(A_1 - A_2)(D_2 - D_1)}{B_1 + B_2} \qquad (3.213)$$

$$D_{eq} = \frac{D_1 B_2 + D_2 B_1}{B_1 + B_2} \qquad (3.214)$$

EXAMPLE 3.13

Assume that the two networks given in Example 3.13 are connected in parallel, as shown in Figure 3.12. Determine the equivalent **A**, **B**, **C**, and **D** constants (Figure 3.36).

Solution

Using the **A**, **B**, **C**, and **D** parameters found previously for networks 1 and 2, that is,

$$\begin{bmatrix} A_1 & B_1 \\ C_1 & D_1 \end{bmatrix} = \begin{bmatrix} 1 & 10\angle 30° \\ 0 & 1 \end{bmatrix}$$

and

$$\begin{bmatrix} A_2 & B_2 \\ C_2 & D_2 \end{bmatrix} = \begin{bmatrix} 1 & 0 \\ 0.025\angle 45° & 1 \end{bmatrix}$$

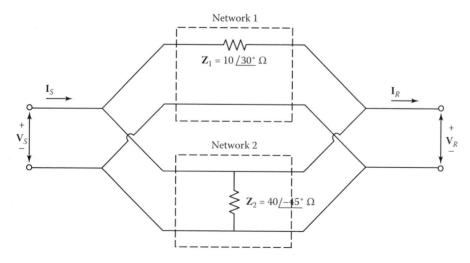

FIGURE 3.36 Transmission networks in parallel for Example 3.13.

the equivalent **A**, **B**, **C**, and **D** constants can be calculated as

$$\mathbf{A}_{eq} = \frac{\mathbf{A}_1\mathbf{B}_2 + \mathbf{A}_2\mathbf{B}_1}{\mathbf{B}_1 + \mathbf{B}_2}$$

$$= \frac{1 \times 0 + 1 \times 10\angle 30°}{10\angle 30° + 0}$$

$$= 1$$

$$\mathbf{B}_{eq} = \frac{\mathbf{B}_1\mathbf{B}_2}{\mathbf{B}_1 + \mathbf{B}_2}$$

$$= \frac{(10\angle 30°) \times 0}{(10\angle 30°) + 0}$$

$$= 0$$

$$\mathbf{C}_{eq} = \mathbf{C}_2 + \mathbf{C}_2 + \frac{(\mathbf{A}_1 - \mathbf{A}_2)(\mathbf{D}_2 - \mathbf{D}_1)}{\mathbf{B}_1 + \mathbf{B}_2}$$

$$= 0 + 0.025\angle 45° + \frac{(1-1)(1-1)}{10\angle 30° - 0}$$

$$= 0.025\angle 45°$$

$$\mathbf{D}_{eq} = \frac{\mathbf{D}_1\mathbf{B}_2 + \mathbf{D}_2\mathbf{B}_1}{\mathbf{B}_1 + \mathbf{B}_2}$$

$$= \frac{1 \times 0 + 1 \times 10\angle 30°}{10\angle 30° + 0}$$

$$= 1$$

Therefore,

$$\begin{bmatrix} \mathbf{A}_{eq} & \mathbf{B}_{eq} \\ \mathbf{C}_{eq} & \mathbf{D}_{eq} \end{bmatrix} = \begin{bmatrix} 1 & 0 \\ 0.025\angle 45° & 1 \end{bmatrix}$$

3.12.7 Terminated Transmission Line

Figure 3.37 shows a four-terminal transmission network connected to (i.e., terminated by) a load $\mathbf{Z}_L$. For the given network,

$$\begin{bmatrix} \mathbf{V}_S \\ \mathbf{I}_S \end{bmatrix} = \begin{bmatrix} \mathbf{A} & \mathbf{B} \\ \mathbf{C} & \mathbf{D} \end{bmatrix} \begin{bmatrix} \mathbf{A}_R \\ \mathbf{I}_R \end{bmatrix} \quad (3.215)$$

or

$$\mathbf{V}_S = \mathbf{A}\mathbf{V}_R + \mathbf{B}\mathbf{I}_R \quad (3.216)$$

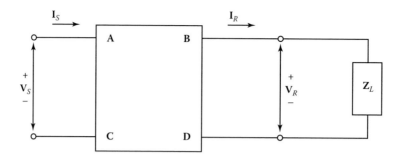

FIGURE 3.37 Terminated transmission line.

and

$$\mathbf{I}_S = \mathbf{CV}_R + \mathbf{DI}_R \tag{3.217}$$

and also

$$\mathbf{V}_R = \mathbf{Z}_L \mathbf{I}_R \tag{3.218}$$

Therefore, the input impedance is

$$\mathbf{Z}_{in} = \frac{\mathbf{V}_S}{\mathbf{I}_S}$$
$$= \frac{\mathbf{AV}_R + \mathbf{BI}_R}{\mathbf{CV}_R + \mathbf{DI}_R} \tag{3.219}$$

or by substituting Equation 3.169 into Equation 3.170,

$$\mathbf{Z}_{in} = \frac{\mathbf{AZ}_L + \mathbf{B}}{\mathbf{CZ}_L + \mathbf{D}} \tag{3.220}$$

Since for the symmetrical and long transmission line,

$$\mathbf{A} = \cosh\sqrt{\mathbf{YZ}} = \cosh\theta$$

$$\mathbf{B} = \sqrt{\frac{\mathbf{Z}}{\mathbf{Y}}}\sinh\sqrt{\mathbf{YZ}} = \mathbf{Z}_c \sinh\theta$$

$$\mathbf{C} = \sqrt{\frac{\mathbf{Y}}{\mathbf{Z}}}\sinh\sqrt{\mathbf{YZ}} = \mathbf{Y}_c \sinh\theta$$

$$\mathbf{D} = \mathbf{A} = \cosh\sqrt{\mathbf{YZ}} = \cosh\theta$$

Steady-State Performance of Transmission Lines

the input impedance, from Equation 3.171, becomes

$$\mathbf{Z}_{in} = \frac{\mathbf{Z}_L \cosh\theta + \mathbf{Z}_c \sinh\theta}{\mathbf{Z}_L \mathbf{Y}_c \sinh\theta + \cosh\theta} \qquad (3.221)$$

or

$$\mathbf{Z}_{in} = \frac{\mathbf{Z}_L[(\mathbf{Z}_c/\mathbf{Z}_L)\sinh\theta + \cosh\theta]}{(\mathbf{Z}_L/\mathbf{Z}_c)\sinh\theta + \cosh\theta} \qquad (3.222)$$

If the load impedance is chosen to be equal to the characteristic impedance, that is,

$$\mathbf{Z}_L = \mathbf{Z}_c \qquad (3.223)$$

the input impedance, from Equation 3.173, becomes

$$\mathbf{Z}_{in} = \mathbf{Z}_c \qquad (3.224)$$

which is independent of θ and the line length. The value of the voltage is constant all along the line.

EXAMPLE 3.14

Figure 3.38 shows a short transmission line that is terminated by a load of 200 kVA at a lagging power factor of 0.866 at 2.4 kV If the line impedance is 2.07 + j0.661 Ω, calculate

(a) Sending-end current
(b) Sending-end voltage
(c) Input impedance
(d) Real and reactive power loss in the line

Solution

(a) From Equation 3.166,

$$\begin{bmatrix} \mathbf{V}_S \\ \mathbf{I}_S \end{bmatrix} = \begin{bmatrix} A & B \\ C & D \end{bmatrix} \begin{bmatrix} \mathbf{V}_R \\ \mathbf{I}_R \end{bmatrix}$$

$$= \begin{bmatrix} 1 & Z \\ 0 & 1 \end{bmatrix} \begin{bmatrix} \mathbf{V}_R \\ \mathbf{I}_R \end{bmatrix}$$

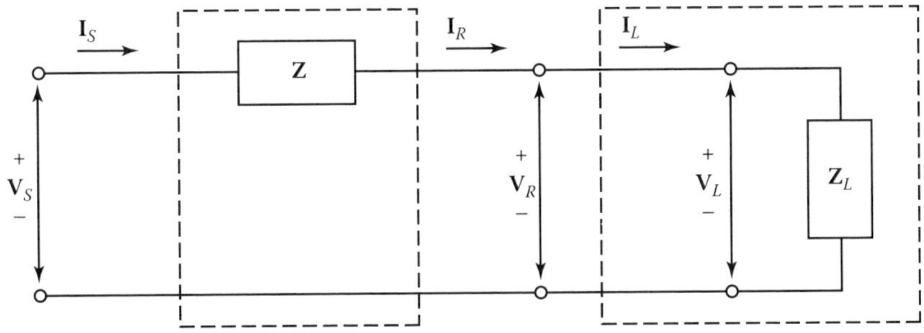

FIGURE 3.38 Transmission system for Example 3.14.

where

$$Z = 2.07 + j0.661 = 2.173 \angle 17.7° \, \Omega$$

$$\mathbf{I}_R = \mathbf{I}_S = \mathbf{I}_L$$

$$\mathbf{V}_R = \mathbf{Z}_L \mathbf{I}_R$$

Since

$$\mathbf{S}_R = 200 \angle 30° = 173.2 + j100 \text{ kVA}$$

and

$$\mathbf{V}_L = 2.4 \angle 0° \text{ kV}$$

then

$$\mathbf{I}_L^* = \frac{\mathbf{S}_R}{\mathbf{V}_L} = \frac{200 \angle 30°}{2.4 \angle 0°} = 83.33 \angle 0° \text{ A}$$

or

$$\mathbf{I}_L = 83.33 \angle -30° \text{ A}$$

hence,

$$\mathbf{I}_S = \mathbf{I}_R = \mathbf{I}_L = 83.33 \angle -30° \text{ A}$$

(b)

$$\mathbf{Z}_L = \frac{\mathbf{V}_L}{\mathbf{I}_L} = \frac{2.4 \times 10^3 \angle 0°}{83.33 \angle -30°} = 28.8 \angle 30° \, \Omega$$

and

$$\mathbf{V}_R = \mathbf{Z}_L \mathbf{I}_R = 28.8 \angle 30° \times 83.33 \angle -30° = 2404 \angle 0° \text{ kV}$$

thus,

$$\mathbf{V}_S = \mathbf{A}\mathbf{V}_R + \mathbf{B}\mathbf{I}_R$$
$$= 2400 \angle 0° + 2.173 \angle 17.7° \times 83.33 \angle -30°$$
$$= 2576.9 - j38.58$$
$$= 2577.2 \angle -0.9° \text{ V}$$

(c) The input impedance is

$$\mathbf{Z}_{in} = \frac{\mathbf{V}_S}{\mathbf{I}_S} = \frac{\mathbf{A}\mathbf{V}_R + \mathbf{B}\mathbf{I}_R}{\mathbf{C}\mathbf{V}_R + \mathbf{D}\mathbf{I}_R}$$

$$= \frac{2577.2 \angle -0.9°}{83.33 \angle -30°} = 30.93 \angle 29.1° \, \Omega$$

Steady-State Performance of Transmission Lines

(d) The real and reactive power loss in the line:

$$\mathbf{S}_L = \mathbf{S}_S - \mathbf{S}_R$$

where

$$\mathbf{S}_S = \mathbf{V}_S \mathbf{I}_S^* = 2.577.2\angle -0.9° \times 83.33\angle 30°$$
$$= 214{,}758\angle 29.1° \text{ VA}$$

or

$$\mathbf{S}_S = \mathbf{I}_S \times \mathbf{Z}_{in} \times \mathbf{I}_S^* = 214{,}758\angle 29.1° \text{ VA}$$

Thus,

$$\mathbf{S}_L = 214{,}758\angle 29.1° - 200{,}000\angle 30°$$
$$= 14{,}444.5 + j4{,}444.4 \text{ VA}$$

that is, the active power loss is 14,444.5 W, and the reactive power loss is 4444.4 vars.

3.12.8 Power Relations Using A, B, C, and D Line Constants

For a given long transmission line, the complex power at the sending and receiving ends are

$$\mathbf{S}_S = P_S + jQ_S = \mathbf{V}_S \mathbf{I}_S^* \tag{3.225}$$

and

$$\mathbf{S}_R = P_R + jQ_R = \mathbf{V}_R \mathbf{I}_R^* \tag{3.226}$$

Also, the sending- and receiving-end voltages and currents can be expressed as

$$\mathbf{V}_S = \mathbf{A}\mathbf{V}_R + \mathbf{B}\mathbf{I}_R \tag{3.227}$$

$$\mathbf{I}_S = \mathbf{C}\mathbf{V}_R + \mathbf{D}\mathbf{I}_R \tag{3.228}$$

$$\mathbf{V}_R = \mathbf{A}\mathbf{V}_S - \mathbf{B}\mathbf{I}_S \tag{3.329}$$

$$\mathbf{I}_R = \mathbf{C}\mathbf{V}_S + \mathbf{D}\mathbf{I}_S \tag{3.230}$$

where

$$\mathbf{A} = A\angle\alpha = \cosh\sqrt{\mathbf{YZ}} \tag{3.231}$$

$$\mathbf{B} = B\angle\beta = \sqrt{\frac{\mathbf{Z}}{\mathbf{Y}}}\sinh\sqrt{\mathbf{YZ}} \tag{3.232}$$

$$\mathbf{C} = C\angle\delta = \sqrt{\frac{\mathbf{Y}}{\mathbf{Z}}} \sinh\sqrt{\mathbf{YZ}} \tag{3.233}$$

$$\mathbf{D} = \mathbf{A} = \cosh\sqrt{\mathbf{YZ}} \tag{3.234}$$

$$\mathbf{V}_R = V_R\angle 0° \tag{3.235}$$

$$\mathbf{V}_S = V_S\angle\delta \tag{3.236}$$

From Equation 3.215,

$$\mathbf{I}_S = \frac{\mathbf{A}}{\mathbf{B}}\mathbf{V}_S - \frac{\mathbf{V}_R}{\mathbf{B}} \tag{3.237}$$

or

$$\mathbf{I}_S = \frac{AV_S}{B}\angle\alpha+\delta-\beta - \frac{V_R\angle-\beta}{B} \tag{3.238}$$

and

$$\mathbf{I}_S^* = \frac{AV_S}{B}\angle-\alpha-\delta+\beta - \frac{V_R\angle\beta}{B} \tag{3.239}$$

and from Equation 3.225

$$\mathbf{I}_R = \frac{\mathbf{V}_S}{\mathbf{B}} - \frac{\mathbf{A}}{\mathbf{B}}\mathbf{V}_R \tag{3.240}$$

or

$$\mathbf{I}_R = \frac{V_S}{B}\angle\delta-\beta - \frac{AV_R}{B}\angle\alpha-\beta \tag{3.241}$$

and

$$\mathbf{I}_R^* = \frac{V_S}{B}\angle-\delta+\beta - \frac{AV_R}{B}\angle-\alpha+\beta \tag{3.242}$$

By substituting Equations 3.237 and 3.240 into Equations 3.223 and 3.224, respectively,

$$\mathbf{S}_S = P_S + jQ_S = \frac{AV_S^2}{B}\angle\beta-\alpha - \frac{V_S V_R}{B}\angle\beta+\alpha \tag{3.243}$$

and

$$\mathbf{S}_R = P_R + jQ_R = \frac{V_S V_R}{B} \angle \beta - \delta - \frac{A V_R^2}{B} \angle \beta - \alpha \qquad (3.244)$$

Therefore, the real and reactive powers at the sending end are

$$P_S = \frac{A V_S^2}{B} \cos(\beta - \alpha) - \frac{V_S V_R}{B} \cos(\beta + \alpha) \qquad (3.245)$$

and

$$Q_S = \frac{A V_S^2}{B} \sin(\beta - \delta) - \frac{V_S V_R}{B} \sin(\beta + \alpha) \qquad (3.246)$$

and the real and reactive powers at the receiving end are

$$P_R = \frac{V_S V_R}{B} \cos(\beta - \delta) - \frac{A V_R^2}{B} \cos(\beta - \alpha) \qquad (3.247)$$

and

$$Q_R = \frac{V_S V_R}{B} \sin(\beta - \delta) - \frac{A V_R^2}{B} \sin(\beta - \alpha) \qquad (3.248)$$

For the constants V_S and V_R, for a given line, the only variable in Equations 3.245 through 3.248 is δ, the power angle. Therefore, treating P_S as a function of δ only in Equation 3.245, P_S is maximum when $\beta + \delta = 180°$. Therefore, the maximum power at the sending end, the maximum input power, can be expressed as

$$P_{S,\max} = \frac{A V_S^2}{B} \cos(\beta - \alpha) + \frac{V_S V_R}{B} \qquad (3.249)$$

and similarly the corresponding reactive power at the sending end, the maximum input vars, is

$$Q_{S,\max} = \frac{A V_S^2}{B} \sin(\beta - \alpha) \qquad (3.250)$$

On the other hand, P_R is maximum when $\delta = \beta$. Therefore, the maximum power obtainable (which is also called the *steady-state power limit*) at the receiving end can be expressed as

$$P_{R,\max} = \frac{V_S V_R}{B} - \frac{A V_R^2}{B} \cos(\beta - \alpha) \qquad (3.251)$$

and similarly, the corresponding reactive power delivered at the receiving end is

$$Q_{R,\max} = -\frac{AV_R^2}{B}\sin(\beta - \alpha) \qquad (3.252)$$

In the above equations, V_S and V_R are the phase (line-to-neutral) voltages whether the system is single phase or three phase. Therefore, the total three-phase power transmitted on the three-phase line is three times the power calculated by using the above equations. If the voltages are given in volts, the power is expressed in watts or vars. Otherwise, if they are given in kilovolts, the power is expressed in megawatts or megavars.

For a given value of γ, the power loss P_L in a long transmission line can be calculated as the difference between the sending- and the receiving-end real powers

$$P_L = P_S - P_R \qquad (3.253)$$

and the lagging vars loss is

$$Q_L = Q_S - Q_R \qquad (3.254)$$

EXAMPLE 3.15

Figure 3.39 shows a three-phase, 345-kV ac transmission line with bundled conductors connecting two buses that are voltage regulated. Assume that the series capacitor and shunt reactor compensations are to be considered. The bundled conductor line has two 795-kcmil ACSR conductors per phase. The subconductors are separated by 18 in, and the phase spacing of the flat configuration is 24, 24, and 48 ft. The resistance, inductive reactance, and susceptance of the line are given as 0.059 Ω/mi per phase, 0.588 Ω/mi per phase, and 7.20 × 10⁻⁶ S phase to neutral per phase per mile, respectively. The total line length is 200 mi, and the line resistance may be neglected because simple calculations and approximate answers will suffice. First, assume that there is no compensation in use; that is, both reactors are disconnected and the series capacitor is bypassed. Determine the following:

(a) Total three-phase SIL of line in megavolt-amperes
(b) Maximum three-phase theoretical steady-state power flow limit in megawatts
(c) Total three-phase magnetizing var generation by line capacitance
(d) Open-circuit receiving-end voltage if the line is open at the receiving end

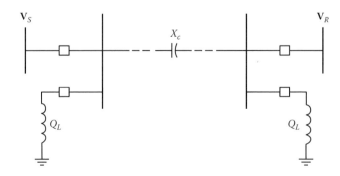

FIGURE 3.39 Compensated line for Example 3.15.

Steady-State Performance of Transmission Lines

Solution

(a) The surge impedance of the line is

$$Z_c = (x_L \times x_c)^{1/2} \tag{3.255}$$

where

$$x_c = \frac{1}{b_c} \ \Omega/\text{mi/phase} \tag{3.256}$$

Thus,

$$Z_c = \left(\frac{x_L}{b_c}\right)^{1/2}$$

$$= \left(\frac{0.588}{7.20 \times 10^{-6}}\right)^{1/2}$$

$$= 285.77 \ \Omega/\text{mi/phase} \tag{3.257}$$

Thus, the total three-phase SIL of the line is

$$\text{SIL} = \frac{|kV_{R(L-L)}|^2}{Z_c}$$

$$= \frac{345^2}{285.77}$$

$$= 416.5 \ \text{MVA/mi}$$

(b) Neglecting the line resistance,

$$P = P_S = P_R$$

or

$$P = \frac{V_S V_R}{X_L} \sin\delta \tag{3.258}$$

When $\delta = 90°$, the maximum three-phase theoretical steady-state power flow limit is

$$P_{max} = \frac{V_S V_R}{X_L}$$

$$= \frac{(345 \ kV)2}{117.6}$$

$$= 1012.1 \ \text{MW} \tag{3.259}$$

(c) Using a nominal π circuit representation, the total three-phase magnetizing var generated by the line capacitance can be expressed as

$$Q_c = V_S^2 \frac{b_c l}{2} + V_R^2 \frac{b_c l}{2}$$

$$= V_S^2 \frac{B_c}{2} + V_R^2 \frac{B_c}{2} \qquad (3.260)$$

Hence,

$$Q_c = (345 \times 10^3)^2 \left[\frac{1}{2} (7.20 \times 10^{-6})200 \right] + (345 \times 10^3)^2 \left[\frac{1}{2} (7.20 \times 10^{-6})200 \right]$$

$$= 171.4 \text{ Mvar}$$

(d) If the line is open at the receiving end, the open-circuit receiving-end voltage can be expressed as

$$\mathbf{V}_S = \mathbf{V}_{R(oc)} \cosh \gamma l \qquad (3.261)$$

or

$$\mathbf{V}_{R(oc)} = \frac{\mathbf{V}_S}{\cosh \gamma l} \qquad (3.262)$$

where

$$\gamma = j\omega\sqrt{LC}$$

$$= j\omega \left(\frac{x_L}{\omega} \frac{1}{\omega x_c} \right)^{1/2}$$

$$= j \left(\frac{x_L}{x_c} \right)^{1/2}$$

$$= j[(0.588)(7.20 \times 10^{-6})]^{1/2} = j0.0021 \text{ rad/mi} \qquad (3.263)$$

and

$$\gamma l = j(0.0021)(200) = j0.4115 \text{ rad}$$

thus,

$$\cosh \gamma l = \cosh(0 + j0.4115)$$
$$= \cosh(0)\cos(0.4115) + j\sinh(0)\sin(0.4115)$$
$$= 0.9165$$

Steady-State Performance of Transmission Lines

therefore,

$$V_{R(oc)} = \frac{345\,kV}{0.9165}$$
$$= 376.43\,kV$$

Alternatively,

$$V_{R(oc)} = V_S \frac{X_c}{X_c + X_L}$$
$$= (345\,kV)\left(\frac{-1388.9}{-1388.9 + 117.6}\right)$$
$$= 376.74\,kV \tag{3.264}$$

EXAMPLE 3.16

Use the data given in Example 3.13 and assume that the shunt compensation is now used. Assume also that the two shunt reactors are connected to absorb 60% of the total three-phase magnetizing var generation by line capacitance and that half of the total reactor capacity is placed at each end of the line. Determine the following:

(a) Total three-phase SIL of the line in megavolt-amperes
(b) Maximum three-phase theoretical steady-state power flow limit in megawatts
(c) Three-phase megavolt-ampere rating of each shunt reactor
(d) Cost of each reactor at $10/kVA
(e) Open-circuit receiving-end voltage if the line is open at the receiving end

Solution

(a) SIL = 416.5, as before, in Example 3.13.
(b) P_{max} = 1012.1 MW, as before
(c) The three-phase megavolt-ampere rating of each shunt reactor is

$$\frac{1}{2}Q_L = \frac{1}{2}0.60Q_c$$
$$= \frac{1}{2}0.60(171.4)$$
$$= 51.42\,MVA$$

(d) The cost of each reactor at $10/kVA is

(51.42 MVA/reactor) ($10/kVA) = $514,200

(e) Since

$$\gamma l = j0.260\,rad$$

and

$$\cosh \gamma l = 0.9663$$

then

$$V_{R(oc)} = \frac{345\,\text{kV}}{0.9663}$$
$$= 357.03\,\text{kV}$$

Alternatively,

$$V_{R(oc)} = V_S \frac{X_c}{X_c + X_L}$$
$$= (345\text{ kV})\left(\frac{-13{,}472}{-13{,}472 + 117.6}\right)$$
$$= 357.1\,\text{kV}$$

Therefore, the inclusion of the shunt reactor causes the receiving-end open-circuit voltage to decrease.

3.13 EHV UNDERGROUND CABLE TRANSMISSION

As discussed in the previous sections, the inductive reactance of an overhead high-voltage ac line is much greater than its capacitive reactance. However, the capacitive reactance of an underground high-voltage ac cable is much greater than its inductive reactance because the three-phase conductors are located very close to each other in the same cable. The approximate values of the resultant vars (reactive power) that can be generated by ac cables operating at the phase-to-phase voltages of 132, 220, and 400 kV are 2000, 5000, and 15,000 kVA/mi, respectively. This var generation, due to the capacitive charging currents, sets a practical limit to the possible noninterrupted length of an underground ac cable.

This situation can be compensated for by installing appropriate inductive shunt reactors along the line. This "critical length" of the cable can be defined as the length of cable line that has a three-phase charging reactive power equal in magnitude to the thermal rating of the cable line. For example, the typical critical lengths of ac cables operating at the phase-to-phase voltages of 132, 200, and 400 kV can be given approximately as 40, 25 and 15 mi, respectively.

The study done by Schifreen and Marble [12] illustrated the limitations in the operation of high-voltage ac cable lines due to the charging current. For example, Figure 3.40 shows that the magnitude of the maximum permissible power output decreases as a result of an increase in cable length [13]. Figure 3.41 shows that increasing lengths of cable line can transmit full-rated current (1.0 pu) only if the load power factor is decreased to resolve lagging values. Note that, the critical length is used as the base length in the figures. Table 3.4 [14] gives characteristics of a 345-kV pipe-type cable. Figure 3.42 shows the permissible variation in per-unit vars delivered to the electric system at each terminal of cable line for a given power transmission.

EXAMPLE 3.17

Consider a high-voltage open-circuit three-phase insulated power cable of length *l* shown in Figure 3.43. Assume that a fixed sending-end voltage is to be supplied; the receiving-end voltage floats,

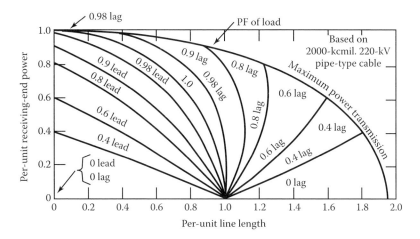

FIGURE 3.40 Power transmission limits of high-voltage ac cable lines. Curved lines: Sending-end current equal to rated or base current of cable. Horizontal lines: Receiving-end current equal to rated or base current of cable. (From Wiseman, R. T., *Trans. Am. Inst. Electr. Eng.*, 26, 803 © 1956 IEEE.)

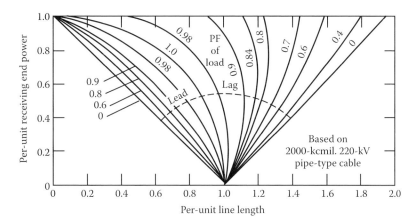

FIGURE 3.41 Receiving-end current limits of high-voltage ac cable lines. Curved lines: Sending-end current equal to rated or base current of cable. (From Wiseman, R. T., *Trans. Am. Inst. Electr. Eng.*, 26, 803 © 1956 IEEE. Used with permission.)

and it is an overvoltage. Furthermore, assume that at some critical length ($l = l_0$), the sending-end current $\mathbf{I}_S$ is equal to the ampacity of the cable circuit, $\mathbf{I}_{I0}$. Therefore, if the cable length is l_0, no load, whatever of 1.0 or the leading power factor can be supplied without overloading the sending end of the cable. Use the general long-transmission-line equations, which are valid for steady-state sinusoidal operation, and verify that the approximate critical length can be expressed as

$$l_0 \cong \frac{I_{I0}}{V_s b}$$

Solution

The long-transmission-line equations can be expressed as

$$\mathbf{V}_S = \mathbf{V}_R \cosh \gamma l + \mathbf{I}_R \mathbf{Z}_c \sinh \gamma l \qquad (3.265)$$

TABLE 3.4
Characteristics of a 345-kV Pipe-Type Cable

Characteristics	Power Factor (%)	Maximum Electric Stress, 300 V/mil				Power Factor (%)	Maximum Electric Stress, 350 V/mil			
Conductor size, kcmil		1,000	1,250	1,500	2,000		1,000	1,230	1,300	2,000
Insulation thickness, mils		1,250	1,173	1,110	1,035		980	915	885	835
E_R kV, 200 kV										
I_T, A	0.5	585	638	680	730	0.3	576	623	636	688
	0.3	653	721	780	860	0.3	637	724	776	847
Rated three-phase mVA	0.5	350	381	406	436	0.3	344	372	392	413
	0.3	390	431	466	516		393	432	463	508
Z, Ω/mi		0.403∠78.0	0.381∠78.0	0.363∠80.0	0.338∠80.0		0.377∠80.0	0.355∠78.5	0.367∠79.4	0.319∠80.0
YS, mi ×10⁻⁴		1.08∠89.7	1.20∠89.7	1.32∠89.7	1.53∠89.7		1.26∠89.7	1.41∠89.7	1.54∠89.7	1.78∠89.7
A numeric mi ×10⁻⁸		6.6∠283.9	6.8∠284.3	6.9∠284.9	7.19∠284.8		6.9∠283.5	7.1∠284.1	7.5∠284.6	7.54∠284.8
Z_0, Ω		61.2∠5.9	56.4∠5.5	52.5∠4.9	67.1∠6.9		54.7∠6.3	50.1∠3.6	48.9∠3.2	42.3∠4.9
I_c A ∠mi		21.6	24.1	26.5	30.5		25.1	28.1	30.8	3.37

		12,900	14,400	13,800	13,800		15,000	16,800	18,000	21,300
3Φ charging, KVA/mi										
S_4, mi	0.5	27.1	26.5	25.7	24.0	0.5	22.9	22.2	21.3	19.4
	0.3	30.3	30.0	29.3	28.2	0.3	26.2	25.7	25.2	23.9
Nominal pipe size, in, 10^{8}/in.										
Earth resistivity, thermal Ω cm		80	80	80	80		80	80	80	80
Conductor temperature, °C		70	70	70	70		70	70	70	70
3Φ dielectric loss, W/ft	0.5	12.2	13.7	14.9	17.3	0.5	14.2	15.8	17.3	20.1
	0.3	7.3	81	9.0	10.3	0.3	8.5	9.5	10.4	12.0
3Φ total loss, W/ft	0.5	29.0	30.8	32.2	34.3	0.5	30.5	32.1	33.5	35.2
	0.3					0.3	29.7	31.5	33.0	34.8
Ratio watts dielectric loss, Φ total loss	0.5	42.2	44.5	46.0	50.5	0.5	46.3	49.3	52.6	37.5
	0.3	38.8	36.9	35.3	30.5	0.3	35.0	33.1	31.7	34.5

Source: Wiseman, R. T., *Trans. Am. Inst. Electr. Eng.*, 26, 803 © 1956 IEEE.

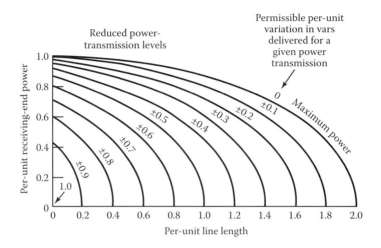

FIGURE 3.42 Permissible variations in per-unit vars delivered to electric system at each terminal of ac cable line for given power transmission. (From Wiseman, R. T., *Trans. Am. Inst. Electr. Eng.*, 26, 803 © 1956 IEEE.)

FIGURE 3.43 Cable system for Example 3.17.

and

$$\mathbf{I}_S = \mathbf{I}_R \cosh \gamma l + \mathbf{V}_R \mathbf{Y}_c \sinh \gamma l \tag{3.266}$$

Since at critical length, $l = l_0$ and

$$\mathbf{I}_R = 0$$

and

$$\mathbf{I}_S = \mathbf{I}_{l_0}$$

from Equation 3.264, the sending-end current can be expressed as

$$\mathbf{I}_{l_0} = \mathbf{V}_R \mathbf{Y}_c \sinh \gamma l_0 \tag{3.267}$$

or

$$\mathbf{I}_{l_0} = \mathbf{V}_R \mathbf{Y}_c \left(\frac{e^{\gamma l_0} - e^{\gamma l_0}}{2} \right) \tag{3.268}$$

$$\mathbf{I}_{l_0} = \mathbf{V}_R \mathbf{Y}_c \left(\frac{[1+\gamma l_0 + (\gamma l_0)^2/2! + \cdots] - [1 - \gamma l_0 + (\gamma l_0)^2/2! - \cdots]}{2} \right)$$

or

$$\mathbf{I}_{l_0} = \mathbf{V}_R \mathbf{Y}_c \left(\gamma l_0 + \frac{(\gamma l_0)^3}{3!} + \cdots \right) \quad (3.269)$$

Neglecting $(\gamma l_0)^3/3!$ and higher powers of γl_0,

$$\mathbf{I}_{l_0} = \mathbf{V}_R \mathbf{Y}_c \gamma l_0 \quad (3.270)$$

Similarly, from Equation 3.263, the sending-end voltage for the critical length can be expressed as

$$\mathbf{V}_S = \mathbf{V}_R \cosh \gamma l_0 \quad (3.271)$$

or

$$\mathbf{V}_S = \mathbf{V}_R \left(\frac{e^{\gamma l_0} + e^{\gamma l_0}}{2} \right) \quad (3.272)$$

or

$$\mathbf{V}_S = \mathbf{V}_R \left(\frac{[1+\gamma l_0 + (\gamma l_0)^2/2! + \cdots] + [1 - \gamma l_0 + (\gamma l_0)^2/2! - \cdots]}{2} \right)$$

or

$$\mathbf{V}_S = \mathbf{V}_R \left(1 + \frac{(\gamma l_0)^2}{2!} + \cdots \right) \quad (3.273)$$

Neglecting higher powers of γl_0,

$$\mathbf{V}_S \cong \mathbf{V}_R \left(1 + \frac{(\gamma l_0)^2}{2!} \right) \quad (3.274)$$

Thus,

$$\mathbf{V}_R = \frac{\mathbf{V}_S}{1 + \frac{(\gamma l_0)^2}{2!} + \cdots} \quad (3.275)$$

Substituting Equation 3.273 into Equation 3.268,

$$\mathbf{I}_{l_0} = \left(\frac{\mathbf{V}_S}{1 + \frac{(\gamma l_0)^2}{2!} + \cdots} \right) \mathbf{Y}_c \gamma l_0 \tag{3.276}$$

or

$$\mathbf{I}_{l_0} = \mathbf{V}_S \mathbf{Y}_c \gamma l_0 \left(1 + \frac{(\gamma l_0)^2}{2!} \right)^{-1}$$

$$= \mathbf{V}_S \mathbf{Y}_c \gamma l_0 \left(1 - \frac{(\gamma l_0)^2}{2!} + \cdots \right) \tag{3.277}$$

or

$$\mathbf{I}_{l_0} = \mathbf{V}_S \mathbf{Y}_c \gamma l_0 - \frac{\mathbf{V}_S \mathbf{Y}_c (\gamma l_0)^3}{2!} \tag{3.278}$$

Neglecting the second term,

$$\mathbf{I}_{l_0} \cong \mathbf{V}_S \mathbf{Y}_c \gamma l_0 \tag{3.279}$$

Therefore, the critical length can be expressed as

$$l_0 \cong \frac{\mathbf{I}_{l_0}}{\mathbf{V}_S \mathbf{Y}_c \gamma} \tag{3.280}$$

where

$$\mathbf{Y}_c = \sqrt{\frac{\mathbf{y}}{\mathbf{z}}}$$

$$\gamma = \sqrt{\mathbf{z} \times \mathbf{y}}$$

Thus,

$$\mathbf{y} = \mathbf{Y}_c \gamma \tag{3.281}$$

or

$$\mathbf{y} = g + jb \tag{3.282}$$

Therefore, the critical length can be expressed as

$$l_0 \cong \frac{\mathbf{I}_{l_0}}{\mathbf{V}_S \times \mathbf{y}} \tag{3.283}$$

or

$$l_0 \cong \frac{\mathbf{I}_{l_0}}{\mathbf{V}_S \times g + jb} \tag{3.284}$$

Steady-State Performance of Transmission Lines

Since, for cables, $g \ll b$,

$$y \cong b \angle 90°$$

and assuming

$$\mathbf{I}_{l_0} \cong I_{l_0} \angle 90°$$

from Equation 5.231, the critical length can be expressed as

$$l_0 \cong \frac{\mathbf{I}_{l_0}}{\mathbf{V}_S \times b} \tag{3.285}$$

EXAMPLE 3.18

Figure 3.44a shows an open-circuit high-voltage insulated ac underground cable circuit. The critical length of uncompensated cable is l_0 for which $\mathbf{I}_S = \mathbf{I}_0$ is equal to cable ampacity rating. Note that, $Q_0 = 3V_S I_0$, where the sending-end voltage $\mathbf{V}_S$ is regulated and the receiving-end voltage $\mathbf{V}_R$ floats. Here, $|\mathbf{V}_R|$ differs little from $|\mathbf{V}_S|$ because of the low series inductive reactance of cables. On the basis of the given information, investigate the performances with $\mathbf{I}_R = 0$ (i.e., *zero load*).

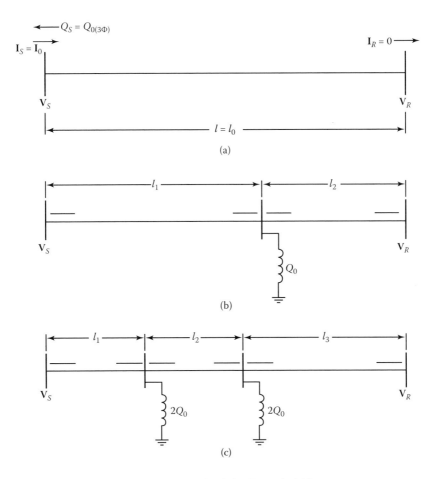

FIGURE 3.44 Insulated HV underground cable circuit for Example 3.18.

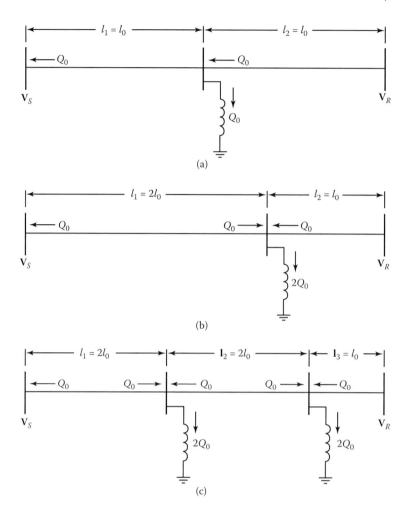

FIGURE 3.45 Solution for Example 3.16.

(a) Assume that one shunt inductive reactor sized to absorb Q_0 magnetizing vars is to be purchased and installed as shown in Figure 3.44b. Locate the reactor by specifying l_1 and l_2 in terms of l_0. Place arrowheads on the four short lines, indicated by a solid line, to show the directions of magnetizing var flows. Also show on each line the amounts of var flow, expressed in terms of Q_0.
(b) Assume that one reactor size $2Q_0$ can be afforded and repeat part (a) on a new diagram.
(c) Assume that two shunt reactors, each of size $2Q_0$, are to be installed, as shown in Figure 3.44c, hoping, as usual, to extend the feasible length of cable. Repeat part (a).

Solution

The answers for parts (a), (b), and (c) are given in Figure 3.45a, b and c, respectively.

3.14 GAS-INSULATED TRANSMISSION LINES

The gas-insulated transmission line (GIL) is a system for transmitting electric power at bulk power ratings over long distances. Its first application took place in 1974 to connect the electric generator

Steady-State Performance of Transmission Lines

of a hydro pump storage plant in Germany. For that application of the GIL, a tunnel was built in the mountain for a 420-kV overhead line. To this day, a total of more than 100 km of GILs have been built worldwide at high-voltage levels ranging from 135 to 550 kV. The GILs have also been built at power plants having adverse environmental conditions. Thus, in situations for which overhead lines are not feasible, the GIL may be an acceptable alternative since it provides a solution for a line without reducing transmission capacity under any kinds of climate conditions. This is because the GIL transmission system is independent of environmental conditions since it is completely sealed inside a metallic enclosure.

Applications of GIL include connecting high-voltage transformers with high-voltage switchgear within power plants, connecting high-voltage transformers inside the cavern power plants to overhead lines on the outside, connecting gas-insulated switchgear (GIS) with overhead lines, and serving as a bus duct within GIS.

At the beginning, the GIL system was only used in special applications owing to its high cost. Today, the second-generation GIL system is used for high-power transmission over long distances owing to the substantial reduction in its cost. This is accomplished not only by its much lower cost but also by the use of an N_2–SF_6 gas mixture for electrical insulation. The advantages of the GIL system include low losses, low magnetic field emissions, greater reliability with high transmission capacity, no negative impacts on the environment or the landscape, and underground laying with a transmission capacity that is equal to an overhead transmission line.

EXAMPLE 3.19

A power utility company is required to build a 500 kV line to serve a nearby town. There are two possible routes for the construction of the necessary power line. Route A is 80 mi long and goes around a lake. It has been estimated that the required overhead transmission line will cost $1 million per mile to build and $500 per mile per year to maintain. Its salvage value will be $2000 per mile at the end of 40 years.

On the other hand, route B is 50 mi long and is an underwater (submarine) line that goes across the lake. It has been estimated that the required underwater line using submarine power cables will cost $4 million to build per mile and $1500 per mile per ear to maintain. Its salvage value will be $6000 per mile at the end of 40 years.

It is also possible to use GIL in route C, which goes across the lake. Route C is 30 mi in length. It has been estimated that the required GIL transmission will cost $7.6 million per mile to build and $200 per mile to maintain. Its salvage value will be $1000 per mile at the end of 40 years. It has also been estimated that if the GIL alternative is elected, the relative savings in power losses will be $17.5 × 10^6 per year in comparison with the other two alternatives.

Assume that the fixed charge rate is 10% and that the annual ad valorem (property) taxes are 3% of the first costs of each alternative. The cost of energy is $0.10 per kWh. Use any engineering economy interest tables* and determine the economically preferable alternative.

Solution

OVERHEAD TRANSMISSION:

The first cost of the 500 kV overhead transmission line is

$$P = (\$1,000,000/\text{mi})(80 \text{ mi}) = \$80,000,000$$

and its estimated salvage value is

$$F = (\$2000/\text{mi})(80 \text{ mi}) = \$160,000$$

---

* For example, see *Engineering Economy for Engineering Managers*, T. Gönen, Wiley, 1990.

The annual equivalent cost of capital invested in the line is

$$A_1 = \$80,000,000(A/P)_{40}^{10\%} - \$100,000(A/F)_{40}^{10\%}$$
$$= \$80,000,000(0.10226) - \$100,000(0.00226)$$
$$= \$8,180,800 - \$266 = \$8,180,534$$

The annual equivalent cost of the tax and maintenance is

$$A_2 = (\$80,000,000)(0.03) + (\$500/\text{mi})(80 \text{ mi}) = \$2,440,000$$

The total annual equivalent cost of the overhead transmission line is

$$A = A_1 + A_2 = \$8,180,534 + \$2,440,000$$
$$= \$10,620,534$$

SUBMARINE TRANSMISSION:

The first cost of the 500 kV submarine power transmission line is

$$P = (\$4,000,000/\text{mi})(50 \text{ mi}) \ \$200,000,000$$

and its estimated salvage value is

$$F = (\$6000/\text{mi})(50 \text{ mi}) \ \$300,000$$

The annual equivalent cost of capital invested in the line is

$$A_1 = \$200,000,000(A/P)_{40}^{10\%} - \$300,000(A/F)_{40}^{10\%}$$
$$= \$200,000,000(0.10296) - \$300,000(0.00296)$$
$$= \$20,591,112$$

The annual equivalent cost of tax and maintenance is

$$A_2 = (\$200,000,000)(0.03) + (\$1500/\text{mi})(50 \text{ mi}) = \$6,075,000$$

The total annual equivalent cost of the overhead transmission line is

$$A = A_1 + A_2 = \$20,591,112 + \$6,075,000$$
$$= \$26,666,112$$

GIL TRANSMISSION:

The first cost of the 500 kV GIL transmission line is

$$P = (\$7,600,000/\text{mi})(30 \text{ mi}) \ \$228,000,000$$

and its estimated salvage value is

$$F = (\$1000/mi)(30\ mi)\ \$30,000$$

The annual equivalent cost of capital invested in the GIL line is

$$A_1 = \$228,000,000(A/P)_{40}^{10\%} - \$30,000(A/F)_{40}^{10\%}$$
$$= \$228,000,000(0.10226) - \$30,000(0.00226)$$
$$= \$23,315,280 - \$67.5 = \$23,315,212.5$$

The annual equivalent cost of the tax and maintenance is

$$A_2 = (\$228,000,000)(0.03) + (\$200/mi)(30\ mi) = \$6,846,000$$

The total annual equivalent cost of the GIL transmission line is

$$A = A_1 + A_2 = \$23,315,212.5 + \$6,846,000$$
$$= \$30,221,212.5$$

Since the relative savings in power losses is $17,500,000, then the total net annual equivalent cost of the GIL transmission is

$$A_{net} = \$30,221,212.5 - \$17,500,000$$
$$= \$12,721,212.5$$

The results show that the use of overhead transmission for this application is the best choice. The next best alternative is the GIL transmission. However, the above example is only a rough and very simplistic estimate. In real applications, there are many other cost factors that need to be included in such comparisons.

EXAMPLE 3.20

Consider transmitting 2100-MVA electric power across 30 km by using an overhead transmission line (OH) versus by using a GIL. The resulting power losses at peak load are 820 and 254 kW/km for the overhead transmission and the GIL, respectively. Assume that the annual load factor and the annual power loss factor are the same and are equal to 0.7 for both alternatives. Also, assume that the cost of electric energy is $0.10 per kWh. Determine the following:

(a) The power loss of the overhead line at peak load
(b) The power loss of the GIL
(c) The total annual energy loss of the overhead transmission line at peak load
(d) The total annual energy loss of the gas insulated transmission line at peak load
(e) The average energy loss of the overhead transmission line
(f) The average energy loss of the GIL at peak load
(g) The average annual cost of losses of the overhead transmission line
(h) The average annual cost of losses of the GIL
(i) The annual resultant savings in losses using the GIL
(j) Find the breakeven (or payback) period when the gas-insulated line alternative is selected, if the investment cost of the gas-insulated line is $200,000,000

Solution

(a) The power loss of the overhead transmission line at peak load is

$$(\text{Power loss})_{\text{OH line}} = (829 \text{ kW/km})30 \text{ km} = 24,870 \text{ kW}$$

(b) The power loss of the GIL transmission line at peak load is

$$(\text{Power loss})_{\text{GIL line}} = (254 \text{ kW/km})30 \text{ km} = 7620 \text{ kW}$$

(c) The total annual energy loss of the overhead transmission line at peak load is

$$(\text{Total annual energy loss})_{\text{at peak}} = (24,870 \text{ kW})(8760 \text{ h/yr})$$
$$= 21,786 \times 10^4 \text{ kWh/yr}$$

(d) The total annual energy loss of the gas-insulated line at peak load is

$$(\text{Total annual energy loss})_{\text{at peak}} = (7620 \text{ kW})(8760 \text{ h/yr})$$
$$= 6675.2 \times 10^4 \text{ kWh/yr}$$

(e) The total annual energy loss of the overhead transmission line at peak load is

$$(\text{Average annual energy loss})_{\text{OH line}} = 0.7(21,786 \times 10^4 \text{ kWh/yr})$$
$$= 15,250.2 \times 10^4 \text{ kWh/yr}$$

(f) The average energy loss of the gas-insulated line at peak load is

$$(\text{Average annual energy loss})_{\text{GIL line}} = 0.7(6675 \text{ kWh/yr})$$
$$= 4672.64 \times 10^4 \text{ kWh/yr}$$

(g) The average annual cost of losses of the overhead transmission line is

$$(\text{Average annual cost of losses})_{\text{OH line}} = (\$0.10/\text{kWh})(15,250.2 \times 10^4 \text{ kWh/yr})$$
$$= \$1525.02 \times 10^3/\text{yr}$$

(h) The average annual cost of losses of the GIL is

$$(\text{Average annual cost of losses})_{\text{GIL line}} = (\$0.10/\text{kWh})(4672.64 \times 10^4 \text{ kWh/yr})$$
$$= \$467.264 \times 10^3/\text{yr}$$

(i) The annual resultant savings in power losses using the GIL is

$$\text{Annual savings in losses} = (\text{Annual cost of losses})_{\text{OH line}} - (\text{Annual cost of losses})_{\text{GIL line}}$$
$$= \$1525 \times 10^3 - \$467.264 \times 10^3$$
$$= \$1057.736 \times 10^3/\text{yr}$$

(j) If the GIL alternative is selected

$$\text{Breakeven period} = \frac{\text{Total investment cost}}{\text{Savings per year}}$$

$$= \frac{\$200{,}000{,}000}{\$1057.736 \times 10^3} \cong 189 \text{ years}$$

3.15 BUNDLED CONDUCTORS

Bundled conductors are used at or above 345 kV. Instead of one large conductor per phase, two or more conductors of approximately the same total cross section are suspended from each insulator string. Therefore, by having two or more conductors per phase in close proximity compared with the spacing between phases, the voltage gradient at the conductor surface is significantly reduced. The bundles used at the extra-high-voltage range usually have two, three, or four subconductors, as shown in Figure 3.46. The bundles used at the ultrahigh-voltage range may also have 8, 12, and even 16 conductors.

Bundle conductors are also called *duplex*, *triplex*, and so on, conductors, referring to the number of subconductors, and are sometimes referred to as grouped or multiple conductors. The advantages derived from the use of *bundled* conductors instead of single conductors per phase are (1) reduced line inductive reactance; (2) reduced voltage gradient; (3) increased corona critical voltage and, therefore, less corona power loss, audible noise, and radio interference; (4) more power may be carried per unit mass of the conductor; and (5) the amplitude and duration of high-frequency vibrations may be reduced. The disadvantages of bundled conductors include (1) increased wind and ice loading; (2) suspension is more complicated and duplex or quadruple insulator strings may be required; (3) increased tendency to gallop; (4) increased cost; (5) increased clearance requirements at structures; and (6) increased charging kilovolt-amperes.

If the subconductors of a bundle are transposed, the current will be divided exactly between the conductors of the bundle. The GMRs of bundled conductors made up of two, three, and four subconductors can be expressed, respectively, as

$$D_s^b = (D_s \times d)^{1/2} \tag{3.286}$$

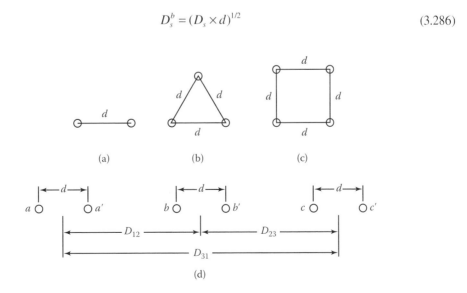

FIGURE 3.46 Bundle arrangements: (a) two-conductor bundle; (b) three-conductor bundle; (c) four-conductor bundle; (d) cross section of bundled-conductor three-phase line with horizontal tower configuration.

$$D_s^b = (D_s \times d^2)^{1/3} \tag{3.287}$$

$$D_s^b = 1.09(D_s \times d_3)^{1/4} \tag{3.288}$$

where
D_s^b = GMR of bundled conductor
D_s = GMR of subconductors
d = distance between two subconductors

Therefore, the *average* inductance per phase is

$$L_a = 2 \times 10^{-7} \ln \frac{D_{eq}}{D_s^b} \quad \text{H/m} \tag{3.289}$$

and the inductive reactance is

$$X_L = 0.1213 \ln \frac{D_{eq}}{D_s^b} \quad \Omega/\text{mi} \tag{3.290}$$

where

$$D_{eq} \triangleq D_m \, (D_{12} \times D_{23} \times D_{31})^{1/3} \tag{3.291}$$

The modified GMRs (to be used in capacitance calculations) of bundled conductors made up of two, three, and four subconductors can be expressed, respectively, as

$$D_{sC}^b = (r \times d)^{1/2} \tag{3.292}$$

$$D_{sC}^b = (r \times d^2)^{1/3} \tag{3.293}$$

$$D_{sC}^b = 1.09(r \times d^3)^{1/4} \tag{3.294}$$

where
D_{sC}^b = modified GMR of bundled conductor
r = outside radius of subconductors
d = distance between two subconductors

Therefore, the line-to-neutral capacitance can be expressed as

$$C_n = \frac{2\pi \times 8.8538 \times 10^{-12}}{\ln\left(\dfrac{D_{eq}}{D_{sC}^b}\right)} \quad \text{F/m} \tag{3.295}$$

or

$$C_n = \frac{55.63 \times 10^{-12}}{\ln\left(\dfrac{D_{eq}}{D_{sC}^b}\right)} \quad \text{F/m} \tag{3.296}$$

For a two-conductor bundle, the maximum voltage gradient at the surface of a subconductor can be expressed as

$$E_0 = \frac{V_0\left(1 + \dfrac{2r}{d}\right)}{2r \ln\left(\dfrac{D}{\sqrt{r \times d}}\right)} \tag{3.297}$$

EXAMPLE 3.21

Consider the bundled-conductor three-phase 200-km line shown in Figure 3.46d. Assume that the power base is 100 MVA and the voltage base is 345 kV. The conductor used is a 1113 kcmil ACSR, and the distance between two subconductors is 12 in. Assume that the distances D_{12}, D_{23}, and D_{31} are 26, 26, and 52 ft, respectively, and determine the following:

(a) Average inductance per phase in henries per meter
(b) Inductive reactance per phase in ohms per kilometer and ohms per mile
(c) Series reactance of line in per units
(d) Line-to-neutral capacitance of line in farads per meter
(e) Capacitive reactance to neutral of line in ohm per kilometers and ohm per miles.

Solution

(a) From Table A.3 in Appendix A, D_s is 0.0435 ft; therefore,

$$D_s^b = (D_s \times d)^{1/2}$$
$$= (0.0435 \times 0.3048 \times 12 \times 0.0254)^{1/2}$$
$$= 0.0636 \, \text{m}$$

$$D_{eq} = (D_{12} \times D_{23} \times D_{31})^{1/3}$$
$$= (26 \times 26 \times 52 \times 0.3048^3)^{1/3}$$
$$= 9.9846 \, \text{m}$$

Thus, from Equation 3.219,

$$L_a = 2\times 10^{-7} \ln \frac{D_{eq}}{D_s^b}$$

$$= 2\times 10^{-7} \ln \left(\frac{9.9846}{0.0636}\right)$$

$$= 1.0112\,\mu H/m$$

(b)

$$X_L = 2\pi f L_a$$

$$= 2\pi 60 \times 1.0112 \times 10^{-6} \times 10^3$$

$$= 0.3812\,\Omega/km$$

and

$$X_L = 0.3812 \times 1.609$$

$$= 0.6134\,\Omega/mi$$

(c)

$$Z_B = \frac{345^2}{100}$$

$$= 1190.25\,\Omega$$

$$X_L = \frac{0.3812 \times 200}{1190.25}$$

$$= 0.0641\,pu$$

(d) From Table A.3, the outside diameter of the subconductor is 1.293 in; therefore, its radius is

$$r = \frac{1.293 \times 0.3048}{2 \times 12}$$

$$= 0.0164\,m$$

$$D_{sC}^b = (r \times d)^{1/2}$$

$$= (0.0164 \times 12 \times 0.0254)^{1/2}$$

$$= 0.0707\,m$$

Thus, the line-to-neutral capacitance of the line is

$$C_n = \frac{55.63 \times 10^{-12}}{\ln\left(D_{eq}/D_{sC}^b\right)}$$

$$= \frac{55.63 \times 10^{-12}}{\ln(9.9846/0.0707)}$$

$$= 11.238 \times 10^{-12} \text{ F/m}$$

(c) The capacitive reactance to the neutral of the line is

$$X_c = \frac{1}{2\pi f C_n}$$

$$= \frac{10^{12} \times 10^{-3}}{2\pi 60 \times 11.238}$$

$$= 0.236 \times 10^6 \text{ }\Omega\text{-km}$$

and

$$X_c = \frac{0.236 \times 10^6}{1.609}$$

$$= 0.147 \times 10^6 \text{ }\Omega \text{ mi}$$

3.16 EFFECT OF GROUND ON CAPACITANCE OF THREE-PHASE LINES

Consider three-phase line conductors and their images below the surface of the ground, as shown in Figure 3.47. Assume that the line is transposed and that conductors a, b, and c have the charges q_a, q_b, and q_c, respectively, and their images have the charges $-q_a$, $-q_b$, and $-q_c$. The line-to-neutral capacitance can be expressed as [4,6]

$$C_n = \frac{2\pi \times 8.8538 \times 10^{-12}}{\ln\left(\dfrac{D_{eq}}{r}\right) - \ln\left(\dfrac{l_{12}l_{23}l_{31}}{h_{11}h_{22}h_{33}}\right)^{1/3}} \text{ F/m} \quad (3.298)$$

If the effect of the ground is not taken into account, the line-to-neutral capacitance is

$$C_n = \frac{2\pi \times 8.8538 \times 10^{-12}}{\ln\left(\dfrac{D_{eq}}{r}\right)} \text{ F/m} \quad (3.299)$$

As one can see, the effect of the ground increases the line capacitance. However, since the conductor heights are much larger than the distances between them, the effect of the ground is usually ignored for three-phase lines.

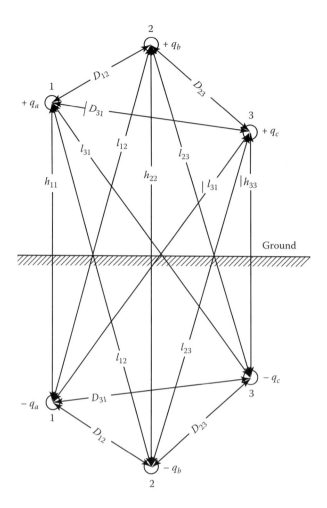

FIGURE 3.47 Three-phase line conductors and their images.

3.17 ENVIRONMENTAL EFFECTS OF OVERHEAD TRANSMISSION LINES

Recently, the importance of minimizing the environmental effects of overhead transmission lines has increased substantially because of the increasing use of greater extra-high- and ultrahigh-voltage levels. The magnitude and effect of radio noise, television interference, audible noise, electric field, and magnetic fields must not only be predicted and analyzed in the line design stage but also measured directly. Measurements of corona-related phenomena must include radio and television station signal strengths, and radio, television, and audible-noise levels.

To determine the effects of transmission line of these quantities, measurements should be taken at three different times: (1) before the construction of the line; (2) after construction, but before energization; and (3) after energization of the line. Noise measurements should be made at several locations along a transmission line. Also, at each location, measurements may be made at several points that might be of particular interest. Such points may include the point of maximum noise, the edge of the right of way, and the point 50 ft from the outermost conductor.

Overhead transmission lines and stations also produce electric and magnetic fields, which have to be taken into account in the design process. The study of field effects (e.g., induced voltages and currents in conducting bodies) is becoming especially crucial as the operating voltage levels of transmission lines have been increasing due to the economics and operational benefits involved. Today, for example, such study at ultrahigh-voltage level involves the following:

1. Calculation and measurement techniques for electric and magnetic fields
2. Calculation and measurement of induced currents and voltages on objects of various shapes for all line voltages and design configurations
3. Calculation and measurement of currents and voltages induced in people as result of various induction mechanisms
4. Investigation of sensitivity of people to various field effects
5. Study of conditions resulting in fuel ignition, corona from grounded objects, and other possible field effects [14]

Measurements of the transmission line electric field must be made laterally at midspan and must extend at least to the edges of the right of way to determine the profile of the field. Further, related electric field effects such as currents and voltages induced in vehicles and fences should also be considered. Magnetic field effects are of much less concern than electric field effects for extra-high- and ultrahigh-voltage transmission because magnetic field levels for normal values of load current are low. The quantity and character of currents induced in the human body by magnetic effects have considerably less impact than those arising from electric induction. For example, the induced current densities in the human body are less than one-tenth those caused by electric field induction. Furthermore, most environmental measurements are highly affected by prevailing weather conditions and transmission line geometry. The weather conditions include temperature, humidity, barometric pressure, precipitation levels, and wind velocity.

REFERENCES

1. Elgerd, O. I., *Electric Energy Systems Theory: An Introduction*. McGraw-Hill, New York, 1971.
2. Neuenswander, J. R., *Modern Power Systems*. International Textbook Company, Scranton, PA, 1971.
3. Stevenson, W. D., Jr., *Elements of Power System Analysis*, 3rd ed. McGraw-Hill, New York, 1975.
4. Anderson, P. M., *Analysis of Faulted Power Systems*. Iowa State Univ. Press, Ames, IA, 1973.
5. Fink, D. G., and Beaty, H. W., *Standard Handbook for Electrical Engineers*, 11th ed. McGraw-Hill, New York, 1978.
6. Wagner, C. F., and Evans, R. D., *Symmetrical Components*. McGraw-Hill, New York, 1933.
7. Weedy, B. M., *Electric Power Systems*, 2nd ed. Wiley, New York, 1972.
8. Concordia, C., and Rusteback, E., Self-excited oscillations in a transmission system using series capacitors. *IEEE Trans. Power Appar. Syst.* PAS-89 (no. 7) 1504–1512 (1970).
9. Elliott, L. C., Kilgore, L. A., and Taylor, E. R., The prediction and control of self-excited oscillations due to series capacitors in power systems. *IEEE Trans. Power Appar. Syst.* PAS-90 (no. 3) 1305–1311 (1971).
10. Kilgore, L., Taylor, E. R., Jr., Ramey, D. G., Farmer, R. G., and Schwalb, A. L, Solutions to the problems of subsynchronous resonance in power systems with series capacitors. *Proc. Am. Power Conf.* 35, 1120–1128 (1973).
11. Bowler, C. E. J., Concordia, C., and Tice, J. B., Subsynchronous torques on generating units feeding series-capacitor compensated lines, *Proc. Am. Power Conf.* 35, 1129–1136 (1973).
12. Schifreen, C. S., and Marble, W. C., Changing current limitations in operation of high-voltage cable lines. *Trans. Am. Inst. Electr. Eng.* 26, 803–817 (1956).
13. Wiseman, R. T., Discussions to charging current limitations in operation of high-voltage cable lines, *Trans. Am. Inst. Electr. Eng.* 26, 803–817 (1956).
14. Electric Power Research Institute, *Transmission Line Reference Book: 345 kV and Above*. EPRI, Palo Alto, CA, 1979.

GENERAL REFERENCES

Bowman, W. I., and McNamee, J. M., Development of equivalent pi and T matrix circuits for long untransposed transmission lines. *IEEE Trans. Power Appar. Syst.* PAS-83, 625–632 (1964).

Clarke, E., *Circuit Analysis of A-C Power Systems*, vol. 1. General Electric Company, Schenectady, NY, 1950.

Cox, K. J., and Clark, E., Performance charts for three-phase transmission circuits under balanced operations. *Trans. Am. Inst. Electr. Eng.* 76, 809–816 (1957).

Electric Power Research Institute, *Transmission Line Reference Book: 115–138 kV Compact Line Design*. EPRI, Palo Alto, CA, 1978.

Gönen, T., *Electric Power Distribution System Engineering*, 2nd ed. CRC Press, Boca Raton, FL, 2008.

Gönen, T., Nowikowski, J., and Brooks, C. L., Electrostatic unbalances of transmission lines with "*N*" overhead ground wires—Part I, *Proc. of Modeling and Simulation Conf.*, Pittsburgh, April 24–25, 1986, vol. 17, pt. 2, pp. 459–464.

Gönen, T., Nowikowski, J., and Brooks, C. L., Electrostatic unbalances of transmission lines with "*N*" overhead ground wires—Part II, *Proc. of Modeling and Simulation Conf.*, Pittsburgh, April 24–25, 1986, vol. 17, pt. 2, pp. 465–470.

Gönen, T., Yousif, S., and Leng, X., Fuzzy logic evaluation of new generation impact on existing transmission system, *Proc. of IEEE Budapest Tech'99 Conf.*, Budapest, Hungary, August 29–September 2, 1999.

Gross, C. A., *Power System Analysis*, Wiley, New York, 1979.

Institute of Electrical and Electronics Engineers, *Graphic Symbols for Electrical and Electronics Diagrams,* IEEE Stand. 315–1971 for American National Standards Institute (ANSI) Y32.2-1971). IEEE, New York, 1971.

Kennelly, A. E., *The Application of Hyperbolic Functions to Electrical Engineering Problems*, 3rd ed. McGraw-Hill, New York, 1925.

Skilling, H. H., *Electrical Engineering Circuits*, 2nd ed. Wiley, New York, 1966.

Travis, I., Per unit quantities, *Trans. Am. Inst. Electr. Eng.* 56, 340–349 (1937).

Woodruf, L. F., *Electrical Power Transmission*. Wiley, New York, 1952.

Zaborsky, J., and Rittenhouse, J. W., *Electric Power Transmission*. Rensselaer Bookstore, Troy, NY, 1969.

PROBLEMS

1. Redraw the phasor diagram shown in Figure 3.14 by using **I** as the reference vector and derive formulas to calculate the
   (a) Sending-end phase voltage, V_s
   (b) Sending-end power-factor angle, Φ_s

2. A three-phase, 60-Hz, a 20-mi-long short transmission line provides 12 MW at a lagging power factor of 0.85 at a line-to-line voltage of 34.5 kV. The line conductors are made of 26-strand 397.5-kcmil ACSR conductors that operate at 50°C and are equilaterally spaced 6 ft apart. Calculate the following:
   (a) Source voltage
   (b) Sending-end power factor
   (c) Transmission efficiency
   (d) Regulation of line

3. Repeat Problem 2 assuming the receiving-end power factor of 0.8 lagging.

4. Repeat Problem 2 assuming the receiving-end power factor of 0.8 leading.

5. A single-phase load is supplied by a 24-kVfeeder whose impedance is 60 + j310 Ω and a 24/2.4-kV transformer whose equivalent impedance is 0.25 + j1.00 Ω referred to its low-voltage side. The load is 210 kW at a leading power factor of 0.9 and 2.3 kV. Calculate the following:
   (a) Sending-end voltage of feeder
   (b) Primary-terminal voltage of transformer
   (c) Real and reactive-power input at sending end of feeder

6. A short three-phase line has the series reactance of 151 Ω per phase. Neglect its series resistance. The load at the receiving end of the transmission line is 15 MW per phase and 12 Mvar lagging per phase. Assume that the receiving-end voltage is given as 115 + j0 kV per phase and calculate
   (a) Sending-end voltage
   (b) Sending-end current

7. A short 40-mi-long three-phase transmission line has a series line impedance of 0.6+/0.95 Ω/mi per phase. The receiving-end line-to-line voltage is 69 kV. It has a full-load receiving-end current of 300∠−30° A. Do the following:
   (a) Calculate the percentage of voltage regulation.
   (b) Calculate the **ABCD** constants of the line.
   (c) Draw the phasor diagram of V_S, V_R, and **I**.
8. Repeat Problem 7 assuming the receiving-end current of 300∠−45° A.
9. A three-phase, 60-Hz, 12-MW load at a lagging power factor of 0.85 is supplied by a three-phase, 138-kV transmission line of 40 mi. Each line conductor has a resistance of 41 Ω/mi and an inductance of 14 mH/mi. Calculate
   (a) Sending-end line-to-line voltage
   (b) Loss of power in transmission line
   (c) Amount of reduction in line power loss if load-power factor were improved to unity
10. A three-phase, 60-Hz transmission line has sending-end voltage of 39 kV and receiving-end voltage of 34.5 kV. If the line impedance per phase is $18 + j57$ Ω, compute the maximum power receivable at the receiving end of the line.
11. A three-phase, 60-Hz, 45-mi-long short line provides 20 MVA at a lagging power factor of 0.85 at a line-to-line voltage of 161 kV. The line conductors are made of 19-strand 4/0 copper conductors that operate at 50°C. The conductors are equilaterally spaced with 4 ft spacing between them.
   (a) Determine the percentage of voltage regulation of the line.
   (b) Determine the sending-end power factor.
   (c) Determine the transmission line efficiency if the line is single phase, assuming the use of the same conductors.
   (d) Repeat part (c) if the line is three phase.
12. A three-phase, 60-Hz, 15-MW load at a lagging power factor of 0.9 is supplied by two parallel connected transmission lines. The sending-end voltage is 71 kV, and the receiving-end voltage on a full load is 69 kV. Assume that the total transmission line efficiency is 98%. If the line length is 10 mi and the impedance of one of the lines is $0.7 + j1.2$ Ω/mi, compute the total impedance per phase of the second line.
13. Verify that $(\cosh \gamma l - 1)/\sinh \gamma l = \tanh(1/2) \gamma l$.
14. Derive Equations 3.78 and 3.79 from Equations 3.76 and 3.77.
15. Find the general circuit parameters for the network shown in Figure P3.1.
16. Find a T equivalent of the circuit shown in Figure P3.1.
17. Assume that the line is a 200-mi-long transmission line and repeat Example 3.6. Use the Y and Z given in Example 3.6 as if they are for the whole 200-mi-long line given in this problem. Use the long-line model.
18. Assume that the line in Example 3.8 is 75 mi long and the load is 100 MVA, and repeat the example. (Use T-model for the medium line.)

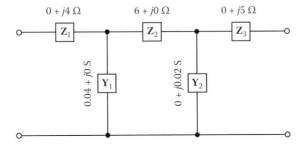

FIGURE P3.1 Network for Problem 15.

19. Develop the equivalent transfer matrix for the network shown in Figure P3.2 by using matrix manipulation.
20. Develop the equivalent transfer matrix for the network shown in Figure P3.3 by using matrix manipulation.
21. Verify Equations 3.202 through 3.205 without using matrix methods.
22. Verify Equations 3.209 through 3.212 without using matrix methods.
23. Assume that the line given in Example 3.6 is a 200-mi-long transmission line. Use the other data given in Example 3.6 accordingly and repeat Example 3.10.
24. Use the data from Problem 23 and repeat Example 3.11.
25. Assume that the shunt compensation of Example 3.16 is to be retained and now 60% series compensation is to be used, that is, the X_c is equal to 60% of the total series inductive reactance per phase of the transmission line. Determine the following:
    (a) Total three-phase SIL of line in megavolt-amperes
    (b) Maximum three-phase theoretical steady-state power flow limit in megawatts
26. Assume that the line given in Problem 25 is designed to carry a contingency peak load of 2 × SIL and that each phase of the series capacitor bank is to be of series and parallel groups of two-bushing, 12-kV, 150-kvar shunt power factor correction capacitors.
    (a) Specify the necessary series-parallel arrangements of capacitors for each phase.
    (b) Such capacitors may cost about $1.50/kvar. Estimate the cost of the capacitors in the entire three-phase series capacitor bank. (Take note that the structure and the switching and protective equipment associated with the capacitor bank will add a great deal more cost.)
27. Use Table 3.4 for a 345-kV, pipe-type, three-phase, 1000-kcmil cable. Assume that the percent power factor cable is 0.5 and maximum electric stress is 300 V/mil and that

$$V_s = \frac{345{,}000}{\sqrt{3}} \angle 0° \text{ V}$$

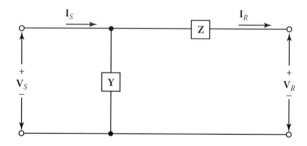

FIGURE P3.2 Network for Problem 19.

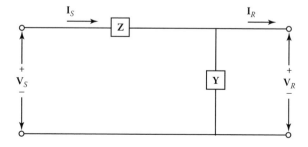

FIGURE P3.3 Network for Problem 20.

Use $I_T = 585$ A for the cable and calculate the following:
(a) Susceptance b of cable
(b) Critical length of cable and compare to value given in Table 4.3

28. Consider the cable given in Problem 27 and use Table 3.4 for the relevant data; determine the value of

$$\mathbf{I}_{lo} = \frac{\mathbf{V}_S}{\mathbf{Z}_c} \tanh \gamma l_0$$

accurately and compare it with the given value of cable ampacity in Table 3.3. (*Hint*: Use the exponential form of the tanh γl_0 function.)

29. Consider Equation 3.52 and verify that the maximum power obtainable (i.e., the steady-state power limit) at the receiving end can be expressed as

$$P_{R,\max} = \frac{|\mathbf{V}_S||\mathbf{V}_R|}{|X|} \sin \gamma$$

30. Repeat Problem 8 assuming that the given power is the sending-end power instead of the receiving-end power.

31. Assume that a three-phase transmission line is constructed of 700 kcmil, 37-strand copper conductors, and the line length is 100 mi. The conductors are spaced horizontally with $D_{ab} = 10$ Ω, $D_{bc} = 8$ Ω, and $D_{ca} = 18$ Ω. Use 60 Hz and 25°C, and determine the following line constants from tables in terms of
    (a) Inductive reactance in ohms per mile
    (b) Capacitive reactance in ohms per mile
    (c) Total line resistance in ohms
    (d) Total inductive reactance in ohms
    (e) Total capacitive reactance in ohms

32. A 60-Hz, single-circuit, three-phase transmission line is 150 mi long. The line is connected to a load of 50 MVA at a logging power factor of 0.85 at 138 kV. The line impedance and admittance are $\mathbf{z} = 0.7688 \angle 77.4°$ Ω/mi and $\mathbf{y} = 4.5239 \times 10^{-6} \angle 90°$ S/mi, respectively. Use the long-line model and determine the following:
    (a) Propagation constant of the line
    (b) Attenuation constant and phase-change constant, per mile, of the line
    (c) Characteristic impedance of the line
    (d) SIL of the line
    (e) Receiving-end current
    (f) Incident voltage at the sending end
    (g) Reflected voltage at the sending end

33. Consider a three-phase transmission and assume that the following values are given:

$\mathbf{V}_{R(L-N)} = 79{,}674.34 \angle 0°$ V, $\mathbf{I}_R = 209.18 \angle -31.8°$ A, $\mathbf{Z}_c = 469.62 \angle 5.37°$ Ω and $\gamma l = 0.0301 + j0.3202$

Determine the following:
(a) Incident and reflected voltages at the receiving end of the line
(b) Incident and reflected voltages at the sending end of the line
(c) Line voltage at the sending end of the line

34. Repeat Example 3.17 but assume that the conductor used is 1431-kcmil ACSR and that the distance between two subconductors is 18 in. Also, assume that the distances D_{12}, D_{23}, and D_{31} are 25, 25, and 50 ft, respectively.

4 Disturbance of Normal Operating Conditions and Other Problems

4.1 INTRODUCTION

The normal operation of a power system may be disturbed or disrupted owing to a system fault when abnormally high currents flow through an abnormal path as a result of the partial or complete failure of the insulation at one or more points of the system. The complete failure of insulation is called a "short circuit" or "fault."

A short circuit occurs on a power system when one or more energized conductors contact other conductors or ground. In the event of insulation failure, it may not be essential for the conductors to be in actual contact. The short circuit can exist by current flowing through an ionized path that may be through air or some other substance that is normally an insulator. At the fault, the voltage between the two parts is reduced to zero in the event of metal-to-metal contacts or to a very low value if the short-circuit path is through an arc. As the result of a fault, currents of abnormally high magnitude flow through the system to the fault point.

The abnormally high current may flow owing to abnormally high voltages (overvoltages) on the system as a result of lightning or switching surges that can puncture through or cause flashover across the surface of insulation. The resulting damage to the insulation or ionization of the surrounding insulation establishes a follow-through power arc. Insulator contamination by moisture with dirt or salt may also cause flashover even during normal voltage conditions.

The switching surges are due to rapid changes in the flow of current that can occur when energizing or deenergizing lines and equipment. Such operations can produce traveling waves that may flash lines or equipment and weak points of insulation. Figures 4.1 and 4.2 show strings of suspension insulators undergoing a lightning impulse flashover test.

Line and apparatus insulation may be subjected to transient overvoltages whenever current is started or stopped. These surges are a component of the "recovery" voltages. The most severe switching surges occur when current that lags or leads the applied voltage by 90° (e.g., fault current or line-charging current) is interrupted.

During unloaded line dropping on a grounded system, the line voltage may go to crest line-to-neutral voltage on the first interruption, three times this value on the first restrike, five times this value on the second strike, etc., as the arc restrikes on the succeeding half-cycles. The magnitude of these switching surges is appreciably greater for systems that are not solidly grounded. Thus, system insulation may be subjected to serious overvoltages with breaker recovery voltages that are still higher when the line-charging current is interrupted.

Since they are mostly constructed of bare conductors, the overhead transmission lines are one of the most vulnerable points in power systems. A considerable amount of faults occur in these overhead transmission lines. Other causes of faults may include wind, sleet, conductor clashing due to conductor galloping, small animals (such as birds, snakes, and squirrels), trees, cranes, airplanes, vandalism, vehicles colliding with poles or towers, line breaks due to excessive ice loading, or other damages to supporting structures.

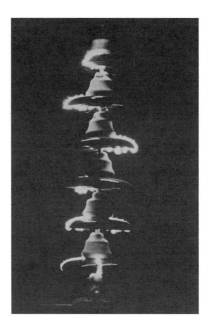

FIGURE 4.1 String of suspension insulators undergoing lightning impulse flashover test. (Courtesy of Ohio Brass Company.)

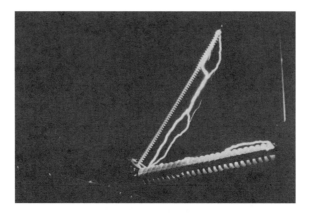

FIGURE 4.2 String of suspension insulators undergoing lightning impulse flashover test. (Courtesy of Ohio Brass Company.)

A fault is not to be considered with an overload. An "overload" simply means that the system carries loads that exceed the normal load for which the system is designed. The voltage at the overload point may drop to a low value but not to zero, as would be the case in the event of the fault. The resulting "undervoltages" condition may reach to the point in the system far from the fault point [1].

It is possible to classify faults as "permanent" and "temporary" faults. Permanent faults damage the equipment. The destruction to the equipment is usually violent owing to sufficient energy flows into the short circuit in a short time to cause an explosion. Such a persistent fault can damage conductors by burning or melting them.

Most of the faults on overhead transmission lines are luckily *temporary* or *transient* in nature. Thus, the service can be restored by isolating and then reclosing the faulty line section very rapidly, allowing only enough time for the air to be deionized after the fault arc has been extinguished so as to prevent restriking. This procedure is called *high-speed reclosing*.

Disturbance of Normal Operating Conditions and Other Problems

Substation buses are designed to withstand the maximum expected faults due to extraordinarily large mechanical forces produced by heavy short-circuit currents. Transformer failures can be due to the insulation deterioration caused by aging or overvoltages as a result of lightning or switching transients. The faults may be external to the transformer, but they can still cause large mechanical forces internal to the transformer. These forces can cause windings or other parts to move and damage insulation or actually cause structural failures.

The types of failures involving the generator include (1) failures that occur in the exciter circuit or control equipment (about 50% of the generator faults); (2) failures that occur in the auxiliary apparatus such as cooling equipment (about 40%); and (3) failures including (a) stator faults due to the breakdown of conductor insulation caused by overvoltage or by overheating as a result of unbalanced currents, (b) rotor faults due to the damaged rotor windings caused by ground faults, (c) overspeed due to sudden loss of load, (d) loss of synchronism caused by an interphase fault or wrong switching, or loss of field, (e) loss of field caused by a pilot-exciter failure or a main-exciter failure, or a field breaker failure, and (f) bearing failure due to the failure of cooling or oil supply.

Two parallel current-carrying line conductors are subject to a force of attraction or repulsion, depending on the current direction. The magnitude of the force affecting each conductor can be expressed as

$$F \propto \frac{I^2}{d} \qquad (4.1)$$

where
I = current in each conductor
d = distance between conductors

In the event that the current flow in each conductor is in the same direction, the force will cause attraction. Otherwise, it will cause repulsion. A recent Electric Power Research Institute (EPRI) study [2] shows that for short-circuit currents, these forces may be enough to cause significant conductor movement, particularly where conductors are closely spaced, for example, in extra-high-voltage conductor bundles or in adjacent phases of a compact line. Of course, the actual movement of a conductor, considering inertia, is a function of both the magnitude of the current and the time it is applied and is therefore dependent on circuit breaker interrupting time.

Other problems during the normal operating conditions include corona, corona losses, and radio noise or radio interference (RI) problems.

4.2 FAULT ANALYSIS AND FAULT TYPES

The purpose of the fault analysis (also called *short-circuit study* or *analysis*) is to calculate the maximum and minimum fault currents and voltages at different locations of the power system for various types of faults so that the appropriate protective schemes, relays, and circuit breakers can be selected in order to rescue the system from the abnormal condition within minimum time.

In practice, to perform the fault analysis, the following *simplifying assumptions* are usually made:

1. The normal loads, line-charging (i.e., shunt) capacitances, shunt elements in transformer equivalent circuits for representing magnetizing reactances or core loss, and other shunt connections to the ground are neglected.
2. All generated (i.e., internal) system voltages are equal (in magnitude) and are in phase.
3. Normally, the series resistances of lines and transformers are neglected if considered small in comparison with their reactances.

4. All the transformers are considered to be at their nominal taps.
5. The generator is represented by a constant-voltage source behind (i.e., in series with) a proper reactance that may be subtransient (X_d''), transient (X_d'), or synchronous, at steady state (X_d) reactance. Usually, the subtransient reactance (X_d'') is selected for the positive-sequence reactance. Therefore, such a representation is sufficient to calculate the magnitudes of the fault currents in the first three to four cycles after the fault takes place.

The first assumption is based on the fact that the fault circuit has predominantly lower impedance than the shunt impedances. Therefore, the saving in computational effort due to this assumption usually justifies the slight loss in accuracy.

The second assumption results from the first assumption. With the first assumption, the power system network becomes open-circuited, and therefore the normal load currents (i.e., prefault currents) are consequently neglected, and thus all the prefault bus voltages will have the same magnitude and phase angle.

Thus, in per-unit analysis, the prefault bus voltages are set equal to $1.0\angle 0°$ pu. In the rare event that taking into account for load current is desirable, this can be done by applying the superposition theorem. The third assumption is usually done for hand calculations and educational purposes. With this assumption, the power system network will contain only reactances, and therefore the system can be represented by its most simplified reactance diagram. However, this assumption is not necessary if the computation will be done using a digital computer.

The fourth assumption neglects the transformer tapings so that the fault analysis can be carried out in a per-unit system. Thus, with this representation, transformers will be out of circuit. As mentioned before, the subtransient reactance, X_d'', is usually selected as the positive-sequence reactance. The value of the negative-sequence reactance is slightly different than the positive-sequence reactance for the salient-type machines.

In the event that the generator is a nonsalient type (i.e., a cylindrical rotor machine) and if the subtransient reactance (X_d'') is selected for the positive-sequence reactance, the negative-sequence reactance becomes identical to the positive-sequence reactance, as shown in Table 4.1. In practice, to calculate the maximum fault currents, it is common to make an assumption that the fault is "bolted," that is, one having no fault impedance ($\mathbf{Z}_f = 0$) resulting from fault arc. (This assumption not only simplifies the fault calculations but also provides a safety factor since the calculated values become larger than the ones calculated using a fault impedance value.)

In the event that the fault has a short-circuit path that is not a metallic path but an arc or a path through the ground, nonlinear impedances are included that tend to inject harmonics into the current and/or voltage. Fault resistance has two components: the resistance of the arc [3] and the resistance of the ground [4]. If the fault is between phases such as line to line, the fault includes only the arc. Thus, the fault resistance includes only the arc resistance. Fault arc resistance is given by Warrington [5] as

$$R_{arc} = \frac{8750 \times \ell}{I^{1.4}} \quad \Omega \tag{4.2}$$

where
ℓ = length of arc in still air in feet
I = fault current in amperes

If time is involved, the arc resistance is calculated from

$$R_{arc} = \frac{8750(d \times 3vt)}{I^{1.4}} \quad \Omega \tag{4.3}$$

where
 d = conductor spacing in feet (from Figure 4.3)
 v = wind velocity in miles per hour
 t = duration in seconds

The relation between the fault current and the arc voltage is given as

$$R_{arc} = \frac{8750(d \times 3vt)}{I^{0.4}} \quad \Omega \tag{4.4}$$

If a high-resistance line-to-ground fault occurs, the important impedances in the fault circuit are the contact to ground and the path through it. Warrington [5] has shown that the resistance of the ground fault is somewhat nonlinear partly because there are small arcs between conducting particles and partly due to the compounds of silicon, carbon, etc., which have nonlinear resistance (see Figure 4.3). It is interesting to note that, in practice, such *fault resistance* is erroneously called the *fault impedance*, and it is assumed to include a fictitious reactive component.

In general, the fault types that may occur in a three-phase power system can be categorized as follows.

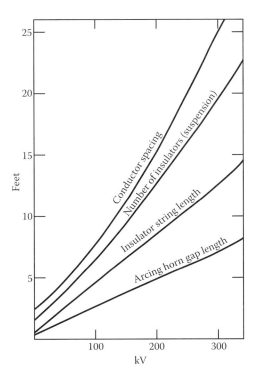

FIGURE 4.3 Minimum arcing distances on overhead lines (based on average tower dimensions in the United Kingdom and the United States). (From Warrington, A. R. van C., *Protective Relays: Their Theory and Practice*, vol. 2. Chapman & Hall, London, 1969.)

A. Shunt faults:
   1. Balanced (also called symmetrical) three-phase faults
      a. Three-phase direct (L–L–L) faults
      b. Three-phase faults through a fault impedance to ground (L–L–L–G)
   2. Unbalanced (also called unsymmetrical) faults
      a. Single line-to-ground (SLG) faults
      b. Line-to-line (L–L) faults
      c. Double line-to-ground (DLG) faults
B. Series faults:
   1. One line open (OLO)
   2. Two lines open (TLO)
   3. Unbalanced series impedance condition
C. Simultaneous faults:
   1. A shunt fault at one fault point and a shunt fault at the other
   2. A shunt fault at one fault point and a series fault at the other
   3. A series fault at one fault point and a series fault at the other
   4. A series fault at one fault point and a shunt fault at the other

Shunt faults are more severe than series faults. Balanced faults are simpler to calculate than unbalanced faults. Simultaneous faults, involving two or more faults that occur simultaneously, are usually considered to be the most difficult fault analysis problem. In this chapter, only balanced faults and series faults are reviewed; unbalanced faults will be reviewed in Chapter 6. The probability of having a simultaneous fault is much less than the shunt fault. Therefore, the discussion of the simultaneous faults is kept beyond the scope of this book. However, for those readers interested in the subject matter, the book by Anderson [6] is highly recommended.

4.3 BALANCED THREE-PHASE FAULTS AT NO LOAD

Consider the per-phase representation of a synchronous generator, as shown in Figure 4.4. Assume that there is a balanced three-phase fault between the points F and N. If the generator voltage is $e(t) = V_m \sin(\omega t + \alpha)$ and the fault occurs at $t = 0$, it can be shown that there will be a transient current $i(t)$ that can be expressed as

$$i(t) = \frac{V_m}{Z} \left[\sin(\omega t + \alpha - \theta) - \sin(\alpha - \theta) e^{-\frac{Rt}{L}} \right] \quad (4.5)$$

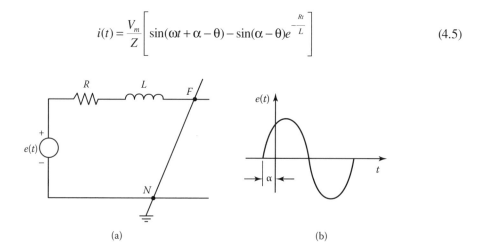

FIGURE 4.4 Synchronous generator with balanced fault: (a) per-phase representation of generator; (b) voltage waveform.

Disturbance of Normal Operating Conditions and Other Problems

where

$$Z = (R^2 + \omega^2 L^2)^{1/2} \quad \theta = \tan^{-1}\left(\frac{\omega L}{R}\right)$$

It can be seen in Equation 4.5 that the first term is a sinusoidal term and its value changes with time and that the second term is a nonperiodic term and its value decreases exponentially with time. The second term is a *unidirectional offset* and is also called the *direct current (dc) component* of the fault current. It will in general exist, and its initial magnitude (i.e., at $t = 0$) can be as large as the magnitude of the steady-state current term, as shown in Figure 4.5a. If the fault occurs at $t = 0$ when the angle $\alpha - \theta = -90°$, the value of the transient current becomes twice the steady-state maximum value and can be expressed as

$$i(t) = \frac{V_m}{2}\left(-\cos\omega t + e^{-\frac{Rt}{L}}\right) \tag{4.6}$$

and is shown in Figure 4.5a. The associated value of α is obtained from

$$\tan\alpha = -\frac{R}{\omega L} \tag{4.7}$$

On the other hand, if $\alpha = 0$, at $t = 0$, the dc offset does not exist, as shown in Figure 4.5b, and the value of the transient current can be expressed as

$$i(t) = \frac{V_m}{Z}\sin\omega t \tag{4.8}$$

Obviously, if a $\alpha - \theta = \pi$ at $t = 0$, the dc offset current again cannot exist. Thus, the value of the transient current depends on the angle α of the voltage wave. However, the time of the fault cannot be predicted in practice, and therefore the value of cannot be known ahead of time. However, the dc component diminishes very fast, usually in 8–10 cycles.

Furthermore, since the voltages generated in the phases of a three-phase synchronous generator are 120° apart from each other, each phase will have, in general, a different offset.

Note that, in the aforementioned discussions, the value of L for the generator is assumed to be constant. In reality, however, the reactance of a synchronous machine varies with time immediately

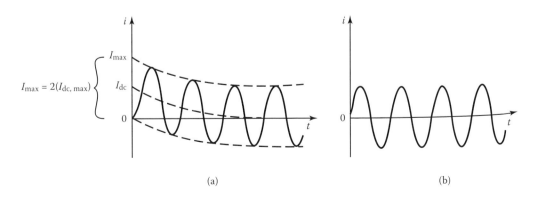

FIGURE 4.5 Balanced fault current wave shapes: (a) $a - \theta = -90°$; (b) $\alpha = \theta$.

after the occurrence of the fault. Thus, it is customary to represent a synchronous generator by a constant driving voltage in series with impedance that varies with time.

This varying impedance consists primarily of reactance since $X \gg R_a$, that is, X is much larger than the armature resistance. Hence, the value of impedance is approximately equal to its reactance. For the purpose of fault current calculations, the variable reactances of a synchronous machine can be represented, as shown in Figure 4.6b, by the following three reactance values:

X_d'' = *subtransient reactance:* determines the fault current during the first cycle after the fault occurs. In about 0.05–0.1 s, this reactance increases to

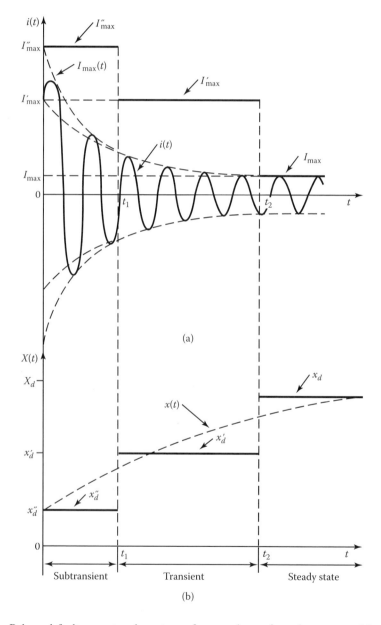

FIGURE 4.6 Balanced fault current and reactance for one phase of synchronous machine: (a) balanced instantaneous fault current without dc offset; (b) reactance $X(t)$ vs. time with stepped approximation.

X'_d = *transient reactance:* determines the fault current after several cycles at 60 Hz. In about 0.2–2 s, it reactance increases to

$X_d = X_s$ = *synchronous reactance:* determines the fault current after a steady-state condition is reached

This representation of the machine reactance by three different reactances is due to the fact that the flux across the air gap of the machine is much greater at the instant the fault occurs than it is a few cycles later. Thus, when a fault occurs at the terminals of a synchronous machine, time is necessary for the decrease in flux across the air gap. As the flux lessens, the armature current lessens since the voltage produced by the air gap flux regulates the current.

Thus, the subtransient reactance X''_d includes the leakage reactance of the stator and rotor windings of the generator, the influences of damper windings and of the solid parts of the rotor body being included in the rotor leakage. The subtransient reactance is also called the *initial reactance*. The transient reactance X''_d includes the leakage reactance of the stator and excitation windings of the generator. It is usually larger than the subtransient reactance.

If, however, the rotor has laminated poles and yokes and no damper windings, the transient reactance is the same as the subtransient reactance. If, however, the rotor has laminated poles and yokes and no damper windings,* the transient reactance is the same as the subtransient reactance.

The synchronous reactance X_d is the total reactance of the armature winding, which includes the stator leakage reactance and the armature reaction reactance of the generator. It is much larger than the transient reactance X''_d. Note that, all three reactances are considered to be the positive-sequence reactance of the synchronous machine.

In a *salient-pole* machine, the index *d* means that the reactances refer to a position of the rotor such that the axis of the rotor winding coincides with the axis of the stator winding so that the flux flows directly into the pole face. Therefore, it is called the *direct axis*, and thus the three reactances are also known as the *direct-axis reactances*.

In addition to these reactances, the generator also has reactances in the corresponding *quadrature axis*, that is, X''_q, X'_q and X_q, due to the flux path between poles, that is, midway between the field poles. The quadrature axis is 90 electrical degrees apart from the adjacent direct axes. However, the quadrature axis reactances are not relevant to the fault calculations.

Note that in a *nonsalient-pole* machine (i.e., *cylindrical-rotor machine*), values of X_d and X_q are basically equal. Therefore, there is no need to differentiate X_d from X_q but only call the synchronous reactance X_s. For the sake of simplification, in this book, all synchronous machines are assumed to have cylindrical rotors.

If the generator is operating at no load before the occurrence of a three-phase fault at its terminals, then the continuously varying symmetrical maximum current, $I_{max}(t)$, and reactance can be approximated with the discrete current levels, as shown in Figure 4.6, so that

$$X''_d = \frac{E_{max}}{I''_{max}} \qquad (4.9)$$

$$X'_d = \frac{E_{max}}{I'_{max}} \qquad (4.10)$$

$$X_d = \frac{E_{max}}{I_{max}} \qquad (4.11)$$

---

* They are located in the pole faces of a generator and are used to reduce the effects of hunting.

where E_{max} is the no-load line-to-neutral maximum voltage of the generator. Alternatively, using the root mean square (rms) values,

$$I'' = \frac{E_g}{X_d''} \tag{4.12}$$

$$I' = \frac{E_g}{X_d'} \tag{4.13}$$

$$I = \frac{E_g}{X_d} \tag{4.14}$$

where
  E_g = no-load line-to-neutral rms voltage
  I'' = subtransient current,* rms value without dc offset
  I' = transient current, rms value without dc offset
  I = steady-state current, rms value

Note that, the importance of the reactances given by Equations 4.9 through 4.11 depends on what percentage they represent of the short-circuit impedance. For example, if the fault occurs right at the terminals of the generator, they are very important; however, if the fault is remote from the generator, their importance is smaller.

The fault current will be lagging in power in a system where $X \gg R$. Table 4.1 gives the typical values of the reactances for synchronous machines.

It is interesting to observe in Figure 4.6a that the total alternating component of armature current consists of the steady-state value and the two components that decay with time constants T_d' and T_d''. It can be expressed as

$$I_{ac} = (I'' - I')\exp\left(-\frac{t}{T_d''}\right) + (I' - I)\exp\left(-\frac{t}{T'}\right) + I \tag{4.15}$$

where all quantities are in rms values and are equal but displaced 120 electrical degrees in the three phases.

4.4 FAULT INTERRUPTION

As most fault-protective devices, such as circuit breakers and fuses, operate well before steady-state conditions are reached, generator synchronous reactance is almost never used in calculating fault currents for application of these devices. As discussed before, to determine the initial symmetrical rms current, the subtransient reactances of the synchronous generators and motors are used.

However, the interrupting capacity of circuit breakers is determined using the subtransient reactance for generators and transient reactance for the synchronous motors. The effects of induction motors are ignored. It is appropriate to use subtransient reactance for synchronous motors if fast-acting circuit breakers are used. For example, modern air blast circuit breakers usually operate in 2.5 cycles of 60 Hz. Older circuit breakers and those used on lower voltages may take eight cycles or more to operate. Note that, thus far, the dc offset (or unidirectional current component) has been excluded in the above discussions. With fast-acting circuit breakers, the actual current to be interrupted is increased by the dc

---

* It is also called the *initial symmetrical rms current*.

component of the fault current, and the initial symmetrical rms current value is increased by a specific factor depending on the speed of the circuit breaker.* For example, if the circuit breaker opening time is eight, three, or two cycles, then the corresponding multiplying factor is 1.0, 1.2, or 1.4, respectively. Therefore, the interrupting capacity (or rating) of a circuit breaker is expressed as

$$S_{\text{interrupting}} = \sqrt{3}(V_{\text{prefault}})(I'')\zeta \times 10^{-6} \text{ MVA} \qquad (4.16)$$

where
V_{prefault} = prefault voltage at point of fault in volts
I'' = initial symmetrical rms current in amperes
ζ = multiplying factor

Note that, Equation 4.16 includes only the ac component. The multiplying factors and the reactance types are given in Table 4.1 [7]. As discussed before, the asymmetrical current wave decays gradually to a symmetrical current, the rate of decay of the dc component being determined by the L/R of the system supplying the current. The time constant for dc component decay can be found from

$$T_{dc} = \text{circuit } L/R \text{ s} \qquad (4.17a)$$

or

$$T_{dc} = \frac{\text{circuit } L/R}{2\pi} \text{ cycles} \qquad (4.17b)$$

The momentary duty (or rating) of a circuit breaker is expressed as

$$S_{\text{momentary}} = \sqrt{3}(V_{\text{prefault}})(I'')(1.6) \times 10^{-6} \text{ MVA} \qquad (4.18)$$

This equation includes the dc component. Thus, the rms momentary current can be expressed as

$$I_{\text{momentary}} = 1.6 \times I'' \text{ A} \qquad (4.19)$$

Here, the $I_{\text{momentary}}$ current is the total rms current that includes ac and dc components, and it is used for oil circuit breakers of 115 kV and above. The circuit breaker must be able to withstand this rms current during the first half-cycle after the fault occurs. Note that, if the I'' is measured in peak amperes, then the peak momentary current is expressed as

$$I_{\text{momentary}} = 2.7 \times I'' \text{ A} \qquad (4.20)$$

In the United States, the ratings of circuit breakers are given in the American National Standards Institute (ANSI) standards [8] based on symmetrical current,† in terms of nominal voltage, rated maximum voltage, rated voltage range factor K, rated continuous current, and rated short-circuit current. The required symmetrical current-interrupting capability is defined as

$$\begin{pmatrix} \text{Required symmetrical} \\ \text{current-interrupting capability} \end{pmatrix} = \begin{pmatrix} \text{Rated short-} \\ \text{circuit current} \end{pmatrix} \begin{pmatrix} \text{Rated maximum voltage} \\ \text{Operating voltage} \end{pmatrix}$$

---

* Note that, the fault megavolt-ampere is often referred to as the *fault level*.
† The ratings of the circuit breakers can also be based on total current, which includes the dc component.

TABLE 4.1
Reactance Quantities and Multiplying Factors for Application of Circuit Breakers

		Reactance Quantity for Use in X_1		
	Multiplying Factor	Synchronous Generators and Condensers	Synchronous Motors	Induction Machines
A. Circuit Breaker Interrupting Duty				
1. General case:				
Eight-cycle or slower circuit breakers[a]	1.0	Subtransient[b]	Transient	Neglect
Five-cycle circuit breakers	1.1			
Three-cycle circuit breakers	1.2			
Two-cycle circuit breakers	1.4			
2. Special case for circuit breakers at generator voltage only. For short-circuit calculations of more than 500,000 kVA (before the application of any multiplying factor) fed predominantly direct from generators, or through current-limiting reactors only:		Subtransient[b]	Transient	Neglect
Eight-cycle or slower circuit breakers[a]	1.1			
Five-cycle circuit breakers	1.2			
Three-cycle circuit breakers	1.3			
Two-cycle circuit breakers	1.5			
3. Air circuit breakers rated 600 V and less	1.25	Subtransient	Subtransient	Subtransient
B. Mechanical Stress and Momentary Duty of Circuits Breakers				
1. General case	1.6	Subtransient	Subtransient	Subtransient
2. At 500 V and below, unless current is fed predominantly by directly connected synchronous machines or through reactors	1.5	Subtransient	Subtransient	Subtransient

Source: Westinghouse Electric Corporation, *Electrical Transmission and Distribution Reference Book*. WEC, Pittsburgh, 1964.

[a] As old circuit breakers are slower that modern ones, it might be expected that a low multiplier could be used with old circuit breakers. However, modern circuit breakers are likely to be more effective than their slower predecessors, and therefore, the application procedure with the older circuit breakers should be more conservative than with modern circuit breakers. Also, there is no assurance that a short circuit will not change its character and initiate a higher current flow through a circuit breaker while it is opening. Consequently, the factors to be used with older and slower circuit breakers well may be the same as for modern eight-cycle circuit breakers.

[b] This is based on the condition that any hydroelectric generators involved have amortisseur (damper) windings. For hydro-electric generators without amortisseur windings, a value of 75% of the transient reactance should be used for this calculation rather than the subtransient value.

The standard dictates that for operating voltages below $1/K$ times the rated maximum voltage, the required symmetrical current-interrupting capability of the circuit breaker is equal to K times the rated short-circuit current. Table 4.2 gives outdoor circuit breaker ratings based on symmetrical current. Note that, the rated voltage factor K is defined as the ratio of the rated maximum voltage to the lower limit of the range of operating voltage in which the required symmetrical and asymmetrical interrupting capabilities vary in inverse proportion to the operating voltage [9]. Therefore,

TABLE 4.2
Outdoor Circuit Breaker Ratings Based on Symmetrical Current

Nominal rms Voltage Class (kV)	Rated Maximum rms Voltage (kV)	Rated Voltage Range Factor, K	Rated Continuous rms Current (kA)	Rated Short-Circuit rms Current (at Rated Maximum kV) (kA)	Rated Interrupting Time (Cycles)	Rated Maximum rms Voltage Divided by K (kV)	Maximum rms Symmetrical Interrupting Capability[a] (kA)
14.4	15.5	2.67	0.6	8.9	5	5.8	24
14.4	25.5	1.29	1.2	18	5	12	23
23	25.8	2.15	1.2	11	5	12	24
34.5	38	1.65	1.2	22	5	23	36
46	48.3	1.21	1.2	17	5	40	21
69	72.5	1.21	1.2	19	5	60	23
115	121	1	1.2	20	3	121	20
115	121	1	1.6	40	3	121	40
⋮	⋮	⋮	⋮	⋮	⋮	⋮	⋮
115	121	1	3	63	3	121	63
138	145	1	1.2	20	3	145	20
138	145	1	1.6	40	3	145	40
138	145	1	2	40	3	145	40
⋮	⋮	⋮	⋮	⋮	⋮	⋮	⋮
138	145	1	3	80	3	145	80
161	169	1	1.2	16	3	169	16
161	169	1	1.6	31.5	3	169	31.5
⋮	⋮	⋮	⋮	⋮	⋮	⋮	⋮
161	169	1	2	50	3	169	50
230	242	1	1.6	31.5	3	242	31.5
230	242	1	2	31.5	3	242	31.5
⋮	⋮	⋮	⋮	⋮	⋮	⋮	⋮
230	242	1	3	63	3	242	63
345	362	1	2	40	3	362	40
345	362	1	3	40	3	362	40
500	550	1	2	40	2	550	40
500	550	1	3	40	2	550	40
700	765	1	2	40	2	765	40
700	765	1	3	40	2	765	40

[a] Equal to K times the rated short-circuit rms current.

general expressions (which take into account the rated voltage range factor K) for the rms and peak momentary currents, respectively, are

$$I_{momentary} = 1.6 \times K \times I'' \text{ A} \qquad (4.21)$$

and

$$I_{momentary} = 2.7 \times K \times I'' \text{ A} \qquad (4.22)$$

Notice that, in Table 4.2, the factor K is 1 for the nominal voltages of 115 kV and above. Therefore, Equations 4.21 and 4.22 become the same as Equations 4.19 and 4.20, respectively.

EXAMPLE 4.1

A circuit breaker has a rated maximum rms voltage of 38 kV and is being operated at 34.5 kV. Determine the following:

(a) The highest symmetrical current-interrupting capability
(b) The operating voltage at the highest symmetrical current capability
(c) The associated rms momentary current rating
(d) The associated peak momentary current rating

Solution

(a) From Table 4.2, the rated voltage range factor K is 1.65, the rated continuous rms current is 1200 A, and the rated short-circuit rms symmetrical current at the rated maximum rms voltage of 38 kV is 22,000 A. However, since the circuit breaker is used at 34.5 kV, its symmetrical current-interrupting capability is

$$(22{,}000 \text{ A})\left(\frac{38 \text{ kV}}{34.5 \text{ kV}}\right) = 24{,}232 \text{ A}$$

The highest symmetrical current-interrupting capability is

$$(22{,}000 \text{ A}) K = 22{,}000 \times 1.65 \cong 36{,}000 \text{ A}$$

(b) Which is possible when the operating voltage is

$$\frac{38 \text{ kV}}{K} = \frac{38}{1.65} \cong 23 \text{ kV}$$

(c) Note that, at lower operating voltages, the highest symmetrical current-interrupting capability of 36,000 A cannot be exceeded. Hence, the associated rms momentary current is

$$I_{momentary} = 1.6 \times K \times I''$$
$$= 1.6(36{,}000 \text{ A})$$
$$= 57{,}600 \text{ A rms}$$

(d) The associated peak momentary current rating is

$$I_{momentary} = 2.7 \times K \times I''$$
$$= 2.7(36{,}000 \text{ A})$$
$$= 97{,}200 \text{ A peak}$$

A simplified procedure for determining the symmetrical fault current is known as the "E/X method" and is described in Section 5.3.1 of ANSI C37.010. This method [6,9] gives results approximating those obtained by more rigorous methods. In using this method, it is necessary first to make an E/X calculation. The method then corrects this calculation to take into account both the dc and ac decay of the current, depending on circuit parameters X/R. The approximation basically results owing to the use of curves.

Disturbance of Normal Operating Conditions and Other Problems

EXAMPLE 4.2

Consider the system shown in Figure 4.7 and assume that the generator is unloaded and running at the rated voltage with the circuit breaker open at bus 3. Assume that the reactance values of the generator are given as $X_d'' = X_1 = X_2 = 0.14$ pu and $X_0 = 0.08$ pu based on its ratings. The transformer impedances are $Z_1 = Z_2 = Z_0 = j0.05$ pu based on its ratings. The transmission line TL_{23} has $Z_1 = Z_2 = j0.04$ pu and $Z_0 = j0.10$ pu. Assume that the fault point is located on bus 1, and determine the subtransient fault current for a three-phase fault in per units and amperes. Select 25 MVA as the megavolt-ampere base and 8.5 and 138 kV as the low-voltage and high-voltage bases.

Solution

$$I_f'' = \frac{E_g}{X_d''}$$

$$= \frac{1.0\angle 0°}{j0.14}$$

$$= j7.143 \text{ pu}$$

The current base for the low-voltage side is

$$I_{B(LV)} = \frac{S_B}{\sqrt{3}V_{B(LV)}}$$

$$= \frac{25,000 \text{ kVA}}{\sqrt{3}(8.5 \text{ kV})}$$

$$= 1698.1 \text{ A}$$

Therefore,

$$I_f'' = |I_f''| = (7.143)(1698.1)$$

$$= 12,129.52 \text{ A}$$

Example 4.3

Use the results of Example 4.2 and determine the following:

(a) The possible value of the maximum of the dc current component
(b) Total maximum instantaneous current

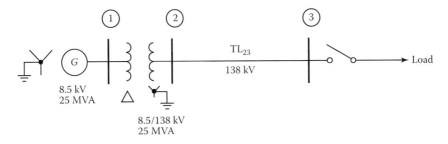

FIGURE 4.7 Transmission system for Example 4.2.

(c) Momentary current
(d) Interrupting capacity of the three-cycle circuit breaker if located at bus 1
(e) Momentary interrupting capacity of the three-cycle circuit breaker if located at bus 1

Solution

(a) Let

$$I_{dc,max} = \text{peak-to-peak amplitude}$$

then

$$I_{dc\,max} = \sqrt{2}\left(I_f''\right)$$
$$= \sqrt{2}(I_{rms})$$
$$= \sqrt{2}(7.143\,pu)$$
$$= 10.1\,pu$$

(b) From Figure 4.5a,

$$I_{max}'' = 2I_{dc,max}$$
$$= 2\left(\sqrt{2}I_f''\right)$$
$$= 20.2\,pu$$

(c) $I_{momentary}$ current represents the summation of I_{ac} and I_{dc}, where I_{dc} is about 50% of I_{ac}. Hence,

$$I_{momentary} = 1.6 \times I_f$$
$$= 1.6(7.143\,pu)$$
$$= 11.43\,pu$$

(d)

$$S_{interrupting} = \sqrt{3}V_f I_f'' \zeta \times 10^{-6}$$
$$= \sqrt{3}(8500)(12{,}129.52)(1.2) \times 10^{-6}$$
$$= 214.3\,MVA$$

(e)

$$S_{momentary} = \sqrt{3}V_f I_f''(1.6) \times 10^{-6}$$
$$= \sqrt{3}(8500)(12{,}129.52)(1.6) \times 10^{-6}$$
$$= 285.7\,MVA$$

4.5 BALANCED THREE-PHASE FAULTS AT FULL LOAD

In the previous section, balanced three-phase faults have been discussed under the assumption that the generator supplies a no-load current. In this section, it will be assumed that before the fault, the generator supplies a load current that is significantly large and therefore cannot be neglected. The calculation of the balanced three-phase fault can be performed by using any of the following two methods:

Method 1: Using Machine Internal Voltage Behind Subtransient Reactance

Consider the one-line diagram shown in Figure 4.8a and assume that the three-phase synchronous generator feeds the three-phase synchronous motor. The load current flowing before the fault, at point F, is $\mathbf{I}_L$ and the voltage at the fault point is $\mathbf{V}_F$. The line impedance between the generator and the motor is $\mathbf{Z}_{12}$. Therefore, the terminal voltage $\mathbf{V}_t$ of the generator and the motor is not the same as

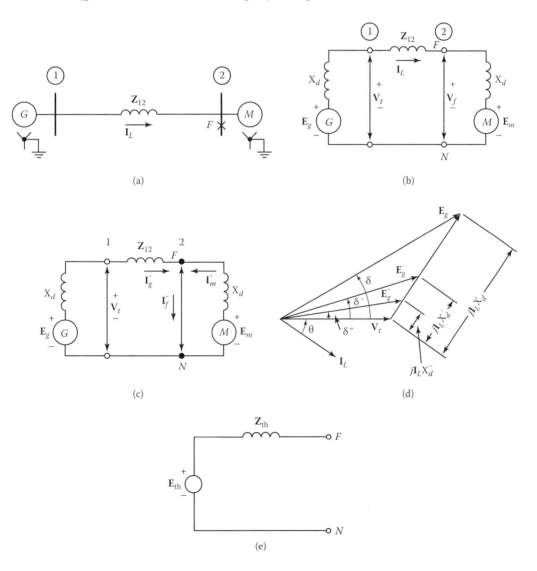

FIGURE 4.8 Calculation of balanced three-phase fault at full load: (a) one-line diagram of system under study; (b) equivalent circuit before fault; (c) equivalent circuit after fault; (d) phasor diagram for generator; (e) Thévenin equivalent of circuit given in (b).

shown in Figure 4.8b. Since there is a prefault load current $\mathbf{I}_L$, the internal voltages of the synchronous generator can be expressed as

$$\mathbf{E}''_g = \mathbf{V}_t + \mathbf{I}_L X''_d \tag{4.23}$$

$$\mathbf{E}'_g = \mathbf{V}_t + \mathbf{I}_L X'_d \tag{4.24}$$

$$\mathbf{E}_g = \mathbf{V}_t + \mathbf{I}_L X_d \tag{4.25}$$

where
$\mathbf{E}''_g$ = voltage behind the subtransient reactance X''_d (also called *subtransient internal voltage*)
$\mathbf{E}'_g$ = voltage behind the transient reactance X'_d (also called *transient internal voltage*)
$\mathbf{E}_g$ = voltage behind the synchronous reactance X_d (also called *steady-state internal voltage*)

Obviously, in the event that the generator is operating at no load, all three internal reactances are the same, thus

$$\mathbf{E}''_g = \mathbf{E}'_g = \mathbf{E}_g \tag{4.26}$$

as used in Equations 4.12 through 4.14. Figure 4.8d shows a phasor diagram depicting the relations between the internal voltages and the terminal voltage V_t for the generator.

Similarly, the internal voltages of the synchronous motor can be expressed as

$$\mathbf{E}''_m = \mathbf{V}_t - \mathbf{I}_L X''_d \tag{4.27}$$

$$\mathbf{E}'_m = \mathbf{V}_t - \mathbf{I}_L X'_d \tag{4.28}$$

$$\mathbf{E}_m = \mathbf{V}_t - \mathbf{I}_L X_d \tag{4.29}$$

The proper selection of internal voltage depends on the time passed since the occurrence of the fault. For example, if the fault current immediately after the fault is required, the subtransient internal voltage $\mathbf{E}''_g$ is used.

Note that, when a synchronous motor is involved in the fault, its field stays energized even though the motor itself receives no energy from the line. Because of the inertia caused by its rotor and connected mechanical load, it rotates for some time. Therefore, the motor starts acting as a generator and contributes current to the fault current, as shown in Figure 4.8c. Therefore, the fault current contributions of the generator and the motor can be expressed as

$$\mathbf{I}''_g = \frac{\mathbf{E}''_g}{X''_d + \mathbf{Z}_{12}} \tag{4.30}$$

and

$$\mathbf{I}''_m = \frac{\mathbf{E}''_m}{X''_d} \tag{4.31}$$

Therefore, the total current, which includes both fault and load currents, can be found from

$$\mathbf{I}''_f = \mathbf{I}''_g + \mathbf{I}''_m \tag{4.32}$$

Method 2: Using the Thévenin Voltage and Impedance at the Fault Point

The subtransient fault current can be determined by using Thévenin's theorem, as shown in Figure 4.8e. Note that, the Thévenin voltage, $\mathbf{E}_{th}$, is the same as the V_f voltage before the fault, as shown in Figure 4.8b, where the Thévenin impedance is

$$\mathbf{Z}_{th} = \frac{(X''_d + X_{12})X''_d}{(X''_d + X_{12}) + X''_d} \tag{4.33}$$

Thus, the subtransient fault current at the fault point F is

$$\mathbf{I}''_f = \frac{\mathbf{E}_{th}}{\mathbf{Z}_{th}}$$
$$= \frac{\mathbf{V}_f}{\mathbf{Z}_{th}} \tag{4.34}$$

By using the current division, the contributions of the generator and motor to the fault current can be found as

$$\mathbf{I}''_{f(g)} = \left[\frac{X''_d}{(X''_d + X_{12}) + X''_d} \right] \mathbf{I}''_f \tag{4.35}$$

and

$$\mathbf{I}''_{f(m)} = \left[\frac{(X''_d + X_{12})}{(X''_d + X_{12}) + X''_d} \right] \mathbf{I}''_f \tag{4.36}$$

To find the total current contributions of the generator and the motor, the prefill load current $\mathbf{I}_L$ has to be taken into account. Thus, the total subtransient currents in the generator and the motor can be expressed as

$$\mathbf{I}''_g = \mathbf{I}''_{f(g)} + \mathbf{I}_L \tag{4.37}$$

and

$$\mathbf{I}''_m = \mathbf{I}''_{f(m)} - \mathbf{I}_L \tag{4.38}$$

and hence, the total fault current can be expressed as

$$\mathbf{I}''_f = \mathbf{I}''_g + \mathbf{I}''_m \tag{4.39}$$

The corresponding fault (or short-circuit) megavolt-ampere can be found as

$$S_f = \sqrt{3}(V_{\text{prefault}})(I_f)10^{-6} \quad \text{MVA} \tag{4.40}$$

where the prefault voltage is the nominal voltage in kilovolts.

In the event that the nominal prefault voltage and fault current are known for a given fault point, the corresponding Thévenin equivalent system reactance can be found as

$$X_{\text{th}} = \frac{V_{\text{prefault}} \times 1000}{\sqrt{3}I_f} \quad \Omega \tag{4.41}$$

Alternatively, finding I_f from Equation 4.40 and substituting into Equation 4.41,

$$X_{\text{th}} = \frac{(V_{\text{prefault}})^2}{S_f} \quad \Omega \tag{4.42}$$

where the prefault voltage is in kilovolts and the fault megavolt-ampere is in megavolt-amperes.

Furthermore, in the event that base voltage and prefault voltage are the same, then the Thévenin equivalent system reactance can be expressed as

$$X_{\text{th}} = \frac{S_B}{S_f} \quad \text{pu} \tag{4.43}$$

or

$$X_{\text{th}} = \frac{I_B}{I_f} \quad \text{pu} \tag{4.44}$$

EXAMPLE 4.4

Consider the system shown in Figure 4.8 and assume that the prefault voltage at the terminals of the motor is $1.0\angle 0°$ pu. The subtransient reactances of the generator and the motor are both 0.07 pu. The load current $\mathbf{I}_L$ is $0.8 + j0.5$ pu. The line reactance X_{12} is 0.09 pu. Use the first method given in Section 4.4 and determine the following:

(a) Subtransient internal voltage of the generator
(b) Subtransient internal voltage of the motor
(c) Fault current contribution of the generator
(d) Fault current contribution of the motor
(e) Total fault current

Solution

(a)

$$\mathbf{I}_L = 0.8 + j0.50 = 0.9434\angle 32° \text{ pu}$$

Disturbance of Normal Operating Conditions and Other Problems

Thus, the prefault terminal voltage of the generator is

$$\mathbf{V}_t = \mathbf{V}_f + \mathbf{I}_L X_{12}$$
$$= 1.0\angle 0° + (0.8 + j0.5)(0.09\angle 90°)$$
$$= 0.955 + j0.072 = 0.9577\angle 4.3° \text{ pu}$$

$$\mathbf{E}''_g = \mathbf{V}_f + \mathbf{I}_L X''_d$$
$$= (0.955 + j0.072) + (0.9434\angle 32°)(0.07\angle 90°)$$
$$= 0.920 + 0.128 \text{ pu}$$

(b)

$$\mathbf{V}_t = \mathbf{V}_f = 1.0\angle 0° \text{ pu}$$

Hence,

$$\mathbf{E}''_m = \mathbf{V}_t - \mathbf{I}_L X''_d$$
$$= 1.0\angle 0° - (0.9434\angle 32°)(0.07\angle 90°)$$
$$= 1.035 - j0.056 \text{ pu}$$

(c)

$$\mathbf{I}''_g = \frac{\mathbf{E}''_g}{X''_d + X_{12}}$$
$$= \frac{0.920 + j0.128}{j0.07 + j0.09}$$
$$= 0.8 - j5.7502 \text{ pu}$$

(d)

$$\mathbf{I}''_m = \frac{\mathbf{E}''_m}{X''_d}$$
$$= \frac{1.035 - j0.056}{j0.07}$$
$$= -0.8 - j14.7855 \text{ pu}$$

(e)

$$\mathbf{I}''_f = \mathbf{I}''_g + \mathbf{I}''_m$$
$$= -j20.5357 \text{ pu}$$

EXAMPLE 4.5

Use the data given in Example 4.4 and determine

(a) Thévenin's impedance at the fault point
(b) Subtransient fault at the fault point
(c) Subtransient fault contribution of the generator
(d) Subtransient fault contribution of the motor
(e) Total current contribution of the generator
(f) Total current contribution of the motor
(g) Total fault current at the fault point

Solution

(a)
$$Z_{th} = \frac{(X_d'' + X_{12})X_d''}{(X_d'' + X_{12}) + X_d''}$$

$$= \frac{(j0.16)j0.07}{j0.16 + j0.07}$$

$$= j0.0487 \text{ pu}$$

(b)
$$I_f'' = \frac{E_{th}}{Z_{th}} = \frac{V_f}{Z_{th}}$$

$$= \frac{1.0\angle 0°}{j0.0487}$$

$$= 20.5357\angle -90° \text{ pu}$$

(c)
$$I_{f(g)}'' = \left[\frac{X_d''}{(X_d'' + X_{12}) + X_d''}\right] I_f''$$

$$= \left[\frac{j0.07}{j0.16 + j0.07}\right] \times 20.5357\angle -90°$$

$$= -j6.25 \text{ pu}$$

(d)
$$I_{f(m)}'' = \left[\frac{(X_d'' + X_{12})}{(X_d'' + X_{12}) + X_d''}\right] I_f''$$

$$= \left[\frac{j0.16}{j0.16 + j0.07}\right](20.5357\angle -90°)$$

$$= -j14.2857 \text{ pu}$$

(e)

$$I_g'' = I_{f(g)}'' + I_L$$
$$= -j6.25 + (0.8 + j0.5)$$
$$= 0.8 - j5.75 \text{ pu}$$

(f)

$$I_f'' = I_{f(m)}'' - I_L$$
$$= -j14.2857 - (0.8 + j0.5)$$
$$= -0.8 - j14.7857 \text{ pu}$$

(g)

$$I_f'' = I_g'' + I_m''$$
$$= (0.8 - j5.57) - 0.8 - j14.7857$$
$$= -j20.5357 \text{ pu}$$

It checks with the previous results.

4.6 APPLICATION OF CURRENT-LIMITING REACTORS

In the event of having a fault, the fault current is limited by the system reactance, which includes the impedance of the generators, transformers, lines, and other components of the system. The reactance of modern generators are large enough to limit the short-circuit megavoltampere at any point to a value the circuit breakers are capable of interrupting. However, if the system is large or some of the generators are old, additional impedance may be required, and it can be provided by reactors.

Current-limiting reactors are coils used to limit current during fault conditions, and to perform this task, it is important that magnetic saturation at high current does not reduce the coil reactance. Reactors can be either air-cored type or iron-cored type. Air-cored-type reactors do not have magnetic saturation, and therefore, their reactances are independent of current. Air-cored reactors are of two types: oil immersed and dry type. Oil-immersed reactors can be cooled by any of the means used for cooling the power transformer. On the other hand, dry-type reactors are usually cooled by natural ventilation but are also designed with forced-air and heat-exchanger auxiliaries.

Reactors are usually built as a single-phase unit. If dry-type reactors are located near metal objects such as I-beams, plates, and channels, magnetic shielding must be used to prevent the reactor flux inducing eddy currents in the objects. Otherwise, the proximity of metal objects will increase the power loss of the reactor and change its reactance.

Because of the required clearances and construction details necessary to minimize corona, these reactors are limited to 34.5 kV as a maximum insulation class. However, oil-immersed reactors can be used at any voltage level, for either indoor or outdoor. They provide high safety against flashover, high thermal capacity, and having no magnetic field outside the tank to cause heating or magnetic forces in surrounding metal objects during short circuits.

Air-cored reactors are designed mechanically (against the great mechanical stresses that take place under short-circuit conditions) and thermally for not more than $33\frac{1}{3}$ times the normal full-load current for 5 s under short-circuit conditions.

Iron-cored reactors are usually built as oil-immersed reactors. Their reactance-to-resistance ratio is much greater than for the air-cored type. They can be designed for any voltage level and are more expensive than air-cored reactors.

In general, the effective resistance of a reactor is negligibly small (its ratio of R/X is about 0.03). The inductive reactance, X, of a reactor can be determined from

$$X_r = \frac{(\%X_r) \times V_r}{100(\sqrt{3}) \times I_r} \quad \Omega/\text{phase} \tag{4.45}$$

or

$$X_r = \frac{(\%X_r) \times V_{2r}}{100 \times P_r} \quad \Omega/\text{phase} \tag{4.46}$$

where
$\%X_r$ = reactor reactance in percent
V_r = reactor-rated voltage in kilovolts
I_r = reactor-rated current in kiloamperes
P_r = reactor power rating in megavolt-amperes

Since the fault levels increase owing to the growth in interconnection of power systems, it may be necessary to increase the system reactance by locating reactors at strategic points in the system [10]. The reactors may be located (1) in series with generators as shown in Figure 4.9a, (2) in series with the lines or feeders as shown in Figure 4.9b, (3) between buses as shown in Figure 4.9c, (4) in a tie–bus arrangement as shown in Figure 4.9d, and (5) in a ring arrangement as shown in Figure 4.9e and f.

The reactors in series with generators are usually not used because modern generators have enough leakage reactance. Also, under steady-state conditions, there is a large voltage drop and power loss in each reactor. Furthermore, in the event of having a fault on or near the buses, the generators may drop out of synchronism owing to the bus undervoltage conditions. Therefore, such an arrangement is only used to protect old generators with low leakage reactances.

The reactors in series with lines are more commonly used than the first method because in the event of having a fault at point F in Figure 4.9b, there will be a large voltage drop in the associated reactor but only a small reduction in the bus voltage. Thus, the synchronism of the generators will be intact, and the fault can be isolated by the loss of the faulted line. The arrangements shown in Figure 4.9c and d tend to limit and localize the disturbances to the faulted bus and generator.

In the ring arrangements shown in Figure 4.9e and f, the current to be transferred between two sections flows through two paths in parallel, whereas in the tie–bus arrangement the current flows through two reactors in series. Therefore, the reactors in the tie–bus arrangement have only one-third of the reactance of ring reactors. However, they carry twice as much current as the ring reactors.

EXAMPLE 4.6

Three 20-kV solidly grounded generators are connected to three reactors in a tie–bus arrangement as shown in Figure 4.9d. The reactances of generators and the reactors are 0.2 and 0.1 pu, respectively, based on a 50-MVA base. If there is a symmetrical three-phase fault at the fault point F, determine the following:

(a) Short-circuit megavolt-amperes
(b) Fault current distribution in system

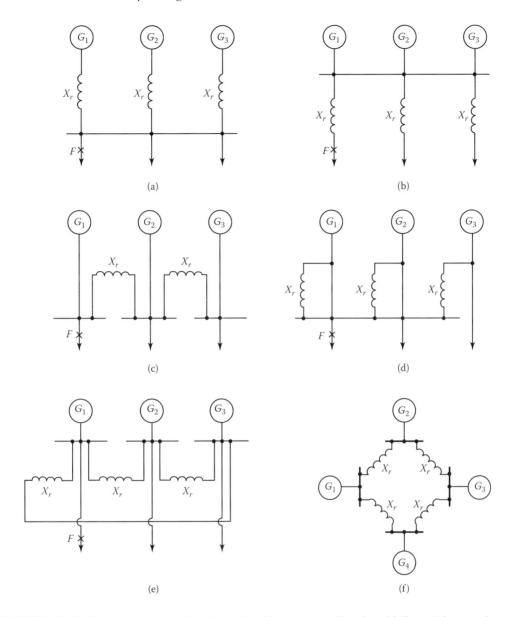

FIGURE 4.9 Various reactor connections: (a) series with generators; (b) series with lines; (c) between buses; (d) tie–bus system; (e) ring system; (f) ring system.

Solution

(a) Figure 4.10a shows the equivalent circuit after the fault. The Thévenin equivalent reactance can be calculated from Figure 4.10b as

$$X_{th} = \frac{(j0.2)(j0.25)}{j0.2 + j0.25}$$

$$= j0.1111 \, \text{pu}$$

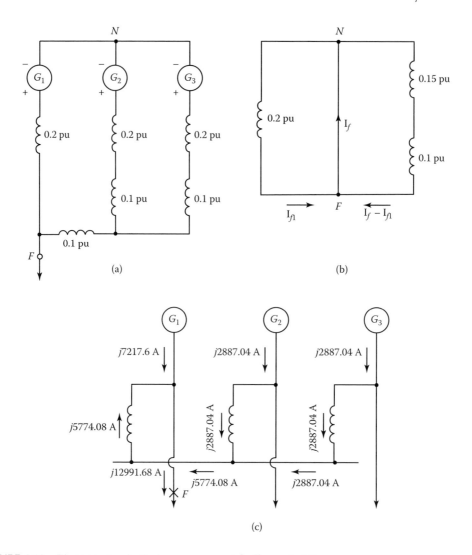

FIGURE 4.10 Three reactors in tie–bus arrangement for Example 4.5.

Therefore, the short-circuit megavolt-amperes can be found as

$$S_f = \frac{S_{B(3\phi)}}{|X_{th}|}$$

$$= \frac{50 \text{ MVA}}{0.1111}$$

$$= 450.05 \text{ MVA}$$

(b) Since the short-circuit current at the fault point F is

$$\mathbf{I}_f = \frac{\mathbf{V}_f}{\mathbf{Z}_{th}}$$

$$= \frac{1.0\angle 0°}{j0.1111}$$

$$= 9\angle -90° \text{ pu}$$

and

$$I_B = \frac{50\,\text{MVA}}{\sqrt{3}(20\,\text{kV})}$$

$$= 1443.4\,\text{A}$$

then

$$\mathbf{I}_f = (9\angle-90°)1443.4$$

$$= 12{,}990.6\angle-90°\,\text{A}$$

Assume that generator G_1 supplies some $\mathbf{I}_{f_1}$ amount of the short-circuit current $\mathbf{I}_f$. Then, from Figure 4.10b,

$$0.2\mathbf{I}_{f_1} = 0.25(12{,}990.6\angle-90°)$$

Thus,

$$\mathbf{I}_{f_1} = 16{,}238.25\angle-90°\,\text{A}$$

The remaining short-circuit current distribution is shown in Figure 4.10c.

4.7 INSULATORS

4.7.1 Types of Insulators

An *insulator* is a material that prevents the flow of an electric current and can be used to support electrical conductors. The function of an insulator is to provide for the necessary clearances between the line conductors, between conductors and ground, and between conductors and the pole or tower. Insulators are made of porcelain, glass, and fiberglass treated with epoxy resins. However, porcelain is still the most common material used for insulators.

The basic types of insulators include (1) pin-type insulators, (2) suspension insulators, and (3) strain insulators. The pin insulator gets its name from the fact that it is supported on a pin. The pin holds the insulator, and the insulator has the conductor tied to it. They may be made in one piece for voltages below 23 kV, in two pieces for voltages from 23 to 46 kV, in three pieces for voltages from 46 to 69 kV, and in four pieces for voltages from 69 to 88 kV. Pin insulators are seldom used on transmission lines having voltages above 44 kV, although some 88-kV lines using pin insulators are in operation.

The glass pin insulator is mainly used on low-voltage circuits. The porcelain pin insulator is used on secondary mains and services, as well as on primary mains, feeders, and transmission lines. Figure 4.11 shows typical pin-type porcelain insulators. A modified version of the pin-insulator is known as the post insulator. The post-type insulators are used on distribution, subtransmission, and transmission lines and are installed on wood, concrete, and steel poles. The line post insulators are constructed for vertical or horizontal mounting. The line post insulators are usually made as one-piece solid porcelain units. Figure 4.12 shows a typical post-type porcelain insulator. Suspension insulators consist of a string of interlinking separate disks made of porcelain. A string may consist of many disks depending on the line voltage.* For example, on average, for 115-kV lines usually seven disks are used; however, for 345-kV lines usually 18 disks are used.

The suspension insulator, as its name implies, is suspended from the cross-arm (or a pole or tower) and has the line conductor fastened to the lower end, as shown in Figures 4.1 and 4.2. Figure 4.13 shows a typical suspension- or strain-type porcelain insulator (with a breakdown strength of

---

* In average practice, the number of units used in an insulator string is approximately proportional to the line voltage, with a slight increase for the highest voltages and with some allowances for the length of the insulator unit. For example, 4 or 5 units have generally been used at 69 kV, 7 or 8 at 115 kV, 8–10 at 138 kV, 9–11 at 161 kV, 14–20 at 230 kV, 15–18 at 345 kV, 24–35 at 500 kV (with the 35-unit insulator strings used at high altitudes), 33–35 at 735 kV (Hydro-Quebec), and 30–35 at 765 kV.

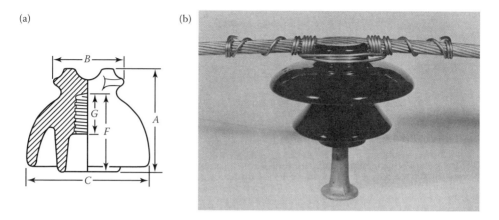

FIGURE 4.11 Typical pin-insulators: (a) one-piece pin insulator; (b) two-piece pin insulator.

between 12 and 28 kV/mm) disks. When there is a dead end of the line or there is corner or a sharp curve, or the line crosses a river, etc., the line will withstand great strain.

The assembly of suspension units arranged to dead-end the conductor of such a structure is called a *dead-end*, or *strain*, insulator. In such an arrangement, suspension insulators are used as strain insulators. The dead-end string is usually protected against damage from arcs by using one to three additional units and installing arcing horns or rings, as shown in Figure 4.14. Such devices are designed to ensure that an arc (e.g., due to lightning impulses) will hold free of the insulator string.

The *arcing horns* protect the insulator string by providing a shorter path for the arc, as shown in Figure 4.14a. The effectiveness of the *arcing ring* (or *grading shield*), shown in Figure 4.14b, is due to its tendency to equalize the voltage gradient over the insulator, causing a more uniform field. Therefore, protection of the insulator is not dependent on simply providing a shorter arcing path, as is the case with horns. Figure 4.14c shows a control ring developed by Ohio Brass company that can be used to "control" the voltage stress at the line end of the insulator strings.

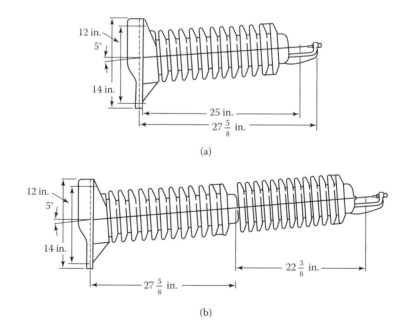

FIGURE 4.12 Typical (side) post-type porcelain insulators used in: (a) 69 kV; (b) 138 kV.

FIGURE 4.13 Typical suspension- or strain-type porcelain insulator disk. (Courtesy of Ohio Brass Company.)

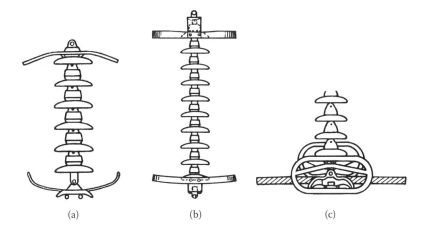

FIGURE 4.14 Devices used to protect insulator strings: (a) suspension string with arcing horns; (b) suspension string with grading shields (or arcing rings); (c) suspension string with control ring. (Courtesy of Ohio Brass Company.)

It has been shown that their use can also reduce the corona formation on the line hardware. Control rings are used on single-conductor high-voltage transmission lines operating above 250 kV. Transmission lines with bundled conductors do not require the use of arcing horns and rings nor control rings, provided that the bundle is not made up of two conductors one above the other.

4.7.2 Testing of Insulators

The operating performance of a transmission line depends largely on the insulation. Therefore, experience has shown that for a satisfactory operation, the dry flashover voltage of the assembled insulator must be equal to three to five times the nominal operating voltage, and its leakage path must be about twice the shortest air gap distance. Thus, insulators used on overhead lines are subjected to tests that can generally be classified as (1) design tests, (2) performance tests, and (3) routine tests.

The *design tests* include the dry flashover test, pollution flashover test, wet flashover test, pollution flashover test, and impulse test. The *flashover voltage* is defined as the voltage at which the insulator surface breaks down (by ionization of the air surrounding the insulator), allowing current to flow on the outside of the insulator between the conductor and the cross-arm.

TABLE 4.3
Flashover Characteristics of Suspension Insulator Strings and Air Gaps

Impulse Air Gap		Impulse Flashover, Positive Critical	No. of Insulator	Wet 60-Hz Flashover	Wet 60-Hz Air Gap	
in.	mm	(kV)	Units[a]	(kV)	mm	in.
8	203	150	1	50	254	10
14	356	255	2	90	305	12
21	533	355	3	130	406	16
26	660	440	4	170	508	20
32	813	525	5	215	660	26
38	965	610	6	255	762	30
43	1092	695	7	295	889	35
49	1245	780	8	335	991	39
55	1397	860	9	375	1118	44
60	1524	945	10	415	1245	49
66	1676	1025	11	455	1346	53
71	1803	1105	12	490	1473	58
77	1956	1185	13	525	1575	62
82	2083	1265	14	565	1676	66
88	2235	1345	15	600	1778	70
93	2362	1425	16	630	1880	74
99	2515	1505	17	660	1981	78
104	2642	1585	18	690	2083	82
110	2794	1665	19	720	2184	86
115	2921	1745	20	750	2286	90
121	3073	1825	21	780	2388	94
126	3200	1905	22	810	2464	97
132	3353	1985	23	840	2565	101
137	3480	2065	24	870	2692	106
143	3632	2145	25	900	2794	110
148	3759	2225	26	930	2921	115
154	3912	2305	27	960	3023	119
159	4039	2385	28	990	3124	123
165	4191	2465	29	1020	3251	128
171	4343	2550	30	1050	3353	132

Source: Transmission Line Design Manual, U.S. Department of the Interior, Denver, Colorado, 1980.

[a] Insulator units are 146×254 mm $\left(5\frac{3}{4} \times 10 \text{ in}\right)$ or 146×267 mm $\left(5\frac{3}{4} \times 10 \text{ in}\right)$.

Whether an insulator breaks down depends not only on the magnitude of the applied voltage but also on the rate at which the voltage increases. Since insulations have to withstand steep-fronted lightning and switching surges when they are in use, their design must provide the flashover voltage* on a steep-fronted impulse waveform that is greater than that on a normal system waveform. The ratio of these voltages is defined as the *impulse ratio*. Thus,

$$\text{impulse ratio} = \frac{\text{impulse flashover voltage}}{\text{power frequency flashover voltage}} \tag{4.47}$$

---

* This phenomenon is studied in the laboratory by subjecting insulators to voltage impulses by means of a "lightning generator."

Disturbance of Normal Operating Conditions and Other Problems

Table 4.3 gives flashover characteristics of suspension insulator strings and air gaps [10]. The performance tests include the puncture test, mechanical test, temperature test, porosity test, and electromechanical test (for suspension insulators only). The event that takes place when the dielectric of the insulator breaks down and allows current to flow inside the insulator between the conductor and the cross-arm is called the *puncture*. Thus, the design must facilitate the occurrence of a flashover at a voltage that is lower than the voltage for puncture. An insulator may survive a flashover without damage but must be replaced when punctured. The test of the glaze on porcelain insulators is called the *porosity test*. The routine tests include the proof-load test, corrosion test, and high-voltage test (for pin insulators only).*

4.7.3 Voltage Distribution over a String of Suspension Insulators

Figure 4.15 shows the voltage distribution along the surface of a single clean insulator disk (known as the cap-and-pin insulator unit) used in suspension insulators. Note that, the highest voltage gradient takes place close to the cap and the pin (which are made up of metal), whereas much lower voltage gradients take place along most of the remaining surfaces.

The underside (i.e., the *inner skirt*) of the insulator has been given the shape, as shown in Figure 4.15, to minimize the effects of moisture and contamination and to provide the longest path possible for the leakage currents that might flow on the surface of the insulator. In the figure, the voltage drop between the cap and the pin has been taken as 100% of the total voltage.

Thus, approximately 24% of this voltage is distributed along the surface of the insulator from the cap to point 1 and only 6% from point 1 to point 9. The remaining 70% of this voltage is distributed between point 9 and the pin. The main problem with suspension insulators having a string of identical insulator disks is the nonuniform distribution voltage over the string.

Each insulator disk with its hardware (i.e., cap and pin) constitutes a capacitor, the hardware acting as the plates or electrodes and the porcelain as the dielectric. Figure 4.16 shows the typical voltage distribution on the surfaces of three clean cap-and-pin insulator units connected in series [11]. The figure clearly illustrates that when several units are connected in series, (1) the voltage on each insulator over the string is not the same, (2) the location of the unit within the insulator string dictates the voltage distribution, and (3) the maximum voltage gradient takes place at the pin of the insulator unit nearest to the line conductor.

As shown in Figure 4.17a, when several insulator units are placed in series, two sets of capacitances take place; the series capacitance C_1 (i.e., the capacitance of each insulator unit) and the shunt capacitance to ground, C_2. Note that, all the charging current I for the series and shunt capacitances flows through the first (with respect to the conductor) of the series capacitance C_1.

The I_1 portion of this current flows through the first shunt capacitance C_2, leaving the remaining $I-I_1$ portion of the current to flow through the second series capacitance, and so on. Therefore, this diminishing current flowing through the series capacitance C_1 results in a diminishing voltage (drop) distribution through them from the conductor end to the ground end (i.e., cross-arm), as illustrated in Figure 4.17b. Thus,

$$V_5 > V_4 > V_3 > V_2 > V_1$$

In summary, the voltage distribution over a string of identical suspension insulator units is not uniform owing to the capacitances formed in the air between each cap/pin junction and the grounded (metal) tower. However, other air capacitances exist between metal parts at different potentials. For example, there are air capacitances between the cap–pin junction of each unit and the line conductor. Figure 4.18 shows the resulting equivalent circuit for the voltage

---

* For further information, see the ANSI Standards C29.1–C29.9.

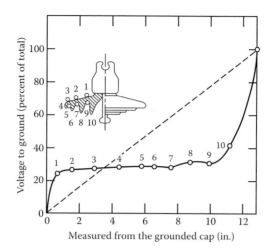

FIGURE 4.15 Voltage distribution along surface of single dean cap-and-pin suspension insulator. (From Edison Electric Institute, *EHV Transmission Line Reference Book*, EEI, New York, 1968.)

distribution along a clean eight-unit insulator string. The voltage distribution on such a string can be expressed as

$$V_k = \frac{V_n}{\beta^2 \sinh \beta n}\left[\frac{C_2}{C_1}\right]\sinh \beta k + \frac{C_3}{C1}\sinh \beta(k-n) + \frac{C_3}{C_1}\sinh \beta n \qquad (4.48)$$

where
  V_k = voltage across k units from ground end
  V_n = voltage across all n units (i.e., applied line-to-ground voltage in volts)

$$\beta = \text{a constant} = \left(\frac{C_2 + C_1}{2}\right)^{1/2} \qquad (4.49)$$

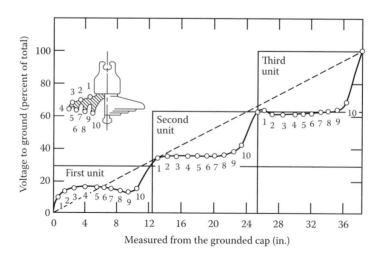

FIGURE 4.16 Typical voltage distribution on surfaces of three clean cap-and-pin suspension insulator units in series. (From Edison Electric Institute, *EHV Transmission Line Reference Book*. EEI, New York, 1968.)

Disturbance of Normal Operating Conditions and Other Problems

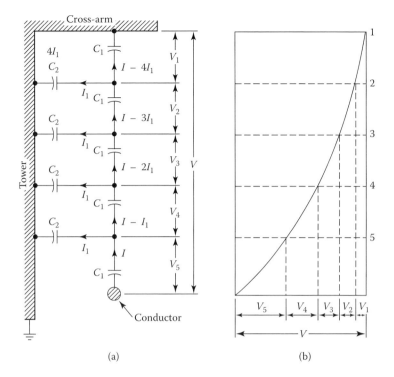

FIGURE 4.17 Voltage distribution among suspension insulator units [11].

C_1 = capacitance between cap and pin of each unit
C_2 = capacitance of one unit to ground
C_3 = capacitance of one unit to line conductor

However, the capacitance C_3 is usually very small, and thus its effect on the voltage distribution can be neglected. Hence, Equation 4.48 can be reexpressed as

$$V_k = V_n \left(\frac{\sinh \alpha k}{\sinh \alpha n} \right) \quad (4.50)$$

where

$$\alpha = \text{a constant} = \left(\frac{C_2}{C_1} \right)^{1/2} \quad (4.51)$$

Figure 4.19 shows how the voltage changes along the eight-unit string of insulators when the ratio C_2/C_1 is about .083333 and the ratio C_3/C_1 is about zero (i.e., $C_3 = 0$). It is interesting to note that a calculation based on Equation 4.48 gives almost the same result. The ratio C_2/C_1 is usually somewhere between 0.1 and 0.2.

Furthermore, there is the air capacitance that exists between the conductor and the tower. However, it has no effect on the voltage distribution over the insulator string, and therefore it can be neglected.

Note that, this method of calculating the voltage distribution across the string is based on the assumption that the insulator units involved are clean and dry, and therefore they act as a purely capacitive voltage divider. In reality, however, the insulator units may not be clean or dry. Thus, in

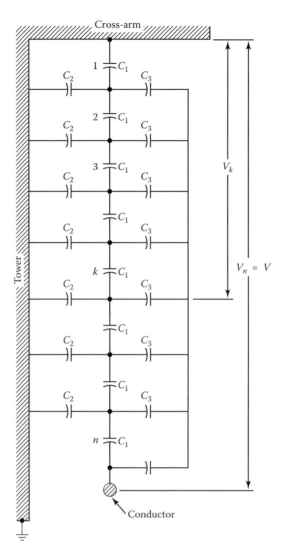

FIGURE 4.18 Equivalent circuit for voltage distribution along clean eight-unit insulator string. (Adopted from Edison Electric Institute, *EHV Transmission Line Reference Book*. EEI, New York, 1968.)

the equivalent circuit of the insulator string, each capacitance C_1 should be shunted by a resistance R representing the leakage resistance.

Such resistance depends on the presence of contamination (i.e., pollution) on the insulator surfaces and is considerably modified by rain and fog. If, however, the units are badly contaminated, the surface leakage (resistance) currents could be greater than the capacitance currents, and the extent of the contamination could vary from unit to unit, causing an unpredictable voltage distribution. It is also interesting to note that if the insulator unit nearest to the line conductor is electrically stressed to its safe operating value, then all the other units are electrically understressed, and consequently, the insulator string as a whole is being inefficiently used. Therefore, the string efficiency (in per units) for an insulator string made up of n series units can be defined as

$$\text{String efficiency} = \frac{\text{voltage across string}}{n(\text{voltage across unit adjacent to line conductor})} \quad (4.52)$$

Disturbance of Normal Operating Conditions and Other Problems

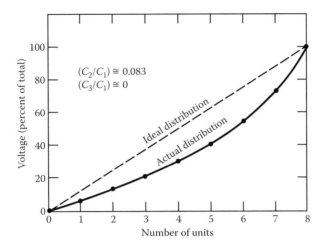

FIGURE 4.19 Voltage distribution along clean eight-unit cap-and-pin insulator string. (From Edison Electric Institute, *EHV Transmission Line Reference Book*. EEI, New York, 1968.)

If the unit adjacent to the line conductor is about to flash over, then the whole string is about to flash over. Therefore, the string efficiency can be reexpressed as

$$\text{String efficiency} = \frac{\text{flashover voltage of string}}{n(\text{flashover voltage of one unit})} \quad (4.53)$$

Note that, the string efficiency decreases as the number of units increases. The methods to improve the *string efficiency* (*grading*) include the following:

1. By grading the insulators so that the top unit has the minimum series capacitance C_1, whereas the bottom unit has the maximum capacitance. This may be done by using different sizes of disks and hardware or by putting metal caps on the disks or by a combination of both methods.* However, this is a rarely used method since it would involve stocking spares of different types of units, which is contrary to the present practice of the utilities to standardize on as few types as possible.
2. By installing a large circular or oval grading shield ring (i.e., an arcing ring) at the line end of the insulator string [13]. This method introduces a capacitance C_3, as shown in Figure 4.18, from the ring to the insulator hardware to neutralize the capacitance C_2 from the hardware to the tower. This method substantially improves the string efficiency. However, it is not usually possible in practice to achieve completely uniform voltage distribution by using the grading shield, especially if the string has a large number of units.
3. By reducing the air (shunt) capacitance C_3, between each unit and the tower (i.e., the ground), by increasing the length of the cross-arms. However, this method is restricted in practice owing to the reduction in cross-arm rigidity and the increase in tower cost.
4. By using a semiconducting (or stabilizing) high-resistance glaze on the insulator units to achieve a resistor voltage divider effect. This method is based on the fact that the string efficiency increases owing to the increase in surface leakage resistance current when the units are wet. Thus, the leakage resistance current becomes the same for all units, and the voltage distribution improves since it does not depend on the capacitance currents only. However, this method is restricted by the risk of thermal instability.

---

* Proposed by Peek [13,14].

4.7.4 Insulator Flashover due to Contamination

An insulator must be capable of enduring extreme and sudden temperature changes such as ice, sleet, and rain as well as environmental contaminants such as smoke, dust, salt, fogs, saltwater sprays, and chemical fumes without deterioration from chemical action, breakage from mechanical stresses, or electrical failure. Further, the insulating material must be thick enough to resist puncture by the combined working voltage of the line and any probable transient whose time lag to spark over is great.

If this thickness is greater than the desirable amount, then two or more pieces are used to achieve the proper thickness. The thickness of a porcelain part must be so related to the distance around it that it will flash over before it will puncture. The ratio of puncture strength to flashover voltage is called the "safety factor" of the part or of the insulator against puncture. This ratio should be high enough to provide sufficient protection for the insulator from puncture by the transients.

The insulating materials mainly used for line insulators are (1) wet-process porcelain, (2) dry-process porcelain, and (3) glass. However, wet-process porcelain is used much more than dry-process porcelain. One of the reasons for this is that wet-process porcelain has greater resistance to impact and is practically incapable of being penetrated by moisture without glazing, whereas dry-process porcelain is not.

However, in general, dry-process porcelain has a somewhat higher crushing strength. Dry-process porcelain is only used for the lowest voltage lines. As a result of recent developments in the technology of glass manufacturing, glass insulators, which are very tough and have low internal resistance, can be produced. Thus, usage of glass insulators is increasing.

To select insulators properly for a given overhead line design, not only the aforementioned factors but also the geographic location of the line needs to be considered. For example, the overhead lines that will be built along the seashore, especially in California, will be subjected to winds blowing in from the ocean, which carry a fine salt vapor that deposits salt crystals on the windward side of the insulator.

On the order hand, if the line is built in areas where rain is seasonal, the insulator surface leakage resistance may become so low during the dry seasons that insulators flash over without warning. Another example is that if the overhead line is going to be built near gravel pits, cement mills, and refineries, its insulators may become so contaminated that extra insulation is required. Contamination flashovers on transmission systems are initiated by airborne particles deposited on the insulators. These particles may be of natural origin or they may be generated by pollution that is mostly a result of industrial, agricultural, or construction activities. Thus, when line insulators are contaminated, many insulator flashovers occur during light fogs unless arcing rings protect the insulators or special fog-type insulators are used.

Table 4.4 lists the types of contaminants causing contamination flashover [14]. The mixed contamination condition is the most common, caused by the combination of industrial pollution and sea salt or by the combination of several industrial pollutions. Table 4.4 also presents the prevailing weather conditions at the time of flashover. Note that fog, dew, drizzle, and mist are common weather conditions, accounting for 72% of the total. In general, a combination of dew and fog is considered as the most severe wetting condition, even though fog is not necessary for the wetting process.

Note that, the surface leakage resistance of an insulator is unaffected by deposits of dry dirt. However, when these contamination deposits become moist or wet, they constitute continuous conducting layers. Leakage current starts to flow in these layers along the surface of the insulators. This leakage current heats the wet contamination, and the water starts to evaporate from those areas where the product of current density and surface resistivity is greater, causing the surface resistivity to further increase. Therefore, the current continues to flow around such a dry spot, causing the current density in the neighboring regions to increase. This, in turn, produces more heat, which evaporates the moisture in these surrounding regions, causing the formation of circular patterns

TABLE 4.4
Numbers of Flashovers Caused by Various Contaminant, Weather, and Atmospheric Conditions

	Weather and Atmospheric Conditions								
Type of Contaminant	Fog	Dew	Drizzle, Mist	Ice	Rain	No Wind	High Wind	Wet Snow	Fair
Sea salt	14	11	22	1	12	3	12	3	—
Cement	12	10	16	2	11	4	1	4	—
Fertilizer	7	5	8	—	1	1	—	4	—
Fly ash	11	6	19	1	6	3	1	3	1
Road salt	8	2	6	—	4	2	—	6	—
Potash	3		3	—	—	—	—	—	—
Cooling tower	2	2	2	—	2	—	—	—	—
Chemicals	9	5	7	1	1	—	—	1	1
Gypsum	2	1	2	—	2	—	—	2	—
Mixed contamination	32	19	37	—	13	1	—	1	—
Limestone	2	1	2	—	4	—	2	2	—
Phosphate and sulfate	4	1	4	—	3	—	—	—	—
Paint	1		1	—	—	1	—	—	—
Paper mill	2	2	4	—	2	—	—	1	—
Drink milk	1	1	1	—	—	1	—	1	—
Acid exhaust	2		3	—	—	—	—	1	—
Bird droppings	2	2	3	—	1	2	—	—	2
Zinc industry	2	1	2	—	1	—	—	1	—
Carbon	5	4	5	—	—	4	3	3	—
Soap	2	2	1	—	—	1	—	—	—
Steel works	6	5	3	2	2	—	—	1	—
Carbide residue	2	1	1	1	—	—	—	1	—
Sulfur	3	2	2	—	—	1	—	1	—
Copper and nickel salt	2	2	2	—	—	2	—	1	—
Wood fiber	1	1	1	—	1	—	—	1	—
Bulldozing dust	2	1	1	—	—	2	—	—	—
Aluminum plant	2	2	1	—	1	—	—	—	—
Sodium plant	1		1	—	—	—	—	—	—
Active dump	1	1	1	—	—	—	—	—	—
Rock crusher	3	3	5	—	1	—	—	—	—

Source: Electric Power Research Institute, *Transmission Line Reference Book: 345 kV and Above*, 2nd ed. EPRI, Palo Alto, CA, 1982.

known as "dry bands" until the leakage current is decreased to a value insufficient to sustain further evaporation, and the voltage builds up across the dry bands.

Further wetting results in further reduction of the resistance, and small flashovers take place on the dry bands on which moisture droplets fall. Since many dry bands on the insulator are in about the same condition, the arcs extend rapidly over the whole surface, forcing all dry bands to discharge in a rapid cascade known as the "flashover" of the insulator.

Figure 4.20 illustrates the phenomenon of insulator flashover due to contamination. Severe contamination may reduce the 60-Hz flashover voltage from approximately 50 kV rms per unit to as low as 6 to 9 kV rms per unit. The condition for such flashover may be developed during the melting of contaminated ice on the insulator by leakage currents. An insulator flashover due to contamination

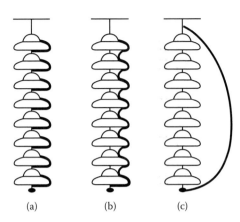

FIGURE 4.20 Changes in channel position of contaminated flashover. (From Edison Electric Institute, *EHV Transmission Line Reference Book*. EEI, New York, 1968.)

is easily distinguished from other types of flashover due to the fact that the arc always begins close to the surface of each insulator unit, as shown in Figure 4.20a. As shown in Figure 4.20c, only in the final stage does the flashover resemble an air strike. Furthermore, since the insulator unit at the conductor end has the greatest voltage, the flashover phenomenon usually starts at that insulator unit.

To prevent insulator flashovers, the insulators of an overhead transmission line may be cleaned simply by washing them, a process that can be done basically either by conventional techniques or by a new technique. In the conventional techniques, the line is deenergized, and its conductors are grounded at each pole or tower where the members of an insulator cleaning crew wash and wipe the insulators by hand.

In the new technique, the line is kept energized while the insulators may be cleaned by high-pressure water jets produced by a truck-mounted high-pressure pump that forces water through a nozzle at 500–850 psi, developing a round solid stream. The water jet strikes the insulator with a high velocity, literally tearing the dirt and other contaminants from the insulator surface. The cost of insulator cleaning per unit is very low by this technique.

Certain lines may need insulator cleaning as often as three times a year. To overcome the problem of surface contamination, some insulators may be covered with a thin film of silicone grease that absorbs the dirt and makes the surface water form into droplets rather than a thin film. This technique is especially effective for spot contamination where maintenance is possible, and it is also used against sea salt contamination.

Finally, specially built semiconducting glazed insulators having a resistive coating are used. The heat produced by the resistive coating keeps the surface dry and provides for a relatively linear potential distribution.

4.7.5 Insulator Flashover on Overhead High-Voltage dc Lines

Even though mechanical considerations are similar for ac and dc lines, electrical characteristics of insulators on dc lines are significantly different from those on ac lines. For example, when conventional ac insulators are used on dc lines, flashover takes place much more frequently than on an ac line of equivalent voltage. This is caused partly by the electrostatic forces of the steady dc field, which increases the deposit of pollution on the insulator surface.

Further, arcs tend to develop into flashovers more readily in the absence of voltage zero. To improve the operating performance and reduce the construction cost of overhead high-voltage dc (HVDC) lines by using new insulating materials and new insulator configurations particularly suited to dc voltages stress, more compact line designs can be produced, therefore saving money on towers and rights of way.

For example, to improve the operating performance and reduce the construction cost of overhead HVDC lines, the EPRI has sponsored the development of a new insulator. One of the more popular designs, the composite insulator, uses a fiberglass rod for mechanical and electrical strength and flexible skirts made of organic materials for improved flashover performance. The composite insulator appears to be especially attractive for use on HVDC lines because it is better able to withstand flashover in all types of contaminated environments, particularly in areas of light and medium contamination.

Furthermore, there are various design measures that may be taken into account to prevent contamination flashovers, for example, overinsulation, installment of V-string insulators, and installment of horizontal string insulators. Overinsulation may be applicable in the areas of heavy contamination.

Up to 345 kV, overinsulation is often achieved by increasing the number of insulators. However, severe contamination may dictate the use of very large leakage distances that may be as large as double the nominal requirements. Thus, electrical, mechanical, and economic restrictions may limit the use of this design measure.

The use of the V-string insulators can prevent the insulation contamination substantially. They self-clean more effectively in rain than vertical string insulators since both sides of each insulator disk are somewhat exposed to rain. They can be used in heavy contamination areas very effectively.

The installment of horizontal insulator strings is the most effective design measure that can be used to prevent contamination flashovers in the very heavy contamination areas. The contaminants are most effectively washed away on such strings. However, they may require a strain tower support depending on the tower type.

Other techniques used include the installation of specially designed and built insulators. For example, the use of fog-type insulators has shown that the contamination flashover can be effectively reduced since most of the flashovers occur in conditions where there is mist, dew, and fog.

4.8 GROUNDING

4.8.1 Electric Shock and Its Effects on Humans

To properly design a grounding* for the high-voltage lines and/or substations, it is important to understand the electrical characteristics of the most important part of the circuit, the human body. In general, shock currents are classified according to the degree of the severity of the shock they cause. For example, currents that produce direct physiological harm are called *primary shock currents*. However, currents that cannot produce direct physiological harm but may cause involuntary muscular reactions are called *secondary shock currents*. These shock currents can be either steady state or transient in nature. In ac power systems, steady-state currents are sustained currents of 60 Hz or its harmonics.

The transient currents, on the other hand, are capacitive discharge currents whose magnitudes diminish rapidly with time. Table 4.5 gives the possible effects of electrical shock currents on humans. Note that, the threshold value for a normally healthy person to be able to feel a current is about 1 mA.† This is the value of current at which a person is just able to detect a slight tingling sensation on the hands or fingers due to current flow [15].

Currents of approximately 10–30 mA can cause lack of muscular control. In most humans, a current of 100 mA will cause ventricular fibrillation. Currents of higher magnitudes can stop the heart completely or cause severe electrical burns.

Ventricular fibrillation is a condition where the heart beats in an abnormal and ineffective manner, with fatal results. Thus, its threshold is the main concern in grounding design. It is defined as very rapid uncoordinated contractions of the ventricles of the heart, resulting in loss of synchronization

---

* It is called the *equipment grounding*.
† Experiments have long ago established the well-known fact that electrical shock effects are due to current, not voltage [15].

TABLE 4.5
Typical Effects of Electrical Shock Current on Humans

60 Hz Current	Effect
0–1 mA	No sensation (not felt)
0–3 mA	Perceptible, mild
3–5 mA	Annoyance, pain, or surprise
5–10 mA	Painful shock
10–15 mA	Local muscle contractions, sufficient to cause "freezing" to circuit for 2.5% of population
15–30 mA	Local muscle contractions, sufficient to cause "freezing" to circuit for 50% of population
30–50 mA	Difficulty in breathing, can cause loss of consciousness
50–100 mA	Possible ventricular fibrillation of the heart
100–200 mA	Certain ventricular fibrillation of the heart
>200 mA	Severe burns and muscular contractions; heart more apt to stop than fibrillated
Over a few amperes	Irreparable damage to body tissue

between heartbeat and pulse beat. IEEE (Institute of Electrical and Electronics Engineers) Standard (Std.) 80-2000 gives the following equation to find the nonfibrillating current of magnitude I_b at a duration ranging from 0.03 to 3.0 s in relation to the energy absorbed by the body as

$$S_b = (I_b)^2 \times t_s \tag{4.54}$$

where

I_b = rms magnitude of the current through the body in amperes
I_s = duration of the current exposure in seconds
S_b = empirical constant related to the electric shock energy tolerated by a certain percent of a given population

A human heart can be seen as a muscle operating rhythmically due to a nerve pulse that provides the heartbeat. Therefore, when a false signal (i.e., the electrical shock) is injected into the heart, it could upset the rhythmic flow of operation of values and other components of the heart. This causes a condition known as ventricular fibrillation. Once this "out-of-phase" rhythm is established, it is extremely difficult to stop. It usually requires the injection of another shock to stop the fibrillation and reestablish the normal rhythm. Obviously, if it takes place in the field or at a remote location, the time delay before medical defibrillation may be too long, causing a fatality.

A fatality may also occur owing to a coronary arrest, that is, the stopping of the heartbeat. Furthermore, the current passing through the body may temporarily paralyze either the nerves or the area of the brain that controls respiration. This may also lead to death by causing cessation of respiration (asphyxia) if the victim has grasped a live conductor and cannot let go.

Currents of 1 mA or greater but less than 6 mA are often defined as the *secondary shock currents* (*let-go currents*). The let-go current is the maximum current level at which a human holding an energized conductor can control his muscles enough to release it. Dalziel's classic experiment [16,17], with 28 women and 134 men, provides data indicating an average let-go current of 10.5 rnA for women and 16 rnA for men, with 6 and 9 rnA as the respective threshold values, as shown in Figure 4.21a.

According to Dalziel, not only the individual's physiological development but also psychological factors can play an important role in limiting both the minimum and maximum values.* In general, currents with magnitudes of 6 mA or greater are known as *primary shock currents*.

---

* "Almost without exception, let-go currents in excess of 18 mA were observed in connection with friendly wagers between students. Thus, it was noted that the highest value was obtained on a student in physiology who boasted that he was as good as any engineering student. Although he made no complaints, it is more than likely that he had a sore arm for at least a week" [17].

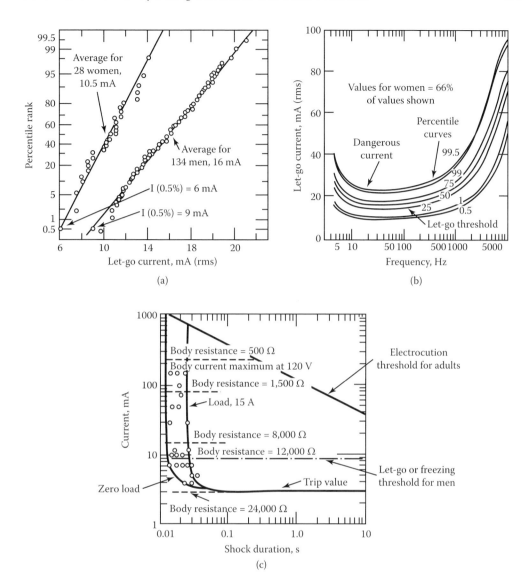

FIGURE 4.21 Effects of shock currents on humans: (a) 60-Hz let-go current distribution curves for 134 men and 28 women; (b) let-go currents vs. frequency; (c) trip current vs. shock duration. (From Dalziel, C. F., and Lee, W. R., *IEEE Spectrum* 6, 44–50 © 1972 IEEE.)

Note that, it is virtually impossible to produce primary shock currents with less than 25 V owing to normal body resistance. Among the possible consequences of primary shock current is ventricular fibrillation. On the basis of the electrocution formula developed by Dalziel [16,17], the 60-Hz minimum required body current leading to possible fatality through ventricular fibrillation can be expressed as

$$I_b = \frac{0.116}{\sqrt{t}} \quad \text{A} \tag{4.55}$$

where t is in seconds, in the range from approximately 8.3 ms to 5 s.

Table 4.6 presents a summary of quantitative effects of electric current on humans based on a report by the IEEE Working Group on Electrostatic Effects of Transmission Lines. It is important to be aware that such tables are developed on the assumption that the individuals involved are 100% fit. However, in reality, not all individuals are 100% fit. Therefore, they are more susceptible to shock hazards. Thus, in these days of the Occupational Safety and Health Administration and lawsuits, utilities have to be very cautious when it comes to grounding. The effects of an electric current passing through the vital parts of a human body depend on the duration, magnitude, and frequency of this current. Experiments have shown that the heart requires about 5 min to return to normal after experiencing a severe shock [16]. Thus, two or more closely spaced shocks (such as those that would take place in systems with automatic reclosing) would tend to have a cumulative effect. Present industry practice considers two closely spaced shocks to be equivalent to a single shock whose duration is the sum of the intervals of the individual shocks. Experiments have also shown that humans are very vulnerable to the effects of electric current at frequencies of 50–60 Hz.

As shown in Figure 4.21b, the human body can tolerate slightly larger currents at 25 Hz and about five times larger at dc current. Similarly, at frequencies of 1000 or 10,000 Hz, even larger currents can be tolerated. In the case of lighting surges, the human body appears able to tolerate very high currents, perhaps on the order of several hundreds of amperes [18]. Figure 4.21c shows the relation between trip current and shock duration for a typical ground fault interrupter (used at 120/240 V level), with electrocution threshold and let-go threshold for adults indicated to provide perspective.

When the human body becomes a part of the electric circuit, the current that passes through it can be found by applying Thévenin's theorem and Kirchhoff's current law, as illustrated in Figure 4.22. For dc and ac currents at 60 Hz, the human body can be substituted by a resistance in the equivalent circuits. The body resistance considered is usually between two extremities, either from one hand to both feet or from one foot to the other one.

Experiments have shown that the body can tolerate much more current flowing from one leg to the other than it can when current flows from one hand to the legs. Figure 4.22a shows a touch contact with current flowing from hand to feet. On the other hand, Figure 4.22b shows a step contact where current flows from one foot to the other. Note that, in each case, the body current I_b is driven by the potential difference between points A and B.

TABLE 4.6
Effect of Electric Current (mA) on Men and Women

Effect	Direct Current		60 Hz rms	
	Men	Women	Men	Women
1. No sensation on hand	1	0.6	0.4	0.3
2. Slight tingling; perception threshold	5.2	3.5	1.1	0.7
3. Shock—not painful but muscular control not lost	9	6	1.8	1.2
4. Painful shock—painful but muscular control not lost	62	41	9	6
5. Painful shock—let-go threshold[a]	76	51	16.0	10.5
6. Painful and severe shock, muscular contractions, breathing difficulty	90	60	23	15
7. Possible ventricular fibrillation from				
Short shocks:				
(a) Shock duration 0.03 s	1300	1300	1000	1000
(b) Shock duration 3.0 s	500	500	100	100
(c) Almost certain ventricular fibrillation (if shock duration over one heartbeat interval)	1375	1375	275	275

Source: IEEE Working Group Report, *IEEE Trans. Power Appar. Syst.* PAS-9, 422–426 © 1972 IEEE.

[a] Threshold for 50% of the males and females tested.

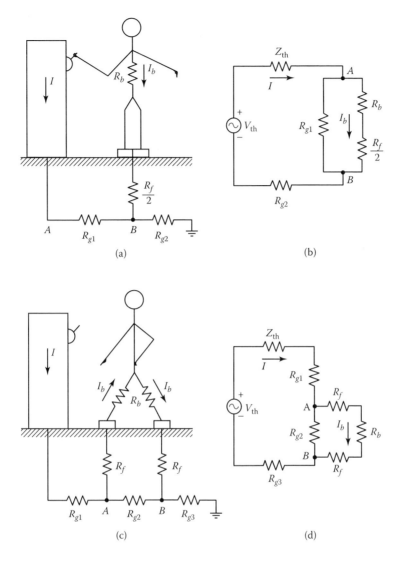

FIGURE 4.22 Typical electric shock hazard situations: (a) touch potential; (b) its equivalent circuit; (e) step potential; (d) its equivalent circuit.

Currents of 1 mA or greater but less than 6 mA are often defined as secondary shock currents (let-go currents). The let-go current is the maximum current level at which a human holding an energized conductor can control his muscles enough to release it. For 99.5% of the population, the 60-Hz minimum required body current, I_b, leading to possible fatality through ventricular fibrillation can be expressed as

$$I_b = \frac{0.116}{\sqrt{t_s}} \text{ A} \quad \text{for 50 kg body weight} \tag{4.56a}$$

or

$$I_b = \frac{0.157}{\sqrt{t_s}} \text{ A} \quad \text{for 70 kg body weight} \tag{4.56b}$$

where t is in seconds in the range from approximately 8.3 ms to 5 s.

The effects of an electric current passing through the vital parts of a human body depend on the duration, magnitude, and frequency of this current. The body resistance considered is usually between two extremities, either from one hand to both feet or from one foot to the other one. Figure 4.23 show five basic situations involving a person and grounded facilities during fault.

Note that, in the figure, the *mesh voltage* is defined by the maximum touch voltage within a mesh of a ground grid. However, the *metal-to-metal touch voltage* defines the difference in potential between metallic objects or structures within the substation site that may be bridged by direct hand-to-hand or hand-to-feet contact. However, the *step voltage* represents the difference in surface potential experienced by a person bridging a distance of 1 m with the feet without contacting any other grounded object.

On the other hand, the *touch voltage* represents the potential difference between the ground potential rise (GPR) and the surface potential at the point where a person is standing while at the same time having a hand in contact with a grounded structure. The *transferred voltage* is a special case of the touch voltage where a voltage is transferred into or out of the substation from or to a remote point external to the substation site [12].

Finally, GPR is the maximum electrical potential that a substation grounding grid may have relative to a distant grounding point assumed to be at the potential of remote earth. This voltage, GPR, is equal to the maximum grid current times the grid resistance. Under normal conditions, the grounded electrical equipment operates at near-zero ground potential. That is, the potential of a grounded neutral conductor is nearly identical to the potential of remote earth. During a ground fault, the portion of fault current that is conducted by substation grounding grid into the earth causes the rise of the grid potential with respect to remote earth.

Exposure to a touch potential normally poses a greater danger than exposure to a step potential. The step potentials are usually smaller in magnitude (due to the greater corresponding body resistance), and the allowable body current is higher than the touch contacts. In either case, the value of the body resistance is difficult to establish.

As said before, experiments have shown that the body can tolerate much more current flowing from one leg to the other than it can when current flows from one hand to the legs. Treating the foot as a circular plate electrode gives an approximate resistance of $3\rho_s$, where ρ_s is the soil resistivity. The resistance of the body itself is usually used as about 2300 Ω hand to hand or 1100 Ω hand to foot.

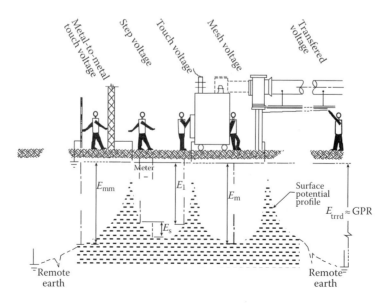

FIGURE 4.23 Possible basic shock situations. (From Keil, R. P., *Substation Grounding*, in *Electric Power Substations Engineering*, McDonald, J. D. (ed.), 2nd ed., CRC Press, Boca Raton, FL, 2007.)

However, IEEE Std. 80-2000 [12] recommends the use of 1000 Ω as a reasonable approximation for body resistance. Therefore, the total branch resistance, for hand-to-foot currents, can be expressed as

$$R_b = 1000 + 1.5\rho_s \text{ Ω for touch voltage} \tag{4.57a}$$

and, for foot-to-foot currents,

$$R_b = 1000 + 6\rho_s \text{ Ω for step voltage} \tag{4.57b}$$

where ρ_s is the soil resistivity in ohm meters. If the surface of the soil is covered with a layer of crushed rock or some other high-resistivity material, its resistivity should be used in Equations 4.56 and 4.57. The touch voltage limit can be determined from

$$V_{touch} = \left(R_b + \frac{R_f}{2}\right)I_b \tag{4.58}$$

and

$$V_{step} = (R_b + 2R_f)I_b \tag{4.59}$$

where

$$R_f = 3C_s\rho_s \tag{4.60}$$

where
- R_b = resistance of human body, typically 1000 Ω for 50 and 60 Hz
- R_f = ground resistance of one foot
- I_b = rms magnitude of the current going through the body in A, per Equations 4.56a and 4.56b
- C_s = surface layer derating factor based on the thickness of the protective surface layer spread above the earth grade at the substation (per IEEE Std. 80-2000, if no protective layer is used, then $C_s = 1$)

Since it is much easier to calculate and measure the potential than the current, the fibrillation threshold, given by Equations 4.56a and 4.56b, are usually given in terms of voltage. Thus, for a person with a body weight of 50 or 70 kg, the maximum allowable (or tolerable) touch voltages, respectively, can be expressed as

$$V_{touch\ 50} = \frac{0.116(1000 + 1.5\rho_s)}{\sqrt{t_s}} \text{ V} \quad \text{for 50 kg body weight} \tag{4.61a}$$

and

$$V_{touch\ 70} = \frac{0.157(1000 + 1.5\rho_s)}{\sqrt{t_s}} \text{ V} \quad \text{for 70 kg body weight} \tag{4.61b}$$

Note that, the above equations are applicable only in the event of no protective surface layer is used. Hence, for the metal-to-metal touch in V, Equations 4.61a and 4.61b become

$$V_{mm\text{-}touch\ 50} = \frac{116}{\sqrt{t_s}} \text{ V} \quad \text{for 50 kg body weight} \tag{4.61c}$$

and

$$V_{\text{mm-touch 70}} = \frac{157}{\sqrt{t_s}} \text{ V} \quad \text{for 70 kg body weight} \qquad (4.61d)$$

The maximum allowable (or tolerable) step voltages, for a person with a body weight of 50 or 70 kg, are given, respectively, as

$$V_{\text{step 50}} = \frac{0.116(1000 + 6C_s\rho_s)}{\sqrt{t_s}} \text{ V} \quad \text{for 50 kg body weight} \qquad (4.62a)$$

and

$$V_{\text{step 70}} = \frac{0.157(1000 + 6C_s\rho_s)}{\sqrt{t_s}} \text{ V} \quad \text{for 70 kg body weight} \qquad (4.62b)$$

$$V_{\text{touch 50}} = \frac{0.116(1000 + 1.5C_s\rho_s)}{\sqrt{t_s}} \text{ V} \quad \text{for 50 kg body weight} \qquad (4.62c)$$

$$V_{\text{touch 70}} = \frac{0.157(1000 + 1.5C_s\rho_s)}{\sqrt{t_s}} \text{ V} \quad \text{for 70 kg body weight} \qquad (4.62d)$$

The above equations are applicable only in the event that a protection surface layer is used. For metal-to-metal contacts, use $\rho_s = 0$ and $C_s = 1$. For more detailed applications, see IEEE Std. 2000 [12]. Also, it is important to note that in using the above equations, it is assumed that they are applicable to 99.5% of the population. There are always exceptions.

4.8.2 Reduction of Factor C_s

Note that, according to IEEE Std. 80-2000, a thin layer of highly resistive protective surface material such as gravel spread across the earth at a substation greatly reduces the possibility of shock situation at that substation. IEEE Std. 80-2000 gives the required equations to determine the ground resistance of one foot on a thin layer of surface material as

$$C_s = 1 + \frac{1.6b}{\rho_s} \sum_{n=1}^{\infty} K^n R_{m(2nh_s)} \qquad (4.63)$$

and

$$C_s = 1 - \frac{0.09\left(1 - \dfrac{\rho}{\rho_s}\right)}{2h_s + 0.09} \qquad (4.64)$$

where

$$K = \frac{\rho - \rho_s}{\rho + \rho_s} \qquad (4.65)$$

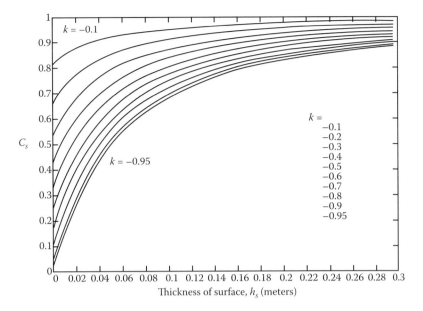

FIGURE 4.24 Surface layer derating factor vs. thickness of surface material in meters. (From Keil, R. P., *Substation Grounding*, in *Electric Power Substations Engineering*, McDonald, J. D. (ed.), 2nd ed., CRC Press, Boca Raton, FL, 2007.)

where
  C_s = surface layer derating factor (It can be considered as a corrective factor to compute the effective foot resistance in the presence of a finite thickness of surface material.) (See Figure 4.24.)
  ρ_s = surface material resistivity, $\Omega \cdot m$
  K = reflection factor between different material resistivities
  ρ = resistivity of earth beneath the substation, $\Omega \cdot m$
  h_s = thickness of the surface material, m
  b = radius of circular metallic disc representing the foot, m
  $R_{m(2nh_s)}$ = mutual ground resistance between two similar, parallel, coaxial plates that are separated by a distance of $(2nh_s)$, $\Omega \cdot m$

Note that, Figure 4.24 gives the exact value of C_s instead of using the empirical Equation 4.64 for it. The empirical equation gives approximate values that are within 5% of the values that can be found with the equation.

Table 4.7 gives typical values for various ground types. However, the resistivity of ground also changes as a function of temperature, moisture, and chemical content. Therefore, in practical applications, the only way to determine the resistivity of soil is by measuring it.

EXAMPLE 4.7

Assume that a human body is part of a 60-Hz electric power circuit for about 0.49 s and that the soil type is average earth. On the basis of IEEE Std. 80-2000, determine the following:

(a) Tolerable touch potential, for 50 kg body weight
(b) Tolerable step potential
(c) Tolerable touch voltage limit for metal-to-metal contact, if the person is 50 kg
(d) Tolerable touch voltage limit for metal-to-metal contact, if the person is 70 kg

TABLE 4.7
Resistivity of Different Soils

Ground Type	Resistivity, ρ_s
Seawater	0.01–1.0
Wet organic soil	10
Moist soil (average earth)	100
Dry soil	1000
Bedrock	10^4
Pure slate	10^7
Sandstone	10^9
Crushed rock	1.5×10^8

Solution

(a) Using Equation 4.61a, for 50 kg body weight

$$V_{\text{touch 50}} = \frac{0.116(1000 + 1.5\rho_s)}{\sqrt{t_s}}$$

$$= \frac{0.116(1000 + 1.5 \times 100)}{\sqrt{0.49}}$$

$$\cong 191\,\text{V}$$

(b) Using Equation 4.61b

$$V_{\text{step 50}} = \frac{0.116(1000 + 6\rho_s)}{\sqrt{t_s}}$$

$$= \frac{0.116(1000 + 6 \times 100)}{\sqrt{0.49}}$$

$$\cong 265\,\text{V}$$

(c) Since $\rho_s = 0$,

$$V_{\text{mm-touch 50}} = \frac{116}{\sqrt{t_s}} = \frac{116}{\sqrt{0.49}} = 165.7\,\text{V for 50 kg body weight}$$

(d) Since $\rho_s = 0$,

$$V_{\text{mm-touch 70}} = \frac{157}{\sqrt{t_s}} = \frac{157}{\sqrt{0.49}} = 224.3\,\text{V for 70 kg body weight}$$

4.8.3 Ground Potential Rise (GPR) and Ground Resistance

The GPR is a function of fault current magnitude, system voltage, and ground (system) resistance. The current through the ground system multiplied by its resistance measured from a point remote from the substation determines the GPR with respect to remote ground.

TABLE 4.8
Effect of Moisture Content on Soil Resistivity

Moisture Content (wt.%)	Resistivity ($\Omega \cdot cm$)	
	Topsoil	Sandy Loam
0	$>10^9$	$>10^9$
2.5	250,000	150,000
5	165,000	43,000
10	53,000	18,500
15	19,000	10,500
20	12,000	6,300
30	6,400	4,200

The ground resistance can be reduced by using electrodes buried in the ground. For example, metal rods or *counterpoise* (i.e., buried conductors) are used for the lines of the grid system are made of copper-stranded cable, on the other hand rods are used for the substations.

The grounding resistance of a buried electrode is a function of (1) the resistance of the electrode itself and the connections to it, (2) the contact resistance between the electrode and the surrounding soil, and (3) the resistance of the surrounding soil, from the electrode surface outward.

The first two resistances are very small with respect to soil resistance and therefore may be neglected in some applications. However, the third one is usually very large depending on the type of soil, chemical ingredients, moisture level, and temperature of the soil surrounding the electrode.

Table 4.8 presents data indicating the effect of moisture contents on soil resistivity. The resistance of the soil can be measured by using the three-electrode method or by using self-contained instruments such as the Biddle Megger ground resistance tester.

If the surface of the soil is covered with a layer of crushed rock or some other high-resistivity material, its resistivity should be used in the previous equations. Table 4.7 gives typical values for various ground types. However, the resistivity of ground also changes as a function of temperature, moisture, and chemical content. Thus, in practical applications, the only way to determine the resistivity of soil is by measuring it.

In general, soil resistivity investigations are required to determine the soil structure. Table 4.7 gives only very rough estimates. The soil resistivity can very substantially with changes in temperature, moisture, and chemical content. To determine the soil resistivity of a specific site, soil resistivity measurements are required to be taken. Since soil resistivity can change both horizontally and vertically, it is necessary to take more than one set of measurements. IEEE Std. 80-2000 [12] describes various measuring techniques in detail. There are commercially available computer programs that use the soil data and calculate the soil resistivity and provide a confidence level based on the test. There is also a graphical method that was developed by Sunde [29] to interpret the test results.

4.8.4 Ground Resistance

Ground is defined as a conducting connection, either intentional or accidental, by which an electric circuit or equipment becomes grounded. Thus, *grounded* means that a given electric system, circuit, or device is connected to the earth or to some other equivalent conducting body of relatively large extent, serving in the place of the former with the purpose of establishing and maintaining the potential of conductors connected to it approximately at the potential of the earth and allowing for conducting electric currents from and to the earth of its equivalent.

Thus, a safe grounding design should provide the following:

1. A means to carry and dissipate electric currents into ground under normal and fault conditions without exceeding any operating and equipment limits or adversely affecting the continuity of service
2. Assurance for such a degree of human safety so that a person working or walking in the vicinity of grounded facilities is not subjected to the danger of critical electrical shock

However, a low ground resistance is not, in itself, a guarantee of safety. For example, about three or four decades ago, a great many people assumed that any object grounded, however crudely, could be safely touched. This misconception probably contributed to many tragic accidents in the past. Thus, since there is no simple relation between the resistance of the ground system as a whole and the maximum shock current to which a person might be exposed, a system or system component (e.g., substation or tower) of relatively low ground resistance may be dangerous under some conditions, but another system component with very high ground resistance may still be safe or can be made safe by careful design.

Table 4.9 gives data showing the effect of temperature on soil resistivity.

Figure 4.25 shows a ground rod driven into the soil and conducting current in all directions. The resistance of the soil has been illustrated in terms of successive shells of the soil of equal thickness. With increased distance from the electrode, the soil shells have greater area and therefore lower resistance. Thus, the shell nearest the rod has the smallest cross section of the soil and therefore the highest resistance. Measurements have shown that 90% of the total resistance surrounding an electrode is usually with a radius of 6–10 ft. Table 4.10 gives formulas to determine resistance to ground of various types of electrodes [25].

The assumptions that have been made in deriving these formulas are that the soil is perfectly homogeneous and the resistivity is of the same known value throughout the soil surrounding the electrode. Of course, these assumptions are seldom true. The only way one can be sure of the resistivity of the soil is by actually measuring it at the actual location of the electrode and at the actual depth.

Figure 4.26 shows the variation of soil resistivity with depth for a soil having uniform moisture content at all depths [25a]. In reality, however, deeper soils have greater moisture content, and the advantage of depth is more visible. Some nonhomogeneous soils can also be modeled by using the two-layer method [26–29].

The resistance of the soil can be measured by using the three-electrode method or by using self-contained instruments such as the Biddle Megger ground resistance tester. Figure 4.27 shows the approximate ground resistivity distribution in the United States.

TABLE 4.9
Effect of Temperature on Soil Resistivity[a]

Temperature (°C)	°F	Resistivity (Ω·cm)
20	68	7,200
10	50	9,900
0 (water)	32	13,800
0 (ice)	32	30,000
−5	23	79,000
−15	14	330,000

[a] Sandy loam with 15.2% moisture.

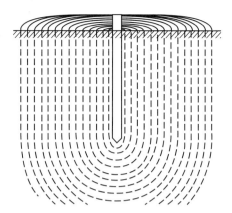

FIGURE 4.25 Resistance of earth surrounding an electrode.

4.8.5 Soil Resistivity Measurements

Table 4.7 gives estimates on soil classification that are only an approximation of the actual resistivity of a given site. Actual resistivity tests therefore are crucial. They should be made at a number of places within the site. In general, substation sites where the soil has uniform resistivity throughout the entire area and to a considerable depth are seldom found.

4.8.5.1 Wenner Four-Pin Method

More often than not, there are several layers, each having a different resistivity. Furthermore, lateral changes also take place, but with respect to the vertical changes; these changes usually are more gradual. Hence, soil resistivity tests should be made to find out if there are any substantial changes in resistivity with depth. If the resistivity varies considerably with depth, it is often desirable to use an increased range of probe spacing to get an estimate of the resistivity of the deeper layers.

IEEE Std. 81-1983 describes a number of measuring techniques. The Wenner four-pin method is the most commonly used technique. Figure 4.28 illustrates this method. In this method, four probes (or pins) are driven into the earth along a straight line, at equal distances apart, driven to a depth b. The voltage between the two inner (i.e., potential) electrodes is then measured and divided by the current between the two outer (i.e., current) electrodes to give a value of resistance R. The apparent resistivity of soil is determined from

$$\rho_a = \frac{4\pi a R}{1 + \frac{2a}{\sqrt{a^2 + 4b^2}} - \frac{a}{\sqrt{a^2 + b^2}}} \tag{4.66}$$

where
    ρ_a = apparent resistivity of the soil in ohm meters
    R = measured resistivity in ohms
    a = distance between adjacent electrodes in meters
    b = depth of the electrodes in meters

In the event that b is small in comparison to a, then

$$\rho_a = 2\pi a R \tag{4.67}$$

TABLE 4.10
Formulas for Calculations of Resistance to Ground

Hemisphere, radius a	$R =$	$\dfrac{\rho}{2\pi a}$
One ground rod, length L, radius a	$R =$	$\dfrac{\rho}{2\pi L}\left(\ln\dfrac{4L}{a} - 1\right)$
Two ground rod, $s > L$; spacing s	$R =$	$\dfrac{\rho}{4\pi L}\left(\ln\dfrac{4L}{a} - 1\right) + \dfrac{\rho}{4\pi s}\left(1 - \dfrac{L^2}{3s^2} + \dfrac{2L^4}{5s^4}\cdots\right)$
Two ground rod, $s > L$; spacing s	$R =$	$\dfrac{\rho}{4\pi L}\left(\ln\dfrac{4L}{a} + \ln\dfrac{4L}{s} - 2 + \dfrac{s}{2L} - \dfrac{s^2}{16L^2} + \dfrac{s^4}{512L^4}\cdots\right)$
Buried horizontal wire, length $2L$, depth $s/2$	$R =$	$\dfrac{\rho}{4\pi L}\left(\ln\dfrac{4L}{a} + \ln\dfrac{4L}{s} - 2 + \dfrac{s}{2L} - \dfrac{s^2}{16L^2} + \dfrac{s^4}{512L^4}\cdots\right)$
Rigth-angle turn of wire, length of arm L, depth $s/2$	$R =$	$\dfrac{\rho}{4\pi L}\left(\ln\dfrac{2L}{a} + \ln\dfrac{2L}{s} - 0.2373 + 0.2146\,\dfrac{s}{L} + 0.1035\,\dfrac{s^2}{L^2} - 0.0424\,\dfrac{s^4}{L^4}\cdots\right)$
Three-point star, length of arm L, depth $s/2$	$R =$	$\dfrac{\rho}{6\pi L}\left(\ln\dfrac{2L}{a} + \ln\dfrac{2L}{s} + 1.071 - 0.209\,\dfrac{s}{L} + 0.238\,\dfrac{s^2}{L^2} - 0.054\,\dfrac{s^4}{L^4}\cdots\right)$
Four-point star, length of arm L, depth $s/2$	$R =$	$\dfrac{\rho}{8\pi L}\left(\ln\dfrac{2L}{a} + \ln\dfrac{2L}{s} + 2.912 - 1.071\,\dfrac{s}{L} + 0.645\,\dfrac{s^2}{L^2} - 0.145\,\dfrac{s^4}{L^4}\cdots\right)$

Disturbance of Normal Operating Conditions and Other Problems

✶	Six-point star, length of arm L, depth $s/2$	$R = \dfrac{\rho}{12\pi L}\left(\ln\dfrac{2L}{a} + \ln\dfrac{2L}{s} + 6.851 - 3.128\,\dfrac{s}{L} + 1.758\,\dfrac{s^2}{L^2} - 0.490\,\dfrac{s^4}{L^4}\cdots\right)$
✸	Eight-point star, length of arm L, depth $s/2$	$R = \dfrac{\rho}{16\pi L}\left(\ln\dfrac{2L}{a} + \ln\dfrac{2L}{s} + 10.98 - 5.51\,\dfrac{s}{L} + 3.26\,\dfrac{s^2}{L^2} - 1.17\,\dfrac{s^4}{L^4}\cdots\right)$
◯	Ring of wire, diameter of ring D, diameter of wire d, depth $s/2$	$R = \dfrac{\rho}{2\pi^2 D}\left(\ln\dfrac{8D}{d} + \ln\dfrac{4D}{s}\right)$
∣	Buried horizontal strip, length $2L$, section a by b, depth $s/2$, $b < a/8$	$R = \dfrac{\rho}{4\pi L}\left(\ln\dfrac{4L}{a} + \dfrac{a^2 - \pi ab}{2(a+b)^2} + \ln\dfrac{4L}{s} - 1 + \dfrac{s}{2L} - \dfrac{s^2}{16L^2} + \dfrac{s^4}{512L^4}\cdots\right)$
⬤	Buried vertical round plate, radius a, depth $s/2$	$R = \dfrac{\rho}{8a} + \dfrac{\rho}{4\pi s}\left(1 - \dfrac{7}{12}\,\dfrac{a^2}{s^2} + \dfrac{33}{40}\,\dfrac{a^4}{s^4}\cdots\right)$
⬤	Buried vertical round plate, radius a, depth $s/2$	$R = \dfrac{\rho}{8a} + \dfrac{\rho}{4\pi s}\left(1 + \dfrac{7}{24}\,\dfrac{a^2}{s^2} + \dfrac{99}{320}\,\dfrac{a^4}{s^4}\cdots\right)$

Source: Dwight, H. B., *Electr. Eng. (Am. Inst. Electr. Eng.)* 55, 1319–1328 © 1936 IEEE.

[a] Approximate formulas, including effects of images. Dimensions must be in centimeters to give resistance in ohms. The symbol ρ is the resistivity of earth in ohm centimeters.

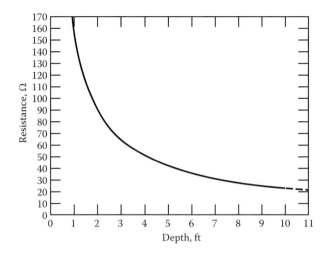

FIGURE 4.26 Variation of soil resistivity with depth for soil having uniform moisture content at all depths. (From National Bureau of Standards Technical Report 108.)

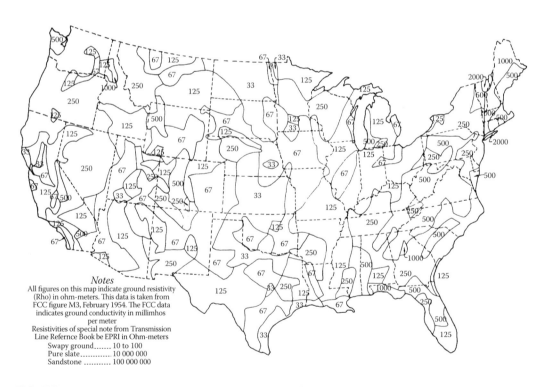

FIGURE 4.27 Approximate ground resistivity distribution in the United States. (From Farr, H. H., *Transmission Line Design Manual*. U.S. Department of the Interior, Water and Power Resources Service, Denver, 1980.)

Disturbance of Normal Operating Conditions and Other Problems

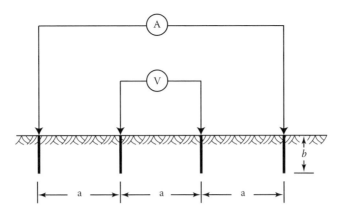

FIGURE 4.28 Wenner four-pin method. (From Gönen, T., *Electric Power Transmission System Engineering*, 2nd ed., CRC Press, Boca Raton, FL, 2009.)

The current tends to flow near the surface for the small probe spacing, whereas more of the current penetrates deeper soils for large spacing. Because of this fact, the previous two equations can be used to determine the apparent resistivity ρ_a at a depth a.

The Wenner four-pin method obtains the soil resistivity data for deeper layers without driving the test pins to those layers. No heavy equipment is needed to do the four-pin test. The results are not greatly affected by the resistance of the test pins or the holes created in driving the test pins into the soil. Because of these advantages, the Wenner method is the most popular method.

4.8.5.2 Three-Pin or Driven-Ground Rod Method

IEEE Std. 81-1983 describes a second method of measuring soil resistivity. It is illustrated in Figure 4.29. In this method, the depth (L_r) of the driven rod located in the soil to be tested is varied. The other two rods are known as *reference rods*. They are driven to a shallow depth in a straight line. The location of the voltage rod is varied between the test rod and the current rod. Alternatively, the voltage rod can be placed on the other side of the driven rod. The apparent resistivity is found from

$$\rho_a = \frac{2\pi L_r R}{\ln\left(\frac{8L_r}{d}\right) - 1} \tag{4.68}$$

where
  L_r = length of the driven rod in meters
  d = diameter of the rod in meters
  R = measured resistivity in ohms

A plot of the measured resistivity value ρ_a vs. the rod length (L_r) provides a visual aid for finding out earth resistivity variations with depth. An advantage of the driven-rod method, even though not related necessarily to the measurements, is the ability to determine to what depth the ground rods can be driven. This knowledge can save the need to redesign the ground grid. Because of hard layers in the soil such as rock and hard clay, it becomes practically impossible to drive the test rod any further, resulting in insufficient data.

A disadvantage of the driven-rod method is that when the test rod is driven deep in the ground, it usually losses contact with the soil owing to the vibration and the larger-diameter couplers resulting

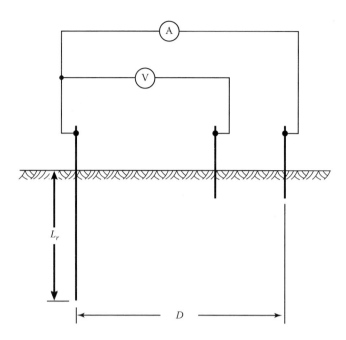

FIGURE 4.29 Circuit diagram for three-pin or driven-ground rod method. (From Gönen, T., *Electric Power Transmission System Engineering*, 2nd ed., CRC Press, Boca Raton, FL, 2009.)

in higher measured resistance values. A ground grid designed with these higher soil resistivity values may be unnecessarily conservative. Thus, this method presents an uncertainty in the resistance value.

4.9 SUBSTATION GROUNDING

Grounding at the substation has paramount importance. Again, the purpose of such a grounding system includes the following:

1. To provide the ground connection for the grounded neutral for transformers, reactors, and capacitors
2. To provide the discharge path for lightning rods, arresters, gaps, and similar devices
3. To ensure safety to operating personnel by limiting potential differences that can exist in a substation
4. To provide a means of discharging and deenergizing equipment in order to proceed with the maintenance of the equipment
5. To provide a sufficiently low resistance path to ground to minimize the rise in ground potential with respect to remote ground

A multigrounded, common neutral conductor used for a primary distribution line is always connected to the substation grounding system where the circuit originates and to all grounds along the length of the circuit. If separate primary and secondary neutral conductors are used, the conductors have to be connected together provided the primary neutral conductor is effectively grounded.

The substation grounding system is connected to every individual equipment, structure, and installation so that it can provide the means by which grounding currents are connected to remote areas. It is extremely important that the substation ground has a low ground resistance, adequate current-carrying capacity, and safety features for personnel. It is crucial to have the substation ground resistance very low so that the total rise of the ground system potential will not reach values that are unsafe for human contact.*

The substation grounding system is normally made of buried horizontal conductors and driven ground rods interconnected (by clamping, welding, or brazing) to form a continuous grid (also called *mat*) network. A continuous cable (usually it is 4/0 bare copper cable buried 12–18 in. below the surface) surrounds the grid perimeter to enclose as much ground as possible and to prevent current concentration and thus high gradients at the ground cable terminals. Inside the grid, cables are buried in parallel lines and with uniform spacing (e.g., about 10 × 20 ft).

All substation equipment and structures are connected to the ground grid with large conductors to minimize the grounding resistance and limit the potential between equipment and the ground surface to a safe value under all conditions. All substation fences are built inside the ground grid and attached to the grid in short intervals to protect the public and personnel. The surface of the substation is usually covered with crushed rock or concrete to reduce the potential gradient when large currents are discharged to ground and to increase the contact resistance to the feet of the personnel in the substation.

IEEE Std. 80-1976 [24] provides a formula for a quick simple calculation of the grid resistance to ground after a minimum design has been completed. It is expressed as

$$R_{grid} = \frac{\rho_s}{4r} + \frac{\rho_s}{L_T} \tag{4.69}$$

where

ρ_s = soil resistivity in ohm meters
L = total length of grid conductors in meters
R = radius of circle with area equal to that of grid in meters

IEEE Std. 80-2000 provides the following equation to determine the grid resistance after a minimum design has been completed:

$$R_{grid} = \frac{\rho_s}{4}\sqrt{\frac{\pi}{A}} \tag{4.70}$$

Also, IEEE Std. 80-2000 provides the following equation to determine the upper limit for grid resistance to ground after a minimum design has been completed:

$$R_{grid} = \frac{\rho_s}{4}\sqrt{\frac{\pi}{A}} + \frac{\rho_s}{L_T} \tag{4.71}$$

where

R_{grid} = grid resistance in *ohms*
ρ = soil resistance in *ohm meters*
A = area of the ground in square meters
L_T = total buried length of conductors in meters

However, Equation 4.71 requires a uniform soil resistivity. Hence, a substantial engineering judgment is necessary for reviewing the soil resistivity measurements to decide the value of soil resistivity. However, it does provide a guideline for the uniform soil resistivity to be used in the

---

* *Mesh voltage* is the worst possible value of a touch voltage to be found within a mesh of a ground grid if standing at or near the center of the mesh.

ground grid design. Alternatively, Sverak et al. [19] provides the following formula for the grid resistance:

$$R_{grid} = \rho_s \left[\frac{1}{L_T} + \frac{1}{\sqrt{20A}} \left(1 + \frac{1}{1 + h\sqrt{\frac{20}{A}}} \right) \right] \quad (4.72)$$

where
R_{grid} = substation ground resistance, Ω
ρ_s = soil resistivity, $\Omega \cdot m$
A = area occupied by the ground grid, m^2
H = depth of the grid, m
L_T = total buried length of conductors, m

IEEE Std. 80-1976 also provides formulas to determine the effects of the grid geometry on the step and mesh voltage in volts. Mesh voltage is the worst possible value of a touch voltage to be found within a mesh of a ground grid if standing at or near the center of the mesh. They can be expressed as

$$E_{step} = \frac{\rho_s \times K_s \times K_i \times I_G}{L_s} \quad (4.73)$$

and

$$E_{mesh} = \frac{\rho_s \times K_m \times K_i \times I_G}{L_m} \quad (4.74)$$

where
ρ_s = average soil resistivity in ohm meters
K_s = step coefficient
K_m = mesh coefficient
K_i = irregularity coefficient
I_G = maximum rms current flowing between ground grid and earth in amperes
L_s = total length of buried conductors, including cross connections, and (optionally) the total effective length of ground rods in meters
L_m = total length of buried conductors, including cross connections, and (optionally) the combined length of ground rods in meters

Many utilities have computer programs for performing grounding grid studies. The number of tedious calculations that must be performed to develop an accurate and sophisticated model of a system is no longer a problem.

In general, in the event of a fault, overhead ground wires, neutral conductors, and directly buried metal pipes and cables conduct a portion of the ground fault current away from the substation ground grid and have to be taken into account when calculating the maximum value of the grid current. On the basis of the associated equivalent circuit and resultant current division, one can determine what portion of the total current flows into the earth and through other ground paths. It can be used to determine the approximate amount of current that did not use the ground as flow path. The fault current division factor (also known as the *split factor*) can be expressed as

$$S_{split} = \frac{I_{grid}}{3I_{a0}} \quad (4.75)$$

where

S_{split} = fault current division factor
I_{grid} = rms symmetrical grid current, A
I_{a0} = zero-sequence fault current, A

The *split factor* is used to determine the approximate amount of current that did not use the ground flow path. Computer programs can determine the split factor easily, but it is also possible to determine the split factor through graphs. With the Y ordinate representing the split factor and the X axis representing the grid resistance, it is obvious that the grid resistance has to be known to determine the split factor. As previously said, the split factor determines the approximate amount of current that uses the earth as a return path. The amount of current that does enter the earth is found from the following equation. Hence, the design value of the *maximum grid current* can be found from

$$I_G = D_f \times I_{grid} \tag{4.76}$$

where

I_G = maximum grid current in amperes
D_f = decrement factor for the entire fault duration of t_f, given in seconds
I_{grid} = rms symmetrical grid current in amperes

Here, Figure 4.30 illustrates the relation between asymmetrical fault current, dc decaying component, and symmetrical fault current, and the relation between the variables I_F, I_f, and D_f for the fault duration t_f.

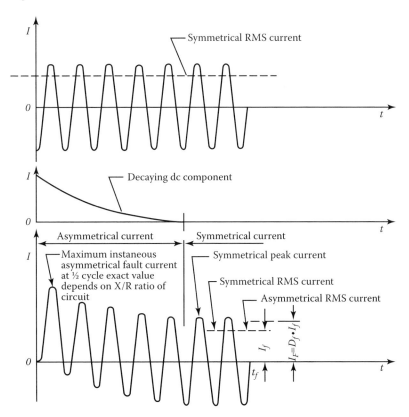

FIGURE 4.30 Relation between asymmetrical fault current, dc decaying component, and symmetrical fault current.

The *decrement factor* is an adjustment factor that is used in conjunction with the symmetrical ground fault current parameter in safety-oriented grounding calculations. It determines the rms equivalent of the asymmetrical current wave for a given fault duration, accounting for the effect of initial dc offset and its attenuation during the fault. The decrement factor can be calculated from

$$D_f = \sqrt{1 + \frac{T_a}{I_f}\left(1 - e^{-\frac{2t_f}{T_a}}\right)} \qquad (4.77)$$

where

t_f = time duration of fault in seconds
$T_a = \dfrac{X}{\omega R}$ = dc offset time constant in seconds

Here, t_f should be chosen as the fastest clearing time, and includes breaker and relay time for transmission substations. It is assumed here that the ac components do not decay with time.

The symmetrical grid current is defined as that portion of the symmetrical ground fault current that flows between the grounding grid and surrounding earth. It can be expressed as

$$I_{\text{grid}} = S_f \times I_f \qquad (4.78)$$

where

I_f = rms value of symmetrical ground fault current in amperes
S_f = fault current division factor

IEEE Std. 80-2000 provides a series of current based on computer simulations for various values of ground grid resistance and system conditions to determine the grid current. On the basis of those split-current curves, one can determine the maximum grid current.

4.10 GROUND CONDUCTOR SIZING FACTORS

The flow of excessive currents will be very dangerous if the right equipment is not used to help dissipate the excessive currents. Ground conductors are means of providing a path for excessive currents from the substation to ground grid. Hence, the ground grid than can spread the current into the ground, creating a zero potential between the substation and the ground. Table 4.11 gives the list of possible conductors that can be used for such conductors. In the United States, there are only two types of conductors, namely, copper and/or copper-clad steel conductors, that are used for this purpose. The copper one is mainly used because of its high conductivity and the high resistance to corrosion. The next step is to determine the size of ground conductor that needs to be buried underground.

Thus, based on the *symmetrical conductor current*, the required conductor size can be found from

$$I_f = A_{\text{mm}^2}\left[\left(\frac{\text{TCAP} \times 10^{-4}}{t_c \times \alpha_r \times \rho_r}\right)\ln\left(\frac{K_0 + T_{\max}}{K_0 + T_{\text{amb}}}\right)\right]^{1/2} \qquad (4.79)$$

if the conductor size needs to be found in square millimeters, the conductor size can be found from

$$A_{\text{mm}^2} = \frac{I_f}{\left[\left(\dfrac{\text{TCAP} \times 10^{-4}}{t_c \times \alpha_r \times \rho_r}\right)\ln\left(\dfrac{K_0 + T_{\max}}{K_0 + T_{\text{amb}}}\right)\right]^{1/2}} \qquad (4.80)$$

TABLE 4.11
Material Constants of the Typical Grounding Material Used

Description	K_f	T_m (°C)	α_r Factor at 20°C (1/°C)	ρ_r 20°C (mΩ·cm)	K_0 at 0°C (0°C)	Fusing Temperature, T_m (0°C)	Material Conducting (%)	TCAP Thermal Capacity [J/cm³ × °C]
Copper annealed soft-drawn	7	1083	0.0393	1.72	234	1083	100	3.42
Copper annealed hard-drawn	1084	1084	0.00381	1.78	242	1084	97	3.42
Copper-clad steel wire	1084	12.06	0.00378	5.86	245	1084	30	3.85
Stainless steel 304	1510	14.72	0.00130	15.86	749	1400	2.4	3.28
Zinc-coated steel rod	28.96	28.96	0.0030	72	293	419	8.6	4.03

Alternatively, in the event that the conductor size needs to be found in kilo-circular mils, since

$$A_{\text{kcmil}} = 1.974 \times A_{\text{mm}^2} \tag{4.81}$$

then Equation 4.72 can be expressed as

$$I_f = 5.07 \times 10^{-3} A_{\text{kcmil}} \left[\left(\frac{\text{TCAP} \times 10^{-4}}{t_c \times \alpha_r \times \rho_r} \right) \ln\left(\frac{K_0 + T_{\max}}{K_0 + T_{\text{amb}}} \right) \right]^{1/2} \tag{4.82}$$

Note that both α_r and ρ_r can be found at the same reference temperature of T_r (°C). Also, note that Equations 4.79 and 4.82 can also be used to determine the short-time temperature rise in a ground conductor. Thus, taking other required conversions into account, the conductor size in kilo-circular mils can be found from

$$A_{\text{kcmil}} = \frac{197.4 \times I_f}{\left[\left(\frac{\text{TCAP} \times 10^{-4}}{t_c \times \alpha_r \times \rho_r} \right) \ln\left(\frac{K_0 + T_{\max}}{K_0 + T_{\text{amb}}} \right) \right]^{1/2}} \tag{4.83}$$

where

I_f = rms current (without dc offset), kA
A_{mm^2} = conductor cross section, mm²
A_{kcmil} = conductor cross section, kcmil
TCAP = thermal capacity per unit volume, J/(cm³·°C). (It is found from Table 4.11, per IEEE Std. 80-2000.)
t_c = duration of current, s
α_r = thermal coefficient of resistivity at reference temperature T_r, 1/°C. (It is found from Table 4.11, per IEEE Std. 80-2000 for 20°C.)
ρ_r = resistivity of the ground conductor at reference temperature T_r, μΩ·cm. (It is found from Table 4.11, per IEEE Std. 80-2000 for 20°C.)

$K_0 = 1/\alpha_0$ or $(1/\alpha_r) - T_r$, °C
T_{max} = maximum allowable temperature, °C
T_{amb} = ambient temperature, °C
I_f = rms current (without dc offset), kA
A_{mm^2} = conductor cross section, mm²
A_{kcmil} = conductor cross section, kcmil

For a given conductor material, once the TCAP is found from Table 4.11 or calculated from

$$\text{TCAP}[J/(cm^3 \times °C)] = 4.184(J/cal) \times SH[(cal/(g \times °C))] \times SW(g/cm^3) \tag{4.84}$$

where
SH = specific heat, in cal/(g × °C) is related to the thermal capacity per unit volume in J/(cm³ × °C)
SW = specific weight, in g/cm³ is related to the thermal capacity per unit volume in J/(cm³ × °C)

Thus, TCAP is defined by

$$\text{TCAP}[J/(cm^3 \times °C)] = 4.184(J/cal) \times SH[(cal/(g \times °C))] \times SW(g/cm^3) \tag{4.85}$$

Asymmetrical fault currents consist of subtransient, transient, and steady-state ac components, and the dc offset current component. To find the asymmetrical fault current (i.e., if the effect of the dc offset is needed to be included in the fault current), the equivalent value of the asymmetrical current I_F is found from

$$I_F = D_f \times I_f \tag{4.86}$$

where I_F represents the rms value of an asymmetrical current integrated over the entire fault duration, t_c, can be found as a function of X/R by using D_f, before using Equations 4.79 or 4.82, and where D_f is the decrement factor and is found from

$$D_f = \left[1 + \frac{T_a}{t_f}\left(1 - e^{\frac{-2t_f}{T_a}}\right)\right]^{1/2} \tag{4.87}$$

where t_f is the time duration of fault in seconds, and T_a is the dc offset time constant in seconds. Note that,

$$T_a = \frac{X}{\omega R} \tag{4.88}$$

and for 60 Hz,

$$T_a = \frac{X}{120\pi R} \tag{4.89}$$

The resulting I_F is always greater than I_f. However, if the X/R ratio is less than 5 and the fault duration is greater than 1 s, the effects of the dc offset are negligible.

Disturbance of Normal Operating Conditions and Other Problems

4.11 MESH VOLTAGE DESIGN CALCULATIONS

If the GPR value exceeds the tolerable touch and step voltages, it is necessary to perform the mesh voltage design calculations to determine whether the design of a substation is safe. If the design is again unsafe, conductors in the form of ground rods are added to the design until the design is considered safe. The mesh voltage is found from

$$E_{mesh} = \frac{\rho \times K_m \times K_i \times I_G}{L_M} \tag{4.90}$$

where
- ρ = soil resistivity, $\Omega \cdot m$
- K_m = mesh coefficient
- K_i = correction factor for grid geometry
- I_G = maximum grid current that flows between ground grid and surrounding earth, A
- L_m = length of $L_c + L_R$ for mesh voltage, m
- L_c = total length of grid conductor, m
- L_R = total length of ground rods, m

The mesh coefficient K_m is determined from

$$K_M = \frac{1}{2\pi}\left[\ln\left\{\frac{D^2}{16h \times d} + \frac{(D+2h)2}{8D \times d} - \frac{h}{4d}\right\} + \frac{K_{ii}}{K_h} \times \ln\left\{\frac{8}{\pi(2n-1)}\right\}\right] \tag{4.91}$$

where
- d = diameter of grid conductors, m
- D = spacing between parallel conductors, m
- K_{ii} = irregularity factor (*corrective weighting factor* that adjusts for the effects of inner conductors on the corner mesh)
- K_h = corrective weighting factor that highlight for the effects of grid depth
- n = geometric factor
- h = depth of ground grid conductors, m

Note that, the value of K_{ii} depends on the following circumstances:

(a) For the grids with ground rods existing in grid corners as well as perimeter:

$$K_{ii} = 1 \tag{4.92}$$

(b) For the grids with no or few ground rods with none existing in corners or perimeter:

$$K_{ii} = \frac{1}{(2n)^{\frac{2}{n}}} \tag{4.93}$$

and

$$K_h = \sqrt{1 + \frac{h}{h_0}} \tag{4.94}$$

where h_0 = grid reference depth = 1 m.

The effective number of parallel conductors (n) given in a given grid are found from

$$n = n_a \times n_b \times n_c \times n_d \tag{4.95}$$

where

$$n_a = \frac{2L_c}{L_p}$$

$n_b = 1$, for square grids
$n_c = 1$, for square and rectangular grids
$n_d = 1$, for square, rectangular, and L-shaped grids

Otherwise, the following equations are used to determine the n_b, n_c, and n_d so that

$$n_b = \sqrt{\frac{L_p}{4\sqrt{A}}} \tag{4.96}$$

$$n_c = \left[\frac{L_X \times L_Y}{A}\right]^{\frac{0.7A}{L_X \times L_Y}} \tag{4.97}$$

$$n_d = \frac{D_m}{\sqrt{L_X^2 + L_Y^2}} \tag{4.98}$$

where
L_p = peripheral length of the grid, m
L_C = total length of the conductor in the horizontal grid, m
A = area of the grid, m²
L_X = maximum length of the grid in the X direction, m
L_Y = maximum length of the grid in the Y direction, m
d = diameter of grid conductors, m
D = spacing between parallel conductors, m
D_m = maximum distance between any two points on the grid, m
h = depth of ground grid conductors, m

Note that, the irregularity factor is determined from

$$K_{ii} = 0.644 + 0.148n \tag{4.99}$$

The effective buried length (L_M) for grids:

(a) With little or no ground rods but none located in the corners or along the perimeter of the grid:

$$L_M = L_C + L_R \tag{4.100}$$

where
L_R = total length of all ground rods, m
L_C = total length of the conductor in the horizontal grid, m

(b) With ground rods in corners and along the perimeter and throughout the grid:

$$L_M = L_C + \left[1.55 + 1.22\left(\frac{L_R}{\sqrt{L_X^2 + L_Y^2}}\right)\right] L_R \qquad (4.101)$$

where L_R = length of each ground rod, m.

4.12 STEP VOLTAGE DESIGN CALCULATIONS

According to IEEE STD. 80-2000, in order for the ground system to be safe, step voltage has to be less than the tolerable step voltage. Furthermore, step voltages within the grid system designed for safe mesh voltages will be well within the tolerable limits, the reason for this is both feet and legs are in series rather than in parallel and the current takes the path from one leg to the other rather than through vital organs. The step voltage is determined from

$$E_{step} = \frac{\rho \times K_s \times K_i \times I_G}{L_S} \qquad (4.102)$$

where
K_s = step coefficient
L_S = buried conductor length, m

Again, for grids with or without ground rods,

$$L_S = 0.75 L_C + 0.85 L_R \qquad (4.103)$$

so that the step coefficient can be found from

$$K_S = \frac{1}{\pi}\left[\frac{1}{2h} + \frac{2}{D+h} + \frac{1}{D}(1 - 0.5^{n-2})\right] \qquad (4.104)$$

where h = depth of ground grid conductors in meters, usually between 0.25 m < h < 2.5 m.

4.13 TYPES OF GROUND FAULTS

In general, it is difficult to determine which fault type and location will result in the greatest flow of current between the ground grid and the surrounding earth because no simple rule applies. IEEE Std. 80-2000 recommends not to consider multiple simultaneous faults since their probability of occurrence is negligibly small. Instead, it recommends investigating single-line-to-ground and line-to-line-to-ground faults.

4.13.1 Line-to-Line-to-Ground Fault

For a line-to-line-to-ground (i.e., double line-to-ground) fault, IEEE Std. 80-2000 gives the following equation to calculate the zero-sequence fault current:

$$I_{a0} = \frac{E(R_2 + jX_2)}{(R_1 + jX_1)[R_0 + R_2 + 3R_f + j(X_0 + X_2)] + (R_2 + jX_2)(R_0 + 3R_f + jX_0)} \qquad (4.105)$$

where

I_{a0} = symmetrical rms value of zero-sequence fault current, A
E = phase-to-neutral voltage, V
R_f = estimated resistance of the fault, Ω (normally it is assumed $R_f = 0$)
R_1 = positive-sequence system resistance, Ω
R_2 = negative-sequence system resistance, Ω
R_0 = zero-sequence system resistance, Ω
X_1 = positive-sequence system reactance (subtransient), Ω
X_2 = negative-sequence system reactance, Ω
X_0 = zero-sequence system reactance, Ω

The values of R_0, R_1, R_2, and X_0, X_1, X_2 are determined by looking into the system from the point of fault. In other words, they are determined from the Thévenin equivalent impedance at the fault point for each sequence.* Often, however, the resistance quantities given in the above equation is negligibly small. Hence,

$$I_{a0} = \frac{E \times X_2}{X_1(X_0 + X_2)(X_0 + X_2)} \qquad (4.106)$$

4.13.2 SINGLE-LINE-TO-GROUND FAULT

For a single-line-to-ground fault, IEEE Std. 80-2000 gives the following equation to calculate the zero-sequence fault current,

$$I_{a0} = \frac{E}{3R_f + R_0 + R_1 + R_2 + j(X_0 + X_1 + X_2)} \qquad (4.107)$$

Often, however, the resistance quantities in the above equation are negligibly small. Hence,

$$I_{a0} = \frac{E}{X_0 + X_1 + X_2} \qquad (4.108)$$

4.14 GROUND POTENTIAL RISE

As said before in Section 4.8.2, the GPR is a function of fault-current magnitude, system voltage, and ground-system resistance. The GPR with respect to remote ground is determined by multiplying the current flowing through the ground system by its resistance measured from a point remote from the substation. Here, the current flowing through the grid is usually taken as the maximum available line-to-ground fault current.

The GPR is a function of fault current magnitude, system voltage, and ground (system) resistance. The current through the ground system multiplied by its resistance measured from a point remote from the substation determines the GPR with respect to remote ground. Hence, GPR can be found from

$$V_{GPR} = I_G \times R_g \qquad (4.109)$$

---

* It is often acceptable to use $X_1 = X_2$, especially if an appreciable percentage of the positive-sequence reactance to the point of fault is that of static equipment and transmission lines.

where
  V_{GPR} = ground potential rise, V
  R_g = ground grid resistance, Ω

For example, if a ground fault current of 20,000 A is flowing into a substation ground grid due to a line-to-ground fault and the ground grid system has a 0.5 Ω resistance to the earth, the resultant IR voltage drop would be 10,000 V. It is clear that such 10,000-V IR voltage drop could cause serious problems to communication lines in and around the substation in the event that the communication equipment and facilities are not properly insulated and/or neutralized.

The ground grid resistance can be found from

$$R_g = \rho \left[\frac{1}{L_T} + \frac{1}{\sqrt{20A}} \left(1 + \frac{1}{1 + h\sqrt{\frac{20}{A}}} \right) \right] \quad (4.110)$$

where
  L_T = total buried length of conductors, m
  h = depth of the grid, m
  A = area of substation ground surface, m²

To aid the substation grounding design engineer, IEEE Std. 80-2000 includes a design procedure that has a 12-step process, as shown in Figure 4.30, in terms of substation grounding design procedure block diagram, based on a preliminary of a somewhat arbitrary area; that is, the standard suggests the grid be approximately the size of the distribution substation. However, some references state a common practice that is to extend the grid 3 m beyond the perimeter of the substation fence.

EXAMPLE 4.8

Let the starting grid be a 10-by-10 ground grid. Design a proper substation grounding to provide safety measures for anyone going near or working on a substation. Hence, use the IEEE 12-step process shown in Figure 4.31, then build a grid large enough to dissipate the ground fault current into the earth. A large grounding grid extending far beyond the substation fence and made of a single copper plate would have the most desirable effect for dispersing fault currents to remote earth and thereby ensure the safety of personnel at the surface. Unfortunately, a copper plate of such size is not an economically viable option. The alternative is to design a grid by using a series of horizontal conductors and vertical ground rods. Of course, the application of conductors and rods depends on the resistivity of the substation ground. Change the variables as necessary in order to meet specifications for grounding of the substation. The variables include the size of the grid, the size of the conductors used, the amount of conductors used, and the spacing of each grounding rod. Use 12,906 A as the maximum value fault current, a maximum clearing time of 0.5 s, and a conductor diameter of 211.6 kcmil, based on the given information. The soil resistivity is 35 Ω·m and the crushed rock resistivity on the surface of the substation is 2000 Ω·m. Assume that the substation has no transmission line shield wires and but four distribution neutrals. Design a grid by using a series of horizontal conductors and vertical ground rods based on the resistivity of the soil in ohm meters.

Solution

STEP 1: FIELD DATA

Assume that the uniform average soil resistivity of the substation ground is measured to be 35 Ω·m.

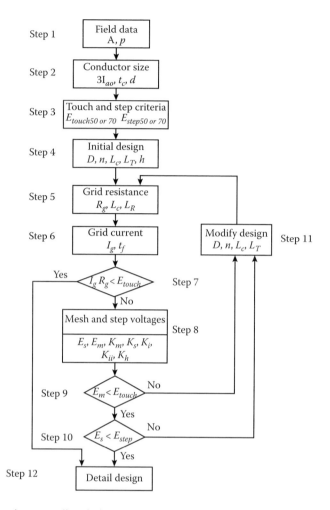

FIGURE 4.31 Substation grounding design procedure block diagram.

STEP 2: CONDUCTOR SIZE

The analysis of the grounding grid should be based on the most conservative fault safety conditions. For example, the fault current $3I_{a0}$ is the assumed maximum value with all current dispersed through the grid, that is, there is no alternative path for ground other than through the grid to remote earth. Since the maximum value of the fault current is given as 12,906 A, the conductor size is selected based on the current-carrying capacity, in addition to the amount of time the fault is going to take place. Thus, use a maximum fault current of 12,906 A, a maximum clearing time of 0.5 s, and a conductor diameter of 211.6 kcmil. The crushed rock resistivity is 2000 Ω·m. The surface derating factor is 0.714. The diameter of the conductor can be found from Table A1.

STEP 3: TOUCH AND STEP VOLTAGE CRITERIA

According to the federal law, all known hazards must be eliminated where the GPR takes place for the safety of workers at a work site. To remove the hazards associated with GPR, a grounding grid is designed to reduce the hazardous potentials at the surface. First, it is necessary to determine what is not hazardous to the body. For two body types, the potential safe step and touch voltages a human could withstand before the fault is cleared need to be determined from Equations 4.62a and 4.62b for touch voltages, and from Equations 4.62c and 4.62d for step voltages as

$$V_{\text{touch } 50} = 516 \text{ V and } V_{\text{touch } 70} = 698 \text{ V}$$

and

$$V_{\text{step 50}} = 1517 \text{ V and } V_{\text{step 70}} = 2126 \text{ V}$$

STEP 4: INITIAL DESIGN

The initial design consists of factors obtained from the general knowledge of the substation. The preliminary size of the grounding grid system is largely based on the size of the substation to include all dimensions within the perimeter of the fence. To establish economic viability, the maximum area is considered and formed in the shape of a square with an area of 100 m². However, the touch voltage has exceeded the limit.

Therefore, an alternative grid size is developed as shown in Figure 4.32, this time in the shape of a rectangle with ground rods. The horizontal distance is called L_X and is measured 24 m, while the vertical distance L_Y is measured 12 m. The area of the grid is 288 m². The horizontal conductors were conservatively spaced at the minimum distance of 3 m apart. The total length of the horizontal conductors L_T is 258 m. A total of 12 grounding rods at a length of 2.5 m each accounted for a total ground rod length L_C of 228 m. The grid burial depth is 0.5 m. From Equation 4.95, the effective number of parallel conductors (n) is found as 6.5.

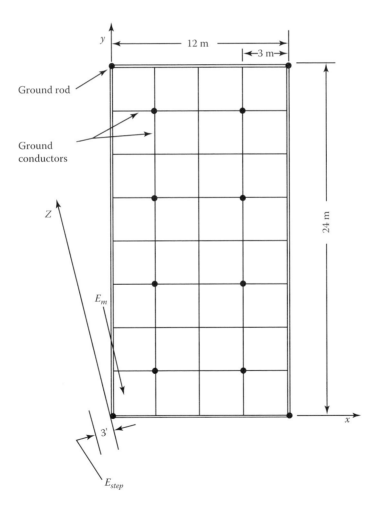

FIGURE 4.32 Preliminary design.

STEP 5: GRID RESISTANCE

A good grounding system provides a low resistance to remote earth in order to minimize the GPR. The next step is to evaluate the grid resistance by using Equation 4.103. All design parameters can be found in Tables 4.12A and 4.12B, and in the preliminary design outline is shown in Figure 4.32. Table 4.13 gives approximate equivalent impedance of transmission line overhead shield wires and distribution feeder neutrals, according to their numbers. From Equation 4.110 for $L_T = 258$ m, a grid area of $A = 288$ m², $\rho = 35$ Ω·m, and $h = 0.5$ m, the grid resistance is

$$R_g = \rho \left[\frac{1}{L_T} + \frac{1}{\sqrt{20A}} \left(1 + \frac{1}{1 + h\sqrt{\frac{20}{A}}} \right) \right]$$

$$= 35 \left[\frac{1}{258} + \frac{1}{\sqrt{20 \times 288}} \left(1 + \frac{1}{1 + 0.5\sqrt{\frac{20}{288}}} \right) \right]$$

$$= 1.0043 \; \Omega$$

STEP 6: GRID CURRENT

From Equation 4.109, the GPR is determined as

$$V_{GPR} = I_G \times R_g$$

This is important because to determine the GPR and compare it to the tolerable touch voltage is the first step to find out whether the grid design is a safe design for the people in and around the substation. The next step is to find the the grid current I_G. However, the split factor should first be determined from the following equation

$$S_f = \left| \frac{Z_{eq}}{Z_{eq} + R_g} \right|$$

Since the substation has no impedance line shield wires and four distribution neutrals, from Table 4.12, the equivalent impedance can be found as $Z_{eq} = 0.322 + j0.242$ Ω. Thus, $R_g = 1.0043$ Ω and a total fault current of $3I_{a0} = 12{,}906$ A, a decrement factor of $D_f = 1.026$. Thus, the current division factor (or the split factor) can be found as

TABLE 4.12A
Initial Design Parameters

ρ	A	L_T	L_C	L_R	L_T	h	L_X	L_Y	D
35 Ω	288 m	258 Ω	228 Ω	30 m	2.5 Ω	0.5 m	24 m	12 m	3 m

TABLE 4.12B
Initial Design Parameters

t_c	h_s	D	$3I_{a0}$	ρ_s	D_f	L_p	n_c	n_d	t_f
0.05 s	0.11 m	0.01 m	12,906 A	2000 Ω	1.03	75 m	1	1	0.5 s

TABLE 4.13
Approximate Equivalent Impedance of Transmission Line Overhead Shield Wires and Distribution Feeder Neutrals

No. of Transmission Lines	No. of Distribution Neutrals	R_{tg} = 15 and $R_{d\phi}$ = 25 $R + jX$ Ω	R_{tg} = 15 and $R_{d\phi}$ = 25 $R + jX$ Ω
1	1	0.91 + j0.485 Ω	3.27 + j0.652 Ω
1	2	0.54 + j0.33 Ω	2.18 + j0.412 Ω
1	4	0.295 + j0.20 Ω	1.32 + j0.244 Ω
4	4	0.23 + j0.12 Ω	0.817 + j0.16 Ω
0	4	0.322 + j0.242 Ω	1.65 + j0.291 Ω

$$S_f = \left| \frac{Z_{eq}}{Z_{eq} + R_g} \right|$$

$$= \left| \frac{(0.322 + j0.242)}{(0.322 + j0.242) + 1.0043} \right|$$

$$\cong 0.2548$$

since

$$I_g = S_f \times 3I_{a0}$$
$$= 0.2548 \times 12{,}906$$
$$= 3288 \text{ A}$$

thus,

$$I_G = D_f \times I_g$$
$$= 1.026 \times 3288$$
$$= 3375 \text{ A}$$

STEP 7: DETERMINATION OF GPR

As said before, the product of I_G and R_g is the GPR. It is necessary to compare the GPR to the tolerable touch voltage, $V_{\text{touch 70}}$. If the GPR is larger than the $V_{\text{touch 70}}$, further design evaluations are necessary and the tolerable touch and step voltages should be compared with the maximum mesh and step voltages. Hence, first determine the GPR as

$$\text{GPR} = I_G \times R_g$$
$$= 3375 \times 1.0043$$
$$= 3390 \text{ V}$$

Check to see whether

$$\text{GPR} > V_{\text{touch70}}$$

Indeed,

$$3390\ V > 698\ V$$

As it can be observed from the results, the GPR is much larger than the step voltage. Therefore, further design considerations are necessary and thus the step and mesh voltages must be calculated and compared with the tolerable touch and step voltage as follows.

STEP 8: MESH AND STEP VOLTAGE CALCULATIONS

(a) **Determination of the Mesh Voltage**

To calculate the mesh equation by using Equation 4.90, it is necessary first to calculate the variables K_m and K_i. Here, K_m can be determined from Equation 4.91. However, again letting $D = 3$ m, $h = 0.5$ m, $d = 0.01$ m, find the following equations as

$$n = n_a \times n_b \times n_c \times n_d$$

where

$$n_a = \frac{2 \times L_C}{L_P}$$

$$= \frac{2 \times 228}{72}$$

$$= 6.33$$

and

$$n_b = \sqrt{\frac{L_P}{4 \times \sqrt{A}}}$$

$$= \sqrt{\frac{72}{4\sqrt{288}}}$$

$$= 1.03$$

$n_c = 1$, for rectangular grid
$n_d = 1$, for rectangular grid
Thus,

$$n = 6.33 \times 1.03 \times 1 \times 1 = 6.52$$

Since
$K_{ii} = 1$, for rectangular and square grids

$$K_h = \sqrt{1 + \frac{h}{h_0}}$$

$$= \sqrt{1 + \frac{0.5}{1}}$$

$$= 1.22$$

So that from Equation 4.91,

$$K_M = \frac{1}{2\pi}\left[\ln\left\{\frac{D^2}{16h \times d} + \frac{(D+2h)2}{8D \times d} - \frac{h}{4d}\right\} + \frac{K_{ii}}{K_h} \times \ln\left\{\frac{8}{\pi(2n-1)}\right\}\right]$$

$$= \frac{1}{2\pi}\left[\ln\left\{\frac{3^2}{16 \times 0.5 \times 0.01} + \frac{(3+2\times 0.5)2}{8 \times 3 \times 0.01} - \frac{0.5}{4 \times 0.01}\right\} + \frac{1}{1.22} \times \ln\left\{\frac{8}{\pi(26.52-1)}\right\}\right]$$

$$= 0.61$$

Also, since

$$K_i = 0.644 + 0.148 \times n$$
$$= 0.644 + 0.148 \times 6.52$$
$$= 1.61$$

Thus, the mesh voltage is determined from Equation 4.90 as

$$E_{mesh} = \frac{\rho \times K_m \times K_i \times I_G}{L_C + \left[1.55 + 1.22\left(\frac{L_R}{\sqrt{L_X^2 + L_Y^2}}\right)\right]L_R}$$

$$= \frac{35 \times 0.61 \times 1.61 \times 3375}{288 + \left[1.55 + 1.22\left(\frac{12}{\sqrt{24^2+12^2}}\right)\right]12}$$

$$= 470\,V$$

(b) Determination of the Step Voltage

For the ground to be safe, the step voltage has to be less than the tolerable step voltage. Also, step voltages within a grid system designed for safe mesh voltages will be well within the tolerable limits. This is because both feet and legs are in series rather than in parallel, and the current takes the path from one leg to the other rather than through vital organs. The step voltages are calculated from Equation 4.102

$$E_{step} = \frac{\rho \times K_s \times K_i \times I_G}{L_S}$$

where
K_s = step coefficient
L_S = buried conductor length, m

For grids with or without ground rods, L_S is determined from Equation 4.103 as

$$L_S = 0.75 L_C + 0.85 L_R$$
$$= 0.75 \times 228 + 0.85 \times 30$$
$$= 196.5\,m$$

The step coefficient is found from Equation 4.104

$$K_s = \frac{1}{\pi}\left[\frac{1}{2h} + \frac{2}{D+h} + \frac{1}{D}(1-0.5^{n-2})\right]$$

$$= \frac{1}{\pi}\left[\frac{1}{2\times 0.5} + \frac{2}{3+0.5} + \frac{1}{3}(1-0.5^{6.52-2})\right]$$

$$= 0.511\,\text{V}$$

where h = depth of ground grid conductors in meters, usually between 0.25 m < h < 2.5 m. All other variables are as defined before.

Thus, the step voltage can be found as

$$E_{step} = \frac{\rho \times K_s \times K_i \times I_G}{L_S}$$

$$= \frac{35 \times 0.511 \times 1.61 \times 3375}{181.2}$$

$$= 536.11\,\text{V}$$

STEP 9: COMPARISON OF E_{MESH} VS. V_{TOUCH}

Here, the mesh voltage that is calculated in Step 8 is compared with the tolerable touch voltages calculated in Step 4. If the calculated mesh voltage E_{mesh} is greater than the tolerable $V_{touch\,70}$, further design evaluations are necessary. If the mesh voltage E_{mesh} is smaller than the $V_{touch\,70}$, then it can be moved to the next step and compare E_{step} with $V_{step\,70}$. Accordingly,

$$470\,\text{V} < 700\,\text{V}$$

$$E_{mesh} < V_{touch\,70}$$

Here, the present grid design passes the second critical criteria. Hence, it can be moved to Step 10 to find out whether the final criterion is met.

STEP 10: COMPARISON OF E_{STEP} VS. $V_{STEP\,70}$

At this step, E_{step} is compared with the calculated tolerable step voltage $V_{step\,70}$. If

$$E_{step} > V_{step\,70}$$

A refinement is of the preliminary design is necessary and can be accomplished by decreasing the total grid resistance, closer grid spacing, adding more ground grid rods, if possible, and/or limiting the total fault current. On the other hand, if

$$E_{step} < V_{step\,70}$$

then the designed grounding grid is considerably safe. Since

$$536.11\,\text{V} < 2126\,\text{V}$$

then for the design

$$E_{step} < V_{step\,70}$$

Disturbance of Normal Operating Conditions and Other Problems   233

In summary, according to the calculations, the calculated mesh and step voltages are smaller than the tolerable touch and step voltages; therefore, in a typical shock situation, humans that become part of the circuit during a fault will have only what is considered a safe amount of current through their bodies.

There are many variables that can be changed in order to meet specifications for grounding a substation. Some variables include the size of the grid, the size of the conductors used, the amount of conductors used, and the spacing of each ground rod. After many processes an engineer has to go through, the project would then be put into construction if it is approved. Designing safe substation grounding is obviously not an easy task; however, there are certain procedures that an engineer can follow to make the designing substation grounding easier.

4.15 TRANSMISSION LINE GROUNDS

High-voltage transmission lines are designed and built to withstand the effects of lightning with minimum damage and interruption of operation. If the lightning strikes an overhead ground wire (also called *static wire*) on a transmission line, the lightning current is conducted to ground through the ground wire installed along the pole or through the metal tower. The top of the line structure is raised in potential to a value determined by the magnitude of the lightning current and the surge impedance of the ground connection.

In the event that the impulse resistance of the ground connection is large, this potential can be in the magnitude of thousands of volts. If the potential is greater than the insulation level of the apparatus, a flashover will take place, causing an arc. The arc, in turn, will start the operation of protective relays, causing the line to be taken out of service. In the event that the transmission structure is well grounded and there is a sufficient coordination between the conductor insulation and the ground resistance, a flashover can generally be avoided.

The transmission line grounds can be designed in various ways to achieve a low ground resistance. For example, a pole butt grounding plate or butt coil can be used on wood poles. A butt coil is a spiral coil of bare copper wire installed at the bottom of a pole. The wire of the coil is extended up the pole as the ground wire lead. In practice, usually one or more ground rods are used instead to achieve the required low ground resistance.

The sizes of the rods used are usually 5/8 or 3/4 in. in diameter and 10 ft in length. The thickness of the rod does not play a major role in reducing the ground resistance as does the length of the rod. Multiple rods are usually used to provide the low ground resistance require by the high-capacity structures. However, if the rods are moderately close to each other, the overall resistance will be more than if the same number of rods were spaced far apart. In other words, adding a second rod does not provide a total resistance of half that of a single rod unless the two are several rod lengths apart (actually infinite distance). Lewis [30] has shown that at 2 ft apart, the resistance of two pipes (used as ground rods) in parallel is about 61% of the resistance of one of them, and at 6 ft apart it is about 55% of the resistance of one pipe.

Where there is bedrock near the surface or where sand is encountered, the soil is usually very dry and therefore has high resistivity. Such situations may require a grounding system known as the counterpoise, made of buried metal (usually galvanized steel wire) strips, wires, or cables. The counterpoise for an overhead transmission line consists of a special grounding terminal that reduces the surge impedance of the ground connection and increases the coupling between the ground wire and the conductors.

The basic types of counterpoises used for transmission lines located in areas with sandy soil or rock close to the surface are the continuous type (also called the *parallel type*) and the radial type (also called the *crowfoot type*), as shown in Figure 4.33. The continuous counterpoise is made of one or more conductors buried under the transmission line for its entire length.

The counterpoise wires are connected to the overhead ground (or *static*) wire at all towers or poles. However, the radial-type counterpoise is made of a number of wires, and extends radially (in some fashion) from the tower legs. The number and length of the wires are determined by the tower

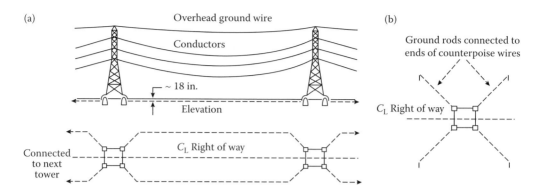

FIGURE 4.33 Two basic types of counterpoises: (a) continuous (parallel), (b) radial. (From Gönen, T., *Electric Power Transmission System Engineering*, 2nd ed., CRC Press, Boca Raton, FL, 2009.)

location and the soil conditions. The counterpoise wires are usually installed with a cable plow at a length of 18 in. or more so that they will not be disturbed by cultivation of the land.

A multigrounded, common neutral conductor used for a primary distribution line is always connected to the substation grounding system where the circuit originates and to all grounds along the length of the circuit. If separate primary and secondary neutral conductors are used, the conductors have to be connected together provided that the primary neutral conductor is effectively grounded. The resistance of a single buried horizontal wire, when it is used as radial counterpoise, can be expressed [16] as

$$R = \frac{\rho}{\pi \ell} \left(\ln \frac{2\ell}{2(ad)^{1/2}} - 1 \right) \quad \text{when } d \ll \ell \qquad (4.111)$$

where
  ρ = ground resistivity in ohm meters
  ℓ = length of wire in meters
  a = radius of wire in meters
  d = burial depth in meters

It is assumed that the potential is uniform over the entire length of the wire. This is only true when the wire has ideal conductivity. If the wire is very long, such as with the radial counterpoise, the potential is not uniform over the entire length of the wire. Hence, Equation 4.83 cannot be used. Instead, the resistance of such a continuous counterpoise when $\ell(r/\rho)^{1/2}$ is large can be expressed as

$$R = (r\rho)^{1/2} \coth\left[\ell\left(\frac{r}{\rho}\right)^{1/2}\right] \qquad (4.112)$$

where r = resistance of wire in ohm meters. If the lightning current flows through a counterpoise, the effective resistance is equal to the surge impedance of the wire. The wire resistance decreases as the surge propagates along the wire. For a given length counterpoise, the transient resistance will diminish to the steady-state resistance if the same wire is used in several shorter radial counterpoises rather than as a continuous counterpoise. Thus, the first 250 ft of counterpoise is most effective when it comes to grounding of lightning currents.

4.16 TYPES OF GROUNDING

In general, transmission and subtransmission systems are solidly grounded. Transmission systems are usually connected grounded wye, but subtransmission systems are often connected in delta. Delta systems may also be grounded through grounding transformers. In most high-voltage systems, the neutrals are solidly grounded, that is, connected directly to the ground. The advantages of such grounding are

1. Voltages to ground are limited to the phase voltage
2. Intermittent ground faults and high voltages due to arcing faults are eliminated
3. Sensitive protective relays operated by ground fault currents clear these faults at an early stage

The grounding transformers used are normally either small distribution transformers (that are connected normally in wye–delta, having their secondaries in delta), or small grounding autotransformers with interconnected wye or "zig-zag" windings, as shown in Figure 4.34. The three-phase autotransformer has a single winding. If there is a ground fault on any line, the ground current flows equally in the three legs of the autotransformer. The interconnection offers the minimum impedance to the flow of the single-phase fault current.

The transformers are only used for grounding and carry little current except during a ground fault. Because of that, they can be fairly small. Their ratings are based on the stipulation that they carry current for no more than 5 min since the relays normally operate long before that. The grounding transformers are connected to the substation ground.

All substation equipment and structures are connected to the ground grid with large conductors to minimize the grounding resistance and limit the potential between equipment and the ground surface to a safe value under all conditions. As shown in Figure 4.34, all substation fences are built inside the ground grid and attached to the grid at short intervals to protect the public and personnel. Furthermore, the surface of the substation is usually covered with crushed rock or concrete to reduce the potential gradient when large currents are discharged to ground and to increase the contact resistance to the feet of the personnel in the substation.

As said before, the substation grounding system is connected to every individual equipment, structure, and installation in order to provide the means by which grounding currents are conducted to remote areas. Thus, it is extremely important that the substation ground has a low ground resistance, adequate current-carrying capacity, and safety features for the personnel.

It is crucial to have the substation ground resistance very low so that the total rise of the grounding system potential will not reach values that are unsafe for human contact. Therefore, the substation

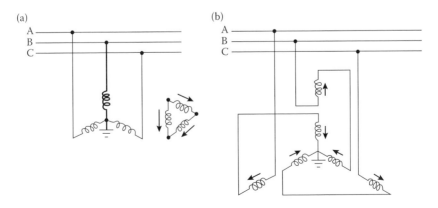

FIGURE 4.34 Grounding transformers used in delta-connected systems: (a) using wye–delta-connected small distribution transformers, or (b) using grounding autotransformers with interconnected wye or "zig-zag" windings. (From Gönen, T., *Electric Power Transmission System Engineering*, 2nd ed., CRC Press, Boca Raton, FL, 2009.)

grounding system normally is made up of buried horizontal conductors and driven ground rods interconnected (by clamping, welding, or brazing) to form a continuous grid (also called mat) network, as shown in Figure 4.35.

Notice that, a continuous cable (usually it is 4/0 bare stranded copper cable buried 12–18 in. below the surface) surrounds the grid perimeter to enclose as much ground as possible and to prevent current concentration and thus high gradients at the ground cable terminals. Inside the grid, cables are buried in parallel lines and with uniform spacing (e.g., about 10 × 20 ft).

All substation equipment and structures are connected to the ground grid with large conductors to minimize the grounding resistance and limit the potential between equipment and the ground surface to a safe value under all conditions. As shown in Figure 4.34, all substation fences are built inside the ground grid and attached to the grid at short intervals to protect the public and personnel.

Furthermore, the surface of the substation is usually covered with crushed rock or concrete to reduce the potential gradient when large currents are discharged to the ground and to increase the contact resistance to the feet of the personnel in the substation.

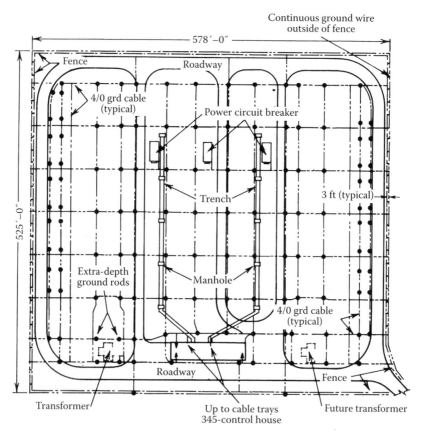

FIGURE 4.35 Typical grounding (grid) system design for 345-kV substation. (From Fink, D. G., and Beaty, H. W., eds., *Standard Handbook for Electrical Engineers*, 11th ed. Used with permission. © 1978 McGraw-Hill.)

Today, many utilities have computer programs for performing grounding grid studies. Thus, the number of tedious calculations that must be performed to develop an accurate and sophisticated model of a system is no longer a problem. For example, Figure 4.36a shows a typical computerized grounding grid design with all relevant soil and system data. Figure 4.36b shows the meshes with hazardous potentials as determined by the computer. Figure 4.36c shows the results of the first refinement in the grid design indicating the hazardous touch potentials. Finally, Figure 4.36d shows the final refinement with no hazardous touch potentials.

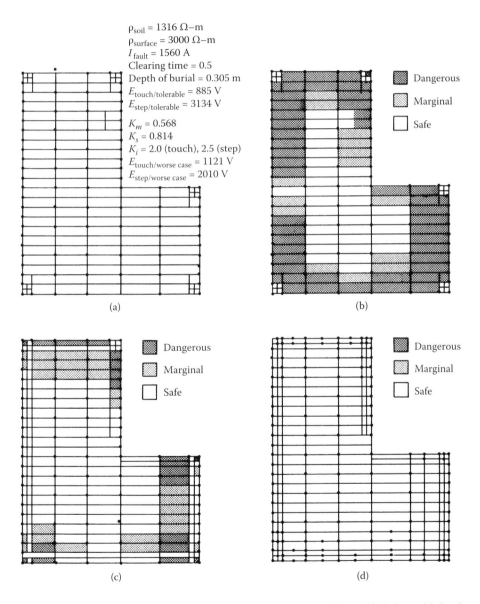

FIGURE 4.36 Computerized grounding grid design: (a) typical grounding grid design with its data; (b) meshes with hazardous potentials as identified by computer; (c) first refinement of design; (d) final refinement of design with no hazardous touch potentials. (From Institute of Electrical and Electronics Engineers, *IEEE Recommended Practice for Industrial and Commercial Power System Analysis*, IEEE Stand. 399-1980 © 1980 IEEE.)

REFERENCES

1. Warrington, A. R. van C., *Protective Relays: Their Theory and Practice*, vol. 2. Chapman & Hall, London, 1969.
2. Electric Power Research Institute, *Transmission Line Reference Book*. EPRI, Palo Alto, CA, 1978.
3. Strom, A. P., Long 60 cycle arcs in air. *Trans. Am. Inst. Electr. Eng.* 65, 113–117 (1946) (discussion, pp. 504–507).
4. Gilkeson, C. L., Jeanne, P. A., and Vaage, E. F., Power system faults to ground. Part II. Fault resistance. *Trans. Am. Inst. Electr. Eng.* 56, 428–433, 474 (1937).
5. Warrington, A. R. van C., Reactance relays negligibly affected by arc impedance. *Electr. World* 98 (12), 502–505 (1931).
6. Aderson, P. M., *Analysis of Faulted Power Systems*, Wiley-Interscience, New York, 1995.
7. Westinghouse Electric Corporation, *Electrical Transmission and Distribution Reference Book*. WEC, Pittsburgh, 1964.
8. American National Standards Institute, *Schedules of Preferred Ratings and Related Required Capabilities for AC High-Voltage Circuit Breakers Rated on a Symmetrical Current Basis*, ANSI C37.06-1971. ANSI, New York, 1971.
9. Gönen, T., *A Practical Guide for Calculation of Short-Circuit Currents and Selection of High Voltage Circuit Breakers*. Black & Veatch Co., Overland Park, KS, 1977.
10. Farr, H. H., *Transmission Line Design Manual*. U.S. Department of the Interior, Water and Power Resources Service, Denver, 1980.
11. Edison Electric Institute, *EHV Transmission Line Reference Book*. EEI, New York, 1968.
12. IEEE Standard. 2000, *IEEE Guide for Safety in AC Substation Grounding*. IEEE Std. 80-2000.
13. Peek, F. W., Jr., Electrical characteristics of the suspension insulator. Part I. *Trans. Am. Inst. Electr. Eng.* 31, 907–930 (1912).
14. Peek, F. W., Jr., Electric characteristics of the suspension insulator. Part II. *Trans. Am. Inst. Electr. Eng.* 39, 1685–1705 (1920).
15. Ferris, L. P. et al., Effects of electrical shock on the heart. *Trans. Am. Inst. Electr. Eng.* 55, 498–515, 1263 (1936).
16. Dalziel, C. F., Electrical shock hazard. *IEEE Spectrum* 9 (2), 41–50 (1972).
17. Dalziel, C. F., and Lee, W. R., Lethal electrical currents. *IEEE Spectrum* 6, 44–50 (1969).
18. IEEE Working Group Report, Electrostatic effects of overhead transmission lines. Part I. Hazards and effects. *IEEE Trans. Power Appar. Syst.* PAS-9 (2), 422–426 (1972).
19. Sverak, J. G. et al., Safe substation grounding. Part I. *IEEE Trans. Power Appar. Syst.* PAS-100 (9), 4281–4290 (1981).
20. IEEE Working Group Report, Electrostatic effects of overhead transmission lines. Part II. Methods of calculation. *IEEE Trans. Power Appar. Syst.* PAS-9t (2), 426–444 (1972).
21. Lee, W. R., Death from electrical shock. *Proc. Inst. Electr. Eng.* 113 (1), 144–148 (1966).
22. Gönen, T., and Bckiroglu, H., Electrical safety in industrial plants: Some considerations. *Hazard Prev.* 13 (5), 4–7 (1977).
23. Institute of Electrical and Electronics Engineers, *IEEE Recommended Practice for Industrial and Commercial Power System Analysis*, IEEE Stand. 399-1980. IEEE, New York, 1980.
24. Institute for Electrical and Electronics Engineers, *IEEE Guide for Safety in AC Substation Grounding*, IEEE Std. 80-1976. IEEE, New York, 1976.
25. Dwight, H. B., Calculation of resistances to ground. *Electr. Eng.* (*Am. Inst. Electr. Eng.*) 55, 1319–1328 (1936).
25a. National Bureau of Standards Technical Report 108.
26. Dawalibi, F., and Mukhedkar, D., Optimum design of substation grounding in two-layer earth structure. Part I. Analytical study. *IEEE Trans. Power Appar. Syst.* PAS-94 (2), 252–261 (1975).
27. Dawalibi, F., and Mukhedkar, D., Optimum design of substation grounding in two-layer earth structure. Part II. Comparison between theoretical and experimental results. *IEEE Trans. Power Appar. Syst.* PAS-94 (2), 262–266 (1975).
28. Dawalibi, F., and Mukhedkar, D., Optimum design of substation grounding in two-layer earth structure. Part III. Study of grounding grids performance and new electrodes configuration. *IEEE Trans. Power Appar. Syst.* PAS-94 (2), 267–272 (1975).
29. Sunde, E. D., *Earth Conduction Effect in Transmission System*. Macmillan, New York, 1968.
30. Lewis. W. W. *The Protection of Transmission Systems against Lighting*. Dover, New York, 1965.

31. Fink, D. G., and Beaty, H. W., eds., *Standard Handbook for Electrical Engineers*, 11th ed. McGraw-Hill, New York, 1978.
32. Electric Power Research Institute, *Transmission Line Reference Book: 345 kV and Above*, 2nd ed. EPRI, Palo Alto, CA, 1982.
33. Gönen, T., *Electric Power Transmission System Engineering*, 2nd ed., CRC Press, Boca Raton, FL, 2009.

GENERAL REFERENCES

Alperöz, N., *Enerji Dagitimi*. Istanbul State Engineering and Architectural Academy Press, Istanbul, Turkey, 1974.

American Institute of Electrical Engineers, *AIEE Lightning Reference Book*. Lightning and Insulator Subcommittee, AIEE, New York, 1937.

Anderson, J. G., Monte Carlo computer calculation of transmission line lightning performance. *Trans. Am. Inst. Electr. Eng.* 80 (Part 3), 414–419 (1961).

Anderson, O. W., Laplacian electrostatic field calculations by finite elements with automatic grid generation. *IEEE Trans. Power Appar. Syst.* PAS-73, 682–689 (1973).

Bellaschi, P. L., Armington, R. E., and Snowden, A. E., Impulse and sixty cycle characteristics of driven grounds. II. *Trans. Am. Inst. Electr. Eng.* 61, 349–363 (1942).

Biegelmeier, U. G., Die Recleatung der Z-Schwelle des Herzkammerflimmerns fur die Festiegun von Beruhrungsspanungs Greuzeu bei den Schutzman Bradhmer Gegen Elektrische Unfallen. *Elektrotech. Maschinenbau* 93 (1), 1–8 (1976).

Biegelmeier, U. G., and Rotter, K., Elektrische Wilderstrande und Strome in Menschlicken Korper. *Elektrotech. Maschinenbau* 89, 104–109 (1971).

Bodier, G., La securite des personnes et la question des mises à la terra dans les postes de distribution. *Bull. Soc. Fr. Electr.* 7 (74), 545–562 (1947).

Elek, A., Hazards of electric shock at stations during fault and methods of reduction. *Ont. Hydro. Res. News* 10 (1), 1–6 (1958).

Gönen, T., *Electric Power Transmission System Engineering*, 2nd ed., CRC Press, Boca Raton, FL, 2009.

Gross, C. A., *Power System Analysis*. Wiley, New York, 1979.

Gross, E. T. B. et al., Grounding grids for high voltage stations. *Trans. Am. Inst. Electr. Eng.* 72, 799-810 (1953).

Heppe, R. J., Step potentials and body currents for near grounds in two-layer earth. *IEEE Trans. Power Appar. Syst.* PAS-98 (1), 45–59 (1979).

IEEE Committee Report, A survey of the problem of insulator contamination in the United States and Canada. I. *IEEE Trans. Power Appar. Syst.* PAS-90, 2577–2585 (1971).

IEEE Committee Report, A survey of the problem of insulator contamination in the United States and Canada. II. *IEEE Trans. Power Appar. Syst.* PAS-91, 1948–1954 (1972).

Institute of Electrical and Electronics Engineers, *IEEE Tutorial Course: Application of Power Circuit Breakers*, Publ. No. 75CH0975-3-PWR. IEEE, New York, 1975.

Institute of Electrical and Electronics Engineers, *IEEE Recommended Practice for Grounding of Industrial and Commercial Power Systems*, IEEE Stand. 142-1982. IEEE, New York, 1982.

Kaminski, J., Jr., Long time mechanical and electrical strength in suspended insulators. *Trans. Am. Inst. Electr. Eng.* 82, 446–452 (1963).

Karady, G., and Lamontagne, G., Electrical and contamination performance of synthetic insulators for 735 kV transmission lines. *IEEE Power Eng. Soc. Summer Meet.* 1976, 502–5 (1976).

Kyser, H., *Elektrikle Energi Nakli* (translated into Turkish by M. Dilege). Istanbul Technical Univ. Press, Istanbul, Turkey, 1952.

Langer, H., Messungen von Erderspannungen in einew 220 kV Umspanwerk. *Elektrotech. Z.* 75 (4), 97–105 (1954).

Looms, J. S. T., and Proctor, F. H., The development of an epoxy-based insulator for UHV. *IEEE Power Eng. Soc. Summer Meet.*, 1976 Pap. No. A76 342-6 (1976).

Maxwell, J. C., *A Treatise on Electricity and Magnetism*. Dover, New York, 1954.

Nasser, E., Zum Problem des Fremdschichtuberschlages an Isolatoren. *F.TZ-A. Elektrotech.* 2 83, 356–365 (1962).

Nasser, E., *Fundamentals of Gaseous Ionization and Plasma Electronics*. Wiley, New York, 1971.

Nasser, E., *An Annotated Bibliography on the Problem of Insulator Contamination of the Electric Energy System*, ERI Proj. Rep. No. 713-S. Engineering Research Institute, Iowa State University, Ames, IA, 1973.

Neuenswander, J. R., *Modern Power Systems*. International Textbook Co., Scranton, Pennsylvania, 1971.

Numajiri, F., Analysis of transmission line lightning voltages by digital computer. *Electr. Eng. Jpn.* 84 (8), 53–63 (1964).

Nunnally, H. N. et al., Computer simulation for determining step and touch potentials resulting from faults or open neutrals in URD cable. *IEEE Trans. Power, Appar. Syst.* PAS-98 (3), 1130–1136 (1979).

Prache, P. M., and James, J., *Uzak Mesafe Yeralti Hatlari Dersteri* (Translated into Turkish by Daarafakioglu). Istanbul Technical Univ. Press. Istanbul. Turkey, 1958.

Qalziel, C. F., A study of the hazards of impulse currents. *Trans. Am. Inst. Electr. Eng.* 72 (Part 2), 1032–1042 (1953)

Roeper, R., *Kurzchlussstrome in Drehstromnetzen*, 5th Ger. ed. (translated as *Short-Circuit Currents in Three-Phase Networks*). Siemens Aktienges, Munich, Germany, 1972.

Rudenberg, R., *Transient Performance of Electric Power Systems*. McGraw-Hill, New York, 1950.

Rudenberg, R., *Electrical Shock Waves in Power Systems*. Harvard Univ. Press, Cambridge, MA, 1968.

Rumeli, A., The mechanism of flashover of polluted insulation. Ph.D. Thesis, University of Strathclyde, Glasgow, Scotland, 1967.

Stevenson, W. D., Jr., *Elements of Power System Analysis*, 4th ed. McGraw-Hill, New York, 1982.

Sverak, J. G., Optimized grounding grid design using variable spacing technique. *IEEE Trans. Power Appar. Syst.* PAS-95 (1), 362–374 (1976).

Udo, T., Minimum phase-to-phase electrical clearances for substations based on switching surges and lightning surges. *IEEE Trans. Power Appar. Syst.* PAS-85, 838–845 (1966).

Warrington, A. R. van C., *Protective Relays: Their Theory and Practice*, vol. 1. Chapman & Hall, London, 1962.

Zaborszky, J., Efficiency of grounding grids with nonuniform soil. *Trans. Am. Inst. Electr. Eng.* 74, 1230–1233 (1955).

PROBLEMS

1. Verify Equation 4.5 by using the
    (a) Classical calculus approach
    (b) Laplace transformation approach
2. Use Equation 4.6 and verify that if a fault occurs at $t = 0$ when the angle $\alpha - \theta = 90°$, then the value of the maximum transient current becomes twice the steady-state maximum value.
3. Repeat Example 4.1 assuming that the fault is located on bus 2.
4. Repeat Example 4.1 assuming that the fault is located on bus 3.
5. Use the results of Problem 3 and repeat Example 4.2.
6. Use the results of Problem 4 and repeat Example 4.2.
7. Consider the system shown in Figure P4.1 and the following data:
    Generator G_1: 15 kV, 50 MVA, $X_1 = X_2 = 0.10$ pu and $X_0 = 0.05$ pu based on its own ratings
    Synchronous motor: 15 kV, 20 MVA, $X_1 = X_2 = 0.20$ pu and $X_0 = 0.07$ pu based on its own ratings
    Transformer T_1: 15/115 kV, 30 MVA, $X_1 = X_2 = X_0 = 0.06$ pu based on its own ratings
    Transformer T_2: *115/15* kV, 25 MVA, $X_1 = X_2 = X_0 = 0.07$ pu based on its own ratings
    Transmission line TL23: $X_1 = X_2 = 0.03$ pu and $X_0 = 0.10$ pu based on its own ratings
    Assume a three-phase fault at bus 1 and determine the fault current. Use 50 MVA as the megavolt-ampere base.
8. Repeat Problem 7 assuming that bus 2 is faulted.

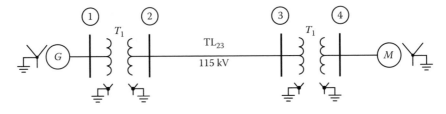

FIGURE P4.1 Two-bus system for Problem 7.

Disturbance of Normal Operating Conditions and Other Problems

9. Repeat Problem 7 assuming that bus 3 is faulted.
10. Repeat Problem 7 assuming that bus 4 is faulted.
11. A generator is connected through an eight-cycle circuit breaker to an unloaded transformer. Its subtransient, transient, and steady-state reactances are given as 8%, 16%, and 100%, respectively. It is operating at no load and rated voltage $1.0\angle 0°$ pu when a three-phase short circuit occurs between the breaker and the transformer. Determine the following:
    (a) Sustained (i.e., steady-state) short-circuit current in breaker in per units
    (b) Initial symmetrical rms current in breaker in per units
    (c) Maximum possible dc component of short circuit in breaker in per units
12. Consider Example 4.5 and assume that the tie–bus arrangement is replaced by an equivalent ring arrangement (as shown in Figure 4.9e), so that the resulting short-circuit megavolt-ampere base is equal to the one found in Example 4.5. Determine the following:
    (a) Necessary per-unit reactance of each reactor
    (b) Fault current distribution
13. Assume that a string of five insulators is used to suspend one line conductor, as shown in Figure P4.2. Consider only the series capacitance C_1 and the shunt capacitance C_2 and define the ratio C_1/C_2 as k. Using the charging currents and Kirchhoff's current and voltage laws, derive an expression to determine the voltage distribution across various units of the suspension insulator in terms of the ratio k. [Do not use Equations 4.48 or 4.50.]
14. Use the results of Problem 13 and assume that the string of five insulators is used to suspend one conductor of a 6–9 kV, three-phase, overhead line. If the ratio of k is 0.1, determine voltages V_1, V_2, V_3, V_4, and V_5.
15. Repeat Problem 13 using the Q charges on the various capacitors. Assume that the potential differences across the capacitors at any instant are as shown in Figure P4.2.

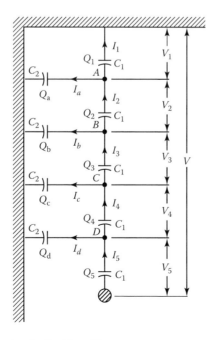

FIGURE P4.2 Five-insulator string for Problem 13.

16. Consider Figure P4.3 and assume that an arcing ring has been installed at the line end of the insulator string.
    (a) Repeat Problem 14
    (b) Determine the sting efficiency
17. Consider Figure P4.4, and in order to achieve uniform voltage distribution over the five insulators, determine the following required values:
    (a) Air capacitance C_a
    (b) Air capacitance C_b
    (c) Air capacitance C_c
    (d) Air capacitance C_d
18. Assume that a post-type insulator consists of three-pin insulators fixed one above another and used to support a bus of one 115 kV three-phase system, as shown in Figure P4.5. Use the results of Problem 13. If the voltage across the top pin insulator is twice that of the voltage across the bottom insulator, determine the voltage across the middle insulator.
19. Assume that a string of eight insulators is used to suspend a line conductor of a 138-kV three-phase system, as shown in Figure 4.5. Assume that $C_2 = 0.1 \times C_1$ and $C_3 = 0.02 \times C_1$ and use Equation 4.48. Determine the following:
    (a) Voltage across three units from the ground end
    (b) Voltage across five units from the ground end
20. Repeat Problem 19 using Equation 4.50.
21. Consider Problem 19 and assume that $C_3 = 0$. Use the results of Problem 13 and repeat Problem 19.
22. Repeat Example 4.6 assuming wet organic soil.
23. Repeat Example 4.6 assuming dry soil.

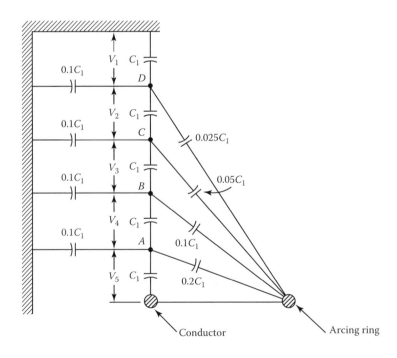

FIGURE P4.3 Five-insulator string with an arcing ring for Problem 16.

24. Consider Example 4.6 and Equations 4.62 and 4.63. Assume that $K_s = 1.4$, $K_m = 7.0$, and $K_i = 2$ and 2.5 for the touch and step potentials, respectively. Determine the required minimum total length of corresponding grid conductors.

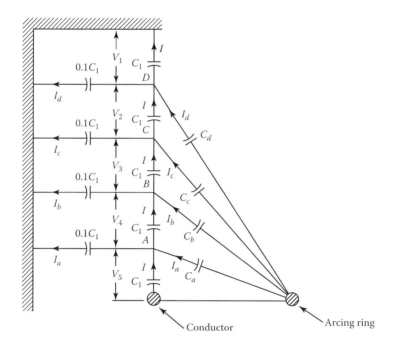

FIGURE P4.4 Five-insulator string for Problem 17.

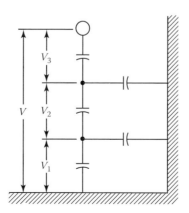

FIGURE P4.5 Post-type insulator for Problem 18.

5 Symmetrical Components and Sequence Impedances

5.1 INTRODUCTION

In general, it can be said that truly balanced three-phase systems exist only in theory. In reality, many systems are very nearly balanced and for practical purposes can be analyzed as if they were truly balanced systems. However, there are also emergency conditions (e.g., unsymmetrical faults, unbalanced loads, open conductors, or unsymmetrical conditions arising in rotating machines) where the degree of unbalance cannot be neglected. To protect the system against such contingencies, it is necessary to size protective devices, such as fuses and circuit breakers, and set the protective relays. Therefore, to achieve this, currents and voltages in the system under such unbalanced operating conditions have to be known (and therefore calculated) in advance.

In 1918, Fortescue [1] proposed a method for resolving an unbalanced set of n related phasors into n sets of balanced phasors called the *symmetrical components* of the original unbalanced set. The phasors of each set are of equal magnitude and spaced 120° or 0° apart. The method is applicable to systems with any number of phases; however, in this book, only three-phase systems will be discussed.

Today, the symmetrical component theory is widely used in studying unbalanced systems. Furthermore, many electrical devices have been developed and are operating based on the concept of symmetrical components. The examples include (1) the negative-sequence relay to detect system faults, (2) the positive-sequence filter to make generator voltage regulators respond to voltage changes in all three phases rather than in one phase alone, and (3) the Westinghouse-type HCB pilot wire relay using positive- and zero-sequence filters to detect faults.

5.2 SYMMETRICAL COMPONENTS

Any unbalanced three-phase system of phasors can be resolved into three balanced systems of phasors: (1) positive-sequence system, (2) negative-sequence system, and (3) zero-sequence system, as illustrated in Figure 5.1.

The *positive-sequence system* is represented by a balanced system of phasors having the same phase sequence (and therefore positive phase rotation) as the original unbalanced system. The phasors of the positive-sequence system are equal in magnitude and displaced from each other by 120°, as shown in Figure 5.1b.

The *negative-sequence system* is represented by a balanced system of phasors having the opposite phase sequence (and therefore negative phase rotation) to the original system. The phasors of the negative-sequence system are also equal in magnitude and displaced from each other by 120°, as shown in Figure 5.1c.

The zero-sequence system is represented by three single phasors that are equal in magnitude and angular displacements, as shown in Figure 5.1d. Note that, in the hook, the subscripts 0, 1, and 2 denote the zero sequence, positive sequence, and negative sequence, respectively. Therefore, three voltage phasors V_a, V_b, and V_c of an unbalanced set, as shown in Figure 5.1a can be expressed in terms of their symmetrical components as

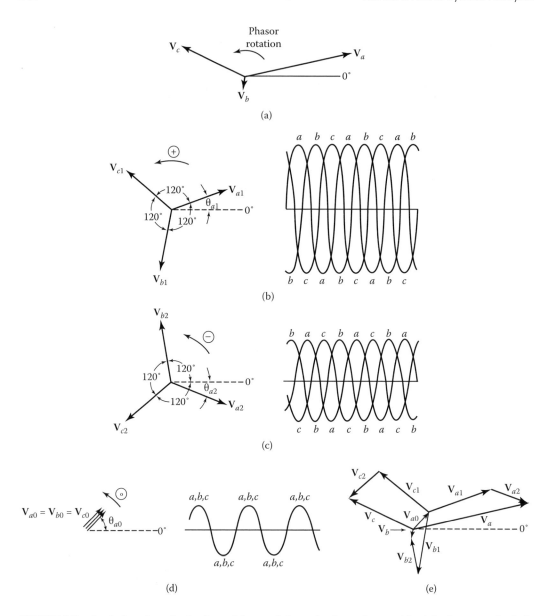

FIGURE 5.1 Analysis and synthesis of set of three unbalanced voltage phasors: (a) original system of unbalanced phasors; (b) positive-sequence components; (c) negative-sequence components; (d) zero-sequence components; (e) graphical addition of phasors to get original unbalanced phasors.

$$\mathbf{V}_a = \mathbf{V}_{a1} + \mathbf{V}_{a2} + \mathbf{V}_{a0} \tag{5.1}$$

$$\mathbf{V}_b = \mathbf{V}_{b1} + \mathbf{V}_{b2} + \mathbf{V}_{b0} \tag{5.2}$$

$$\mathbf{V}_c = \mathbf{V}_{c1} + \mathbf{V}_{c2} + \mathbf{V}_{c0} \tag{5.3}$$

Symmetrical Components and Sequence Impedances

Figure 5.1e shows the graphical additions of the symmetrical components of Figures 5.1b through 5.1d to obtain the original three unbalanced phasors shown in Figure 5.1a.

5.3 OPERATOR a

Because of the application of the symmetrical components theory to three-phase systems, there is a need for a *unit phasor* (or *operator*) that will rotate another phasor by 120° in the counterclockwise direction (i.e., it will add 120° to the phase angle of the phasor) but leave its magnitude unchanged when it is multiplied by the phasor (see Figure 5.2). Such an operator is a complex number of unit magnitude with an angle of 120° and is defined by

$$\begin{aligned}
\mathbf{a} &= 1\angle 120° \\
&= 1e^{j\left(\frac{2\pi}{3}\right)} \\
&= 1(\cos 120° + j\sin 120°) \\
&= -0.5 + j0.866
\end{aligned}$$

where

$$j = \sqrt{-1}$$

It is clear that if operator **a** is designated as

$$\mathbf{a} = 1\angle 120°$$

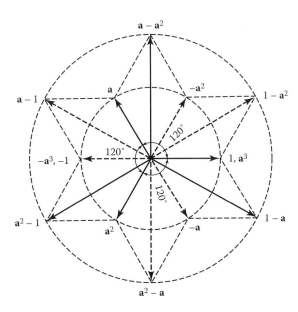

FIGURE 5.2 Phasor diagram of various powers and functions of operator **a**.

then

$$\begin{aligned}\mathbf{a}^2 &= \mathbf{a} \times \mathbf{a} \\ &= (1\angle 120°)(1\angle 120°) = 1\angle 240° = 1\angle{-}120° \\ \mathbf{a}^3 &= \mathbf{a}^2 \times \mathbf{a} \\ &= (1\angle 240°)(1\angle 120°) = 1\angle 360° = 1\angle 0° \\ \mathbf{a}^4 &= \mathbf{a}^3 \times \mathbf{a} \\ &= (1\angle 0°)(1\angle 120°) = 1\angle 120° = \mathbf{a} \\ \mathbf{a}^5 &= \mathbf{a}^3 \times \mathbf{a}^2 \\ &= (1\angle 0°)(1\angle 240°) = 1\angle 240° = \mathbf{a}^2 \\ \mathbf{a}^6 &= \mathbf{a}^3 \times \mathbf{a}^3 \\ &= (1\angle 0°)(1\angle 0°) = 1\angle 0° = \mathbf{a}^3 \\ &\vdots \\ \mathbf{a}^{n+3} &= \mathbf{a}^n \times \mathbf{a}^3 = \mathbf{a}^n\end{aligned}$$

Figure 5.2 shows a phasor diagram of the various powers and functions of operator **a**. Various combinations of operator **a** are given in Table 5.1. In manipulating quantities involving operator **a**, it is useful to remember that

$$1 + \mathbf{a} + \mathbf{a}^2 = 0 \qquad (5.4)$$

5.4 RESOLUTION OF THREE-PHASE UNBALANCED SYSTEM OF PHASORS INTO ITS SYMMETRICAL COMPONENTS

In the application of the symmetrical component, it is customary to let phase **a** be the reference phase. Therefore, using operator **a**, the symmetrical components of the positive-, negative-, and zero-sequence components can be expressed as

TABLE 5.1
Powers and Functions of Operator a

Power or Function	In Polar Form	In Rectangular Form
$\mathbf{a}$	$1\angle 120°$	$-0.5 + j0.866$
$\mathbf{a}^2$	$1\angle 240° = 1\angle{-}120°$	$-0.5 - j0.866$
$\mathbf{a}^3$	$1\angle 360° = 1\angle 0°$	$1.0 + j0.0$
$\mathbf{a}^4$	$1\angle 120°$	$-0.5 + j0.866$
$1 + \mathbf{a} = -\mathbf{a}^2$	$1\angle 60°$	$0.5 + j0.866$
$1 - \mathbf{a}$	$\sqrt{3}\angle{-}30°$	$1.5 - j0.866$
$1 + \mathbf{a}^2 = -\mathbf{a}$	$1\angle{-}60°$	$0.5 - j0.866$
$1 - \mathbf{a}^2$	$\sqrt{3}\angle 30°$	$1.5 + j0.866$
$\mathbf{a} - 1$	$\sqrt{3}\angle 150°$	$-1.5 + j0.866$
$\mathbf{a} + \mathbf{a}^2$	$1\angle 180°$	$-1.0 + j0.0$
$\mathbf{a} - \mathbf{a}^2$	$\sqrt{3}\angle 90°$	$0.0 + j1.732$
$\mathbf{a}^2 - \mathbf{a}$	$\sqrt{3}\angle{-}90°$	$0.0 - j1.732$
$\mathbf{a}^2 - 1$	$\sqrt{3}\angle{-}150°$	$-1.5 - j0.866$
$1 + \mathbf{a} + \mathbf{a}^2$	$0\angle 0°$	$0.0 + j0.0$

Symmetrical Components and Sequence Impedances

$$\mathbf{V}_{b1} = \mathbf{a}^2 \mathbf{V}_{a1} \tag{5.5}$$

$$\mathbf{V}_{c1} = \mathbf{a} \mathbf{V}_{a1} \tag{5.6}$$

$$\mathbf{V}_{b2} = \mathbf{a} \mathbf{V}_{a2} \tag{5.7}$$

$$\mathbf{V}_{c2} = \mathbf{a}^2 \mathbf{V}_{a2} \tag{5.8}$$

$$\mathbf{V}_{b0} = \mathbf{V}_{c0} = \mathbf{V}_{a0} \tag{5.9}$$

Substituting the above equations into Equations 5.2 and 5.3, as appropriate, the phase voltages can be expressed in terms of the sequence voltages as

$$\mathbf{V}_a = \mathbf{V}_{a1} + \mathbf{V}_{a2} + \mathbf{V}_{a0} \tag{5.10}$$

$$\mathbf{V}_b = \mathbf{a}^2 \mathbf{V}_{a1} + \mathbf{a} \mathbf{V}_{a2} + \mathbf{V}_{a0} \tag{5.11}$$

$$\mathbf{V}_c = \mathbf{a} \mathbf{V}_{a1} + \mathbf{a}^2 \mathbf{V}_{a2} + \mathbf{V}_{a0} \tag{5.12}$$

Equations 5.10 through 5.12 are known as the *synthesis equations*. Therefore, it can be shown that the sequence voltages can be expressed in terms of phase voltages as

$$\mathbf{V}_{a0} = \frac{1}{3} \left(\mathbf{V}_a + \mathbf{V}_b + \mathbf{V}_c \right) \tag{5.13}$$

$$\mathbf{V}_{a1} = \frac{1}{3} \left(\mathbf{V}_a + \mathbf{a} \mathbf{V}_b + \mathbf{a}^2 \mathbf{V}_c \right) \tag{5.14}$$

$$\mathbf{V}_{a2} = \frac{1}{3} \left(\mathbf{V}_a + \mathbf{a}^2 \mathbf{V}_b + \mathbf{a} \mathbf{V}_c \right) \tag{5.15}$$

which are known as the *analysis equations*. Alternatively, the synthesis and analysis equations can be written, respectively, in matrix form as

$$\begin{bmatrix} \mathbf{V}_a \\ \mathbf{V}_b \\ \mathbf{V}_c \end{bmatrix} = \begin{bmatrix} 1 & 1 & 1 \\ 1 & \mathbf{a}^2 & \mathbf{a} \\ 1 & \mathbf{a} & \mathbf{a}^2 \end{bmatrix} \begin{bmatrix} \mathbf{V}_{a0} \\ \mathbf{V}_{a1} \\ \mathbf{V}_{a2} \end{bmatrix} \tag{5.16}$$

and

$$\begin{bmatrix} \mathbf{V}_{a0} \\ \mathbf{V}_{a1} \\ \mathbf{V}_{a2} \end{bmatrix} = \frac{1}{3} \begin{bmatrix} 1 & 1 & 1 \\ 1 & \mathbf{a} & \mathbf{a}^2 \\ 1 & \mathbf{a}^2 & \mathbf{a} \end{bmatrix} \begin{bmatrix} \mathbf{V}_a \\ \mathbf{V}_b \\ \mathbf{V}_c \end{bmatrix} \tag{5.17}$$

or

$$[\mathbf{V}_{abc}] = [\mathbf{A}][\mathbf{V}_{012}] \tag{5.18}$$

and

$$[\mathbf{V}_{012}] = [\mathbf{A}]^{-1}[\mathbf{V}_{abc}] \tag{5.19}$$

where

$$[\mathbf{A}] = \begin{bmatrix} 1 & 1 & 1 \\ 1 & \mathbf{a}^2 & \mathbf{a} \\ 1 & \mathbf{a} & \mathbf{a}^2 \end{bmatrix} \tag{5.20}$$

$$[\mathbf{A}]^{-1} = \frac{1}{3}\begin{bmatrix} 1 & 1 & 1 \\ 1 & \mathbf{a} & \mathbf{a}^2 \\ 1 & \mathbf{a}^2 & \mathbf{a} \end{bmatrix} \tag{5.21}$$

$$[\mathbf{V}_{abc}] = \begin{bmatrix} \mathbf{V}_a \\ \mathbf{V}_b \\ \mathbf{V}_c \end{bmatrix} \tag{5.22}$$

$$[\mathbf{V}_{012}] = \begin{bmatrix} \mathbf{V}_{a0} \\ \mathbf{V}_{a1} \\ \mathbf{V}_{a2} \end{bmatrix} \tag{5.23}$$

The synthesis and analysis equations in terms of phase and sequence currents can be expressed as

$$\begin{bmatrix} \mathbf{I}_a \\ \mathbf{I}_b \\ \mathbf{I}_c \end{bmatrix} = \begin{bmatrix} 1 & 1 & 1 \\ 1 & \mathbf{a}^2 & \mathbf{a} \\ 1 & \mathbf{a} & \mathbf{a}^2 \end{bmatrix} \begin{bmatrix} \mathbf{I}_{a0} \\ \mathbf{I}_{a1} \\ \mathbf{I}_{a2} \end{bmatrix} \tag{5.24}$$

and

$$\begin{bmatrix} \mathbf{I}_{a0} \\ \mathbf{I}_{a1} \\ \mathbf{I}_{a2} \end{bmatrix} = \frac{1}{3}\begin{bmatrix} 1 & 1 & 1 \\ 1 & \mathbf{a} & \mathbf{a}^2 \\ 1 & \mathbf{a}^2 & \mathbf{a} \end{bmatrix} \begin{bmatrix} \mathbf{I}_a \\ \mathbf{I}_b \\ \mathbf{I}_c \end{bmatrix} \tag{5.25}$$

Symmetrical Components and Sequence Impedances

or

$$[\mathbf{I}_{abc}] = [\mathbf{A}][\mathbf{I}_{012}] \quad (5.26)$$

and

$$[\mathbf{I}_{012}] = [\mathbf{A}]^{-1}[\mathbf{I}_{abc}] \quad (5.27)$$

EXAMPLE 5.1

Determine the symmetrical components for the phase voltages of $\mathbf{V}_a = 7.3\angle 12.5$, $\mathbf{V}_b = 0.4\angle -100°$, and $\mathbf{V}_c = 4.4\angle 154°$ V

Solution

$$\mathbf{V}_{a0} = \frac{1}{3}(\mathbf{V}_a + \mathbf{V}_b + \mathbf{V}_c)$$
$$= \frac{1}{3}(7.3\angle 12.5° + 0.4\angle -100° + 4.4\angle 154°)$$
$$= 1.47\angle 45.1° \text{ V}$$

$$\mathbf{V}_{a1} = \frac{1}{3}(\mathbf{V}_a + \mathbf{a}\mathbf{V}_b + \mathbf{a}^2\mathbf{V}_c)$$
$$= \frac{1}{3}\left[7.3\angle 12.5° + (1\angle 120°)(0.4\angle -100°) + (1\angle 240°)(4.4\angle 154°)\right]$$
$$= 3.97\angle 20.5° \text{ V}$$

$$\mathbf{V}_{a2} = \frac{1}{3}(\mathbf{V}_a + \mathbf{a}^2\mathbf{V}_b + \mathbf{a}\mathbf{V}_c)$$
$$= 3\left[7.3\angle 12.5° + (1\angle 240°)(0.4\angle -100°) + (1\angle 20°)(4.4\angle 154°)\right]$$
$$= 2.52\angle -19.7° \text{ V}$$

$$\mathbf{V}_{b0} = \mathbf{V}_{a0}$$
$$= 1.47\angle 45.1° \text{ V}$$

$$\mathbf{V}_{b1} = \mathbf{a}^2\mathbf{V}_{a1}$$
$$= (1\angle 240°)(3.97\angle 20.5°)$$
$$= 3.97\angle 260.5° \text{ V}$$

$$\mathbf{V}_{b2} = \mathbf{a}\mathbf{V}_{a2}$$
$$= (1\angle 120°)(2.52\angle -19.7°)$$
$$= 2.52\angle 100.3° \text{ V}$$

$$\mathbf{V}_{c0} = \mathbf{V}_{a0}$$
$$= 1.47 \angle 45.1° \text{ V}$$

$$\mathbf{V}_{c1} = \mathbf{a}\mathbf{V}_{a1}$$
$$= (1\angle 120°)(3.97\angle 20.5°)$$
$$= 3.97 \angle 140.5° \text{ V}$$

$$\mathbf{V}_{c2} = \mathbf{a}^2\mathbf{V}_{a2}$$
$$= (1\angle 240°)(2.52\angle -19.7°)$$
$$= 2.52 \angle 220.3° \text{ V}$$

Note that, the resulting values for the symmetrical components can be checked numerically (e.g., using Equation 5.11) or graphically, as shown in Figure 5.1e.

5.5 POWER IN SYMMETRICAL COMPONENTS

The three-phase complex power at any point of a three-phase system can be expressed as the sum of the individual complex powers of each phase so that

$$\mathbf{S}_{3\phi} = P_{3\phi} + jQ_{3\phi}$$
$$= \mathbf{S}_a + \mathbf{S}_b + \mathbf{S}_c \qquad (5.28)$$
$$= \mathbf{V}_a\mathbf{I}_a^* + \mathbf{V}_b\mathbf{I}_b^* + \mathbf{V}_c\mathbf{I}_c^*$$

or, in matrix notation,

$$\mathbf{S}_{3\phi} = \begin{bmatrix} \mathbf{V}_a & \mathbf{V}_b & \mathbf{V}_c \end{bmatrix} \begin{bmatrix} \mathbf{I}_a \\ \mathbf{I}_b \\ \mathbf{I}_c \end{bmatrix}^*$$
$$= \begin{bmatrix} \mathbf{V}_a \\ \mathbf{V}_b \\ \mathbf{V}_c \end{bmatrix}^t \begin{bmatrix} \mathbf{I}_a \\ \mathbf{I}_b \\ \mathbf{I}_c \end{bmatrix}^* \qquad (5.29)$$

or

$$\mathbf{S}_{3\phi} = [\mathbf{V}_{abc}]^t [\mathbf{I}_{abc}]^* \qquad (5.30)$$

Symmetrical Components and Sequence Impedances

where

$$[\mathbf{V}_{abc}] = [\mathbf{A}][\mathbf{V}_{012}]$$

$$[\mathbf{I}_{abc}] = [\mathbf{A}][\mathbf{I}_{012}]$$

and therefore,

$$[\mathbf{V}_{abc}]^t = [\mathbf{V}_{012}]^t [\mathbf{A}]^t \tag{5.31}$$

$$[\mathbf{I}_{abc}]^* = [\mathbf{A}]^* [\mathbf{I}_{012}]^* \tag{5.32}$$

Substituting Equations 5.31 and 5.32 into Equation 5.30,

$$\mathbf{S}_{3\phi} = [\mathbf{V}_{012}]^t [\mathbf{A}]^t [\mathbf{A}]^* [\mathbf{I}_{012}]^* \tag{5.33}$$

where

$$[\mathbf{A}]^t [\mathbf{A}]^* =$$

$$= \begin{bmatrix} 1 & 1 & 1 \\ 1 & a^2 & a \\ 1 & a & a^2 \end{bmatrix} \begin{bmatrix} 1 & 1 & 1 \\ 1 & a & a^2 \\ 1 & a^2 & a \end{bmatrix}$$

$$= \begin{bmatrix} 3 & 0 & 0 \\ 0 & 3 & 0 \\ 0 & 0 & 3 \end{bmatrix}$$

$$= 3 \begin{bmatrix} 1 & 0 & 0 \\ 0 & 1 & 0 \\ 0 & 0 & 1 \end{bmatrix}$$

Therefore,

$$\mathbf{S}_{3\phi} = 3 \left[\mathbf{V}_{012} \right]^t \left[\mathbf{I}_{012} \right]^*$$

$$= 3 \begin{bmatrix} \mathbf{V}_{a0} & \mathbf{V}_{a1} & \mathbf{V}_{a2} \end{bmatrix} \begin{bmatrix} \mathbf{I}_{a0} \\ \mathbf{I}_{a1} \\ \mathbf{I}_{a2} \end{bmatrix}^* \tag{5.34a}$$

or

$$\mathbf{S}_{3\phi} = 3 \left[\mathbf{V}_{a0} \mathbf{I}_{a0}^* + \mathbf{V}_{a1} \mathbf{I}_{a1}^* + \mathbf{V}_{a2} \mathbf{I}_{a2}^* \right] \tag{5.34b}$$

Note that, there are no cross terms (e.g., $\mathbf{V}_{a0}\mathbf{I}_{a1}^*$ or $\mathbf{V}_{a1}\mathbf{I}_{a0}^*$) in this equation, which indicates that there is no coupling of power among the three sequences. Also, note that the symmetrical components of voltage and current belong to the same phase.

EXAMPLE 5.2

Assume that the phase voltages and currents of a three-phase system are given as

$$[\mathbf{V}_{abc}] = \begin{bmatrix} 0 \\ 50 \\ -50 \end{bmatrix} \text{ and } [\mathbf{I}_{abc}] = \begin{bmatrix} -5 \\ j5 \\ -5 \end{bmatrix}$$

and determine the following:

(a) Three-phase complex power using Equation 5.30
(b) Sequence voltage and current matrices, that is, $[\mathbf{V}_{012}]$ and $[\mathbf{I}_{012}]$
(c) Three-phase complex power using Equation 5.34

Solution

(a)

$$\mathbf{S}_{3\phi} = [\mathbf{V}_{abc}]^t [\mathbf{I}_{abc}]^*$$

$$= \begin{bmatrix} 0 & 50 & -50 \end{bmatrix} \begin{bmatrix} 5 \\ -j5 \\ +5 \end{bmatrix}$$

$$= -250 - j250$$

$$= 353.5534 \angle 45° \text{ VA}$$

(b)

$$[\mathbf{V}_{012}] = [\mathbf{A}]^{-1}[\mathbf{V}_{abc}]$$

$$= \frac{1}{3}\begin{bmatrix} 1 & 1 & 1 \\ 1 & a & a^2 \\ 1 & a^2 & a \end{bmatrix} \begin{bmatrix} 0 \\ 50 \\ -50 \end{bmatrix}$$

$$= \begin{bmatrix} 0.0 \angle 0° \\ 28.8675 \angle 90° \\ 28.8675 \angle -90° \end{bmatrix} \text{ V}$$

$$[\mathbf{I}_{012}] = [\mathbf{A}]^{-1}[\mathbf{I}_{abc}]$$

$$= \frac{1}{3}\begin{bmatrix} 1 & 1 & 1 \\ 1 & a & a^2 \\ 1 & a^2 & a \end{bmatrix} \begin{bmatrix} -5 \\ j5 \\ -5 \end{bmatrix}$$

$$= \begin{bmatrix} 3.7268 \angle 153.4° \\ 2.3570 \angle 165° \\ 2.3570 \angle -75° \end{bmatrix}$$

(c)

$$\mathbf{S}_{3\phi} = 3\left[\mathbf{V}_{a0}\mathbf{I}_{a0}^* + \mathbf{V}_{a1}\mathbf{I}_{a1}^* + \mathbf{V}_{a2}\mathbf{I}_{a2}^*\right]$$
$$= 353.5534\angle -45° \text{ VA}$$

5.6 SEQUENCE IMPEDANCES OF TRANSMISSION LINES

5.6.1 SEQUENCE IMPEDANCES OF UNTRANSPOSED LINES

Figure 5.3a shows a circuit representation of an untransposed transmission line with unequal self-impedances and unequal mutual impedances. Here,

$$[\mathbf{V}_{abc}] = [\mathbf{Z}_{abc}][\mathbf{I}_{abc}] \tag{5.35}$$

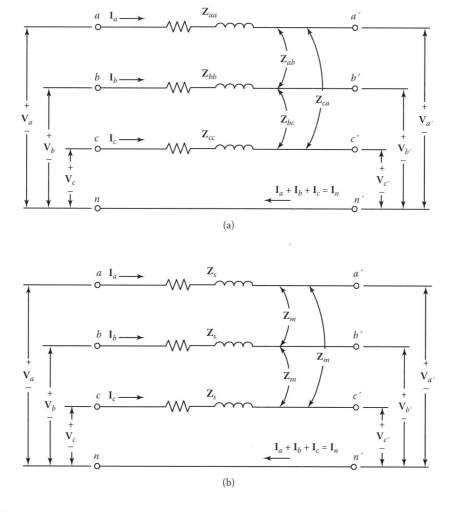

FIGURE 5.3 Transmission line circuit diagrams: (a) with unequal series and unequal impedances; (b) with equal series and equal mutual impedances.

where

$$[\mathbf{Z}_{abc}] = \begin{bmatrix} \mathbf{Z}_{aa} & \mathbf{Z}_{ab} & \mathbf{Z}_{ac} \\ \mathbf{Z}_{ba} & \mathbf{Z}_{bb} & \mathbf{Z}_{bc} \\ \mathbf{Z}_{ca} & \mathbf{Z}_{cb} & \mathbf{Z}_{cc} \end{bmatrix} \qquad (5.36)$$

in which the self-impedances are

$$\mathbf{Z}_{aa} \neq \mathbf{Z}_{bb} \neq \mathbf{Z}_{cc}$$

and the mutual impedances are

$$\mathbf{Z}_{ab} \neq \mathbf{Z}_{bc} \neq \mathbf{Z}_{ca}$$

Multiplying both sides of Equation 5.35 by $[\mathbf{A}]^{-1}$ and also substituting Equation 5.26 into Equation 5.35,

$$[\mathbf{A}]^{-1}[\mathbf{V}_{abc}] = [\mathbf{A}]^{-1}[\mathbf{Z}_{abc}][\mathbf{A}][\mathbf{I}_{012}] \qquad (5.37)$$

where the similarity transformation is defined as

$$[\mathbf{Z}_{012}] \triangleq [\mathbf{A}]^{-1}[\mathbf{Z}_{abc}][\mathbf{A}] \qquad (5.38)$$

Therefore, the sequence impedance matrix of an untransposed transmission line can be calculated using Equation 5.38 and can be expressed as

$$[\mathbf{Z}_{012}] = \begin{bmatrix} \mathbf{Z}_{00} & \mathbf{Z}_{01} & \mathbf{Z}_{02} \\ \mathbf{Z}_{10} & \mathbf{Z}_{11} & \mathbf{Z}_{12} \\ \mathbf{Z}_{20} & \mathbf{Z}_{21} & \mathbf{Z}_{22} \end{bmatrix} \qquad (5.39)$$

or

$$[\mathbf{Z}_{012}] = \begin{bmatrix} (\mathbf{Z}_{s0} + 2\mathbf{Z}_{m0}) & (\mathbf{Z}_{s2} - \mathbf{Z}_{m2}) & (\mathbf{Z}_{s1} - \mathbf{Z}_{m1}) \\ (\mathbf{Z}_{s1} - \mathbf{Z}_{m1}) & (\mathbf{Z}_{s0} - \mathbf{Z}_{m0}) & (\mathbf{Z}_{s2} + 2\mathbf{Z}_{m2}) \\ (\mathbf{Z}_{s2} - \mathbf{Z}_{m2}) & (\mathbf{Z}_{s1} + 2\mathbf{Z}_{m1}) & (\mathbf{Z}_{s0} - \mathbf{Z}_{m0}) \end{bmatrix} \qquad (5.40)$$

where, by definition,

$$\mathbf{Z}_{s0} = \text{zero-sequence self-impedance}$$
$$\triangleq \frac{1}{3}(\mathbf{Z}_{aa} + \mathbf{Z}_{bb} + \mathbf{Z}_{cc}) \qquad (5.41)$$

Symmetrical Components and Sequence Impedances

$\mathbf{Z}_{s1}$ = positive-sequence self-impedance

$$\triangleq \frac{1}{3}\left(\mathbf{Z}_{aa} + \mathbf{a}\mathbf{Z}_{bb} + \mathbf{a}^2\mathbf{Z}_{cc}\right) \tag{5.42}$$

$\mathbf{Z}_{s2}$ = negative-sequence self-impedance

$$\triangleq \frac{1}{3}\left(\mathbf{Z}_{aa} + \mathbf{a}^2\mathbf{Z}_{bb} + \mathbf{a}\mathbf{Z}_{cc}\right) \tag{5.43}$$

$\mathbf{Z}_{m0}$ = zero-sequence mutual impedance

$$\triangleq \frac{1}{3}\left(\mathbf{Z}_{bc} + \mathbf{Z}_{ca} + \mathbf{Z}_{ab}\right) \tag{5.44}$$

$\mathbf{Z}_{m1}$ = positive-sequence mutual impedance

$$\triangleq \frac{1}{3}\left(\mathbf{Z}_{bc} + \mathbf{a}\mathbf{Z}_{ca} + \mathbf{a}^2\mathbf{Z}_{ab}\right) \tag{5.45}$$

$\mathbf{Z}_{m2}$ = negative-sequence mutual impedance

$$\triangleq \frac{1}{3}\left(\mathbf{Z}_{bc} + \mathbf{a}^2\mathbf{Z}_{ca} + \mathbf{a}\mathbf{Z}_{ab}\right) \tag{5.46}$$

Therefore,

$$[\mathbf{V}_{012}] = [\mathbf{Z}_{012}][\mathbf{I}_{012}] \tag{5.47}$$

Note that the matrix in Equation 5.40 is not a symmetrical matrix, and therefore, the application of Equation 5.47 will show that there is a mutual coupling among the three sequences, which is not a desirable result.

5.6.2 Sequence Impedances of Transposed Lines

The remedy is either to completely transpose the line or to place the conductors with equilateral spacing among them so that the resulting mutual impedances* are equal to each other, that is, $\mathbf{Z}_{ab} = \mathbf{Z}_{bc} = \mathbf{Z}_{ca} = \mathbf{Z}_m$, as shown in Figure 5.3b. Furthermore, if the self-impedances of conductors are equal to each other, that is, $\mathbf{Z}_{aa} = \mathbf{Z}_{bb} = \mathbf{Z}_{cc} = \mathbf{Z}_s$, Equation 5.36 can be expressed as

$$[\mathbf{Z}_{abc}] = \begin{bmatrix} \mathbf{Z}_s & \mathbf{Z}_m & \mathbf{Z}_m \\ \mathbf{Z}_m & \mathbf{Z}_s & \mathbf{Z}_m \\ \mathbf{Z}_m & \mathbf{Z}_m & \mathbf{Z}_s \end{bmatrix} \tag{5.48}$$

---

* In passive networks $\mathbf{Z}_{ab} = \mathbf{Z}_{ba}$, $\mathbf{Z}_{bc} = \mathbf{Z}_{cb}$, etc.

where

$$\mathbf{Z}_s = \left[(r_a + r_e) + j0.1213 \ln \frac{D_e}{D_s}\right] l \quad \Omega \tag{5.49}$$

$$\mathbf{Z}_m = \left[r_e + j0.1213 \ln \frac{D_e}{D_{eq}}\right] l \quad \Omega \tag{5.50}$$

$$D_{eq} \triangleq D_m = (D_{ab} \times D_{bc} \times D_{ca})^{1/3}$$

r_a = resistance of a single conductor a.

r_e is the resistance of Carson's [2] equivalent (and fictitious) earth return conductor. It is a function of frequency and can be expressed as

$$r_e = 1.588 \times 10^{-3} f \ \Omega/\text{mi} \tag{5.51}$$

or

$$r_e = 9.869 \times 10^{-4} f \ \Omega/\text{km} \tag{5.52}$$

At 60 Hz, $r_e = 0.09528$ Ω/mi. The quantity D_e is a function of both the earth resistivity ρ and the frequency f and can be expressed as

$$D_e = 2160 \left(\frac{\rho}{f}\right)^{1/2} \text{ft} \tag{5.53}$$

where ρ is the earth resistivity and is given in Table 5.2 for various earth types. If the actual earth resistivity is unknown, it is customary to use an average value of 100 Ω/m for ρ. Therefore, at 60 Hz, $D_e = 2788.55$ ft. D_s is the GMR of the phase conductor as before. Therefore, by applying Equation 5.38,

TABLE 5.2
Resistivity of Different Soils

Ground Type (Ω/m)	Resistivity ρ
Seawater	0.01–1.0
Wet organic soil	10
Moist soil (average earth)	100
Dry soil	1000
Bedrock	10^4
Pure slate	10^7
Sandstone	10^9
Crushed rock	1.5×10^8

Symmetrical Components and Sequence Impedances

$$[\mathbf{Z}_{012}] = \begin{bmatrix} (\mathbf{Z}_s + 2\mathbf{Z}_m) & 0 & 0 \\ 0 & (\mathbf{Z}_s - \mathbf{Z}_m) & 0 \\ 0 & 0 & (\mathbf{Z}_s - \mathbf{Z}_m) \end{bmatrix} \quad (5.54)$$

where, by definition,

$$\mathbf{Z}_0 = \text{zero-sequence impedance at 60 Hz}$$
$$\triangleq \mathbf{Z}_{00} = \mathbf{Z}_s + 2\mathbf{Z}_m \quad (5.55a)$$

$$= \left[(r_a + 3r_e) + j0.1213 \ln \frac{D_e^3}{D_s \times D_{eq}^2} \right] l \quad \Omega \quad (5.55b)$$

$$\mathbf{Z}_1 = \text{positive-sequence impedance at 60 Hz}$$
$$\triangleq \mathbf{Z}_{11} = \mathbf{Z}_s - \mathbf{Z}_m \quad (5.56a)$$

$$= \left[r_a + j0.1213 \ln \frac{D_{eq}}{D_s} \right] l \quad \Omega \quad (5.56b)$$

$$\mathbf{Z}_2 = \text{negative-sequence impedance at 60 Hz}$$
$$\triangleq \mathbf{Z}_{22} = \mathbf{Z}_s - \mathbf{Z}_m \quad (5.57a)$$

$$= \left[r_a + j0.1213 \ln \frac{D_{eq}}{D_s} \right] l \quad \Omega \quad (5.57b)$$

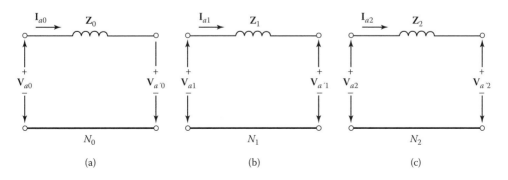

FIGURE 5.4 Sequence networks of a transmission line: (a) zero-sequence network; (b) positive-sequence network; (c) negative-sequence network.

Thus, Equation 5.54 can be expressed* as

$$[\mathbf{Z}_{012}] = \begin{bmatrix} \mathbf{Z}_0 & 0 & 0 \\ 0 & \mathbf{Z}_1 & 0 \\ 0 & 0 & \mathbf{Z}_2 \end{bmatrix} \quad (5.58)$$

Both Equations 5.54 and 5.58 indicate that there is no mutual coupling among the three sequences, which is the desirable result. Therefore, the zero-, positive-, and negative-sequence currents cause voltage drops only in the zero-, positive-, and negative-sequence networks, respectively, of the transmission line. Also note, in Equation 5.54, that the positive- and negative-sequence impedances of the transmission line are equal to each other but they are far less than the zero-sequence impedance of the line. Figure 5.4 shows the sequence networks of a transmission line.

5.6.3 Electromagnetic Unbalances due to Untransposed Lines

If the line is neither transposed nor its conductors equilaterally spaced, Equation 5.48 cannot be used. Instead, use the following equation:

$$[\mathbf{Z}_{abc}] = \begin{bmatrix} \mathbf{Z}_{aa} & \mathbf{Z}_{ab} & \mathbf{Z}_{ac} \\ \mathbf{Z}_{ba} & \mathbf{Z}_{bb} & \mathbf{Z}_{bc} \\ \mathbf{Z}_{ca} & \mathbf{Z}_{cb} & \mathbf{Z}_{cc} \end{bmatrix} \quad (5.59)$$

where

$$\mathbf{Z}_{aa} = \mathbf{Z}_{bb} = \mathbf{Z}_{cc} = \left[(r_a + r_e) + j0.1213 \ln \frac{D_e}{D_s} \right] l \quad (5.60)$$

---

* Equations 5.55 and 5.57 can easily be modified so that they can give approximate sequence impedances at other frequencies. For example, their expressions at 50 Hz are given as

$$\mathbf{Z}_0 = \text{zero-sequence impedance at 50 Hz}$$
$$\triangleq \mathbf{Z}_{00} = \mathbf{Z}_s + 2\mathbf{Z}_m$$
$$= \left[(r_a + 3r_e) + j0.1213 \left(\frac{50 \text{ Hz}}{60 \text{ Hz}} \right) \ln \frac{D_e^3}{D_s \times D_{eq}^2} \right] l \quad \Omega$$
$$= \left[(r_a + 3r_e) + j0.10108 \ln \frac{D_e^3}{D_s \times D_{eq}^2} \right] l \quad \Omega$$

$$\mathbf{Z}_1 = \text{positive-sequence impedance at 50 Hz}$$
$$\triangleq \mathbf{Z}_{11} = \mathbf{Z}_s - \mathbf{Z}_m$$
$$= \left[r_a + j0.10108 \ln \frac{D_{eq}}{D_s} \right] l \quad \Omega$$

$$\mathbf{Z}_2 = \text{negative-sequence impedance at 50 Hz}$$
$$\triangleq \mathbf{Z}_{22} = \mathbf{Z}_s - \mathbf{Z}_m$$
$$= \left[r_a + j0.10108 \ln \frac{D_{eq}}{D_s} \right] l \quad \Omega$$

Symmetrical Components and Sequence Impedances

$$\mathbf{Z}_{ab} = \mathbf{Z}_{ba} = \left[r_e + j0.1213\ln\frac{D_e}{D_{ab}}\right]l \quad (5.61)$$

$$\mathbf{Z}_{ac} = \mathbf{Z}_{ca} = \left[r_e + j0.1213\ln\frac{D_e}{D_{ac}}\right]l \quad (5.62)$$

$$\mathbf{Z}_{bc} = \mathbf{Z}_{cb} = \left[r_e + j0.1213\ln\frac{D_e}{D_{bc}}\right]l \quad (5.63)$$

The corresponding sequence impedance matrix can be found from Equation 5.38 as before. Therefore, the associated sequence admittance matrix can be found as

$$[\mathbf{Y}_{012}] = [\mathbf{Z}_{012}]^{-1} \quad (5.64a)$$

$$= \begin{bmatrix} \mathbf{Y}_{00} & \mathbf{Y}_{01} & \mathbf{Y}_{02} \\ \mathbf{Y}_{10} & \mathbf{Y}_{11} & \mathbf{Y}_{12} \\ \mathbf{Y}_{20} & \mathbf{Y}_{21} & \mathbf{Y}_{22} \end{bmatrix} \quad (5.64b)$$

Therefore,

$$[\mathbf{I}_{012}] = [\mathbf{Y}_{012}][\mathbf{V}_{012}] \quad (5.65)$$

Since the line is neither transposed nor its conductors equilaterally spaced, there is an electromagnetic unbalance in the system. Such unbalance is determined from Equation 5.65 with only positive-sequence voltage applied. Therefore,

$$\begin{bmatrix} \mathbf{I}_{a0} \\ \mathbf{I}_{a1} \\ \mathbf{I}_{a2} \end{bmatrix} = \begin{bmatrix} \mathbf{Y}_{00} & \mathbf{Y}_{01} & \mathbf{Y}_{02} \\ \mathbf{Y}_{10} & \mathbf{Y}_{11} & \mathbf{Y}_{12} \\ \mathbf{Y}_{20} & \mathbf{Y}_{21} & \mathbf{Y}_{22} \end{bmatrix} \begin{bmatrix} 0 \\ \mathbf{V}_{a1} \\ 0 \end{bmatrix} \quad (5.66a)$$

$$= \begin{bmatrix} \mathbf{Y}_{01} \\ \mathbf{Y}_{11} \\ \mathbf{Y}_{21} \end{bmatrix} \mathbf{V}_{a1} \quad (5.66b)$$

According to Gross and Hesse [3], the per-unit unbalances for zero sequence and negative sequence can be expressed, respectively, as

$$\mathbf{m}_0 \triangleq \frac{\mathbf{I}_{a0}}{\mathbf{I}_{a1}} \text{ pu} \quad (5.67a)$$

$$= \frac{\mathbf{Y}_{01}}{\mathbf{Y}_{11}} \text{ pu} \quad (5.67b)$$

and

$$\mathbf{m}_2 \triangleq \frac{\mathbf{I}_{a2}}{\mathbf{I}_{a1}} \quad \text{pu} \qquad (5.68a)$$

$$= \frac{\mathbf{Y}_{21}}{\mathbf{Y}_{11}} \quad \text{pu} \qquad (5.68b)$$

Since, in physical systems [3],

$$\mathbf{Z}_{22} \gg \mathbf{Z}_{02} \text{ or } \mathbf{Z}_{21}$$

and

$$\mathbf{Z}_{00} \gg \mathbf{Z}_{20} \text{ or } \mathbf{Z}_{01}$$

the approximate values of the per-unit unbalances for zero and negative sequences can be expressed, respectively, as

$$\mathbf{m}_0 \cong -\frac{\mathbf{Z}_{01}}{\mathbf{Z}_{00}} \quad \text{pu} \qquad (5.69a)$$

and

$$\mathbf{m}_2 \cong -\frac{\mathbf{Z}_{21}}{\mathbf{Z}_{22}} \quad \text{pu} \qquad (5.69b)$$

EXAMPLE 5.3

Consider the compact-line configuration shown in Figure 5.5. The phase conductors used are made up of 500-kcmil, 30/7-strand ACSR (aluminum conductor steel reinforced conductor) conductor. The line length is 40 mi and the line is not transposed. Use 50°C and 60 Hz. Ignore the overhead ground wire. If the earth has an average resistivity, determine the following:

(a) Line impedance matrix
(b) Sequence impedance matrix of line

Solution

(a) The conductor parameters can be found from Table A.3 (Appendix A) as

$$r_a = r_b = r_c = 0.206 \ \Omega/\text{mi}$$

$$D_s = D_{sa} = D_{sb} = D_{sc} = 0.0311 \ \text{ft}$$

Symmetrical Components and Sequence Impedances

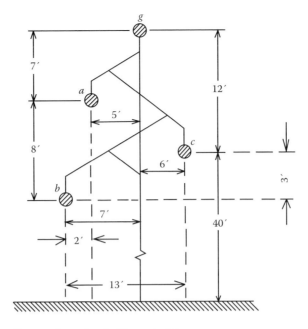

FIGURE 5.5 Compact-line configuration for Example 5.3.

$$D_{ab} = (2^2 + 8^2)^{1/2} = 8.2462 \text{ ft}$$

$$D_{bc} = (3^2 + 13^2)^{1/2} = 13.3417 \text{ ft}$$

$$D_{ac} = (5^2 + 11^2)^{1/2} = 12.0830 \text{ ft}$$

Since the earth has an average resistivity, $D_e = 2788.5$ ft. At 60 Hz, $r_e = 0.09528$ Ω/mi. From Equation 5.60, the self-impedances of the line conductors are

$$\begin{aligned}
\mathbf{Z}_{aa} = \mathbf{Z}_{bb} = \mathbf{Z}_{cc} &= \left[(r_a + r_e) + j0.1213\ln\frac{D_e}{D_s}\right]l \\
&= \left[(0.206 + 0.09528) + j0.1213\ln\frac{2788.5}{0.0311}\right] \times 40 \\
&= 12.0512 + j55.3495 \text{ Ω}
\end{aligned}$$

The mutual impedances calculated from Equations 5.61 through 5.63 are

$$\begin{aligned}
\mathbf{Z}_{ab} = \mathbf{Z}_{ba} &= \left[r_e + j0.1213\ln\frac{D_e}{D_{ab}}\right]l \\
&= \left[0.09528 + j0.1213\ln\frac{2788.5}{0.2462}\right] \times 40 \\
&= 3.8112 + j28.2650 \text{ Ω}
\end{aligned}$$

$$\mathbf{Z}_{bc} = \mathbf{Z}_{cb} = \left[r_e + j0.1213 \ln \frac{D_e}{D_{bc}} \right] l$$

$$= \left[0.09528 + j0.1213 \ln \frac{2788.5}{13.3417} \right] \times 40$$

$$= 3.8112 + j25.9297 \; \Omega$$

$$\mathbf{Z}_{ac} = \mathbf{Z}_{ca} = \left[r_e + j0.1213 \ln \frac{D_e}{D_{ac}} \right] l$$

$$= \left[0.09528 + j0.1213 \ln \frac{2788.5}{12.0830} \right] \times 40$$

$$= 3.8112 + j26.4107 \; \Omega$$

Therefore,

$$[\mathbf{Z}_{abc}] = \begin{bmatrix} (12.0512 + j55.3495) & (3.8112 + j26.4107) & (3.8112 + j26.4107) \\ (3.8112 + j28.2650) & (12.0512 + j55.3495) & (3.8112 + j25.9297) \\ (3.8112 + j26.4107) & (3.8112 + j25.9297) & (12.0512 + j55.3495) \end{bmatrix}$$

(b) Thus, the sequence impedance matrix of the line can be found from Equation 5.38 as

$$[\mathbf{Z}_{012}] = [\mathbf{A}]^{-1}[\mathbf{Z}_{abc}][\mathbf{A}]$$

$$= \begin{bmatrix} (19.67 + j109.09) & (0.54 + j0.47) & (-0.54 + j0.47) \\ (-0.54 + j0.47) & (8.24 + j28.48) & (-1.07 - j0.94) \\ (0.54 + j0.47) & (1.07 - j0.94) & (8.24 + j28.48) \end{bmatrix}$$

EXAMPLE 5.4

Repeat Example 5.3 assuming that the line is *completely transposed*.

Solution

(a) From Equation 5.49,

$$\mathbf{Z}_s = \left[(r_a + r_e) + j0.1213 \ln \frac{D_e}{D_s} \right] l \; \Omega$$

$$= 12.0512 + j53.3495 \; \Omega$$

as before

Symmetrical Components and Sequence Impedances

From Equation 5.50,

$$\mathbf{Z}_m = \left[r_e + j0.1213\ln\frac{D_e}{D_{eq}} \right] l \quad \Omega$$

where

$$D_{eq} = (8.2462 \times 13.3417 \times 12.0830)^{1/3} = 11 \text{ ft}$$

Thus,

$$\mathbf{Z}_m = \left[0.09528 + j0.1213\ln\frac{2788.5}{11} \right] \times 40$$
$$= 3.8112 + j26.8684 \ \Omega$$

Therefore,

$$[\mathbf{Z}_{abc}] = \begin{bmatrix} \mathbf{Z}_s & \mathbf{Z}_m & \mathbf{Z}_m \\ \mathbf{Z}_m & \mathbf{Z}_s & \mathbf{Z}_m \\ \mathbf{Z}_m & \mathbf{Z}_m & \mathbf{Z}_s \end{bmatrix}$$
$$= \begin{bmatrix} (12.0512 + j55.3495) & (3.8112 + j26.8684) & (3.8112 + j26.8684) \\ (3.8112 + j26.8684) & (12.0512 + j55.3495) & (3.8112 + j26.8684) \\ (3.8112 + j26.8684) & (3.8112 + j26.8684) & (12.0512 + j55.3495) \end{bmatrix}$$

(b) From Equation 5.54

$$[\mathbf{Z}_{012}] = \begin{bmatrix} (\mathbf{Z}_s + 2\mathbf{Z}_m) & 0 & 0 \\ 0 & (\mathbf{Z}_s - \mathbf{Z}_m) & 0 \\ 0 & 0 & (\mathbf{Z}_s - \mathbf{Z}_m) \end{bmatrix}$$
$$= \begin{bmatrix} 19.6736 + j109.086 & 0 & 0 \\ 0 & 8.2400 + j28.4811 & 0 \\ 0 & 0 & 8.2400 + j28.4811 \end{bmatrix}$$

or, by substituting Equations 5.55b and 5.56b into Equation 5.58,

$$[\mathbf{Z}_{012}] = \begin{bmatrix} 19.6736 + j109.086 & 0 & 0 \\ 0 & 8.2400 + j28.4811 & 0 \\ 0 & 0 & 8.2400 + j28.4811 \end{bmatrix}$$

EXAMPLE 5.5

Consider the results of Example 5.3 and determine the following:

(a) Per-unit electromagnetic unbalance for zero sequence
(b) Approximate value of per-unit electromagnetic unbalance for zero sequence
(c) Per-unit electromagnetic unbalance for negative sequence
(d) Approximate value of per-unit electromagnetic unbalance for negative sequence

Solution

The sequence admittance of the line can be found as

$$[Y_{012}] = [Z_{012}]^{-1}$$

$$= \begin{bmatrix} (1.60 \times 10^{-3} - j8.88 \times 10^{-3}) & (7.57 \times 10^{-5} + j1.93 \times 10^{-4}) & (-2.01 \times 10^{-4} + j6.15 \times 10^{-5}) \\ (-2.01 \times 10^{-4} + j6.15 \times 10^{-5}) & (9.44 \times 10^{-3} - j3.25 \times 10^{-2}) & (-4.55 \times 10^{-4} - j1.55 \times 10^{-3}) \\ (7.57 \times 10^{-5} + j1.93 \times 10^{-4}) & (1.60 \times 10^{-3} - j2.54 \times 10^{-4}) & (9.44 \times 10^{-3} - j3.25 \times 10^{-2}) \end{bmatrix}$$

(a) From Equation 5.67b,

$$m_0 = \frac{Y_{01}}{Y_{11}}$$

$$= \frac{7.57 \times 10^{-5} + j1.93 \times 10^{-4}}{9.44 \times 10^{-3} - j3.25 \times 10^{-2}}$$

$$= 0.61 \angle 142.4° \%$$

(b) From Equation 5.69a,

$$m_0 \cong -\frac{Z_{01}}{Z_{11}}$$

$$= -\frac{0.54 + j0.47}{19.67 + j109.09}$$

$$= 0.64 \angle 141.3° \%$$

(c) From Equation 5.68b,

$$m_2 = \frac{Y_{21}}{Y_{11}}$$

$$= \frac{160 \times 10^{-3} - j2.54 \times 10^{-4}}{9.44 \times 10^{-3} - j3.25 \times 10^{-2}}$$

$$= 4.79 \angle 64.8° \%$$

(d) From Equation 5.69b,

$$\mathbf{m}_2 \cong -\frac{\mathbf{Z}_{21}}{\mathbf{Z}_{22}}$$

$$= \frac{1.07 - j0.94}{8.24 + j28.48}$$

$$= 4.8\angle 64.8°\ \%$$

5.6.4 Sequence Impedances of Untransposed Line with Overhead Ground Wire

Assume that the untransposed line shown in Figure 5.5 is *shielded* against direct lightning strikes by the overhead ground wire u (used instead of g).

Therefore,

$$[\mathbf{V}_{abcu}] = [\mathbf{Z}_{abcu}][\mathbf{I}_{abcu}] \tag{5.70}$$

but since for the ground wire $\mathbf{V}_a = 0$,

$$\begin{bmatrix} \mathbf{V}_a \\ \mathbf{V}_b \\ \mathbf{V}_c \\ \hline 0 \end{bmatrix} = \left[\begin{array}{ccc|c} \mathbf{Z}_{aa} & \mathbf{Z}_{ab} & \mathbf{Z}_{ac} & \mathbf{Z}_{au} \\ \mathbf{Z}_{ba} & \mathbf{Z}_{bb} & \mathbf{Z}_{bc} & \mathbf{Z}_{bu} \\ \mathbf{Z}_{ca} & \mathbf{Z}_{cb} & \mathbf{Z}_{cc} & \mathbf{Z}_{cu} \\ \hline \mathbf{Z}_{ua} & \mathbf{Z}_{ub} & \mathbf{Z}_{uc} & \mathbf{Z}_{uu} \end{array} \right] \begin{bmatrix} \mathbf{I}_a \\ \mathbf{I}_b \\ \mathbf{I}_c \\ \hline \mathbf{I}_u \end{bmatrix} \tag{5.71}$$

The matrix $[\mathbf{Z}_{abcu}]$ can be determined using Equations 5.59 through 5.63, as before, and also using the following equations:

$$\mathbf{Z}_{au} = \mathbf{Z}_{ua} = \left[r_e + j0.1213 \ln \frac{D_e}{D_{au}} \right] l \tag{5.72}$$

$$\mathbf{Z}_{bu} = \mathbf{Z}_{ub} = \left[r_e + j0.1213 \ln \frac{D_e}{D_{bu}} \right] l \tag{5.73}$$

$$\mathbf{Z}_{cu} = \mathbf{Z}_{uc} = \left[r_e + j0.1213 \ln \frac{D_e}{D_{cu}} \right] l \tag{5.74}$$

$$\mathbf{Z}_{uu} = \mathbf{Z}_{uu} = \left[r_e + j0.1213 \ln \frac{D_e}{D_{uu}} \right] l \tag{5.75}$$

where r_u and D_{uu} are the resistance and GMR of the overhead ground wire, respectively.

The matrix $[\mathbf{Z}_{abcu}]$ given in Equation 5.71 can be reduced to $[\mathbf{Z}_{abc}]$ by using *the Kron reduction technique*. Therefore, Equation 5.71 can be reexpressed as

$$\begin{bmatrix} \mathbf{V}_{abc} \\ 0 \end{bmatrix} = \left[\begin{array}{c|c} \mathbf{Z}_1 & \mathbf{Z}_2 \\ \hline \mathbf{Z}_3 & \mathbf{Z}_4 \end{array} \right] \begin{bmatrix} \mathbf{I}_{abc} \\ 0 \end{bmatrix} \tag{5.76}$$

where the submatrices $[Z_1]$, $[Z_2]$, $[Z_3]$, and $[Z_4]$ are specified in the partitioned matrix $[Z_{abcu}]$ in Equation 5.71. Therefore, after the reduction,

$$[V_{abc}] = [Z_{abc}][I_{abc}] \tag{5.77}$$

where

$$[Z_{abc}] \triangleq [Z_1] - [Z_2][Z_4]^{-1}[Z_3] \tag{5.78}$$

Therefore, the sequence impedance matrix can be found from

$$[Z_{012}] = [A]^{-1}[Z_{abc}][A] \tag{5.79}$$

Thus, the sequence admittance matrix becomes

$$[Y_{012}] = [Z_{012}]^{-1} \tag{5.80}$$

5.7 SEQUENCE CAPACITANCES OF TRANSMISSION LINE

5.7.1 THREE-PHASE TRANSMISSION LINE WITHOUT OVERHEAD GROUND WIRE

Consider Figure 3.47 and assume that the three-phase conductors are charged. Therefore, for sinusoidal steady-state analysis, both voltage and charge density can be represented by phasors. Thus,

$$[V_{abc}] = [P_{abc}][Q_{abc}] \tag{5.81}$$

or

$$\begin{bmatrix} V_a \\ V_b \\ V_c \end{bmatrix} = \begin{bmatrix} P_{aa} & P_{ab} & P_{ac} \\ P_{ba} & P_{bb} & P_{bc} \\ P_{ca} & P_{cb} & P_{cc} \end{bmatrix} \begin{bmatrix} q_a \\ q_b \\ q_c \end{bmatrix} \tag{5.82}$$

where
 $[P_{abc}]$ = matrix of potential coefficients, based on Figure 3.47
where

$$p_{aa} = \frac{1}{2\pi\varepsilon} \ln \frac{h_{11}}{r_a} \quad \text{F}^{-1}\text{m} \tag{5.83}$$

$$p_{bb} = \frac{1}{2\pi\varepsilon} \ln \frac{h_{22}}{r_b} \quad \text{F}^{-1}\text{m} \tag{5.84}$$

$$p_{cc} = \frac{1}{2\pi\varepsilon} \ln \frac{h_{33}}{r_c} \quad \text{F}^{-1}\text{m} \tag{5.85}$$

$$p_{ab} = p_{ba} = \frac{1}{2\pi\varepsilon} \ln \frac{l_{12}}{D_{12}} \quad \text{F}^{-1}\text{m} \tag{5.86}$$

Symmetrical Components and Sequence Impedances

$$p_{bc} = p_{cb} = \frac{1}{2\pi\varepsilon} \ln \frac{l_{23}}{D_{23}} \quad \text{F}^{-1}\text{m} \tag{5.87}$$

$$p_{ac} = p_{ca} = \frac{1}{2\pi\varepsilon} \ln \frac{l_{31}}{D_{31}} \quad \text{F}^{-1}\text{m} \tag{5.88}$$

Therefore, from Equation 5.81,

$$[Q_{abc}] = [P_{abc}]^{-1} [V_{abc}] \quad \text{C/m} \tag{5.89a}$$

$$= [C_{abc}] [V_{abc}] \quad \text{C/m} \tag{5.89b}$$

since

$$[C_{abc}] = [P_{abc}]^{-1} \quad \text{F/m} \tag{5.90}$$

or

$$[C_{abc}] = \begin{bmatrix} C_{aa} & -C_{ab} & C_{ac} \\ -C_{ba} & C_{bb} & -C_{bc} \\ -C_{ca} & -C_{cb} & C_{cc} \end{bmatrix} \quad \text{F/m} \tag{5.91}$$

where $[C_{abc}]$ is the *matrix of Maxwell's coefficients*, the diagonal terms are *Maxwell's (or capacitance) coefficients*, and the off-diagonal terms are *electrostatic induction coefficients*.

Therefore, the sequence capacitances can be found by using the similarity transformation as

$$[C_{012}] \triangleq [A]^{-1} [C_{abc}] [A] \quad \text{F/m} \tag{5.92a}$$

$$= \begin{bmatrix} C_{00} & C_{01} & C_{02} \\ C_{10} & C_{11} & C_{12} \\ C_{20} & C_{21} & C_{22} \end{bmatrix} \quad \text{F/m} \tag{5.92b}$$

Note that, if the line is *transposed*, the matrix of potential coefficients can be expressed in terms of self- and mutual-potential coefficients as

$$[P_{abc}] = \begin{bmatrix} p_s & p_m & p_m \\ p_m & p_s & p_m \\ p_m & p_m & p_s \end{bmatrix} \tag{5.93}$$

Therefore, using the similarity transformation,

$$[P_{012}] \triangleq [A]^{-1} [P_{abc}] [A] \tag{5.94a}$$

$$= \begin{bmatrix} p_0 & 0 & 0 \\ 0 & p_1 & 0 \\ 0 & 0 & p_2 \end{bmatrix} \quad (5.94b)$$

Thus,

$$[C_{012}] \triangleq [P_{012}]^{-1} \quad (5.95a)$$

$$= \begin{bmatrix} 1/p_0 & 0 & 0 \\ 0 & 1/p_1 & 0 \\ 0 & 0 & 1/p_2 \end{bmatrix} \quad (5.95b)$$

$$= \begin{bmatrix} C_0 & 0 & 0 \\ 0 & C_1 & 0 \\ 0 & 0 & C_2 \end{bmatrix} \quad (5.95c)$$

Alternatively, the sequence capacitances can approximately be calculated without using matrix algebra. For example, the zero-sequence capacitance can be calculated [4] from

$$C_0 = \frac{29.842}{\ln\left(\dfrac{H_{aa}}{D_{aa}}\right)} \text{ nF/mi} \quad (5.96)$$

where
H_{aa} = GMD between three conductors and their images

$$= \left[h_{11} \times h_{22} \times h_{33} \left(l_{12} \times l_{23} \times l_{31} \right)^2 \right]^{1/9} \quad (5.97)$$

D_{aa} = self-GMD of overhead conductors as composite group but with D_s of each conductor taken as its radius

$$= \left[r_a \times r_b \times r_c \left(D_{12} \times D_{23} \times D_{31} \right)^2 \right]^{1/9} \quad (5.98)$$

Note that, D_s has been replaced by the conductor radius since all charge on a conductor resides on its surface. The positive- and negative-sequence capacitances of a line are the same owing to the fact that the physical parameters do not vary with a change in sequence of the applied voltage. Therefore, they are the same as the line-to-neutral capacitance C_n and can be calculated from Equation 3.298 or 3.299.

Symmetrical Components and Sequence Impedances

Note that, the mutual capacitances of the line can be found from Equation 5.91. The capacitances to ground can be expressed as

$$\begin{bmatrix} C_{ag} \\ C_{bg} \\ C_{cg} \end{bmatrix} = \begin{bmatrix} C_{aa} & C_{ab} & C_{ac} \\ -C_{ab} & -C_{bb} & C_{bc} \\ -C_{ac} & C_{bc} & -C_{cc} \end{bmatrix} \begin{bmatrix} 1 \\ -1 \\ -1 \end{bmatrix} \quad (5.99)$$

If the line is transposed, the capacitance to ground is an average value that can be determined from

$$C_{g,avg} = \frac{1}{3}\left(C_{ag} + C_{bg} + C_{cg}\right) \quad (5.100)$$

Also, note that the shunt admittance matrix of the line is

$$[\mathbf{Y}_{abc}] = j\omega\, [C_{abc}] \quad (5.101)$$

Therefore,

$$[\mathbf{Y}_{012}] = [\mathbf{A}]^{-1}\, [\mathbf{Y}_{abc}]\, [\mathbf{A}] \quad (5.102)$$

Thus,

$$\left[C_{012}\right] = \frac{\left[\mathbf{Y}_{012}\right]}{j\omega} \quad (5.103)$$

Hence,

$$[\mathbf{I}_{012}] = j\omega\, [C_{012}]\, [\mathbf{V}_{012}] = j\,[B_{012}]\, [\mathbf{V}_{012}] \quad (5.104)$$

and

$$[\mathbf{I}_{abc}] = [\mathbf{A}]\, [\mathbf{I}_{012}] \quad (5.105)$$

or

$$\begin{aligned}\left[\mathbf{I}_{abc}\right] &= j\omega\left[C_{abc}\right]\left[\mathbf{V}_{abc}\right] \\ &= j\left[B_{abc}\right]\left[\mathbf{V}_{abc}\right]\end{aligned} \quad (5.106)$$

5.7.2 Three-Phase Transmission Line with Overhead Ground Wire

Consider Figure 5.6a and assume that the line is transposed and that the overhead ground wire is denoted by u and that there are nine capacitances involved. The voltages and charge densities involved can be represented by phasors. Therefore,

$$[\mathbf{V}_{abcu}] = [P_{abcu}]\, [Q_{abcu}] \quad (5.107)$$

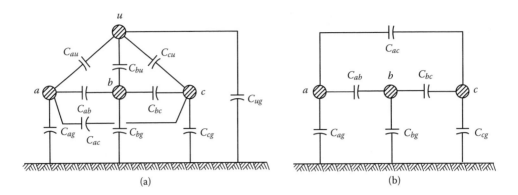

FIGURE 5.6 Three-phase line with one overhead ground wire u: (a) equivalent circuit showing ground wire; (b) equivalent circuit without showing ground wire.

but since, for the ground wire, $V_u = 0$,

$$\begin{bmatrix} V_a \\ V_b \\ V_c \\ \hline 0 \end{bmatrix} = \begin{bmatrix} p_{aa} & p_{ab} & p_{ac} & | & p_{au} \\ p_{ba} & p_{bb} & p_{bc} & | & p_{bu} \\ p_{ca} & p_{cb} & p_{cc} & | & p_{cu} \\ \hline p_{ua} & p_{ub} & p_{uc} & | & p_{uu} \end{bmatrix} \begin{bmatrix} q_a \\ q_b \\ q_c \\ \hline q_u \end{bmatrix} \quad (5.108)$$

The matrix $[P_{abcu}]$ can be calculated as before. The corresponding matrix of the Maxwell coefficients can be found as

$$[C_{abcu}] = [P_{abcu}]^{-1} \quad (5.109)$$

The corresponding equivalent circuit is shown in Figure 5.6a. Such equivalent circuit representation is convenient to study switching transients, traveling waves, overvoltages, etc.

The matrix $[P_{abcu}]$ given in Equation 5.108 can be reduced to $[P_{abc}]$ by using the Kron reduction technique. Therefore, Equation 5.97 can be reexpressed as

$$\begin{bmatrix} V_{abc} \\ \hline 0 \end{bmatrix} = \begin{bmatrix} P_1 & | & P_2 \\ \hline P_3 & | & P_4 \end{bmatrix} \begin{bmatrix} Q_{abc} \\ \hline Q_u \end{bmatrix} \quad (5.110)$$

Where the submatrices $[P_1]$, $[P_2]$, $[P_3]$, and $[P_4]$ are specified in the partitioned matrix $[P_{abcu}]$ in Equation 5.108. Thus, after the reduction,

$$[V_{abc}] = [P_{abc}][Q_{abc}] \quad (5.111)$$

where

$$[P_{abc}] \triangleq [P_1] - [P_2][P_4]^{-1}[P_3] \quad (5.112)$$

Thus, the corresponding matrix of the Maxwell coefficients can be found as

$$[C_{abc}] = [P_{abc}]^{-1} \quad (5.113)$$

as before. The corresponding equivalent circuit is shown in Figure 5.6b, and such representation is convenient to study a load-flow problem. Of course, the average capacitances to ground can be found as before.

Alternatively, the sequence capacitances can approximately be calculated without using the matrix algebra. For example, the zero-sequence capacitance can be calculated [4] from

$$C_0 = \frac{29.842 \ln\left(\dfrac{h_{gg}}{D_{gg}}\right)}{\ln\left(\dfrac{H_{aa}}{D_{aa}}\right) \times \ln\left(\dfrac{H_{gg}}{D_{gg}}\right) - \left[\ln\left(\dfrac{H_{ag}}{D_{ag}}\right)\right]^2} \quad \text{nF/mi} \qquad (5.114)$$

where
- H_{aa} = given by Equation 5.97
- D_{aa} = given by Equation 5.98
- h_{gg} = GMD between ground wires and their images
- D_{gg} = self-GMD of ground wires with $D_s = r_g$
- H_{ag} = GMD between phase conductors and images of ground wires
- D_{ag} = GMD between phase conductors and ground wires

If the transmission line is untransposed, both electrostatic and electromagnetic unbalances exist in the system. If the system neutral is (*solidly*) grounded, in the event of an electrostatic unbalance, there will be a neutral residual current flow in the system due to the unbalance in the charging currents of the line. Such residual current flow is continuous and independent of the load. Since the neutral is grounded, $\mathbf{V}_n = \mathbf{V}_{a0} = 0$, and the *zero-sequence displacement* or *unbalance* is

$$\mathbf{d}_0 \triangleq \frac{\mathbf{C}_{01}}{\mathbf{C}_{11}} \qquad (5.115)$$

and the *negative-sequence unbalance* is

$$\mathbf{d}_2 \triangleq \frac{\mathbf{C}_{21}}{\mathbf{C}_{11}} \qquad (5.116)$$

If the system neutral is not grounded, there will be the neutral voltage $\mathbf{V}_n \neq 0$, and therefore, the neutral point will be shifted. Such zero-sequence neutral displacement or unbalance is defined as

$$\mathbf{d}_0 \triangleq \frac{\mathbf{C}_{01}}{\mathbf{C}_{00}} \qquad (5.117)$$

EXAMPLE 5.6

Consider the line configuration shown in Figure 5.5. Assume that the 115-kV line is not transposed and its conductors are made up of 500-kcmil, 30/7-strand ACSR conductors. Ignore the overhead ground wire and determine the following:

(a) Matrix of potential coefficients
(b) Matrix of Maxwell's coefficients

(c) Matrix of sequence capacitances
(d) Zero- and negative-sequence electrostatic unbalances, assuming that the system neutral is solidly grounded

Solution

(a) The corresponding potential coefficients arc calculated using Equations 5.83 through 5.88. For example,

$$p_{aa} = \frac{1}{2\pi\varepsilon} \ln\left(\frac{h_{11}}{r_a}\right)$$

$$= 11.185 \ln\left(\frac{90}{0.037667}\right)$$

$$= 87.0058 \text{ F}^{-1}\text{m}$$

$$p_{ab} = \frac{1}{2\pi\varepsilon} \ln\left(\frac{l_{12}}{D_{12}}\right)$$

$$= 11.185 \ln\left(\frac{82.0244}{8.2462}\right)$$

$$= 53.6949 \text{ F}^{-1}\text{m}$$

where

$$l_{12} = \left[22 + (45+37)^2\right]^{1/2}$$

$$= 82.0244 \text{ ft}$$

$$D_{12} = \left(2^2 + 8^2\right)^{1/2}$$

$$= 8.2462 \text{ ft}$$

The others can also be found similarly. Therefore,

$$[P_{abc}] = \begin{bmatrix} 87.0058 & 25.6949 & 21.9131 \\ 25.6949 & 84.8164 & 19.7635 \\ 21.9131 & 19.7635 & 85.6884 \end{bmatrix}$$

(b)

$$[C_{abc}] = [P_{abc}]^{-1}$$

$$= \begin{bmatrix} 1.31\times10^{-2} & -3.38\times10^{-3} & -2.58\times10^{-3} \\ -3.38\times10^{-3} & 1.33\times10^{-2} & -2.21\times10^{-3} \\ -2.58\times10^{-3} & -2.21\times10^{-3} & 1.28\times10^{-2} \end{bmatrix}$$

(c)

$$[\mathbf{C}_{012}] = [\mathbf{A}]^{-1}[\mathbf{C}_{abc}][\mathbf{A}]$$

$$= \begin{bmatrix} 7.666 \times 10^{-3} + j3.1 \times 10^{-2} & -2.38 \times 10^{-4} + j8.94 \times 10^{-5} & -2.38 \times 10^{-4} + j8.94 \times 10^{-5} \\ -2.38 \times 10^{-4} - j8.94 \times 10^{-5} & 1.58 \times 10^{-2} - j4.37 \times 10^{-19} & 5.33 \times 10^{-4} + j6.02 \times 10^{-4} \\ -2.38 \times 10^{-4} + j8.94 \times 10^{-5} & 5.33 \times 10^{-4} + j6.02 \times 10^{-4} & 1.58 \times 10^{-2} - j1.30 \times 10^{-18} \end{bmatrix}$$

(d) From Equation 5.115,

$$d_0 = \frac{C_{01}}{C_{11}}$$

$$= \frac{-2.38 \times 10^{-4} + j8.94 \times 10^{-5}}{1.58 \times 10^{-2}}$$

$$= 0.0160 \angle 159° \text{ or } 1.60\%$$

and from Equation 5.116,

$$d_2 = \frac{C_{21}}{C_{11}}$$

$$= -\frac{5.32 \times 10^{-4} + j6.02 \times 10^{-4}}{1.58 \times 10^{-2}}$$

$$= 0.0508 \angle 228.5° \text{ or } 5.08\%$$

5.8 SEQUENCE IMPEDANCES OF SYNCHRONOUS MACHINES

In general, the impedances to positive-, negative-, and zero-sequence currents in synchronous machines (as well as other rotating machines) have different values. The positive-sequence impedance of the synchronous machine can be selected to be its *subtransient* (X''_d), *transient* (X'_d), or synchronous* (X_d) reactance depending on the time assumed to elapse from the instant of fault initiation to the instant at which values are desired (e.g., for relay response, breaker opening, or sustained fault conditions). Usually, however, in fault studies, the subtransient reactance is taken as the positive-sequence reactance of the synchronous machine.

The negative-sequence impedance of a synchronous machine is usually determined from

$$\mathbf{Z}_2 = jX_2 = j\left(\frac{X''_d + X''_q}{2}\right) \tag{5.118}$$

In a cylindrical-rotor synchronous machine, the subtransient and negative-sequence reactances are the same, as shown in Table 5.3.

The zero-sequence impedance of a synchronous machine varies widely and depends on the pitch of the armature coils. It is much smaller than the corresponding positive- and negative-sequence

---

* It is also called the *direct-axis synchronous reactance*. It is also denoted by X_s.

TABLE 5.3
Typical Reactances of Three-Phase Synchronous Machines

	Turbine Generators						Salient-Pole Generators						Synchronous Condensers					
	Two Pole			Four Pole			With Dampers			With Dampers			Air Cooled			Hydrogen Cooled		
	Low	Avg.	High	Low	Avg.	High	Low	Avg.	High	Low	Avg.	High	Low	Avg.	High	Low	Avg.	High
X_d	0.95	1.2	1.45	1.00	1.2	1.45	0.6	1.25	1.5	0.6	1.25	1.5	1.25	1.85	2.2	1.5	2.2	2.65
X'_d	0.12	0.15	0.21	0.2	0.23	0.28	0.2	0.3	0.5	0.2	0.3	0.5	0.3	0.4	0.5	0.36	0.48	0.6
X''_d	0.07	0.09	0.14	0.12	0.14	0.17	0.13	0.2	0.32	0.2	0.3	0.5	0.19	0.27	0.3	0.23	0.32	0.36
X_q	0.92	1.16	1.42	0.92	1.16	1.42	0.4	0.7	0.8	0.4	0.7	0.8	0.95	1.15	1.3	1.1	1.35	1.55
X_2	0.07	0.09	0.14	0.12	0.14	0.17	0.13	0.2	0.32	0.35	0.48	0.65	0.18	0.26	0.4	0.22	0.31	0.48
X_0	0.01	0.03	0.08	0.015	0.08	0.14	0.03	0.18	0.23	0.03	0.19	0.24	0.025	0.12	0.15	0.03	0.14	0.18

Source: Westinghouse Electric Corporation, *Electrical Transmission and Distribution Reference Book.* WEC, East Pittsburgh, 1964.

reactances. It can be measured by connecting the three armature windings in series and applying a single-phase voltage. The ratio of the terminal voltage of one phase winding to the current is the zero-sequence reactance. It is approximately equal to the zero-sequence reactance. Table 5.3 [5] gives typical reactance values of three-phase synchronous machines. Note that, in the above discussion, the resistance values are ignored because they are much smaller than the corresponding reactance values.

Figure 5.7 shows the equivalent circuit of a cylindrical-rotor synchronous machine with constant field current. Since the coil groups of the three-phase stator armature windings are displaced from each other by 120 electrical degrees, balanced three-phase sinusoidal voltages are induced in the stator windings. Furthermore, each of the three self-impedances and mutual impedances are equal to each other, respectively, owing to the machine symmetry. Therefore, taking into account the neutral impedance $\mathbf{Z}_n$ and applying Kirchhoff's voltage law it can be shown that

$$\mathbf{E}_a = (R_\phi + jX_s + \mathbf{Z}_n)\mathbf{I}_a + (jX_m + \mathbf{Z}_n)\mathbf{I}_b + (jX_m + \mathbf{Z}_n)\mathbf{I}_c + \mathbf{V}_a \quad (5.119)$$

$$\mathbf{E}_b = (jX_m + \mathbf{Z}_n)\mathbf{I}_a + (R_\phi + jX_s + \mathbf{Z}_n)\mathbf{I}_b + (jX_m + \mathbf{Z}_n)\mathbf{I}_c + \mathbf{V}_b \quad (5.120)$$

$$\mathbf{E}_c = (jX_m + \mathbf{Z}_n)\mathbf{I}_a + (jX_m + \mathbf{Z}_n)\mathbf{I}_b + (R_\phi + jX_s + \mathbf{Z}_n)\mathbf{I}_c + \mathbf{V}_c \quad (5.121)$$

or, in matrix form,

$$\begin{bmatrix} \mathbf{E}_a \\ \mathbf{E}_b \\ \mathbf{E}_c \end{bmatrix} = \begin{bmatrix} \mathbf{Z}_s & \mathbf{Z}_m & \mathbf{Z}_s \\ \mathbf{Z}_m & \mathbf{Z}_s & \mathbf{Z}_m \\ \mathbf{Z}_m & \mathbf{Z}_m & \mathbf{Z}_s \end{bmatrix} \begin{bmatrix} \mathbf{I}_a \\ \mathbf{I}_b \\ \mathbf{I}_c \end{bmatrix} + \begin{bmatrix} \mathbf{V}_a \\ \mathbf{V}_b \\ \mathbf{V}_c \end{bmatrix} \quad (5.122)$$

where

$$\mathbf{Z}_s = R_\phi + jX_s + \mathbf{Z}_n \quad (5.123)$$

$$\mathbf{Z}_m = jX_m + \mathbf{Z}_n \quad (5.124)$$

$$\mathbf{E}_a = \mathbf{E}_a \quad (5.125)$$

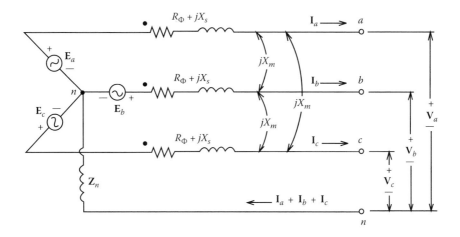

FIGURE 5.7 Equivalent circuit of cylindrical-rotor synchronous machine.

$$\mathbf{E}_b = \mathbf{a}^2 \mathbf{E}_a \quad (5.126)$$

$$\mathbf{E}_c = \mathbf{a}\mathbf{E}_a \quad (5.127)$$

Alternatively, Equation 5.122 can be written in shorthand matrix notation as

$$[\mathbf{E}_{abc}] = [\mathbf{Z}_{abc}][\mathbf{I}_{abc}] + [\mathbf{V}_{abc}] \quad (5.128)$$

Multiplying both sides of this equation by $[\mathbf{A}]^{-1}$ and also substituting Equation 5.26 into it,

$$[\mathbf{A}]^{-1}[\mathbf{E}_{abc}] = [\mathbf{A}]^{-1}[\mathbf{Z}_{abc}][\mathbf{A}][\mathbf{I}_{012}] + [\mathbf{A}]^{-1}[\mathbf{V}_{abc}] \quad (5.129)$$

where

$$[\mathbf{A}]^{-1}[\mathbf{E}_{abc}] = \frac{1}{3}\begin{bmatrix} 1 & 1 & 1 \\ 1 & \mathbf{a} & \mathbf{a}^2 \\ 1 & \mathbf{a}^2 & \mathbf{a} \end{bmatrix}\begin{bmatrix} \mathbf{E} \\ \mathbf{a}^2\mathbf{E} \\ \mathbf{a}\mathbf{E} \end{bmatrix}$$

$$= \begin{bmatrix} 0 \\ \mathbf{E} \\ 0 \end{bmatrix} \quad (5.130)$$

$$[\mathbf{Z}_{012}] \triangleq [\mathbf{A}]^{-1}[\mathbf{Z}_{abc}][\mathbf{A}] \quad (5.38)$$

$$[\mathbf{V}_{012}] = [\mathbf{A}]^{-1}[\mathbf{V}_{abc}] \quad (5.19)$$

Also, due to the symmetry of the machine,

$$[\mathbf{Z}_{012}] = \begin{bmatrix} \mathbf{Z}_s + 2\mathbf{Z}_m & 0 & 0 \\ 0 & \mathbf{Z}_s - \mathbf{Z}_m & 0 \\ 0 & 0 & \mathbf{Z}_s - \mathbf{Z}_m \end{bmatrix} \quad (5.131)$$

or

$$[\mathbf{Z}_{012}] = \begin{bmatrix} \mathbf{Z}_{00} & 0 & 0 \\ 0 & \mathbf{Z}_{11} & 0 \\ 0 & 0 & \mathbf{Z}_{22} \end{bmatrix} \quad (5.132)$$

where

$$\mathbf{Z}_{00} = \mathbf{Z}_s + 2\mathbf{Z}_m = R_\phi + j(X_s + 2X_m) + 3\mathbf{Z}_n \quad (5.133)$$

$$\mathbf{Z}_{11} = \mathbf{Z}_s - \mathbf{Z}_m = R_\phi + j(X_s - X_m) \quad (5.134)$$

$$\mathbf{Z}_{22} = \mathbf{Z}_s - \mathbf{Z}_m = R_\phi + j(X_s - X_m) \quad (5.135)$$

Symmetrical Components and Sequence Impedances

Therefore, Equation 5.128 in terms of the symmetrical components can be expressed as

$$\begin{bmatrix} 0 \\ \mathbf{E}_a \\ 0 \end{bmatrix} = \begin{bmatrix} \mathbf{Z}_{00} & 0 & 0 \\ 0 & \mathbf{Z}_{11} & 0 \\ 0 & 0 & \mathbf{Z}_{22} \end{bmatrix} \begin{bmatrix} \mathbf{I}_{a0} \\ \mathbf{I}_{a1} \\ \mathbf{I}_{a2} \end{bmatrix} + \begin{bmatrix} \mathbf{V}_{a0} \\ \mathbf{V}_{a1} \\ \mathbf{V}_{a2} \end{bmatrix} \quad (5.136)$$

or, in shorthand matrix notation,

$$[\mathbf{E}] = [\mathbf{Z}_{012}][\mathbf{I}_{012}] + [\mathbf{V}_{012}] \quad (5.137)$$

Similarly,

$$\begin{bmatrix} \mathbf{V}_{a0} \\ \mathbf{V}_{a1} \\ \mathbf{V}_{a2} \end{bmatrix} = \begin{bmatrix} 0 \\ \mathbf{E}_a \\ 0 \end{bmatrix} - \begin{bmatrix} \mathbf{Z}_{00} & 0 & 0 \\ 0 & \mathbf{Z}_{11} & 0 \\ 0 & 0 & \mathbf{Z}_{22} \end{bmatrix} \begin{bmatrix} \mathbf{I}_{a0} \\ \mathbf{I}_{a1} \\ \mathbf{I}_{a2} \end{bmatrix} \quad (5.138)$$

or

$$[\mathbf{V}_{012}] = [\mathbf{E}] - [\mathbf{Z}_{012}][\mathbf{I}_{012}] \quad (5.139)$$

Note that, the machine sequence impedances in the above equations are

$$\mathbf{Z}_0 \triangleq \mathbf{Z}_{00} - 3\mathbf{Z}_n \quad (5.140)$$

$$\mathbf{Z}_1 \triangleq \mathbf{Z}_{11} \quad (5.141)$$

$$\mathbf{Z}_2 \triangleq \mathbf{Z}_{22} \quad (5.142)$$

The expression given in Equation 5.140 is due to the fact that the impedance $\mathbf{Z}_n$ is external to the machine. Figure 5.8 shows the sequence networks of a synchronous machine.

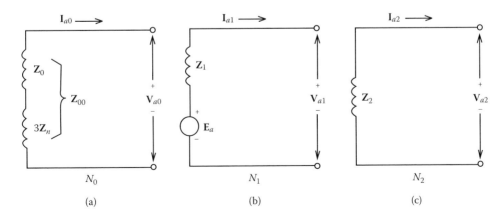

FIGURE 5.8 Sequence networks of synchronous machine: (a) zero-sequence network; (b) positive-sequence network; (c) negative-sequence network.

5.9 ZERO-SEQUENCE NETWORKS

It is important to note that the zero-sequence system, in a sense, is not a three-phase system but a single-phase system. This is because the zero-sequence currents and voltages are equal in magnitude and in phase at any point in all the phases of the system. However, the zero-sequence currents can only exist in a circuit if there is a complete path for their flow. Therefore, if there is no

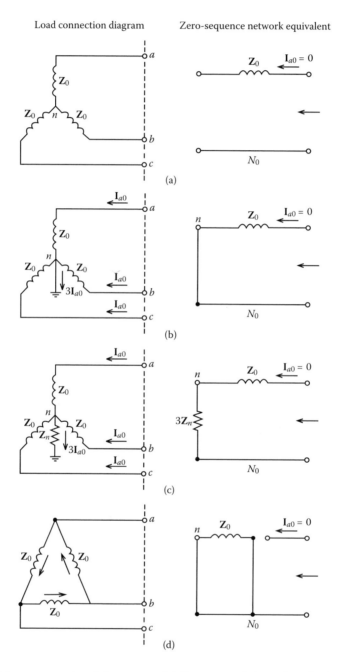

FIGURE 5.9 Zero-sequence network for wye- and delta-connected three-phase loads: (a) wye-connected load with undergrounded neutral; (b) wye-connected load with grounded neutral; (c) wye-connected load grounded through neutral impedance; (d) delta-connected load.

complete path for zero-sequence currents in a circuit, the zero-sequence impedance is infinite. In a zero-sequence network drawing, this infinite impedance is indicated by an open circuit.

Figure 5.9 shows zero-sequence networks for wye- and delta-connected three-phase loads. Note that, a wye-connected load with an ungrounded neutral has infinite impedance to zero-sequence currents since there is no return path through the ground or a neutral conductor, as shown in Figure 5.9a. On the other hand, a wye-connected load with solidly grounded neutral, as shown in Figure 5.9b, provides a return path for the zero-sequence currents flowing through the three phases and their sum, $3\mathbf{I}_{a0}$, flowing through the ground. If the neutral is grounded through some impedance $\mathbf{Z}_n$ as shown in Figure 5.9c, an impedance of $3\mathbf{Z}_n$ should be inserted between the neutral point n and the zero-potential bus N_0 in the zero-sequence network. The reason for this is that a current of $3\mathbf{I}_{a0}$ produces a zero-sequence voltage drop of $3\mathbf{I}_{a0}\mathbf{Z}_n$ between the neutral point n and the ground. Therefore, to reflect this voltage drop in the zero-sequence network, where the zero-sequence current $3\mathbf{I}_{a0}$ flows, the neutral impedance should be $3\mathbf{Z}_n$. A delta-connected load, as shown in Figure 5.9d, provides no path for zero-sequence currents flowing in the line. Therefore, its zero-sequence impedance, as seen from its terminals, is infinite. Yet, it is possible to have zero-sequence currents circulating within the delta circuit. However, they have to be produced in the delta by zero-sequence voltages or by induction from an outside source.

5.10 SEQUENCE IMPEDANCES OF TRANSFORMERS

A three-phase transformer may be made up of three identical single-phase transformers. If this is the case, it is called a three-phase *transformer bank*. Alternatively, it may be built as a three-phase transformer having a single common core (either with shell-type or core-type design) and a tank. For the sake of simplicity, here only the three-phase transformer banks will be reviewed. The impedance of a transformer to both positive- and negative-sequence currents is the same. Even though the zero-sequence series impedances of three-phase units are little different than the positive- and negative-sequence series impedances, it is often assumed in practice that series impedances of all sequences are the same without paying attention to the transformer type

$$\mathbf{Z}_0 = \mathbf{Z}_1 = \mathbf{Z}_2 = \mathbf{Z}_{trf} \tag{5.143}$$

If the flow of zero-sequence current is prevented by the transformer connection, $\mathbf{Z}_0$ is infinite.

Figure 5.10 shows zero-sequence network equivalents of three-phase transformer banks made up of three identical single-phase transformers having two windings with excitation currents neglected. The possible paths for the flow of zero-sequence current are indicated on the connection diagrams, as shown in Figure 5.10a, 5.10c, and 5.10e. If there is no path shown on the connection diagram, this means that the transformer connection prevents the flow of the zero-sequence current by not providing a path for it, as indicated in Figure 5.10b, 5.10d, and 5.10f.

Note that even though the delta–delta bank can have zero-sequence currents circulating within its delta windings, it also prevents the flow of the zero-sequence current outside the delta windings by not providing a return path for it, as shown in Figure 5.10e.

Also, note that if the neutral point n of the wye winding (shown in Figure 5.10a or c) is grounded through $\mathbf{Z}_n$, the corresponding zero-sequence impedance $\mathbf{Z}_0$ should be replaced by $\mathbf{Z}_0 + 3\mathbf{Z}_n$.

If the wye winding is solidly grounded, the $\mathbf{Z}_n$ is zero, and therefore $3\mathbf{Z}_n$ should be replaced by a short circuit. On the other hand, if the connection is ungrounded, the $\mathbf{Z}_n$ is infinite, and therefore $3\mathbf{Z}_n$ should be replaced with an open circuit. It is interesting to observe that the type of grounding only affects the zero-sequence network, not the positive- and negative-sequence networks.

It is interesting to note that there is no path for the flow of zero-sequence current in a wye-grounded–wye-connected three-phase transformer bank, as shown in Figure 5.10b. This is because there is no zero-sequence current in any given winding on the wye side of the transformer bank since it has an ungrounded wye connection. Therefore, because of the lack of equal and opposite ampere turns in

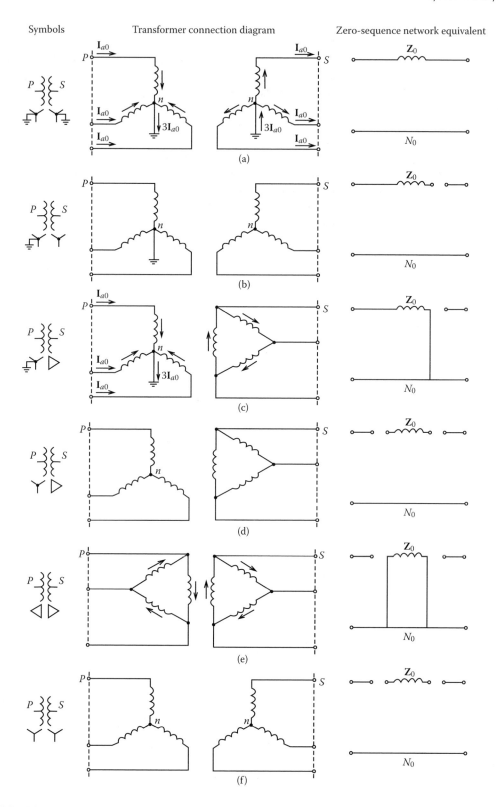

FIGURE 5.10 Zero-sequence network equivalents of three-phase transformer banks made of three identical single-phase transformers with two windings.

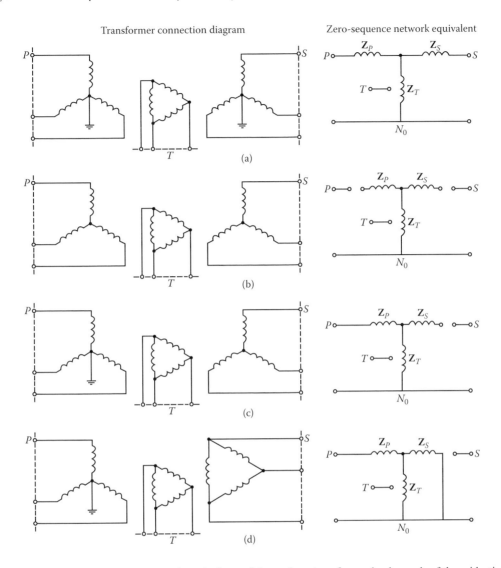

FIGURE 5.11 Zero-sequence network equivalents of three-phase transformer banks made of three identical single-phase transformers with three windings.

the wye side of the transformer bank, there cannot be any zero-sequence current in the corresponding winding on the wye-grounded side of the transformer, with the exception of a negligible small magnetizing current.

Figure 5.11 shows zero-sequence network equivalents of three-phase transformer banks made of three identical single-phase transformers with three windings. The impedances of the three-winding transformer between primary, secondary, and tertiary terminals, indicated by P, S, and T, respectively, taken two at a time with the other winding open, are $\mathbf{Z}_{PS}$, $\mathbf{Z}_{PT}$, and $\mathbf{Z}_{ST}$, the subscripts indicating the terminals between which the impedances are measured. Note that, only the wye–wye connection with delta tertiary, shown in Figure 5.11a, permits zero-sequence current to flow in from either wye line (as long as the neutrals are grounded).

EXAMPLE 5.7

Consider the power system shown in Figure 5.12 and the associated data given in Table 5.4. Assume that each three-phase transformer bank is made of three single-phase transformers. Do the following:

(a) Draw the corresponding positive-sequence network
(b) Draw the corresponding negative-sequence network
(c) Draw the corresponding zero-sequence network

Solution

(a) The positive-sequence network is shown in Figure 5.13a
(b) The negative-sequence network is shown in Figure 5.13b
(c) The zero-sequence network is shown in Figure 5.13c

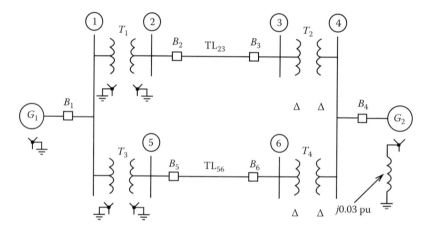

FIGURE 5.12 Power system for Example 5.7.

TABLE 5.4
System Data for Example 5.7

Network Component	MVA Rating	Voltage Rating (kV)	X_1 (pu)	X_2 (pu)	X_0 (pu)
G_1	200	20	0.2	0.14	0.06
G_2	200	13.2	0.2	0.14	0.06
T_1	200	20/230	0.2	0.2	0.2
T_2	200	13.2/230	0.3	0.3	0.3
T_3	200	20/230	0.25	0.25	0.25
T_4	200	13.2/230	0.35	0.35	0.35
TL_{23}	200	230	0.15	0.15	0.3
TL_{56}	200	230	0.22	0.22	0.5

Symmetrical Components and Sequence Impedances

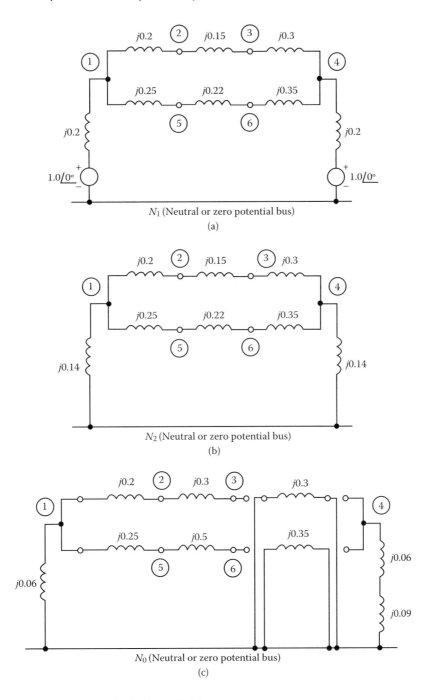

FIGURE 5.13 Sequence networks for Example 5.7.

EXAMPLE 5.8

Consider the power system given in Example 5.7 and assume that there is a fault on bus 3. Reduce the sequence networks drawn in Example 5.7 to their Thévenin equivalents "looking in" at bus 3.

(a) Show the steps of the positive-sequence network reduction
(b) Show the steps of the negative-sequence network reduction
(c) Show the steps of the zero-sequence network reduction

Solution

(a) Figure 5.14 shows the steps of the positive-sequence network reduction. Note that the delta that exits between nodes 1, 3, and 4, as shown in Figure 5.14a, must be replaced by its equivalent wye configuration, as shown in Figure 5.14b, by performing the following calculations:

$$\mathbf{Z}_1 = j\frac{0.35 \times 0.82}{0.35 + 0.82 + 0.3}$$
$$= j0.1952 \text{ pu}$$

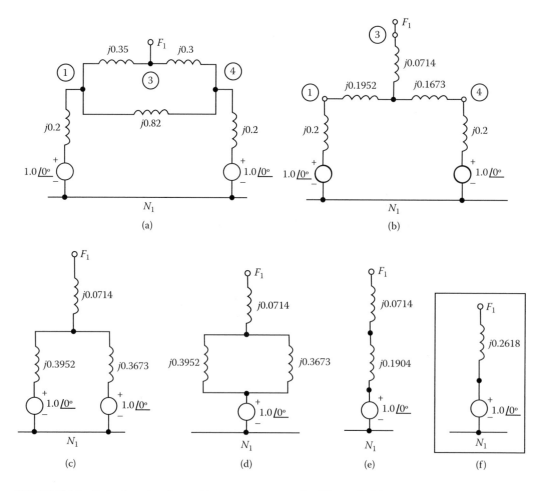

FIGURE 5.14 Reduction steps for positive-sequence network of Example 5.8.

$$Z_2 = j\frac{0.3 \times 0.82}{0.35 + 0.82 + 0.3}$$
$$= j0.1673 \text{ pu}$$

$$Z_3 = j\frac{0.35 \times 0.3}{0.35 + 0.3 + 0.82}$$
$$= j0.0714 \text{ pu}$$

(b) Figure 5.15 shows the steps of the negative-sequence network reduction. Note that, the delta that exits between nodes 1, 3, and 4, as shown in Figure 5.15a, must be replaced by its equivalent wye configuration, as shown in Figure 5.15b, by performing the calculations as in part (a) above.

(c) Figure 5.16 shows the steps of the zero-sequence network reduction.

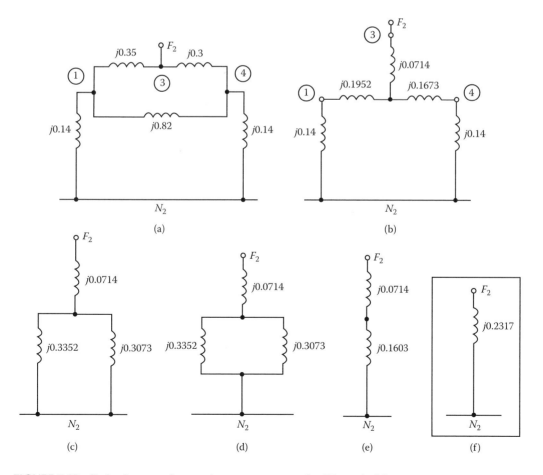

FIGURE 5.15 Reduction steps for negative-sequence network of Example 5.8.

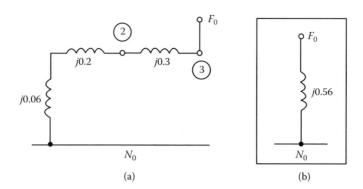

FIGURE 5.16 Reduction steps for zero-sequence network of Example 5.8.

REFERENCES

1. Fortescue, C. L., Method of symmetrical coordinates applied to the solution of polyphase networks. *Trans. Am. Inst. Electr. Eng.* 37, 1027–1140 (1918).
2. Carson, J. R., Wave propagation in overhead wires with ground return. *Bell Syst. Tech. J.* 5 539–554 (1926).
3. Gross, E. T. B., and Hesse, M. H., Electromagnetic unbalance of untransposed lines, *Trans. Am. Inst. Electr. Eng.* 72 (Pt. 3) 1323–1336 (1953).
4. Anderson. P. M., *Analysis of Faulted Power Systems*. Iowa State Univ. Press, Ames, IA, 1973.
5. Westinghouse Electric Corporation, *Electrical Transmission and Distribution Reference Book*. WEC, East Pittsburgh, 1964.

GENERAL REFERENCES

Atabekov, G. I., *The Relay Protection of High Voltage Networks*. Pergamon Press, New York, 1960.
Brown, H. E., *Solution of Large Networks by Matrix Methods*. Wiley, New York, 1975.
Calabrese, G. O., *Symmetrical Components Applied to Electric Power Network*. Ronald Press, New York, 1959.
Clarke, E., Simultaneous faults on three-phase systems, *Trans. Am. Inst. Electr. Eng.* 50, 919–941 (1931).
Clarke, E., *Circuit Analysis of A-C Power Systems*, vol. 1. General Electric Co., Schenectady, New York, 1960.
Clarke, E., *Circuit Analysis of A-C Power Systems*, vol. 2. General Electric Co., Schenectady, New York, 1960.
Clem, J. E., Reactance of transmission lines with ground return. *Trans. Am. Inst. Electr. Eng.* 50, 901–918 (1931).
Dawalibi, F., and Niles, G. B., Measurements and computations of fault current distribution of overhead transmission lines. *IEEE Trans. Power Appar. Syst.* 3, 553–560 (1984).
Duesterhoeft, W. C., Schutz, M. W., Jr., and Clarke, E., Determination of instantaneous currents and voltages by means of alpha, beta, and zero components. *Trans. Am. Inst. Electr. Eng.* 70 (Pt. 3) 1248–1255 (1951).
Elgerd, O. I., *Electric Energy Systems Theory: An Introduction*. McGraw-Hill, New York, 1971.
Ferguson, W. H., Symmetrical component network connections for the solution of phase interchange faults. *Trans. Am. Inst. Electr. Eng.* 78 (Pt 3) 948–950 (1959).
Garin, A. N., Zero-phase-sequence characteristics of transformers, parts I and II. *Gen. Electr. Rev.* 43, 131–136, 174–170 (1940).
Gönen, T., *Electric Power Distribution System Engineering*. McGraw-Hill, New York, 1986.
Gönen, T., Nowikowski, J., and Brooks, C. L., Electrostatic unbalances of transmission Lines with 'N' overhead ground wires, Part I. *Proc. Model. Simul. Conf.* 17 (Pt. 2) 459–464 (1986).
Gönen, T., Nowikowski, J., and Brooks, C. L., Electrostatic unbalances of transmission lines with 'N' overhead ground wires, Part I. *Proc. Model. Simul. Conf.* 17 (Pt. 2) 465–470 (1986).
Gönen, T., *Modern Power System Analysis*, Wiley, New York, 1987.
Gönen, T., and Haj-mohamadi, M. S., Electromagnetic unbalances of six-phase transmission lines. *Electr. Power Energy Syst.* 11 (2) 78–84 (1989).
Gönen, T., *Electric Power Distribution System Engineering*. CRC Press, Boca Raton, FL, 2008.

Harder, E. L., Sequence network connections for unbalanced load and fault conditions. *Electr. J.* 34 (12), 481–488, 1937.

Hobson, J. E., and Whitehead, D. L., Symmetrical components, in *Electrical Transmission and Distribution Reference Book*, Chapter 2. WEC, East Pittsburgh, Pennsylvania, 1964.

Lyle, A. G., *Major Faults on Power Systems*. Chapman & Hall, London, 1952.

Lyon, W. V., *Applications of the Method of Symmetrical Components*. McGraw-Hill, New York, 1937.

Neuenswander, J. R., *Modern Power Systems*. International Textbook Co., New York, 1971.

Roper, R., *Kurzchlussstrime in Drehstromnetzen*, 5th Ger. ed. (translated as *Short-Circuit Currents in Three-Phase Networks*). Siemens Aktienges, Munich, Germany, 1972.

Stevenson, W. D., Jr., *Elements of Power System Analysis*, 4th ed. McGraw-Hill, New York, 1982.

Wagner, C. F., and Evans, R. D., *Symmetrical Components*, McGraw-Hill, New York, 1933.

Wagner, C. F., and Evans, R. D., *Symmetrical Components*. McGraw-Hill, New York, 1941.

Weedy, B. M., *Electric Power Systems*, 3rd ed. Wiley, New York, 1979.

PROBLEMS

1. Determine the symmetrical components for the phase currents of $\mathbf{I}_a = 125\angle 20°$, $\mathbf{I}_b = 175\angle -100°$, and, $\mathbf{I}_c = 95\angle 155°$ A.
2. Assume that the unbalanced phase currents are $\mathbf{I}_a = 100\angle 180°$, $\mathbf{I}_b = 100\angle 0°$, and, $\mathbf{I}_c = 10\angle 20°$ A.
   (a) Determine the symmetrical components.
   (b) Draw a phasor diagram showing $\mathbf{I}_{a0}$, $\mathbf{I}_{a1}$, $\mathbf{I}_{a2}$, $\mathbf{I}_{b0}$, $\mathbf{I}_{b1}$, $\mathbf{I}_{b2}$, $\mathbf{I}_{c0}$, $\mathbf{I}_{c1}$, and $\mathbf{1}_{c2}$ (i.e., the positive-, negative-, and zero-sequence currents for each phase).
3. Assume that $\mathbf{V}_{a1} = 180\angle 0°$, $\mathbf{V}_{a2} = 100\angle 100°$, and, $\mathbf{V}_{a0} = 250\angle -40°$ V.
   (a) Draw a phasor diagram showing all the nine symmetrical components.
   (b) Find the phase voltages $[\mathbf{V}_{abc}]$ using the equation

   $$[\mathbf{V}_{abc}] = [\mathbf{A}][\mathbf{V}_{012}]$$

   (c) Find the phase voltages $[\mathbf{V}_{abc}]$ graphically and check the results against the ones found in part b.
4. Repeat Example 5.2 assuming that the phase voltages and currents are given as

   $$[\mathbf{V}_{abc}] = \begin{bmatrix} 100\angle 0° \\ 100\angle 60° \\ 100\angle -60° \end{bmatrix} \text{ and } [\mathbf{I}_{abc}] = \begin{bmatrix} 10\angle -30° \\ 10\angle 30° \\ 10\angle -90° \end{bmatrix}$$

5. Determine the symmetrical components for the phase currents of $\mathbf{I}_a = 100\angle 20°$, $\mathbf{I}_b = 50\angle -20°$, and, $\mathbf{I}_c = 150\angle 180°$ A. Draw a phasor diagram showing all the nine symmetrical components.
6. Assume that $\mathbf{I}_{a0} = 50 - j86.6$, $\mathbf{I}_{a1} = 200\angle 0°$, and, $\mathbf{I}_a = 400\angle 0°$ A. Determine the following:
   (a) The negative sequence current $\mathbf{I}_{a2}$
   (b) The faulted phase b current $\mathbf{I}_b$
   (c) The faulted phase c current $\mathbf{I}_c$
7. Determine the symmetrical components for the phase currents of $\mathbf{I}_a = 200\angle 0°$, $\mathbf{I}_b = 175\angle -90°$, and, $\mathbf{I}_c = 100\angle 90°$ A.
8. Use the symmetrical components for the phase voltages and verify the following line-to-line voltage equations:
   (a) $\mathbf{V}_{ab} = \sqrt{3}\left(\mathbf{V}_{a1} \angle 30° + \mathbf{V}_{a2} \angle -30°\right)$
   (b) $\mathbf{V}_{bc} = \sqrt{3}\left(\mathbf{V}_{a1} \angle -90° + \mathbf{V}_{a2} \angle 90°\right)$
   (c) $\mathbf{V}_{ca} = \sqrt{3}\left(\mathbf{V}_{a1} \angle 150° + \mathbf{V}_{a2} \angle -150°\right)$

9. Consider Example 5.3 and assume that the voltage applied at the sending end of the line is $69\angle 0°$ kV Determine the phase current matrix from Equation 5.35.
10. Consider a three-phase horizontal line configuration and assume that the phase spacings are $D_{ab} = 30$ ft, $D_{bc} = 30$ ft, and $D_{ca} = 60$ ft. The line conductors are made of 500 kcmil, 37-strand copper conductors. Assume that the 100-mi-long untransposed transmission line operates at 50°C, 60 Hz. If the earth has an average resistivity, determine the following:
    (a) Self-impedances of line conductors in ohms per mile
    (b) Mutual impedances of line conductors in ohms per mile
    (c) Phase impedance matrix of line in ohms
11. Consider a 50-mi-long completely transposed transmission line operating at 25°C, 50 Hz, and having 500-kcmil ACSR conductors. The three-phase conductors have a triangular configuration with spacings of $D_{ab} = 6$ ft, $D_{bc} = 10$ ft, and $D_{ca} = 8$ ft. If the earth is considered to be dry earth, determine the following:
    (a) Zero-sequence impedance of line
    (b) Positive-sequence impedance of line
    (c) Negative-sequence impedance of line
12. Consider a three-phase, vertical pole-top conductor configuration. Use 50°C, 60 Hz and assume that the phase spacings are $D_{ab} = 72$ in., $D_{bc} = 72$ in., and $D_{ca} = 144$ in. The line conductors are made of 795-kcmil, 30/19-strand ACSR. If the line is 100 mi long and not transposed, determine the following:
    (a) Phase impedance matrix of line
    (b) Phase admittance matrix of line
    (c) Sequence impedance matrix of line
    (d) Sequence admittance matrix of line
13. Repeat Problem 12 assuming that the phase spacings are $D_{ab} = 144$ in., $D_{bc} = 144$ in., and $D_{ca} = 288$ in.
14. Repeat Problem 12 assuming that the conductor is 795-kcmil, 61% conductivity, 37-strand, hard-drawn aluminum.
15. Repeat Problem 12 assuming that the conductor is 750-kcmil, 97.3% conductivity, 37-strand, hard-drawn copper conductor.
16. Consider the line configuration shown in Figure 5.5. Assume that the 115-kV line is transposed and its conductors are made up of 500-kcmil, 30/7-strand ACSR conductors. Ignore the overhead ground wire but consider the heights of the conductors and determine the zero-sequence capacitance of the line in nanofarads per mile and nanofarads per kilometer.
17. Solve Problem 16 taking into account the overhead ground wire. Assume that the overhead ground wire is made of 3/8-in. E.B.B. steel conductor.
18. Repeat Example 5.6 without ignoring the overhead ground wire. Assume that the overhead ground wire is made of 3/8-in. E.B.B. steel conductor.
19. Consider the line configuration shown in Figure 5.5. Assume that the 115-kV line is transposed and its conductors are made of 500-kcmil, 30/7-strand ACSR conductors. Ignore the effects of conductor heights and overhead ground wire and determine the following:
    (a) Positive- and negative-sequence capacitances to ground of line in nanofarads per mile
    (b) The 60-Hz susceptance of line in microsiemens per mile
    (c) Charging kilovolt-ampers per phase per mile of line
    (d) Three-phase charging kilovolt-amperes per mile of line
20. Repeat Problem 19 without ignoring the effects of conductor heights.
21. Consider the untransposed line shown in Figure P5.1. Assume that the 50-mi-long line has an overhead ground wire of 3/0 ACSR and that the phase conductors are of 556.5-kcmil, 30/7 strand, ACSR. Use a frequency of 60 Hz, an ambient temperature of 50°C, and average earth resistivity and determine the following:
    (a) Phase impedance matrix of line

(b) Sequence impedance matrix of line
(c) Sequence admittance matrix of line
(d) Electrostatic zero- and negative-sequence unbalance factors of line

22. Repeat Problem 21 assuming that there are two overhead ground wires, as shown in Figure P5.2.

23. Consider the power system given in Example 5.7 and assume that transformers T_1 and T_3 are connected as delta–wye grounded, and T_2 and T_4 are connected as wye grounded–delta, respectively. Assume that there is a fault on bus 3 and do the following:
    (a) Draw the corresponding zero-sequence network.
    (b) Reduce the zero-sequence network to its Thévenin equivalent looking in at bus 3.

24. Consider the power system given in Example 5.7 and assume that all four transformers are connected as wye grounded–wye grounded. Assume there is a fault on bus 3 and do the following:
    (a) Draw the corresponding zero-sequence network.
    (b) Reduce the zero-sequence network to its Thévenin equivalent looking in at bus 3.

25. Consider the power system given in Problem 10. Use 25 MVA as the megavolt-ampere base and draw the positive-, negative-, and zero-sequence networks (but do not reduce them). Assume that the two three-phase transformer bank connections are:
    (a) Both wye-grounded
    (b) Delta–wye grounded for transformer T_1 and wye grounded–delta for transformer T_2
    (c) Wye grounded–wye for transformer T_1 and delta–wye for transformer T_2

26. Assume that a three-phase, 45-MVA, 34.5/115-kV transformer bank of three single-phase transformers, with nameplate impedances of 7.5%, is connected wye–delta with the high-voltage side delta. Determine the zero-sequence equivalent circuit (in per-unit values) under the following conditions:
    (a) If neutral is ungrounded
    (b) If neutral is solidly grounded
    (c) If neutral is grounded through 10-Ω resistor
    (d) If neutral is grounded through 4000-μF capacitor

27. Consider the system shown in Figure P5.3. Assume that the following data are given based on 20 MVA and the line-to-line base voltages as shown in Figure P5.3.

    Generator G_1: $X_1 = 0.25$ pu, $X_2 = 0.15$ pu, $X_0 = 0.05$ pu
    Generator G_2: $X_2 = 0.90$ pu, $X_2 = 0.60$ pu, $X_0 = 0.05$ pu
    Transformer T_1: $X_1 = X_2 = X_0 = 0.10$ pu

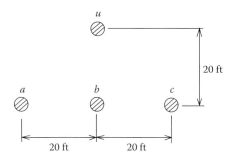

FIGURE P5.1 System for Problem 21.

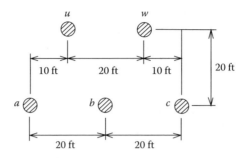

FIGURE P5.2 System for Problem 22.

Transformer T_2: $X_1 = X_2 = 0.10$ pu, $X_0 = \infty$
Transformer T_3: $X_1 = X_2 = X_0 = 0.50$ pu
Transformer T_4: $X_1 = X_2 = 0.30$ pu, $X_0 = \infty$
Transmission line TL_{23}: $X_1 = X_2 = 0.15$ pu, $X_0 = 0.50$ pu
Transmission line TL_{35}: $X_1 = X_2 = 0.30$ pu, $X_0 = 1.00$ pu
Transmission line TL_{57}: $X_1 = X_2 = 0.30$ pu, $X_0 = 1.00$ pu

(a) Draw the corresponding positive-sequence network
(b) Draw the corresponding negative-sequence network
(c) Draw the corresponding zero-sequence network

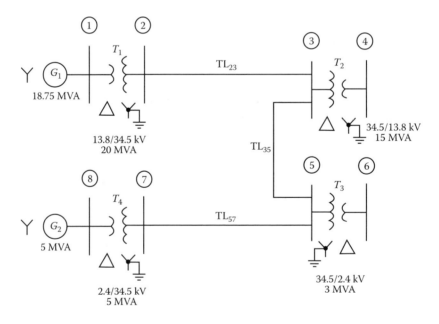

FIGURE P5.3 System for Problem 27.

6 Analysis of Unbalanced Faults

6.1 INTRODUCTION

Most of the faults that occur on power systems are not the balanced (i.e., *symmetrical*) three-phase faults but the unbalanced (i.e., *unsymmetrical*) faults, specifically the single line-to-ground (SLG) faults. For example, Reference [5] gives the typical frequency of occurrence for the three-phase, SLG, line-to-line, and double line-to-ground (DLG) faults as 5%, 70%, 15%, and 10%, respectively.

In general, the three-phase fault is considered to be the most severe one. However, it is possible that the SLG fault may be more severe than the three-phase fault under two circumstances: (1) the generators involved in the fault have solidly grounded neutrals or low-impedance neutral impedances and (2) it occurs on the wye-grounded side of delta–wye-grounded transformer banks. The line-to-line fault current is about 86.6% of the three-phase fault current.

Faults can be categorized as shunt faults (short circuits), series faults (open conductor), and simultaneous faults (having more than one fault occurring at the same time). Unbalanced faults can be easily solved by using the symmetrical components of an unbalanced system of currents or voltages. Therefore, an unbalanced system can be converted to three fictitious networks: the positive-sequence (the only one that has a driving voltage), the negative-sequence, and the zero-sequence networks interconnected to each other in a particular fashion depending on the fault type involved. In this book, only shunt faults are reviewed.

6.2 SHUNT FAULTS

The voltage to ground of phase a at the fault point F before the fault occurred is $\mathbf{V}_F$, and it is usually selected as $1.0\angle 0°$ pu. However, it is possible to have a $\mathbf{V}_F$ value that is not $1.0\angle 0°$ pu. If so, Table 6.1 [8] gives formulas to calculate the fault currents and voltages at the fault point F and their corresponding symmetrical components for various types of faults. Note that, the positive-, negative-, and zero-sequence impedances are viewed from the fault point as $\mathbf{Z}_1$, $\mathbf{Z}_2$, and $\mathbf{Z}_0$, respectively. In the table, $\mathbf{Z}_f$ is the fault impedance and $\mathbf{Z}_{eq}$ is the equivalent impedance to replace the fault in the positive-sequence network. Also, note that the value of the impedance $\mathbf{Z}_g$ is zero in Table 6.1.

6.2.1 SLG Fault

In general, the SLG fault on a transmission system occurs when one conductor falls to ground or contacts the neutral wire. Figure 6.1a shows the general representation of an SLG fault at a fault point F with a fault impedance $\mathbf{Z}_f$.* Usually, the fault impedance $\mathbf{Z}_f$ is ignored in fault studies. Figure 6.1b shows the interconnection of the resulting sequence networks. For the sake of simplicity in fault calculations, the faulted phase is usually assumed to be phase a, as shown in Figure 6.1b.

However, if the faulted phase in reality is other than phase a (e.g., phase b), the phases of the system can simply be relabeled (i.e., a, b, c becomes c, a, b) [4]. A second method involves the use

---

* The fault impedance $\mathbf{Z}_f$ may be thought of as the impedances in the arc (in the event of having a flashover between the line and a tower), the tower, and the tower footing.

TABLE 6.1
Fault Currents and Voltages at Fault Point F and Their Corresponding Symmetrical Components for Various Types of Faults

	Three-Phase Fault through Three-Phase Fault Impedance, Z_f	Line-to-Line Phases b and c Shorted through Fault Impedance, Z_f	Line-to-Ground Fault, Phase a Grounded through Fault Impedance, Z_f	DLG Fault, Phases b and c Shorted, then Grounded through Fault Impedance, Z_f
$\mathbf{I}_{a1}$	$\mathbf{I}_{a1} = \dfrac{\mathbf{V}_f}{\mathbf{Z}_1 + \mathbf{Z}_f}$	$\mathbf{I}_{a1} = -\mathbf{I}_{a2} = \dfrac{\mathbf{V}_f}{\mathbf{Z}_1 + \mathbf{Z}_2 + \mathbf{Z}_f}$	$\mathbf{I}_{a1} = \mathbf{I}_{a2} = \mathbf{I}_{a0}$ $= \dfrac{\mathbf{V}_f}{\mathbf{Z}_0 + \mathbf{Z}_1 + \mathbf{Z}_2 + 3\mathbf{Z}_f}$	$\mathbf{I}_{a1} = -(\mathbf{I}_{a2} + \mathbf{I}_{a0})$ $= \dfrac{\mathbf{V}_f}{\mathbf{Z}_1 + \dfrac{\mathbf{Z}_2(\mathbf{Z}_0 + 3\mathbf{Z}_f)}{\mathbf{Z}_2 + \mathbf{Z}_0 + 3\mathbf{Z}_f}}$
$\mathbf{I}_{a2}$	$\mathbf{I}_{a2} = 0$	$\mathbf{I}_{a2} = -\mathbf{I}_{a1}$	$\mathbf{I}_{a2} = \mathbf{I}_{a1}$	$\mathbf{I}_{a2} = -\mathbf{I}_{a1} \dfrac{\mathbf{Z}_0 + 3\mathbf{Z}_f}{\mathbf{Z}_2 + \mathbf{Z}_0 + 3\mathbf{Z}_f}$
$\mathbf{I}_{a0}$	$\mathbf{I}_{a0} = 0$	$\mathbf{I}_{a0} = 0$	$\mathbf{I}_{a0} = \mathbf{I}_{a1}$	$\mathbf{I}_{a0} = -\mathbf{I}_{a1} \dfrac{\mathbf{Z}_2}{\mathbf{Z}_2 + \mathbf{Z}_0 + 3\mathbf{Z}_f}$
$\mathbf{V}_{a1}$	$\mathbf{V}_{a1} = \mathbf{I}_{a1} \mathbf{Z}_f$	$\mathbf{V}_{a1} = \mathbf{V}_{a2} + \mathbf{I}_{a1}\mathbf{Z}_f$ $= \mathbf{I}_{a1}(\mathbf{Z}_2 + \mathbf{Z}_f)$	$\mathbf{V}_{a1} = -(\mathbf{V}_{a2} = \mathbf{V}_{a2} + \mathbf{V}_{a0}) + \mathbf{I}_{a1}(3\mathbf{Z}_f)$ $= \mathbf{I}_{a1}(\mathbf{Z}_0 + \mathbf{Z}_2 + 3\mathbf{Z}_f)$	$\mathbf{V}_{a1} = \mathbf{V}_{a2} = \mathbf{V}_{a0} - 3\mathbf{I}_{a0}\mathbf{Z}_f$ $= \mathbf{I}_{a1} \dfrac{\mathbf{Z}_2 + (\mathbf{Z}_0 + 3\mathbf{Z}_f)}{\mathbf{Z}_2 + \mathbf{Z}_0 + 3\mathbf{Z}_f}$
$\mathbf{V}_{a2}$	$\mathbf{V}_{a2} = 0$	$\mathbf{V}_{a2} = -\mathbf{I}_{a2}\mathbf{Z}_2 = \mathbf{I}_{a1}\mathbf{Z}_2$	$\mathbf{V}_{a2} = -\mathbf{I}_{a2}\mathbf{Z}_2 = -\mathbf{I}_{a1}\mathbf{Z}_2$	$\mathbf{V}_{a2} = \mathbf{I}_{a2}\mathbf{Z}_2$ $= \mathbf{I}_{a1} \dfrac{\mathbf{Z}_2(\mathbf{Z}_0 + 3\mathbf{Z}_f)}{\mathbf{Z}_2 + \mathbf{Z}_0 + 3\mathbf{Z}_f}$
$\mathbf{V}_{a0}$	$\mathbf{V}_{a0} = 0$	$\mathbf{V}_{a0} = 0$	$\mathbf{V}_{a0} = -\mathbf{I}_{a0}\mathbf{Z}_0$ $= -\mathbf{I}_{a1}\mathbf{Z}_0$	$\mathbf{V}_{a0} = -\mathbf{I}_{a0}\mathbf{Z}_0$ $= \mathbf{I}_{a1} \dfrac{\mathbf{Z}_0 \mathbf{Z}_2}{\mathbf{Z}_2 + \mathbf{Z}_0 + 3\mathbf{Z}_f}$
$\mathbf{Z}_{eq}$	$\mathbf{Z}_{eq} = \mathbf{Z}_f$	$\mathbf{Z}_{eq} = \mathbf{Z}_2 + \mathbf{Z}_f$	$\mathbf{Z}_{eq} = \mathbf{Z}_0 + \mathbf{Z}_2 + 3\mathbf{Z}_f$	$\mathbf{Z}_{eq} = \dfrac{\mathbf{Z}_2(\mathbf{Z}_0 + 3\mathbf{Z}_f)}{\mathbf{Z}_2 + \mathbf{Z}_0 + 3\mathbf{Z}_f}$
$\mathbf{I}_{af}$	$\dfrac{\mathbf{V}_f}{\mathbf{Z}_1 + \mathbf{Z}_f}$	0	$\dfrac{3\mathbf{V}_f}{\mathbf{Z}_0 + \mathbf{Z}_1 + \mathbf{Z}_2 + 3\mathbf{Z}_f}$	0
$\mathbf{I}_{bf}$	$\dfrac{a^2 \mathbf{V}_f}{\mathbf{Z}_1 + \mathbf{Z}_f}$	$-j\sqrt{3} \dfrac{\mathbf{V}_f}{\mathbf{Z}_1 + \mathbf{Z}_2 + \mathbf{Z}_f}$	0	$-j\sqrt{3}\mathbf{V}_f \dfrac{\mathbf{Z}_0 + 3\mathbf{Z}_f - a\mathbf{Z}_2}{\mathbf{Z}_1 \mathbf{Z}_2 + (\mathbf{Z}_1 + \mathbf{Z}_2)(\mathbf{Z}_0 + 3\mathbf{Z}_f)}$
$\mathbf{I}_{cf}$	$\dfrac{a\mathbf{V}_f}{\mathbf{Z}_1 + \mathbf{Z}_f}$	$j\sqrt{3} \dfrac{\mathbf{V}_f}{\mathbf{Z}_1 + \mathbf{Z}_2 + \mathbf{Z}_f}$	0	$j\sqrt{3}\mathbf{V}_f \dfrac{\mathbf{Z}_0 + 3\mathbf{Z}_f - a^2 \mathbf{Z}_2}{\mathbf{Z}_1 \mathbf{Z}_2 + (\mathbf{Z}_1 + \mathbf{Z}_2)(\mathbf{Z}_0 + 3\mathbf{Z}_f)}$
$\mathbf{V}_{af}$	$\mathbf{V}_f \dfrac{\mathbf{Z}_f}{\mathbf{Z}_1 + \mathbf{Z}_f}$	$\mathbf{V}_f \dfrac{2\mathbf{Z}_2 + \mathbf{Z}_f}{\mathbf{Z}_1 + \mathbf{Z}_2 + \mathbf{Z}_f}$	$\mathbf{V}_f \dfrac{3\mathbf{Z}_f}{\mathbf{Z}_0 + \mathbf{Z}_1 + \mathbf{Z}_2 + 3\mathbf{Z}_f}$	$\mathbf{V}_f \dfrac{3\mathbf{Z}_2(\mathbf{Z}_0 + 2\mathbf{Z}_f)}{\mathbf{Z}_1 \mathbf{Z}_2 + (\mathbf{Z}_1 + \mathbf{Z}_2)(\mathbf{Z}_0 + 3\mathbf{Z}_f)}$
$\mathbf{V}_{bf}$	$\mathbf{V}_f \dfrac{a^2 \mathbf{Z}_f}{\mathbf{Z}_1 + \mathbf{Z}_f}$	$\mathbf{V}_f \dfrac{a^2 \mathbf{Z}_f - \mathbf{Z}_2}{\mathbf{Z}_1 + \mathbf{Z}_2 + \mathbf{Z}_f}$	$\mathbf{V}_f \dfrac{3a^2 \mathbf{Z}_f - j\sqrt{3}(\mathbf{Z}_2 - a\mathbf{Z}_0)}{\mathbf{Z}_0 + \mathbf{Z}_1 + \mathbf{Z}_2 + 3\mathbf{Z}_f}$	$\mathbf{V}_f \dfrac{-3\mathbf{Z}_f \mathbf{Z}_2}{\mathbf{Z}_1 \mathbf{Z}_2 + (\mathbf{Z}_1 + \mathbf{Z}_2)(\mathbf{Z}_0 + 3\mathbf{Z}_f)}$
$\mathbf{V}_{cf}$	$\mathbf{V}_f \dfrac{a\mathbf{Z}_f}{\mathbf{Z}_1 + \mathbf{Z}_f}$	$\mathbf{V}_f \dfrac{a\mathbf{Z}_f - \mathbf{Z}_2}{\mathbf{Z}_1 + \mathbf{Z}_2 + \mathbf{Z}_f}$	$\mathbf{V}_f \dfrac{3a\mathbf{Z}_f + j\sqrt{3}(\mathbf{Z}_2 - a^2\mathbf{Z}_0)}{\mathbf{Z}_0 + \mathbf{Z}_1 + \mathbf{Z}_2 + 3\mathbf{Z}_f}$	$\mathbf{V}_f \dfrac{-3\mathbf{Z}_f \mathbf{Z}_2}{\mathbf{Z}_1 \mathbf{Z}_2 + (\mathbf{Z}_1 + \mathbf{Z}_2)(\mathbf{Z}_0 + 3\mathbf{Z}_f)}$
$\mathbf{V}_{bc}$	$j\sqrt{3}\mathbf{V}_f \dfrac{\mathbf{Z}_f}{\mathbf{Z}_1 + \mathbf{Z}_f}$	$j\sqrt{3}\mathbf{V}_f \dfrac{\mathbf{Z}_f}{\mathbf{Z}_1 + \mathbf{Z}_2 + \mathbf{Z}_f}$	$j\sqrt{3}\mathbf{V}_f \dfrac{3\mathbf{Z}_f + \mathbf{Z}_0 + 2\mathbf{Z}_2}{\mathbf{Z}_0 + \mathbf{Z}_1 + \mathbf{Z}_2 + 3\mathbf{Z}_f}$	0

(continued)

Analysis of Unbalanced Faults

TABLE 6.1 (Continued)
Fault Currents and Voltages at Fault Point *F* and Their Corresponding Symmetrical Components for Various Types of Faults

	Three-Phase Fault through Three-Phase Fault Impedance, Z_f	Line-to-Line Phases *b* and *c* Shorted through Fault Impedance, Z_f	Line-to-Ground Fault, Phase *a* Grounded through Fault Impedance, Z_f	DLG Fault, Phases *b* and *c* Shorted, then Grounded through Fault Impedance, Z_f
V_{ca}	$j\sqrt{3}V_f \dfrac{a^2 Z_f}{Z_1 + Z_f}$	$j\sqrt{3}V_f \dfrac{a^2 Z_f - j\sqrt{3}Z_2}{Z_1 + Z_2 + Z_f}$	$j\sqrt{3}V_f \dfrac{a^2(3Z_f + Z_0) - Z_2}{Z_0 + Z_1 + Z_2 + 3Z_f}$	$\sqrt{3}V_f \dfrac{\sqrt{3}Z_2(Z_0 + 3Z_f)}{Z_1 Z_2 + (Z_1 + Z_2)(Z_0 + 3Z_f)}$
V_{ab}	$j\sqrt{3}V_f \dfrac{a Z_f}{Z_1 + Z_f}$	$j\sqrt{3}V_f \dfrac{a Z_f - j\sqrt{3}Z_2}{Z_1 + Z_2 + Z_f}$	$j\sqrt{3}V_f \dfrac{a(3Z_f + Z_0) - Z_2}{Z_0 + Z_1 + Z_2 + 3Z_f}$	$-\sqrt{3}V_f \dfrac{\sqrt{3}Z_2(Z_0 + 3Z_f)}{Z_1 Z_2 + (Z_1 + Z_2)(Z_0 + 3Z_f)}$

Source: Clarke, E., *Circuit Analysis of A-C Power Systems*, Vol. 1. General Electric Co., Schenectady, New York, 1960.

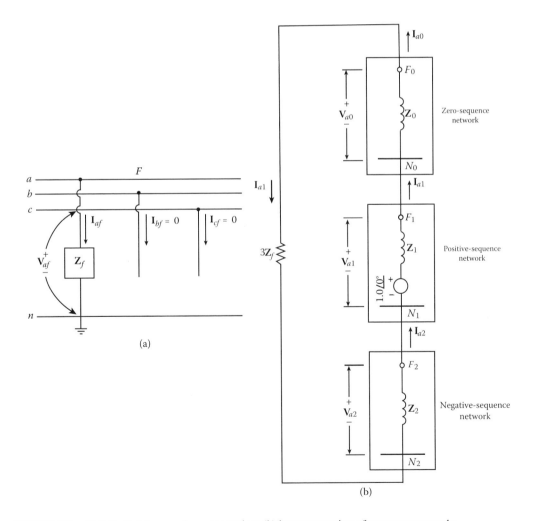

FIGURE 6.1 SLG fault: (a) general representation; (b) interconnection of sequence networks.

of the "generalized fault diagram" of Atabekov [4], further developed by Anderson [3]. From Figure 6.1b, it can be observed that the zero-, positive-, and negative-sequence currents are equal to each other. Therefore,

$$\mathbf{I}_{a0} = \mathbf{I}_{a1} = \mathbf{I}_{a2} = \frac{1.0\angle 0°}{\mathbf{Z}_0 + \mathbf{Z}_1 + \mathbf{Z}_2 + 3\mathbf{Z}_f} \tag{6.1}$$

Since

$$\begin{bmatrix} \mathbf{I}_{af} \\ \mathbf{I}_{bf} \\ \mathbf{I}_{cf} \end{bmatrix} = \begin{bmatrix} 1 & 1 & 1 \\ 1 & \mathbf{a}^2 & \mathbf{a} \\ 1 & \mathbf{a} & \mathbf{a}^2 \end{bmatrix} \begin{bmatrix} \mathbf{I}_{a0} \\ \mathbf{I}_{a1} \\ \mathbf{I}_{a2} \end{bmatrix} \tag{6.2}$$

the fault current for phase a can be found as

$$\mathbf{I}_{af} = \mathbf{I}_{a0} + \mathbf{I}_{a1} + \mathbf{I}_{a2}$$

or

$$\mathbf{I}_{af} = 3\mathbf{I}_{a0} = 3\mathbf{I}_{a1} = 3\mathbf{I}_{a2} \tag{6.3}$$

From Figure 6.1a,

$$\mathbf{V}_{af} = \mathbf{Z}_f \mathbf{I}_{af} \tag{6.4}$$

Substituting Equation 6.3 into Equation 6.4, the voltage at faulted phase a can be expressed as

$$\mathbf{V}_{af} = 3\mathbf{Z}_f \mathbf{I}_{a1} \tag{6.5}$$

But,

$$\mathbf{V}_{af} = \mathbf{V}_{a0} + \mathbf{V}_{a1} + \mathbf{V}_{a2} \tag{6.6}$$

Therefore,

$$\mathbf{V}_{a0} + \mathbf{V}_{a1} + \mathbf{V}_{a2} = 3\mathbf{Z}_f \mathbf{I}_{a1} \tag{6.7}$$

which justifies the interconnection of sequence networks in series, as shown in Figure 6.1b.

Once the sequence currents are found, the zero-, positive-, and negative-sequence voltages can be found from

$$\begin{bmatrix} \mathbf{V}_{a0} \\ \mathbf{V}_{a1} \\ \mathbf{V}_{a2} \end{bmatrix} = \begin{bmatrix} 0 \\ 1.0\angle 0° \\ 0 \end{bmatrix} - \begin{bmatrix} \mathbf{Z}_0 & 0 & 0 \\ 0 & \mathbf{Z}_1 & 0 \\ 0 & 0 & \mathbf{Z}_2 \end{bmatrix} \begin{bmatrix} \mathbf{I}_{a0} \\ \mathbf{I}_{a1} \\ \mathbf{I}_{a2} \end{bmatrix} \tag{6.8}$$

as

$$\mathbf{V}_{a0} = -\mathbf{Z}_0 \mathbf{I}_{a0} \tag{6.9}$$

$$\mathbf{V}_{a1} = 1.0 - \mathbf{Z}_1 \mathbf{I}_{a1} \tag{6.10}$$

$$\mathbf{V}_{a2} = -\mathbf{Z}_2 \mathbf{I}_{a2} \tag{6.11}$$

Analysis of Unbalanced Faults

In the event of having an SLG fault on phase b or c, the voltages related to the known phase a voltage components can be found from

$$\begin{bmatrix} \mathbf{V}_{af} \\ \mathbf{V}_{bf} \\ \mathbf{V}_{cf} \end{bmatrix} = \begin{bmatrix} 1 & 1 & 1 \\ 1 & a^2 & a \\ 1 & a & a^2 \end{bmatrix} \begin{bmatrix} \mathbf{V}_{a0} \\ \mathbf{V}_{a1} \\ \mathbf{V}_{a2} \end{bmatrix} \quad (6.12)$$

as

$$\mathbf{V}_{bf} = \mathbf{V}_{a0} + \mathbf{a}^2 \mathbf{V}_{a1} + \mathbf{a} \mathbf{V}_{a2} \quad (6.13)$$

and

$$\mathbf{V}_{cf} = \mathbf{V}_{a0} + \mathbf{a} \mathbf{V}_{a1} + \mathbf{a}^2 \mathbf{V}_{a2} \quad (6.14)$$

EXAMPLE 6.1

Consider the system described in Examples 5.7 and 5.8 and assume that there is an SLG fault, involving phase a, and that the fault impedance is $5 + j0$ Ω. Also, assume that $\mathbf{Z}_0$ and $\mathbf{Z}_2$ are $j0.56$ and $j0.3619$ Ω, respectively.

(a) Show the interconnection of the corresponding equivalent sequence networks.
(b) Determine the sequence and phase currents.
(c) Determine the sequence and phase voltages.
(d) Determine the line-to-line voltages.

Solution

(a) Figure 6.2 shows the interconnection of the resulting equivalent sequence networks.
(b) The impedance base on the 230-kV line is

$$\mathbf{Z}_B = \frac{230^2}{200}$$

$$= 264.5 \text{ Ω}$$

Therefore,

$$\mathbf{Z}_f = \frac{5 \text{ Ω}}{264.5 \text{ Ω}}$$

$$= 0.0189 \text{ pu Ω}$$

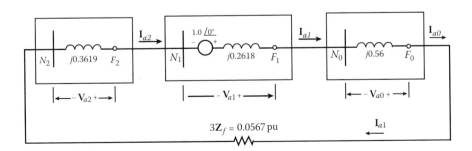

FIGURE 6.2 Interconnection of resultant equivalent sequence networks of Example 6.2.

Thus, the sequence currents and the phase currents are

$$I_{a0} = I_{a1} = I_{a2} = \frac{1.0\angle 0°}{Z_0 + Z_1 + Z_2 + 3Z_f}$$

$$= \frac{1.0\angle 0°}{j0.56 + j0.2618 + j0.3619 + 0.0567}$$

$$= 0.8438\angle -87.3° \quad \text{pu A}$$

and

$$\begin{bmatrix} I_{af} \\ I_{bf} \\ I_{cf} \end{bmatrix} = \begin{bmatrix} 1 & 1 & 1 \\ 1 & a^2 & a \\ 1 & a & a^2 \end{bmatrix} \begin{bmatrix} 0.8438\angle -87.3° \\ 0.8438\angle -87.3° \\ 0.8438\angle -87.3° \end{bmatrix}$$

$$= \begin{bmatrix} 2.5314\angle -87.3° \\ 0 \\ 0 \end{bmatrix} \quad \text{pu A}$$

(c) The sequence and phase voltages are

$$\begin{bmatrix} V_{a0} \\ V_{a1} \\ V_{a2} \end{bmatrix} = \begin{bmatrix} 0 \\ 1.0\angle 0° \\ 0 \end{bmatrix} - \begin{bmatrix} j0.56 & 0 & 0 \\ 0 & j0.2618 & 0 \\ 0 & 0 & j0.3619 \end{bmatrix} \begin{bmatrix} 0.8438\angle -87.3° \\ 0.8438\angle -87.3° \\ 0.8438\angle -87.3° \end{bmatrix}$$

$$= \begin{bmatrix} 0.4725\angle -177.7° \\ 0.7794\angle -0.8° \\ 0.3054\angle -177.7° \end{bmatrix} \quad \text{pu V}$$

and

$$\begin{bmatrix} V_{af} \\ V_{bf} \\ V_{cf} \end{bmatrix} = \begin{bmatrix} 1 & 1 & 1 \\ 1 & a^2 & a \\ 1 & a & a^2 \end{bmatrix} \begin{bmatrix} 0.4725\angle -177.7° \\ 0.7794\angle -0.8° \\ 0.3054\angle -177.7° \end{bmatrix}$$

$$= \begin{bmatrix} 0.0479\angle 86.26° \\ 1.823\angle -127° \\ 1.1709\angle 127.5° \end{bmatrix} \quad \text{pu V}$$

(d) The fine-to-line voltages at the fault point are

$$V_{abf} = V_{af} - V_{bf}$$

$$= 0.0479\angle 87.26° - 1.1823\angle 207.7°$$

$$= 1.146\angle 51.85° \quad \text{pu V}$$

$$V_{bcf} = V_{bf} - V_{cf}$$

$$= 1.823\angle -127° - 1.1709\angle 127.5°$$

$$= 1.878\angle -89.79° \quad \text{pu V}$$

Analysis of Unbalanced Faults

$$V_{caf} = V_{cf} - V_{af}$$
$$= 1.709\angle 121.5° - 0.0479\angle 86.26°$$
$$= 1.2106\angle 126.2° \quad \text{pu V}$$

EXAMPLE 6.2

Consider the system given in Figure 6.3a and assume that the given impedance values are based on the same megavolt-ampere value. The two three-phase transformer banks are made of three single-phase transformers. Assume that there is an SLG fault, involving phase a, at the middle of the transmission line TL_{23}, as shown in the figure.

(a) Draw the corresponding positive-, negative-, and zero-sequence networks, without reducing them, and their corresponding interconnections.
(b) Determine the sequence currents at fault point F.
(c) Determine the sequence currents at the terminals of generator G_1.
(d) Determine the phase currents at the terminals of generator G_1.
(e) Determine the sequence voltages at the terminals of generator G_1.
(f) Determine the phase voltages at the terminals of generator G_1.
(g) Repeat parts (c) through (f) for generator G_2.

Solution

(a) Figure 6.3b shows the corresponding sequence networks.
(b) The sequence currents at fault point F are

$$I_{a0} = I_{a1} = I_{a2} = \frac{1.0\angle 0°}{Z_0 + Z_1 + Z_2}$$
$$= \frac{1.0\angle 0°}{j0.2619 + j0.25 + j0.25}$$
$$= -j1.3125 \quad \text{pu A}$$

(c) Therefore, the sequence current contributions of generator G_1 can be found by symmetry as

$$I_{a1,G_1} = \frac{1}{2} \times I_{a1}$$
$$= -j0.6563 \quad \text{pu A}$$

and

$$I_{a2,G_1} = \frac{1}{2} \times I_{a2}$$
$$= -j0.6563 \quad \text{pu A}$$

and by current division,

$$I_{a0,G_1} = \frac{0.5}{0.55 + 0.5} \times I_{a0}$$
$$= -j0.6250 \quad \text{pu A}$$

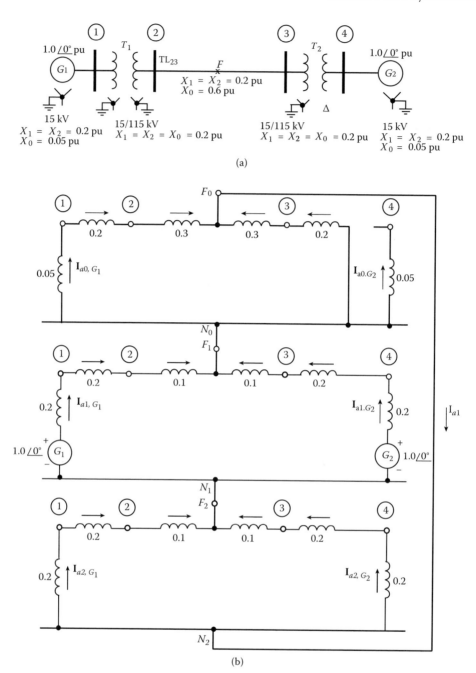

FIGURE 6.3 The system and the solution for Example 6.2.

Analysis of Unbalanced Faults

(d) The phase currents at the terminals of generator G_1 are

$$\begin{bmatrix} I_{af} \\ I_{bf} \\ I_{cf} \end{bmatrix} = \begin{bmatrix} 1 & 1 & 1 \\ 1 & a^2 & a \\ 1 & a & a^2 \end{bmatrix} \begin{bmatrix} 0.6250\angle-90° \\ 0.6563\angle-90° \\ 0.6563\angle-90° \end{bmatrix}$$

$$= \begin{bmatrix} 1.9376\angle-90° \\ 0.0313\angle 90° \\ 0.0313\angle 90° \end{bmatrix} \text{ pu A}$$

(e) The sequence voltages at the terminals of generator G_1 are

$$\begin{bmatrix} V_{a0} \\ V_{a1} \\ V_{a2} \end{bmatrix} = \begin{bmatrix} 0 \\ 1.0\angle 0° \\ 0 \end{bmatrix} - \begin{bmatrix} j0.2619 & 0 & 0 \\ 0 & j0.25 & 0 \\ 0 & 0 & j0.25 \end{bmatrix} \begin{bmatrix} 0.6250\angle-90° \\ 0.6563\angle-90° \\ 0.6563\angle-90° \end{bmatrix}$$

$$= \begin{bmatrix} 0.1637\angle 180° \\ 0.8360\angle 0° \\ 0.1641\angle 180° \end{bmatrix} \text{ pu V}$$

(f) Therefore, the phase voltages are

$$\begin{bmatrix} V_{af} \\ V_{bf} \\ V_{cf} \end{bmatrix} = \begin{bmatrix} 1 & 1 & 1 \\ 1 & a^2 & a \\ 1 & a & a^2 \end{bmatrix} \begin{bmatrix} 0.1637\angle 180° \\ 0.8360\angle 0° \\ 0.1641\angle 180° \end{bmatrix}$$

$$= \begin{bmatrix} 0.5082\angle 0° \\ 0.9998\angle 240° \\ 0.9998\angle 120° \end{bmatrix} \text{ pu V}$$

(g) Similarly, for generator G_2, by symmetry,

$$I_{a1,G_2} = \frac{1}{2} \times I_{a1}$$

$$= -j0.6563 \text{ pu A}$$

and

$$I_{a2,G_2} = \frac{1}{2} \times I_{a2}$$

$$= -j0.6563 \text{ pu A}$$

and by inspection

$$I_{a0,G_2} = 0$$

However, since transformer T_2 has wye–delta connections and the U.S. Standard terminal markings provide that $V_{a1(HV)}$ leads $V_{a1(LV)}$ by 30° and $V_{a2(HV)}$ lags $V_{a2(LV)}$ by 30°, regardless of which side has the delta-connected windings, taking into account the 30° phase shifts,

$$I_{a1,G_2} = 0.6563\angle-90°-30°$$

$$= 0.6563\angle-120° \text{ pu A}$$

and

$$\mathbf{I}_{a2,G_2} = 0.6563\angle -90° + 30°$$
$$= 0.6563\angle -60° \quad \text{pu A}$$

This is because generator G_2 is on the low-voltage side of the transformer. Therefore,

$$\begin{bmatrix} \mathbf{I}_{af} \\ \mathbf{I}_{bf} \\ \mathbf{I}_{cf} \end{bmatrix} = \begin{bmatrix} 1 & 1 & 1 \\ 1 & a^2 & a \\ 1 & a & a^2 \end{bmatrix} \begin{bmatrix} 0 \\ 0.6563\angle -120° \\ 0.6563\angle -60° \end{bmatrix}$$

$$= \begin{bmatrix} 1.1368\angle -90° \\ 1.1368\angle 90° \\ 0 \end{bmatrix} \quad \text{pu A}$$

The positive- and negative-sequence voltages on the G_2 side are the same as on the G_1 side. Thus,

$$\mathbf{V}_{a1} = 0.8434\angle 0° \quad \text{pu V}$$

$$\mathbf{V}_{a2} = 0.1641\angle 180° \quad \text{pu V}$$

Again, taking into account the 30° phase shifts,

$$\mathbf{V}_{a1} = 0.8434\angle 0° - 30°$$
$$= 0.8434\angle -30° \quad \text{pu V}$$

$$\mathbf{V}_{a2} = 0.1641\angle 180° + 30°$$
$$= 0.1641\angle 210° \quad \text{pu V}$$

Obviously

$$\mathbf{V}_{a0} = 0$$

Therefore, the phase voltages at the terminals of generator G_2 are

$$\begin{bmatrix} \mathbf{V}_{af} \\ \mathbf{V}_{bf} \\ \mathbf{V}_{cf} \end{bmatrix} = \begin{bmatrix} 1 & 1 & 1 \\ 1 & a^2 & a \\ 1 & a & a^2 \end{bmatrix} \begin{bmatrix} 0 \\ 0.8434\angle -30° \\ 0.1641\angle 210° \end{bmatrix}$$

$$= \begin{bmatrix} 0.7745\angle -40.6° \\ 0.7745\angle 220.6° \\ 1.0775\angle 90° \end{bmatrix} \quad \text{pu V}$$

6.2.2 Line-to-Line Fault

In general, a line-to-line (L-L) fault on a transmission system occurs when two conductors are short-circuited.* Figure 6.4a shows the general representation of a line-to-line fault at fault point F with a fault impedance $\mathbf{Z}_f$. Figure 6.4b shows the interconnection of resulting sequence networks. It is

---

* Note that $|\mathbf{I}_{f,\text{L-L}}| = 0.866 \times |\mathbf{I}_{f,3\phi}|$. Therefore, if the magnitude of the three-phase fault current is known, the magnitude of the line-to-line fault current can readily be found.

Analysis of Unbalanced Faults

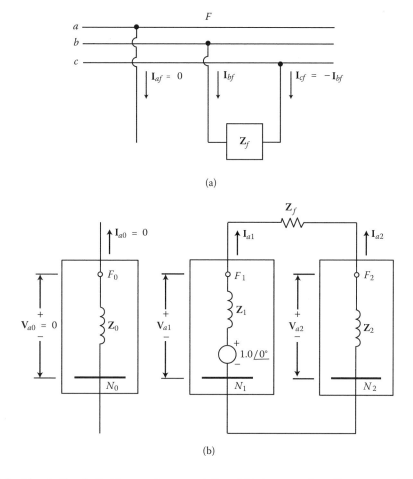

FIGURE 6.4 Line-to-line fault: (a) general representation; (b) interconnection of sequence networks.

assumed, for the sake of symmetry, that the line-to-line fault is between phases *b* and *c*. It can be observed from Figure 6.4a that

$$\mathbf{I}_{af} = 0 \tag{6.15}$$

$$\mathbf{I}_{bf} = -\mathbf{I}_{cf} \tag{6.16}$$

$$\mathbf{V}_{bc} = \mathbf{V}_b - \mathbf{V}_c = \mathbf{Z}_f \mathbf{I}_{bf} \tag{6.17}$$

From Figure 6.4b, the sequence currents can be found as

$$\mathbf{I}_{a0} = 0 \tag{6.18}$$

$$\mathbf{I}_{a1} = -\mathbf{I}_{a2} = \frac{1.0\angle 0°}{\mathbf{Z}_1 + \mathbf{Z}_2 + \mathbf{Z}_f} \tag{6.19}$$

If $\mathbf{Z}_f = 0$,

$$\mathbf{I}_{a1} = -\mathbf{I}_{a2} = \frac{1.0\angle 0°}{\mathbf{Z}_1 + \mathbf{Z}_2} \tag{6.20}$$

Substituting Equations 6.18 and 6.19 into Equation 6.2, the fault currents for phases a and b can be found as

$$\mathbf{I}_{cf} = -\mathbf{I}_{cf} = \sqrt{3}\mathbf{I}_{a1}\angle -90° \tag{6.21}$$

Similarly, substituting Equations 6.18 and 6.19 into Equation 6.8, the sequence voltages can be found as

$$\mathbf{V}_{a0} = 0 \tag{6.22}$$

$$\mathbf{V}_{a1} = 1.0 - \mathbf{Z}_1\mathbf{I}_{a1} \tag{6.23}$$

$$\mathbf{V}_{a2} = -\mathbf{Z}_2\mathbf{I}_{a2} = \mathbf{Z}_2\mathbf{I}_{a1} \tag{6.24}$$

Also, substituting Equations 6.22 through 6.24 into Equation 6.12,

$$\mathbf{V}_{af} = \mathbf{V}_{a1} + \mathbf{V}_{a2} \tag{6.25}$$

or

$$\mathbf{V}_{af} = 1.0 + \mathbf{I}_{a1}(\mathbf{Z}_2 - \mathbf{Z}_1) \tag{6.26}$$

and

$$\mathbf{V}_{bf} = \mathbf{a}^2\mathbf{V}_{a1} + \mathbf{a}\mathbf{V}_{a2} \tag{6.27}$$

or

$$\mathbf{V}_{bf} = \mathbf{a}^2 + \mathbf{I}_{a1}(\mathbf{a}\mathbf{Z}_2 - \mathbf{a}^2\mathbf{Z}_1) \tag{6.28}$$

and

$$\mathbf{V}_{cf} = \mathbf{a}\mathbf{V}_{a1} + \mathbf{a}^2\mathbf{V}_{a2} \tag{6.29}$$

or

$$\mathbf{V}_{cf} = \mathbf{a} + \mathbf{I}_{a1}(\mathbf{a}^2\mathbf{Z}_2 - \mathbf{a}\mathbf{Z}_1) \tag{6.30}$$

Thus, the line-to-line voltages can be expressed as

$$\mathbf{V}_{ab} = \mathbf{V}_{af} - \mathbf{V}_{bf} \tag{6.31}$$

or

$$\mathbf{V}_{ab} = \sqrt{3}(\mathbf{V}_{a1}\angle 30° + \mathbf{V}_{a2}\angle -30°) \tag{6.32}$$

and

$$\mathbf{V}_{bc} = \mathbf{V}_{bf} - \mathbf{V}_{cf} \tag{6.33}$$

or

$$\mathbf{V}_{bc} = \sqrt{3}(\mathbf{V}_{a1}\angle -90° + \mathbf{V}_{a2}\angle 90°) \qquad (6.34)$$

and

$$\mathbf{V}_{ca} = \mathbf{V}_{cf} - \mathbf{V}_{af} \qquad (6.35)$$

or

$$\mathbf{V}_{ca} = \sqrt{3}(\mathbf{V}_{a1}\angle 150° + \mathbf{V}_{a2}\angle -150°) \qquad (6.36)$$

EXAMPLE 6.3

Repeat Example 6.1 assuming that there is a line-to-line fault, involving phases *b* and *c*, at bus 3.

Solution

(a) Figure 6.5 shows the interconnection of the resulting equivalent sequence networks.
(b) The sequence and the phase currents are

$$\mathbf{I}_{a0} = 0$$

$$\mathbf{I}_{a1} = -\mathbf{I}_{a2} = \frac{1.0\angle 0°}{\mathbf{Z}_1 + \mathbf{Z}_2 + \mathbf{Z}_f}$$

$$= \frac{1.0\angle 0°}{j0.2618 + j0.3619 + 0.0189}$$

$$= 1.6026\angle -88.3° \quad \text{pu A}$$

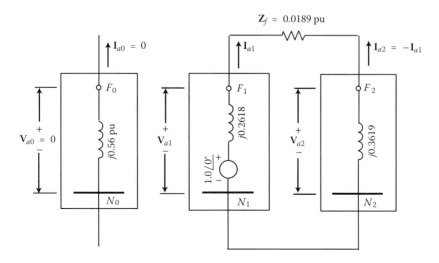

FIGURE 6.5 Interconnection of resultant equivalent sequence networks of Example 6.3.

and

$$\begin{bmatrix} \mathbf{I}_{af} \\ \mathbf{I}_{bf} \\ \mathbf{I}_{cf} \end{bmatrix} = \begin{bmatrix} 1 & 1 & 1 \\ 1 & \mathbf{a}^2 & \mathbf{a} \\ 1 & \mathbf{a} & \mathbf{a}^2 \end{bmatrix} \begin{bmatrix} 0 \\ 1.6026\angle -88.3° \\ 1.6026\angle 91.7° \end{bmatrix}$$

$$= \begin{bmatrix} 0 \\ 2.7758\angle -178.3° \\ 2.7758\angle 1.7° \end{bmatrix} \text{ pu A}$$

(c) The sequence and phase voltages are

$$\begin{bmatrix} \mathbf{V}_{a0} \\ \mathbf{V}_{a1} \\ \mathbf{V}_{a2} \end{bmatrix} = \begin{bmatrix} 0 \\ 1.0\angle 0° \\ 0 \end{bmatrix} - \begin{bmatrix} j0.56 & 0 & 0 \\ 0 & j0.2618 & 0 \\ 0 & 0 & j0.3619 \end{bmatrix} \begin{bmatrix} 0 \\ 1.6026\angle -88.3° \\ 1.6026\angle 91.7° \end{bmatrix}$$

$$= \begin{bmatrix} 0 \\ 0.5808\angle -1.2° \\ 0.5800\angle 1.7° \end{bmatrix} \text{ pu V}$$

and

$$\begin{bmatrix} \mathbf{V}_{af} \\ \mathbf{V}_{bf} \\ \mathbf{V}_{cf} \end{bmatrix} = \begin{bmatrix} 1 & 1 & 1 \\ 1 & \mathbf{a}^2 & \mathbf{a} \\ 1 & \mathbf{a} & \mathbf{a}^2 \end{bmatrix} \begin{bmatrix} 0 \\ 0.5808\angle -1.3° \\ 0.5800\angle 1.7° \end{bmatrix}$$

$$= \begin{bmatrix} 1.1604\angle 0.2° \\ 0.6061\angle -179.7° \\ 0.5540\angle -171.8° \end{bmatrix} \text{ pu V}$$

(d) The line-to-line voltages at the fault point are

$$\mathbf{V}_{abf} = \mathbf{V}_{af} - \mathbf{V}_{bf}$$
$$= 1.7667 + j0.0008$$
$$= 1.7668\angle 0.3° \text{ pu V}$$

$$\mathbf{V}_{bcf} = \mathbf{V}_{bf} - \mathbf{V}_{cf}$$
$$= -0.0524 - j0.0016$$
$$= 0.0525\angle -178.3° \text{ pu V}$$

$$\mathbf{V}_{caf} = \mathbf{V}_{cf} - \mathbf{V}_{af}$$
$$= -1.7143 - j0.0065$$
$$= 1.7143\angle -179.8° \text{ pu V}$$

6.2.3 DLG Fault

In general, the DLG fault on a transmission system occurs when two conductors fall and are connected through ground or when two conductors contact the neutral of a three-phase grounded system. Figure 6.6a shows the general representation of a DLG fault at a fault point F with a fault impedance $\mathbf{Z}_f$ and the impedance from line to ground $\mathbf{Z}_g$ (which can be equal to zero or infinity). Figure 6.6b shows the interconnection of resultant sequence networks. As before, it is assumed, for the sake of symmetry, that the DLG fault is between phases b and c. It can be observed from Figure 6.6a that

$$\mathbf{I}_{af} = 0 \tag{6.37}$$

$$\mathbf{V}_{bf} = (\mathbf{Z}_f + \mathbf{Z}_g)\mathbf{I}_{bf} + \mathbf{Z}_g\mathbf{I}_{cf} \tag{6.38}$$

$$\mathbf{V}_{cf} = (\mathbf{Z}_f + \mathbf{Z}_g)\mathbf{I}_{cf} + \mathbf{Z}_g\mathbf{I}_{bf} \tag{6.39}$$

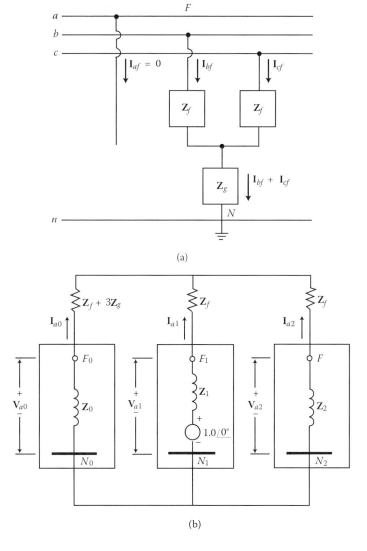

FIGURE 6.6 DLG fault: (a) general representation; (b) interconnection of sequence networks.

From Figure 6.6b, the positive-sequence currents can be found as

$$\mathbf{I}_{a1} = \frac{1.0\angle 0°}{(\mathbf{Z}_1+\mathbf{Z}_f) + \dfrac{(\mathbf{Z}_2+\mathbf{Z}_f)(\mathbf{Z}_0+\mathbf{Z}_f+3\mathbf{Z}_g)}{(\mathbf{Z}_2+\mathbf{Z}_f)+(\mathbf{Z}_0+\mathbf{Z}_f+3\mathbf{Z}_g)}} \quad (6.40\text{a})$$

$$= \frac{1.0\angle 0°}{(\mathbf{Z}_1+\mathbf{Z}_f) + \dfrac{(\mathbf{Z}_2+\mathbf{Z}_f)(\mathbf{Z}_0+\mathbf{Z}_f+3\mathbf{Z}_g)}{\mathbf{Z}_0+\mathbf{Z}_2+2\mathbf{Z}_f+3\mathbf{Z}_g}} \quad (6.40\text{b})$$

The negative- and zero-sequence currents can be found, by using current division, as

$$\mathbf{I}_{a2} = -\left[\frac{(\mathbf{Z}_0+\mathbf{Z}_f+3\mathbf{Z}_g)}{(\mathbf{Z}_2+\mathbf{Z}_f)+(\mathbf{Z}_0+\mathbf{Z}_f+3\mathbf{Z}_g)}\right]\mathbf{I}_{a1} \quad (6.41)$$

and

$$\mathbf{I}_{a0} = -\left[\frac{(\mathbf{Z}_2+\mathbf{Z}_f)}{(\mathbf{Z}_2+\mathbf{Z}_f)+(\mathbf{Z}_0+\mathbf{Z}_f+3\mathbf{Z}_g)}\right]\mathbf{I}_{a1} \quad (6.42)$$

or as an alternative method, since

$$\mathbf{I}_{af} = 0 = \mathbf{I}_{a0} + \mathbf{I}_{a1} + \mathbf{I}_{a2}$$

then if $\mathbf{I}_{a1}$ and $\mathbf{I}_{a2}$ are known,

$$\mathbf{I}_{a0} = -(\mathbf{I}_{a1} + \mathbf{I}_{a2}) \quad (6.43)$$

Note that, in the event of having $\mathbf{Z}_f = 0$ and $\mathbf{Z}_g = 0$, the positive-, negative-, and zero-sequences can be expressed as

$$\mathbf{I}_{a1} = \frac{1.0\angle 0°}{\mathbf{Z}_1 + \dfrac{\mathbf{Z}_0 \times \mathbf{Z}_2}{\mathbf{Z}_0+\mathbf{Z}_2}} \quad (6.44)$$

and by current division

$$\mathbf{I}_{a2} = -\left[\frac{\mathbf{Z}_0}{\mathbf{Z}_0+\mathbf{Z}_2}\right]\mathbf{I}_{a1} \quad (6.45)$$

$$\mathbf{I}_{a0} = -\left[\frac{\mathbf{Z}_2}{\mathbf{Z}_0+\mathbf{Z}_2}\right]\mathbf{I}_{a1} \quad (6.46)$$

Note that, the fault current for phase a is already known to be

$$\mathbf{I}_{af} = 0$$

the fault currents for phases a and b can be found by substituting Equations 6.40 through 6.42 into Equation 6.2 so that

$$\mathbf{I}_{bf} = \mathbf{I}_{a0} + \mathbf{a}^2\mathbf{I}_{a1} + \mathbf{a}\mathbf{I}_{a2} \tag{6.47}$$

and

$$\mathbf{I}_{cf} = \mathbf{I}_{a0} + \mathbf{a}\mathbf{I}_{a1} + \mathbf{a}^2\mathbf{I}_{a2} \tag{6.48}$$

It can be shown that the total fault current flowing into the neutral is

$$\mathbf{I}_n = \mathbf{I}_{bf} + \mathbf{I}_{cf} + 3\mathbf{I}_{a0} \tag{6.49}$$

The sequence voltages can be found from Equation 6.8 as

$$\mathbf{V}_{a0} = -\mathbf{Z}_0\mathbf{I}_{a0} \tag{6.50}$$

$$\mathbf{V}_{a1} = 1.0 - \mathbf{Z}_1\mathbf{I}_{a1} \tag{6.51}$$

$$\mathbf{V}_{a2} = -\mathbf{Z}_2\mathbf{I}_{a2} \tag{6.52}$$

Similarly, the phase voltages can be found from Equation 6.12 as

$$\mathbf{V}_{af} = \mathbf{V}_{a0} + \mathbf{V}_{a1} + \mathbf{V}_{a2} \tag{6.53}$$

$$\mathbf{V}_{bf} = \mathbf{V}_{a0} + \mathbf{a}^2\mathbf{V}_{a1} + \mathbf{a}\mathbf{V}_{a2} \tag{6.54}$$

$$\mathbf{V}_{cf} = \mathbf{V}_{a0} + \mathbf{a}\mathbf{V}_{a1} + \mathbf{a}^2\mathbf{V}_{a2} \tag{6.55}$$

or, alternatively, the phase voltages $\mathbf{V}_{bf}$ and $\mathbf{V}_{cf}$ can be determined from Equations 6.38 and 6.39. As before, the line-to-line voltages can be found from

$$\mathbf{V}_{ab} = \mathbf{V}_{af} - \mathbf{V}_{bf} \tag{6.56}$$

$$\mathbf{V}_{bc} = \mathbf{V}_{bf} - \mathbf{V}_{cf} \tag{6.57}$$

$$\mathbf{V}_{ca} = \mathbf{V}_{cf} - \mathbf{V}_{af} \tag{6.58}$$

Note that, in the event of having $\mathbf{Z}_f = 0$ and $\mathbf{Z}_g = 0$, the sequence voltages become

$$\mathbf{V}_{a0} = \mathbf{V}_{a1} = \mathbf{V}_{a2} = 1.0 - \mathbf{Z}_1\mathbf{I}_{a1} \tag{6.59}$$

where the positive-sequence current is found by using Equation 6.44. Once the sequence voltages are determined from Equation 6.59, the negative- and zero-sequence currents can be determined from

$$\mathbf{I}_{a2} = -\frac{\mathbf{V}_{a2}}{\mathbf{Z}_2} \quad (6.60)$$

and

$$\mathbf{I}_{a0} = -\frac{\mathbf{V}_{a0}}{\mathbf{Z}_0} \quad (6.61)$$

Using the relationship given in Equation 6.59, the resultant phase voltages can be expressed as

$$\mathbf{V}_{af} = \mathbf{V}_{a0} + \mathbf{V}_{a1} + \mathbf{V}_{a2} = 3\mathbf{V}_{a1} \quad (6.62)$$

$$\mathbf{V}_{bf} = \mathbf{V}_{cf} = 0 \quad (6.63)$$

Therefore, the line-to-line voltages become

$$\mathbf{V}_{abf} = \mathbf{V}_{af} - \mathbf{V}_{bf} = \mathbf{V}_{af} \quad (6.64)$$

$$\mathbf{V}_{bcf} = \mathbf{V}_{bf} - \mathbf{V}_{cf} = 0 \quad (6.65)$$

$$\mathbf{V}_{caf} = \mathbf{V}_{cf} - \mathbf{V}_{af} = -\mathbf{V}_{af} \quad (6.66)$$

EXAMPLE 6.4

Repeat Example 6.1 assuming that there is a DLG fault with $Z_f = 5\ \Omega$ and $Z_g = 5\ \Omega$, involving phases b and c, at bus 3.

Solution

(a) Figure 6.7 shows the interconnection of the resulting equivalent sequence networks.
(b) Since

$$Z_f + 3Z_g = \frac{5+30}{264.5}$$

$$= 0.1323 \quad \text{pu}\ \Omega$$

the sequence currents are

$$\mathbf{I}_{a1} = \frac{1.0\angle 0°}{(\mathbf{Z}_1 + \mathbf{Z}_f) + \frac{(\mathbf{Z}_2 + \mathbf{Z}_f)(\mathbf{Z}_0 + \mathbf{Z}_f + 3\mathbf{Z}_g)}{(\mathbf{Z}_2 + \mathbf{Z}_f) + (\mathbf{Z}_0 + \mathbf{Z}_f + 3\mathbf{Z}_g)}}$$

$$= \frac{1.0\angle 0°}{(j0.2618 + 0.0189) + \frac{(j0.3619 + 0.0189)(j0.56 + 0.1323)}{j0.3619 + 0.0189 + j0.56 + 0.1323}}$$

$$= 2.0597\angle -84.5° \quad \text{pu}\ \Omega$$

Analysis of Unbalanced Faults

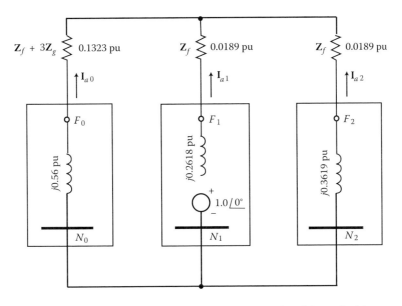

FIGURE 6.7 Interconnection of resultant equivalent sequence networks of Example 6.4.

$$\mathbf{I}_{a2} = -\left[\frac{(\mathbf{Z}_0 + \mathbf{Z}_f + 3\mathbf{Z}_g)}{(\mathbf{Z}_2 + \mathbf{Z}_f) + (\mathbf{Z}_0 + \mathbf{Z}_f + 3\mathbf{Z}_g)}\right]\mathbf{I}_{a1}$$

$$= -\left[\frac{0.5754\angle 76.7°}{0.9342\angle 80.7°}\right](2.0597\angle -84.5°)$$

$$= -1.2686\angle -88.5° \quad \text{pu}\,\Omega$$

$$\mathbf{I}_{a0} = -\left[\frac{(\mathbf{Z}_2 + \mathbf{Z}_f)}{(\mathbf{Z}_2 + \mathbf{Z}_f) + (\mathbf{Z}_0 + \mathbf{Z}_f + 3\mathbf{Z}_g)}\right]\mathbf{I}_{a1}$$

$$= -\left[\frac{0.3624\angle 87°}{0.9342\angle 80.7°}\right](2.0597\angle -84.5°)$$

$$= -0.799\angle -78.2° \quad \text{pu}\,\Omega$$

and the phase currents are

$$\begin{bmatrix} \mathbf{I}_{af} \\ \mathbf{I}_{bf} \\ \mathbf{I}_{cf} \end{bmatrix} = \begin{bmatrix} 1 & 1 & 1 \\ 1 & \mathbf{a}^2 & \mathbf{a} \\ 1 & \mathbf{a} & \mathbf{a}^2 \end{bmatrix} \begin{bmatrix} -0.799\angle -78.2° \\ 2.0597\angle -84.5° \\ -1.2686\angle -88.5° \end{bmatrix}$$

$$= \begin{bmatrix} 0 \\ 3.2677\angle 162.7° \\ 2.9653\angle 27.6° \end{bmatrix} \text{pu A}$$

(c) The sequence and phase voltages are

$$\begin{bmatrix} \mathbf{V}_{a0} \\ \mathbf{V}_{a1} \\ \mathbf{V}_{a2} \end{bmatrix} = \begin{bmatrix} 0 \\ 1.0\angle 0° \\ 0 \end{bmatrix} - \begin{bmatrix} j0.56 & 0 & 0 \\ 0 & j0.2618 & 0 \\ 0 & 0 & j0.3619 \end{bmatrix} \begin{bmatrix} -0.799\angle -78.2° \\ 2.0597\angle -84.5° \\ -1.2686\angle -88.5° \end{bmatrix}$$

$$= \begin{bmatrix} 0.4474\angle 11.8° \\ 0.4662\angle -6.4° \\ 0.4591\angle 1.5° \end{bmatrix} \text{ pu V}$$

and

$$\begin{bmatrix} \mathbf{V}_{af} \\ \mathbf{V}_{bf} \\ \mathbf{V}_{cf} \end{bmatrix} = \begin{bmatrix} 1 & 1 & 1 \\ 1 & \mathbf{a}^2 & \mathbf{a} \\ 1 & \mathbf{a} & \mathbf{a}^2 \end{bmatrix} \begin{bmatrix} 0.4474\angle 11.8° \\ 0.4662\angle -6.4° \\ 0.4591\angle 1.5° \end{bmatrix}$$

$$= \begin{bmatrix} 1.3611\angle 2.2° \\ 0.1333\angle 126.1° \\ 0.1198\angle 74.4° \end{bmatrix} \text{ pu V}$$

(d) The line-to-line voltages at the fault point are

$$\mathbf{V}_{abf} = \mathbf{V}_{af} - \mathbf{V}_{bf}$$
$$= 1.4386 - j0.0555$$
$$= 1.4397\angle -2.2° \text{ pu } \Omega$$

$$\mathbf{V}_{abf} = \mathbf{V}_{af} - \mathbf{V}_{bf}$$
$$= 1.4386 - 0.0555$$
$$= 1.4397 - 2.2° \text{ pu } \Omega$$

$$\mathbf{V}_{bcf} = \mathbf{V}_{bf} - \mathbf{V}_{cf}$$
$$= -0.1107 - j0.0077$$
$$= 0.111\angle 184° \text{ pu } \Omega$$

$$\mathbf{V}_{caf} = \mathbf{V}_{cf} - \mathbf{V}_{af}$$
$$= -1.3279 + j0.0632$$
$$= 1.3294\angle 177.3° \text{ pu } \Omega$$

6.2.4 Symmetrical Three-Phase Faults

In general, the three-phase (3ϕ) fault is not an unbalanced (i.e., unsymmetrical) fault. Instead, the three-phase fault is a balanced (i.e., symmetrical) fault that could also be analyzed using symmetrical components. Figure 6.8a shows the general representation of a balanced three-phase fault at a fault point F with impedances $\mathbf{Z}_f$ and $\mathbf{Z}_g$. Figure 6.8b shows the lack of interconnection of resulting

Analysis of Unbalanced Faults

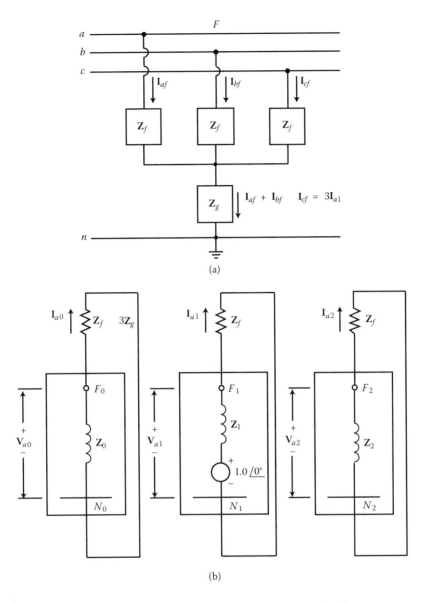

FIGURE 6.8 Three-phase fault: (a) general representation; (b) interconnection of sequence networks.

sequence networks. Instead, the sequence networks are short-circuited over their own fault impedances and are therefore isolated from each other. Since only the positive-sequence network is considered to have internal voltage source, the positive-, negative-, and zero-sequence currents can be expressed as

$$\mathbf{I}_{a0} = 0 \tag{6.67}$$

$$\mathbf{I}_{a2} = 0 \tag{6.68}$$

$$\mathbf{I}_{a1} = \frac{1.0\angle 0°}{\mathbf{Z}_1 + \mathbf{Z}_f} \tag{6.69}$$

If the fault impedance $\mathbf{Z}_f$ is zero,

$$\mathbf{I}_{a1} = \frac{1.0\angle 0°}{\mathbf{Z}_1} \tag{6.70}$$

Substituting Equations 6.67 and 6.68 into Equation 6.2,

$$\begin{bmatrix} \mathbf{I}_{af} \\ \mathbf{I}_{bf} \\ \mathbf{I}_{cf} \end{bmatrix} = \begin{bmatrix} 1 & 1 & 1 \\ 1 & \mathbf{a}^2 & \mathbf{a} \\ 1 & \mathbf{a} & \mathbf{a}^2 \end{bmatrix} \begin{bmatrix} 0 \\ \mathbf{I}_{a1} \\ 0 \end{bmatrix} \tag{6.71}$$

from which

$$\mathbf{I}_{af} = \mathbf{I}_{a1} = \frac{1.0\angle 0°}{\mathbf{Z}_1 + \mathbf{Z}_f} \tag{6.72}$$

$$\mathbf{I}_{bf} = \mathbf{a}^2 \mathbf{I}_{a1} = \frac{1.0\angle 240°}{\mathbf{Z}_1 + \mathbf{Z}_f} \tag{6.73}$$

$$\mathbf{I}_{cf} = \mathbf{a}\mathbf{I}_{a1} = \frac{1.0\angle 120°}{\mathbf{Z}_1 + \mathbf{Z}_f} \tag{6.74}$$

Since the sequence networks are short-circuited over their own fault impedances,

$$\mathbf{V}_{a0} = 0 \tag{6.75}$$

$$\mathbf{V}_{a1} = \mathbf{Z}_f \mathbf{I}_{a1} \tag{6.76}$$

$$\mathbf{V}_{a2} = 0 \tag{6.77}$$

Therefore, substituting Equations 6.75 through 6.77 into Equation 6.12,

$$\begin{bmatrix} \mathbf{V}_{af} \\ \mathbf{V}_{bf} \\ \mathbf{V}_{cf} \end{bmatrix} = \begin{bmatrix} 1 & 1 & 1 \\ 1 & \mathbf{a}^2 & \mathbf{a} \\ 1 & \mathbf{a} & \mathbf{a}^2 \end{bmatrix} \begin{bmatrix} 0 \\ \mathbf{V}_{a1} \\ 0 \end{bmatrix} \tag{6.78}$$

Thus,

$$\mathbf{V}_{af} = \mathbf{V}_{a1} = \mathbf{Z}_f \mathbf{I}_{a1} \tag{6.79}$$

$$\mathbf{V}_{bf} = \mathbf{a}^2 \mathbf{V}_{a1} = \mathbf{Z}_f \mathbf{I}_{a1} \angle 240° \tag{6.80}$$

$$\mathbf{V}_{cf} = \mathbf{a}\mathbf{V}_{a1} = \mathbf{Z}_f \mathbf{I}_{a1} \angle 120° \tag{6.81}$$

Analysis of Unbalanced Faults

Hence, the line-to-line voltages become

$$\mathbf{V}_{ab} = \mathbf{V}_{af} - \mathbf{V}_{bf} = \mathbf{V}_{a1}(1-\mathbf{a}^2) = \sqrt{3}\mathbf{Z}_f\mathbf{I}_{a1}\angle 30° \qquad (6.82)$$

$$\mathbf{V}_{bc} = \mathbf{V}_{bf} - \mathbf{V}_{cf} = \mathbf{V}_{a1}(\mathbf{a}^2-\mathbf{a}) = \sqrt{3}\mathbf{Z}_f\mathbf{I}_{a1}\angle -90° \qquad (6.83)$$

$$\mathbf{V}_{ca} = \mathbf{V}_{cf} - \mathbf{V}_{af} = \mathbf{V}_{a1}(\mathbf{a}-1) = \sqrt{3}\mathbf{Z}_f\mathbf{I}_{a1}\angle 150° \qquad (6.84)$$

Note that, in the event of having $\mathbf{Z}_f = 0$,

$$\mathbf{I}_{af} = \frac{1.0\angle 0°}{\mathbf{Z}_1} \qquad (6.85)$$

$$\mathbf{I}_{bf} = \frac{1.0\angle 240°}{\mathbf{Z}_1} \qquad (6.86)$$

$$\mathbf{I}_{cf} = \frac{1.0\angle 120°}{\mathbf{Z}_1} \qquad (6.87)$$

and

$$\mathbf{V}_{af} = 0 \qquad (6.88)$$

$$\mathbf{V}_{bf} = 0 \qquad (6.89)$$

$$\mathbf{V}_{cf} = 0 \qquad (6.90)$$

and, of course,

$$\mathbf{V}_{a0} = 0 \qquad (6.91)$$

$$\mathbf{V}_{a1} = 0 \qquad (6.92)$$

$$\mathbf{V}_{a2} = 0 \qquad (6.93)$$

EXAMPLE 6.5

Repeat Example 6.1 assuming that there is a symmetrical three-phase fault with $\mathbf{Z}_f = 5\ \Omega$ and $\mathbf{Z}_g = 10\ \Omega$ at bus 3. Let $\mathbf{Z}_0 = j0.5$ pu. Use $\mathbf{Z}_B = 264.5\ \Omega$.

Solution

(a) Figure 6.9 shows the interconnection of the resulting equivalent sequence networks.
(b) The sequence and phase currents are

$$\mathbf{I}_{a0} = \mathbf{I}_{a2} = 0$$

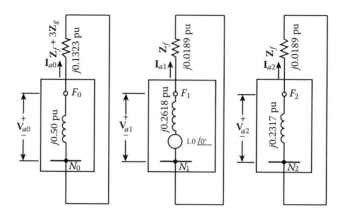

FIGURE 6.9 Interconnection of sequence networks of Example 6.5.

$$I_{a1} = \frac{1.0\angle 0°}{Z_1 + Z_f}$$

$$= \frac{1.0\angle 0°}{j0.2618 + 0.0189}$$

$$= 3.8098\angle -85.9° \quad \text{pu A}$$

and

$$\begin{bmatrix} I_{af} \\ I_{bf} \\ I_{cf} \end{bmatrix} = \begin{bmatrix} 1 & 1 & 1 \\ 1 & a^2 & a \\ 1 & a & a^2 \end{bmatrix} \begin{bmatrix} 0 \\ 3.8098\angle -85.9° \\ 0 \end{bmatrix}$$

$$= \begin{bmatrix} 3.8098\angle -85.9° \\ 3.8098\angle 154.1° \\ 3.8098\angle 34.1° \end{bmatrix} \quad \text{pu A}$$

(c) The sequence and phase voltages are

$$\begin{bmatrix} V_{a0} \\ V_{a1} \\ V_{a2} \end{bmatrix} = \begin{bmatrix} 0 \\ 1.0\angle 0° \\ 0 \end{bmatrix} - \begin{bmatrix} j0.5 & 0 & 0 \\ 0 & j0.2618 & 0 \\ 0 & 0 & j0.2317 \end{bmatrix} \begin{bmatrix} 0 \\ 3.8098\angle -85.9° \\ 0 \end{bmatrix}$$

$$= \begin{bmatrix} 0 \\ 0.0720\angle -85.9° \\ 0 \end{bmatrix} \quad \text{pu V}$$

Analysis of Unbalanced Faults

and

$$\begin{bmatrix} \mathbf{V}_{af} \\ \mathbf{V}_{bf} \\ \mathbf{V}_{cf} \end{bmatrix} = \begin{bmatrix} 1 & 1 & 1 \\ 1 & \mathbf{a}^2 & \mathbf{a} \\ 1 & \mathbf{a} & \mathbf{a}^2 \end{bmatrix} \begin{bmatrix} 0 \\ 0.0720\angle -85.9° \\ 0 \end{bmatrix}$$

$$= \begin{bmatrix} 0.0720\angle -85.9° \\ 0.0720\angle 154.9° \\ 0.0720\angle 34.1° \end{bmatrix} \text{ pu V}$$

(d) The line-to-line voltages at the fault point are

$$\mathbf{V}_{abf} = \mathbf{V}_{af} - \mathbf{V}_{bf}$$
$$= 0.0601 - j0.1023$$
$$= 0.1247\angle -55.9° \text{ pu V}$$

$$\mathbf{V}_{bcf} = \mathbf{V}_{bf} - \mathbf{V}_{cf}$$
$$= -0.1248 - j0.0099$$
$$= 0.1252\angle 184.5° \text{ pu } \Omega$$

$$\mathbf{V}_{caf} = \mathbf{V}_{cf} - \mathbf{V}_{af}$$
$$= 0.0545 + j0.1122$$
$$= 0.1247\angle 64.1° \text{ pu V}$$

6.2.5 Unsymmetrical Three-Phase Faults

However, it is forseeable that not all three-phase faults are symmetrical. There are also unsymmetrical faults, as shown in Figures 6.10–6.14. Figure 6.10 shows an unsymmetrical three-phase fault made up of a SLG fault and a line-to-line fault, both at the same fault point F. Figure 6.11 shows an unsymmetrical three-phase fault with a fault impedance $\mathbf{Z}_f$ connected between two lines. Figures 6.12 shows an unsymmetrical three-phase fault with a fault impedance on each phase where $\mathbf{Z}_{f1} \neq \mathbf{Z}_{f2}$. Figure 6.13 shows an unsymmetrical three-phase fault to ground. Figure 6.14 shows an unsymmetrical three-phase fault with delta-connected fault impedances where $\mathbf{Z}_{f1} \neq \mathbf{Z}_{f2}$. The derivation of the equations that are necessary to calculate fault currents has been left to the interested reader.

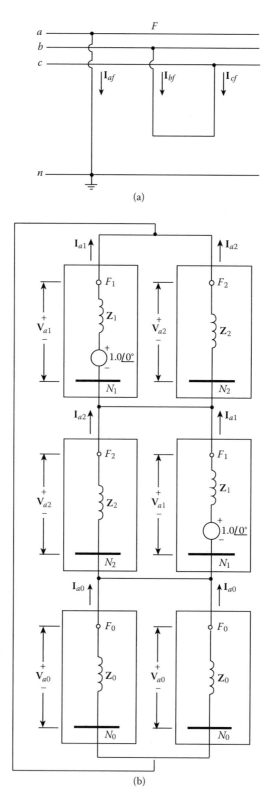

FIGURE 6.10 (a) Representation of an unsymmetrical three-phase fault and (b) interconnection of resultant equivalent sequence networks of Example 6.5.

Analysis of Unbalanced Faults

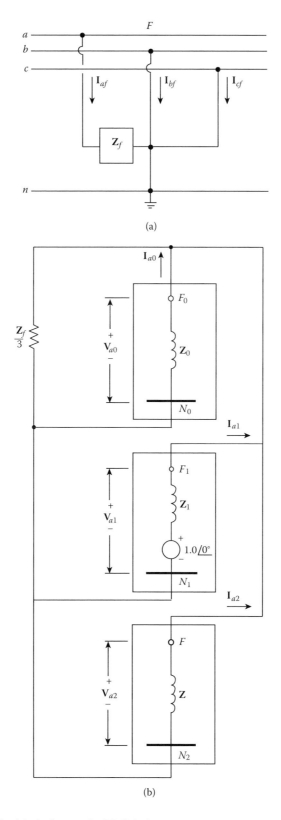

FIGURE 6.11 Generalized fault diagram for DLG fault.

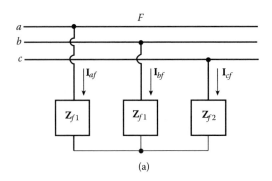

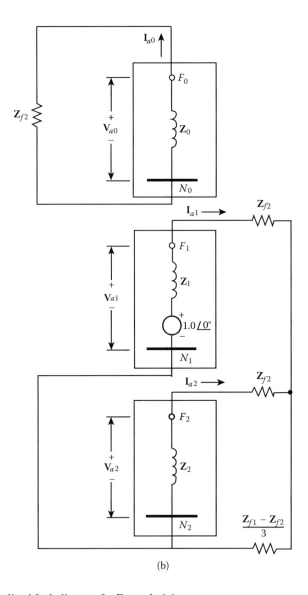

FIGURE 6.12 Generalized fault diagram for Example 6.6.

Analysis of Unbalanced Faults

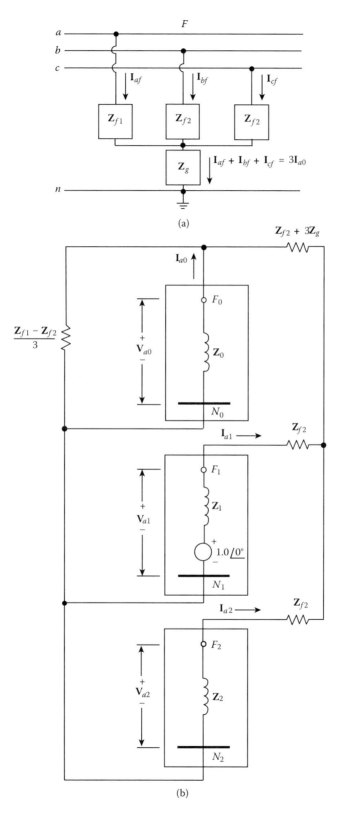

FIGURE 6.13 Generalized fault diagram for Example 6.7.

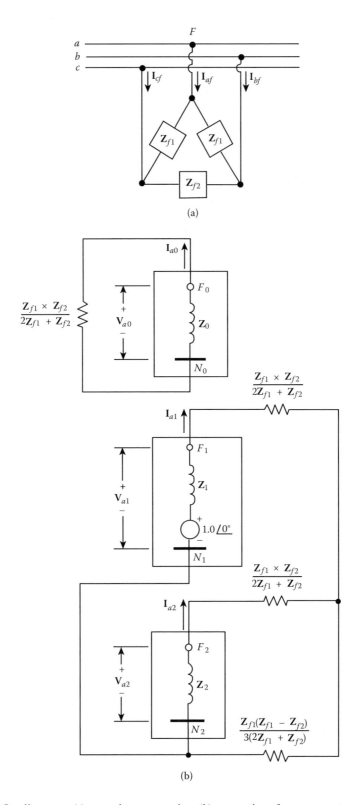

FIGURE 6.14 One line open: (a) general representation; (b) connection of sequence networks.

6.3 GENERALIZED FAULT DIAGRAMS FOR SHUNT FAULTS

In the event that an SLG fault occurs on phase b or c instead of on phase a, the calculation of the fault current can be determined by using one of the following two methods. The first method is analytical and involves the use of Kron's famous primitive network concept [5]. It will not be discussed in this book. However, the interested reader is urged to see an excellent review of the subject given by Anderson [3]. The second method involves the use of the generalized fault diagram of Atabekov [4], which in turn is based on Harder [6] and Hobson and Whitehead [7]. This method will be briefly reviewed in this section and later in Section 6.6. Again, the interested reader is highly recommended to see the fine review and modern application of this subject given by Anderson [3].

Figure 6.15 shows the generalized fault diagram for an SLG fault on any given phase. The resulting positive-, negative-, and zero-sequence networks are coupled by means of ideal transformers, or phase shifters, with complex turns ratios of 1, a, or a^2. Note that, such setup provides the currents and voltages at the output of the ideal transformers and lead the currents and voltages at the input by

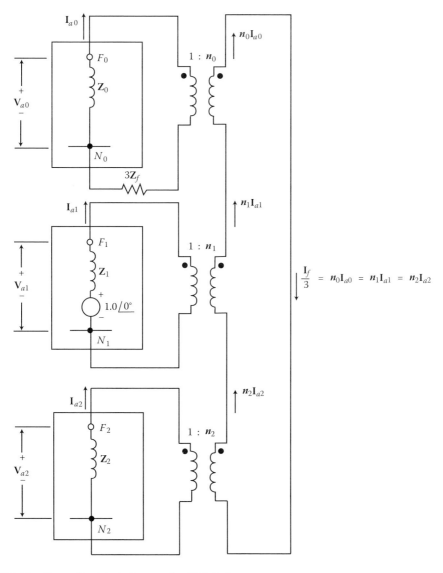

FIGURE 6.15 Generalized fault diagram for SLG fault.

TABLE 6.2
Complex Turns Ratios of Phase Shifters for SLG and GLG Faults

		Phase Shift		
SLG Fault on Phase	DLG Fault on Phases	n_0	n_1	n_2
a	b – c	1	1	1
b	c – a	1	$\mathbf{a}^2$	a
c	a – b	1	a	$\mathbf{a}^2$

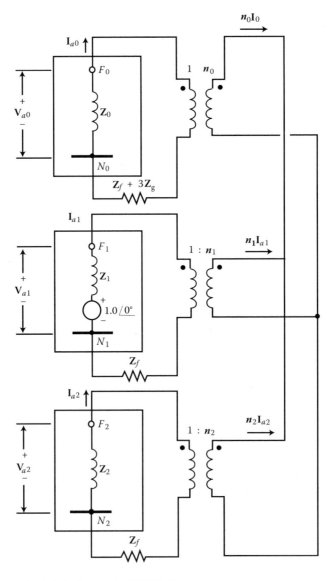

FIGURE 6.16 Generalized fault diagram for DLG fault.

Analysis of Unbalanced Faults

phase angles of 0, 120°, or 240°, respectively, without changing their magnitudes. There is no need here to consider the ideal transformer as a physical apparatus but merely as a symbolic phase shifter with a phase shift of $n = je^{j\theta}$. The appropriate phase shifts for a given SLG fault can be obtained from Table 6.2. For Example, if the SLG fault is on phase b, then the proper phase shifts are $n_0 = 1$, $n_1 = a^2$, and $n_2 = a$.

Figure 6.16 shows the generalized fault diagram for a DLG fault involving any two phases. Again, the appropriate phase shifts for a given DLG fault can be obtained from Table 6.2. For example, if the DLG fault involves phases a and b, then the proper phase shifts are $n_0 = 1$, $n_1 = a$, and $n_2 = a^2$.

Example 6.6

Consider the system described in Example 6.1 and assume that the SLG fault involves phase b.

(a) Draw the generalized fault diagram.
(b) Determine the sequence currents.
(c) Determine the phase currents.

Solution

(a) Figure 6.17 shows the resulting generalized fault diagram.
(b) Since

$$I_{a0} = a^2 I_{a1} = a I_{a2}$$

then

$$I_a = 0 \text{ and } I_c = 0$$

Thus, as before,

$$I_{a1} = \frac{1.0\angle 0°}{Z_0 + Z_1 + Z_2 + 3Z_f}$$

$$= 0.8438\angle -87.3° \text{ pu}$$

but

$$I_{a0} = a^2 I_{a1}$$

$$= 1.0\angle 240°(0.8438\angle -87.3°)$$

$$= 0.8438\angle 152.7° \text{ pu}$$

and

$$I_{a2} = \frac{I_{a0}}{a}$$

$$= \frac{0.8438\angle 152.7°}{1.0\angle 120°}$$

$$= 0.8438\angle 32.7° \text{ pu}$$

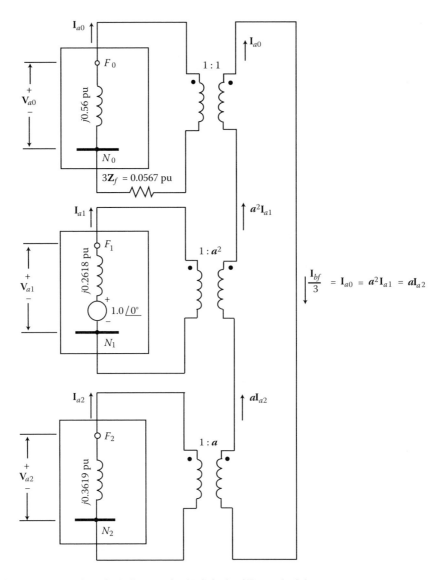

FIGURE 6.17 Generalized fault diagram for SLG fault of Example 6.6.

(c) Therefore,

$$\mathbf{I}_a = \mathbf{I}_c = 0$$

hence,

$$\begin{bmatrix} \mathbf{I}_{af} \\ \mathbf{I}_{bf} \\ \mathbf{I}_{cf} \end{bmatrix} = \begin{bmatrix} 1 & 1 & 1 \\ 1 & a^2 & a \\ 1 & a & a^2 \end{bmatrix} \begin{bmatrix} 0.8438\angle 152.7° \\ 0.8438\angle -87.3° \\ 0.8438\angle 32.7° \end{bmatrix}$$

$$= \begin{bmatrix} 0 \\ 2.5314\angle 152.7° \\ 0 \end{bmatrix} \text{pu}$$

Example 6.7

Consider the system described in Example 6.4 and assume that the DLG fault involves phases a and b.

(a) Draw the generalized fault diagram.
(b) Determine the sequence currents.
(c) Determine the phase currents.

Solution

(a) Figure 6.18 shows the resulting generalized fault diagram.
(b) From Example 6.4

$$\mathbf{I}_{a1} = 2.0597\angle -84.5° \text{ pu}$$

hence,

$$\mathbf{I}_{a0} = -\left(\frac{j0.3808}{1.0731}\right)a\mathbf{I}_{a1}$$

$$= -0.3549(1\angle 120°)(2.0597\angle -84.5°)$$

$$= 0.7309\angle 215.5° \text{ pu}$$

and

$$\mathbf{I}_{a2} = -\left(\frac{j0.6923}{1.0731}\right)\frac{\mathbf{I}_{a1}}{a}$$

$$= -0.6451(1\angle -120°)(2.0597\angle -84.5°)$$

$$= 1.3288\angle -24.5° \text{ pu}$$

(c)

$$\mathbf{I}_f = \mathbf{I}_a + \mathbf{I}_b = 3\mathbf{I}_{a0}$$

$$= 3(0.7309\angle 215.5°)$$

$$= 2.1927\angle 215.5° \text{ pu}$$

$$\mathbf{I}_a = \mathbf{I}_{a0} + \mathbf{I}_{a1} + \mathbf{I}_{a2}$$

$$= 0.7309\angle 215.5° + 2.0597\angle -84.5° + 1.3288\angle -24.5°$$

$$= 0.8116 - j3.0256$$

$$= 3.1326\angle -75° \text{ pu}$$

$$\mathbf{I}_b = \mathbf{I}_{a0} + a^2\mathbf{I}_{a1} + a\mathbf{I}_{a2}$$

$$= 0.7309\angle 215.5° + (1\angle 240°)(2.0597\angle -84.5°) + (1\angle 120°)(1.3288\angle -24.5°)$$

$$= 2.5966 + j1.7524$$

$$= 2.1326\angle 146° \text{ pu}$$

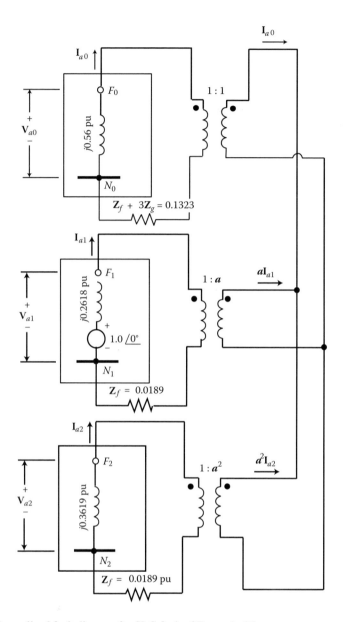

FIGURE 6.18 Generalized fault diagram for SLG fault of Example 6.7.

$$I_c \triangleq 0$$

Also, as a check,

$$I_f = I_a + I_b$$
$$= 3.1326\angle -75° + 3.1326\angle 146°$$
$$= 2.1927\angle 215.5° \text{ pu}$$

6.4 SERIES FAULTS

In general, series (longitudinal) faults are due to an unbalanced series impedance condition of the lines. One or two broken lines, or an impedance inserted in one or two lines, may be considered as *series faults*. In practice, a series fault is encountered, for example, when line (or circuits) are controlled by circuit breakers (or by fuses) or any device that does not open all three phases; one or two phases of the line (or the circuit) may be open while the other phases or phase is closed.

Figure 6.19 shows a series fault due to one line (phase a) being open, which causes a series unbalance. In a series fault, contrary to a shunt fault, there are two fault points, F and F', one

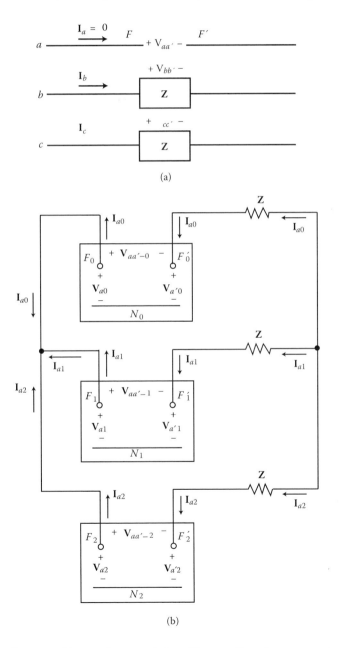

FIGURE 6.19 One line open: (a) general representation; (b) connection of sequence networks.

on either side of the unbalance. The series line impedances **Z**'s can take any values between zero and infinity (in this case, obviously, the line impedance between points F and F' of phase a is infinity).

The sequence networks include the symmetrical portions of the system, looking back to the left of F and to the right of F'. Since in series faults, there is no connection between lines or between line(s) and neutral, only the sequence voltages of $\mathbf{V}_{aa'-0}$, $\mathbf{V}_{aa'-1}$, and $\mathbf{V}_{aa'-2}$ are of interest, not the sequence voltages of $\mathbf{V}_{a0}$, $\mathbf{V}_{a1}$, $\mathbf{V}_{a2}$, etc. (as it was the case with the shunt faults).

6.4.1 One Line Open

From Figure 6.19, it can be observed that the line impedance for the open-line conductor in phase a is infinity, whereas the line impedances for the other two phases have some finite values. Hence, the positive-, negative-, and zero-sequence currents can be expressed as

$$\mathbf{I}_{a1} = \frac{\mathbf{V}_F}{\mathbf{Z} + \mathbf{Z}_1 + (\mathbf{Z} + \mathbf{Z}_0)(\mathbf{Z} + \mathbf{Z}_2)/(2\mathbf{Z} + \mathbf{Z}_0 + \mathbf{Z}_2)} \tag{6.94}$$

and by current division,

$$\mathbf{I}_{a2} = \left(-\frac{\mathbf{Z} + \mathbf{Z}_0}{2\mathbf{Z} + \mathbf{Z}_0 + \mathbf{Z}_2}\right)\mathbf{I}_{a1} \tag{6.95}$$

and

$$\mathbf{I}_{a0} = \left(-\frac{\mathbf{Z} + \mathbf{Z}_2}{2\mathbf{Z} + \mathbf{Z}_0 + \mathbf{Z}_2}\right)\mathbf{I}_{a1} \tag{6.96}$$

or simply

$$\mathbf{I}_{a0} = -(\mathbf{I}_{a1} + \mathbf{I}_{a2}) \tag{6.97}$$

6.4.2 Two Lines Open

If two lines are open as shown in Figure 6.20, then the line impedances for one line open (OLO) in phases b and c are infinity, whereas the line impedance of phase a has some finite value. Thus,

$$\mathbf{I}_b = \mathbf{I}_c = 0 \tag{6.98}$$

and

$$\mathbf{V}_{aa'} = \mathbf{Z}\mathbf{I}_a \tag{6.99}$$

Analysis of Unbalanced Faults

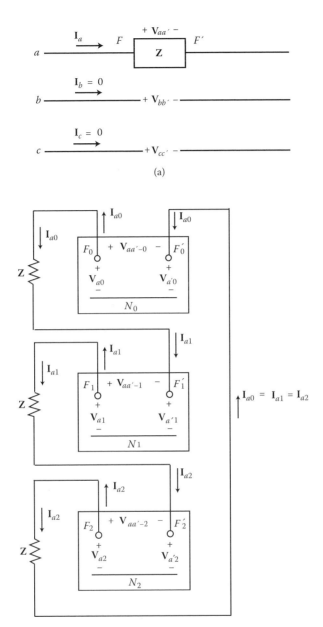

FIGURE 6.20 Two-lines open: (a) general representation; (b) interconnection of sequence networks.

By inspection of Figure 6.20, the positive-, negative-, and zero-sequence currents can be expressed as

$$\mathbf{I}_{a1} = \mathbf{I}_{a2} = \mathbf{I}_{a0} = \frac{\mathbf{V}_F}{\mathbf{Z}_0 + \mathbf{Z}_1 + \mathbf{Z}_2 + 3\mathbf{Z}_f} \tag{6.100}$$

6.5 DETERMINATION OF SEQUENCE NETWORK EQUIVALENTS FOR SERIES FAULTS

Since the series faults have two fault pints (i.e., F and F'), contrary to the shunt faults having only one fault point, the direct application of Thévenin's theorem is not possible. Instead, what is needed is a two-port Thévenin equivalent of the sequence networks as suggested by Anderson [3,8].

6.5.1 Brief Review of Two-Port Theory

Figure 6.21 shows a general two-port network for which it can be written that

$$\begin{bmatrix} V_1 \\ V_2 \end{bmatrix} = \begin{bmatrix} Z_{11} & Z_{12} \\ Z_{21} & Z_{22} \end{bmatrix} \begin{bmatrix} I_1 \\ I_2 \end{bmatrix} \qquad (6.101)$$

where the *open-circuit impedance parameters* can be determined by leaving the ports open and expressed in terms of voltage and current as

$$Z_{11} = \left. \frac{V_1}{I_1} \right|_{I_2=0} \qquad (6.102)$$

$$Z_{12} = \left. \frac{V_1}{I_2} \right|_{I_1=0} \qquad (6.103)$$

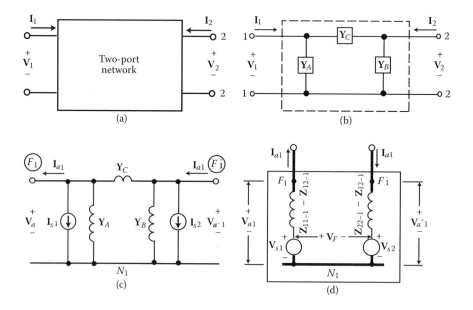

FIGURE 6.21 Application of two-port network theory for determining equivalent positive-sequence network for series faults: (a) general two-port network; (b) general π-equivalent positive sequence network; (c) equivalent positive-sequence network; and (d) uncoupled positive-sequence network.

Analysis of Unbalanced Faults

$$\mathbf{Z}_{21} = \left.\frac{\mathbf{V}_2}{\mathbf{I}_1}\right|_{\mathbf{I}_2=0} \tag{6.104}$$

$$\mathbf{Z}_{22} = \left.\frac{\mathbf{V}_2}{\mathbf{I}_2}\right|_{\mathbf{I}_1=0} \tag{6.105}$$

Alternatively, it can be observed that

$$\begin{bmatrix} \mathbf{I}_1 \\ \mathbf{I}_2 \end{bmatrix} = \begin{bmatrix} \mathbf{Y}_{11} & \mathbf{Y}_{12} \\ \mathbf{Y}_{21} & \mathbf{Y}_{22} \end{bmatrix} \begin{bmatrix} \mathbf{V}_1 \\ \mathbf{V}_2 \end{bmatrix} \tag{6.106}$$

where the *short-circuit admittance parameters* can be determined (by short-circuiting the ports) from

$$\mathbf{Y}_{11} = \left.\frac{\mathbf{I}_1}{\mathbf{V}_1}\right|_{\mathbf{V}_2=0} \tag{6.107}$$

$$\mathbf{Y}_{21} = \left.\frac{\mathbf{I}_2}{\mathbf{V}_1}\right|_{\mathbf{V}_2=0} \tag{6.108}$$

$$\mathbf{Y}_{12} = \left.\frac{\mathbf{I}_1}{\mathbf{V}_2}\right|_{\mathbf{V}_1=0} \tag{6.109}$$

$$\mathbf{Y}_{22} = \left.\frac{\mathbf{I}_2}{\mathbf{V}_2}\right|_{\mathbf{V}_1=0} \tag{6.110}$$

Figure 6.21b shows a general π-equivalent of a two-port network in terms of admittances. The $\mathbf{Y}_A$, $\mathbf{Y}_B$, and $\mathbf{Y}_C$ admittances can be found from

$$\mathbf{Y}_A = \mathbf{Y}_{11} + \mathbf{Y}_{12} \tag{6.111}$$

$$\mathbf{Y}_B = \mathbf{Y}_{22} + \mathbf{Y}_{12} \tag{6.112}$$

$$\mathbf{Y}_C = -\mathbf{Y}_{12} \tag{6.113}$$

6.5.2 Equivalent Zero-Sequence Networks

By comparing the zero-sequence network shown in Figure 6.19 with the general two-port network shown in Figure 6.21a, it can be observed that

$$\mathbf{I}_1 = -\mathbf{I}_{a0} \tag{6.114}$$

$$\mathbf{I}_2 = \mathbf{I}_{a0} \quad (6.115)$$

$$\mathbf{V}_1 = \mathbf{V}_{a0} \quad (6.116)$$

$$\mathbf{V}_2 = \mathbf{V}_{a'0} \quad (6.117)$$

Hence, substituting Equations 6.114 through 6.117 into Equation 6.106, it can be expressed for the Thévenin equivalent of the zero-sequence network that

$$\begin{bmatrix} -\mathbf{I}_{a0} \\ \mathbf{I}_{a0} \end{bmatrix} = \begin{bmatrix} \mathbf{Y}_{11-0} & \mathbf{Y}_{12-0} \\ \mathbf{Y}_{21-0} & \mathbf{Y}_{22-0} \end{bmatrix} \begin{bmatrix} \mathbf{V}_{a0} \\ \mathbf{V}_{a'0} \end{bmatrix} \quad (6.118)$$

6.5.3 Equivalent Positive- and Negative-Sequence Networks

Figure 6.16c shows the equivalent positive-sequence network as an active two-port network with internal sources. Thus, it can be expressed for the two-port Thévenin equivalent of the positive-sequence network that

$$\begin{bmatrix} -\mathbf{I}_{a1} \\ \mathbf{I}_{a1} \end{bmatrix} = \begin{bmatrix} \mathbf{Y}_{11-1} & \mathbf{Y}_{12-1} \\ \mathbf{Y}_{21-1} & \mathbf{Y}_{22-1} \end{bmatrix} \begin{bmatrix} \mathbf{V}_{a1} \\ \mathbf{V}_{a'1} \end{bmatrix} + \begin{bmatrix} \mathbf{I}_{s1} \\ \mathbf{I}_{s2} \end{bmatrix} \quad (6.119)$$

or, alternatively,

$$\begin{bmatrix} \mathbf{V}_{a1} \\ \mathbf{V}_{a'1} \end{bmatrix} = \begin{bmatrix} \mathbf{Z}_{11-1} & \mathbf{Z}_{12-1} \\ \mathbf{Z}_{21-1} & \mathbf{Z}_{22-1} \end{bmatrix} \begin{bmatrix} -\mathbf{I}_{a1} \\ \mathbf{I}_{a1} \end{bmatrix} + \begin{bmatrix} \mathbf{V}_{s1} \\ \mathbf{V}_{s2} \end{bmatrix} \quad (6.120)$$

where $\mathbf{I}_{s1}$, $\mathbf{V}_{s1}$ and $\mathbf{I}_{s2}$, $\mathbf{V}_{s2}$ represent internal sources 1 and 2, respectively. As before, the admittances $\mathbf{Y}_A$, $\mathbf{Y}_B$, and $\mathbf{Y}_C$ can be determined from

$$\mathbf{Y}_A = \mathbf{Y}_{11-1} + \mathbf{Y}_{12-1} \quad (6.121)$$

$$\mathbf{Y}_B = \mathbf{Y}_{22-1} + \mathbf{Y}_{12-1} \quad (6.122)$$

$$\mathbf{Y}_C = -\mathbf{Y}_{12-1} \quad (6.123)$$

The two-port Thévenin equivalent of the negative-sequence network would be the same as the one shown in Figure 6.16c but without the internal sources.

Anderson [3] shows that Equation 6.120 can be simplified as

$$\begin{bmatrix} \mathbf{V}_{a1} \\ \mathbf{V}_{a'1} \end{bmatrix} = \begin{bmatrix} \mathbf{V}_{s1} \\ \mathbf{V}_{s2} \end{bmatrix} - \begin{bmatrix} (\mathbf{Z}_{11-1} - \mathbf{Z}_{12-1})\mathbf{I}_{a1} \\ -(\mathbf{Z}_{22-1} - \mathbf{Z}_{12-1})\mathbf{I}_{a1} \end{bmatrix} \quad (6.124)$$

Analysis of Unbalanced Faults

due to the fact that I_{a1} leaves the network at fault point F and enters at fault point F' due to external connection. This facilitates the voltage V_{a1} to be expressed in terms of the equivalent impedance Z_{s1} and the current I_{a1} to be expressed as it has been done for the shunt faults, where

$$V_{a1} = V_F - Z_1 I_{a1} \quad (6.125)$$

Therefore, it can be concluded that the port of the positive-sequence network are completely uncoupled and that the resulting uncoupled positive-sequence network can be shown as in Figure 6.16d. The voltages V_{a1} and $V_{a'1}$ can be found from Equation 6.120 as

$$\begin{bmatrix} V_{a1} \\ V_{a'1} \end{bmatrix} = \frac{1}{\Delta_y} \begin{bmatrix} (Y_{12-1} I_{s2} - Y_{22-1} I_{s1}) - (Y_{12-1} + Y_{22-1}) I_{a1} \\ (Y_{12-1} I_{s1} - Y_{11-1} I_{s2}) + (Y_{12-1} + Y_{11-1}) I_{a1} \end{bmatrix} \quad (6.126)$$

where

$$\Delta_y = \det \begin{bmatrix} Y_{11-1} & Y_{12-1} \\ Y_{21-1} & Y_{22-1} \end{bmatrix} \quad (6.127)$$

or

$$\Delta_y = Y_{11-1} Y_{22-1} - Y_{12-1}^2 \quad (6.128)$$

EXAMPLE 6.8

Consider the system shown in Figure 6.22 and assume that there is a series fault at fault point A, which is located at the middle of the transmission line TL_{AB}, as shown in the figure, and determine the following:

(a) Admittance matrix associated with the positive-sequence network
(b) Two-port Thévenin equivalent of the positive-sequence network
(c) Two-port Thévenin equivalent of the negative-sequence network

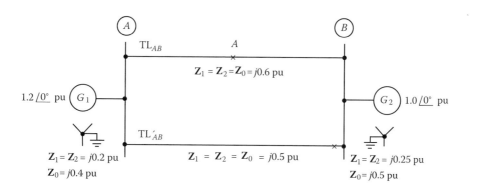

FIGURE 6.22 System diagram for Example 6.8.

Solution

Figure 6.18 shows the steps that are necessary to determine the positive- and negative-sequence network equivalents. Figure 6.18a shows the impedance diagram of the system for the positive sequence. Figure 6.18b shows the resulting two-port equivalent with input and output currents and voltages.

(a) To determine the elements of the admittance matrix $\mathbf{Y}$, it is necessary to remove the internal voltage sources, shown in Figure 6.18b, by short-circuiting them. Then, with $\mathbf{V}_2 = 0$ (i.e., by short circuiting the terminals of the second port), apply $\mathbf{V}_1 = 1.0\angle 0°$ pu and determine the parameter

$$Y_{11} = \frac{I_1}{V_1}\bigg|_{V_2=0} = I_1 = \frac{1.0\angle 0°}{0.4522\angle 90°} = -j2.2115 \text{ pu}$$

and

$$Y_{21} = \frac{I_2}{V_1}\bigg|_{V_2=0} = I_2 = -0.1087 I_1 = j0.2404 \text{ pu}$$

Now, with $\mathbf{V}_1$ and $\mathbf{V}_2 = 1.0\angle 0°$ pu, determine

$$Y_{22} = \frac{I_2}{V_2}\bigg|_{V_1=0} = I_2 = \frac{1.0\angle 0°}{0.4782\angle 90°} = -j2.0912 \text{ pu}$$

and

$$Y_{12} = Y_{21} = j0.2404 \text{ pu}$$

Hence,

$$\mathbf{Y} = \begin{bmatrix} Y_{11-1} & Y_{12-1} \\ Y_{21-1} & Y_{22-1} \end{bmatrix}$$

$$= \begin{bmatrix} -j2.2115 & j0.2404 \\ j0.2404 & -2.0912 \end{bmatrix}$$

(b) To find the source currents $\mathbf{I}_{s1}$ and $\mathbf{I}_{s2}$, short circuit both F and F' to neutral and use the superposition theorem, so that

$$\mathbf{I}_{s1} = \mathbf{I}_{s1(1.2)} + \mathbf{I}_{s1(1.0)}$$

$$= 2.0193\angle 90° + 0.7212\angle 90°$$

$$= 2.7405\angle 90° \text{ pu}$$

Analysis of Unbalanced Faults

and

$$\mathbf{I}_{s2} = \mathbf{I}_{s2(1.2)} + \mathbf{I}_{s2(1.0)}$$
$$= 0.4326\angle 90° + 1.4904\angle 90°$$
$$= 1.9230\angle 90° \text{ pu}$$

Figure 6.23c shows the resulting two-port Thévenin equivalent of the positive-sequence network. Figure 6.23d shows the corresponding coupled positive-sequence network.

(c) Figure 6.23e shows the resulting two-port Thévenin equivalent of the negative-sequence network. Notice that it is the same as the one for the positive-sequence network but with-

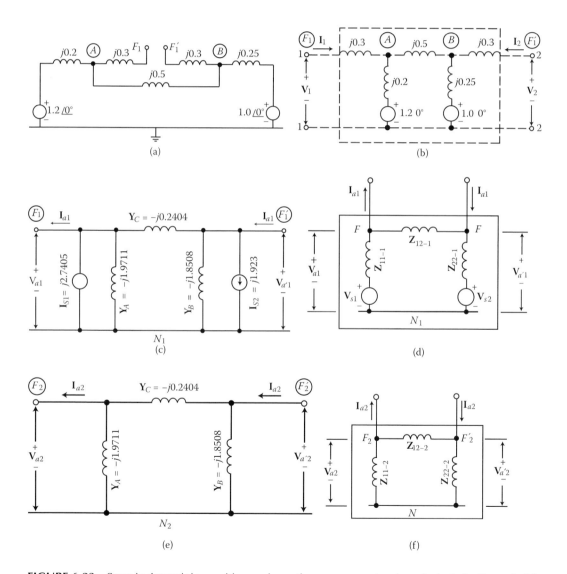

FIGURE 6.23 Steps in determining positive- and negative-sequence network equivalents for Example 6.8: (a) system diagram; (b) resulting two-port equivalent with input and output currents and voltages; (c) two-port Thévenin equivalent of positive-sequence network; (d) resulting coupled positive-sequence network; (e) two-port Thévenin equivalent of negative-sequence network; (f) resulting coupled negative-sequence network.

out its current sources. Figure 6.23f shows the corresponding coupled negative-sequence network.

EXAMPLE 6.9

Consider the solution of Example 6.8 and determine the following:

(a) Uncoupled positive-sequence network
(b) Uncoupled negative-sequence network

Solution

(a) From Example 6.8,

$$\mathbf{Y} = \begin{bmatrix} -j2.2115 & j0.2404 \\ j0.2404 & -2.0912 \end{bmatrix}$$

where

$$\Delta_y = -4.6247 - (-0.0578)$$
$$= -4.5669$$

Since

$$\begin{bmatrix} \mathbf{V}_{a1} \\ \mathbf{V}_{a'1} \end{bmatrix} = \frac{1}{\Delta_y} \begin{bmatrix} (\mathbf{Y}_{12-1}\mathbf{I}_{s2} - \mathbf{Y}_{22-1}\mathbf{I}_{s1}) - (\mathbf{Y}_{12-1} + \mathbf{Y}_{22-1})\mathbf{I}_{a1} \\ (\mathbf{Y}_{12-1}\mathbf{I}_{s1} - \mathbf{Y}_{11-1}\mathbf{I}_{s2}) + (\mathbf{Y}_{12-1} + \mathbf{Y}_{11-1})\mathbf{I}_{a1} \end{bmatrix}$$

where

$$(\mathbf{Y}_{12-1}\mathbf{I}_{s2} - \mathbf{Y}_{22-1}\mathbf{I}_{s1}) = j0.2404(j1.9230) - (-j2.0912)j2.7405$$
$$= -6.1932$$

$$(\mathbf{Y}_{12-1} + \mathbf{Y}_{22-1})\mathbf{I}_{a1} = (j0.2404 - j2.0912)\mathbf{I}_{a1}$$
$$= -j1.8508\mathbf{I}_{a1}$$

$$(\mathbf{Y}_{12-1}\mathbf{I}_{s1} - \mathbf{Y}_{11-1}\mathbf{I}_{s2}) = j0.2404(j2.7405) - (-j2.2115)j1.923$$
$$= -4.9115$$

$$(\mathbf{Y}_{12-1} + \mathbf{Y}_{11-1})\mathbf{I}_{a1} = (j0.2404 - j2.2115))\mathbf{I}_{a1}$$
$$= -j1.9711\mathbf{I}_{a1}$$

Analysis of Unbalanced Faults

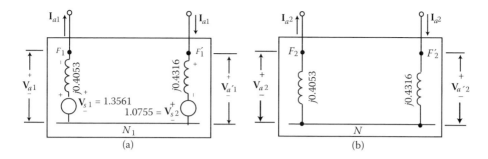

FIGURE 6.24 Uncoupled sequence networks: (a) positive-sequence network; (b) negative-sequence network.

Therefore,

$$\begin{bmatrix} V_{a1} \\ V_{a'1} \end{bmatrix} = \frac{1}{-4.5669} \begin{bmatrix} (-6.1932) + j1.8508 I_{a1} \\ (-4.9115) - j1.9711 I_{a1} \end{bmatrix}$$

$$= \begin{bmatrix} 1.3561 - j0.4053 I_{a1} \\ 1.0755 + j0.4316 I_{a1} \end{bmatrix}$$

(b) Figure 6.24b shows the corresponding uncoupled negative-sequence network.

6.6 GENERALIZED FAULT DIAGRAM FOR SERIES FAULTS

Figure 6.25 shows the generalized fault diagram for OLO involving any given phase. As before, the resulting positive-, negative-, and zero-sequence networks are coupled by means of ideal transformers. The appropriate phase shifts for a given line can be obtained from Table 6.3. For example, if the open line is in phase b, then the proper phase shifts are $n_0 = 1$, $n_1 = a^2$, and $n_2 = a$.

Figure 6.26 shows the generalized fault diagram for a two-line-open (TLO) fault involving any two phases. Again, the appropriate phase shifts for a specific TLO fault can be obtained from Table 6.3. For example, if the TLO fault involves phases b and c, then the proper phase shifts are $n_0 = 1$, $n_1 = 1$, and $n_2 = 1$.

EXAMPLE 6.10

Assume that a given power system has two generators connected to each other over a transmission line and that there is a TLO series fault on the transmission line at some location involving phases a and b. Figure 6.27 shows the resulting two-port Thévenin equivalents of the sequence networks at fault points F and F'. If the remaining line (i.e., line c) has a fault impedance of 0.1 pu, do the following:

(a) Draw the generalized fault diagram.
(b) Determine the positive-sequence current.
(c) Determine the negative-sequence current.
(d) Determine the negative-sequence current.

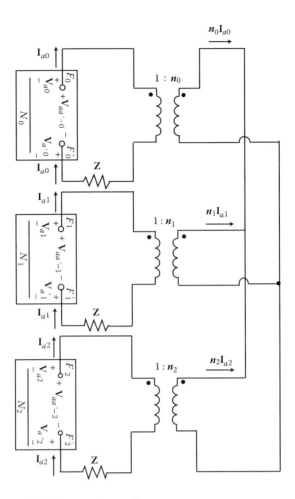

FIGURE 6.25 Generalized fault diagram for one line open.

TABLE 6.3
Complex Turns Ratios of Phase Shifters for OLO and TLO Faults

		Phase Shift		
One Line Open	Two Lines Open	n_0	n_1	n_2
a	$b - c$	1	1	1
b	$c - a$	1	a^2	a
c	$a - b$	1	a	a^2

(e) Determine the line current for phase a.
(f) Determine the line current for phase b.
(g) Determine the line current for phase c.

Solution

(a) Figure 6.28 shows the resulting generalized fault diagram.

Analysis of Unbalanced Faults

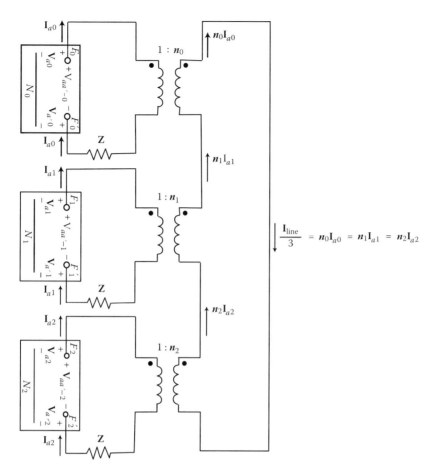

FIGURE 6.26 Generalized fault diagram for two lines open.

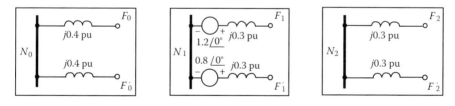

FIGURE 6.27 Resulting two-port Thévenin equivalents of sequence networks for Example 6.10.

(b) From Equation 6.100,

$$\mathbf{I}_{a1} = \frac{V_f}{\mathbf{Z}_0 + \mathbf{Z}_1 + \mathbf{Z}_2 + 3\mathbf{Z}_f}$$

$$= \frac{1.2\angle 0° - 0.8\angle 0°}{j0.8 + j0.6 + j0.6 + 0.3}$$

$$= \frac{0.4\angle 0°}{0.3 + j2.0}$$

$$= 0.1978\angle -81.5° \text{ pu}$$

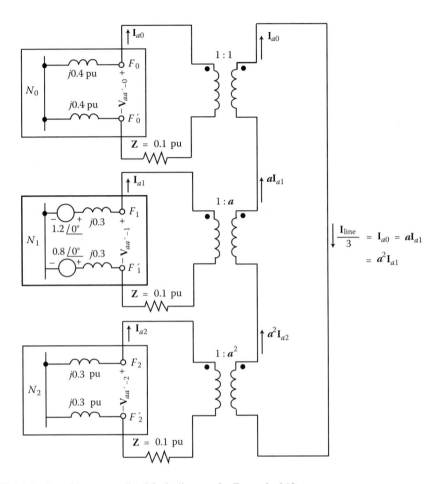

FIGURE 6.28 Resulting generalized fault diagram for Example 6.10.

(c) From Figure 6.28,

$$\mathbf{I}_{a0} = \mathbf{a}\mathbf{I}_{a1} = \mathbf{a}^2\mathbf{I}_{a2}$$

Therefore,

$$\mathbf{I}_{a2} = \frac{\mathbf{I}_{a1}}{\mathbf{a}}$$

$$= \frac{0.1978\angle -81.5°}{1\angle 120°}$$

$$= 0.1978\angle -201.5°$$

$$= 0.1978\angle 158.5° \text{ pu}$$

(d) From part (c),

$$\mathbf{I}_{a0} = \mathbf{a}\mathbf{I}_{a1}$$

$$= (1\angle 120°)(0.1978\angle -81.5°)$$

$$= 0.1978\angle 38.5° \text{ pu}$$

Analysis of Unbalanced Faults

(e) By definition,

$$I_a \triangleq 0$$

but as a check,

$$I_a = I_{a0} + I_{a1} + I_{a2}$$
$$= 0.1978\angle 38.5° + 0.1978\angle -81.5° + 0.1978\angle 158.5° = 0$$

(f) Again by definition, $I_b \triangleq 0$; but as a check,

$$I_b = I_{a0} + a^2 I_{a1} + a I_{a2}$$
$$= 0.1978\angle 38.5° + (1\angle 240°)(0.1978\angle -81.5°) + (1\angle 120°)(0.1978\angle 158.5°)$$
$$= 0.1978\angle 38.5° + 0.1978\angle 158.5° + 0.1978\angle 278.5° = 0$$

(g)

$$I_c = I_{a0} + a I_{a1} + a^2 I_{a2}$$
$$= 0.1978\angle 38.5° + (1\angle 120°)(0.1978\angle -81.5°) + (1\angle 240°)(0.1978\angle 158.5°)$$
$$= 0.1978\angle 38.5° + 0.1978\angle 38.5° + 0.1978\angle 38.5°$$
$$= 0.5934\angle 38.5°$$

As a check,

$$I_f = I_{line} \triangleq 3 I_{a0} = 3(0.1978\angle 38.5°)$$
$$= 0.5934\angle 38.5° \text{ pu}$$

6.7 SYSTEM GROUNDING

A *system neutral ground* is connected to ground from the neutral point or points of a system or rotating machine or transformer. Thus, a *grounded system* is a system that has at least one neutral point that is intentionally grounded, either solidly or through a current-limiting device. For example, most transformer neutrals in transmission systems are solidly grounded.

However, generator neutrals are usually grounded through some type of current-limiting device to limit the ground fault current. Figure 6.29 shows various neutral grounding methods used with generators and resulting zero-sequence networks. Hence, the methods of grounding the system neutral include

1. Ungrounded
2. Solidly grounded
3. Resistance grounded
4. Reactance grounded
5. Peterson coil grounded

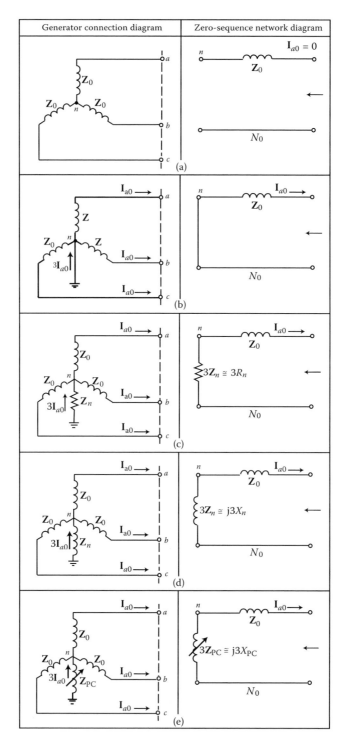

FIGURE 6.29 Various system (neutral) grounding methods used with generators and resulting zero-sequence network: (a) ungrounded; (b) solidly grounded; (c) resistance grounded; (d) reactance grounded; (e) grounded through Peterson coil.

Analysis of Unbalanced Faults

The last four methods above provide grounded neutrals, whereas the first provides an ungrounded (also called *isolated* or *free*) neutral system.

In an ungrounded system, there is no intentional connection between the neutral point or neutral points of the system and the ground, as shown in Figure 6.30a. The line conductors have distributed capacitances between one another (not shown in the figure) and to ground due to capacitive coupling.

Under balanced conditions (assuming a perfectly transposed line), each conductor has the same capacitance to ground. Thus, the charging current of each phase is the same in Figure 6.30b. Hence, the potential of the neutral is the same as the ground potential, as illustrated in Figure 6.30a. The charging currents $\mathbf{I}_{a1}$, $\mathbf{I}_{b1}$, and $\mathbf{I}_{c1}$ lead their respective phase voltages by 90°. Thus,

$$\left|\mathbf{I}_{a1}\right| = \left|\mathbf{I}_{b1}\right| = \left|\mathbf{I}_{c1}\right| = \frac{V_{L-N}}{X_c} \tag{6.129}$$

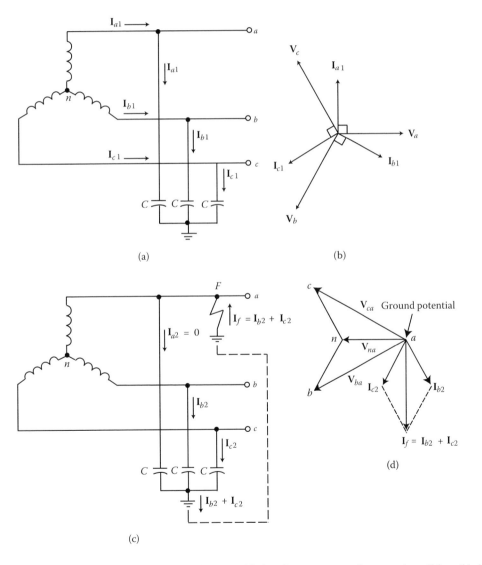

FIGURE 6.30 Representation of ungrounded system: (a) charging currents under normal condition, (b) phasor diagram under normal condition, (c) charging currents during SLG fault, (d) resulting phasor diagram.

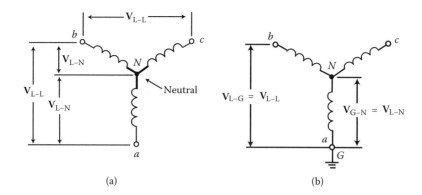

FIGURE 6.31 Voltage diagrams of ungrounded system: (a) before SLG fault; (b) after SLG fault.

where X_c is the capacitive reactance of the line to the ground. These phasor currents are in balance, as shown in Figure 6.30a.

Now assume that there is a line-to-ground fault involving phase a, as shown in Figure 6.30c. As a result of this SLG fault, the potential of phase a becomes equal to the ground potential, and hence no charging current flows in this phase. Therefore, the neutral point shifts from the ground potential position to the position shown in Figure 6.30d. It is also illustrated in Figure 6.31b. A charging current of three times the normal per-phase charging current flows in the faulted phase because the phase voltage of each of the two healthy phases increases by three times its normal phase voltage. Therefore,

$$\mathbf{I}_f = 3\mathbf{I}_{b1} = 3\mathbf{I}_{c1} \qquad (6.130)$$

The insulation of all apparatus connected to the lines is subjected to this high voltage. If it exists for very short periods, the insulation may be adequate to withstand it. However, it will eventually cause the failure of insulation owing to a cumulative weakening action. For operating the protective devices, it is crucial that the magnitude of the current applied should be sufficient to operate them.

However, in the event of an SLG fault on an ungrounded neutral system, the resultant capacitive current is usually not large enough to actuate the protective devices. Furthermore, a current of such magnitude (over 4 or 5 A) flowing through the fault might be sufficient to maintain an arc in the ionized part of the fault. It is possible that such a current may exist even after the SLG fault is cleared.

The phenomenon of persistent arc is called the *arcing ground*. Under such conditions, the system capacity will be charged and discharged in cyclic order, due to which high-frequency oscillations are superimposed on the system. These high-frequency oscillations produce surge voltages as high as six times the normal value that may damage the insulation at any point of the system.*

Neutral grounding is effective in reducing such transient voltage buildup from such intermittent ground faults by reducing neutral displacement from the ground potential and the destructiveness of any high-frequency voltage oscillations following each arc initiation or restrike. Because of these problems with ungrounded neutral systems, in most of the modern high-voltage systems, the neutral systems are grounded.

The advantages of neutral grounding include the following:

1. Voltages of phases are restricted to the line-to-ground voltages since the neutral point is not shifted in this system.
2. The ground relays can be used to protect against the ground faults.
3. The high voltages caused by arcing grounds or transient SLG faults are eliminated.

---

* The condition necessary for producing these overvoltages require that the dielectric strength of the arc path build up at a higher rate, after extinction of the arc, than at the preceding extinction.

4. The overvoltages caused by lightning are easily eliminated contrary to the case of the isolated neutral systems.
5. The induced static charges do not cause any disturbance since they are conducted to ground immediately.
6. It provides a reliable system.
7. It provides a reduction in operating and maintenance expenses.

A power system is solidly (i.e., directly) grounded when a generator, power transformer, or grounding transformer neutral is connected directly to the ground, as shown in Figure 6.32a. When there is an SLG fault on any phase (e.g., phase *a*), the line-to-ground voltage of that phase becomes zero. However, the remaining two phases will still have the same voltages as before, since the neutral point remains unshifted, as shown in Figure 6.32b. Note that, in this system, in addition to the charging currents, the power source also feeds the fault current $\mathbf{I}_f$. To keep the system stable, solid grounding is usually used where the circuit impedance is high enough so that the fault current can be kept within limits.

The comparison of the magnitude of the SLG fault current with the system three-phase current determines how solidly the system should be grounded. The higher the ground fault current in relation to the three-phase current, the more solidly the system is grounded. Most equipment rated 230 kV and above is designed to operate only on an effectively grounded system.*

As a rule of thumb, an effectively grounded system is one in which the ratio of the coefficient of grounding does not exceed 0.80. Here, the *coefficient of grounding* is defined as the ratio of the maximum sustained line-to-ground voltage during faults to the maximum operating line-to-line voltage. At higher voltage levels, insulation is more expensive and therefore, more economy can be achieved from the insulation reduction. However, solid grounding of a generator without external impedance may cause the SLG fault current from the generator to exceed the maximum three-phase fault current the generator can deliver and to exceed the short-circuit current for which its windings are braces. Thus, generators are usually grounded through a resistance, reactance, or Peterson coil to limit the fault current to a value that is less than three-phase fault currents.

In a *resistance-grounded system*, the system neutral is connected to ground through one or more resistors. A system that is properly grounded by resistance is not subject to destructive transient overvoltages. Resistance grounding reduces the arcing hazards and permits ground fault protection.

In a *reactance grounded system*, the system neutral is connected to ground through a reactor. Since the ground fault current that may flow in a reactance-grounded system is a function of the neutral reactance, the magnitude of the reactance in the neutral circuit determines how "solidly" the system is grounded. In fact, whether a system is solidly grounded or reactance grounded depends on the ratio of zero-sequence reactance X_0 to positive-sequence reactance X_1. Hence, a system is reactance grounded if the ratio X_0/X_1 is greater than 3.0. Otherwise, if the ratio is less than 3.0, the system is solidly grounded. The system neutral grounding using Peterson coils will be reviewed in Section 6.8. Characteristics of various methods of system neural grounding are summarized in Table 6.4 [9].

The best way to obtain the system neutral for grounding purposes is to use source transformers or generators with wye-connected windings. The neutral is then readily available.

If the system neutral is not available for some reason, for example, when an existing system is delta connected, the grounding can be done using a zigzag grounding transformer with no secondary winding or a wye–delta grounding transformer. In this case, the delta side must be closed to provide a path for zero-sequence current. The wye winding must be of the same voltage rating as the circuit that is to be grounded. On the other hand, the delta voltage rating can be selected at any standard voltage level.

---

* A system is defined as "effectively grounded" when $R_0 \ll X_1$ and $X_0 \ll 3X_1$, and such relationships exist at any point in the system for any condition of operation and for any amount of generator capacity.

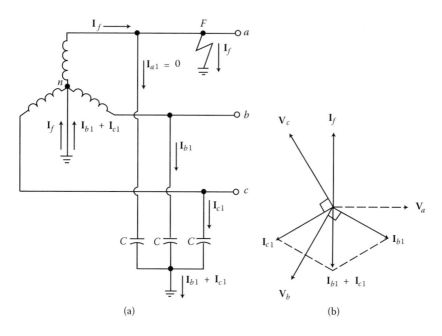

FIGURE 6.32 Representation of SLG fault on solidly grounded system: (a) solidly grounded system; (b) phasor diagram.

TABLE 6.4
System Characteristics with Various Grounding Methods

	Ungrounded	Essentially Solid Grounding		Reactance Grounding High Value Reactor	Ground-Fault Neutralizer	Resistance Grounding Resistance
		Low-Value Solid	Low-Value Reactor			
Current for phase-to ground fault in percent of three-phase fault current	Less than 1%	Varies, may be 100% or greater	Usually designed to produce 25–100%	5–25%	Nearly zero fault current	5–20%
Transient overvoltages	Very high	Not excessive	Not excessive	Very high	Not excessive	Not excessive
Automatic segregation of faulty zone	No	Yes	Yes	Yes	No	Yes
Lightning arresters	Ungrounded-neutral type	Grounded-neutral type	Grounded-neutral type	Ungrounded-neutral type	Ungrounded-neutral type	Ungrounded-neutral type
Remarks	Not recommended owing to overvoltages and nonsegregation of fault	Generally used on systems (1) ≤600 V and (2) >15 kV	Generally used on systems (1) ≤600 V and (2) >15 kV	Not used due to excessive overvoltages	Best suited for high voltage overhead may be self-healing	Generally used on industrial systems of 2.4–15 kV

Source: Beeman, D., ed., *Industrial Power System Handbook.* McGraw-Hill, New York, 1955.

6.8 ELIMINATION OF SLG FAULT CURRENT BY USING PETERSON COILS

In the event that the reactance of a neutral reactor is increased until it is equal to the system capacitance to ground, the system zero-sequence network is in parallel resonance for SLG faults. As a result, a fault current flows through the neutral reactor to ground. A current of approximately equal magnitude, and about 180° out of phase with the reactor current, flows through the system capacitance to ground. These two currents neutralize each other, except for a small resistance component, as they flow through the fault. Such a reactor is called a *ground fault neutralizer*, or an *arc suppression coil*, or a *Peterson coil*. It is basically an iron core reactor that is adjustable by means of taps on the winding.

Resonant grounding is an effective means to clear both transient, due to lightning, small animals, or tree branches, and sustained SLG faults. Other advantages of Peterson coils include extinguishing arcs and reduction of voltage dips owing to SLG faults. The disadvantages of Peterson coils include the need for retuning after any network modification or line-switching operation, the need for the lines to be transposed, and the increase in corona and radio interference under DLG fault conditions.

EXAMPLE 6.11

Consider the subtransmission system shown in Figure 6.33. Assume that loads are connected to buses 2 and 3 and are supplied from bus 1 through 69-kV lines of TL_{12}, TL_{13}, and TL_{23}. The line lengths are 5, 10, and 5 mi for lines TL_{12}, TL_{13}, and TL_{23}, respectively. The lines are transposed and made of three 500-kcmil, 30/7-strand ACSR conductors and there are no ground wires. The geometric mean distance (GMD) between the three conductors and their images (i.e., H_{aa}) is 81.5 ft. The self-GMD of the overhead conductors as a composite group (i.e., D_{aa}) is 1.658 ft. To reduce the SLG faults, a Peterson coil is to be installed between the neutral of the wye-connected secondary of the supply transformer T_1 and ground. The transformer T_1 has a leakage reactance of 5% based on its 25-MVA rating. Do the following:

(a) Determine the total zero-sequence capacitance and susceptance per phase of the system at 60 Hz.
(b) Draw the zero-sequence network of the system.
(c) Determine the continues-current rating of the Peterson coil.
(d) Determine the required reactance value for the Peterson coil.
(e) Determine the inductance value of the Peterson coil.
(f) Determine the continuous kVA rating for the Peterson coil.
(g) Determine the continuous-voltage rating for the Peterson coil.

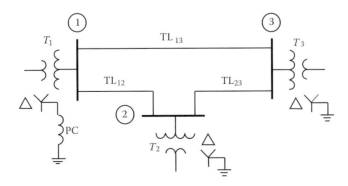

FIGURE 6.33 Subtransmission system for Example 6.11.

Solution

(a)

$$C_0 = \frac{29.842}{\ln(H_{aa}/D_{aa})}$$

$$= \frac{29.842}{\ln(81.5/1.658)}$$

$$= 7.6616 \text{ nF/mi}$$

Therefore,

$$b_0 = \omega C_0$$

$$= 2.8884 \text{ }\mu\text{S/mi}$$

and for the total system,

$$B_0 = b_0 l$$

$$= 2.8884 \times 20$$

$$= 57.7671 \mu\text{S}$$

The total zero-sequence reactance is

$$\sum X_{c0} = \frac{1}{B_0} = \frac{10^6}{57.7671}$$

$$= 17{,}310.8915 \text{ }\Omega$$

and the total zero-sequence capacitance of the system is

$$\sum C_0 = \frac{B_0}{\omega} = \frac{57.7671 \times 10^{-6}}{377}$$

$$= 0.1532 \text{ }\mu\text{F}$$

(b) The resulting zero-sequence network is shown in Figure 6.34.
(c) Since the leakage reactance of transformer T_1 is

$$X_1 = X_2 = X_0 = 0.05 \text{ pu}$$

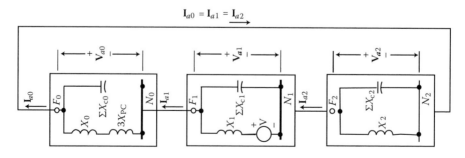

FIGURE 6.34 Interconnection of sequence networks for Example 6.11.

Analysis of Unbalanced Faults

or since

$$Z_B = \frac{kV_B^2}{MVA_B}$$

$$= \frac{69^2}{25}$$

$$= 190.44 \, \Omega$$

To have a zero SLG current,

$$\mathbf{I}_{a0} = \mathbf{I}_{a1} = \mathbf{I}_{a2} = 0$$

Thus, it is required that

$$\mathbf{V}_{a0} = -\mathbf{V}_f$$

where

$$V_F = \frac{69 \times 10^3}{\sqrt{3}}$$

$$= 39{,}837.17 \, V$$

Since $\sum X_{c1} \gg X_1$ and $\sum X_{c2} \gg X_2$, the zero-sequence current component flowing through the Peterson coil (PC) can be expressed as

$$\mathbf{I}_{a0(PC)} = \frac{-\mathbf{V}_{a0}}{j(X_0 + 3X_{PC})}$$

$$= \frac{\mathbf{V}_F}{j(X_0 + 3X_{PC})}$$

or

$$\mathbf{I}_{a0(PC)} = \frac{39{,}837.17 \angle 0°}{j17{,}310.8915}$$

$$= -j2.3013 \, A$$

Therefore, the continuous-current rating for the Peterson coil is

$$\mathbf{I}_{PC} = 3\mathbf{I}_{a0(PC)} = 6.9038 \, A$$

(d) Since

$$3X_{PC} + X_0 = 17{,}310.8915 \, \Omega$$

where

$$X_0 = 9.522 \, \Omega$$

therefore,

$$3X_{PC} = 17{,}310.8915 - 9.522$$
$$= 17{,}301.3695\,\Omega$$

and thus, the required reactance value for the Peterson coil is

$$X_{PC} = \frac{17{,}301.3695\,\Omega}{3}$$
$$= 5767.1232\,\Omega$$

(e) Hence, its inductance is

$$L_{PC} = \frac{X_{PC}}{\omega}$$
$$= \frac{5767.1232}{377}$$
$$= 15.2928\,\text{H}$$

(f) Its continuous kVA rating is

$$S_{PC} = I_{PC}^2 X_{PC}$$
$$= (6.9030)^2 (5767.1232)$$
$$= 274.88\,\text{kVA}$$

(g) The voltage across the Peterson coil is

$$V_{PC} = I_{PC} X_{PC}$$
$$= (6.9030)(5767.1232)$$
$$= 39{,}815.07\,\text{V}$$

which is approximately equal to the line-to-neutral voltage.

6.9 SIX-PHASE SYSTEMS

The six-phase transmission lines are proposed because of their ability to increase power transfer over existing lines and reduce electrical environmental impacts. For example, in six-phase transmission lines, voltage gradients of the conductors are lower, which in turn reduces both audible noise and electrostatic effects without requiring additional insulation.

In multiphase transmission lines, if the line-to-ground voltage is fixed, then the line-to-line voltage decreases as the number of phases increases. Consequently, this enables the line-to-line insulation distance to be reduced.

In such systems, the symmetrical components analysis can also be used to determine the unbalance factors that are caused by the unsymmetrical tower-top configurations.

Analysis of Unbalanced Faults

6.9.1 APPLICATION OF SYMMETRICAL COMPONENTS

A set of unbalanced six-phase currents (or voltages) can be decomposed into six sets of balanced currents (or voltages) that are called the *symmetrical components*. Figure 6.35 shows the balanced voltage sequence sets of a six-phase system [17].

The first set, that is, the first-order positive sequence components, are equal in magnitude and have a 60° phase shift. They are arranged in *abcdef* phase sequence. The remaining sets are the first-order negative sequence, the second-order positive sequence, the second-order negative sequence, the odd sequence, and finally the zero sequence components.

After denoting the phase sequence as *abcdef*, their sequence components can be expressed as

$$\left.\begin{aligned} V_a &= V_{a0^+} + V_{a1^+} + V_{a2^+} + V_{a0^-} + V_{a2^-} + V_{a1^-} \\ V_b &= V_{b0^+} + V_{b1^+} + V_{b2^+} + V_{b0^-} + V_{b2^-} + V_{b1^-} \\ V_c &= V_{c0^+} + V_{c1^+} + V_{c2^+} + V_{c0^-} + V_{c2^-} + V_{c1^-} \\ V_d &= V_{d0^+} + V_{d1^+} + V_{d2^+} + V_{d0^-} + V_{d2^-} + V_{d1^-} \\ V_e &= V_{e0^+} + V_{e1^+} + V_{e2^+} + V_{e0^-} + V_{e2^-} + V_{e1^-} \\ V_f &= V_{f0^+} + V_{f1^+} + V_{f2^+} + V_{f0^-} + V_{f2^-} + V_{f1^-} \end{aligned}\right\} \quad (6.131)$$

6.9.2 TRANSFORMATIONS

By taking phase *a* as the reference phase as usual, the set of voltages given in Equation 6.131 can be expressed in matrix form as

$$\begin{bmatrix} V_a \\ V_b \\ V_c \\ V_d \\ V_e \\ V_f \end{bmatrix} = \begin{bmatrix} 1 & 1 & 1 & 1 & 1 & 1 \\ 1 & b^5 & b^4 & b^3 & b^2 & b \\ 1 & b^4 & b^2 & 1 & b^4 & b^2 \\ 1 & b^3 & 1 & b^3 & 1 & b^3 \\ 1 & b^2 & b^4 & 1 & b^2 & b^4 \\ 1 & b & b^2 & b^3 & b^4 & b^5 \end{bmatrix} \begin{bmatrix} V_{a0^+} \\ V_{a1^+} \\ V_{a2^+} \\ V_{a0^-} \\ V_{a2^-} \\ V_{a1^-} \end{bmatrix} \quad (6.132)$$

or in short-hand matrix notation,

$$[\mathbf{V}_\phi] = [\mathbf{T}_6][\mathbf{V}_s] \quad (6.133)$$

where
  $[\mathbf{V}_\phi]$ = the matrix of unbalanced phase voltages
  $[\mathbf{V}_s]$ = the matrix of balanced sequence voltages
  $[\mathbf{T}_6]$ = the six-phase symmetrical transformation matrix

Similar to the definition of the *a* operator in three-phase systems, it is possible to define a six-phase operator *b* as

$$\mathbf{b} = 1.0 \angle 60° \quad (6.134)$$

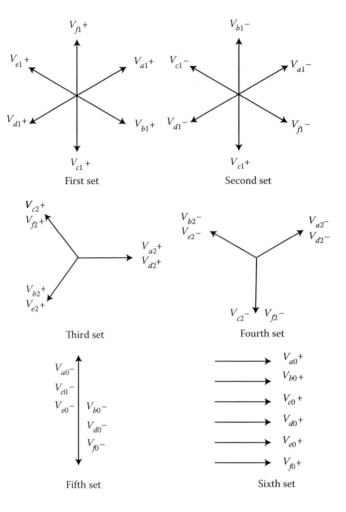

FIGURE 6.35 Balanced voltage-sequence sets of a six-phase system. (From Gönen, T., *Electric Power Transmission System Engineering*, 2nd ed. CRC Press, Roca Baton, FL, 2009.)

or

$$\mathbf{b} = \exp(j\pi/3)$$
$$= 0.5 + j0.866 \qquad (6.135)$$

It can be shown that

$$\mathbf{b} = -\mathbf{a}^2$$
$$= -1(1.0\angle 120°)^2$$
$$= -(1.0\angle 240°) \qquad (6.136)$$
$$= 1.0\angle 60°$$

The relation between the sequence components and the unbalanced phase voltages can be expressed as

$$[\mathbf{V}_s] = [\mathbf{T}_6]^{-1}[\mathbf{V}_\phi] \qquad (6.137)$$

Similar equations can be written for the phase currents and their sequence components as

$$[\mathbf{I}_\phi] = [\mathbf{T}_6][\mathbf{I}_s] \tag{6.138}$$

and

$$[\mathbf{I}_s] = [\mathbf{T}_6]^{-1}[\mathbf{I}_\phi] \tag{6.139}$$

The sequence impedance matrix $[\mathbf{Z}_s]$ can be determined from the phase impedance matrix $[\mathbf{Z}_\phi]$ by applying KVL. Hence,

$$[\mathbf{V}_\phi] = [\mathbf{Z}_\phi][\mathbf{I}_\phi] \tag{6.140}$$

and

$$[\mathbf{V}_s] = [\mathbf{Z}_s][\mathbf{I}_s] \tag{6.141}$$

where
$[\mathbf{Z}_\phi]$ = phase impedance matrix of the line in 6×6
$[\mathbf{Z}_s]$ = sequence impedance matrix of the line in 6×6

After multiplying both sides of Equation 6.141 by $[\mathbf{Z}_s]^{-1}$,

$$\begin{aligned}[\mathbf{I}_s] &= [\mathbf{Z}_s]^{-1}[\mathbf{V}_s] \\ &= [\mathbf{Y}_s][\mathbf{V}_s] \end{aligned} \tag{6.142}$$

where $[\mathbf{Y}_s]$ is the sequence admittance matrix.

Since the unbalanced factors are to be determined after having only the first-order positive sequence voltage applied, the above equation can be reexpressed as

$$\begin{bmatrix} I_{a0^+} \\ I_{a1^+} \\ I_{a2^+} \\ I_{a0^-} \\ I_{a2^-} \\ I_{a1^-} \end{bmatrix} = \begin{bmatrix} Y_{0^+0^+} & Y_{0^+0^+} & Y_{0^+0^+} & Y_{0^+0^-} & Y_{0^+0^-} & Y_{0^+0^-} \\ Y_{0^+0^+} & Y_{0^+0^+} & Y_{0^+0^+} & Y_{0^+0^-} & Y_{0^+0^-} & Y_{0^+0^-} \\ Y_{0^+0^+} & Y_{0^+0^+} & Y_{0^+0^+} & Y_{0^+0^-} & Y_{0^+0^-} & Y_{0^+0^-} \\ Y_{0^-0^+} & Y_{0^-0^+} & Y_{0^-0^+} & Y_{0^-0^-} & Y_{0^-0^-} & Y_{0^-0^-} \\ Y_{0^-0^+} & Y_{0^-0^+} & Y_{0^-0^+} & Y_{0^-0^-} & Y_{0^-0^-} & Y_{0^-0^-} \\ Y_{0^-0^+} & Y_{0^-0^+} & Y_{0^-0^+} & Y_{0^-0^-} & Y_{0^-0^-} & Y_{0^-0^-} \end{bmatrix} \begin{bmatrix} 0 \\ V_{a1^+} \\ 0 \\ 0 \\ 0 \\ 0 \end{bmatrix} \tag{6.143}$$

6.9.3 Electromagnetic Unbalance Factors

For a six-phase transmission line, there are five electromagnetic sequence unbalanced factors. They are called zero, second-order positive, odd, second-order negative, and first-order negative sequence factors. Each of them is computed as the ratio of the corresponding current to the first-order positive sequence current. For example, the zero-sequence unbalanced factor is

$$\mathbf{m}_{0^+} = \frac{I_{a0^+}}{I_{a1^+}} \tag{6.144a}$$

or

$$\mathbf{m}_{0^+} = \frac{(y_{0^+1^+})(V_{a1^+})}{(y_{1^+1^+})(V_{a1^+})}$$

$$= \frac{y_{0^+1^+}}{y_{1^+1^+}} \quad (6.144b)$$

The second-order positive unbalance factor is

$$\mathbf{m}_{2^+} = \frac{I_{a2^+}}{I_{a1^+}} \quad (6.145a)$$

or

$$\mathbf{m}_{2^+} = \frac{(y_{2^+1^+})(V_{a1^+})}{(y_{1^+1^+})(V_{a1^+})}$$

$$= \frac{y_{2^+1^+}}{y_{1^+1^+}} \quad (6.145b)$$

The odd unbalance factor is

$$\mathbf{m}_{0^-} = \frac{I_{a0^-}}{I_{a1^+}} \quad (6.146a)$$

or

$$\mathbf{m}_{0^-} = \frac{(y_{0^-1^+})(V_{a1^+})}{(y_{1^+1^+})(V_{a1^+})}$$

$$= \frac{y_{0^-1^+}}{y_{1^+1^+}} \quad (6.146b)$$

The second-order negative unbalance factor is

$$\mathbf{m}_{2^-} = \frac{I_{a2^-}}{I_{a1^+}} \quad (6.147a)$$

or

$$\mathbf{m}_{2^-} = \frac{(y_{2^-1^+})(V_{a1^+})}{(y_{1^+1^+})(V_{a1^+})}$$

$$= \frac{y_{2^-1^+}}{y_{1^+1^+}} \quad (6.147b)$$

Analysis of Unbalanced Faults

The first-order negative unbalance factor is

$$\mathbf{m}_{1^-} = \frac{I_{a1^-}}{I_{a1^+}} \tag{6.148a}$$

or

$$\mathbf{m}_{1^-} = \frac{(y_{1^-1^+})(V_{a1^+})}{(y_{1^+1^+})(V_{a1^+})}$$

$$= \frac{y_{1^-1^+}}{y_{1^+1^+}} \tag{6.148b}$$

As a result of the unsymmetrical configuration, circulating residual (i.e., zero sequence) currents will flow in the high-voltage system with solidly grounded neutrals. Such currents affect the proper operation of very sensitive elements in ground relays. In addition to the zero-sequence unbalance, negative-sequence charging currents are also produced. They are caused by the capacitive unbalances that will cause the currents to flow through the lines and windings of the transformers and rotating machines in the system, causing additional power losses in the rotating machines and transformers.

6.9.4 Transposition on the Six-Phase Lines

The six-phase transmission lines can be transposed in *complete transposition*, *cyclic transposition*, and *reciprocal transposition*. In a complete transposition, every conductor assumes every possible position with respect to every other conductor over an equal length. Thus, the resultant impedance matrix for a complete transposition can be expressed as

$$[\mathbf{Z}_\phi] = \begin{bmatrix} Z_s & Z_m & Z_m & Z_m & Z_m & Z_m \\ Z_m & Z_s & Z_m & Z_m & Z_m & Z_m \\ Z_m & Z_m & Z_s & Z_m & Z_m & Z_m \\ Z_m & Z_m & Z_m & Z_s & Z_m & Z_m \\ Z_m & Z_m & Z_m & Z_m & Z_s & Z_m \\ Z_m & Z_m & Z_m & Z_m & Z_m & Z_s \end{bmatrix} \tag{6.149}$$

In a six-phase transmission line, it is difficult to achieve a complete transposition. Also, it is not of interest owing to the differences in the line-to-line voltages. Therefore, it is more efficient to implement cyclic transposition or reciprocal cyclic transposition. The impedance matrix for a cyclically transposed line can be expressed as

$$[\mathbf{Z}_\phi] = \begin{bmatrix} Z_s & Z_{m1} & Z_{m2} & Z_{m3} & Z_{m4} & Z_{m5} \\ Z_{m5} & Z_s & Z_{m1} & Z_{m2} & Z_{m3} & Z_{m4} \\ Z_{m4} & Z_{m5} & Z_s & Z_{m1} & Z_{m2} & Z_{m3} \\ Z_{m3} & Z_{m4} & Z_{m5} & Z_s & Z_{m1} & Z_{m2} \\ Z_{m2} & Z_{m3} & Z_{m4} & Z_{m5} & Z_s & Z_{m1} \\ Z_{m1} & Z_{m2} & Z_{m3} & Z_{m4} & Z_{m5} & Z_s \end{bmatrix} \tag{6.150}$$

Similarly, the impedance matrix for a reciprocal cyclically transposed line can be given as

$$[\mathbf{Z}_\phi] = \begin{bmatrix} Z_s & Z_{m1} & Z_{m2} & Z_{m3} & Z_{m2} & Z_{m1} \\ Z_{m1} & Z_s & Z_{m1} & Z_{m2} & Z_{m3} & Z_{m2} \\ Z_{m2} & Z_{m1} & Z_s & Z_{m1} & Z_{m2} & Z_{m3} \\ Z_{m3} & Z_{m2} & Z_{m1} & Z_s & Z_{m1} & Z_{m2} \\ Z_{m2} & Z_{m3} & Z_{m2} & Z_{m1} & Z_s & Z_{m1} \\ Z_{m1} & Z_{m2} & Z_{m3} & Z_{m2} & Z_{m1} & Z_s \end{bmatrix} \quad (6.151)$$

where Z_s is the self-impedance and Z_m is the mutual impedance.

6.9.5 Phase Arrangements

The values of the electromagnetic and electrostatic unbalances will change by changing the phase conductors. However, there is a phase configuration that has the minimum amount of electromagnetic unbalances. The circulating current unbalances can become very large under some phasing arrangements.

6.9.6 Overhead Ground Wires

Overhead ground wires are installed to protect the transmission lines against lightning. However, the overhead ground wires affect both the self and mutual impedances. The resistances of the self and mutual impedances increase slightly while their reactance decrease significantly. The overhead ground wires can increase or decrease some or all of the unbalances depending on the type and size of the configuration. Kron reduction can be used to compare the equivalent impedance matrix for the transmission lines.

6.9.7 Double-Circuit Transmission Lines

The voltage equation of a double-circuit line is given by

$$\begin{bmatrix} \sum \mathbf{V}_{ckt1} \\ \sum \mathbf{V}_{ckt2} \end{bmatrix} = \begin{bmatrix} \mathbf{Z}_{ckt1} & \mathbf{Z}_{ckt1\,ckt2} \\ \mathbf{Z}_{ckt2\,ckt1} & \mathbf{Z}_{ckt2} \end{bmatrix} \begin{bmatrix} \mathbf{I}_{ckt1} \\ \mathbf{I}_{ckt2} \end{bmatrix} \quad (6.152)$$

where $\sum \mathbf{V}_{ckt1}, \sum \mathbf{V}_{ckt2}$, and $\mathbf{I}_{ckt1}, \mathbf{I}_{ckt2}$ are the phase voltages and currents, respectively. Each of them is a column vector of size 6×1. Each [$\mathbf{Z}$] impedance matrix has dimensions of 6×6. The above matrix is solved for the currents, in order to express the unbalance factors in terms of sequence currents. Hence,

$$\begin{bmatrix} \mathbf{I}_{ckt1} \\ \mathbf{I}_{ckt2} \end{bmatrix} = [\mathbf{Y}_{line}] \begin{bmatrix} \sum \mathbf{V}_{ckt1} \\ \sum \mathbf{V}_{ckt2} \end{bmatrix} \quad (6.153)$$

where [$\mathbf{Y}_{line}$] is the admittance matrix of the line having a size 12×12.

Analysis of Unbalanced Faults

To determine the sequence admittance matrix of the line, the appropriate transformation matrix needs to be defined as

$$[\mathbf{T}_{12}] = \left[\begin{array}{c|c} \mathbf{T}_6 & 0 \\ \hline 0 & \mathbf{T}_6 \end{array}\right] \quad (6.154)$$

Premultiplying Equation 6.153 by the transformation matrix $[\mathbf{T}_{12}]$ and postmultiplying by $[\mathbf{T}_{12}]^{-1}$ and also inserting the unity matrix of $[\mathbf{T}_{12}]^{-1}[\mathbf{T}_{12}] = [\mathbf{U}]$ into the right-hand side of the equation,

$$\left[\begin{array}{c} \mathbf{I}_{seq1} \\ \hline \mathbf{I}_{seq2} \end{array}\right] = [\mathbf{T}_{12}][\mathbf{Y}_{line}][\mathbf{T}_{12}]^{-1} \left[\begin{array}{c} \sum \mathbf{V}_{seq1} \\ \hline \sum \mathbf{V}_{seq2} \end{array}\right] \quad (6.155)$$

where $\sum \mathbf{V}_{seq1}$, $\sum \mathbf{V}_{seq2}$, $\mathbf{I}_{seq1}$, and $\mathbf{I}_{seq2}$ are the sequence voltage drops and currents for the first and second circuits, respectively. The sequence admittance matrix is

$$[\mathbf{Y}_{seq}] = [\mathbf{T}_{12}][\mathbf{Y}_{line}][\mathbf{T}_{12}]^{-1} \quad (6.156)$$

The unbalance factors are to be determined with only the first-order positive sequence voltage applied. There are two different unbalances, that is, the *net-through* and the *net-circulating* unbalances. The sequence matrix is found by expanding the above equation to the full 12 × 12 matrix as

$$\begin{bmatrix} I_{a0^+} \\ I_{a1^+} \\ I_{a2^+} \\ I_{a0^-} \\ I_{a2^-} \\ I_{a1^-} \\ \hline I'_{a0^+} \\ I'_{a1^+} \\ I'_{a2^+} \\ I'_{a0^-} \\ I'_{a2^-} \\ I'_{a1^-} \end{bmatrix} = \begin{bmatrix} y_{0^+0^+} & y_{0^+1^+} & y_{0^+2^+} & y_{0^+0^-} & y_{0^+2^-} & y_{0^+1^-} & y_{0^+0'^+} & y_{0^+1'^+} & y_{0^+2'^+} & y_{0^+0'^-} & y_{0^+2'^-} & y_{0^+1'^-} \\ y_{1^+0^+} & y_{1^+1^+} & y_{1^+2^+} & y_{1^+0^-} & y_{1^+2^-} & y_{1^+1^-} & y_{1^+0'^+} & y_{1^+1'^+} & y_{0^+2'^+} & y_{1^+0'^-} & y_{1^+2'^-} & y_{1^+1'^-} \\ y_{2^+0^+} & y_{2^+1^+} & y_{2^+2^+} & y_{2^+0^-} & y_{2^+2^-} & y_{2^+1^-} & y_{2^+0'^+} & y_{2^+1'^+} & y_{0^+2'^+} & y_{2^+0'^-} & y_{2^+2'^-} & y_{2^+1'^-} \\ y_{0^-0^+} & y_{0^-1^+} & y_{0^-2^+} & y_{0^-0^-} & y_{0^-2^-} & y_{0^-1^-} & y_{0^-0'^+} & y_{0^-1'^+} & y_{0^-2'^+} & y_{0^-0'^-} & y_{0^-2'^-} & y_{0^-1'^-} \\ y_{2^-0^+} & y_{2^-1^+} & y_{2^-2^+} & y_{2^-0^-} & y_{2^-2^-} & y_{2^-1^-} & y_{2^-0'^+} & y_{2^-1'^+} & y_{0^-2'^+} & y_{2^-0'^-} & y_{2^-2'^-} & y_{2^-1'^-} \\ y_{1^-0^+} & y_{1^-1^+} & y_{1^-2^+} & y_{1^-0^-} & y_{1^-2^-} & y_{1^-1^-} & y_{1^-0'^+} & y_{1^-1'^+} & y_{0^-2'^+} & y_{1^-0'^-} & y_{1^-2'^-} & y_{1^-1'^-} \\ \hline y_{0'^+0^+} & y_{0'^+1^+} & y_{0'^+2^+} & y_{0'^+0^-} & y_{0'^+2^-} & y_{0'^+1^-} & y_{0'^+0'^+} & y_{0'^+1'^+} & y_{0'^+2'^+} & y_{0'^+0'^-} & y_{0'^+2'^-} & y_{0'^+1'^-} \\ y_{1'^+0^+} & y_{1'^+1^+} & y_{1'^+2^+} & y_{1'^+0^-} & y_{1'^+2^-} & y_{1'^+1^-} & y_{1'^+0'^+} & y_{1'^+1'^+} & y_{1'^+2'^+} & y_{1'^+0'^-} & y_{1'^+2'^-} & y_{1'^+1'^-} \\ y_{2'^+0^+} & y_{2'^+1^+} & y_{2'^+2^+} & y_{2'^+0^-} & y_{2'^+2^-} & y_{2'^+1^-} & y_{2'^+0'^+} & y_{2'^+1'^+} & y_{2'^+2'^+} & y_{2'^+0'^-} & y_{2'^+2'^-} & y_{2'^+1'^-} \\ y_{0'^-0^+} & y_{0'^-1^+} & y_{0'^-2^+} & y_{0'^-0^-} & y_{0'^-2^-} & y_{0'^-1^-} & y_{0'^-0'^+} & y_{0'^-1'^+} & y_{0'^-2'^+} & y_{0'^-0'^-} & y_{0'^-2'^-} & y_{0'^-1'^-} \\ y_{2'^-0^+} & y_{2'^-1^+} & y_{2'^-2^+} & y_{2'^-0^-} & y_{2'^-2^-} & y_{2'^-1^-} & y_{2'^-0'^+} & y_{2'^-1'^+} & y_{2'^-2'^+} & y_{2'^-0'^-} & y_{2'^-2'^-} & y_{2'^-1'^-} \\ y_{1'^-0^+} & y_{1'^-1^+} & y_{1'^-2^+} & y_{1'^-0^-} & y_{1'^-2^-} & y_{1'^-1^-} & y_{1'^-0'^+} & y_{1'^-1'^+} & y_{1'^-2'^+} & y_{1'^-0'^-} & y_{1'^-2'^-} & y_{1'^-1'^-} \end{bmatrix} \begin{bmatrix} 0 \\ \sum V_{a1^+} \\ 0 \\ 0 \\ 0 \\ 0 \\ \hline 0 \\ \sum V_{a1^+} \\ 0 \\ 0 \\ 0 \\ 0 \end{bmatrix}$$

(6.157)

The net-through unbalance factors are defined as

$$\mathbf{m}_{0^+t} \triangleq \frac{I_{a0^+} + I_{a'0^+}}{I_{a1^+} + I_{a'1^+}}$$

$$= \frac{(y_{0^+1^+} + y_{0^+1'^+} + y_{0'^+1^+} + y_{0'^+1'^+})\sum V_{a1+}}{(y_{1^+1^+} + y_{1^+1'^+} + y_{1'^+1^+} + y_{1'^+1'^+})\sum V_{a1+}} \quad (6.158)$$

$$= \frac{y_{0^+1^+} + y_{0^+1'^+} + y_{0'^+1^+} + y_{0'^+1'^+}}{y_k}$$

where

$$y_k = y_{1^+1^+} + y_{1^+1'^+} + y_{1'^+1^+} + y_{1'^+1'^+}$$

$$\mathbf{m}_{2^+t} = \frac{y_{2^+1^+} + y_{2^+1'^+} + y_{2'^+1^+} + y_{2'^+1'^+}}{y_k} \quad (6.159)$$

$$\mathbf{m}_{0^-t} = \frac{y_{0^-1^+} + y_{0^-1'^+} + y_{2'^-1^+} + y_{2'^-1'^+}}{y_k} \quad (6.160)$$

$$\mathbf{m}_{2^-t} = \frac{y_{2^-1^+} + y_{2^-1'^+} + y_{2'^-1^+} + y_{2'^-1'^+}}{y_k} \quad (6.161)$$

$$\mathbf{m}_{1^-t} = \frac{y_{1^-1^+} + y_{1^-1'^+} + y_{1'^-1^+} + y_{1'^-1'^+}}{y_k} \quad (6.162)$$

The net-circulating unbalances are defined as

$$\mathbf{m}_{0^+c} \triangleq \frac{I_{a0^+} + I_{a'0^+}}{I_{a1^+} + I_{a'1^+}}$$

$$= \frac{y_{0^+1^+} + y_{0^+1'^+} - y_{0'^+1^+} - y_{0'^+1'^+}}{y_k} \quad (6.163)$$

$$\mathbf{m}_{2^+c} = \frac{y_{2^+1^+} + y_{2^+1'^+} - y_{2'^+1^+} - y_{2'^+1'^+}}{y_k} \quad (6.164)$$

$$\mathbf{m}_{0^-c} = \frac{y_{0^-1^+} + y_{0^-1'^+} - y_{0'^-1^+} - y_{0'^-1'^+}}{y_k} \quad (6.165)$$

$$\mathbf{m}_{2-c} = \frac{y_{2^-1^+} + y_{2^-1'^+} - y_{2'^-1^+} - y_{2'^-1'^+}}{y_k} \quad (6.166)$$

$$\mathbf{m}_{1-c} = \frac{y_{1^-1^+} + y_{1^-1'^+} - y_{1'^-1^+} - y_{1'^-1'^+}}{y_k} \quad (6.167)$$

REFERENCES

1. Westinghouse Electric Corporation, *Electrical Transmission and Distribution Reference Book*. WEC, East Pittsburgh, 1964.
2. Clarke, E., *Circuit Analysis of A-C Power Systems*, Vol. 1. General Electric Co., Schenectady, New York, 1960.
3. Anderson, P. M., *Analysis of Faulted Power Systems*. Iowa State Univ. Press, Ames, IA, 1973.
4. Atabekov, G. I., *The Relay Protection of High Voltage Networks*. Pergamon Press, New York, 1960.
5. Kron, G., *Tensor Analysis of Networks*. Wiley, New York, 1939.
6. Harder, E. L., Sequence network connections for unbalanced load and fault conditions. *Electr. J.* 34 (12), 481–488 (1977).
7. Hobson, J. E., and Whitehead, D. L., Symmetrical components. *Electrical Transmission and Distribution Reference Book*, Chapter 2. Westinghouse Electric Corp., East Pittsburgh, 1964.
8. Anderson, P. M., Analysis of simultaneous faults by two-port network theory. *IEEE Trans. Power Appar. Syst.* PAS-90 (5), 2199–2205 (1971).
9. Beeman, D., ed., *Industrial Power System Handbook*. McGraw-Hill, New York, 1955.
10. Gönen, T., Haj-mohamadi, M. S., Electromagnetic unbalances of six-phase transmission lines. *Electr. Power Energy Syst.* 11 (2) 78–84 (1989).
11. Gönen, T., *Electric Power Transmission System Engineering*, 2nd ed. CRC Press, Roca Baton, FL, 2009.

GENERAL REFERENCES

AIEE Committee Report, Report on survey of unbalanced charging currents on transmission lines as affecting ground-fault neutralizers. *Trans. Am. Inst. Electr. Eng.* 68, 1328–1329 (1949).
Brown, H. E., *Solution of Large Networks by Matrix Methods*. Wiley, New York, 1975.
Brown, H. H., and Gross, E. T. B., Practical experiences with resonant grounding in a large 34.5 kV system. *Trans. Am. Inst. Electr. Eng.* 69, 1401–1408 (1950).
Calabrese, G. O., *Symmetrical Components Applied to Electric Power Networks*. Ronald Press, New York, 1959.
Clarke, E., *Circuit Analysis of A-C Power Systems*, Vol. 1. General Electric Co., Schenectady, New York, 1961.
Dawalibi, F., and Niles, G. B., Measurements and computations of fault current distribution of overhead transmission lines. *IEEE Trans. Power Appar. Syst.* PAS-I03 (3), 553–560 (1984).
Elgerd, O. I., *Electric Energy Systems Theory: An Introduction*. McGraw-Hill, New York, 1971.
Ferguson, W. H., Symmetrical component network connections for the solution of phase interchange faults. *Trans. Am. Inst. Electr. Eng., Part* 3 78 (44), 948–950 (1959).
Garin, A. N., Zero-phase-sequence characteristics of transformers. Parts I and II. *Gen. Elect. Rev.* 43, 131–136, 174–179 (1940).
Gönen, T., *Electric Power Distribution System Engineering*, 2nd ed. CRC Press, Roca Baton, FL, 2008.
Gönen, T., Nowikowski, J., and Brooks, C. L., Electrostatic unbalances of transmission lines with 'N' overhead ground wires. Part I. *Proc. Model. Simul. Conf.* 17 (Pt. 2) 459–464 (1986).
Gönen, T., Nowikowski, J., and Brooks, C. L., Electrostatic unbalances of transmission lines with 'N' overhead ground wires. Part II. *Proc. Model. Simul. Conf.* 17 (Pt. 2) 465–470 (1986).
Gross, C. A., *Power System Analysis*. Wiley, New York, 1979.
Gross, E. T. B., and Atherton, E. W., Application of resonant grounding in power systems in the United States. *Trans. Am. Inst. Electr. Eng.* 70, 389–397 (1951).
Guile, A. E., and Paterson, W., *Electrical Power Systems*, Vol. 1. Pergamon Press, New York, 1978.
Hardaway, W. D., and Lewis, W. W., Test and operation of Petersen coil on 100-kV system of public service company of Colorado. *Trans. Am. Inst. Electr. Eng.* 57, 295–306 (1938).

Kimbark, E. W., Suppression of ground-fault arcs on single-pale-switched EHV lines by shunt reactors. *IEEE Trans. Power Appar. Syst.* PAS-83, 285–290 (1964).

Lyle, A. G., *Major Faults on Power Systems*. Chapman & Hall, London, 1952.

Lyon, J. A. M., The electrostatic unbalance of transmission lines and its effect on the application of Petersen coils. *Electr. Eng. Am. Inst. Electr. Eng.* 58, 107–111 (1939).

Lyon. W. V., *Applications of the Method of Symmetrical Components*. McGraw-Hill, New York, 1937.

Matsushita, K. et al., Applications of mutually coupled reactor. *IEEE Trans. Power Appar. Syst.* PAS-I03 (3) 530–535 (1984).

Mortlock, J. R., Davies, M. W. H., and Jackson, W., *Power System Analysis*. Chapman & Hall, London, 1952.

Neuenswander, J. R., *Modern Power Systems*. International Textbook Co., Scranton, PA, 1971.

North, J. R., von Voigtlander, F., Halperin, H., and Hunter, E. M., Discussions on some engineering features of Petersen coils and their application. *Trans. Am. Inst. Electr. Eng.* 57, 289–291 (1938).

Roeper, R., *Kurzschlussströme in Drehstromnetzen*, 5th Ger. ed. (translated as *Short Circuit Currents in Three-Phase Networks*). Siemens Aktienges., Munich, Germany, 1972.

Rüdenberg, R., *Transient Performance of Electric Power Systems-Phenomena in Lumped Networks*, 1st ed. McGraw-Hill, New York, 1950.

Stevenson, W. D., Jr., *Elements of Power System Analysis*, 4th ed. McGraw-Hill, New York, 1982.

Stigant, S. A., *Mathematical and Geometrical Techniques for Symmetrical Component Faults Studies*. MacDonald & Co., London, 1965.

Tomlinson, H. R., Ground-fault neutralizer grounding of unit-connected generators. *Trans. Am. Inst. Electr. Eng.* 72 (8), 953–961 (1953).

Wagner, C. F., and Evans, R. D., *Symmetrical Components*. McGraw-Hill, New York, 1933.

Weedy, B. M., *Electric Power Systems*, 3rd ed. Wiley, New York, 1979.

PROBLEMS

1. Consider the system shown in Figure P6.1. Assume that the following data are given based on 20 MVA and the line-to-line base voltages as shown in Figure P6.1.

   Generator G_1: $X_1 = 0.25$ pu, $X_2 = 0.15$ pu, $X_0 = 0.05$ pu
   Generator G_2: $X_1 = 0.90$ pu, $X_2 = 0.60$ pu, $X_0 = 0.05$ pu
   Transformer T_1: $X_1 = X_2 = X_0 = 0.10$ pu
   Transformer T_2: $X_1 = X_2 = 0.10$ pu, $X_0 = \infty$

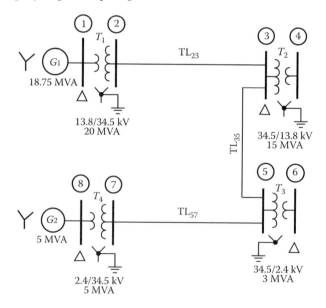

FIGURE P6.1 Eight-bus system for Problem 1.

Transformer T_3: $X_1 = X_2 = X_0 = 0.50$ pu
Transformer T_4: $X_1 = X_2 = 0.30$ pu, $X_0 = \infty$
Transmission line TL_{23}: $X_1 = X_2 = 0.15$ pu, $X_0 = 0.50$ pu
Transmission line TL_{35}: $X_1 = X_2 = 0.30$ pu, $X_0 = 1.00$ pu
Transmission line TL_{57}: $X_1 = X_2 = 0.30$ pu, $X_0 = 1.00$ pu
  (a) Draw the corresponding positive-sequence network.
  (b) Draw the corresponding negative-sequence network.
  (c) Draw the corresponding zero-sequence network.
2. Use the system and its data from Problem 7 (Chapter 4) and assume an SLG fault at bus 4. Assume that $\mathbf{Z}_f$ is $j0.1$ pu based on 50 MVA. Determine the fault current in per units and amperes.
3. Consider the system given in Problem 7 (Chapter 4) and assume that there is a line-to-line fault at bus 3 involving phases b and c. Determine the fault currents for both phases in per units and amperes.
4. Consider the system given in Problem 7 (Chapter 4) and assume that there is a DLG fault at bus 2, involving phases b and c. Assume that $\mathbf{Z}_f$ is $j0.1$ pu and $\mathbf{Z}_g$ is $j0.2$ pu (where $\mathbf{Z}_g$ is the neutral-to-ground impedance) both based on 50 MVA.
5. Consider the system given in Example 4.1 and determine the following:
  (a) Line-to-ground fault (Also, find the ratio of this line-to-ground fault current to the three-phase fault current found in Example 4.1.)
  (b) Line-to-line fault (Also, find the ratio of this line-to-line fault current to the previously calculated three-phase fault current).
  (c) DLG fault
6. Repeat Problem 5 assuming that the fault is located on bus 2.
7. Repeat Problem 5 assuming that the fault is located on bus 3.
8. Consider the system shown in Figure P6.8a Assume that loads, line capacitance, and transformer-magnetizing currents are neglected and that the following data is given based on 20 MVA and the line-to-line voltages as shown in Figure P6.2a. Do not neglect the resistance of the transmission line TL_{23}. The prefault positive-sequence voltage at bus 3 is $\mathbf{V}_{an} = 1.0\angle 0°$ pu, as shown in Figure P6.2b.
Generator: $X_1 = 0.20$ pu, $X_2 = 0.10$ pu, $X_0 = 0.05$ pu
Transformer T_1: $X_1 = X_2 = 0.05$ pu, $X_0 = X_1$ (looking into high-voltage side)

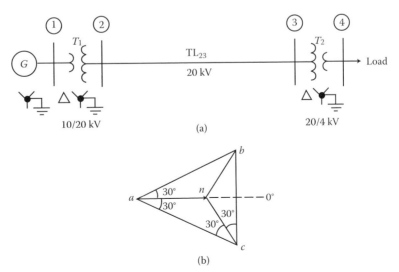

FIGURE P6.2 System for Problem 8.

Transformer T_2: $X_1 = X_2 = 0.05$ pu, $X_0 = \infty$ (looking into high-voltage side)
Transmission line: $\mathbf{Z}_1 = \mathbf{Z}_2 = 0.2 + j0.2$ pu, $\mathbf{Z}_0 = 0.6 + j0.6$ pu

Assume that there is a bolted (i.e., with zero fault impedance) line-to-line fault on phases b and c at bus 3 and determine the following:

(a) Fault current $\mathbf{I}_{bf}$ in per units and amperes
(b) Phase voltages $\mathbf{V}_a$, $\mathbf{V}_b$, and $\mathbf{V}_c$ at bus 2 in per units and kilovolts
(c) Line-to-line voltages $\mathbf{V}_{ab}$, $\mathbf{V}_{bc}$, and $\mathbf{V}_{ca}$ at bus 2 in kilovolts
(d) Generator line currents $\mathbf{I}_a$, $\mathbf{I}_b$, and $\mathbf{I}_c$

Given: per-unit positive-sequence currents on the low-voltage side of the transformer bank lag positive-sequence currents on the high-voltage side by 30° and similarly for negative-sequence currents excepting that the low-voltage currents lead the high-voltage by 30°.

9. Consider Figure P6.3 and assume that the generator ratings are 2.40/4.16Y kV, 15 MW (3ϕ), 18.75 MVA (3ϕ), 80% power factor, 2 poles, 3600 rpm. The generator reactances are $X_1 = X_2 = 0.10$ pu and $X_0 = 0.05$ pu, all based on generator ratings. Note that, the given value of X_1 is subtransient reactance X'', one of several different positive-sequence reactances of a synchronous machine. The subtransient reactance corresponds to the initial symmetrical fault current (the transient dc component not included) that occurs before demagnetizing armature magnetomotive force begins to weaken the net field excitation. If manufactured in accordance with U.S. standards, the coils of a synchronous generator will withstand the mechanical forces that accompany a three-phase fault current, but not more. Assume that this generator is to supply a four-wire, wye-connected distribution. Therefore, the neutral grounding reactor X_n should have the smallest possible reactance. Consider both SLG and DLG faults. Assume the prefault positive-sequence internal voltage of phase a is $2500\angle0°$ or $1.042\angle0°$ pu and determine the following:

(a) Specify X_n in ohms and in per units.
(b) Specify the minimum allowable momentary symmetrical current rating of the reactor in amperes.
(c) Find the initial symmetrical voltage across the reactor, $\mathbf{V}_n$, when a bolted SLG fault occurs on the oil circuit breaker terminal in volts.

10. Consider the system shown in Figure P6.4 and the following data:
Generator G: $X_1 = X_2 = 0.10$ pu and $X_0 = 0.05$ pu based on its ratings
Motor: $X_1 = X_2 = 0.10$ pu and $X_0 = 0.05$ pu based on its ratings
Transformer T_1: $X_1 = X_2 = X_0 = 0.05$ pu based on its ratings
Transformer T_2: $X_1 = X_2 = X_0 = 0.10$ pu based on its ratings
Transmission line TL_{23}: $X_1 = X_2 = X_0 = 0.09$ pu based on 25 MVA
Assume that bus 2 is faulted and determine the faulted phase currents.

(a) Determine the three-phase fault.
(b) Determine the line-to-ground fault involving phase a.

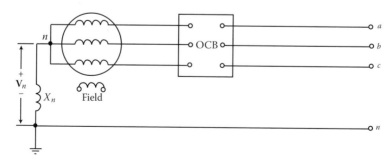

FIGURE P6.3 Generator system for Problem 9.

Analysis of Unbalanced Faults

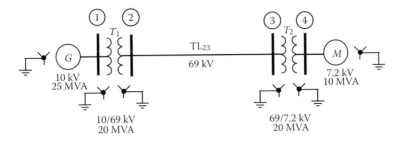

FIGURE P6.4 Transmission system for Problem 10.

    (c) Use the results of part (a) and calculate the line-to-neutral phase voltages at the fault point.
11. Consider the system given in Problem 10 and assume a line-to-line fault, involving phases b and c, at bus 2 and determine the faulted phase currents.
12. Consider the system shown in Figure P6.5 and assume that the associated data is given in Table P6.1 and is based on a 100-MVA base and referred to nominal system voltages.
    Assume that there is a three-phase fault at bus 6. Ignore the prefault currents and determine the following:
    (a) Fault current in per units at faulted bus 6
    (b) Fault current in per units in transmission line TL_{25}
13. Use the results of Problem 12 and calculate the line-to-neutral phase voltages at the faulted bus 6.
14. Repeat Problem 12 assuming a line-to-ground fault, with $\mathbf{Z}_f = 0$ pu, at bus 6.
15. Use the results of Problem 14 and calculate the line-to-neutral phase voltages at the following buses:
    (a) Bus 6
    (b) Bus 2
16. Repeat Problem 12 assuming a line-to-line fault at bus 6.
17. Repeat Problem 12 assuming a DLG fault, with $\mathbf{Z}_f = 0$ and $\mathbf{Z}_g = 0$, at bus 6.

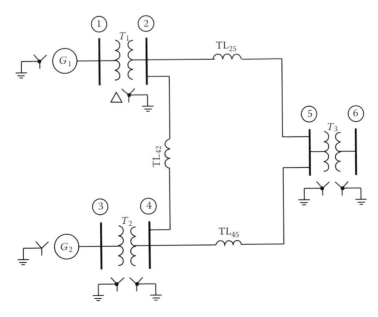

FIGURE P6.5 Transmission system for Problem 12.

TABLE P6.1
Data for Problem 6.12

Network Component	X_1 (pu)	X_2 (pu)	X_0 (pu)
G_1	0.35	0.35	0.09
G_2	0.35	0.35	0.09
T_1	0.10	0.10	0.10
T_2	0.10	0.10	0.10
T_3	0.05	0.05	0.05
TL_{42}	0.45	0.45	1.80
TL_{25}	0.35	0.35	1.15
TL_{45}	0.35	0.35	1.15

18. Consider the system shown in Figure P6.6 and data given in Table P6.2. Assume that there is a fault at bus 2. After drawing the corresponding sequence networks, reduce them to their Thévenin equivalents "looking in" at bus 2 for
    (a) Positive-sequence network
    (b) Negative-sequence network
    (c) Zero-sequence network
19. Use the solution of Problem 18 and calculate the fault currents for the following faults and draw the corresponding interconnected sequence networks.
    (a) SLG fault at bus 2 assuming that the faulted phase is phase a
    (b) DLG fault at bus 2 involving phases b and c
    (c) Three-phase fault at bus 2
20. Use the solution of Problem 18 and calculate the fault currents for the following faults and draw the corresponding interconnected sequence networks and calculate the fault currents, assuming that
    (a) SLG fault at bus 2 involves phase b
    (b) DLG fault at bus 2 involves phases c and a

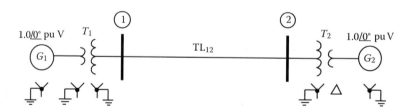

FIGURE P6.6 Transmission system for Problem 18.

TABLE P6.2
Table for Problem 18

Network Component	Base MVA	X_1 (pu)	X_2 (pu)	X_0 (pu)
G_1	100	0.2	0.15	0.05
G_2	100	0.3	0.2	0.05
T_1	100	0.2	0.2	0.2
T_2	100	0.15	0.15	0.15
TL_{12}	100	0.6	0.6	0.9

Analysis of Unbalanced Faults

21. Repeat parts (a) and (b) of Problem 19 assuming that
    (a) SLG fault at bus 2 involves phase c
    (b) DLG fault at bus 2 involves phases a and b
22. Repeat Example 6.6 assuming that the fault impedance is zero.
23. Repeat Example 6.6 assuming that the fault involves phase c.
24. Repeat Example 6.7 assuming that the $\mathbf{Z}_f$ and $\mathbf{Z}_g$, are zero.
25. Repeat Example 6.7 assuming that the fault involves phases c and a.
26. Consider the system shown in Figure P6.7 and data given in Table P6.3. Assume that there is an SLG fault at bus 3. Do the following:
    (a) Determine the Thévenin equivalent positive-sequence impedance.
    (b) Determine the Thévenin equivalent negative-sequence impedance.
    (c) Determine the Thévenin equivalent zero-sequence impedance.
    (d) Determine the positive-, negative-, and zero-sequence currents.
    (e) Determine the phase currents in per units and amperes.
    (f) Determine the positive-, negative-, and zero-sequence voltages.
    (g) Determine the phase voltages in per units and kilovolts.
    (h) Determine the line-to-line voltages in per units and kilovolts.
    (i) Draw a voltage phasor diagram using before-the-fault line-to-neutral and line-to-line voltage values.
    (j) Draw a voltage phasor diagram using the resultant after-the-fault line-to-neutral and line-to-line voltage values.

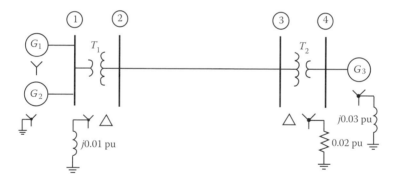

FIGURE P6.7 Transmission system for Problem 26.

TABLE P6.3
Table for Problem 26

Network Component	Base MVA	Voltage Rating (kV)	X_1 (pu)	X_2 (pu)	X_0 (pu)
G_1	100	13.8	0.15	0.15	0.05
G_2	100	13.8	0.15	0.15	0.05
G_3	100	13.8	0.15	0.15	0.05
T_1	100	13.8/115	0.20	0.20	0.20
T_2	100	115/13.8	0.18	0.18	0.18
TL_{23}	100	115	0.30	0.30	0.90

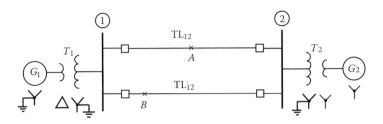

FIGURE P6.8 Tarnsmission system for Problem 27.

27. Consider the system shown in Figure P6.8 and assume that the following data on the same base are given:
    Generator G_1: $X_1 = 0.15$ pu, $X_2 = 0.10$ pu, $X_0 = 0.05$ pu
    Generator G_2: $X_1 = 0.30$ pu, $X_2 = 0.20$ pu, $X_0 = 0.10$ pu
    Transformer T_1: $X_1 = X_2 = X_0 = 0.10$ pu
    Transformer T_2: $X_1 = X_2 = X_0 = 0.15$ pu
    Transmission line TL_{12}: $X_1 = X_2 = 0.30$ pu, $X_0 = 0.60$ pu
    Transmission line TL_{23}: $X_1 = X_2 = 0.30$ pu, $X_0 = 0.60$ pu
    Assume that fault point A is located at the middle of the top transmission line, as shown in the figure, and determine the fault current(s) in per units for the following faults:
    (a) SLG fault (involving phase a)
    (b) DLG fault (involving phases b and c)
    (c) Three-phase fault
28. Repeat Problem 27 assuming that the fault point is n and is located at the beginning of the bottom line.
29. Consider the system shown in Figure P6.9 and assume that the following data on the same base are given:
    Generator G_1: $X_1 = 0.15$ pu, $X_2 = 0.10$ pu, $X_0 = 0.05$ pu
    Generator G_2: $X_1 = 0.15$ pu, $X_2 = 0.10$ pu, $X_0 = 0.05$ pu
    Transformer T_1: $X_1 = X_2 = X_0 = 0.10$ pu
    Transformer T_2: $X_1 = X_2 = X_0 = 0.15$ pu
    Transmission lines: $X_1 = X_2 = 0.30$ pu, $X_0 = 0.60$ (all three are identical)
    Assume that the fault point A is located at the middle of the bottom line, as shown in the figure, and determine the fault current(s) in per units for the following faults:
    (a) SLG fault (involving phase a)
    (b) SLG fault (involving phases b and c)
    (c) DLG fault (involving phases b and c)
    (d) Three-phase fault
30. Repeat Problem 29 assuming that the faulted point is B and is located at the end of the bottom line.

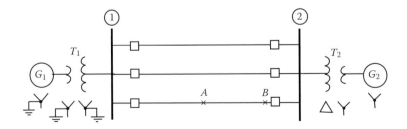

FIGURE P6.9 Transmission system for Problem 29.

31. Consider the system shown in Figure P6.10 and its data given in Table P6.4. Assume that there is an SLG fault involving phase a at fault point F.
    (a) Draw the corresponding equivalent positive-sequence network.
    (b) Draw the corresponding equivalent negative-sequence network.
    (c) Draw the corresponding equivalent zero-sequence network.
32. Use the results of Problem 31 and determine the interior sequence currents flowing in each of the four transmission lines.
    (a) Positive-sequence currents
    (b) Negative-sequence currents
    (c) Zero-sequence currents

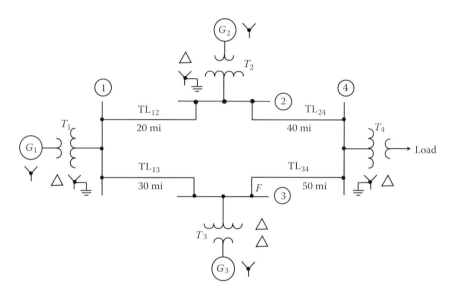

FIGURE P6.10 Four-bus system for Problem 31.

TABLE P6.4
Table for Problem 31

Network Component	Base MVA	Base kV$_{(L-L)}$	X_1 (pu)	X_2 (pu)	X_0 (pu)
G_1	100	230	0.15	0.15	
G_2	100	230	0.20	0.20	
G_3	100	230	0.25	0.25	
T_1	100	230	0.10	0.10	0.10
T_2	100	230	0.09	0.09	0.09
T_3	100	230	0.08	0.08	0.08
T_4	100	230	0.11	0.11	0.11
TL$_{12}$	100	230	0.10	0.10	0.36
TL$_{13}$	100	230	0.20	0.20	0.60
TL$_{24}$	100	230	0.35	0.35	1.05
TL$_{34}$	100	230	0.40	0.40	1.20

33. Use the results of Problem 32 and determine the interior phase currents in each of the four transmission lines.
    (a) Phase a currents
    (b) Phase b currents
    (c) Phase c currents
34. Use the results of Problems 32 and 33 and draw a three-line diagram of the given system. Show the phase and sequence currents on it.
    (a) Determine the SLG fault current.
    (b) Is the fault current equal to the sum of the zero-sequence currents (i.e., $\mathbf{I}_{f(\text{SLG})} = \sum 3\mathbf{I}_{a0}$)?
35. Repeat Example 6.9 assuming that there is a series fault at the fault point B of the bottom line (i.e., TL'_{AB}).
36. Repeat Example 6.10 using the results of Problem 35.
37. Consider the system given in Example 6.8 and determine the following:
    (a) Admittance matrix associated with zero-sequence network
    (b) Two-port Thévenin equivalent of zero-sequence network
38. Use the results of Problem 37 and determine the uncoupled zero-sequence network.
39. Consider the system shown in Figure P6.11 and assume that the equivalent of a large system is shown with two buses of interest and two interconnecting lines. Assume that one conductor of the top line becomes open.
    (a) Draw the corresponding positive-, negative-, and zero-sequence networks, without reducing them, and their interconnections.
    (b) Determine the uncoupled positive-sequence Thévenin equivalent.
    (c) Determine the uncoupled negative-sequence Thévenin equivalent.
    (d) Determine the uncoupled zero-sequence Thévenin equivalent.
    (e) Using the uncoupled sequence equivalents found in parts (b) through (d), repeat part (a).
40. Repeat Example 6.10 assuming an OLO fault involving phase a and $\mathbf{Z} = j0.1$ pu.
41. Use the solutions of Example 6.9 and Problem 38, and assume that the series fault is an OLO fault on phase b, $Z = j0.1$ pu.
    (a) Draw the generalized fault diagram.
    (b) Determine the positive-sequence current.
    (c) Determine the negative-sequence current.
    (d) Determine the zero-sequence current.
    (e) Determine the line current for phase a.
    (f) Determine the line current for phase b.
    (g) Determine the line current for phase c.
42. Repeat Problem 41 assuming that the open line is phase c.
43. Repeat Problem 41 assuming that there is a TLO series fault involving phases c and a.
44. Consider the OLO series fault representation given in Figure 6.19. Assume that there is a $\mathbf{Z}_f$ impedance between the fault points F and F' instead of being open and that the impedances ($\mathbf{Z}$'s) shown on lines b and c are zero. Redraw Figure 6.19 to reflect these changes and mathematically verify the resulting interconnection of the sequence networks.

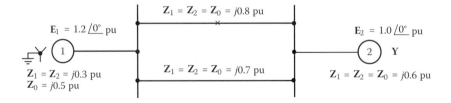

FIGURE P6.11 Transmission system for Problem 39.

Analysis of Unbalanced Faults

45. Consider Figure 6.19 and assume that there is no OLO series fault on phase a but the previous fault points F and F' are short-circuited with a line impedance $\mathbf{Z}$ of zero. However, the line impedances ($\mathbf{Z}$'s) on phases b and c are not zero. Redraw Figure 6.19 to reflect these changes and mathematically verify the resulting interconnection of the sequence networks.

46. Consider Figure 6.19 and assume that, between the points F and F', phases a, b, and c have the impedances $\mathbf{Z}_f$, $\mathbf{Z}$, and Z, respectively. Redraw Figure 6.19 to reflect these changes and mathematically verify the resulting interconnection of the sequence networks.

47. Use the result of Problem 6.19c and assume that the prefault load currents in line TL_{12} where $\mathbf{I}_{a0} = \mathbf{I}_{a2} = 0$ and $\mathbf{I}_{a1} = 0.6\angle-30°$ pu.

48. Assume that there is an OLO fault and an SLG fault with fault impedance $\mathbf{Z}_f$ both on phase a at a given fault point.
    (a) Determine the general representation diagram of the simultaneous fault.
    (b) Show the interconnection of sequence networks.
    (c) Mathematically verify the interconnection drawn in part (b).

49. Consider Figure 6.35a and c and verify Equation 6.130.

50. Repeat Example 6.11 assuming that all three lines of the transmission system have ground wires. The self-GMD of ground wires, D_{gg}, is 0.03125 ft. The GMD between phase conductors and ground wires, D_{ag}, is 13.0628 ft. The GMD between ground wires and their images, H_{gg}, is 104 ft. The GMD between phase conductors and images of ground wires, H_{ag}, is 92.8102 ft.

7 System Protection

7.1 INTRODUCTION

A power system can be thought of as a chain, the links of which are the generators, the power transformers, the switchgear, the transmission lines, the distribution circuits, and the utilization apparatus. The failure of any link destroys the capacity of the chain to perform the function for which it was intended.

The continuity of the chain can be secured by providing alternate links. For example, the transmission lines, being exposed to the natural elements, are much more vulnerable to faults than the power transformers and switchgear. Therefore, alternate transmission lines may be economically justified, whereas alternates for the power transformers and the switchgear would not.

Note that, *switchgear* is a general term covering switching or interrupting devices and their combination with associated control, instrumentation, metering, protective, and regulating devices as well as assemblies of these devices with associated interconnections, accessories, and supporting structures. In power systems, the switchgears are usually located in generating plant switchyards, transmission substations, bulk power substations, and distribution substations, as shown in Figures 7.1 through 7.4.

Figure 7.5 shows a typical 345-kV single-circuit transmission line with steel tower structure and bundled conductors. The transmission lines between the electrical power systems of separate utility companies are known as interconnections or tie lines. They provide the links for the exchange of electric power, contributing to increased efficiency and higher continuity of service. The advantages of interconnections include (1) economical interchange, (2) sharing of generation reserves, (3) utilization of large and more efficient generation units, (4) sharing large investments required by nuclear power plants, and (5) system support during emergencies.

Thus, it is required that interconnections have high capacity and therefore operate at voltages of 230 kV and up. Most commonly, 345 and 500 kV are used, with the trend to higher voltages and in some cases to high-voltage direct current (dc). It is required that these interconnections must be highly reliable. Therefore, their design and protection are developed with utmost precaution. For example, high-speed tripping is essential for all internal faults, and it is crucial that the link be maintained during external faults and system disturbances.

In general, the protection must provide coverage for most contingencies than might be justified in other system areas. High-speed simultaneous tripping of all terminals for all line or internal faults (1) minimizes line damage, (2) improves transient stability of the power systems, and (3) permits high-speed reclosing. In general, less than 10% of all faults are permanent.

Therefore, immediate reclosing and restoration of the line can improve transient stability and continue the advantage of energy interchange for over 90% of the faults if there are other interconnections between the two systems. This would require that the protective relays of each terminal communicate with each other to determine if the fault is internal or external. This is known as *pilot relaying*, and it requires a channel between the terminals.

On high-voltage lines, a single instantaneous reclosure is used and is usually within 12 cycles depending on the time necessary to dissipate the ionized air at the fault. Reclosing limits the phase separation of synchronous machines while the breaker is open, and therefore reduces the power oscillation that follows only the faulted phase so that power is never completely cut off. On low-voltage lines, the reclosure operation is usually repeated three times at intervals of between 15 and 20 s.

FIGURE 7.1 Callaway nuclear power plant switchyard. (Courtesy of Union Electric Company.)

FIGURE 7.2 Typical 345/138-kV transmission bulk-power station. (Courtesy of Union Electric Company.)

System Protection

FIGURE 7.3 Typical 138/34.5-kV bulk power substation. (Courtesy of Union Electric Company.)

FIGURE 7.4 Typical 34.5/4.16-kV distribution substation. (Courtesy of Union Electric Company.)

FIGURE 7.5 Typical 345-kV single-circuit transmission line with steel-tower structure and bundled conductor. (Courtesy of Union Electric Company.)

In the event that the breaker reopens after the third reclosure, the relay equipment locks it open, and it becomes necessary to reclose by hand. Figures 7.6 and 7.7 show typical automatic circuit reclosures. Figure 7.8 shows a typical three-phase oil circuit breaker (CB).

The task of protective relays is operate the correct CBs so as to disconnect only the faulty apparatus from the system as fast as possible, thus minimizing the interruption and damage that can be caused by a sustained fault may include (1) damage to the equipment causing destruction and fire, (2) explosions in equipment containing insulating oil, (3) overheating of system equipment, (4) causing undervoltages or overvoltages in the vicinity of the fault in the system, (5) blocking power flow, (6) causing reduction in stability margins, (7) causing improper operation of equipment

FIGURE 7.6 Typical three-phase automatic circuit recloser. (Courtesy of Union Electric Company.)

System Protection

FIGURE 7.7 Typical three-phase installed recloser. (Courtesy of Union Electric Company.)

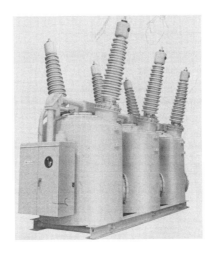

FIGURE 7.8 Typical oil circuit breaker. (Courtesy of McGraw-Edison.)

due to system margins, and (8) causing the system to become unbalanced and "break up" (i.e., *loss synchronism*) by an event known as *cascading*.

7.2 BASIC DEFINITIONS AND STANDARD DEVICE NUMBERS

Auxiliary relay. A relay that operates in response to the opening or closing of its operating circuit to assist another relay in the performance of its function. It may be instantaneous or may have a time lag, and may operate within large limits of the characteristic quantity.

Blocking. Preventing the relay from tripping, either due to its own characteristic or to an additional relay.

Burden. The loading imposed by the circuits of the relay on the energizing input power source or sources. In other words, the relay burden is the power required to operate the relay. The relay burden is usually given as volt-amperes (VA) at current transformer (CT)–rated current or impedance at rated current. Therefore, in a sense, the term *burden*, like the term *load*, is not precisely defined and can mean current, power (volt-amperes for ac and watts for dc), or impedance, depending on context.

Characteristic angle. The phase angle at which the performance of the relay is declared. It is usually the angle at which maximum sensitivity occurs.

Characteristic quantity. The quantity, the value of which characterizes the operation of the relay, for example, current for an overcurrent relay, voltage for a voltage relay, phase angle for a directional relay, time for an independent time delay relay, and impedance for an impedance relay. Some relays have a calibrated response to one or more quantities.

Characteristics (of a relay in steady state). The locus of the pickup or reset when drawn on a graph. In some relays, the two curves are coincident and become the locus, of balance or zero torque.

Dependent time-delay relay. A time-delay relay in which the time delay varies with the value of the energizing quantity.

Dropout or reset. A relay drops out when it moves from the energized position to the unenergized position.

Energizing quantity. The electrical quantity, that is, current or voltage either alone or in combination with other electrical quantities required for the functioning of the relay.

Independent time-delay relay. A time-delay relay in which the time delay is independent of the energizing quantity.

Instantaneous relay. A relay that operates and resets with no intentional time delay.

Inverse time-delay relay. A dependent time-delay relay having an operating time that is an inverse function of the electrical characteristic quantity.

Inverse time-delay relay with definite minimum. A relay in which the time delay varies inversely with the characteristic quantity up to a certain value, after which the time delay becomes substantially independent.

Knee-point emf. That sinusoidal electromotive force (EMF) applied to the secondary terminals of a CT, which, when increased by 10%, causes the exciting current to increase by 50%.

Overshoot time. The time during which stored operating energy dissipated after the characteristic quantity has been suddenly restored from a specified value to the value it had at the initial position of the relay.

Pickup. A relay is said to pick up when it changes from the unenergized position to the energized position (by closing its contacts).

Pilot channel. A means of interconnection between relaying points for the purpose of protection.

Protective gear. The apparatus, including protective relays, transformers, and auxiliary equipment, for use in a protective system.

Protective relay. An electrical device designed to initiate isolation of a part of an electrical system, or to operate an alarm signal in the case of a fault or other abnormal condition.

Protective scheme. The coordinated arrangements for the protection of a power system. It may include several protective systems.

Protective system. A combination of protective gears designed to secure, under predetermined conditions, usually abnormal, the disconnection of an element of a power system, or to give an alarm signal, or both.

Reach. A distance relay operates whenever the impedance seen by the relay is less than a prescribed value. This impedance or the corresponding distance is known as the reach of the relay.

Resetting value. The maximum value of the energizing quantity that is insufficient to hold the relay contacts closed after operation.

Setting. The actual value of the energizing or characteristic quantity at which the relay is designed to operate under given conditions. Such values are usually marked on the relay and may be expressed as direct values, percentages of rated values, or multiples.

System Protection

Stability. The quality whereby a protective system remains inoperative under all conditions other than those for which it is specifically designed to operate. For example, a protective system is said to be stable when it will restrain from tripping for a large external fault current due to a fault occurring outside the protected zone.

Time-delay relay. A relay having an international delaying device. The various circuit devices including relays have been given identifying numbers sometimes with appropriate suffix letters for use on schematic and writing diagrams. A selected list of the standard device numbers are given in Table 7.1

Underreach. The tendency of the relay to restraint at impedances larger than its setting. In other words, it is due to error in relay measurement resulting in wrong operation.

Unit or element. A self-contained relay unit that in conjunction with one or more other relay units performs a complex relay function, for example, a directional unit combined with an overcurrent unit gives a directional overcurrent relay.

TABLE 7.1
Standard Device Numbers

Device No.	Definition
2	Time-delay starting, or closing, relay
21	Distance relay
25	Synchronizing, or synchronism-check, device
27	Undervoltage relay
30	Annunciator relay
32	Directional power relay
37	Undercurrent or underpower relay
46	Reverse-phase or phase balance current relay
49	Machine, or transformer, thermal relay
50	Instantaneous overcurrent relay
51	ac time overcurrent relay
52	ac circuit breakers
55	Power factor relay
59	Overvoltage relay
60	Voltage balance relay
61	Current balance relay
64	Ground fault protective relay
67	ac directional overcurrent relay
68	Blocking relay
76	dc overcurrent relay
78	Phase-angle-measuring, or out-of-step, protective relay
79	ac reclosing relay
81	Frequency relay
83	Automatic selective control, or transfer, relay
85	Carrier, or pilot-wire, receiver relay
86	Locking-out relay
87	Differential protective relay
92	Voltage and power directional relay

7.3 FACTORS AFFECTING PROTECTIVE SYSTEM DESIGN

There are numerous factors affecting the design of a protective system. In general, they can be classified as

1. Economics in terms of initial investment and life-cycle costs, including the operating and maintenance costs.
2. Compliance with operating practices of the utility industry in terms of standards and accepted practices to permit efficient system operation and flexibility for the future.
3. Taking into account past experiences in terms of previous and anticipated problems within the system.
4. Available measures of faults or troubles in terms of fault magnitudes and location of CTs and voltage transformers (VTs).

Of course, there are many other factors affecting the selection of a protective system. Some of them are shown in Figure 7.9 [1].

The basic information necessary for developing a protective system may be summarized as (1) system configuration in terms of a one-line diagram; (2) present system protection scheme (if there is any) and associated problems; (3) existing operating practices and standards; (4) degree of protection necessary; (5) impedance data of the lines and transformers; (6) possible future expansion considerations; (7) fault study in terms of minimum and maximum fault currents; (8) ratios and locations of CTs; (9) ratios, locations, and connections of potential transformers; and (10) minimum and maximum system loads.

7.4 DESIGN CRITERIA FOR PROTECTIVE SYSTEMS

The design criteria for a protective system are

1. **Reliability.** It is a measure of the degree that the protective system will function properly in terms of both *dependability* (i.e., performing correctly when required) and *security* (i.e., avoiding unnecessary operation).
2. **Selectivity (or Discrimination).** The quality whereby a protective system distinguishes between those conditions for which it is intended to operate and those for which it must not operate. In other words, the selectivity of a protective system is its ability to recognize a fault and trip a minimum number of CBs to clear the fault. A well-designed protective system should provide maximum continuity of service with minimum system disconnection.
3. **Speed.** It is the ability of the protective system to disconnect a faulty system element as quickly as possible with minimum fault time and equipment damage. Therefore, a protective relay must operate at the required speed. It should neither be too slow, which may result in damage to the equipment, nor should it be too fast, which may result in undesired operation during transient faults. The speed of operation also has direct effect on the general stability of the power system. The shorter the time for which a fault is allowed to persist on the system, the more load can be transferred between given points on the power system without loss of synchronism. Figure 7.10 shows the curves that represent the power that can be transmitted as a function of fault-clearing time for various types of faults. Obviously, a fast fault-clearing time t_1 permits a higher power transfer than a longer clearing time t_2. Currently, the fault-clearing times on bulk power systems are in the $t1$ region (about three cycles on a 60-Hz base), and thus, power transfers are almost at a maximum. Also, it can be observed that the most severe fault is the three-phase fault, and the least severe fault is the line-to-ground fault in terms of transmission of power.

System Protection

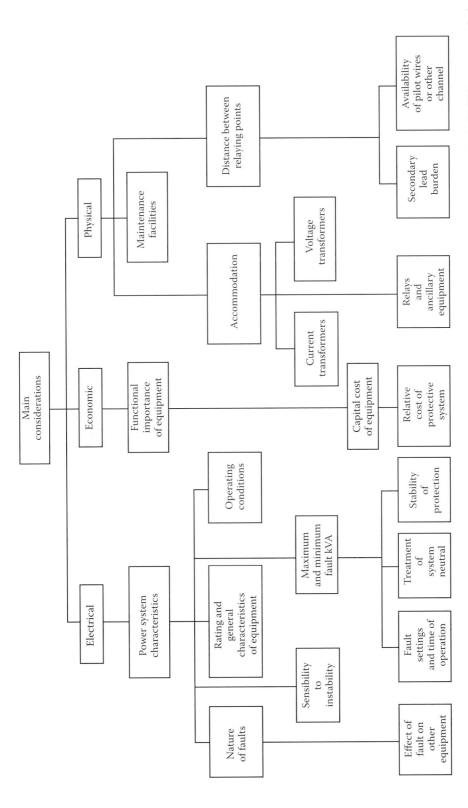

FIGURE 7.9 Factors affecting selection of protective system. (From GEC Measurements Ltd., *Protective Relays Application Guide*, 2nd ed. GEC Measurements Ltd., Stafford, UK, 1975.)

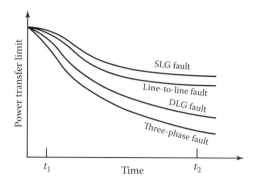

FIGURE 7.10 Typical values of power that can be transmitted as function of fault clearance time.

4. *Simplicity.* It is the sign of a good design in terms of minimum equipment and circuitry. However, the simplest protective system may not always be the most economical one even though it may be the most reliable one owing to fewer elements that can malfunction.
5. *Economics.* It dictates to achieve the maximum protection possible at minimum cost. It is possible to design a very reliable protective system but at a very high cost. Therefore, high reliability should not be pursued as an end in itself, regardless of cost, but should rather be balanced against economy, taking all factors into account.

Protection is not needed when the system is operating normally. It is only needed when the system is not operating normally. Therefore, in that sense, protection is a form of insurance against any failures of the system. Its premium is its capital and maintenance costs, and its return is the possible prevention of loss of system stability and the minimization of any possible damages. The cost of protection is generally extremely small compared with the cost of equipment protected. The art of protective relaying is constantly changing and advancing. However, the basic principles of relay operation and application remain the same. Thus, the purpose of this chapter is to review these fundamental principles and then show their applications to the protection of particular system elements. However, the emphasis will be on the transmission system.

7.5 PRIMARY AND BACKUP PROTECTION

Protection is the art or science of continuously monitoring the power system and detecting the presence of a fault and initiating the correct tripping of the CBs. However, the CBs alone are not sufficient to clear faults. They must be supplemented by protective relays.

The relays are required to detect the existence of faults and, if one does exist, to determine which breakers should be opened to clear it. Therefore, a protective system includes CTs, VTs, and protective relays with their associated wiring [known as the alternating current (ac) part of the system] and the relay contacts that close a circuit from the station battery to the CB's trip coil (known as the dc part of the system and is supplied by the station's batteries).

In general, all protective relays have two positions: (1) the normal position, usually with their contact circuit open, and (2) the fault position, usually with their contact circuit closed. Figure 7.11 shows the basic connections of a protective relay. Note that, after the breaker has tripped, its auxiliary switch (indicated by S in the figure) opens the highly inductive trip coil circuit and the relay can reset when deenergized by the opening of the breaker.

A relay has to be sufficiently sensitive to operate under minimum fault conditions for a fault within its own zone while remaining stable under maximum load or through fault conditions. Therefore, a relay should be able to differentiate a fault current from an overload current. For example, in the

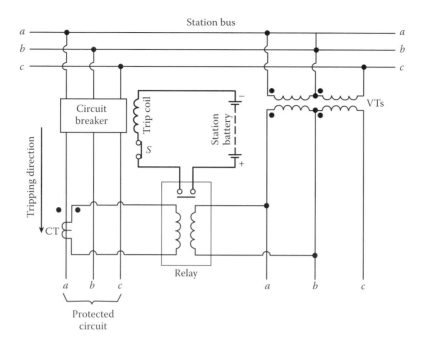

FIGURE 7.11 Typical ac relay connection showing one phase only.

event that a transformer is protected, the relay should not operate for the inrush of magnetizing current (known as inrush current), which may be at a magnitude of five to seven times the full load current. Similarly, the possible power swings that may take place in interconnected systems should also be ignored by the relay.

As shown in Figure 7.12, to be adequately protected with minimum interruptions, a power system can be divided into protective zones for (1) generators (or generator-transformer), (2) transformers, (3) buses, (4) transmission lines, and (5) motors. Note that, each protective zone has its own protective relays for detecting the existence of a fault in that zone and its own CBs for disconnecting that zone from the rest of the system.

Thus, a *protected zone* can be defined as the portion of a power system protected by a given protective system or a part of that protective system. In a well-designed primary protective system, any failure occurring within a given zone should cause the "tripping" (i.e., opening) of all CBs within that zone and only those breakers.

Therefore, *primary protection* can be defined as the protective system that is normally expected to operate in response to a fault in the protected zone. As shown in Figure 7.12, each zone is overlapped in order to prevent the possibility of unprotected (blind) areas. The principle of overlapping protection around a CB by means of CT connections has been illustrated in Figures 7.12 and 7.13. Figure 7.13a and b show the connections for "dead-tank" and "live-tank" breakers, respectively. Both type connections are commonly used in extra-high-voltage (EHV) transmission systems.

Any fault that exists between the CTs will operate both zone 1 and zone 2 relays and trip all CBs in the two zones. For example, if a fault occurs at fault point F_1 circuit breakers B_8 and B_7 should be opened. If a fault occurs at fault point F_2, circuit breakers B_5, B_7, and B_8 should be opened.

For the possible event of failure of the primary protection, for example, due to malfunctioning of a primary relay or a CB's failure to open when needed, a backup protection must be provided to remove the fault part from the system. Therefore, *backup protection* can be defined as the protective system intended to supplement the primary protection in case the latter should be ineffective or to deal with faults in those parts of the power system that are not readily included in the operating

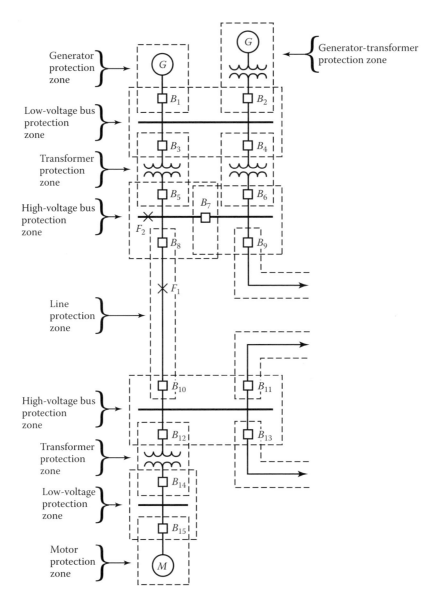

FIGURE 7.12 Typical zones of protection in power system.

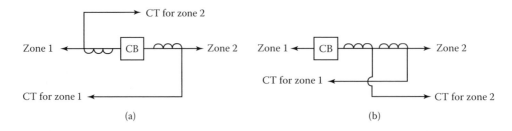

FIGURE 7.13 Principle of overlapping protection around circuit breaker.

zones of the primary protection. The necessary "backup relaying" can be located on another element of the power system, possibly the next adjacent station. This will avoid the simultaneous failure of both the primary and backup relaying.

When so located, this type of relaying is called the *remote backup relaying*. The remote backup is slow and usually disconnects greater portion of the power system than is necessary to remove the faulty part. For example, if circuit breaker B_8 should fail to open for a fault at the fault point F_2, circuit breaker B_{10} should be opened instead. In any case, the opening of the backup breaker(s) must be delayed long enough to give the proper breakers a chance to open first.

In general, backup protection is provided for possible failures in the primary relaying system and CBs. The causes of a relay failure may include failure of the relay itself, failure of auxiliary devices, loss of dc control supply, and failure in CTs or VTs and their circuits. On the other hand, the causes of a CB failure may include failure of the main contacts to interrupt, mechanical failure of the tripping mechanism, open- or short-circuit trip coil, and loss of dc supply.

Thus, any backup protection must provide both relay backup as well as CB backup. In general, there are three kinds of backup relays: those that trip the same CB if the main relay fails (relay backup); those that open the next nearest CBs on the same bus, if one of the local breakers fails to open (CB backup); and those that operate from the next station, in the direction toward the source, to back up both the relays and CBs (remote backup).

If faster backup relaying is necessary, the backup relaying may be located in the same location as the primary relaying to provide "local backup." Therefore, in local backup relaying, fault is cleared locally in the same station where it has taken place.

Today, systems are increasingly becoming complex as a result of larger numbers of interconnection of more and more generating stations, an increasing number of fault current paths, and greater line loadings. Although it may be possible to increase the sensitivity or reach settings of remote relays so that they can recognize the faults with in-feed, there may be cases where such settings are infeasible due to required sensitivity or reach can cause the relay to operate on load currents or on less important synchronizing power swings.

Thus, the local backup relaying can effectively be used at such troublesome locations. However, the local backup relaying should be as completely separated from the primary relaying as is possible.

As said before, the remote backup is inherently slow. For example, when distance relays are used for line protection in the previous example, backup clearing times for faults near circuit breaker B_8 will usually be somewhere between 0.25 and 0.5 s, whereas the clearing times for faults near circuit breaker B_{10} will be between 1 and 3 s.

When backup is provided with time overcurrent relays, the backup clearing times will usually be greater. The delay time that is necessary to coordinate the primary and the backup protection system is called the *coordination time delay*. Today, due to the complexity involved, the coordination of protective relays, especially for the interconnected systems, are done by using various computer programs that are available in the utility industry.

7.6 RELAYS

As defined previously, a protective relay is an electrical device designed to initiate isolation of a part of an electric system, or to operate an alarm signal, in the case of a fault or other abnormal condition. Basically, a protective relay consists of an operating element and a set of contacts. The operating element receives input from the instrument transformers in the form of currents, voltages, or a combination of currents and voltages (e.g., impedance and power).

The relay may respond to (1) a change in magnitude in the input quantity, (2) the phase angle between two quantities, (3) the sum (or difference) of two quantities, or (4) the ratio of the quantities. In any case, the relay performs a measuring (or comparison) operation based on the input and translates the result into a motion of contacts. Therefore, for example, the output state of an electromechanical relay is either "trip" (with its contacts closed) or "block or block to trip" (with its contacts open). When they close, the contacts either actuate a warning signal or complete the trip circuit of a CB, which in turn isolates the faulty part by interrupting the flow of current into that part.

In general, protective relays can be classified by their constructions, functions, or applications. By construction, they can be either electromechanical or solid-state (or static) types. In general, the electromechanical relays are robust, inexpensive, and relatively immune to the harsh environment of a substation. However, they require regular maintenance by skilled personnel. Furthermore, their design is somewhat limited in terms of available characteristics, tap settings, and burden capability.

On the other hand, solid-state relays consist of analog circuits in addition to logic gates and are capable of producing any desired relay characteristics. Therefore, today, solid-state relays have been primarily used in areas where the application of conventional methods is difficult or impossible (e.g., high-voltage or EHV transmission line protection by phase comparison). The relays using transistors for phase or amplitude comparison can be made smaller, cheaper, faster, and more reliable than electromechanical relays. They can be made shock-proof and require very little maintenance. Furthermore, their great sensitivity allows smaller CTs to be used and more sophisticated characteristics to be obtained.

Contrary to electromechanical relays, solid-state relays provide switching action, without any physical motion of any contacts, by changing its state from nonconducting to conducting or vice versa. Electromechanical relays can be classified as magnetic attraction, magnetic induction, D'Arsonval, and thermal units. The most widely used types of magnetic attraction relays include plunger (solenoid), clapper, and polar.

The *plunger type* of construction consists of a cylindrical coil with an external magnetic structure and a center plunger (*armature*), as shown in Figure 7.14a. When the current or voltage applied

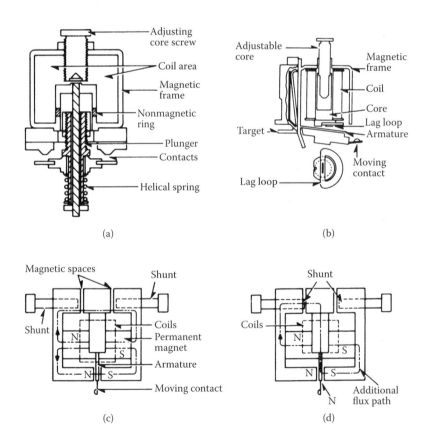

FIGURE 7.14 Various types of magnetic attraction relays: (a) plunger type; (b) clapper type; (c) polar-type unit with balanced air gaps; (d) polar-type unit with unbalanced air gaps. (Courtesy of Westinghouse Electric Corporation.)

to the coil is more than the pickup value, the plunger moves upward to operate a set of contacts. The force required to move the plunger is proportional to the square of the current in the coil. Plunger relays are instantaneous with typical operating times of 5–50 ms, with the longer times occurring near the threshold values of pickup. The plunger-type relay shown in Figure 7.14a is used as a high drop-out instantaneous overcurrent relay.

The *clapper type* (also called the *hinged armature type*) of construction consists of a U-shaped magnetic frame with a movable armature across the open end. The armature is hinged at one side and spring restrained at the other, as shown in Figure 7.14b. When the electrical coil is energized, the armature meets a fixed contact, opening or closing a set of contacts with a torque proportional to the square of the coil current. The pickup and drop-out values of clapper relays are less accurate than those of plunger relays. They are basically used as auxiliary and "go/no-go" relays. The unit shown in the figure operates as an instantaneous overcurrent or instantaneous trip unit.

Polar-type relays operate from a dc applied to a coil wound around the hinged armature in the center of the magnetic structure. A permanent magnet across the structure polarizes the armature gap poles, as shown in Figure 7.14c and d. Two nonmagnetic spacers located at the rear of the magnetic frame are bridged by two adjustable magnetic shunts. This arrangement facilitates the magnetic flux paths to be adjusted for pickup and contact action. With balanced air gaps, as shown in Figure 7.14c, the flux paths are indicated and the armature floats in the center with the coil deenergized.

On the other hand, with the gaps unbalanced, as shown in Figure 7.14d, some of the flux is shunted through the armature. Therefore, the resulting polarization holds the armature against one pole with the coil deenergized. Current in the coil magnetizes the armature either north or south, increasing or decreasing any prior polarization of the armature. This polarization can be fast or gradual depending on design and adjustments. The left gap adjustment, shown in Figure 7.14d, controls the pickup value; the right gap adjustment controls the reset current value.

The *magnetic-induction-type* relays can be classified into two basic types: induction disk and cylinder units. The *induction disk* unit consists of a metallic disk of copper or aluminum that rotates between the pole faces of an electromagnet. It operates by the torque derived from the interaction of fluxes produced by an electromagnet with those from induced currents in the plane of the rotatable disk. The induction unit shown in Figure 7.15a has three poles on one side of the disk and a common magnetic keeper on the opposite side. The main coil is on the center leg. Current I in the main coil produces flux Φ, which passes through the air gap and disk to the keeper. Flux Φ is divided into Φ_L through the left-hand leg and Φ_R through the right-hand leg. A short-circuit lagging coil on the left leg causes Φ_L to lag both Φ_R and Φ, producing a split-phase motor action. Flux Φ_L induces voltages V_s, and current I_s flows, basically in phase, in the shorted lag coil. Flux Φ_T is the total flux produced by the main coil current I. The three fluxes cross the air gap and induce eddy currents in the disk. As a result, these eddy currents set up counter fluxes, and the interaction of the two sets of fluxes produces the torque that rotates the disk. The induction disk unit is always used as a time-delay unit due to the inertia of the moving disk [2].

The *cylinder-type magnetic induction relay* consists of a metallic cylinder with one end closed with a cup, which rotates in an annular air gap between the pole faces of electromagnets and a central core. Since its operation is similar to that of an induction motor with salient poles for the stator windings, it is also called the *induction cup-type unit*. Figure 7.15b shows the basic unit used for relays that has an inner steel core at the center of the square electromagnet, with a thin-walled aluminum cylinder rotating in the air gap. Cylinder travel is limited to a few degrees by the contact and the associated stops, and a spiral spring provides reset torque. Operating torque is a function of the product of the two operating quantities and the cosine of the angle between them. The torque can be expressed as

$$T = k_c \mathbf{I}_1 \mathbf{I}_2 \cos(\theta_{12} - \Phi) - k_s \quad (7.1)$$

where

k_c = design constant
Φ = design constant
I_1 = current flowing through coil 1
I_2 = current flowing through coil 2
k_s = restraining spring torque

Since the rotating parts of the induction cup unit are of low inertia, it is capable of high-speed operation. Therefore, it can be used for functions requiring instantaneous operation. The multiplicity of poles also permits measurement of more than one electrical quantity.

Figure 7.15c shows the *D'Arsonval unit*, which has a magnetic structure and an inner permanent magnet form a two-pole cylindrical core. A moving coil loop in the air gap is energized by dc, which reacts with the air gap flux to produce rotational torque. The *thermal unit* consists of two layers of different metals welded together to form a *bimetallic strip* or coil that has one end fixed and the other end free. As the temperature changes, the different coefficients of thermal expansion of the two metals cause the free end of the coil or strip to move, operating a contact structure for relay applications.

Relays of this type are mostly used for overload protection. Figure 7.16 shows an electromagnetic-attraction *balanced-beam-type* relay unit. It is an overcurrent type of current-balanced relay unit. It is a special type of clapper construction. The beam (armature) is attracted by electromagnets that

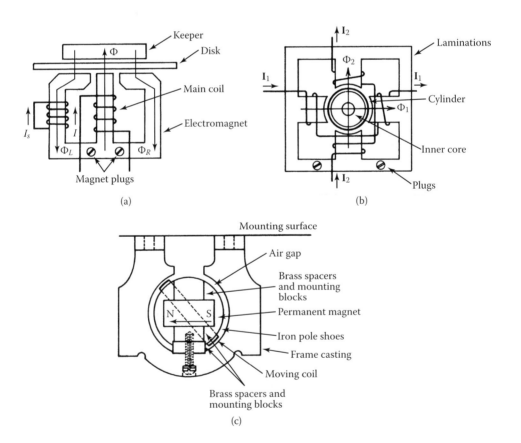

FIGURE 7.15 Various types of magnetic induction relays: (a) induction disk type; (b) cylinder type; (c) D'Arsonval type. (Courtesy of Westinghouse Electric Corporation.)

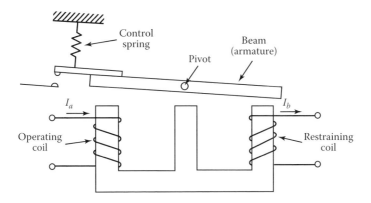

FIGURE 7.16 Balanced-beam relay unit.

are operated by the appropriate parameters, usually two currents or a current and a voltage. A slight mechanical bias is built into the unit by having a control spring in order to keep the contacts open, except when operation is required. Some units have a separate armature at the end of the beam that is drawn into the fixed-position operating coil when current flows into the coil.

Also, various contacts and spring arrangements are possible. The resetting value of a balanced-beam unit is low compared with its operating value because the magnetic gap is small under one pole in the normal position and large under the other. However, in the operating position, this is reversed. The force at each end of the beam is proportional to the square of the gap flux, as in the attracted armature relay.

The gap flux is proportional to the current and decreases inversely approximately as the square of the total air gap length in the magnetic circuit. Another commonly used structure is an induction-type relay having two overcurrent elements acting in opposition on a rotor. In the event that the negative-torque effect on the control spring is ignored, the torque equation of either type can be expressed as

$$T = k_a I_a^2 - k_b I_b^2 \tag{7.2}$$

where k_a and k_b are design constants. When the relay is on the threshold of operating, the net torque becomes zero, and therefore,

$$k_a I_a^2 = k_b I_b^2$$

Thus, the operating characteristic can be expressed as

$$\frac{I_a}{I_b} = \left(\frac{k_a}{k_b}\right)^{1/2} = K \tag{7.3}$$

where K is a constant. If the operating quantities are taken as voltage V and current I, then the pulls on the armatures by the electromagnets are equal to $k_a V^2$ and $k_b I^2$, and therefore the condition for the operation (closing its contacts) of the unit is

$$k_a V^2 > k_b I^2 \tag{7.4}$$

or

$$\left(\frac{k_b}{k_a}\right)^{1/2} > \frac{V}{I} \qquad (7.5)$$

that is,

$$\left(\frac{k_b}{k_a}\right)^{1/2} > Z \qquad (7.6)$$

indicating that the relay will operate when the impedance it "sees" is less than a predetermined value.

The *solid-state relays* (or *static relays*) are extremely fast in their operation because they have no moving parts and have very quick response times. Today, the static relays are very reliable primarily due to high-reliability performance of modern silicon planar transistors. Appropriate circuits are designed to make detection involving phase angles, current and voltage magnitudes, timing, and others. Figure 7.17 illustrates how a static relay could measure the phase angle between a voltage and a current. The voltage and current sine waves, shown in Figures 7.17a and 7.14b, are supplied to separate squaring amplifiers whose function is to convert the sine wave to a square wave that is zero during the negative half-cycle, as shown in Figure 7.17c and d.

These square waves are commonly called "*blocks*" and can be supplied to a comparator circuit in such a way that an output is received only when both signals are present. The duration of their overlap or the duration of the comparator output, shown in Figure 7.17e, is the complement of the phase angle between the current and voltage.

For example, to receive an output when the voltage lags or leads current by 90°, the half-cycle current and voltage blocks must overlap or be coincident for 120°. Figure 7.17e shows the duration of overlap or comparator output for the case where the current leads the voltage by 90°. In actual practice, it is usually the complement of the angle that is measured.

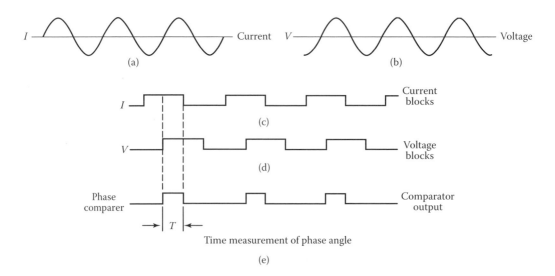

FIGURE 7.17 Typical waveforms used in static relay-measuring operations.

System Protection

The logic circuits used in the static relays can be classified into (1) fault-sensing and data-processing logic units; (2) amplification logic units; and (3) auxiliary logic units. The first type of logic circuits use comparators to detect faults. Magnitude comparators (i.e., comparison logic units) are employed to detect both instantaneous and time overcurrent faults. As an example, Figure 7.18a shows the dements used in a single-phase definite time overcurrent relay.

Note that, an ac is converted to a proportional direct voltage and compared with a fixed dc level. When it exceeds the reference level, a timer is initiated. After the set time delay, the second-level detector operates to activate the output element. The input circuit consists of CT, the secondary current of which is rectified and supplied into a resistive shunt. The current setting of the relay can be changed by means of taps on the CT or by varying the value of the secondary shunt.

A time-delay setting call be achieved by changing the resistance value in the resistor–capacitor (RC) delay circuit with a calibrated potentiometer. Instantaneous operation above a set level is obtained by by-passing the time-delay element to a second-level detector. The circuit of a simplified level detector is shown in Figure 7.18b. It includes an output stage driving the coil of an attracted armature relay. Note that, all transistors are biased off until the input voltage exceeds that at the emitter of TR1, set by potentiometer chain R1 and R2. When this voltage is exceeded, both transistors turn on, and the output relay is activated. This circuit has the advantages of drawing no current when in the nonoperated stage and giving a dropout level of almost 100% of the pickup level. Timing can be performed by connecting an RC time-delay network, shown dashed in Figure 7.18b, at the input to the second-level detector monitoring the voltage across the capacitor.

A variable reference magnitude comparator is used for ground distance relays. A phase angle comparator gives an output when the phase angle between two quantities exceeds the pickup level. These logic circuits are used in phase, distance, and directional relays for high-voltage and EHV transmission line protection. Figure 7.19 shows a sophisticated system that uses distance functions to provide protection for faults of all types.

The basic system provides a standard step distance relaying scheme, or it can be used with a communication channel to provide the most commonly used pilot relaying schemes. The main measuring function is controlled by phase selectors (often called starters) that select the faulted phase or phase pair to which the measuring function must be connected during a fault. Three phase selectors are provided, each connected on a per-phase basis.

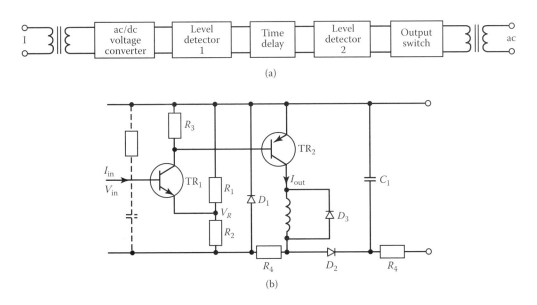

FIGURE 7.18 Definite time overcurrent relay: (a) block diagram; (b) circuit of level detector.

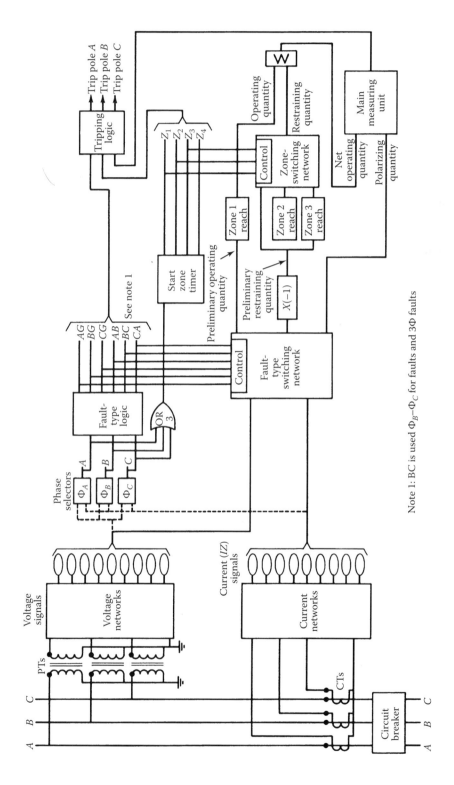

FIGURE 7.19 Basic system of (SLS 1000) modular transmission line protection. (Courtesy of General Electric Company.)

System Protection

Typically, the measuring function would be set with a zone 1 reach, and tripping would be initiated in zone 1 one time following the operation of the phase selectors provided the fault fell within the reach of the main measuring function. The phase selectors also start zone timing functions that extend the reach of the main measuring function as time progresses to provide stepped distance protection.

Up to three zones of protection are thus provided via this single measuring function. A fourth zone can be connected to initiate tripping through the phase selector. Figure 7.20 shows the main measuring function of the system. There can be up to eight distance measuring functions in the SLS 1000: three phase selectors; the main measuring function; a permissive zone measuring function for use in pilot schemes and out-of-step detection; and three blocking functions, which are required only in blocking-type schemes or when an additional zone of time-delayed protection is required.

In this system, the phase selectors must perform the following functions:

1. Detect the presence of a fault on the portion of the system to which they are applied.
2. Identify the faulted phase during single line-to-ground faults.
3. Identify the faulted phases during multiphase faults.
4. Initiate zone timing to extend the reach of the main measuring function.
5. Operate correctly during and immediately following the open-pole period following a single-pole trip when such tripping is employed.
6. Not operate on load current. Provisions are included to make the function lenticular when load is of concern.

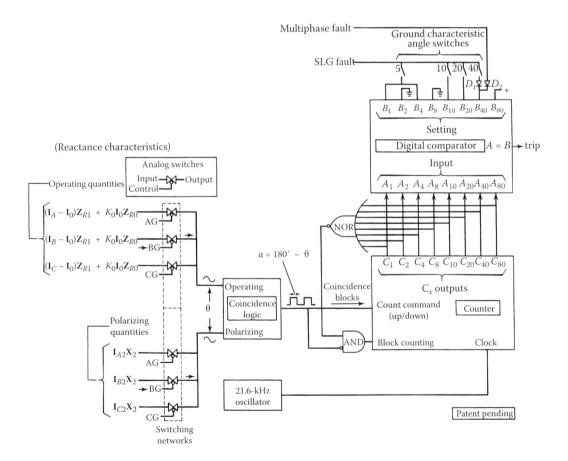

FIGURE 7.20 Main measuring function of (SLS 1000 system) modular transmission line protection. (Courtesy of General Electric Company.)

If there were no load current in the system, or if fault currents were always above load current, the simplest type of phase selector would be an overcurrent function. Unfortunately, in many cases, fault currents are less than load current, thus precluding the use of overcurrent functions. For this reason, distance-type function are preferred and usually employed. The distance function used in this system is a variation of a positive-sequence polarized mho ground distance function.*

7.7 SEQUENCE FILTERS

Sequence filters (also known as *sequence networks*) are used in a three-phase system to measure (and therefore to indicate the presence of) symmetrical components of current and voltage. In a perfectly balanced power system, as mentioned in Chapter 5, there are no zero- and negative-sequence current and voltage quantities, but only positive-sequence quantities exist. Therefore, for example, the presence of a zero-sequence current usually indicates a ground fault on the system, but it could also be due to unbalanced leakage currents over the surface of contaminated insulators or to unbalanced line-to-ground charging currents. Zero-sequence voltage is used to polarize directional ground fault relays.

As said before, positive-sequence power flows toward the fault; negative- and zero-sequence power flows away from the fault, since the fault is the source of negative- and zero-sequence voltage. Thus, negative-sequence current does not occur in a balanced load and, when present on a faulted line, always flows away from the fault.

Negative-sequence currents cause excessive heating of alternator rotors. The need for a response to three-phase faults means that sensitivity to the positive-sequence component is also required. It is necessary to proportion these two quantities so as to ensure that no system condition occurring in practice gives rise to a "blind spot" similar to those possible with summation transformers.

Positive-sequence voltage is sometimes used to operate the automatic voltage regulator controlling the excitation of an ac generator. In summary, currents and voltages may be separated from the corresponding line currents and voltages by segregating networks called filters and are used for carrying out the necessary protective operations (e.g., the activation of relays, control of transmitters, and receivers). In this section, a few of the many available filters will be discussed.†

The simplest sequence filters are the zero-sequence filters, as shown in Figure 7.21. Figure 7.21a shows a zero-sequence current filter that consists of three CT secondaries connected in parallel. Note that, the capital and lowercase subscripts are used for primary (system) and secondary (relaying) quantities, respectively. The CTs basically act as current sources so that the filter output is

$$\mathbf{I}_a + \mathbf{I}_b + \mathbf{I}_c = 3\mathbf{I}_{a0} \tag{7.7}$$

Similarly, Figure 7.21b shows a zero-sequence voltage filter made up of three potential transformers connected in series with the primary in grounded wye. The potential transformers act as voltage sources so that the filter output is

$$\mathbf{V}_a + \mathbf{V}_b + \mathbf{V}_c = 3\mathbf{V}_{a0} \tag{7.8}$$

The current polarity markings on all transformers are important. In the event of having a minus sign error, there will not be any zero-sequence current or voltages, neither in magnitude nor in phase.

---

* The interested reader should see Reference 3 for the detailed information about this system.
† The interested reader should read References 2 and 4 and especially Reference 5 for further information about filters.

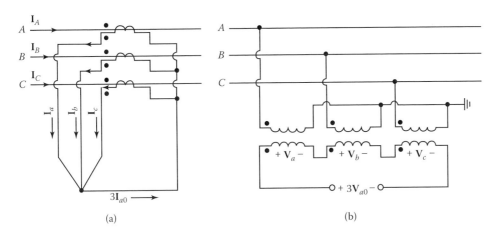

FIGURE 7.21 Zero sequence filters: (a) current filter; (b) voltage filter.

Figure 7.22 shows two composite sequence current filters. Figure 7.22a shows a positive- and zero-sequence filter consisting of three coupled inductors with mutual reactance X_m and self-impedance Z_s and two external resistors R_1 and R_0 Therefore, the open-circuit voltage can be expressed as

$$\mathbf{V}_F = \mathbf{I}_a R_1 + \mathbf{I}_n R_0 + jX_m (\mathbf{I}_c - \mathbf{I}_b) \tag{7.9}$$

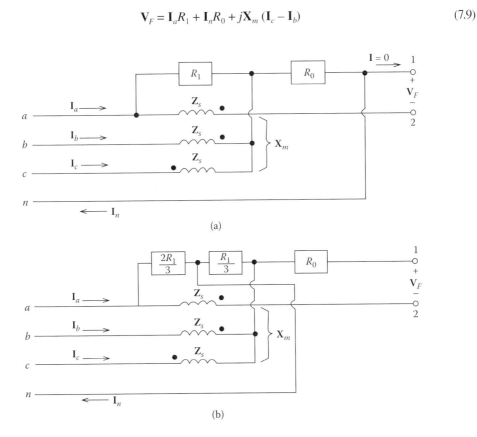

FIGURE 7.22 Composite sequence current filters: (a) positive- and zero-sequence current filter; (b) positive- or negative-sequence current filter.

where

$$\mathbf{I}_n = \mathbf{I}_a + \mathbf{I}_b + \mathbf{I}_c \qquad (7.10)$$

Thus, it can be expressed in terms of the sequence currents as

$$\mathbf{V}_F = \left(R_1 + 3R_0\right)\mathbf{I}_{a0} + \left(R_1 - \sqrt{3}X_m\right)\mathbf{I}_{a1} + \left(R_1 + \sqrt{3}X_m\right)\mathbf{I}_{a2} \qquad (7.11)$$

Changing the values of R_1, R_0, and X_m and modifying the connections by interchanging $\mathbf{I}_b$ and $\mathbf{I}_c$ in the circuit can produce different output characteristics. For example, when R_1 is selected as $X_m = R_1/\sqrt{3}$ and substituted into Equation 7.11, the result becomes

$$\mathbf{V}_F = (R_1 + 3R_0)\mathbf{I}_{a0} + 2R_1\mathbf{I}_{a1} \qquad (7.12)$$

Thus, a relay connected to the output terminals will react to zero- and positive-sequence currents.

Figure 7.22b shows a similar sequence network that can be used as a positive- or negative-sequence current filter. The open-circuit voltage can be expressed as

$$\mathbf{V}_F = \left(\frac{2R_1}{3}\right)\mathbf{I}_a + jX_m\left(\mathbf{I}_c - \mathbf{I}_b\right) - \frac{R_1}{3}\left(\mathbf{I}_b + \mathbf{I}_c\right) \qquad (7.13)$$

which can be reexpressed in terms of the sequence currents as

$$\mathbf{V}_F = \left(R_1 - \sqrt{3}X_m\right)\mathbf{I}_{a1} + \left(R_1 + \sqrt{3}X_m\right)\mathbf{I}_{a2} \qquad (7.14)$$

Therefore, if R_1 is selected such that $X_m = R_1/\sqrt{3}$ and substituted into Equation 7.14, the result becomes

$$\mathbf{V}_F = 2R_1\mathbf{I}_{a2} \qquad (7.15)$$

Thus, a relay connected to the output terminals will react to a negative-sequence current. On the other hand, current $\mathbf{I}_b$ and $\mathbf{I}_c$ can be interchanged so that

$$\mathbf{V}_F = \left(R_1 + \sqrt{3}X_m\right)\mathbf{I}_{a1} + \left(R_1 - \sqrt{3}X_m\right)\mathbf{I}_{a2} \qquad (7.16)$$

Furthermore, if R_1 is selected such that $X_m = R_1/\sqrt{3}$, before, the result becomes

$$\mathbf{V}_F = 2R_1\mathbf{I}_{a1} \qquad (7.17)$$

Hence, a relay connected to the output terminals will respond to positive-sequence current.

7.8 INSTRUMENT TRANSFORMERS

Instrument transformers are used both to provide safety for the operator and equipment from high voltage and to permit proper insulation levels and current-carrying capacity in relays, meters, and other instruments. In the United States, the standard instruments and relays are rated at 5 A and/or

System Protection

120 V, 60 Hz. The basic instrument transformers are of two types: CTs and VTs (formerly called potential transformers).

In either case, the external load applied to the secondary of an instrument transformer is referred to as its "burden." The term *burden* usually describes the impedance connected to the transformer secondary winding but may specify the volt-amperes supplied to the load. For example, a transformer supplying 5 A to a resistive burden of 0.5 Ω may also be said to have a burden of 12.5 VA at 5 A.

7.8.1 Current Transformers

Current transformers can be constructed in various ways. For example, a CT may have two separate windings on a magnetic steel core, similar to that of the potential transformers. However, it differs in that the primary winding consists of a few turns of heavy wire capable of carrying the full-load current, whereas the secondary winding consists of many turns of smaller wire with a current-carrying capacity of from 5 to 20 A, depending on the design. It is called a *wound-type current transformer* owing to its wound primary coil.

Another very common type of construction is the so-called window-, through-, or donut-type CT in which the core has an opening through which the conductor carrying the primary-load current is passed. This *bushing-type current transformer* is made up of an annular-shaped core with secondary winding. It may be built into various types of apparatus such as CBs, power transformers, generators, or switchgear, the core being arranged to encircle an insulating bushing through which a power conductor passes.

A CT ratio is selected on the basis of the continuous-current ratings of the connected apparatus (e.g., relays, measuring instruments, and auxiliary CTs) and of the secondary winding of the CT itself. It is usually selected so that the secondary current is about 5 A at maximum primary-load current. If delta-connected CTs are used, the factor of $\sqrt{3}$ must be taken into account. Some of the standard CT ratios are 50:5, 100:5, 150:5, 200:5, 250:5, 300:5, 400:5, 450:5, 500:5, 600:5, 800:5, 900:, 1000:5, and 1200:5.

When CTs are interconnected, the relative polarities of primary and secondary terminals become important. Figure 7.21a shows a wye connection of CTs. Figure 7.23 shows two possible connections for delta-connected CTs. The output currents of the delta connection shown in Figure 7.23a can be expressed as

$$\mathbf{I}_a - \mathbf{I}_b = (1 - a^2)\mathbf{I}_{a1} + (1 - a)\mathbf{I}_{a2} \tag{7.18}$$

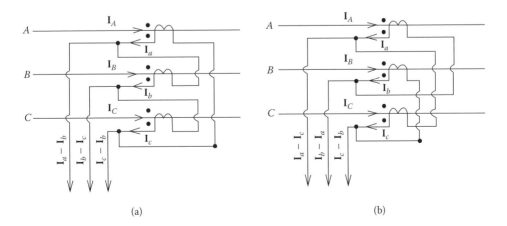

FIGURE 7.23 Delta connections of CTs.

$$\mathbf{I}_b - \mathbf{I}_c = (a^2 - a)\mathbf{I}_{a1} + (a - a^2)\mathbf{I}_{a2} \tag{7.19}$$

$$\mathbf{I}_c - \mathbf{I}_a = (a - 1)\mathbf{I}_{a1} + (a^2 - 1)\mathbf{I}_{a2} \tag{7.20}$$

Similarly, the output currents of the delta connection in Figure 7.23b can be expressed as

$$\mathbf{I}_a - \mathbf{I}_c = (1 - a)\mathbf{I}_{a1} + (1 - a^2)\mathbf{I}_{a2} \tag{7.21}$$

$$\mathbf{I}_b - \mathbf{I}_a = (a^2 - 1)\mathbf{I}_{a1} + (a - 1)\mathbf{I}_{a2} \tag{7.22}$$

$$\mathbf{I}_c - \mathbf{I}_b = (a^2 - a)\mathbf{I}_{a1} + (a - a^2)\mathbf{I}_{a2} \tag{7.23}$$

which shows that the connection shown in Figure 7.23b is merely the reverse of the connection shown in Figure 7.23a.

Also, note that the zero-sequence currents are not present in the output circuits; they merely circulate in the delta connection. In the event that there is a three-phase fault, only positive-sequence currents will exist. Therefore, the output currents of the delta connection given by Equations 7.18 through 7.20 can be reexpressed as

$$\mathbf{I}_a - \mathbf{I}_b = (1 - a^2)\mathbf{I}_{a1} \tag{7.24}$$

$$\mathbf{I}_b - \mathbf{I}_c = (a^2 - a)\mathbf{I}_{a1} \tag{7.25}$$

$$\mathbf{I}_c - \mathbf{I}_a = (a - 1)\mathbf{I}_{a1} \tag{7.26}$$

In the event that there is a line-to-line fault involving phases b and c, the same output currents can be reexpressed as

$$\mathbf{I}_a - \mathbf{I}_b = (a - a^2)\mathbf{I}_{a1} \tag{7.27}$$

$$\mathbf{I}_b - \mathbf{I}_c = (a^2 - a)\mathbf{I}_{a1} \tag{7.28}$$

$$\mathbf{I}_c - \mathbf{I}_a = (a - a^2)\mathbf{I}_{a1} \tag{7.29}$$

since $\mathbf{I}_{a2} = -\mathbf{I}_{a1}$. Similarly, in the event that there is a line-to-ground fault involving phase a, the same output currents can be expressed as

$$\mathbf{I}_a - \mathbf{I}_b = 3\mathbf{I}_{a1} \tag{7.30}$$

$$\mathbf{I}_b - \mathbf{I}_c = 0 \tag{7.31}$$

$$\mathbf{I}_c - \mathbf{I}_a = -3\mathbf{I}_{a1} \tag{7.32}$$

since

$$\mathbf{I}_{a1} = \mathbf{I}_{a2}$$

Figure 7.24 shows the equivalent circuit of an ideal CT. Note that, the primary-side impedance $\mathbf{Z}_p$ has been neglected since it does not affect the perfectly transformed current $\mathbf{I}_s$ (which is equal to $N_p\mathbf{I}_p/N_s$) or the voltage across X_e, where X_e is the transformer excitation reactance referred to as the secondary and is saturable.

The term $\mathbf{Z}_s$ is the transformer leakage reactance referred to as the secondary, $\mathbf{Z}_B$ is the connected external impedance (or burden) in terms of secondary leads, relay, instruments, etc., and $\mathbf{I}_L$

System Protection

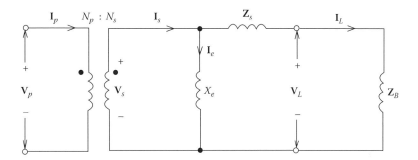

FIGURE 7.24 Equivalent circuit of ideal CT.

is the maximum secondary current and can be estimated by dividing the known maximum fault current by the selected current by the selected CT ratio. Therefore, it can be shown that

$$\mathbf{I}_s = \left(\frac{N_p}{N_s}\right)\mathbf{I}_p \tag{7.33}$$

$$\mathbf{V}_s = \left(\frac{N_s}{N_p}\right)\mathbf{V}_p \tag{7.34}$$

$$\mathbf{I}_L = \mathbf{I}_s - \mathbf{I}_e \tag{7.35}$$

$$\mathbf{V}_s = \mathbf{I}_L (\mathbf{Z}_s + \mathbf{Z}_B) \tag{7.36}$$

By substituting Equation 7.33 into Equation 7.35 and in turn substituting the resulting equation into Equation 7.36,

$$\mathbf{V}_s = (\mathbf{Z}_s + \mathbf{Z}_B)\left[\left(\frac{N_p}{N_s}\right)\mathbf{I}_p - \mathbf{I}_e\right] \tag{7.37}$$

The relationship between $\mathbf{V}_s$ and $\mathbf{I}_e$ is not a linear one owing to the saturation of the CT core. Therefore, CTs have ratio errors. The deviation of $\mathbf{I}_L$ from $\mathbf{I}_e$ is called the *CT error* and can be expressed as a percentage,

$$\text{CT error} = \frac{\mathbf{I}_s - \mathbf{I}_L}{\mathbf{I}_s} \times 100 \tag{7.38}$$

or using the excitation current $\mathbf{I}_e$,

$$\text{CT error} = \frac{\mathbf{I}_e}{\mathbf{I}_s} \times 100 \tag{7.39}$$

The CT error can be very high if the impedance burden is too large. However, the CT error can be reduced to an acceptable level by choosing the CT properly with respect to the burden. The acceptable CT error* can be determined by calculation for some types and by testing for other types. Therefore, the performance of a CT can be estimated by (1) the formula method, (2) the CT

---

* Note that, the CT error consists of separate ratio and phase angle errors, and that the latter is important in revenue-metering installations.

saturation curve method, and (3) the ANSI transformer relaying accuracy class method. Here, only the first two methods will be discussed.

7.8.1.1 Method 1. The Formula Method

This method is based on the known values of the maximum core flux density. For example, it is known that the silicon steels used in transformers saturate from 1.2013 to 1.9375 T (with an average value of 1.55 T), with newer units having the greater values. In this method, the secondary rms voltage V_s is first determined in volts from Equation 7.36, and it is substituted into the following equation to determine the maximum care flux density:

$$\mathbf{V}_s = 4.44 f \times A \times N_s \times \beta_m \text{ V} \tag{7.40}$$

where
- f = frequency in hertz
- A = cross-sectional area of iron core in square meters
- N_s = number of turns in secondary winding
- β_m = maximum core flux density in teslas

If the calculated value of the maximum core flux density β_m does not exceed its known value, it can be assumed that the error in the secondary current (and therefore the CT error) due to saturation is not significant.

EXAMPLE 7.1

Assume that a CT has a rated current ratio of 500: 5 A, $\mathbf{Z}_s = 0.242$ Ω, and $\mathbf{Z}_B = 0.351$ Ω. The core area is 3 in.² (or 1.9356×10^{-3} m²). The CT must operate at a maximum primary current of 10,000 A. If the core is built in silicon steel, determine whether the CT will saturate.

Solution

The turns ratio is 500/5 = 100. In the event that the CT does not saturate, the maximum secondary current is

$$\mathbf{I}_L = \frac{10,000}{100} = 100 \text{ A}$$

From Equation 7.36, the secondary rms voltage is

$$\begin{aligned}\mathbf{V}_s &= \mathbf{I}_L(\mathbf{Z}_s + \mathbf{Z}_B) \\ &= 100(0.242 + 0.351) \\ &= 59.2991 \text{ V}\end{aligned}$$

From Equation 7.40,

$$59.2991 = (4.44 \times 60)(1.9356 \times 10^{-3})(100)\beta_m$$

Then,

$$\beta_m = 1.15 \text{ T}$$

Since 1.15 T is less than even the lower limit of 1.2013 T, the CT will not saturate.

7.8.1.2 Method 2. The Saturation Curve Method

In this method, the excitation current I_e for a given value of secondary voltage V_s is determined from a saturation curve. Figure 7.25 shows various saturation curves for a multiratio bushing-type CT. Such curves are developed by applying rms voltage to the CT secondary while measuring the rms currents, with the primary-side open-circuited.

EXAMPLE 7.2

Consider Example 7.1 and determine the CT error using the saturation curve method.

Solution

In Example 7.1, the secondary rms voltage V_s has been determined to be 59.2991 V. Therefore, the corresponding excitation current I_e can be found from Figure 7.25 as 0.1 A. Since $I_s \cong I_L = 100$ A, the CT error is

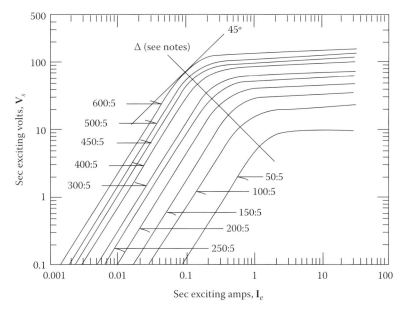

Current ratio	Turn ratio	Secondary resistance Ω^1
50 : 5	10 : 1	0.061
100 : 5	20 : 1	0.082
150 : 5	30 : 1	0.104
200 : 5	40 : 1	0.125
250 : 5	50 : 1	0.146
300 : 5	60 : 1	0.168
400 : 5	80 : 1	0.211
450 : 5	90 : 1	0.230
500 : 5	100 : 1	0.242
600 : 5	120 : 1	0.296

Notes:
1. Above the line: voltage for given exciting current will not be less than 95% of curve value.
2. Below line: exciting current for given voltage will not exceed curve value by more than 25%.

FIGURE 7.25 Saturation curves for multiratio bushing-type CT with an ANSI accuracy classification of C100. (From Westinghouse Electric Corporation, *Applied Protective Relaying*. Relay-Instrument Division, WEC, Newark, NJ, 1976.)

$$\text{CT error} = \frac{I_e}{I_s} \times 100$$
$$= \frac{0.1}{100} \times 100$$
$$= 0.1\%$$

7.8.2 Voltage Transformers

Voltage transformers are connected across the points at which the voltage is to be measured. Therefore, they are similar to small power transformers, differing only in details of design that control ratio accuracy over the specified range of output. In general, a very high accuracy is not demanded from VTs in the protection system applications. Therefore, a VT can be represented as an ideal transformer so that

$$\mathbf{V}_s = \frac{N_s}{N_p} \mathbf{V}_p$$

The rated output of a VT seldom exceeds a few hundred volt-amperes. The secondary voltage of a VT is usually 120 V line to line or 69.3 V line to neutral.

For certain low-voltage applications, the secondary voltage of a VT may be 115 V line to line or 66.4 V line to neutral. Some of the standard voltage ratios are 1:1, 2:1, 2.5:1, 4:1, 5:1, 20:1, 40:1, 60:1, 100:1, 200:1, 300:1, 400:1, 600:1, 800:1, 1000:1, 2000:1, 3000:1, and 4500:1.

A VT must be insulated to withstand overvoltages, including impulse voltages, of a level equal to the withstand value of the switchgear with which it is associated and the high-voltage system. VTs designed for medium-voltage circuits have dry-type insulation, but for high-voltage and EHV systems oil-immersed units are general.

At higher voltages, their cost may become the prohibiting factor. Therefore, for voltage at high-voltage and EHV levels, capacitor voltage transformers (CVTs) are used instead. Figure 7.26 shows

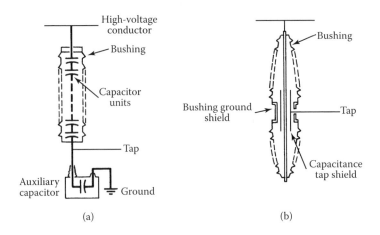

FIGURE 7.26 Capacitor voltage transformers: (a) coupling-capacitor voltage divider; (b) capacitance-bushing voltage divider. (Mason, C. R.: *The Art and Science of Protective Relaying*. 1956. Copyright Wiley-VCH Verlag GmbH & Co. KGaA. Reproduced with permission.)

System Protection

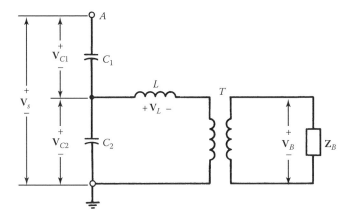

FIGURE 7.27 Equivalent circuit of capacitor voltage transformer.

two different types of CVTs. The coupling-capacitor voltage divider shown in Figure 7.26a is made up of a series stack of capacitors with the secondary tap taken from the last unit, which is called the *auxiliary capacitor* [6]. The *capacitance-bushing voltage divider* shown in Figure 7.26b uses capacitance coupling of a specially constructed bushing of a CB or power transformer. As shown in Figure 7.26b, a particular level of the bushing is tapped for a secondary voltage. Figure 7.27 shows the equivalent circuit of a CVT. The inductance L may be a separate unit or it may be included in the form of a leakage reactance in the transformer T. The tuning inductance L is adjusted so that the burden voltage $\mathbf{V}_B$ is in phase with the line-to-neutral system voltage $\mathbf{V}_s$ between the phase A conductor and the ground. Therefore,

$$X_L = \frac{X_{C1} \times X_{C2}}{X_{C1} + X_{C2}} \tag{7.41}$$

in series resonance. Values of C_1 and C_2 are so chosen that a major part of the voltage drops across C_1 (i.e., $C_1 \ll C_2$ so that $\omega L \cong 1/\omega C_2$). At normal frequency, when C and L are in resonance and therefore cancel each other, the circuit acts in the same way as in a conventional VT. At the other frequencies, however, a reactive component exists that modifies the errors.

In general, however, VTs are far more accurate than the CTs. The VTs are usually connected in wye–wye, delta–delta, or delta–wye as required by the specific application. The open-delta connection can be used in some applications. Obviously, it is less expensive since it requires only two VTs.

Figure 7.28 shows various VT connections. Figure 7.29a shows a typical application of VTs on the low-voltage side of a wye–delta-connected power transformer in order to supply voltage to distance relays. Figure 7.29b shows a VT connection used for distance and ground relays. Note that, the wye-broken delta auxiliary VTs shown in Figure 7.29b are used to provide zero-sequence polarizing voltage for directional ground relays.

7.9 R–X DIAGRAM

The characteristics of a relay (e.g., a distance relay) and a system can be graphically represented in terms of only two variables (i.e., R and X or $|\mathbf{Z}|$ and θ), rather than the three variables (i.e., $\mathbf{V}$, $\mathbf{I}$, and θ) [7]. The R–X diagram is also called the *impedance diagram* (or the *Z-plane* or simply the *complex plane*).

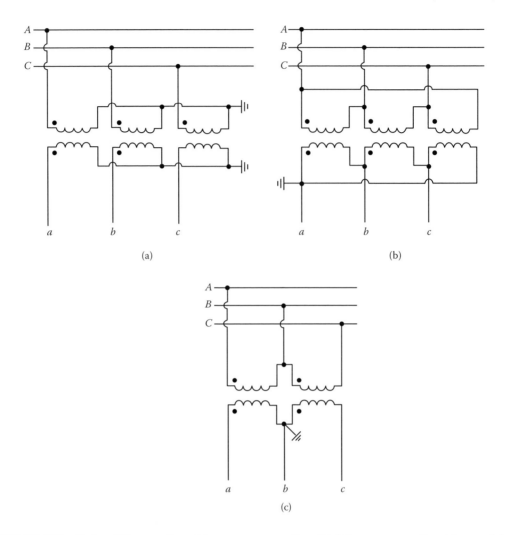

FIGURE 7.28 Various VT connections: (a) wye–wye connection; (b) delta–wye connection; (c) open–delta connection.

The complex variable **Z** is determined by dividing the rms magnitude of voltage by the rms magnitude of current. The resulting **Z** can be expressed in rectangular or polar form as

$$\mathbf{Z} = R + jX = |\mathbf{Z}|\, e^{j\theta}$$

where the values of R and X represent the coordinates of a point on the R–X diagram which, as shown in Figure 7.30, was originally given by a combination of the three variables, **V**, **I**, and θ. The values of R and X can be positive or negative, whereas **Z** must always be positive. Any negative values of **Z** must be neglected since they have no significance.

In addition to the plot of the operating characteristic of a given relay, the system conditions affecting the operation of this relay can be superimposed on the same R–X diagram so that the response of the relay can be determined. To achieve this relay operation, the system characteristic has to be within the operating region of the relay characteristic.

For example, if the relay involved is a distance relay, the signs of R and X are positive, according to the convention, when real power and lagging reactive power flow are in the tripping

System Protection

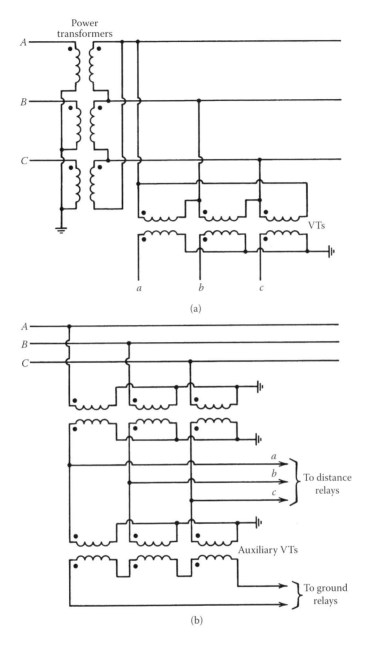

FIGURE 7.29 Typical VT applications: (a) used at low-voltage side of wye–delta connected power transformer for use with distance relays; (b) used for distance and ground relays.

direction of the relay under balanced three-phase conditions, as shown in Figure 7.31. Table 7.2 gives the conventional signs of R and X depending on the power flow direction in the example system. Note that, the superimposed system and relay characteristics have to be in terms of the same phase quantities and the same scale. The quantities are usually given in per units even through they can also be in actual ohms. When ohms are used, both system and relay characteristics have to be in terms of the same phase quantities and the same scale. The quantities are usually given in per units even though they can also be in actual ohms. When ohms are used,

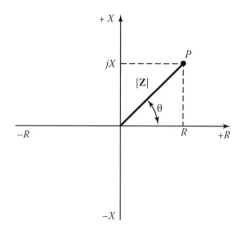

FIGURE 7.30 R–X diagram.

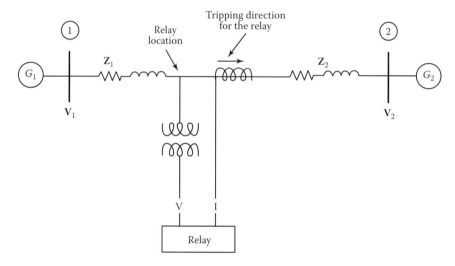

FIGURE 7.31 Example system for illustrating convention for relating relay and system characteristics on R–X diagram.

TABLE 7.2
Conventional Signs of *R* and *X*

Condition	Sign of *R*	Sign of *X*
Real power from bus 1 toward bus 2	+	
Real power from bus 2 toward bus 1	−	
Lagging reactive power from bus 1 toward bus 2		+
Lagging reactive power from bus 2 toward bus 1		−
Leading reactive power from bus 1 toward bus 2		−
Leading reactive power from bus 2 toward bus 1		+

System Protection

both system and relay characteristics have to be on either a primary or a secondary basis using the following relationship:

$$\text{Secondary ohms} = (\text{primary ohms})\left(\frac{\text{CT ratio}}{\text{VT ratio}}\right) \quad (7.42)$$

In the event that a short transmission line has a series impedance of $\mathbf{Z}_L$ and negligible shunt admittance, the sending-end voltage $\mathbf{V}_S$ can be expressed in terms of the line voltage drop and the receiving-end voltage $\mathbf{V}_R$ as

$$\mathbf{V}_S = \mathbf{IZ}_L + \mathbf{V}_R$$

Therefore, the sending-end impedance $\mathbf{Z}_S$ can be expressed in terms of the receiving-end impedance $\mathbf{Z}_R$ and the line impedance $\mathbf{Z}_L$ as

$$\mathbf{Z}_S = \mathbf{Z}_L + \mathbf{Z}_R \quad (7.43)$$

where

$$\mathbf{Z}_S = \frac{\mathbf{V}_S}{\mathbf{I}}$$

$$\mathbf{Z}_R = \frac{\mathbf{V}_R}{\mathbf{I}}$$

The receiving-end load impedance can also be expressed as

$$\mathbf{Z}_R = R_R + jX_R$$

where

$$R_R = \frac{|\mathbf{V}|^2 \times P}{P^2 + Q^2} \quad (7.44)$$

$$X_R = \frac{|\mathbf{V}|^2 \times Q}{P^2 + Q^2} \quad (7.45)$$

EXAMPLE 7.3

Assume that a short transmission line has the receiving-end load S_R and voltage magnitude $|\mathbf{V}_R|$ of 2.5 + j0.9 and 1.0 in per units, respectively. If the line series impedance $|\mathbf{Z}_L|$ is 0.1 + j0.25 per unit, determine the following:

(a) Receiving-end impedance $\mathbf{Z}_R$
(b) Sending-end impedance $\mathbf{Z}_S$
(c) Impedance diagram shown in R–X diagram
(d) Power angle δ

Solution

(a) From Equations 7.44 and 7.45,

$$R_R = \frac{|\mathbf{V}|^2 \times P}{P^2 + Q^2}$$

$$= \frac{|1.0|^2 \times (2.5)}{2.5^2 + 0.9^2}$$

$$= 0.3541 \, \text{pu}$$

and

$$X_R = \frac{|\mathbf{V}|^2 \times Q}{P^2 + Q^2}$$

$$= \frac{|1.0|^2 \times (0.9)}{2.5^2 + 0.9^2}$$

$$= 0.1275 \, \text{pu}$$

Therefore,

$$\mathbf{Z}_R = 0.3451 + j0.1275$$
$$= 0.3764 \angle 19.8° \, \text{pu}$$

(b) From Equation 7.43,

$$\mathbf{Z}_S = (0.1 + j0.25)(0.3451 + j0.1275)$$
$$= 0.4541 + j0.3775$$
$$= 0.5905 \angle 39.7° \, \text{pu}$$

(c) The impedance diagram is shown in Figure 7.32.

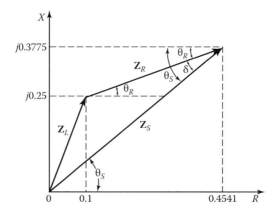

FIGURE 7.32 The R–X diagram for Example 7.2.

(d) The power angle is

$$\delta = \delta_S - \theta_R$$
$$= 39.7° - 19.8°$$
$$= 19.9°$$

7.10 RELAYS AS COMPARATORS

In general, the task of a relay is to detect the change between normal and faulted conditions and send a signal when a fault occurs. The relay achieves this task by measuring different functions of applied quantities and by making a comparison between two or more different inputs, or simply comparing one measured quantity with a standard (or predetermined) value. Therefore, the relay intelligence itself is a "comparator."

There are various types of comparators. The two most common types of comparators involve the use of amplitude and phase comparison. The amplitude and phase relation depends on the system conditions and for a predetermined value of this relation, indicative of a particular type and location of fault, the relay operates. With the exception of certain relay types (e.g., overcurrent relays), where only one electrical quantity overcomes a mechanical quantity such as the restraint from a spring, it is usual to compare two electrical quantities.

7.11 DUALITY BETWEEN PHASE AND AMPLITUDE COMPARATORS

Consider two phasor quantities **A** and **B** such that*

$$\mathbf{A} = |\mathbf{A}|e^{j\theta} = \mathbf{A}_p + j\mathbf{A}_q \tag{7.46}$$

$$\mathbf{B} = |\mathbf{B}|e^{j\Phi} = \mathbf{B}_p + j\mathbf{B}_q \tag{7.47}$$

and assume that they are the input signals of an amplitude comparator and the relay operates when

$$|\mathbf{A}| > |\mathbf{B}|$$

If the input signals are altered to $|\mathbf{A} + \mathbf{B}|$ and $|\mathbf{A} - \mathbf{B}|$ so that the relay operates when

$$|\mathbf{A} + \mathbf{B}| > |\mathbf{A} - \mathbf{B}|$$

then it acts as a phase comparator, as shown in Figure 7.33. This is due to the fact that both **A** and **B** must have the same polarity in order to satisfy the equation and to operate the relay.

Similarly, a phase comparator with inputs **A** and **B** operates when **A** and **B** have the same directional sense or polarity. Therefore, in the event that these inputs are altered to $|\mathbf{A} + \mathbf{B}|$ and $|\mathbf{A} - \mathbf{B}|$, as shown in Figure 7.34, then the relay acts as an amplitude comparator when $|\mathbf{A} + \mathbf{B}|$ and $|\mathbf{A} - \mathbf{B}|$ have the same directional sense, that is, $|\mathbf{A}| > |\mathbf{B}|$. Thus, in the event that the input quantities are altered to the sum and difference of the original two input quantities, an amplitude comparator can be used as a phase comparator or vice versa.

---

* Note that, these are not the **A** and **B** of the **ABCD** transmission line constants.

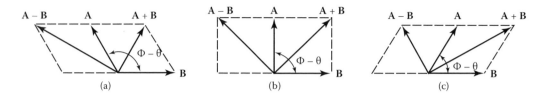

FIGURE 7.33 Phase comparison using an amplitude comparator: (a) when $|A + B| < |A - B|$, $\Phi - \theta > 90°$; (b) when $|A + B| = |A - B|$, $\Phi - \theta = 90°$; (c) when $|A + B| > |A - B|$, $\Phi - \theta > 90°$.

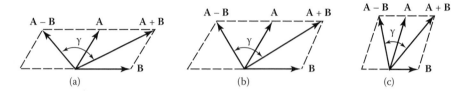

FIGURE 7.34 Amplitude comparison using phase comparator: (a) when $|A| < |B|$, $\delta > 90°$; (b) when $|A| < |B|$, $\delta = 90°$; (c) when $|A| > |B|$, $\delta < 90°$.

7.12 COMPLEX PLANES

Consider the two phasor quantities given by

$$\mathbf{A} = |\mathbf{A}|e^{j\theta} = \mathbf{A}_p + j\mathbf{A}_q \tag{7.46}$$

$$\mathbf{B} = |\mathbf{B}|e^{j\Phi} = \mathbf{B}_p + j\mathbf{B}_q \tag{7.47}$$

from which

$$\frac{\mathbf{A}}{\mathbf{B}} = \left|\frac{\mathbf{A}}{\mathbf{B}}\right|\cos i(\theta - \Phi) + j\left|\frac{\mathbf{A}}{\mathbf{B}}\right|\sin(\theta - \Phi) \tag{7.48}$$

If **B** is taken as the reference phasor, then

$$\frac{\mathbf{A}}{\mathbf{B}} = \left|\frac{\mathbf{A}}{\mathbf{B}}\right|\cos\theta + j\left|\frac{\mathbf{A}}{\mathbf{B}}\right|\sin\theta \tag{7.49}$$

or in terms of the real and imaginary (or quadrature) components

$$\frac{\mathbf{A}}{\mathbf{B}} = \left|\frac{\mathbf{A}}{\mathbf{B}}\right|_p + j\left|\frac{\mathbf{A}}{\mathbf{B}}\right|_q \tag{7.50}$$

If phasors **A** and **B** represent voltage and current, respectively, then the real and imaginary components of **A/B** symbolize resistance R and reactance X. Therefore, as already discussed in Section 7.9, it can be plotted in the **Z**-plane. Thus, the resultant diagram is called the *impedance diagram*. Similarly, if phasor **B/A** is plotted, its real and imaginary components represent conductance G and susceptance B. Thus, phasor **B/A** can be plotted in the *Y*-plane. Therefore, the resultant diagram is called the *admittance diagram*.

On the other hand, if phasors **A** and **B** both represent currents or voltages, then the real and imaginary components of **A/B** and **B/A** symbolize quantities that are defined by Warrington [8,9] as the "alpha" and "beta" quantities, respectively. Therefore,

$$\alpha = \frac{\mathbf{A}}{\mathbf{B}} = \left|\frac{\mathbf{A}}{\mathbf{B}}\right| e^{j(\theta-\Phi)}$$
$$= \left|\frac{\mathbf{A}}{\mathbf{B}}\right|_p + j\left|\frac{\mathbf{A}}{\mathbf{B}}\right|_q \quad (7.51)$$
$$= \alpha_p + j\alpha_q$$

Thus, such phasor can be plotted in the α-plane and the resultant diagram is called the *alpha diagram*. Similarly,

$$\beta = \frac{\mathbf{B}}{\mathbf{A}} = \left|\frac{\mathbf{B}}{\mathbf{A}}\right| e^{j(\Phi-\theta)}$$
$$= \left|\frac{\mathbf{B}}{\mathbf{A}}\right|_p + j\left|\frac{\mathbf{B}}{\mathbf{A}}\right|_q \quad (7.52)$$
$$= \beta_p + j\beta_q$$

Therefore, such phasor can be plotted in the β-plane, and the resultant diagram is called the *beta diagram*.

In general, the operating characteristic of a relay is plotted on a polar graph whose ordinates are the real and imaginary components of *A/B* or *B/A*, where **A** and **B** are the two quantities compared by the comparator.

The equations representing such operating characteristics are of the second-order type and therefore represent circles, sectors of circles, or straight lines.

Thus, the operating characteristics represent the threshold (or boundary) of operation where the comparator has zero input; therefore, the output is *positive* (*tripping*) on one side of the characteristic and *negative* (*blocking*) on the other side.

In the event that the relay characteristic is a circle passing through the origin when plotted on the α-plane, as shown in Figure 7.35a, it becomes a straight line outside the origin when plotted on

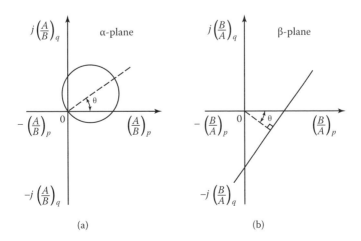

FIGURE 7.35 Typical relay-operating characteristic: (a) plotted on α-plane, (b) plotted on β-plane.

the β-plane, as shown in Figure 7.35b. The opposite of this is also true. Furthermore, if the circular characteristics are not going through the origin, they are orthogonal in the two planes.

7.13 GENERAL EQUATION OF COMPARATORS

To develop the general equation of comparators, assume that the logic configuration of a relay is made up of two weighted summers and a comparator, as shown in Figure 7.36a. Assume also, that $\mathbf{S}_1$ and $\mathbf{S}_2$ are the two input signals such that when the phase relationship or amplitude relationship complies with predetermined threshold conditions, tripping is initiated. Let the two input signals be

$$\mathbf{S}_1 = k_1\mathbf{A} + k_2\mathbf{B} \tag{7.53}$$

$$\mathbf{S}_2 = k_3\mathbf{A} + k_4\mathbf{B} \tag{7.54}$$

where k_1, k_2, k_3, and k_4 are design constants. Alternatively,

$$\mathbf{S}_1 = (k_1|\mathbf{A}|\cos\theta + k_2|\mathbf{B}|\cos\Phi) + j(k_1|\mathbf{A}|\sin\theta + k_2|B|\sin\Phi) \tag{7.55}$$

$$\mathbf{S}_2 = (k_3|\mathbf{A}|\cos\theta + k_4|\mathbf{B}|\cos\Phi) + j(k_3|\mathbf{A}|\sin\theta + k_4|B|\sin\Phi) \tag{7.56}$$

Figure 7.36b shows the phasor diagram. Both $\mathbf{S}_1$ and $\mathbf{S}_2$ are input to the comparator, which produces a trip (operate) signal whenever $|\mathbf{S}_1| > |\mathbf{S}_2|$ in an amplitude comparison mode.

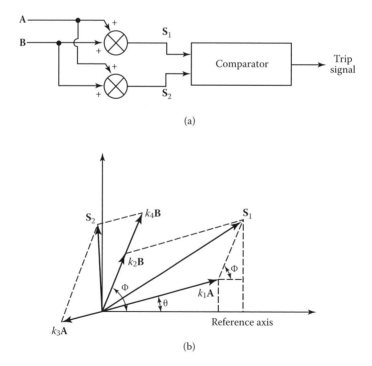

FIGURE 7.36 General comparator representation: (a) block diagram; (b) phasor diagram.

System Protection

7.14 AMPLITUDE COMPARATOR

The trip signal is produced for an amplitude comparator when $|S_2| > |S_1|$. Therefore, their moduli will be equal at the threshold of operation, that is, $|S_1| = |S_2|$ for any phase angle between them (the locus of which represents the relay-tripping characteristic). Thus, at the threshold of operation

$$(k_1 |A| \cos \theta + k_2 |B| \cos \Phi)^2 + (k_1 |A| \sin \theta + k_2 |B| \sin \Phi)^2 = \\ (k_3 |A| \cos \theta + k_4 |B| \cos \Phi)^2 + (k_3 |A| \sin \theta + k_4 |B| \sin \Phi)^2 \quad (7.57)$$

Rearranging the terms,

$$\left(k_1^2 - k_3^2\right)|A|^2 + 2(k_1 k_2 - k_3 k_4)|A||B|\cos(\Phi - \theta) + \left(k_2^2 - k_4^2\right)|B|^2 = 0 \quad (7.58)$$

Dividing by $\left(k_2^2 - k_4^2\right)|A|^2$,

$$\left|\frac{B}{A}\right|^2 + 2\left|\frac{B}{A}\right|\left(\frac{k_1 k_2 - k_3 k_4}{k_2^2 - k_4^2}\right)\cos(\Phi - \theta) + \left(\frac{k_1^2 - k_3^2}{k_2^2 - k_4^2}\right) = 0 \quad (7.59)$$

which can be reexpressed as

$$|\boldsymbol{\beta}|^2 + 2|\boldsymbol{\beta}||\boldsymbol{\beta}_0|\cos(\Phi - \theta) + \boldsymbol{\beta}_0^2 = r^2 \quad (7.60)$$

where

$$\boldsymbol{\beta} \triangleq \left|\frac{B}{A}\right| e^{j(\Phi - \theta)} \quad (7.61)$$

$$\boldsymbol{\beta}_0 = \frac{k_1 k_2 - k_3 k_4}{k_2^2 - k_4^2} = |\boldsymbol{\beta}_0| e^{j\theta} \quad (7.62)$$

$$r = \frac{k_1 k_4 - k_2 k_3}{k_2^2 - k_4^2} \quad (7.63)$$

Equation 7.60 represents the equation of a circle of radius r, and center located at $\boldsymbol{\beta}_0$ on the β-plane, as shown in Figure 7.37, having $|\boldsymbol{\beta}_0| \cos \theta$ and $j|\boldsymbol{\beta}_0| \sin \theta$ as coordinates represented by β_p and $j\beta_q$.

Similarly, it is possible to express Equation 7.58 in terms of the alpha quantities so that

$$|\boldsymbol{\alpha}|^2 + 2|\boldsymbol{\alpha}||\boldsymbol{\alpha}_0| \cos (\Phi - \theta) + |\boldsymbol{\alpha}_0|^2 = r^2 \quad (7.64)$$

where

$$\boldsymbol{\alpha} \triangleq \left|\frac{A}{B}\right| e^{j(\theta - \Phi)} \quad (7.65)$$

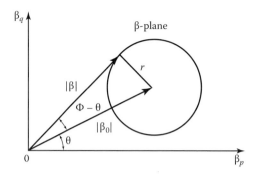

FIGURE 7.37 Threshold characteristics of comparator plotted in β-plane.

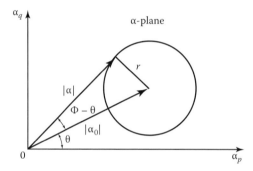

FIGURE 7.38 Threshold characteristics of comparator plotted in α-plane.

$$\boldsymbol{\alpha}_0 = \frac{k_1 k_2 - k_3 k_4}{k_1^2 - k_3^2} \tag{7.66}$$

$$r = \frac{k_1 k_4 - k_2 k_3}{k_1^2 - k_3^2} \tag{7.67}$$

Equation 7.64 represents the equation of a circle of radius r and center located at $\boldsymbol{\alpha}_0$ on the α-plane, as shown in Figure 7.38, having $|\boldsymbol{\alpha}_0| \cos \theta$ and $j|\boldsymbol{\alpha}_0| \sin \theta$ as coordinates represented by α_p and $j\alpha_q$.

7.15 PHASE COMPARATOR

The trip signal is produced for a phase comparator when the product of the two signals $\mathbf{S}_1$ and $\mathbf{S}_2$ is positive. The signals $\mathbf{S}_1$ and $\mathbf{S}_2$ are given by Equations 7.55 and 7.56, respectively, as before. Therefore, they can be expressed as

$$\mathbf{S}_1 = |S_1| e^{j\theta_1} \tag{7.68}$$

$$\mathbf{S}_2 = |S_2| e^{j\theta_2} \tag{7.69}$$

System Protection

The product of $\mathbf{S}_1$ and $\mathbf{S}_2$ is maximum when the two phasors are in phase. Therefore, at the threshold of operation

$$\theta_1 - \theta_2 = \pm 90° \tag{7.70}$$

by taking the tangent of both sides

$$\tan(\theta_1 - \theta_2) = \pm\infty \tag{7.71}$$

or

$$\frac{\tan\theta_1 - \tan\theta_2}{1 + \tan\theta_1 \tan\theta_2} = \pm\infty \tag{7.72}$$

which is true when

$$1 + \tan\theta_1 \tan\theta_2 = 0 \tag{7.73}$$

or

$$\tan\theta_1 = \frac{1}{\tan\theta_2} \tag{7.74}$$

The quantities $\tan\theta_1$ and $\tan\theta_2$ can be determined from Equations 7.55 and 7.56, respectively, as

$$\tan\theta_1 = \frac{k_1|\mathbf{A}|\sin\theta + k_2|\mathbf{B}|\sin\Phi}{k_1|\mathbf{A}|\cos\theta + k_2|\mathbf{B}|\cos\Phi} \tag{7.75}$$

$$\tan\theta_2 = \frac{k_3|\mathbf{A}|\sin\theta + k_4|\mathbf{B}|\sin\Phi}{k_3|\mathbf{A}|\cos\theta + k_4|\mathbf{B}|\cos\Phi} \tag{7.76}$$

Substituting Equations 7.75 and 7.76 into Equation 7.74,

$$\frac{k_1|\mathbf{A}|\sin\theta + k_2|\mathbf{B}|\sin\Phi}{k_1|\mathbf{A}|\cos\theta + k_2|\mathbf{B}|\cos\Phi} = \frac{-k_1|\mathbf{A}|\cos\theta + k_4|\mathbf{B}|\cos\Phi}{k_3|\mathbf{A}|\sin\theta + k_4|\mathbf{B}|\sin\Phi} \tag{7.77}$$

or

$$k_1 k_3 |\mathbf{A}|^2 + k_2 k_4 |\mathbf{B}|^2 + (k_1 k_4 + k_2 k_3)|\mathbf{A}||\mathbf{B}|\cos(\Phi - \theta) = 0 \tag{7.78}$$

Dividing by $k_2 k_4 |\mathbf{A}|^2$

$$\left|\frac{\mathbf{B}}{\mathbf{A}}\right|^2 + \left|\frac{\mathbf{B}}{\mathbf{A}}\right|\left(\frac{k_1 k_4 + k_2 k_3}{k_2 k_4}\right)\cos(\Phi - \theta) + \frac{k_1 k_3}{k_2 k_4} = 0 \tag{7.79}$$

which can be reexpressed as

$$|\boldsymbol{\beta}|^2 - 2|\boldsymbol{\beta}||\boldsymbol{\beta}_0|\cos(\Phi-\theta) + \boldsymbol{\beta}_0^2 = r^2 \tag{7.80}$$

where

$$\boldsymbol{\beta}_0 = -\frac{k_1 k_4 + k_2 k_3}{2 k_2 k_4} \tag{7.81}$$

$$r = \frac{k_1 k_4 + k_2 k_3}{2 k_2 k_4} \tag{7.82}$$

Equation 7.80 represents the equation of a circle of radius r and center located at $\boldsymbol{\beta}_0$ on the β-plane, having $|\boldsymbol{\beta}_0|\cos\theta$ and $j|\boldsymbol{\beta}_0|\sin\theta$ as coordinates represented by β_p and $j|\beta_q|$.

Similarly, it is possible to express the equation in terms of the alpha quantities so that

$$|\boldsymbol{\alpha}|^2 - 2|\boldsymbol{\alpha}||\boldsymbol{\alpha}_0|\cos(\Phi-\theta) + |\boldsymbol{\alpha}_0|^2 = r^2 \tag{7.83}$$

where

$$\boldsymbol{\alpha}_0 = -\frac{k_1 k_4 + k_2 k_3}{2 k_1 k_3} \tag{7.84}$$

$$r = \frac{k_1 k_4 + k_2 k_3}{2 k_1 k_4} \tag{7.85}$$

Equation 7.83 represents the equation of a circle of radius r and center located at $\boldsymbol{\alpha}_0$ on the α-plane having $|\boldsymbol{\alpha}_0|\cos\theta$ and $j|\boldsymbol{\alpha}_0|\sin\theta$ as coordinates represented by α_p and $j\alpha_q$.

Table 7.3 gives the values of r and $\boldsymbol{\beta}_0$ for amplitude and phase comparators for the β-plane. Similarly, Table 7.4 gives the values of r and $\boldsymbol{\alpha}_0$ for amplitude and phase comparators for the α-plane. The electromagnetic relays that operate inherently as amplitude comparators include balanced beam, hinged armature, plunger, and induction disk relays with shaded pole-driving magnets.

TABLE 7.3
Values of r and β_0 for Amplitude and Phase Comparators on β-Plane

Quantity	Amplitude Comparator	Phase Comparator
r	$\dfrac{k_1 k_4 - k_2 k_3}{k_2^2 - k_4^2}$	$\dfrac{k_1 k_4 - k_2 k_3}{2 k_2 k_4}$
β_0	$\dfrac{k_1 k_2 - k_3 k_4}{k_2^2 k_4^2}$	$-\dfrac{k_1 k_4 + k_2 k_3}{2 k_2 k_4}$

System Protection

TABLE 7.4
Values of r and α_0 for Amplitude and Phase Comparators on α-Plane

Quantity	Amplitude Comparator	Phase Comparators
r	$\dfrac{k_1 k_4 - k_2 k_3}{k_1^2 - k_3^2}$	$\dfrac{k_1 k_4 - k_2 k_3}{2 k_1 k_3}$
α_0	$\dfrac{k_1 k_2 - k_3 k_4}{k_1^2 - k_3^2}$	$-\dfrac{k_1 k_4 + k_2 k_3}{2 k_1 k_3}$

On the other hand, the electromagnetic relays that operate inherently as phase comparators include induction cup, induction dynamometer, and induction magnets. The solid-state relays are capable of producing any desired relay characteristics using phase or amplitude comparison.

EXAMPLE 7.4

Assume that an amplitude comparator has a threshold characteristic whose center and radius are given by Equations 7.62 and 7.63. Determine the properties of this characteristic for the following values of the constants.

(a) When $k_1 = k_3$
(b) When $k_1 \neq k_3$
(c) When $k_1 = k_3$ and $k_2 = -k_4$

Solution

(a) From Equations 7.63 and 7.62, the radius of the circle and the location of its center can be found as

$$r = -\frac{k_1 k_4 + k_2 k_3}{k_2^2 - k_4^2}$$
$$= \frac{-k_3(k_2 - k_4)}{(k_2 + k_4)(k_2 - k_4)}$$
$$= \frac{-k_3}{(k_2 + k_4)}$$

and

$$c = -\frac{k_1 k_2 + k_3 k_4}{k_2^2 - k_4^2}$$
$$= \frac{-k_3(k_2 - k_4)}{(k_2 + k_4)(k_2 - k_4)}$$
$$= \frac{-k_3}{(k_2 + k_4)}$$

Thus, the circle passes through the origin in the β-plane.

(b) However, when $k_1 \neq k_3$, the circle is obviously offset.
(c) When $k_1 = k_3$ and $k_2 = -k_4$,

$$r = -\frac{k_1 k_4 + k_2 k_3}{k_2^2 - k_4^2}$$

$$= \frac{k_3(k_4 - k_2)}{(k_2 + k_4)(k_2 - k_4)}$$

$$= \frac{-k_3}{0}$$

$$= \infty$$

and

$$c = -\frac{k_1 k_2 + k_3 k_4}{k_2^2 - k_4^2}$$

$$= \frac{k_3}{0}$$

$$= \infty$$

Therefore, the characteristic is a straight line passing through the origin, that is, a directional characteristic.

7.16 GENERAL EQUATION OF RELAYS

In most relays, at least one of the design constants k_1, k_2, k_3, and k_4 is zero and two of them are often equal. Hence, the practical case becomes relatively simple. For example, if there are no more than two quantities involved, it can be shown that the operation of most relays can be predicted by the use of the following general relay (torque) equations:

$$T = k_a|\mathbf{A}|^2 - k_b|\mathbf{B}|^2 + k_c|\mathbf{A}||\mathbf{B}|\cos(\Phi - \theta) - k_s \qquad (7.86)$$

where

k_a, k_b, k_c = scalar constants
k_s = adjustable spring constant representing mechanical restraining torque
$|\mathbf{A}|, |\mathbf{B}|$ = two electrical quantities being compared
Φ = phase angle between A and B
θ = relay characteristic angle (value of maximum torque in electromagnetic relay or maximum output in static relay)

At the threshold of operation under steady-state conditions,

$$k_a|\mathbf{A}|^2 - k_b|\mathbf{B}|^2 + k_c|\mathbf{A}||\mathbf{B}|\cos(\Phi - \theta) - k_s = 0 \qquad (7.87)$$

It represents all the circular and straight-line characteristics that can be obtained from any two-input relay. If the two quantities are current $\mathbf{I}$ and voltage $\mathbf{V}$ then

$$k_a|\mathbf{I}|^2 - k_b|\mathbf{V}|^2 + k_c|\mathbf{I}||\mathbf{V}|\cos(\Phi - \theta) - k_s = 0 \qquad (7.88)$$

Here, the current winding produces a torque $k_a|\mathbf{I}|^2$ and the potential winding a torque $k_b|\mathbf{V}|^2$, whereas the torque due to interaction of current and potential windings will be $VI\cos(\Phi - \theta)$.

System Protection

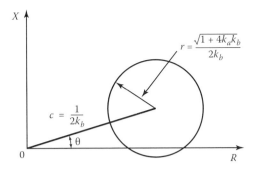

FIGURE 7.39 Characteristics of two-input relay.

In single-quantity relays, k_s is a constant and is used as a level indicator. Whereas in relays using two input quantities, its value is reduced to zero. Therefore, setting $k_s = 0$ and dividing Equation 7.88 by $k_b|\mathbf{I}|^2$,

$$\frac{k_a}{k_b} - \left|\frac{\mathbf{V}}{\mathbf{I}}\right|^2 + \left|\frac{\mathbf{V}}{\mathbf{I}}\right|\frac{k_c \cos(\Phi - \theta)}{k_b} = 0 \tag{7.89}$$

If $k_c = 1$,

$$\left|\frac{\mathbf{V}}{\mathbf{I}}\right|^2 - \left|\frac{\mathbf{V}}{\mathbf{I}}\right|\frac{\cos(\Phi - \theta)}{k_b} = \left|\frac{k_a}{k_b}\right|^2 \tag{7.90}$$

adding $|1/2k_b|^2$ on both sides of the equation,

$$\left|\frac{\mathbf{V}}{\mathbf{I}}\right|^2 - \left|\frac{\mathbf{V}}{\mathbf{I}}\right|\frac{\cos(\Phi - \theta)}{k_b} + \left|\frac{1}{2k_b}\right|^2 = \left|\frac{k_a}{k_b}\right|^2 + \left|\frac{1}{2k_b}\right|^2 \tag{7.91}$$

It represents a circle on the R–X complex plane (polar diagram) having $|\mathbf{V}/\mathbf{I}| \cos \Phi$ and $j|\mathbf{V}/\mathbf{I}| \sin \Phi$ as coordinates, as shown in Figure 7.39. The radius of the circle is

$$r = \frac{\sqrt{1 + 4k_a k_b}}{2k_b} \tag{7.92}$$

and its center is located at $1/2\, k_b$ from the origin at an angle θ from the reference axis.

7.17 DISTANCE RELAYS

A distance relay responds to input quantities as a function of the electrical circuit distance between the relay location and the point of faults. Since the impedance of a transmission line is proportional to its length, for distance measurement it is suitable to employ a distance relay that is capable of measuring the impedance of a line up to a given point.

It is designed to operate only for faults taking place between the relay location and the selected point. Thus, it discriminates the faults that may take place between different line sections by comparing the current and voltage of the power system to determine whether a fault exists within or outside its operating zone.

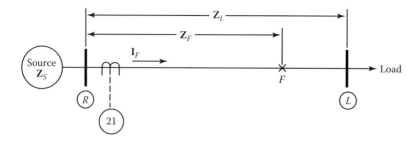

FIGURE 7.40 Representation of line section between two buses.

Consider the transmission line shown in Figure 7.40 and assume that a balanced-beam-type distance relay (indicated in the figure by its standard device number, 21) is located at bus R and receives a secondary current equal to the primary fault current and a secondary voltage equal to the product of the fault current and the impedance of the line up to the fault point F (i.e., $\mathbf{V}_F = \mathbf{I}_F\mathbf{Z}_F$). Since the relay is designed so that its operating torque is proportional to the current and its restraining torque is proportional to the voltage, then in conformity with the relative number of ampere turns applied to each coil, there is a definite ratio at which the two torques are equal (called the *balance point of the relay*).

Thus, any increase in the current coil ampere turns without an associated increase in the voltage coil ampere turns causes the relay to become unbalanced. Hence, below a given ratio of voltage to current, the operating torque is greater than the restraining torque and the relay closes its contacts. On the other hand, above a given ratio of voltage to current, the restraining torque is larger than the operating torque, the relay restrains, and the contacts stay open.

The ohmic setting of such relays can be adjusted by altering the relationship of the ampere turns of the operating coil to those of the restraint coil, making it possible to select a setting suitable for the length of the line to be protected. The locus of points where the operating and restraining torques are equal is the threshold characteristic.

It depends on the ratio of voltage to current and the phase angle between them, and it may be plotted on an R–X diagram. If the locus of power system impedances such as those of faults, power swings, and loads are plotted on the same diagram, the relay performance under such system disturbances can be observed.

Figure 7.41a shows a single-line diagram of a three-phase system. Assume that there is a fault at the end of the line so that it can be represented by a simple impedance loop to which a voltage V has

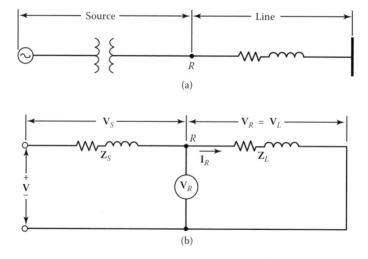

FIGURE 7.41 Power system arrangement to study relationship between source-to-line ratio and relay voltage.

System Protection

been applied, as shown in Figure 7.41b. The voltage V may be either a wye or a delta open-circuit voltage of the power system, depending on the fault type. The distance relay is located at the point R, and the current $\mathbf{I}_R$ and voltage and voltage $\mathbf{V}_R$ are applied to it. The impedances $\mathbf{Z}_S$ and $\mathbf{Z}_L$ are the source and line impedances, respectively, due to their positions with respect to the relay location. It is interesting to note that the source impedance $\mathbf{Z}_S$ is a measure of the fault level at the relaying point. If the fault involves ground, the $\mathbf{Z}_S$ is also affected by the method of system grounding behind at the relaying point. If the fault involves ground, the $\mathbf{Z}_S$ is also affected by the method of system grounding behind the relaying point.

The line impedance $\mathbf{Z}_L$ is a measure of the impedance of the protection section. Thus, the voltage $\mathbf{V}_R$ applied to the relay is equal to $\mathbf{I}_R \mathbf{Z}_L$ for a fault at the reach point. Thus,

$$\mathbf{V}_R = \mathbf{I}_R \mathbf{Z}_L \tag{7.93}$$

but

$$\mathbf{I}_R = \frac{\mathbf{V}}{\mathbf{Z}_S + \mathbf{Z}_L} \tag{7.94}$$

Hence,

$$\mathbf{V}_R = \left[\frac{\mathbf{Z}_L}{\mathbf{Z}_S + \mathbf{Z}_L}\right] \mathbf{V} \tag{7.95}$$

or

$$\mathbf{V}_R = \left[\frac{1}{(\mathbf{Z}_S/\mathbf{Z}_L) + 1}\right] \mathbf{V} \tag{7.96}$$

Figure 7.41b illustrates this relationship between the relay voltage $\mathbf{V}_R$ and the ratio $\mathbf{Z}_S/\mathbf{Z}_L$. It is applicable for all types of short circuits.

However, for phase faults, the voltage V is the line-to-line voltage and $\mathbf{Z}_S/\mathbf{Z}_L$ is the ratio of the positive-sequence source impedance to the positive-sequence line impedance. Therefore,

$$\mathbf{V}_R = \left[\frac{1}{(\mathbf{Z}_{S1}/\mathbf{Z}_{L1}) + 1}\right] \mathbf{V}_{(L-L)} \tag{7.97}$$

On the other hand, for ground faults, the voltage V is the line-to-neutral voltage and $\mathbf{Z}_S/\mathbf{Z}_L$ is a composite ratio taking into account the positive-, negative-, and zero-sequence impedances. Thus,

$$\mathbf{V}_R = \left[\frac{1}{(\mathbf{Z}_S/\mathbf{Z}_L) + 1}\right] \mathbf{V}_{(L-N)} \tag{7.98}$$

where

$$\mathbf{Z}_S = \mathbf{Z}_{S1} + \mathbf{Z}_{S2} + \mathbf{Z}_{S0} \tag{7.99}$$

$$\mathbf{Z}_L = \mathbf{Z}_{L1} + \mathbf{Z}_{L2} + \mathbf{Z}_{l0} \tag{7.100}$$

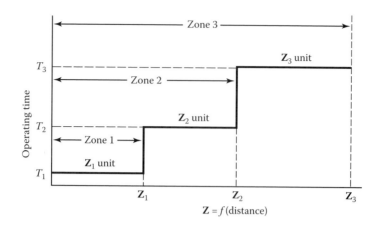

FIGURE 7.42 Operating time versus impedance for impedance-type distance relay.

It is interesting to note that, under normal circumstances, the relay measures a greater impedance than that of the line because it also measures the impedance of the load that is connected to the line. In the event that having a solid fault, the fault causes the load to be short-circuited, the relay measures only the impedance of the line.

On the other hand, in the event that the fault is not a solid fault (e.g., an arcing fault), its impedance, which is in parallel with that of the load and the lines to the load, is added to that of the faulted line section. Thus, the fault appears to be more far off than it actually is. Also, it shrinks the reach of the relay.

Hence, partly for this reason, and partly because the relay cannot be built with a hundred percent accuracy, a second distance-measuring unit is used to provide protection for the far end of the line near the next bus. Also, a third unit is typically used to provide backup protection for the first two units in the next line section, as shown in Figure 7.42.

The term *distance* is used for a family of relays that respond to a ratio of voltage to current and therefore to impedance or a component of impedance. The distance relays are classified according to their polar characteristics, the number of inputs they have, and the method by which the comparison is made. The common types compare two input quantities in either magnitude or phase to obtain characteristics that are either straight lines or circles when plotted on an R–X diagram. The basic distance relay types are (1) impedance, (2) reactance, (3) admittance (or mho), (4) offset mho (or modified impedance), and (5) ohm relays.

7.17.1 Impedance Relay

Consider an amplitude comparator (e.g., balanced beam structure), and assume that it is at the threshold of operation so that Equation 7.58 is applicable. Therefore,

$$\left(k_1^2 - k_3^2\right)|\mathbf{A}|^2 + 2(k_1 k_2 - k_3 k_4)|\mathbf{A}||\mathbf{B}|\cos(\Phi - \theta) + \left(k_2^2 - k_4^2\right)|\mathbf{B}|^2 = 0 \tag{7.58}$$

if the constants are so adjusted so that the input signals are

$$\mathbf{S}_1 = k_1 \mathbf{V} \tag{7.101}$$

$$\mathbf{S}_2 = k_4 \mathbf{V} \tag{7.102}$$

System Protection

that is, $k_2 = k_3 = 0$, $\mathbf{A} = \mathbf{V}$, and $\mathbf{B} = \mathbf{I}$. Substituting Equations 7.101 and 7.102 into Equation 7.58

$$k_1^2 |\mathbf{V}|^2 = k_4^2 |\mathbf{I}|^2$$

or

$$\frac{\mathbf{V}}{\mathbf{I}} = \frac{k_4}{k_1} = \text{constant } k \qquad (7.103)$$

that is,

$$\mathbf{Z} = \text{constant } k \qquad (7.104)$$

Therefore, the impedance relay does not consider the phase angle between the current and the voltage applied to it.

Because of this, its impedance characteristic when plotted on an R–X diagram is a circle with its center at the origin of the coordinates and of radius equal to its setting in ohms, as shown in Figure 7.43b. The relay operates for all impedance values that are less than its setting, that is, for all points within the cross-hatched circle. It restrains for all points outside the circle. The relay, shown in Figure 7.43, is located on bus A and is nondirectional, that is, it operates for all faults along the vector AB and also for all faults behind the bus A up to an impedance AC. Note that, the line ABC represents the angle by which the fault current lags the driving voltage and that A is the relaying point.

The characteristic shown in Figure 7.43 indicates that the relay is not directional. To make the relay nonresponsive to bus and line faults behind it, directional control is essential. This can be achieved by the addition of a directional unit. Figure 7.44 shows a typical per-phase arrangement for a three-zone distance relay with directional unit used for transmission line protection [10].

The impedance characteristic of the directional relay is a straight line on the R–X diagram. The combined characteristic of the directional and impedance relays is the cross-hatched semicircle shown in Figure 7.45. The three impedance units are labeled $\mathbf{Z}_1$, $\mathbf{Z}_2$, and $\mathbf{Z}_3$.

The associated regions are called *zones of protection*. It is usually a standard practice to set the zone 1 relay for about 80% reach (to cover 80% of the line impedance) and instantaneous operation.

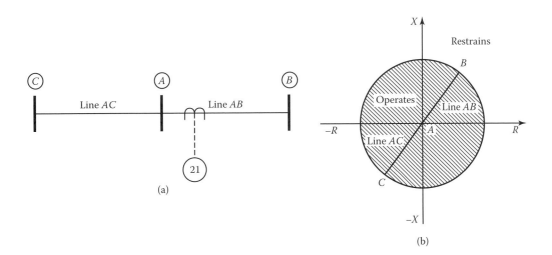

FIGURE 7.43 Characteristics of impedance relay.

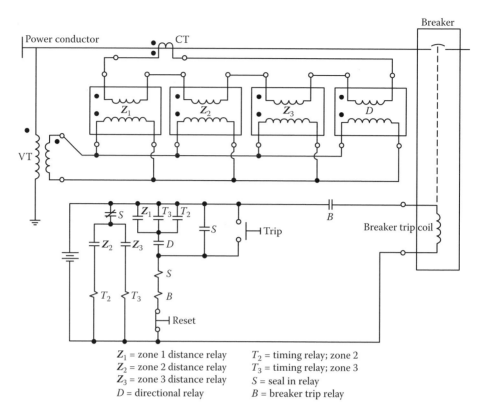

FIGURE 7.44 Typical per-phase arrangement for three-zone distance relay with directional unit. (Gross, C. A.: *Power System Analysis*. 1979. Copyright Wiley-VCH Verlag GmbH & Co. KGaA. Reproduced with permission.)

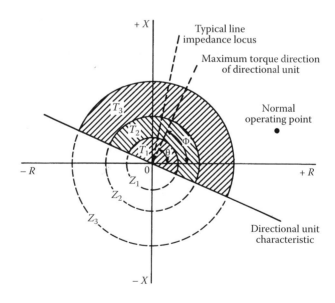

FIGURE 7.45 Operating and time-delay characteristics of impedance-type distance relay with directional restraint.

System Protection

Thus, the resulting shortened zone 1 is called as an *underreaching zone*. The zones 2 and 3 relay units are typically set for longer reaches and time delays. For example, the reach of the zone 2 relay extends beyond the line terminal well into the lines connected to the remote bus. Therefore, it is called as an *overreaching zone*. Typically, the zone 2 relay is set for 120% reach and operates at a time delay T_2 of 12–18 cycles.

At any event, it should be long enough to provide selectivity with the slowest of (1) line relays of adjoining line sections, (2) bus differential relays of the bus at the other end of the line, or (3) transformer differential relays of transformers on the bus at the other end of the line. On the other hand, the zone 3 relay provides backup protection for faults in adjoining line sections. Therefore, its reach should extend beyond the end of the longest adjoining line section under the conditions that cause the maximum amount of underreach, namely, arcs and intermediate current sources. Typically, the zone 3 relay is set for 250% reach on transmission lines and operates at a time delay T_3 of 60 cycles.*

In Figure 7.45, Φ is the angle of maximum torque of directional unit, and θ is the line impedance angle. This origin represents the relay location. If there is a metallic (or solid) fault in the tripping direction, the impedance lies on the line impedance locus indicated by this angle θ. The $\mathbf{Z}_1$ and $\mathbf{Z}_2$ units provide the primary protection for a given transmission line section, whereas $\mathbf{Z}_2$ and $\mathbf{Z}_3$ provide backup protection for adjoining buses and line sections. The disadvantages of the impedance relays are

1. It is nondirectional. Therefore, it will see faults both in front of and behind the relaying point. Thus, it requires a directional unit to provide it with correct discrimination.
2. It is affected by arc resistance during arcing faults on the protected line.
3. It is highly sensitive to power swings, due to the large area covered by its impedance circle.

If the impedance relays operate owing to the changes in line-to-line voltages (known as *delta voltages*, e.g., $\mathbf{V}_b - \mathbf{V}_c$) and the difference between line currents (known as *delta currents*, e.g., $\mathbf{I}_b - \mathbf{I}_c$), they are called *phase relays*. Figure 7.46b shows such "phase" arrangement, which is also called the *delta connection*. The phase relays can detect

$$\mathbf{Z}_{ab} = \frac{\mathbf{V}_{ab}}{\mathbf{I}_{ab}}$$

or

$$\mathbf{Z}_{bc} = \frac{\mathbf{V}_{bc}}{\mathbf{I}_{bc}}$$

or

$$\mathbf{Z}_{ca} = \frac{\mathbf{V}_{ca}}{\mathbf{I}_{ca}}$$

where the delta voltages must be determined from

$$\mathbf{V}_{ab} = \frac{1}{\sqrt{3}}\left(\mathbf{V}_{an} - \mathbf{V}_{bn}\right)$$

---

* In certain applications such as at EHV levels, it may be extremely difficult to coordinate the zone 3 relay with the zone 2 relays of the neighboring lines. Because of the difficulty, the remote backup protection using the zone 3 units are often neglected.

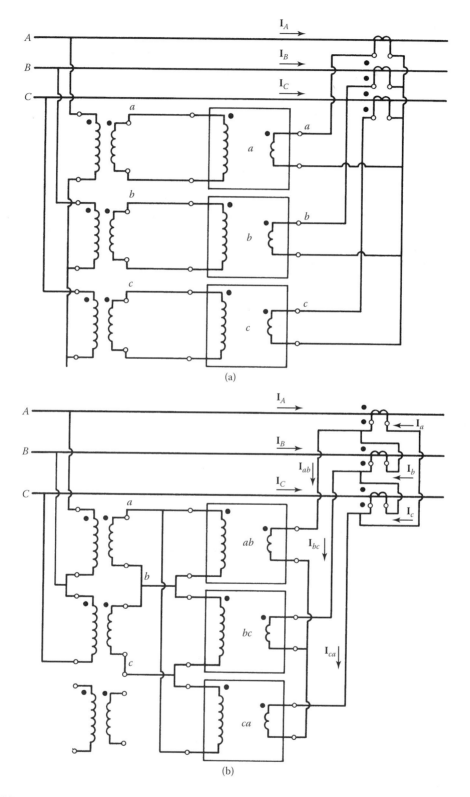

FIGURE 7.46 Distance relay connections: (a) ground relay connections; (b) phase relay connections. (Gross, C. A.: *Power System Analysis*. 1979. Copyright Wiley-VCH Verlag GmbH & Co. KGaA. Reproduced with permission.)

System Protection

$$V_{ab} = \frac{1}{\sqrt{3}}\left(V_{an} - V_{bn}\right)$$

$$V_{ab} = \frac{1}{\sqrt{3}}\left(V_{an} - V_{bn}\right)$$

and the delta currents must be determined from

$$I_{ab} = \frac{1}{\sqrt{3}}\left(I_a - I_b\right)$$

$$I_{ab} = \frac{1}{\sqrt{3}}\left(I_a - I_b\right)$$

$$I_{ca} = \frac{1}{\sqrt{3}}\left(I_c - I_a\right)$$

Thus, the phase relays operate properly on three-phase faults, all line-to-line faults, and double line-to-ground faults. However, they do not operate properly on single line-to-ground faults. For such faults, three additional relays, which use line-to-neutral voltages, line currents, and the zero-sequence currents, are needed. These relays are called *ground relays*. Figure 7.46a shows such "ground" arrangement, which is also called *wye connection*. The ground relays can detect

$$Z_a = \frac{V_a}{I_a}$$

or

$$Z_b = \frac{V_b}{I_b}$$

or

$$Z_c = \frac{V_c}{I_c}$$

where the wye values are readily available from the fault studies using symmetrical components. The ground relays operate properly not only on single line-to-ground faults but also on double line-to-ground faults and three-phase faults. However, they do not operate properly on line-to-line faults.

7.17.2 Reactance Relay

Consider a phase comparator (e.g., the induction cylinder or double-induction-loop structure), and assume that it is at the threshold of operation so that Equation 7.78 is applicable. Therefore,

$$k_1 k_3 |\mathbf{A}|^2 + k_2 k_4 |\mathbf{B}|^2 + (k_1 k_4 + k_2 k_3)|\mathbf{A}||\mathbf{B}|\cos(\Phi - \theta) = 0 \quad (7.78)$$

if the input signals are

$$\mathbf{S}_1 = -k_a \mathbf{V} + k_b \mathbf{I} \angle \Phi - \theta \quad (7.105)$$

$$\mathbf{S}_2 = k_b \mathbf{I} \angle \Phi - \theta \quad (7.106)$$

that is, $k_1 = -k_a$, $k_2 = k_4 = k_b \angle \Phi$, $k_3 = 0$, $A = V$, and $B = V$. Substituting Equations 7.105 and 7.106 into Equation 7.78,

$$k_b^2 |\mathbf{I}|^2 = -k_a k_b |\mathbf{I}||\mathbf{V}|\cos(\Phi - \theta) = 0 \quad (7.107)$$

or

$$\mathbf{Z}\cos(\Phi - \theta) = \frac{k_b}{k_a} \quad (7.108)$$

If $\Phi = \dfrac{\pi}{2}$, then

$$\mathbf{Z}\sin\theta = \frac{k_b}{k_a} \quad (7.109)$$

or

$$X = \frac{k_b}{k_a} = \text{constant } k \quad (7.110)$$

that is,

$$X = \text{constant } k$$

Thus, the characteristic of this relay on the R–X diagram is represented by a straight line parallel to the horizontal axis R, as shown in Figure 7.47a. Note that, the resistance component of the impedance does not affect the operation of the relay and that the relay reacts only to the reactance component. Thus, the relay operates for any point below for the operating characteristic (regardless of whether its location is above or below the R axis). If $\Phi \neq \dfrac{\pi}{2}$ in Equation 7.108, then the straight-line characteristic will not be parallel to the R axis. Such a relay is called the *angle-impedance* relay.

A reactance relay is not affected by the presence of a fault arc resistance since it responds only to the reactive component of the system impedance. However, when fault arc resistance is of such a high value that load and fault current magnitudes are of the same order, the reach of the relay is altered by the value of the load and its power factor and may either overreach or underreach.

System Protection

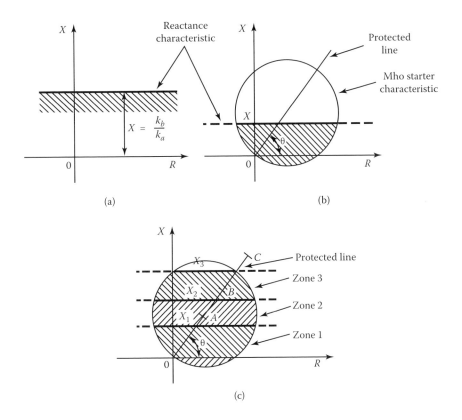

FIGURE 7.47 Characteristic of reactance relay: (a) on R–X plane; (b) combination of mho and reactance characteristics; (c) application for zone protection.

Thus, to provide directional response and to prevent operation under normal-load conditions, a voltage-restraining unit (e.g., mho relay) is added to the relay. Such a modified reactance relay is called the *starting relay*. Figure 7.47b shows such a combination of mho and reactance characteristic. Figure 7.47c shows the application of a reactance relay for zone protection. In this figure, 0 is the relay location, OA is the first line section, AB is the second line section, BC is the third line section, and θ is the line impedance angle.

7.17.3 Admittance (Mho) Relay

Consider a phase comparator and assume that it is at the threshold of operation so that Equation 7.78 is applicable. Therefore,

$$k_1 k_3 |\mathbf{A}|^2 + k_2 k_4 |\mathbf{B}|^2 + (k_1 k_4 + k_2 k_3)|\mathbf{A}||\mathbf{B}|\cos(\Phi - \theta) = 0 \tag{7.78}$$

if the input signals are

$$\mathbf{S}_1 = -k_a \mathbf{V} + k_b \mathbf{I} \angle \Phi - \theta \tag{7.111}$$

$$\mathbf{S}_2 = k_a \mathbf{V} \tag{7.112}$$

that is, $k_1 = -k_a$, $k_2 = k_b \angle \Phi$, $k_3 = k_a$, $k_4 = 0$, $\mathbf{A} = \mathbf{V}$, and $\mathbf{B} = \mathbf{I}$. $\mathbf{A} = \mathbf{V}$ and $\mathbf{B} = \mathbf{V}$. Substituting Equations 7.111 and 7.112 into Equation 7.78,

$$-k_a^2 |\mathbf{V}|^2 + k_a k_b |\mathbf{V}||\mathbf{I}|\cos(\Phi - \theta) = 0 \tag{7.113}$$

or

$$\frac{\mathbf{I}}{\mathbf{V}} \cos(\Phi - \theta) = \frac{k_a}{k_b} \tag{7.114}$$

or

$$\mathbf{Y} \cos(\Phi - \theta) = \frac{k_a}{k_b} \tag{7.115}$$

This represents the admittance (mho) relay characteristic. If it is plotted on the R–X diagram, it is a circle passing through the origin, as shown in Figure 7.48a. If it is plotted on the admittance (i.e., the G–B) diagram, it is a straight line, as shown in Figure 7.48b. The circle passing through the origin makes the relay inherently directional.

Thus, with such a characteristic, the relay measures distances in one direction only. Thus, it does not require a separate distance unit. The reach point setting of a mho relay changes with the fault angle owing to the fact that the impedance measurement is not constant for all angles. The fault angle is therefore dependent on the relative value of R and X. If there is an arcing fault condition, the value of R will increase by the amount of the resistance of arc R_{arc} causing the fault angle to change. This, in turn, causes the relay to underreach if it has a characteristic angle that is equal to the line angle.

Hence, it is good practice to set the relay with its characteristic angle leading the line angle, so that an arc resistance can be taken into account without causing any underreach, as is illustrated in Figure 7.48c. The approximate value of arc resistance arc can be determined from the empirical formula derived by Warrington [8].

$$R_{arc} = \frac{8750\ell}{I^{1.4}} \, \Omega$$

where
ℓ = length of the arc in still air in feet
I = current in arc in amperes

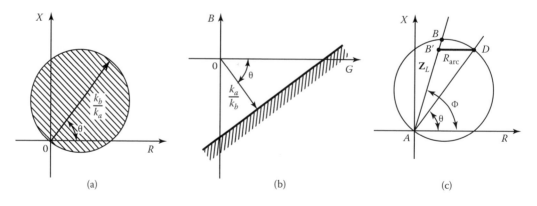

FIGURE 7.48 Characteristic of mho relay: (a) on R–X plane; (b) on G–B plane; (c) reduction of relay reach by fault (arc) resistance.

System Protection

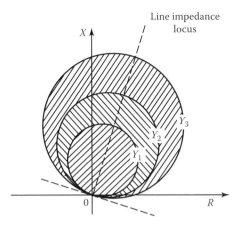

FIGURE 7.49 Operating characteristic of three-zone mho relay.

The length of the arc is initially equal to conductor spacing in the event of line-to-line faults and the distance from conductor to tower in case of line-to-ground faults. With cross wind, when there is a time delay in fault clearance such as zones 2 and 3, the arc is extended considerably, and therefore its resistance is increased. Thus, especially for zones 2 and 3, high-velocity winds may cause the relay to underreach seriously. Such arc resistance can be determined from Equation 4.3.

The effect of the arc resistance is most significant on short lines and with fault currents less than 2000 A during off-peak hours. It is not significant on long-transmission lines, which have overhead ground wires and steel towers. However, it is very significant on long-transmission lines built with wooden poles and without ground wires. Figure 7.49 shows the application of a mho relay for zone protection.

It is interesting to note that the critical arc location is just short of the point on a line at which a distance relay's operation changes from high-speed to intermediate time or from intermediate time to backup time. In other words, an arc within the high-speed zone (i.e., zone 1) will cause the relay to operate in intermediate time (i.e., zone 2); an arc within the intermediate zone will cause the relay operate in backup time (i.e., zone 3); or an arc within the backup zone will prevent relay operation altogether.

If a fault takes place within the instantaneous operation zone (i.e., zone 1), the distance relay will operate instantaneously before the arc can stretch significantly and therefore increase its resistance. Thus, the important thing is the initial amount of arc resistance.

7.17.4 Offset Mho (Modified Impedance) Relay

Assume that the input signals to a phase comparator are given as

$$+\mathbf{S}_1 = -k_a\mathbf{V} + k_2\mathbf{I}\angle\Phi - \theta \tag{7.116}$$

$$\mathbf{S}_2 = -k_a\mathbf{V} + k_4\mathbf{I}\angle\Phi - \theta \tag{7.117}$$

that is $k_1 = -k_a$, $k_2 = k_2\angle\Phi$, $k_3 = k_a$, $k_4 = k_4\angle\Phi$, $A = V$, and $B = I$. Substituting Equations 7.116 and 7.117 into Equation 7.78,

$$-k_a^2V^2 + k_2k_4I^2 + k_a(k_2 - k_4)VI\cos(\Phi - \theta) = 0 \tag{7.118}$$

dividing by I^2,

$$-k_a^2 Z^2 + k_2 k_4 + k_a Z(k_2 - k_4)\cos(\Phi - \theta) = 0 \qquad (7.119)$$

since $Z^2 = R^2 + X^2$ and dividing by $-k_a^2$,

$$R^2 + X^2 - \frac{k_2 k_4}{k_a^2} - \frac{k_2 - k_4}{k_a}(R\cos\Phi + X\sin\Phi) = 0 \qquad (7.120)$$

$$\left[R - \frac{(k_2 - k_4)\cos\Phi}{2k_a}\right]^2 + \left[X - \frac{(k_2 - k_4)\sin\Phi}{2k_a}\right]^2 \le \left[\frac{k_2 + k_4}{2k_a}\right]^2 \qquad (7.121)$$

Thus, the characteristic of this relay on the R–X diagram is represented by a circle with its center located at $(k_2 - k_4)/2k_a \angle \Phi$ and of radius equal to $(k_2 + k_4)/2k_a$, as shown in Figure 7.50a. Under close-up fault conditions, when the voltage applied to a mho relay is at or near zero, the relay may fail to operate unless a "current bias" is introduced into the voltage circuit. Hence, the mho characteristic is shifted to encircle the origin, as shown in Figure 7.50a.

The offset mho unit is used in conjunction with mho measuring units as a fault detector and third-zone measuring unit. Thus, with the backward reach arranged to extend into the bus zone, the offset mho unit will provide backup for bus faults, as shown in Figure 7.50b. However, this cannot be provided in conjunction with reactance measuring units since they will operate instantaneously for bus faults causing the discrimination between primary protection zones to disappear.

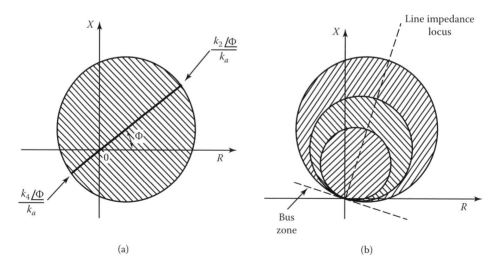

FIGURE 7.50 Offset mho characteristic: (a) typical characteristic; (b) application for zone backup protection.

System Protection

7.17.5 Ohm Relay

The ohm relay characteristic is represented by Equation 7.108, and it is a straight line when plotted on the R–X diagram. Thus, a reactance relay is a special case of an ohm relay. It is also called the *angle impedance relay* and is used as a "blinder" to prevent distance relays from tripping on very severe power swings on long lines to avoid cascade tripping. During severe power swing conditions from which the system is not likely to recover, normal service can be provided only if the swinging sources are separated.

To minimize the system disturbance, an out-of-step tripping scheme having ohm units is used. The scheme usually has two ohm units, the characteristics of which are arranged to be parallel to and on either side of the line impedance vector, as shown in Figure 7.51. As the impedance changes during a power swing, the point representing the impedance moves along the power swing locus, entering into the zone between the two blinders provided by the ohm units O_1 and O_2, after which the ohm units operate. Since no condition other than power swing can cause the impedance vector to move in this way, the scheme operates only during power swings and not during other disturbances (e.g., faults).

EXAMPLE 7.5

Consider the one-line diagram of a 345-kV transmission line shown in Figure 7.52a. The equivalent systems behind buses A and B are represented by the equivalent system impedances in series with constant bus voltages, respectively. Assume that power flow direction is from bus A to bus B. Consider the directional distance relay located at A whose forward direction is in the direction from bus A to bus B. Assume that zone-type distance relays have two units, namely, three phase and phase to phase. Thus, for three-phase faults, the mho characteristic is directional and only operates for faults in the forward direction on line AB. For line-to-line faults (i.e., a–b, b–c, and c–a), the phase-to-phase unit operates. Also, assume that all double line-to-ground faults are protected by an overlap of the two distance units. All other line-to-ground faults are protected by ground distance relays and are not included in this example. Consider only zones 1 and 2 protection for the distance relay located at bus A and do the following:

(a) Draw the locus of the line impedance $\mathbf{Z}_L$ on the R–X diagram.
(b) Draw the two-zone mho characteristics on the R–X diagram for the three-phase units.
(c) Draw the two-zone characteristics on the R–X diagram for the phase-to-phase units.
(d) Indicate the approximate vicinity for the possible location of the normal operation point for power flow from bus A to bus B.

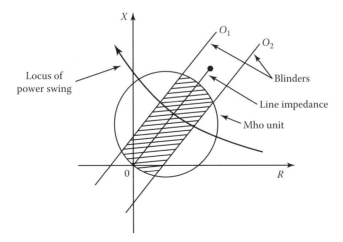

FIGURE 7.51 Use of ohm relay units as blinders to narrow angular range in which tripping can occur during swings.

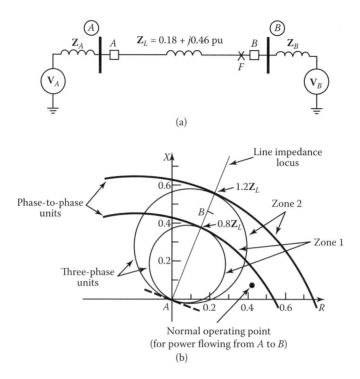

FIGURE 7.52 Typical application of distance relays for protection of transmission line: (a) system one-line diagram; (b) distance relays of A looking toward B.

Solution

The solution is shown in Figure 7.52b.

EXAMPLE 7.6

Assume that Figure 7.52a is the one-line diagram of a 138-kV subtransmission line with a line impedance of 0.2 + j0.7 pu. Consider the directional distance (mho) relay located at A whose forward direction is in the direction from bus A to bus B. Assume that there is a line-to-line fault at the fault point F, which is located at 0.7 pu distance away from bus A. The magnitude of the fault current is 1.2 pu. Assume that the line spacing of 10.3 ft is equal to the arc length. The base quantities for power, voltage, current, and impedance are given as 100 MVA, 138 kV, 418.4 A, and 190.4 Ω, respectively. Consider only zones 1 and 2 protection and determine the following:

(a) Value of arc resistance at fault point in ohms and per units
(b) Value of line impedance including the arc resistance
(c) Line impedance angle without and with arc resistance
(d) Graphically, whether relay will clear the fault instantaneously

Solution

(a) Since the current in the arc is 1.2 pu
or

$$I = 1.2 \times 418.4$$
$$= 502.08 \text{ A}$$

the arc resistance can be found from Equation 4.2 as

$$R_{arc} = \frac{8750\ell}{I^{1.4}}$$

$$= \frac{8750 \times 10.3}{502.08^{1.4}}$$

$$= 14.92 \, \Omega \text{ or } 0.0784 \text{ pu}$$

(b) The impedance seen by the relay is

$$\mathbf{Z}_L + R_{arc} = (0.2 + j0.7)0.7 + 0.0784$$
$$= 0.2184 + j0.49 \text{ pu}$$

(c) The line impedance angle without the arc resistance is

$$\tan^{-1}\left(\frac{0.49}{0.14}\right) = 74.05°$$

and with the arc resistance is

$$\tan^{-1}\left(\frac{0.49}{0.2184}\right) = 65.98°$$

(d) Figure 7.53 shows that even after the addition of the arc resistance, the fault point F moved to point F′, which is still within zone 1. Therefore, the fault will be cleared instantaneously.

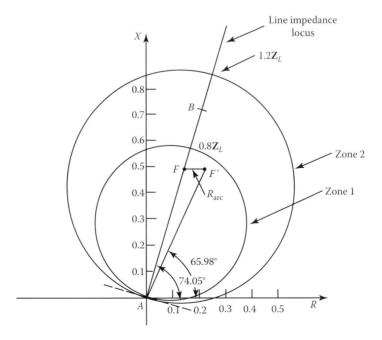

FIGURE 7.53 Graphical determination of fault clearance.

EXAMPLE 7.7

Consider a transmission line TL_{12} protected by the directional distant relays R_{12} and R_{21}, as shown in Figure 7.54a. Determine the following:

(a) Zones of protection for relay R_{12}
(b) Coordination of distance relays R_{12} and R_{23} in terms of operating time versus impedance

Solution

(a) The zone of protection for relay R_{12} is shown in Figure 7.54b.
(b) The coordination of the distance relays R_{12} and R_{23} in terms of operating time versus impedance is illustrated in Figure 7.54c. Note that, zone 3 provides backup protection for the neighboring protection system.

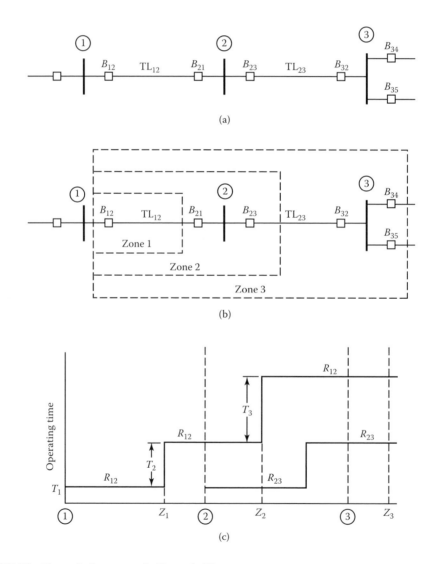

FIGURE 7.54 Transmission system for Example 7.7.

System Protection

EXAMPLE 7.8

Consider the 230-kV transmission system shown in Figure 7.54a. Assume that the positive-sequence impedances of the lines TL_{12} and TL_{23} are $2 + j20$ and $2.5 + j25\ \Omega$, respectively. If the maximum peak load supplied by the line TL_{12} is 100 MVA with a lagging power factor of 0.9, design a three-zone distance-relaying system for the R_{12} impedance relay by determining the following:

(a) Maximum load current
(b) CT ratio
(c) VT ratio
(d) Impedance measured by relay
(e) Load impedance based on secondary ohms
(f) Zone 1 setting of relay R_{12}
(g) Zone 2 setting of relay R_{12}
(h) Zone 3 setting of relay R_{12}

Solution

(a) The maximum load current is

$$I_{max} = \frac{100 \times 10^6}{\sqrt{3}(230 \times 10^3)} = 251.02\ \text{A}$$

(b) Therefore, the CT ratio is 250:5, which gives about 5 A in the secondary winding under the maximum loading.
(c) Since the system voltage to neutral is $(230/\sqrt{3}) = 132.79$ kV and selecting a secondary voltage of 69 V line to neutral, the VT ratio is calculated as

$$\frac{132.79 \times 10^3}{69} = \frac{1924.5}{1}$$

(f) The impedance measured by the relay is

$$\frac{V_p/1924.5}{I_p/50} = 0.026\ Z_{line}$$

Therefore, the impedances of lines TL_{12} and TL_{23} as seen by the relay are approximately $0.052 + j0.5196$ and $0.065 + j0.6495\ \Omega$, respectively.

(g) The load impedance based on secondary ohms is

$$\mathbf{Z}_{load} = \frac{69}{251.02(5/250)}(0.9 + j0.4359)$$
$$= 12.37 + j5.99\ \Omega\ \text{(secondary)}$$

(h) The zone 1 setting of relay R_{12} is

$$\mathbf{Z}_r = 0.80(0.052 + j0.5196)$$
$$= 0.0416 + j0.4157 \ \Omega \ \text{(secondary)}$$

(i) The zone 2 setting of relay R_{12} is

$$\mathbf{Z}_r = 1.20(0.052 + j0.5196)$$
$$= 0.0624 + j0.6235 \ \Omega \ \text{(secondary)}$$

(j) Since the zone 3 setting must reach beyond the longest line connected to bus 2, it is

$$\mathbf{Z}_r = 0.052 + j0.5196 + 1.20(0.065 + j0.6495)$$
$$= 0.130 + j1.299 \ \Omega \ \text{(secondary)}$$

EXAMPLE 7.9

Assume that the R_{12} relay of Example 7.8 is a mho relay and that the relay characteristic angle may be either 30° or 45°. If the 30° characteristic angle is used, the relay ohmic settings can be determined by dividing the required zone reach impedance, in secondary ohms, by $\cos(\theta - 30°)$, where θ is the line angle. Use the 30° characteristic angle and determine the following:

(a) Zone 1 setting of mho relay R_{12}
(b) Zone 2 setting of mho relay R_{12}
(c) Zone 3 setting of mho relay R_{12}

Solution

(a) From Example 7.8, the required zone 1 setting was

$$\mathbf{Z}_r = 0.0416 + j0.4157$$
$$= 0.4178\angle 84.3° \ \Omega \ \text{(secondary)}$$

Therefore,

$$\text{mho relay zone 1 setting} = \frac{0.4178}{\cos(84.30° - 30°)}$$
$$= 0.7157 \ \Omega \ \text{(secondary)}$$

(b) The required zone 2 setting was

$$\mathbf{Z}_r = 0.0624 + j0.6235$$
$$= 0.6266\angle 84.3° \ \Omega \ \text{(secondary)}$$

System Protection

Thus,

$$\text{mho relay zone 2 setting} = \frac{0.6266}{\cos(84.30° - 30°)}$$
$$= 1.0734 \, \Omega \text{ (secondary)}$$

(c) The required zone 3 setting was

$$\mathbf{Z}_r = 0.130 + j1.299$$
$$= 1.3055 \angle 84.3° \, \Omega \text{ (secondary)}$$

Therefore,

$$\text{mho relay zone 3 setting} = \frac{1.3055}{\cos(84.3° - 30°)}$$
$$= 2.2364 \, \Omega \text{ (secondary)}$$

that is, 312.5% of the zone 1 setting.

7.18 OVERCURRENT RELAYS

Lines are protected by overcurrent, distance, or pilot relays, depending on the requirements. Overcurrent relaying is the simplest and cheapest, the most difficult to apply, and the quickest to require readjustment or even "overload" protection, which normally utilizes relays that operate in a time related in some degree to the thermal capability of the plant to be protected.

On the other hand, overcurrent protection is directed totally to the clearance of faults, even though with the settings usually adopted some degree of overload protection is achieved. Therefore, the maximum load currents must be known to determine whether the ratio of the minimum fault current to maximum load current is high enough to enable simple overcurrent-operated relays to be used successfully. Figure 7.55a shows the operating characteristic of a time overcurrent relay.

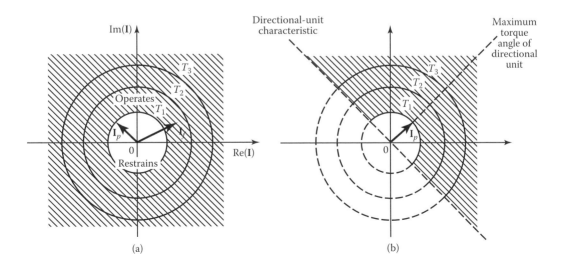

FIGURE 7.55 Operating characteristics of time-overcurrent relays: (a) time-overcurrent relay; (b) directional time-overcurrent relay.

If the magnitude of the fault current I_f is greater than the relay pickup current I_p, the relay operates. Otherwise, it restrains, that is, does not operate. The pickup current is the current to just cause the disk to move and close the associated contact and it is adjusted by current coil taps. Note that, the phase angle of the fault current I_f can be anywhere between 0° and 360° due to the fact that the reference phasor is arbitrary. In Figure 7.55a, the circles represent the time settings, where time T_3 is earlier than time T_2 and T_1, and time T_2 is earlier than time T_1.

The time settings are adjusted by time dial or lever by moving the contact position when the relay is reset, and it changes the time of operation at a given tap and current magnitude. Figure 7.56 shows a typical time-overcurrent relay. When fault current can flow in both directions through the relay location, it is necessary to make the response of the relay directional by the addition or directional control units. The directional unit is basically a power-measuring device in which the system voltage is used as a reference for establishing the relative direction or phase of the fault current. Figure 7.55b shows the operating characteristics of such a directional time-overcurrent relay.

There are two basic forms of overcurrent relays: the instantaneous type and the time-delay type. The instantaneous overcurrent relay is designed to operate with no intentional time delay when the current exceeds the relay setting. The time of operation of such relays is anywhere between 0.016 and 0.1 s. The time-overcurrent relay has an operating characteristic such that its operating time changes inversely as the current flowing in the relay. Figure 7.57 [11] shows the three most often used time-overcurrent characteristics: inverse, very inverse, and extremely inverse, as well as the instaneous time curve. Figure 7.58 shows a comparison of co-type curve shapes. Both the instaneous and the time-delay-type overcurrent relays and are inherently nonselective, that is, they can detect fault conditions not only in their own protected apparatus but also in nearby apparatus.

Thus, in practice, the required selectivity between overcurrent relays protecting different system components has to be achieved on the basis of sensitivity, or operating time, or a combination of both.

Figure 7.56 shows a typical IAC time-overcurrent relay. Figure 7.57 shows the three most often used time–current characteristics of overcurrent relays: inverse, very inverse, and extremely inverse,

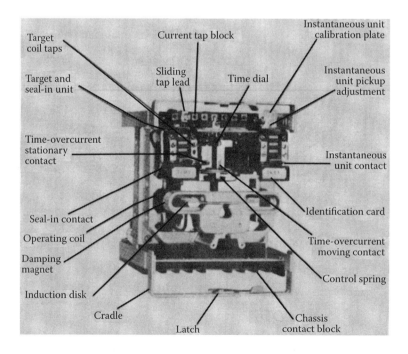

FIGURE 7.56 Typical IAC time-overcurrent relay. (Courtesy of General Electric Company.)

System Protection

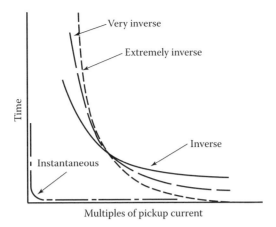

FIGURE 7.57 Time–current characteristics of overcurrent relays.

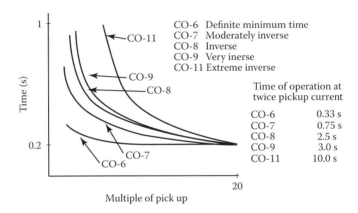

FIGURE 7.58 Comparison of CO-type curve shapes. (Courtesy of Westinghouse Electric Corporation.)

as well as the instantaneous time curve.* Figure 7.58 shows a comparison of CO-type curve shapes. Both the instantaneous and the time-type overcurrent relays are inherently nonselective, that is, they can detect fault conditions not only in their own protected apparatus but also in nearby apparatus.

Thus, in practice, the required selectivity between overcurrent relays protecting different system components has to be achieved on the basis of sensitivity, or operating time, or a combination of both. Figures 7.59 and 7.60 show the time–current curves of types CO-7 and CO-8 overcurrent relays, respectively. Figure 7.61 shows the time–current curves of an IAC overcurrent relay. As said before, the pickup setting I_p of a relay can be adjusted by using the taps on its current (input) winding. For example, the relay CO-7 (its time–current characteristics shown in Figure 7.59) has current-tap settings (CTSs) of 4, 5, 6, 7, 8, 10, and 12 A.

The time–current curves are usually given with multiples of pickup amperes as the abscissa and operating time as the ordinate. Here, the multiples of pickup amperes is the ratio of relay fault current to pickup current, that is, ($|I_f| : |I_p|$). The time–current curves can be shifted up or down by an adjustment called the *time dial setting* (TDS). Note that, in Figure 7.59, the fastest and slowest operations of the relay can be obtained by the TDSs of 1/2 and 11, respectively.

---

* Note that Figure 7.57 may be misleading since the pickup current ordinarily needed for the instantaneous trip unit in several multiples of the current needed for the time-delay unit.

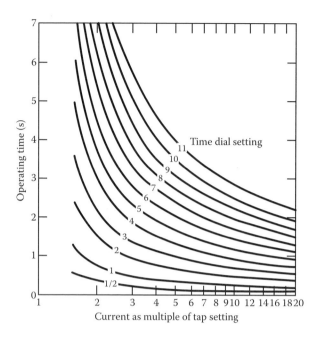

FIGURE 7.59 Time–current curves of type CO-7 overcurrent relays (50–60 cycles) with moderately inverse characteristics (CTSs: 4, 5, 6, 7, 8, 10, and 12 A). (Courtesy of Westinghouse Electric Corporation.)

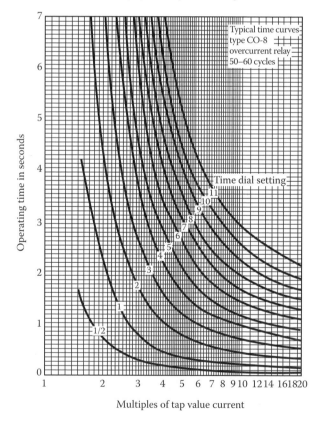

FIGURE 7.60 Time–current curves of type CO-8 overcurrent relays with inverse characteristics. (Courtesy of Westinghouse Electric Corporation.)

System Protection

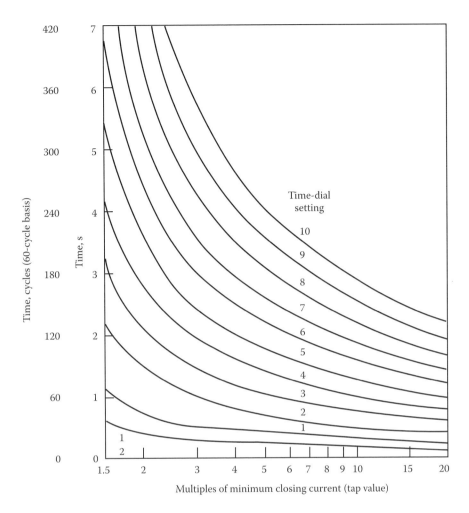

FIGURE 7.61 Time–current curves of IAC overcurrent relays with inverse characteristics. (Courtesy of General Electric Company).

Even though the TDSs are given in integers, the intermediate values can be obtained by interpolating between the given discrete curves. At low values of operating current, the shape of the curve is determined by the effect of the restraining force of the control springs, whereas at high values the effect of saturation dominates in induction overcurrent relays. Thus, a different TDS is achieved by varying the travel of the disk or cup required to close the contacts of the relay.

All time-delay overcurrent induction relays will start to move and will eventually close their contacts on current equal to their current-tap (pickup) setting, assuming that they are in good operating condition and free from dust, etc. However, because of the manufacturing tolerances allowed and the low operating torque available from such small currents, it is desirable to select a CTS such that the relay will not be expected to give accurate time–current performance below approximately 1.5 multiples of minimum pickup. In general, overcurrent relays are used to protect subtransmission lines and distribution circuits since faults in such circuits do not usually affect the system stability and, therefore, do not dictate the use of high-speed relaying.

Occasionally, they are used on transmission lines for primary ground fault protection where distance relays are employed and for ground backup protection on most lines having pilot relaying for primary protection. However, as the demand increases for faster fault-clearing times, distance relaying for ground fault primary and backup protection of transmission lines is slowly replacing overcurrent relaying. Even though the application of overcurrent relaying on radial circuits is relatively simple, its application on the loop and/or interconnected circuits becomes most difficult since it requires readjustment as the system configuration changes. Furthermore, overcurrent relays cannot discriminate between load and fault current, and therefore, when they are used for phase fault protection, they are only applicable when the minimum fault current is larger than the maximum full-load current. Figure 7.62 shows the connections for a ground overcurrent relay together with phase overcurrent relays. It is usually the practice to use a set of two or three overcurrent relays for protection against phase-to-phase faults and a separate overcurrent relay for single line-to-ground faults. Figure 7.63 shows various ways to connect ground relays with respect to phase relays, indicated by the standard devices numbers 64 and 51, respectively.

Under balanced or no-fault conditions, ideally there will be no current flowing through the ground relay.* The relay operations are coordinated with respect to each other in order to provide the desired selectivity and cause the least service interruption while isolating the fault. This procedure is called *relay coordination*. The basic rules of thumb of relay coordination are

1. Whenever possible, relays with the same operating characteristics are to be used in series with each other.
2. The relay farthest from the source must have current settings equal to or less than the relays behind it, that is, the primary current that is necessary to operate the relay in front is always equal to or less than the primary current necessary to operate the relay behind it.
3. The relay settings are first selected to provide the shortest operating times at maximum fault current levels and then checked to determine whether they are satisfactory at minimum fault current levels.

Among the various possible methods to achieve correct relay coordination are (1) discrimination by time that is, time grading; (2) discrimination by current, that is, current grading; and (3) discrimination by both time and current, that is, time–current grading.

In the time grading method, the selectivity is achieved on the basis of the time operation of the relays. The time of operation of the relays at various locations is so adjusted that the relay farthest from the source will have minimum time of operation, and as it is approached toward the source, the operating time increases. The basic disadvantage of this method is that the longest fault clearance time takes place for faults in the section closest to the source where the faults are most severe.

The current grading method is based on the fact that the fault current along the length of the protected circuit decreases as the distance from the source to the fault location increases. Therefore, the relays controlling the various CBs are set to operate at properly tapered values such that only the relay nearest to the fault trips its breaker. However, in practice, it is difficult to determine the magnitude of the current accurately and also the accuracy of the relays under transient conditions will probably suffer. Furthermore, the relay cannot differentiate between faults that are very close to each other (e.g., on each side of a bus) because the difference in the current would be extremely small. Because of the aforementioned difficulties involved, usually a combination of the two gradings (i.e., time–current grading) is used.

In the time–current grading method, the time–current grading is achieved with the help of relays that have inverse time-overcurrent relay characteristics. With this characteristic, the time of operation is inversely proportional to the fault current level and the actual characteristic is a function of both time and current settings of the relay. Figures 7.59 through 7.61 show some typical inverse-time relay characteristics. Hence, there are basic adjustable settings on all inverse-time relays: the CTSs

---

* Sometimes, in practice, there is trouble with the residual ground relay connections of Figures 7.62 and 7.63. For example, false ground tripping sometimes occurs because of unequal CT ratios, transients, harmonics, etc., if a sensitive ground relay pickup is attempted.

System Protection

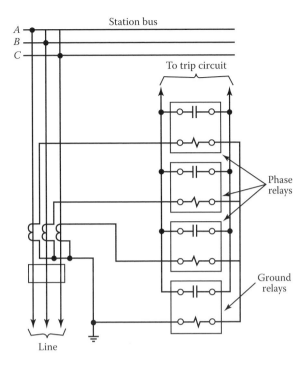

FIGURE 7.62 Phase and ground protection of circuit using induction-type overcurrent relays.

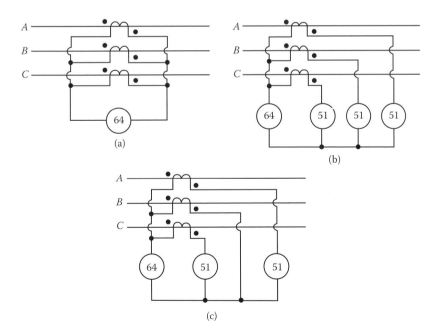

FIGURE 7.63 Various types of ground relay connections.

and the TDSs. Thus, the pickup current is determined by adjusting the CTSs. Note that, the pickup current is the current that causes the relay to operate and close the contact. To determine the CDS settings, the maximum fault current that can flow at each relay location on the given system has to be calculated.

A three-phase fault under maximum generation causes the maximum fault current and a line-to-line fault, whereas under minimum generation it causes minimum fault current. Thus, the relay must respond to the fault currents between these two extreme values. On a radial system, the lowest CDS setting has to be at the farthest end. The settings are increased for the succeeding relays toward the source.

As said before, the reset position of the moving contact of an inverse-time relay is the time dial. It changes the operation time of the relay for a given tap setting and current magnitude. To have a proper coordination between various relays on a radial system, the relay farthest from the source must be set to operate in the minimum possible time. The time settings are increased for the succeeding relays toward the source. For inverse-time overcurrent relays, the time setting is determined on the basis of the maximum fault current. In the event that the relay has correct discrimination at maximum fault current, it will automatically have a greater discrimination at the minimum fault current because the characteristic curve is more inverse on a lower current region.

The time interval that is necessary between two adjacent relays is called the coordination delay time (CDT). It is the minimum interval that permits a relay and its CB to clear a fault in its operating zone. With modern CBs, it is possible to use a CDT of 0.4 s.

It depends on (1) the fault current interrupting time of the CB (it is about 0.1 s or six cycles for the modern CBs), (2) the overtravel of the relay (when the relay is deenergized, operation may continue for a little bit longer until any stored energy has been dissipated), (3) errors due to relay and CT tolerances (causing departure of actual relay operation from published characteristics) and errors introduced by calculation approximations, and (4) some extra allowance (safety margin) that is necessary to ensure that a satisfactory contact gap remains. The sum of factors 2–4 is called the *error margin* and is typically given as 0.3 s.

Figure 7.64 shows the application of time-overcurrent relays to a series of radial lines. It illustrates how time coordination is achieved between inverse time-overcurrent relays at each breaker location. A vertical line drawn through any assumed fault location will intersect the operating time curves of various relays and will thereby shown the time at which each relay would operate if the fault current continued to flow for that length of time.

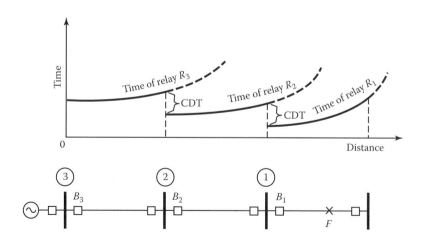

FIGURE 7.64 Operating time of overcurrent relays with inverse-time characteristics.

System Protection

For the fault shown, the relay tripping breaker B_1 operates quickly at time T_1, followed by the relays controlling B_2 and B_3 so that B_1 operates before B_2 and B_2 before B_3. Thus, the operating time T_2 of the relay at bus 2 can be expressed as

$$T_2 = T_1 + \text{CDT} \tag{7.122}$$

where

$$\text{CDT} = (\text{operating time of breaker } B_1) + (\text{error margin}) \tag{7.123}$$

Similarly, the operating time T_3 of the relay at bus 3 can be expressed as

$$T_3 = T_2 + \text{CDT} \tag{7.124}$$

EXAMPLE 7.10

Figure 7.65 shows the one-line diagram of a 13.2-kV radial system. Assume that all loads have the same power factor and that the CT ratios are 300:5, 400:5, and 600:5, as shown. The maximum (three-phase) fault currents at buses 1, 2, and 3 are 2400, 2700, and 3000 A, respectively. The operating time of each relay is six cycles. Use three CO-7 relays per breaker and determine the CTs and TDS of each delay.

Solution

The load currents at each of the three buses are

$$I_{L1} = \frac{4.5 \times 10^6}{\sqrt{3}(13.2 \times 10^3)}$$
$$= 196.82 \text{ A}$$

$$I_{L2} = \frac{7 \times 10^6}{\sqrt{3}(13.2 \times 10^3)}$$
$$= 306.17 \text{ A}$$

$$I_{L3} = \frac{14.5 \times 10^6}{\sqrt{3}(13.2 \times 10^3)}$$
$$= 634.21 \text{ A}$$

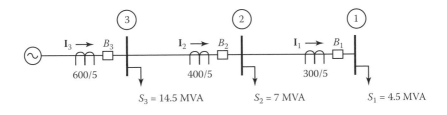

FIGURE 7.65 One-line diagram of radial system.

Using the given CT ratios, the corresponding relay currents due to load currents can be found as

$$I_{L,R1} = \frac{196.82}{300/5}$$
$$= 3.28 \text{ A}$$

$$I_{L,R2} = \frac{306.17}{400/5}$$
$$= 3.83 \text{ A}$$

$$I_{L,R3} = \frac{634.21}{600/5}$$
$$= 5.29 \text{ A}$$

The available CTSs for the CO-7 relay are given in Figure 7.59 as 4, 5, 6, 7, 8, 10, and 12 A. Therefore, the CTSs for the relays R_1, R_2, and R_3 are selected as

$$CTS_1 = 4A$$

$$CTS_2 = 4A$$

$$CTS_3 = 6A$$

The next task that is necessary in this coordination process is the selection of the TDS settings for each relay using the maximum fault currents. Relay R_1 is at the end of the radial system, and therefore no coordination is necessary. The fault current as seen by relay R_1 can be determined as

$$I_{f,R1} = \frac{2,400}{300/5}$$
$$= 40 \text{ A}$$

or, as a multiple of the selected CTS (or the pickup value),

$$\frac{I_{f,R1}}{CTS_1} = \frac{40}{4}$$
$$= 10$$

Since the fastest possible operation is desirable, the smallest TDS is selected. Therefore, the TDS for relay R_1 is

$$TDS_1 = \frac{1}{2}$$

The operating time for relay R_1 can be read from the associated curve in Figure 7.59 as

$$T_1 = 0.15 \text{ s}$$

System Protection

To set relay R_2 as a backup relay, to respond to the balanced three-phase fault at bus 1, it is assumed that the error margin is 0.3. Therefore, its operating time T_2 can be found from Equations 7.122 and 7.123 as

$$T_2 = T_1 + 0.1 + 0.3$$
$$= 0.55 \text{ s}$$

The fault current for a fault at bus 1 as a multiple of the CTS at bus 2 can be found from

$$\frac{I_{f,R1}}{CTS_2} = \frac{40}{4}$$
$$= 10$$

Thus, from the characteristics given in Figure 7.59, for relay R_2 for 0.55 s operating time and 10 ratio, the TDS can be determined as

$$TDS_2 \cong 2.3$$

The next step is to determine the settings for relay R_3. A three-phase fault at bus 2 produces a fault current of 2700 A. Therefore,

$$I_{f,R2} = \frac{2700}{400/5}$$
$$= 33.75 \text{ A}$$

and

$$\frac{I_{f,R2}}{CTS_2} = \frac{33.75}{4}$$
$$= 8.44$$

A new curve for the TDS of 2.3 can be drawn in Figure 7.59 between the two curves shown for the TDSs of 2 and 3. Then, the operating time of relay R_2 can be found from this new curve for the associated multiple of the CTS as 0.60 s. Therefore, permitting the same CDT for relay R_3 to respond to a fault at bus 2 as for relay R_2 responding to a fault at bus 1,

$$T_3 = 0.60 + 0.1 + 0.3$$
$$= 1.0 \text{ s}$$

The corresponding current in the relay can be given as a multiple of the relay pickup current. Thus,

$$\frac{I_{f,R3}}{CTS_3} = \frac{3,000}{(600/5)6}$$
$$= 4.17$$

Therefore, for relay R_3 for a 1.0 s operating time and a 4.17 ratio, the TDS can be determined as

$$TDS_3 \cong 2.8$$

7.19 DIFFERENTIAL PROTECTION

A differential relay can be defined as the relay that operates when the phasor difference of two or more similar electrical quantities exceeds a predetermined amount. Therefore, almost any type of relay, when connected in a particular way, can be made to function as a differential relay provided that (1) it has two or more similar electrical input quantities and (2) these quantities have phase displacement (normally approximately 180°). Thus, it can be said that differential relaying is the most selective relaying principle that detects faults by comparing electrical quantities at all of the terminals of a system element.

Most differential relays are of the "current differential" type. In current differential relaying, CTs are placed in all of the connections to the system element to be protected and their secondaries are connected in parallel with the operating winding of the relay to form a circulating current system, as shown in Figure 7.66a. The dotted line represents the protected zone, which may be a generator, a transformer, a bus, a line segment, etc.

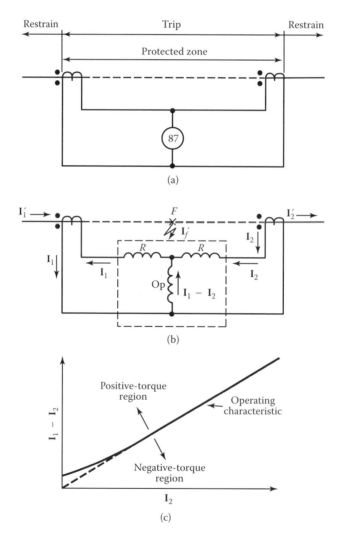

FIGURE 7.66 Protection by using differential relays: (a) principle of differential protection; (b) wiring diagram for differential protection; (c) operating characteristics of a percentage differential relay.

The relay-operating winding is connected between the midpoints (i.e., equipotential points) of the pilot wire. The voltage induced in the secondary of the CTs will circulate a current through the combined impedance of the pilot wires and the CTs, as shown in Figure 7.66b. Normally, the CT secondary currents merely circulate between the CTs and no current flows through the relay operating (indicated with Op in the figure) winding, and therefore, $\mathbf{I}_1 - \mathbf{I}_2 = 0$ which is also the case for the faults occurring outside of the protected zone (i.e., for external faults).

However, should a fault occur at a point F in the protected zone (i.e., internal fault), a difference current of $\mathbf{I}_1 - \mathbf{I}_2 = \mathbf{I}_f$ will flow in the operating winding of the relay, and if it is greater than a predetermined value of current $|\mathbf{I}_p|$, it will cause the relay to trip all of the CBs in the circuits connected to the faulty system element. Therefore, the relay trips when

$$|\mathbf{I}_1 - \mathbf{I}_2| > |\mathbf{I}_p| \tag{7.125}$$

and it restrains when

$$|\mathbf{I}_1 - \mathbf{I}_2| < |\mathbf{I}_p| \tag{7.126}$$

This form of protection is known as Merz–Price protection.

Note that, the relay employed in the wiring diagram of Figure 7.66b is a percentage differential relay. It has an operating winding and two restraining windings. The task of the restraining windings is to prevent undesired relay operation should a current flow in the operating winding due to CT errors during an external fault. The differential current in the operating winding is proportional to $\mathbf{I}_1 - \mathbf{I}_2$, and the equivalent current in the restraining winding is proportional to $\frac{1}{2}(\mathbf{I}_1 + \mathbf{I}_2)$ due to the fact that the two restraining windings are identical. Therefore, the ratio of the differential operating current to the average restraining current is a fixed "percentage."

However, on the basis of the percentage differential relay definition given by the ANSI, the term "through" current is often used to refer to $\mathbf{I}_2$ instead of $\frac{1}{2}(\mathbf{I}_1 + \mathbf{I}_2)$. Therefore, the operating winding current $\mathbf{I}_1 - \mathbf{I}_2$ has to be greater than a certain percentage of the through current $\mathbf{I}_2$ for the relay to operate. Figure 7.66c shows the operating characteristic of such a percentage differential relay. The relay operates in the positive-torque region and restrains in the negative-torque region.

Note that, the purpose of the slope in the operating characteristic of the differential relay is to prevent incorrect operation due to CT error currents that might flow in the differential relay circuit during a severe external fault. These error currents occur because no two CTs will function exactly alike even though they are made to the same specifications and material.

Since they are not absolutely alike, they saturate unequally when high currents flow through them during external faults and their ratio breaks down unequally. When this occurs, the unbalanced current flows in the relay circuit, and the relay has no way of determining whether the current it sees signifies an internal fault or a "mistake" on the part of the CTs, which the relay should ignore.

The percentage differential relays are of two types. One operates on a constant-percentage difference in currents in the two CTs and the other functions on a percentage difference that increases quickly as the fault current increases. The constant-percentage differential relay operates on a 10% slope in order to allow a plus or minus 10% margin for CT errors because of unequal characteristics and saturation. Therefore, the CTs should be selected carefully so that the difference in secondary current output of $\mathbf{I}_1$ and $\mathbf{I}_2$ CTs will not be greater than 5% under maximum fault conditions, which leaves a safety margin of 5% without exceeding the 10% margin built into the relay.

On the other hand, with the increasing-slope differential relay, the margin permitted for CT errors increases quickly as the magnitude of the fault current increases. Thus, the relay operates on a 10% current differential on low-magnitude faults when there is no danger of CT errors and still not function incorrectly during severe external faults.

Power transformers require protection against internal faults and against overheating caused by overloading or prolonged external faults. The effect of prolonged faults on system stability, as well as the possibility of considerable damage to the apparatus, make high-speed relaying essential in most cases. Figure 7.67 shows the wiring diagram for differential protection of a two-winding power transformer. It can be shown that

$$\frac{I'_1}{I'_2} = \frac{N_2}{N_1} \tag{7.127}$$

where N_1 and N_2 are the turns of the primary and secondary windings of the power transformer, respectively. Note that, at the two CTs

$$I_1 = \frac{I'_1}{n_1} \tag{7.128}$$

and

$$I_2 = \frac{I'_2}{n_2} \tag{7.129}$$

where n_1 and n_2 are the turns ratios of the two CTs located on the primary and secondary sides of the power transformer, respectively. Since under normal conditions or during the external faults,

$$I_1 - I_2 = 0 \tag{7.130}$$

or

$$\frac{I'_1}{I'_2} = \frac{n_1}{n_2} \tag{7.131}$$

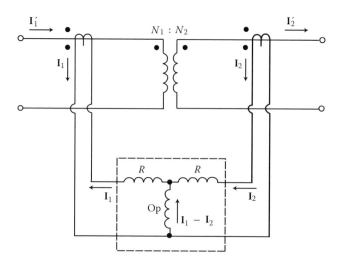

FIGURE 7.67 Differential protection of transformer.

System Protection

substituting Equation 7.131 into Equation 7.127,

$$\frac{N_2}{N_1} = \frac{n_1}{n_2} \qquad (7.132)$$

If there is an internal fault located on the primary side of the power transformer, as shown in Figure 7.67, there will be a current of

$$I_1 - I_2 = \frac{I'_f}{n_1} \qquad (7.133)$$

flowing in the operating winding of the relay. Whereas, if such a fault is located on the secondary side of the power transformer,

$$I_1 - I_2 = \frac{I'_f}{n_2} \qquad (7.134)$$

The differential relays used for transformer protection are of the constant percentage type. However, if the power transformer has a tapping range enabling its ratio to be varied, this must be allowed for in the differential system. Otherwise, there will be a differential current flowing in the operating winding of the relay under the normal-load conditions.

If the three-phase power transformer windings are connected delta–wye or wye–delta, the high-voltage side line current leads the low-voltage side line current by 30°, regardless of whether the wye or the delta winding is on the high-voltage side.* Thus, the balanced three-phase through current goes through a phase change of 30°, which has to be corrected in the CT secondary terminals by appropriate connection of the CT secondary windings.

Furthermore, zero-sequence current flowing on the wye side of the power transformer will not produce current outside the delta on the other side. Thus, the zero-sequence current has to be eliminated from the wye side by connecting the secondary windings of the CTs in delta on the wye-connected side of the power transformer and in wye on the delta-connected side in order to provide the 30° phase shift.

The reason for such CT connections can be further explained by the fact that each line current on the delta-connected side of the power transformer bank is the difference of the currents in two windings of the power transformer, whereas the line current on the wye-connected side is the same as the current in one winding of the transformer.

Hence, the line current on the delta side should be compared with the difference of two line currents on the wye side by connecting the secondary windings of the CTs in delta. If the power transformer were connected in wye–wye, the CTs on both sides would need to be connected in delta.

When CTs are connected in delta, their secondary ratings must be reduced to $1/\sqrt{3}$ times the secondary rating of wye-connected CTs in order that the currents outside the delta may balance with the secondary currents of the wye-connected CTs. Differential relays connected in the aforementioned manner provide protection against turn-to-turn faults as well as against ground faults, phase-to-phase faults, and faults in the leads to the CBs.

Similarly, a three-winding transformer bank can be protected by use of percentage differential relays having three restraining windings one for each transformer winding. However, such a transformer bank can be protected by use of percentage differential relays having only two restraining windings if the third transformer winding is only a delta-connected tertiary with no connections brought out.

---

* Only if ANSI Standard connections are in use.

In the aforementioned discussions, the magnetizing current of the power transformer has not been taken into account. When the transformer is in the normal operation mode, this current is very small. However, when the transformer is first energized, there will be a transient "magnetizing-inrush current" whose magnitude can be 8–12 times the rated load current. It may decay with a time constant, being from perhaps 0.1 s for a 100-kVA transformer up to 1.0 s for a large unit.

The maximum inrush current is produced when the transformer is energized at the zero point of the voltage wave. Residual flux can increase the current still further. Since the inrush current appears in only one primary winding of the transformer, it will cause false operation of the differential relay. Switching at other instants of the voltage wave produces lower values of transient current.

There is a similar, but lesser, inrush when the voltage across a transformer recovers after the clearing of an external fault. In this case, the transformer is partially energized so the "recovery" inrush is less than the "initial" inrush. Also, when a transformer bank is paralleled with a second energized bank, the energized bank can have a sympathetic inrush.

The offset inrush current of the transformer bank switched on the line will find a parallel path in the energized bank, and the dc component may saturate the iron core, causing an "apparent" inrush. Again, this inrush is less than that of the initial inrush, and its magnitude depends on the relative value of the transformer impedance to that of the rest of the system, which forms an additional parallel circuit.

The magnetizing inrush phenomenon produces current input to the primary winding, which has no equivalent on the secondary side. Thus, it appears as unbalanced and cannot be set apart from the internal fault. However, since it is transient in nature, stability can be achieved by providing a small time delay. The standard method for preventing relay operation during the energization period is to utilize the high level of harmonics that are present in the inrush current. The harmonic components in the inrush current are filtered out and introduced into a circuit that restrains the relay from operating. Such relay is called *a harmonic restrained-percentage differential relay*.

Other methods include (1) desensitization, (2) using a voltage-operated automatic tripping suppressor unit in conjunction with the differential relay, and (3) using a differential relay with reduced sensitivity to the inrush wave in terms of having a higher pickup to the offset wave in addition to time delay to override the high initial peaks.

There are other types of relays that are used for transformer protection against internal faults. One such relay is the *Buchholz relay*. It is a gas pressure–actuated relay and is used for alarm and tripping for various fault conditions, including hot spots on the core due to short circuit of lamination insulation and core insulation failure. The Buchholz relay is often called "a sudden pressure relay." Figure 7.68 shows unit generator–transformer protection using differential relays. Note that, there is no generator breaker in this unit protection method. The step-up transformer is usually connected grounded wye–delta, and the neutral of the wye-connected generator is high-resistance grounded through a distribution transformer.

Figure 7.69 shows the differential protection of a bus or a switchgear.* In this scheme, discrimination between internal and external faults is made on the basis of the voltage magnitude that appears across the relay. When an external fault takes place, there is substantially less voltage produced across the relay circuit due to the fact that the secondary current is only opposed, at the most, by the CT terminal resistance and the internal secondary impedance of the CT in the faulted circuit when it is completely saturated.

On the other hand, when internal faults take place, the secondary current is opposed by the magnetizing impedance of the CTs and by the high impedance of the relay circuit, thus producing a large operating voltage on the relay and causing its operation. In this scheme, false relay operations

---

* Switchgear is the power substation apparatus that is used to direct the "how" of power and to isolate power equipment or circuits. It includes circuit breakers, disconnect switches, buses, connections, and the structures on which they are mounted.

System Protection

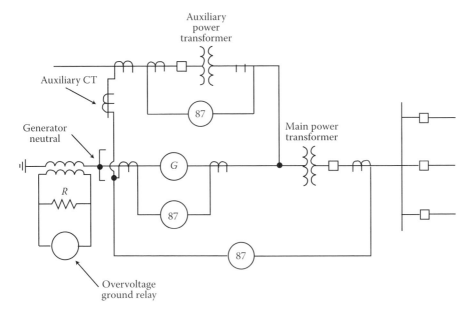

FIGURE 7.68 Unit generator–transformer protection.

can be prevented by adjusting the minimum pickup of the relay well above the maximum voltage produced on an external fault.

It is interesting to note that the problem of CT saturation can be eliminated at its source by using air core CTs called *linear couplers*. The number of secondary turns of such a linear coupler is much greater than the one of a bushing CT. They can be employed without damage with their secondaries open-circuited. Their secondary-excitation characteristic is a straight line with a slope of approximately 5 V per 1000 AT.

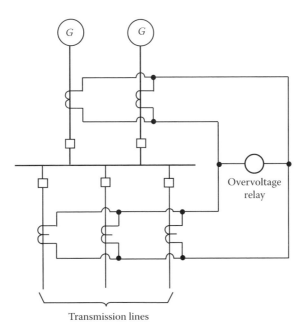

FIGURE 7.69 Differential protection of bus.

EXAMPLE 7.11

Consider a three-phase delta–wye-connected two-winding transformer bank that is to be protected by using percentage differential relays. Assume that the high-voltage side is connected in delta and sketch the necessary wiring diagram for connecting three differential relays one for each phase.

Solution

The solution is self-explanatory, as shown in Figure 7.70.

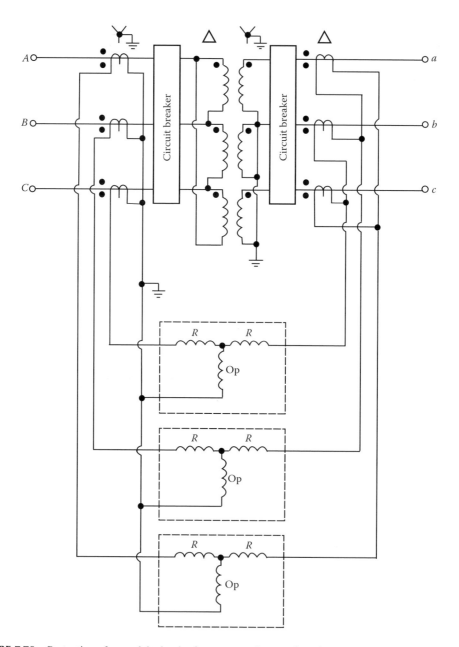

FIGURE 7.70 Protection of wye–delta bank of power transformers by using percentage differential relays.

EXAMPLE 7.12

Consider a three-phase wye–delta–wye-connected three-winding transformer bank that is to be protected by using percentage differential relays. Assume that the tertiary windings are connected in delta and sketch the necessary wiring diagram for connecting three differential relays (each with three restraining windings) one for each phase.

Solution

The solution is self-explanatory, as shown in Figure 7.71.

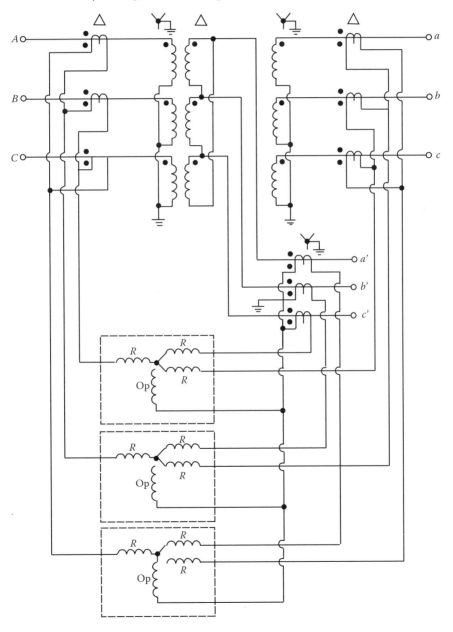

FIGURE 7.71 Protection of wye–delta bank of a three-winding power transformer by using percentage differential relays.

EXAMPLE 7.13

Assume that a three-phase delta–wye-connected, 30-MVA, 69/34.5-kV power transformer is protected by the use of percentage differential relays, as shown in Figure 7.70. If the CTs located on the delta and wye sides are of 300/5 and 1200/5 A, respectively, determine the following:

(a) Output currents of both CTs at full load
(b) Relay current at full load
(c) Minimum relay current setting to permit 25% overload

Solution

(a) The high-voltage side line current is

$$I_{HV} = \frac{30 \times 10^6}{\sqrt{3}(69 \times 10^3)} = 251.02 \text{ A}$$

The low-voltage side line current is

$$I_{LV} = \frac{30 \times 10^6}{\sqrt{3}(34.5 \times 10^3)} = 502.04 \text{ A}$$

Therefore, the output current of the CT located on the high-voltage side is

$$251.02 \left(\frac{5}{300}\right) = 4.1837 \text{ A}$$

and the output current of the CT located on the low-voltage side is

$$502.04 \left(\frac{5}{1200}\right)\sqrt{3} = 3.6232 \text{ A}$$

Note that, the winding current of the delta-connected cr is multiplied by $\sqrt{3}$ to obtain its line current.

(b) The relay current at full load is

$$4.1837 - 3.6232 = 0.5605 \text{ A}$$

(c) Thus, the minimum relay current setting to permit 25% overload is

$$(1.25)(0.5605) = 0.7007 \text{ A}$$

Note that, the output current of the CT located on the low-voltage side is 3.6232 A, contrary to the output current of 4.1837 A of the CT located on the high voltage side. Therefore, it cannot balance the 4.1837 A produced by the high-voltage side under balanced normal-load conditions. Therefore, a "mismatch" condition has been created. Such current mismatch should always be checked to ensure that the relay taps selected have an adequate safety margin. Where necessary, current mismatch values can be reduced by using a set of three auxiliary CTs with turns ratios of

$$\frac{4.1837}{3.6232} = 1.1547$$

or simply by changing CT taps or the tap settings on relay coils themselves. Transformer differential relays ordinarily have abundant restraint taps available so that auxiliary CTs usually are not necessary.

7.20 PILOT RELAYING

Pilot relaying, in a sense, is a means of remote controlling the CBs. Here, the term "pilot" implies that there is some type of channel (or medium) that interconnects the ends of a transmission line over which information on currents and voltages, from each end of the line to the other, can be transmitted. Such systems employ high-speed protective relays at the line terminals in order to ascertain in as short a time as possible whether a fault is within the protected line or external to it.

If the fault is internal to the protected line, all terminals are tripped in high speed. If the fault is external to the protected line, tripping is blocked (i.e., prevented). The location of the fault is pointed out either by the presence or the absence of a pilot signal. The advantages of such high-speed simultaneous clearing of faults at all line terminals by opening CBs include (1) minimum damage to equipment, (2) minimum (transient) stability problems, and (3) automatic reclosing.

Pilot relaying, being a modified form of differential relaying, is the best protection that can be provided for transmission lines. It is inherently selective, suitable for high-speed operation, and capable of good sensitivity. It is usually used to provide primary protection for a line. Backup protection may be provided by a separate set of relays (step-distance relaying), or the relays employed in the pilot may also be used to provide a backup function. The types of pilot channels available for protective relaying include

1. *Separate wire circuits*, called *pilot wires*, operating at power system frequency, audio-frequency tones, or in dc. They can be made up of telephone lines either privately owned or leased. Refer to Figure 7.72b.
2. *Power line carriers*, which use the protected transmission line itself to provide the channel medium for transmission of signals at frequencies of 30–300 kHz. These are the most widely used "pilots" for protective relaying. The carrier transmitter–receivers are connected to the transmission line by coupling capacitor devices that are also used for line voltage measurement. Line traps tuned to the carrier frequency are located at the line terminals, as shown in Figure 7.73. They prevent an external fault behind the relays from shorting out the channel by showing a high impedance to the carrier frequency and a low impedance to the power frequency. The radiofrequency choke acts as a low impedance to 60 Hz power frequency but as a high impedance to the carrier frequency. Therefore, it protects the apparatus from high voltage at the power frequency and, simultaneously, limits the attenuation of the carrier frequency.
3. *Microwave channel*, which uses beamed radio signals, usually in the range of 2–12 GHz, between line-of-sight antennas located at the terminals. This channel can also simultaneously be used for other functions. A continuous tone of one frequency, called the *guard frequency*, is transmitted under normal (or *nonfault*) conditions. When there is an internal fault, the audio tone transmitter is keyed by the protective relaying scheme so that its output is shifted from the guard frequency to a trip frequency.

Pilot-relaying systems use either phase comparison or directional comparison to detect faults. In the phase comparison, the phase position of the power system frequency current at the terminals is compared. Amplitude modulation is used in a phase comparison system. The phase of the modulating signal waveform is not affected by signal attenuation. Identical equipment at each end of the line

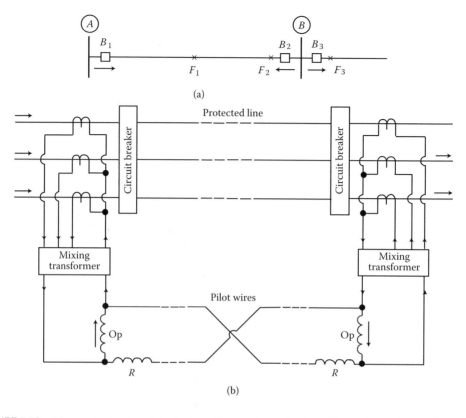

FIGURE 7.72 Line protection by pilot relaying: (a) example application; (b) one form of pilot-wire relaying application.

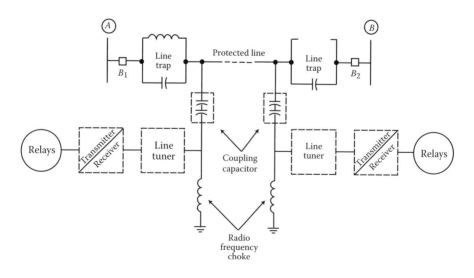

FIGURE 7.73 One-line diagram of power line carrier for pilot relay system.

are modulated in phase during an internal fault and in antiphase when a through-fault current flows (due to an external fault).

Hence, current flow through the line to an external fault is considered 1800 out of phase, and tripping is blocked. If the currents are relatively in phase, an internal fault is indicated, and the line is tripped. Thus, modulation is of the all-or-nothing type, producing half-cycle pulses of carrier signals interspersed with half-periods of zero signals, as shown in Figure 7.74.

Note that, during an external fault, the out-of-phase modulation results in transmission of the carrier signal to the line alternately from each end. Therefore, the transmission from one end fills in the gaps in the signals from the other and vice versa, providing a continuous signal on the line. The presence of the signal is used to block the tripping function. On the other hand, when there is an internal fault, the resulting signal on the line has half-period gaps during which the tripping function, initiated by the relay, is completed.

A pilot-relaying system can also use directional comparison to detect faults. In this case, the fault-detecting relays compare the direction of power flow at the line terminals. Power flow into the line at the terminals points out an internal fault, and the line is tripped. If power flows into the line at one end and out at the other, the fault is considered external, and tripping is not allowed.

Consider the line shown in Figure 7.72a. Assume that directional relays are used and high-speed protection is provided for the entire line (instead of the middle 60%) by pilot relaying. Therefore, both faults F_1 and F_2 are detected as internal faults by the relays located at B_1 and B_2 and are therefore cleared at a high speed.

Note that, both relays see the fault current flowing in the forward direction. Thus, when this information is impressed on the signal by modulation and transmitted to the remote ends over a pilot channel, it is confirmed that the faults are indeed on the protected line. Now assume that there is a

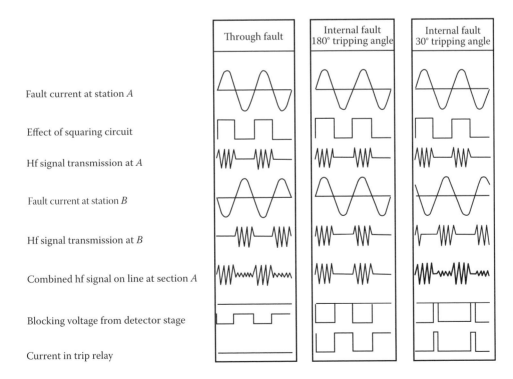

FIGURE 7.74 Carrier current phase comparison: key to operation. (From GEC Measurements Ltd., *Protective Relays Application Guide*, 2nd ed. GEC Measurements Ltd., Stafford, UK, 1975.)

fault at F_3. The relay at B_2 sees it as an external fault and the relays at B_1 and B_3 see it as an internal fault. Upon receiving this directional information at B_1, that relay will be able to block tripping for the fault at F_3.

7.21 COMPUTER APPLICATIONS IN PROTECTIVE RELAYING

Computers have been widely used in the electric power engineering field since the 1950s. The applications include a variety of off-line or on-line tasks. Examples of off-line tasks include fault studies, load-flow studies, transient stability calculations, unit commitment, and relay settings and coordination studies.

Examples of on-line tasks are economic generation scheduling and dispatching, load frequency control, supervisory control and data acquisition (SCADA), sequence-of-events monitoring, sectionalizing, and load management. The applications to computers in protective relaying have been primarily in relay settings and coordination studies and computer relaying.

7.21.1 COMPUTER APPLICATIONS IN RELAY SETTINGS AND COORDINATION

Today, there are various commercially available computer programs that are being used in the power industry to set and coordinate protective equipment. Advantages of using such programs include (1) sparing the relay engineer from routine, tedious, and time-consuming work; (2) facilitating system-wide studies; (3) providing consistent relaying practices throughout the system; and (4) providing complete and updated results of changes in system protection.

In 1960, the Westinghouse Electric Corporation developed its well-known protective device coordination program. Today, it is one of the most comprehensive and complete programs for applying, setting, and checking the coordination of various types of protective relays, fuses, and reclosers. Figure 7.75 shows the block diagram of the program.

Note that, the user must specify the input data for the "data check study" block in terms of both device type and settings for each relay, fuse, or recloser. The program then evaluates the effectiveness of these devices and settings within the existing system and, if necessary, recommends alternative protective devices. Whereas in the "coordination study" block, the user specifies the protective device with no settings, or permits the program to select a device. The program then establishes settings within the ranges specified or it selects a device and settings.

The settings and/or devices are chosen to optimize coordination. The "final coordination study" block shows how the system will behave with the revised settings, which can then be issued by the relay engineer [2]. Of course, no computer program can replace the relay engineer. Such a program is simply a tool to aid the engineer by indicating possible problems in the design and their solutions. The engineer has to use engineering judgment, past experience, and skill in determining the best protection of the system.

7.21.2 COMPUTER RELAYING

Computer hardware technology has advanced considerably since the early 1960s. Newer generations of mini- and microcomputers tend to make digital computer relaying a viable alternative to the traditional relaying systems. Indeed, it appears that a simultaneous change is taking place in traditional relaying systems, which are beginning to use solid-state analog and digital subsystems as their building blocks.

The use of digital computers for protection of power system equipment, however, is of relatively recent origin. The first serious proposals appeared in the late 1960s. For example, in 1966, Last and Stalewski [12] suggested that digital computers should be used in an on-line mode for protection of power systems. Since then, many authors have developed digital computer techniques for protection of lines, transformers, and generators. Significant contributions have been made in the area of line

System Protection 463

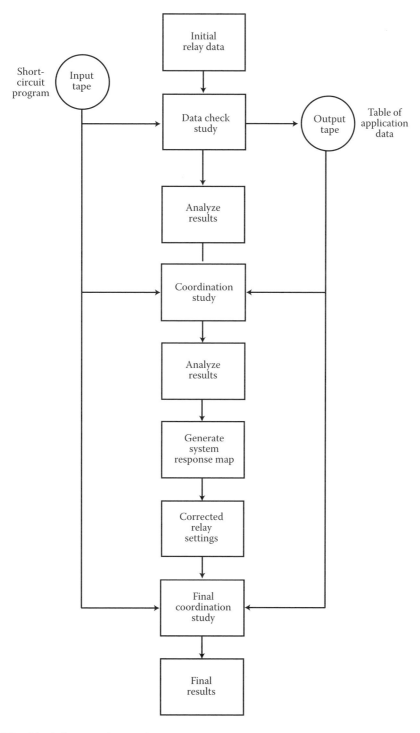

FIGURE 7.75 Block diagram of protective device protection program of Westinghouse Electric Corporation. (From Westinghouse Electric Corporation, *Applied Protective Relaying*. Relay-Instrument Division, WEC, Newark, NJ, 1976.)

protection [13–16]. However, protection of transformers [17,18] and generators [19] using digital computers has somewhat received less attention. In 1969, the feasibility of protecting a substation by a digital computer was investigated by Rockefeller [20]. It examines protection of all types of station equipment into one unified system.

Researchers suggested the use of minicomputers in power system substations for control, data acquisition, and protection against faults and other abnormal operating conditions. This has become known as *computer relaying* and several papers have been written on various techniques of performing relaying functions. Many of these techniques basically use the three-phase lines to duplicate existing relaying characteristics. A few field installations have been made to demonstrate computer relaying techniques and to show that computers can survive in the harsh substation environment. However, today, many utility companies have installations or plans for future implementations of computers as SCADA remotes.

These substation control computers receive data from the transmit data to the central dispatch control computer. It is interesting to note that most of the papers written on computer relaying have been under the auspices of universities. These tend to focus on algorithms and software or on models that can be tested on multipurpose minicomputers or on special-purpose circuits and hardware that are laboratory oriented. For example, a great deal of research has been undertaken at the University of Missouri at Columbia (UMC), since 1968, to develop a computer relaying system that would permit the computer to perform relaying as well as other substation functions [13,21–24]. In 1974, in cooperation with the University of Saskatchewan, the first computer relaying short course was held at the UMC. Since that time, the short course has been offered many times.*

However, the "real-world" test installations have been the result of cooperation between utilities and manufacturers and are mainly concerned with line protection using an impedance algorithm [25]. For example, in 1971, Westinghouse installed the PRODAR 70 computer in the Pacific Gas and Electric Company's Tesla Substation for protection of the Bellota 230-kV line. Also, in 1971, a computer relaying project was initiated by American Electric Power and later joined by IBM Service Corporation. The purpose of the project was to install an IBM System 7 in a substation to perform protective relaying and few data-logging functions [15].

General Electric initiated a joint project with Philadelphia Electric Company to install computer relaying equipment on a 500-kV line. Therefore, it can be said that when the minicomputer became available, the industry realized the potential of the relatively low-cost computer and tried various applications. However, the extremely high costs of software programs to implement specific functions have played an inhibitive role. With the advent of microcomputers, the hardware costs can be further reduced. This permits software simplification since a microprocessor can be dedicated to a specific function.

REFERENCES

1. GEC Measurements Ltd., *Protective Relays Application Guide*, 2nd ed. GEC Measurements Ltd., Stafford, UK, 1975.
2. Westinghouse Electric Corporation, *Applied Protective Relaying*. Relay-Instrument Division, WEC, Newark, NJ, 1976.
3. General Electric Company, *SLS 1000 Transmission Line Protection*, Appl. Manual GET-6749. General Electric Co., Schenectady, New York, 1984.
4. Westinghouse Electric Corporation, *Electrical Transmission and Distribution Reference Book*. WEC, East Pittsburgh, Pennsylvania, 1964.
5. Atabekov, G. I., *The Relay Protection of High Voltage Networks*. Pergamon Press, New York. 1960.
6. Mason, C. R., *The Art and Science of Protective Relaying*. Wiley, New York, 1956.

---

* In September 1969, the IEEE relays Committee formed the Computer Relaying Subcommittee with Dr. James R. Tudor as its chairman and the Substation Computer Working Group with Dr. Lewis N. Walker as its chairman.

7. Neher, J. H., A comprehensive method of determining the performance of distance relays. *Trans. Am. Inst. Electr. Eng.* 56, 833–844, 1515 (1937).
8. Warrington, A. R. van C., *Protective Relays: Their Theory and Practice*, Vol. l. Chapman & Hall, London, 1976.
9. Warrington, A. R. van C., *Protective Relays: Theory and Practice*, 2nd ed., Vol. 2. Chapman & Hall, London, 1974.
10. Gross, C. A., *Power System Analysis*. Wiley, New York, 1979.
11. General Electric Company, *Distribution System Feeder Overcurrent Protection*, Appl. Manual GET-6450. General Electrical Co., Schenectady, New York, 1979.
12. Last, F. H., and Stalewski, A., Protective gear as a part of automatic power system control. *IEEE Conf. Publ.* 16, Part I, 337–343 (1966).
13. Walker, L. N., Ogden, A. D., Ott, G. E., and Tudor J. R., Special purpose digital computer requirements for power system substation needs. *IEEE Power Eng. Soc. Winter Power Meet.*, 1970, Pap. No. 70 CP 142-PWR (1970).
14. Hope, G. S., and Umamaheswaran, V. S., Sampling for computer protection of transmission lines. *IEEE Trans. Power Appar. Syst.* PAS-93 (5), 1524–1534 (1974).
15. Phadke, A. G., Hlibka, T., and Ibrahim, M., A digital computer system for EHV substations: Analysis and field tests. *IEEE Power Eng. Soc. Summer Meet.*, 1975, Pap. No. F75 543-9 (1975).
16. Mann, B. J., and Morrison, I. F., Digital calculation of impedance for transmission line protection. *IEEE Trans. Power Appar. Syst.* PAS-91 (3), 1266–1272 (1972).
17. Sykes. J. A., and Morrison, I. F., A proposed method of harmonic-restraint differential protection of transformers by digital computer. *IEEE Trans. Power Appar. Syst.* PAS-91 (3), 1266–1272 (1972).
18. Sykes, J. A., A new technique for high-speed transformer fault protection suitable for digital computer implementation. *IEEE Power Eng. Soc. Summer Meet.*, 1972, Pap. No. C72 429-9 (1972).
19. Sachdev, M. S., and Wind, D. W., Generator differential protection using a hybrid computer. *IEEE Trans. Power Appar. Syst.* PAS-92 (6), 2063–2072 (1973).
20. Rockefeller, G. D., Fault protection with a digital computer. *IEEE Trans. Power Appar. Syst.* PAS-88, 438–462 (1969).
22. Walker, L. N., Ogden, A. D., Ott, G. E., and Tudor, J. R., Implementation of high frequency transient fault detector. *IEEE Power Eng. Soc. Winter Power Meet.*, 1970, Pap. No. 70CP 140-PWR (1970).
23. Walker, L. N., Ott, G. E., and Tudor, J. R., Simulated power transmission substation. *SWIEEECO Rec. Tech. Pap.*, 153–162 (1970).
24. Walker, L. N., Analysis, design, and simulation of a power transmission substation control system. Ph.D. Dissertation, University of Missouri-Columbia, Columbia, 1970.
25. Horowitz, S. H., ed., *Protective Relaying for Power Systems*. IEEE Press, New York, 1980.

GENERAL REFERENCES

Anderson, P. M., Reliability modeling of protective system. *IEEE Trans. Power Appar. Syst.* PAS·t03 (8), 2207–2214 (1984).
Baird, T. C., Walker, L. N., and Tudor, J. R., Computer relaying: An update. *Mo. Valley Elect. Assoc. Conf.* 1976 (1976).
Gönen, T., *Electric Power Distribution System Engineering*. CRC Press, Boca Raton, FL, 2008.
IEEE Working Group Report, Central computer control and protection functions. *IEEE Trans. Power Appar. Syst.* PAS-97 (1), 7–16 (1978).
Kurihara, T., *Computerized Protection of a Three-Phase Transmission Line*, Power Affiliates Rep. No. 73184. Iowa State University, Ames, IA, 1973.
Stevenson, W. D., Jr., *Elements of Power System Analysis*, 4th ed. McGraw-Hill, New York, 1982.

PROBLEMS

1. Consider the system shown in Figure P7.1 and do the following:
   (a) Sketch the zones of protection.
   (b) Describe the backup protection necessary for a fault at fault point F_1.
   (c) Repeat part (b) for a fault at F_2.
   (d) Repeat part (b) for a fault at F_3.

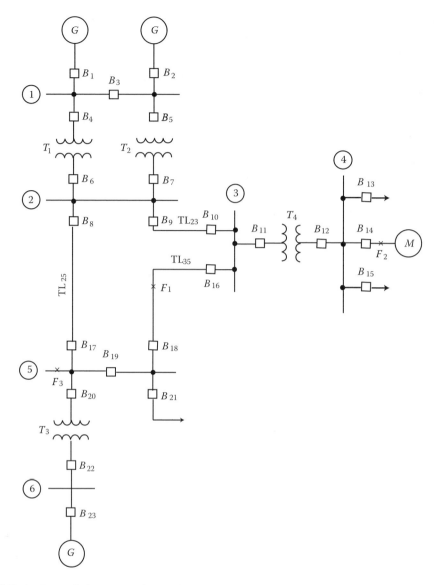

FIGURE P7.1 Transmission system for Problem 1.

2. Consider the system shown in Figure P7.2 and repeat Problem 1.
3. Consider the system, with overlapped primary protective zones, shown in Figure P7.3 and assume that there is a fault at fault point F. Describe the necessary primary and backup protection with their possible implications.
4. Consider the system shown in Figure P7.4 and determine the locations of the necessary backup relays in the event of having a fault at the following locations:
   (a) Fault point F_1
   (b) Fault point F_2
5. Verify Equation 7.11 using Equation 7.9.
6. Verify Equation 7.12 using Equation 7.9.

System Protection

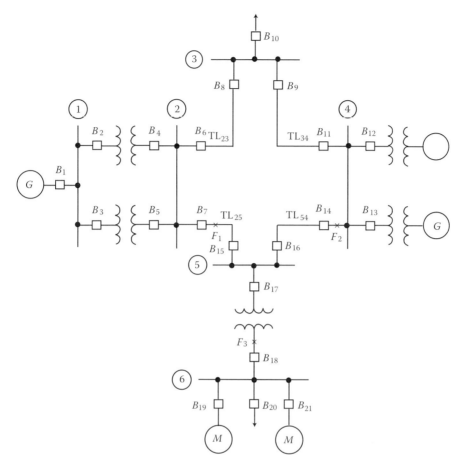

FIGURE P7.2 Transmission system for Problem 2.

7. Assume that a zero-sequence current filter is required to be used to protect a three-phase 4.16-kV, 1.7-MVA load against more than 10% zero-sequence currents. Design a zero-sequence filter to detect this condition.
8. Consider Figure 7.22b and Equation 7.13 and assume that VF is the open-circuit voltage at the output terminals 1 and 2.
   (a) Find the Thévenin's equivalent impedance at terminals 1 and 2.
   (b) If a relay with a resistance R_L is connected between terminals 1 and 2 and $R_l = \sqrt{3}X_m$, verify that the current in relay is

$$I_L = \frac{2R_1 I_{a2}}{R_0 + R_1 + Z_S + R_L}$$

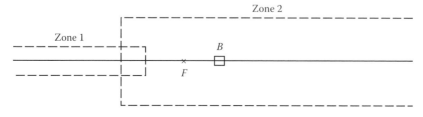

FIGURE P7.3 Protected system for Problem 3.

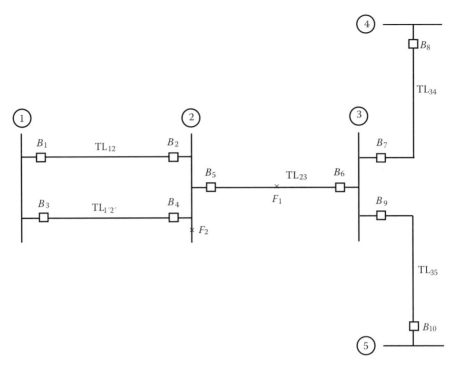

FIGURE P7.4 System for Problem 4.

9. Consider the current equation given in Problem 8b and assume that $R_0 = R_L = 02\ \Omega$, $Z_s = 0.1\ \Omega$, $I_L = 5$ A, and $I_{a2} = 15$ A. Determine the following:
   (a) Required value of R_1
   (b) Required value of X_m
10. Repeat Example 7.1 assuming that $Z_B = 1.0\ \Omega$.
11. Repeat parts (a), (b), and (d) of Example 7.3 assuming that the load is reduced by half.
12. Derive Equation 7.60 from Equation 7.59.
13. Derive Equation 7.64 from Equation 7.58.
14. Derive Equation 7.78 from Equation 7.77.
15. Derive Equation 7.80 from Equation 7.79.
16. Consider the two input signals S_1 and S_2 given by Equations 7.68 and 7.69 and assume that the threshold characteristics for the phase comparator is $|S_1 S_2^*|$ so that the phasor S_1 has a positive projection of S_2. Show mathematically that the above threshold of operation will be satisfied as long as $\theta_1 - \theta_2 \leq \dfrac{\pi}{2}$.
17. Repeat Example 7.5 for the directional distance relay located at B whose forward direction is in the direction from bus B to bus A. Assume that the power flow direction is from bus B to bus A.
18. Repeat Example 7.6 assuming that the fault point F is located at 0.78-pu distance away from bus A.
19. Repeat Example 7.6 assuming that the arc resistance is increased by a 75-mph wind and that the zone 2 relay unit operates at a time delay of 18 cycles.
20. Repeat Example 7.8 assuming that the transmission system is being operated at 138 kV line-to-line voltage and at a maximum peak load of 50 MVA at a lagging power factor of 0.85.
21. Repeat Example 7.9 using the results of Problem 20 and a 45° mho relay characteristic.

System Protection

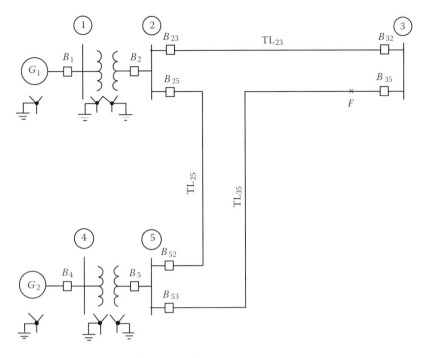

FIGURE P7.5 Transmission system for Problem 22.

22. Consider the 345-kV transmission system shown in Figure P7.5. Assume that all three lines are identical with positive-sequence impedance of 0.02 + j0.2 pu and that the megavolt-ampere base is 200 MVA. Assume also, that all six line breakers are commanded by directional impedance distance relays and considerer only three-phase faults. Set the settings of zones 1, 2, and 3 for 80%, 120%, and 250%, respectively. Determine the following:
   (a) Relay settings for all zones in per units.
   (b) Relay setting for all zones in ohms, if the VTs are rated $(345 \times 10^3/\sqrt{3}):69$ V and the CTs are rated 400:5 A.
   (c) If there is a fault at fault point F located on the line TL_{35} at a 0.15-pu distance away from bus 3, explain the resulting relay operations.
23. Consider the transmission line shown in Figure P7.6. Assume that the line is compensated by series capacitors in order to improve stability limits and voltage regulation and to maximize the load-carrying capability of the system. Assume that the series capacitors are located at the terminals due to economics and that X_{CC} is equal to X_{CB}. Furthermore, assume that the line is protected by directional mho-type distance relays located at band C. Do the following:
   (a) Determine whether the series capacitors present any problem for the relays. If so, what are they?
   (b) Sketch the possible locus of the line impedance on the R–X diagram.

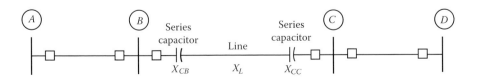

FIGURE P7.6 Transmission system for Problem 23.

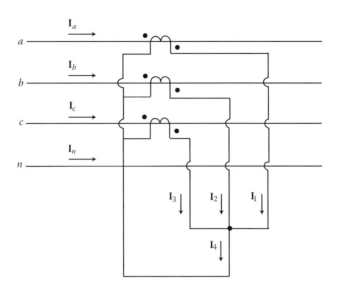

FIGURE P7.7 CT connection for Problem 24.

(c) Sketch the operating characteristic of the distance relay located at B and set to protect the line BC.

(d) Sketch the operating characteristic of the distance relay located at B and set to protect line BA.

24. Consider the CT connection shown in Figure P7.7 and assume that the currents $\mathbf{I}_1$, $\mathbf{I}_2$, and $\mathbf{I}_3$ are measured to be $5\angle 90°$, $3\angle 0°$, and $4\angle -110°$ A, respectively. If each CT has a ratio of 500: 5A, determine the following:
    (a) Neutral current $\mathbf{I}_n$
    (b) Current $\mathbf{I}_4$
    (c) Relationship between currents $\mathbf{I}_n$, $\mathbf{I}_4$, and $\mathbf{I}_{a0}$

25. Repeat Example 7.10 assuming that it is a 11-kV radial system and that the CT ratios are 200:5, 300:5, and 500:5 at buses 1, 2, and 3, respectively.

26. Consider the percentage differential relay protection scheme shown in Figure 7.66b. Assume that the relay has a 0.1 A minimum pickup and 10% slope of characteristic $\mathbf{I}_1 - \mathbf{I}_2$ versus $\dfrac{\mathbf{I}_1 - \mathbf{I}_2}{2}$. The current ratio of each CT is 300:5 A. Assume that a high-resistance ground fault occurred at fault point F and the currents $\mathbf{I}'_1$ and $\mathbf{I}'_2$ are $330 + j0$ and $316 + j0$ A, respectively. Determine whether the relay will operate to trip its CB.

27. Repeat Problem 26 assuming that the currents $\mathbf{I}'_1$ and $\mathbf{I}'_2$ are $330 + j0$ A and $30 + j0$ A, respectively.

28. Consider the protection of a wye–delta bank of power transformers by using percentage differential relays, as shown in Figure 7.70. Assume that the high-voltage side is the primary side of the power transformer. Redraw the protection scheme showing the directions of the currents flowing in all circuits. Clearly identify each current.

8 Power Flow Analysis

8.1 INTRODUCTION

Power flow* (or load flow) is the solution for the normal balanced three-phase steady-state operating conditions of an electric power system. In general, power flow calculations are performed for power system planning and operational planning, and in connection with system operation and control. The data obtained from power flow studies are used for the studies of normal operating mode, contingency analysis, outage security assessment, and optimal dispatching and stability. Note that, normally it is understood that a "contingency" is the loss of a major transmission element or a large generating unit. When a system is able to withstand any single major contingency, it is now called "N-1 secure" according to the new terminology suggested by National Electric Reliability Council (NERC). A "double contingency" is the loss of two transmission lines, or two generators, or a line and a generator [1]. In this case, it can be said that such system is "N-2 secure."

As has been succinctly put by Shipley [2], "the difficulties and the importance of the power flow problem have fascinated mathematicians and engineers throughout the world for a number of years. Many people have devoted a large proportion of their professional life to the solution of the problem. It has received more attention than all the other power system problems combined. The amount of effort devoted to the problem has resulted in an enormous amount of technical publications. The nature of the problem probably precludes the development of a perfect procedure. Therefore, it is likely that progress will continue to be made on improved solutions for a long time."

Before 1929, all *power flow* (or *load flow*, as it was then called) calculations were made by hand. In 1929, network calculators (of Westinghouse) or network analyzers (of General Electric) were employed to perform power flow calculations. The first paper describing the first digital method to solve the power flow problem was published in 1954 [3]. However, the first really successful digital method was developed by Ward and Hale in 1956 [4]. Most of the early iterative methods were based on the Y-matrix of the Gauss–Seidel method [4–10]. It requires minimum computer storage and needs only a small number of iterations for small networks. Unfortunately, as the size of the network is increased, the number of iterations required increases dramatically for large systems. In some cases, the method does not provide a solution at all.

Therefore, the slowly converging behavior of the Gauss–Seidel method and its frequent failure to converge in ill-conditioned situations caused the development of the Z-matrix methods [11–25]. Even though these methods have considerably better converging characteristics, they also have the disadvantage of needing a significantly larger computer storage memory owing to the fact that the Z-matrix is full, contrary to the Y-matrix, which is sparse.

These difficulties encountered in load flow studies led to the development of the Newton–Raphson method. The method was originally developed by Van Ness and Griffin [16–18] and later by others [19–24]. The method is based on the Newton–Raphson algorithm to solve the simultaneous quadratic equations of the power network. Contrary to the Gauss–Seidel algorithm, it needs a larger time per iteration, but only a few iterations, and is significantly independent of the network size.

---

* The term *load* in power system literature is somewhat ambiguous. For example, it may be used to describe current, power, or impedance depending on context. It is interesting to note that, in this country until 1929, a power flow study has been called as *power flow study*. However, after 1929, the name has been changed to *load flow study*. However, recently, IEEE suggested again the use of "power flow study" instead of "load flow study." This author is also biased toward the use of the name power flow study.

Therefore, most of the power flow problems that could not be solved by the Gauss–Seidel method (e.g., systems with negative impedances) are solved with no difficulty by this method. However, it was not computationally competitive on large systems because of the rapid increase of computer time and storage requirement with problem size.

However, the development of a very efficient sparsity-programmed ordered elimination technique by Tinney and others [24a,24b] to solve the simultaneous equations has enhanced the efficiency of the Newton–Raphson method, in terms of speed and storage requirements, and has made it the most widely used power flow method. The method has been further improved by the addition of automatic controls and adjustments (e.g., program-controlled in-phase tap changes, phase-angle regulators, and area interchange control). Since system planning studies and system operations may require multiple-case power flow solutions in some situations, the recent research efforts have been concentrated on the development of the decoupled Newton–Raphson methods [25–30].

These methods are based on the fact that in any power transmission network operating in the steady state, the coupling (i.e., interdependence) between P–θ (i.e., active powers and bus voltage angles) and Q–V (i.e., reactive powers and bus voltage magnitudes) is relatively weak, contrary to the strong coupling between P and θ and between Q and V. Therefore, these methods solve the load flow problem by "decoupling" (i.e., solving separately) the P–θ and Q–V problems. Thus, the solutions are obtained by applying approximations to the Newton–Raphson method. They have adequate accuracy and very fast speed and therefore can be used for online applications and contingency security assessments.

A given power flow algorithm must be reliable and must provide a fast convergence. Today, however, there is no such method that is adequately fast and can provide a feasible solution in every case study. Therefore, research efforts have been overwhelming in the development of numerical methods and programming techniques that can be used to solve power flow problems in an optimum manner [24a,24b,31–64].

In addition, the power systems and therefore the sizes of problems to be solved are continuously increasing at a faster rate than the development of computers with greater capacity. Thus, the theory of network tearing, or *diakoptics*, as proposed by Gabriel Kron and later developed by Happ and others [52,65–70] for power flow, has been receiving recent attention. According to diakoptics, a given large network system can be actually broken into pieces by a technique known as "tearing" so that each piece can be taken into account separately, and the solution attained can be subjected to a transformation that provides the solution to the original problem.

Another interesting development in the power flow problem has been that of probabilistic load flow in which the power inputs and outputs of a network are taken as random variables [50,51,71,72]. This is because the power flow problem is stochastic in nature, that is, some of its input data is subject to uncertainty, and therefore its output data (branch flows) can be expressed as a set of possible values with corresponding frequencies of occurrence, or a probability distribution function.

The uncertainties associated with the input data are due to (1) load forecast uncertainties caused by economics, conservation, etc.; (2) load variability due to weather conditions and load management practices; (3) unavailability (e.g., due to the forced outage and/or maintenance) of installed generator and transmission facilities; (4) delays in the installation of new generation and transmission facilities; (5) sudden changes in fuel prices and fuel availability; and (6) new techniques in power generation and transmission.

The present-day trends are toward the development of interactive power flow programs [46] where the planning engineer can modify data on a computer, either in a dialog mode with the program or on graphic displays, and then direct the program for the solution of the problem. The program algorithm may allow the planner to choose any portion of the output data to be displayed after the solution. It is foreseeable that the future power flow programs will provide improved interactive capabilities and faster algorithms to minimize the difficulties encountered in arriving at solutions within proper limits of time and effort.

8.2 POWER FLOW PROBLEM

The power flow problem can be defined as the calculation of the real and reactive powers flowing in each line, and the magnitude and phase angle of the voltage at each bus of a given transmission system for specified generation and load conditions. The information obtained from the power flow studies can be used to test the system's capability to transfer energy from generation to load without overloading lines and to determine the adequacy of voltage regulation by shunt capacitors, shunt reactors, tap-changing transformers, and the var-supplying capability of rotating machines.

It is possible to define the bus power in terms of generated power, load power, and transmitted power at a given bus. For example, the *bus power* (i.e., the net power) of the ith bus of an n bus power system can be expressed as

$$\begin{aligned} \mathbf{S}_i &= P_i + jQ_i \\ &= (P_{Gi} - P_{Li} - P_{Ti}) + j(Q_{Gi} - Q_{Li} - Q_{Ti}) \end{aligned} \quad (8.1)$$

where
- $\mathbf{S}_i$ = three-phase complex bus power at ith bus
- P_i = three-phase real bus power at ith bus
- Q_i = three-phase reactive bus power at ith bus
- P_{Gi} = three-phase real generated power flowing into ith bus
- P_{Li} = three-phase real load power flowing out of ith bus
- P_{Ti} = three-phase real transmitted power flowing out of ith bus
- Q_{Gi} = three-phase reactive generated power flowing into ith bus
- Q_{Li} = three-phase reactive load power flowing out of ith bus
- Q_{Ti} = three-phase reactive transmitted power flowing out of ith bus

In load flow studies, the basic assumption is that the given power system is a balanced three-phase system operating in its steady state with a constant 60-Hz frequency.* Therefore, the system can be represented by its single-phase positive-sequence network with a lumped series and shunt branches. The power flow problem can be solved either by using the bus admittance matrix ($\mathbf{Y}_{bus}$) or the bus impedance matrix ($\mathbf{Z}_{bus}$) representation of the given network. It is customary to use the nodal analysis approach. Thus, if the bus voltages are known, the bus currents can be expressed as

$$[\mathbf{I}_{bus}] = [\mathbf{Y}_{bus}][\mathbf{V}_{bus}] \quad (8.2)$$

or in its inverse form,

$$\begin{aligned}\left[\mathbf{V}_{bus}\right] &= \left[\mathbf{Y}_{bus}\right]^{-1}\left[\mathbf{I}_{bus}\right] \\ &= \left[\mathbf{Z}_{bus}\right]\left[\mathbf{I}_{bus}\right]\end{aligned} \quad (8.3)$$

If the bus voltages are known, then the bus currents can be calculated from Equation 8.2. Otherwise, if the bus currents are known, then the bus voltages can be found from Equation 8.3.

However, in a given power flow problem, nodal active and reactive powers are the independent variables and nodal voltages are the dependent variables. Therefore, the determination of the nodal voltages, which initially seems to be simple, given that the nodal currents are known, becomes a

---

* In general, power systems operate with a maximum frequency variation of ±0.05 Hz.

TABLE 8.1
Summary of Bus Classification

Bus Type	Known Quantities	Unknown Quantities
Slack	$|V| = 1.0$	P, Q
	$\theta = 0$	
Generator (*PV* bus)	$P, |V|$	Q, θ
Load (*PQ* bus)	P, Q	$|V|, \theta$

nonlinear problem dictating an iterative* solution method since it is nodal power rather than current that is known.

Each bus of a network has four variable quantities associated with it: the real and reactive power, the (line-to-ground) voltage magnitude, and voltage phase angle. Any two of the four may be the independent variables and are specified, whereas the other two remain to be determined. Because of the physical characteristics of generation and load, the electrical conditions at each bus are defined in terms of active and reactive power rather than by bus current. Since the complex power flowing into the *i*th bus can be expressed as

$$\mathbf{V}_i \mathbf{I}_i^* = P_i + jQ_i \tag{8.4}$$

the bus current is related to these variables as

$$\mathbf{I}_i = \frac{P_i - jQ_i}{\mathbf{V}_i^*} \tag{8.5}$$

In general, there are three types of buses in a load flow problem. Each with its own specified variables: (1) *slack* (generator) *bus*,† (2) generator buses, and (3) load buses. Since transmission losses in a given system are associated with the bus voltage profile, until a solution is obtained, the total power generation requirement cannot be determined. Therefore, the generator at the slack bus is used to supply the additional active and reactive power necessary owing to the transmission losses.

Thus, at the slack bus, the magnitude and phase angle of the voltage are known, and the real and reactive power generated are the quantities to be determined. It is only after a solution is converged, that is, all bus voltages are known, that the real and reactive power generation requirements at the slack bus can be determined. In other words, the losses are not known in advance, and consequently the power at the slack bus cannot be specified.

To define the power flow problem to be solved, it is necessary to specify the real power and the voltage magnitude at each generator bus. This is because these quantities are controllable through the governor and excitation controls, respectively. The generator bus is also known as the *PV* bus. Since an overexcited synchronous generator supplies current at a lagging power factor, the reactive power Q of a generator is not required to be specified.

The load bus is also known as the *PQ bus*. This is because the real and reactive powers are specified at a given load bus. Table 8.1 gives the bus types with corresponding known and unknown variables. Note that, the slack bus voltage is set to 1.0 pu with a phase angle of 0° (i.e., the slack bus voltage is used as a reference voltage).

---

* Iteration is a more sophisticated title for the trial-and-error method.
† It is also called the *swing bus*, *floating bus*, or *reference bus*. It is a fictitious concept, created by the load flow analyst.

Power Flow Analysis

It is possible that some load buses may have transformers capable of lap-changing and phase-shifting operations. These types of load buses are known as the voltage-controlled load buses. At the voltage-controlled load buses, the known quantities are the voltage magnitude in addition to the real and reactive powers, and the unknown quantities are the voltage phase angle and the turns ratio.

8.3 SIGN OF REAL AND REACTIVE POWERS

In performing a power flow study, one must remember that the lagging reactive power is a positive reactive power due to the inductive current and the leading reactive power is a negative power due to the capacitive current and that the positive bus current is in the direction that flows toward the bus.

Since the generator current flows toward the bus and the load current flows away from the bus, the sign of power is positive for the generator bus and negative for the load bus. Therefore, the following observations can be made:

1. The real and reactive powers associated with the *inductive load bus* (i.e., the lagging power factor load bus) are both negative.
2. The real and reactive powers associated with the *capacitive load bus* (i.e., the leading power factor load bus) are negative and positive, respectively.
3. The real and reactive powers associated with the *inductive generator bus* (i.e., the bus with a generator operating in lagging power factor mode) are both positive.
4. The real and reactive powers associated with the *capacitive generator bus* (i.e., the bus with a generator operating in leading power factor mode) are positive and negative, respectively.
5. The reactive power of a *shunt capacitive compensation apparatus* located at a bus is positive.

For example, if a load bus is connected to a load that withdraws 5 pu W and 3 pu inductive vars, then the bus current, from Equation 8.5, can be expressed as

$$\mathbf{I}_i = \frac{P_i - jQ_i}{\mathbf{V}_i^*}$$

$$= \frac{-5 - j(-3)}{\mathbf{V}_i^*}$$

$$= \frac{-5 - j3}{\mathbf{V}_i^*} \text{ pu A}$$

However, if a generator bus is connected to a generator that operates in a lagging power factor mode and supplies 5 pu W and 3 pu vars, then the bus current becomes

$$\mathbf{I}_i = \frac{P_i - jQ_i}{\mathbf{V}_i^*}$$

$$= \frac{5 - j(+3)}{\mathbf{V}_i^*}$$

$$= \frac{5 - j3}{\mathbf{V}_i^*} \text{ pu A}$$

8.4 GAUSS ITERATIVE METHOD

Assume that a set of simultaneous linear equations with n unknowns ($x_1, x_2, x_3, \cdots, x_n$ independent variables) is given as

$$a_{11}x_1 + a_{12}x_2 + a_{13}x_3 + \cdots + a_{1n}x_n = b_1$$
$$a_{21}x_1 + a_{22}x_2 + a_{23}x_3 + \cdots + a_{2n}x_n = b_2$$
$$\vdots \qquad (8.6)$$
$$a_{n1}x_1 + a_{n2}x_2 + a_{n3}x_3 + \cdots + a_{nn}x_n = b_n$$

where the a coefficients and the b dependent variables are known. The given Equation set 8.6 can be reexpressed as

$$x_1 = \frac{1}{a_{11}}(b_1 - a_{12}x_2 - a_{23}x_3 - \cdots - a_{1n}x_n)$$
$$x_2 = \frac{1}{a_{22}}(b_2 - a_{21}x_1 - a_{23}x_3 - \cdots - a_{2n}x_n)$$
$$\vdots \qquad (8.7)$$
$$x_n = \frac{1}{a_{nn}}(b_n - a_{n1}x_1 - a_{n2}x_2 - \cdots - a_{n,n-1}x_{n-1})$$

Assume that the initial approximation values of the independent variables are $x_1^{(0)}, x_2^{(0)}, x_3^{(0)}, \ldots, x_n^{(0)}$. Thus, after the substitution, the Equation set 8.7 can be written as

$$x_1^{(1)} = \frac{1}{a_{11}}(b_1 - a_{12}x_2^{(0)} - a_{23}x_3^{(0)} - \cdots - a_{1n}x_n^{(0)})$$
$$x_2^{(1)} = \frac{1}{a_{22}}(b_2 - a_{21}x_1^{0} - a_{23}x_3^{(0)} - \cdots - a_{2n}x_n^{(0)})$$
$$\vdots \qquad (8.8)$$
$$x_n^{(1)} = \frac{1}{a_{nn}}(b_n - a_{n1}x_1^{(0)} - a_{n2}x_2^{(0)} - \cdots - a_{n,n-1}x_{n-1}^{(0)})$$

Power Flow Analysis

where the initial values* are usually selected as

$$x_1^{(0)} = \frac{b_1}{a_{11}}$$

$$x_1^{(0)} = \frac{b_1}{a_{11}}$$

.
.
.

$$x_1^{(0)} = \frac{b_1}{a_{11}}$$

(8.9)

If the results obtained from Equation 8.8 match the initial values within a predetermined tolerance, a convergence (i.e., solution) has been achieved. Otherwise, the new corrected values of the independent variables (i.e., $x_1^{(1)}$, $x_2^{(1)}$, $x_3^{(1)}$, ..., $x_n^{(1)}$) are substituted into the next iteration. Therefore, after the $k + 1$ iteration,

$$x_1^{(k+1)} = \frac{1}{a_{11}}(b_1 - a_{12}x_2^{(k)} - a_{13}x_3^{(k)} - \cdots - a_{1n}x_n^{(k)})$$

$$x_2^{(k+1)} = \frac{1}{a_{22}}(b_2 - a_{21}x_1^{(k)} - a_{23}x_3^{(k)} - \cdots - a_{2n}x_n^{(k)})$$

.
.
.

$$x_n^{(k+1)} = \frac{1}{a_{nn}}(b_n - a_{n1}x_1^{(k)} - a_{n3}x_3^{(k)} - \cdots - a_{nn}x_{n-1}^{(k)})$$

(8.10)

8.5 GAUSS–SEIDEL ITERATIVE METHOD

The Gauss–Seidel iterative method is based on the Gauss iterative method. The only difference is that in the Gauss–Seidel iterative method, a more efficient substitution technique is used. In the iterations, therefore, the newly computed values of x are immediately used in the right sides of the following equations as soon as they are obtained. Therefore,

---

* Note that, the superscript (*i*) indicates the *i*th approximation for x. In other words, it denotes the iteration cycle number and not a power.

$$x_1^{(k+1)} = \frac{1}{a_{11}}(b_1 - a_{12}x_2^{(k)} - a_{13}x_3^{(k)} - \cdots - a_{1n}x_n^{(k)})$$

$$x_2^{(k+1)} = \frac{1}{a_{22}}(b_2 - a_{21}x_1^{(k+1)} - a_{23}x_3^{(k)} - \cdots - a_{2n}x_n^{(k)})$$

$$x_3^{(k+1)} = \frac{1}{a_{22}}(b_3 - a_{31}x_1^{(k+1)} - a_{32}x_2^{(k+1)} - \cdots - a_{3n}x_n^{(k)})$$

$$\vdots$$

$$x_n^{(k+1)} = \frac{1}{a_{nn}}(b_n - a_{n1}x_1^{(k+1)} - a_{n3}x_3^{(k+1)} - \cdots - a_{nn}x_{n-1}^{(k+1)})$$

(8.11)

Note that, the superscript (*i*) indicates the *i*th approximation for *x*. In other words, it denotes the iteration cycle number and not a power.

In Equation 8.11, the circled values are the immediately substituted values, which are determined in preceding steps of the $(k + 1)$th iteration.

8.6 APPLICATION OF GAUSS–SEIDEL METHOD: Y_{bus}

Assume that the neutral of an *n*-bus network system is taken as reference; then the *n* current equations can be expressed in terms of the *n* unknown voltages,

$$[\mathbf{I}_{bus}] = [\mathbf{Y}_{bus}][\mathbf{V}_{bus}]$$

or

$$\mathbf{I}_1 = \mathbf{Y}_{11}\mathbf{V}_1 + \mathbf{Y}_{12}\mathbf{V}_2 + \mathbf{Y}_{13}\mathbf{V}_3 + \cdots + \mathbf{Y}_{1n}\mathbf{V}_n$$
$$\mathbf{I}_2 = \mathbf{Y}_{21}\mathbf{V}_1 + \mathbf{Y}_{22}\mathbf{V}_2 + \mathbf{Y}_{23}\mathbf{V}_3 + \cdots + \mathbf{Y}_{2n}\mathbf{V}_n$$
$$\mathbf{I}_3 = \mathbf{Y}_{31}\mathbf{V}_1 + \mathbf{Y}_{32}\mathbf{V}_2 + \mathbf{Y}_{33}\mathbf{V}_3 + \cdots + \mathbf{Y}_{3n}\mathbf{V}_n$$
$$\vdots$$
$$\mathbf{I}_n = \mathbf{Y}_{n1}\mathbf{V}_1 + \mathbf{Y}_{n2}\mathbf{V}_2 + \mathbf{Y}_{n3}\mathbf{V}_3 + \cdots + \mathbf{Y}_{nn}\mathbf{V}_n$$

(8.12)

Power Flow Analysis

or, in matrix form,

$$\begin{bmatrix} \mathbf{I}_1 \\ \mathbf{I}_2 \\ \mathbf{I}_2 \\ \cdot \\ \cdot \\ \cdot \\ \mathbf{I}_n \end{bmatrix} = \begin{bmatrix} \mathbf{Y}_{11} & \mathbf{Y}_{12} & \mathbf{Y}_{13} & \cdots & \mathbf{Y}_{1n} \\ \mathbf{Y}_{21} & \mathbf{Y}_{22} & \mathbf{Y}_{23} & \cdots & \mathbf{Y}_{2n} \\ \mathbf{Y}_{31} & \mathbf{Y}_{32} & \mathbf{Y}_{33} & \cdots & \mathbf{Y}_{3n} \\ \cdot & \cdot & \cdot & \cdots & \cdot \\ \cdot & \cdot & \cdot & \cdots & \cdot \\ \cdot & \cdot & \cdot & \cdots & \cdot \\ \mathbf{Y}_{n1} & \mathbf{Y}_{n2} & \mathbf{Y}_{n3} & \cdots & \mathbf{Y}_{nn} \end{bmatrix} \begin{bmatrix} \mathbf{V}_1 \\ \mathbf{V}_2 \\ \mathbf{V}_3 \\ \cdot \\ \cdot \\ \cdot \\ \mathbf{V}_n \end{bmatrix} \quad (8.13)$$

Note that, the diagonal element $\mathbf{Y}_{ii}$ is obtained as the algebraic sum of all primitive admittances incident to node i and that the off-diagonal elements $\mathbf{Y}_{ij} = \mathbf{Y}_{ji}$ are obtained as the negative of the primitive (branch) admittance connection nodes i and j, that is, $\mathbf{Y}_{ij} = -\mathbf{y}_{ij}$. In the case of a π representation of transmission circuits, the shunt susceptance would be included in the nodal admittance term, as would the admittance of shunt reactors or the shunt elements of equivalent circuits as those used for representing tapped transformers.

Therefore, the bus voltages for the $(k + 1)$th iteration can be determined from Equation 8.12 when $\mathbf{V}_i^{(k)}$ and $\mathbf{I}_i^{(k)}$ are found after the kth iteration. Thus,

$$\begin{aligned}
\mathbf{V}_1^{(k+1)} &= \frac{1}{\mathbf{Y}_{11}} \left(\mathbf{I}_1^{(k)} - \mathbf{Y}_{12}\mathbf{V}_2^{(k)} - \mathbf{Y}_{13}\mathbf{V}_3^{(k)} - \cdots - \mathbf{Y}_{1n}\mathbf{V}_n^{(k)} \right) \\
\mathbf{V}_2^{(k+1)} &= \frac{1}{\mathbf{Y}_{22}} \left(\mathbf{I}_2^{(k)} - \mathbf{Y}_{21}\mathbf{V}_2^{(k+1)} - \mathbf{Y}_{23}\mathbf{V}_3^{(k)} - \cdots - \mathbf{Y}_{2n}\mathbf{V}_n^{(k)} \right) \\
\mathbf{V}_3^{(k+1)} &= \frac{1}{\mathbf{Y}_{33}} \left(\mathbf{I}_3^{(k)} - \mathbf{Y}_{31}\mathbf{V}_2^{(k+1)} - \mathbf{Y}_{32}\mathbf{V}_3^{(k+1)} - \cdots - \mathbf{Y}_{3n}\mathbf{V}_n^{(k)} \right) \\
\mathbf{V}_n^{(k+1)} &= \frac{1}{\mathbf{Y}_{nn}} \left(\mathbf{I}_n^{(k)} - \mathbf{Y}_{n1}\mathbf{V}_1^{(k+1)} - \mathbf{Y}_{n2}\mathbf{V}_2^{(k+1)} - \cdots - \mathbf{Y}_{n,n-1}\mathbf{V}_{n-1}^{(k+1)} \right)
\end{aligned} \quad (8.14)$$

Even though currents in Equation 8.14 are unknown, they can be expressed in terms of P, Q, and $\mathbf{V}$ as

$$\mathbf{I}_i = \frac{P_i - jQ_i}{\mathbf{V}_i^*} \quad (8.5)$$

A general formula to determine the bus voltage at the ith (PQ) bus can be developed by substituting Equation 8.5 into Equation 8.14 so that

$$\mathbf{V}_i^{(k+1)} = \frac{1}{\mathbf{Y}_{ii}} \left(\frac{P_i - jQ_i}{\mathbf{V}_i^{(k)*}} - \sum_{\substack{j=1 \\ j \neq i}}^{n} \mathbf{Y}_{ij}\mathbf{V}_j^{(k)} \right) \quad \text{for } i = 2,\ldots,n \quad (8.15)$$

Note that, bus 1 is designated as the slack bus with known voltage magnitude and phase angle. Therefore, the bus voltage calculations start with bus 2.

If the ith bus is a *PV* bus where real power and voltage magnitude, rather than reactive power, are given, then the unknown reactive power has to be determined first before each iteration. Thus, for the generator bus i,

$$\mathbf{I}_{gen} = \frac{P_i - jQ_i}{\mathbf{V}_i^*} = \mathbf{Y}_{i1}\mathbf{V}_1 + \mathbf{Y}_{i2}\mathbf{V}_2 + \mathbf{Y}_{i3}\mathbf{V}_3 + \cdots + \mathbf{Y}_{in}\mathbf{V}_n \tag{8.16}$$

$$P_i - jQ_i = \mathbf{V}_i^{(k)} \left[\sum_{j=1}^{n} \mathbf{Y}_{ij} \mathbf{V}_j^{(k)} \right] \tag{8.17}$$

Taking the imaginary part of Equation 8.11,

$$Q_i = -\operatorname{Im}\left[\mathbf{V}_i^{(k)*} \left(\sum_{j=1}^{n} \mathbf{Y}_{ij} \mathbf{V}_i^{(k)} \right) \right] \tag{8.18}$$

Note that the best values of voltages are used in calculating the reactive power Q_i. Once Q_i is found, it is used in Equation 8.15 to determine the new $\mathbf{V}_i$ at the generator bus. Usually, a limit on maximum and/or minimum $Q_{i,\text{spec}}$ may be specified. If the calculated reactive power, $Q_{i,\text{calc}}$, should exceed the limits $Q_{i,\text{spec}}$, then the limiting value of $Q_{i,\text{spec}}$ should be used in place of $Q_{i,\text{calc}}$ in Equation 8.15.

If the magnitude of the new calculated voltage is larger than the magnitude of the specified (original) voltage, the new voltage is corrected by multiplying by the ratio of the specified voltage magnitude to the calculated voltage magnitude, keeping the new phase angle of the calculated voltage. In other words, it is only the voltage magnitude that is being corrected.

In summary, the iteration process in the Gauss–Seidel method starts by assuming initial phasor values for the unknown bus voltages (except for the slack bus) and computing their new values, that is, the corrected voltages. At each bus, the corrected voltage is substituted back into Equation 8.15 as the estimated value for $\mathbf{V}_i^*$ to calculate the new value $\mathbf{V}_i$.

This process is continued for a predetermined number of iterations, usually twice. As the corrected voltage value is determined at each bus, it is employed in finding the corrected voltage at the next. The process is repeated at each bus successively for the rest of the buses to finish the first iteration. This iteration process for the network system is repeated until the voltage correction required for each bus is less than a specified precision index, that is, tolerance.

After the bus voltages $\mathbf{V}_2, \mathbf{V}_3, \mathbf{V}_4, \ldots, \mathbf{V}_n$ are thus solved, the bus power at the slack bus can be determined from

$$\frac{P_1 - jQ_1}{\mathbf{V}_i^*} = \mathbf{Y}_{11}\mathbf{V}_1 + \mathbf{Y}_{12}\mathbf{V}_2 + \mathbf{Y}_{13}\mathbf{V}_3 + \cdots + \mathbf{Y}_{1n}\mathbf{V}_n \tag{8.19}$$

or

$$P_1 - jQ_1 = \mathbf{Y}_{11}\mathbf{V}_1\mathbf{V}_1^* + \mathbf{Y}_{12}\mathbf{V}_2\mathbf{V}_1^* + \mathbf{Y}_{13}\mathbf{V}_3\mathbf{V}_1^* + \cdots + \mathbf{Y}_{1n}\mathbf{V}_n\mathbf{V}_1^* \tag{8.20}$$

As the final step, once all the bus voltages are known, the power flows in all transmission lines can be determined to complete the power flow study. For example, consider the line connecting buses i and j, as shown in Figure 8.1, and assume that the line current $\mathbf{I}_{ij}$ can be expressed as

$$\mathbf{I}_{ij} = \mathbf{I}_{\text{series}} + \mathbf{I}_{\text{shunt}} \tag{8.21}$$

Power Flow Analysis

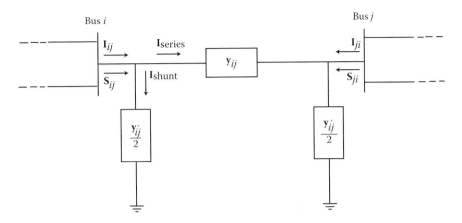

FIGURE 8.1 Line data used in load flow calculation.

where

$$\mathbf{I}_{series} = (\mathbf{V}_i - \mathbf{V}_j)\mathbf{y}_{ij} \qquad (8.22)$$

thus,

$$\mathbf{I}_{ij} = (\mathbf{V}_i - \mathbf{V}_j)\mathbf{y}_{ij} + \mathbf{V}_i \frac{\mathbf{y}'_{ij}}{2} \qquad (8.23)$$

where
$\mathbf{y}_{ij}$ = admittance of line ij
$\mathbf{y}'_{ij}$ = total line-charging admittance

The real and reactive power flow from bus i to bus j can be expressed as

$$\mathbf{S}_{ij} = P_{ij} + jQ_{ij} = \mathbf{V}_i \mathbf{I}^*_{ij} \qquad (8.24)$$

By substituting Equation 8.23 into Equation 8.24,

$$\mathbf{S}_{ij} = P_{ij} + jQ_{ij} = \mathbf{V}_i(\mathbf{V}^*_i - \mathbf{V}^*_j)\mathbf{y}^*_{ij} + \mathbf{V}_i\mathbf{V}^*_i \left(\frac{\mathbf{y}'_{ij}}{2}\right)^* \qquad (8.25)$$

Alternatively, the real and reactive power flow from bus j to bus i can be expressed as

$$\mathbf{S}_{ji} = P_{ji} + jQ_{ji} = \mathbf{V}_j(\mathbf{V}^*_j - \mathbf{V}^*_i)\mathbf{y}^*_{ij} + \mathbf{V}_j\mathbf{V}^*_j \left(\frac{\mathbf{y}'_{ij}}{2}\right)^* \qquad (8.26)$$

Note that, the line plus transformers can also be represented by the series and the two shunt admittances.

8.7 APPLICATION OF ACCELERATION FACTORS

Sometimes the number of iterations required to converge can be significantly reduced by application of a so-called *acceleration factor*. The correction in voltage from $\mathbf{V}_i^{(k)}$ to $\mathbf{V}_i^{(k+1)}$ is multiplied by such a factor in order to bring the new voltage closer to its final value. Therefore, the accelerated new voltage can be expressed as

$$\mathbf{V}_{i(\text{acceleration})}^{(k+1)} = \mathbf{V}_i^{(k)} + \alpha(\mathbf{V}_i^{(k+1)} - \mathbf{V}_i^{(k)})$$
$$= \mathbf{V}_i^{(k)} + \alpha \cdot \Delta \mathbf{V}_i^{(k)} \tag{8.27}$$

However, in the selection of a factor, there is nothing to guarantee a fast convergence. However, at the same time, numerous studies have been made to determine the optimum values of acceleration factors [5,7,10,60,73]. For example, the results of one study is shown in Figure 8.2. The actual value of α depends on the method of solution and the nature of the network system. However, an a value of about 1.6 is usually used in the $\mathbf{Y}_{\text{bus}}$ Gauss–Seidel iterations.

8.8 SPECIAL FEATURES

Today, most of the commercial power flow programs are equipped with automatic adjustment features either to permit the use of off-nominal bus quantities or to simulate the real-life operation on the power system. For example, a transformer provided with off-nominal taps can be used to regulate the voltage on a given bus to a specified level whether or not that transformer physically is capable of such automatic operation. In general, the most commonly used automatic adjustment features are

1. Automatic load-tap-changing (LTC) transformers for voltage control
2. Automatic phase-shifting transformers
3. Area power interchange control

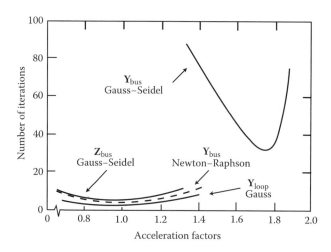

FIGURE 8.2 Effects of acceleration factors on rate of convergence for power flow solutions. (From Stagg, G. W., and El-Abiad, A. H., *Computer Methods in Power System Analysis*. McGraw-Hill, New York, 1968.)

8.8.1 LTC Transformers

The purpose of tap changing is to control the voltage magnitude at a given bus so that it is constant or within certain limits. Assume that an automatic LTC transformer is connected to a particular bus to keep load voltage constant. It is possible to run the power flow program employing one tap setting and without mentioning the magnitude of the load voltage.

If the voltage magnitude determined by the power flow program run exceeds the given limits, a new tap setting can be selected for the next run. In general, when the automatic tap-changing feature is used to represent a manual tap-changing transformer, the output of the power flow program will specify the tap setting that gives the required bus voltage. It is important to note that changing the tap setting (or ratio) will change the system admittance/impedance matrix. Therefore, after each tap ratio adjustment, the $\mathbf{Y}_{bus}$ admittance matrix or the $\mathbf{Z}_{bus}$ impedance matrix has to be adjusted.

Another means of taking into account the LTC transformer is to represent it by its impedance, or admittance, connected in series with an ideal autotransformer, as shown in Figure 8.3a. An equivalent π circuit, as shown in Figure 8.3b, can be developed from this representation to be used in power flow studies.

8.8.2 Phase-Shifting Transformers

Automatic phase-shifting transformers can be employed to control the power in a circuit that is connected in series with the transformer. However, in the event that such apparatus is not present in a system, the automatic phase-shifting feature of the power flow program can be used to determine the phase shift requirement for a given power flow in the series circuit. Note that, the effect of a phase-shifting transformer can be found by replacing a by $\mathbf{a}e^{j\Phi}$ where $|a|$ is unity.

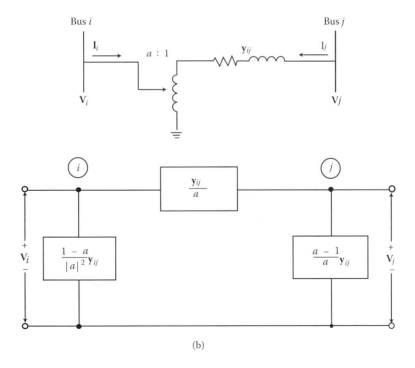

FIGURE 8.3 LTC transformer representations: (a) equivalent circuit; (b) equivalent circuit.

8.8.3 Area Power Interchange Control

If the load flow study is performed for some interconnected power systems or areas, the load flow results must meet the requirements of a specified net power interchange for each area. For example, power flow studies for normal operating conditions often require a certain amount of power transmitted out of an area. Power flow programs with area interchange control have provisions for keeping the power out of the given area constant by controlling the power at one generator, which is called the *regulating generator*.

After each power flow solution, the regulating-generator real-power output is adjusted. The actual net power interchange is calculated by adding tie-line flows algebraically. The regulating generator output is adjusted according to

$$P_{\text{reg}}^{(k+1)} = P_{\text{reg}}^{(k)} + \Delta P_T^{(k)} \tag{8.28}$$

where

$$\Delta P_T^{(k)} = P_{T(\text{sch})} - \Delta P_T^{(k)} \tag{8.29}$$

Thus,

$$P_{\text{reg}}^{(k+1)} = P_{\text{reg}}^{(k)} + \left[P_{T(\text{sch})} - P_T^{(k)} \right] \tag{8.30}$$

where
- $P_{\text{reg}}^{(k+1)}$ = new estimate of power output for regulating generator for the $(k + 1)$th iteration
- $P_{\text{reg}}^{(k)}$ = power output of regulating generator adjusted for the kth iteration
- $P_{T(\text{sch})}$ = scheduled net tie-line flow (or power interchange)
- $P_T^{(k)}$ = actual net power interchange
- $\Delta P_T^{(k)}$ = difference between actual and scheduled power interchanges

The final solution (after several iterations) is obtained when $\Delta P_T^{(k)}$ becomes less than the specified tolerance limit.

EXAMPLE 8.1

Consider the one-line diagram of the IEEE five-bus power system as shown in Figure 8.4. Assume that generator bus 1 is used as the slack bus with a constant voltage magnitude and angle of $1.02\angle 0°$ pu and that generator bus 4 has a constant-voltage magnitude of 1.05 pu and a specified real power of 1.0 pu based on 1.0 MVA as the megavolt-ampere base. The rest of the buses are connected to inductive loads. Note that, the load quantities in terms of P and Q are all negative to indicate the inductive loads. The initial voltage values for all load buses are given as $1.0\angle 0°$, as shown in Figure 8.4. A capacitor bank is connected to bus 4. Use the Gauss–Seidel method and determine the following:

(a) Bus admittance matrix of system
(b) Value of voltage $\mathbf{V}_2$ for the first iteration

Power Flow Analysis

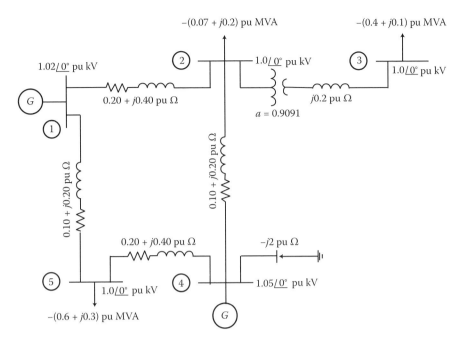

FIGURE 8.4 Five-bus system for Example 8.1.

Solution

(a) The primitive branch admittance can be found as

$$y_{12} = \frac{1}{z_{12}}$$
$$= \frac{1}{0.20 + j0.40}$$
$$= 1.0 - j2.0 \text{ pu}$$

$$y_{15} = \frac{1}{z_{15}}$$
$$= \frac{1}{0.10 + j0.20}$$
$$= 2.0 - j4.0 \text{ pu}$$

$$y_{24} = \frac{1}{z_{24}}$$
$$= \frac{1}{0.10 + j0.20}$$
$$= 2.0 - j4.0 \text{ pu}$$

$$y_{45} = \frac{1}{z_{45}}$$

$$= \frac{1}{0.20 + j0.40}$$

$$= 1.0 - j2.0 \text{ pu}$$

$$y_{12} = \frac{1}{j0.2} \times \frac{1}{(0.9091)^2}$$

$$= 0 + j5.5 \text{ pu}$$

Therefore, the off-diagonal elements of the bus admittance matrix can be expressed as

$$\mathbf{Y}_{12} = \mathbf{Y}_{21}$$
$$= -\mathbf{y}_{12}$$
$$= -1.0 + j2.0 \text{ pu}$$

$$\mathbf{Y}_{15} = \mathbf{Y}_{51}$$
$$= -\mathbf{y}_{15}$$
$$= -2.0 + j4.0 \text{ pu}$$

$$\mathbf{Y}_{23} = \mathbf{Y}_{32}$$
$$= -\mathbf{y}_{23}$$
$$= 0 + j5.5 \text{ pu}$$

$$\mathbf{Y}_{24} = \mathbf{Y}_{42}$$
$$= -\mathbf{y}_{24}$$
$$= -2.0 + j4.0 \text{ pu}$$

$$\mathbf{Y}_{45} = \mathbf{Y}_{54}$$
$$= -\mathbf{y}_{45}$$
$$= -1.0 + j2.0 \text{ pu}$$

The diagonal elements of the bus admittance matrix can be found as

$$\mathbf{Y}_{11} = \mathbf{Y}_{12} + \mathbf{Y}_{15}$$
$$= 3.0 - j6.0$$

$$\mathbf{Y}_{22} = (1.0 - j2.0) + \frac{1}{j0.2} \times \frac{1}{0.9091^2} + (2.0 - j4.0)$$

$$= 3.0 - j12.05$$

Power Flow Analysis

$$Y_{33} = \frac{1}{j0.2}$$
$$= -j5.0$$

$$Y_{44} = (1.0 - j2.0) + (2.0 - j4.0) + \frac{1}{-j2}$$
$$= 3.0 - j5.5$$

$$Y_{55} = (2.0 - j4.0) + (1.0 - j2.0)$$
$$= 3.0 - j6.0$$

Therefore, the bus admittance matrix of the system can be expressed as

$$[Y_{bus}] = \begin{bmatrix} Y_{11} & Y_{12} & Y_{13} & Y_{14} & Y_{15} \\ Y_{21} & Y_{22} & Y_{23} & Y_{24} & Y_{25} \\ Y_{31} & Y_{32} & Y_{33} & Y_{34} & Y_{35} \\ Y_{41} & Y_{42} & Y_{43} & Y_{44} & Y_{45} \\ Y_{51} & Y_{52} & Y_{53} & Y_{54} & Y_{55} \end{bmatrix}$$

$$= \begin{bmatrix} (3-j6) & (-1+j2) & (0+j0) & (0+j0) & (-2+j4) \\ (-1+j2) & (3-j12.05) & (0+j5.5) & (-2+j4) & (0+j0) \\ (0+j0) & (0+j5.5) & (0-j5) & (0+j0) & (0+j0) \\ (0+j0) & (-2+j4) & (0+j0) & (3-j5.5) & (-1+j2) \\ (-2+j4) & (0+j0) & (0+j0) & (-1+j2) & (3-j6) \end{bmatrix}$$

(b) From Equation 8.15, the value of voltage V_2 for the first iteration can be found as

$$V_2^{(1)} = \frac{1}{Y_{22}} \left(\frac{P_2 - jQ_2}{V_2^{(0)*}} - \sum_{\substack{j=1 \\ j \neq 2}}^{5} Y_{2j} V_j^0 \right)$$

$$= \frac{1}{Y_{22}} \left(\frac{P_2 - jQ_2}{V_2^{(0)*}} - \left\{ V_1^{(0)} Y_{21} + V_3^{(0)} Y_{23} + V_4^{(0)} Y_{24} + V_5^{(0)} Y_{25} \right\} \right)$$

$$= \frac{1}{3 - j12.05} \left[\frac{-0.7 + j(0.2)}{1.0} - \{1.02(-1+j2) + 1.0(0+j5.5) + 1.05(-2+j4) + 1.0(0+j0)\} \right]$$

$$= \frac{1}{3 - j12.05} \left[\frac{-0.7 + j(0.2)}{1.0} - (-3.12 + j11.74) \right]$$

$$= \frac{1}{3 - j12.05} [2.42 - j11.54]$$

$$= 0.9495 \angle -2.1366°$$

$$= 0.94886 - j0.035 \, pu$$

It is better to recompute $\mathbf{V}_2$ from Equation 8.15 by using the corrected value of $\mathbf{V}_2^{(1)*}$. Therefore,

$$\mathbf{V}_2^{(1)*} = 0.9495 \angle -2.1366°$$

$$= 0.94886 - j0.035 \text{ pu}$$

Thus,

$$\mathbf{V}_2^{(2)} = \frac{1}{\mathbf{Y}_{22}} \left(\frac{P_2 - jQ_2}{\mathbf{V}_2^{(1)*}} - \left\{ \mathbf{V}_1^{(0)} \mathbf{Y}_{21} + \mathbf{V}_3^{(0)} \mathbf{Y}_{23} + \mathbf{V}_4^{(0)} \mathbf{Y}_{24} + \mathbf{V}_5^{(0)} \mathbf{Y}_{25} \right\} \right)$$

$$= \frac{1}{3 - j12.05} \left[\frac{-0.7 + j(0.2)}{0.9495 \angle 2.1366°} - \{-3.12 + j11.74\} \right]$$

$$= \frac{1}{3 - j12.05} [2.3911 - j11.502]$$

$$\cong 0.946 \angle -2.2364°$$

$$\cong 0.94533 - j0.0369 \text{ pu}$$

8.9 APPLICATION OF GAUSS–SEIDEL METHOD: Z_{bus}

The Gauss–Seidel iterative method can also be applied for the solution of the power flow problem by the bus impedance matrix rather than the bus admittance matrix [11,13]. In this method, Equation 8.3 can be solved directly for the bus voltages in terms of the bus currents using the inverse of the bus admittance matrix. Therefore,

$$[\mathbf{V}_{bus}] = [\mathbf{Y}_{bus}]^{-1} [\mathbf{I}_{bus}]$$
$$= [\mathbf{Z}_{bus}][\mathbf{I}_{bus}] \quad (8.31)$$

where the bus currents can be obtained from Equation 8.5. Therefore, when the bus impedance matrix is known, then for any current vector $[\mathbf{I}^{(k)}]$, it is possible to determine a new voltage vector $[\mathbf{V}^{(k)}]$. After finding the new voltage vector, the associated new current vector can be computed, which in turn provides a new voltage vector.

This iteration process continues until the voltage correction required for each bus is less than a predetermined tolerance limit. To improve the convergence characteristics, it is useful to include, at each load bus, a fixed and artificial impedance to ground to approximate the load on the basis of the estimated voltage at this bus. Otherwise, with no paths to neutral (ground),* the resultant bus admittance is singular, and therefore its inverse will not exist.

An alternative approach eliminates the equation for the slack bus from Equation 8.31 before inverting the bus admittance matrix. In other words, the swing bus is taken as the reference in determining the bus admittance and impedance matrices. Therefore, the new equation can be expressed as

$$[\mathbf{V}_{bus}] = [\mathbf{Z}_{bus}][\mathbf{I}_{bus}] + [\mathbf{V}_s] \quad (8.32)$$

---

* Note that, this is because in power flow studies, the generations and loads are assumed to be external to the network and are excluded. Because of this, there are only high-impedance line-to-ground paths, e.g., line capacitance to ground, static capacitors, and shunt impedance due to transformer magnetizing current.

Power Flow Analysis

where the matrix $\mathbf{V}_s$ is the column vector whose elements are the voltage of the swing bus. Note that, the matrix $\mathbf{Z}$ is different than the one in Equation 8.31 and matrices $\mathbf{V}$ and $\mathbf{I}$ do not have $\mathbf{V}_s$ and $\mathbf{I}_s$, respectively. The iteration process involved in the Gauss–Seidel $\mathbf{Z}_{bus}$ method is similar to the one in the Gauss–Seidel $\mathbf{Y}_{bus}$ method. However, the fundamental Equations 8.5, 8.15, and 8.18 are, respectively, modified as

$$\mathbf{I}_i = \frac{P_i - jQ_i}{\mathbf{V}_i^*} - \mathbf{y}\mathbf{V}_i \text{ for } i = 1,2,\ldots,n, \quad i \neq s \tag{8.33}$$

$$\mathbf{V}_i^{(k+1)} = \sum_{\substack{j=1 \\ j \neq i}}^{n} \mathbf{Z}_{ij}\left(\frac{P_i - jQ_i}{\mathbf{V}_j^{(k)*}} - \mathbf{y}_i \mathbf{V}_i^{(k)}\right) + \mathbf{V}_s \quad \text{for } i = 2,3,\ldots,n \tag{8.34}$$

$$Q_i = -\text{Im}\left[\frac{\mathbf{V}_i^{(k)*}}{\mathbf{Z}_{ii}}\left(\mathbf{V}_i^{(k)} - \mathbf{V}_s - \sum_{\substack{j=1 \\ j \neq i}}^{n} \mathbf{Z}_{ij}\frac{P_j - jQ_j}{\mathbf{V}_j^{(k)*}}\right)\right] \tag{8.35}$$

The method has reliable convergence characteristics but does not have the advantage of the iterative bus admittance methods in terms of storage and speed with respect to larger systems.

8.10 NEWTON–RAPHSON METHOD

Assume that a single-variable equation has been given as

$$f(x) = 0 \tag{8.36}$$

Since any function of x can be expressed as a power series, the given function can be expanded by Taylor's series about a particular point x_0 as

$$f(x) = f(x_0) + \frac{1}{1!}\frac{df(x_o)}{dx}(x - x_0) + \frac{1}{2!}\frac{df^2(x_0)}{dx^2}(x - x_0)^2 + \cdots \\ + \frac{1}{n!}\frac{df^n(x_0)}{dx^n}(x - x_0)^n = 0 \tag{8.37}$$

If the terms beyond the first derivative are dropped (i.e., assuming convergence after the first two terms), a linear approximation results as

$$f(x) = f(x_0) + \frac{df(x_o)}{dx}(x - x_0) = 0 \tag{8.38}$$

from which

$$x_1 = x_0 - \frac{f(x_0)}{df(x_0)/dx} \tag{8.39}$$

However, to prevent any confusion with respect to notation, it is customary to reexpress Equation 8.39 as

$$x^{(1)} = x^{(0)} - \frac{f(x^{(0)})}{df(x^{(0)})/dx} \qquad (8.40)$$

where
$x^{(0)} =$ initial approximation (or estimate)
$x^{(1)} =$ first approximation

Therefore, a recursion formula can be developed so that at the end of the $(k + 1)$th iteration,

$$x^{(k+1)} = x^{(k)} - \frac{f(x^{(k)})}{df(x^{(k)})/dx} \qquad (8.41)$$

or, alternatively,

$$x^{(k+1)} = x^{(k)} - \frac{f(x^{(k)})}{df'(x^{(k)})} \qquad (8.42)$$

Thus,

$$\Delta x = -\frac{f(x^{(k)})}{f'(x^{(k)})} \qquad (8.43)$$

where

$$\Delta x = x^{(k+1)} - x^{(k)} \qquad (8.44)$$

Therefore, Δx (the amount by which $x^{(k)}$ needs to be modified) can be determined by substituting the value of $x^{(k)}$ into $f(x)$ and $f'(x)$, as shown in Equation 8.43. Thus, an iterative solution process is established.

The method can easily be extended to multivariable nonlinear equations in the following manner. Assume that a set of n nonlinear equations with n unknowns is given as

$$f_1(x_1, x_2, x_3, \cdots, x_n) = 0$$
$$f_2(x_1, x_2, x_3, \cdots, x_n) = 0 \qquad (8.45)$$
$$f_n(x_1, x_2, x_3, \cdots, x_n) = 0$$

or, in matrix notation,

$$F(x) = 0 \qquad (8.46)$$

where

$$[x] = \begin{bmatrix} x_1 \\ x_2 \\ x_3 \\ \vdots \\ x_n \end{bmatrix} \quad (8.47)$$

Here, the objective is to solve for the x values expressed in Equation 8.47 so that

$$f(x) = [0] \quad (8.48)$$

As has been done previously, if the given functions are expanded according to Taylor's series and the terms beyond the first derivatives are dropped,

$$F(x) = F(x^{(x)}) + [J(x^{(0)})][x - x^{(0)}] = 0 \quad (8.49)$$

where the coefficient matrix in Equation 8.49 is called the *Jacobian matrix* and can be expressed as

$$[J(x)] \triangleq \begin{bmatrix} \dfrac{\partial f_1}{\partial x_1} & \dfrac{\partial f_1}{\partial x_2} & \cdots & \dfrac{\partial f_1}{\partial x_n} \\ \dfrac{\partial f_2}{\partial x_1} & \dfrac{\partial f_2}{\partial x_2} & \cdots & \dfrac{\partial f_2}{\partial x_n} \\ \vdots & \vdots & & \vdots \\ \dfrac{\partial f_n}{\partial x_1} & \dfrac{\partial f_n}{\partial x_2} & \cdots & \dfrac{\partial f_n}{\partial x_n} \end{bmatrix} \quad (8.50)$$

Therefore, from Equation 8.49,

$$[x^{(k+1)}] = [x^{(k)}] - [J(x^{(x)})]^{-1}[F(x^{(k)})] \quad (8.51)$$

or

$$\begin{aligned}{}[\Delta x] &= [x^{(k+1)}] - [x^{(k)}] \\ &= -[J(x^{(x)})]^{-1}[F(x^{(k)})] \end{aligned} \quad (8.52)$$

EXAMPLE 8.2

Assume as given the nonlinear equations

$$f_1(x_1, x_2) = x_1^2 + 3x_1x_2 - 4 = 0$$

and

$$f_2(x_1, x_2) = x_1x_2 - 2x_2^2 + 5 = 0$$

Determine the values of x_1 and x_2 by using the Newton–Raphson method.

Solution

The Jacobian matrix is

$$[J(x)] = \begin{bmatrix} \dfrac{\partial f_1}{\partial x_1} & \dfrac{\partial f_1}{\partial x_2} \\ \dfrac{\partial f_2}{\partial x_1} & \dfrac{\partial f_2}{\partial x_2} \end{bmatrix}$$

$$= \begin{bmatrix} 2x_1 + 3x_2 & 3x_1 \\ x_2 & x_1 - 3x_2 \end{bmatrix}$$

As the initial approximation, let

$$[x^{(0)}] = \begin{bmatrix} x_1^{(0)} \\ x_2^{(0)} \end{bmatrix}$$

$$= \begin{bmatrix} 1 \\ 2 \end{bmatrix}$$

Therefore, the first iteration can be performed as

$$[x^{(1)}] = [x^{(0)}] - \begin{bmatrix} 8 & 3 \\ 2 & -7 \end{bmatrix}^{-1} [f(x^{(0)})]$$

$$= \begin{bmatrix} 1 \\ 2 \end{bmatrix} - \begin{bmatrix} \dfrac{7}{62} & \dfrac{3}{62} \\ \dfrac{2}{62} & -\dfrac{8}{62} \end{bmatrix} \begin{bmatrix} 3 \\ -1 \end{bmatrix}$$

$$= \begin{bmatrix} 0.7741 \\ 1.7742 \end{bmatrix}$$

Thus, after substituting the new values for the second iteration,

$$[x^{(2)}] = \begin{bmatrix} 0.7097 \\ 1.7742 \end{bmatrix} - \begin{bmatrix} 6.7419 & 2.1290 \\ 1.7742 & -6.3871 \end{bmatrix}^{-1} \begin{bmatrix} 0.2810 \\ -0.0364 \end{bmatrix}$$

$$= \begin{bmatrix} 0.6730 \\ 1.7583 \end{bmatrix}$$

Similarly, after substituting the new values for the third iteration,

$$[x^{(3)}] = \begin{bmatrix} 0.6730 \\ 1.7583 \end{bmatrix} - \begin{bmatrix} 6.6210 & 2.0191 \\ 1.7383 & -6.3602 \end{bmatrix}^{-1} \begin{bmatrix} 0.0031 \\ 0.0001 \end{bmatrix}$$

$$= \begin{bmatrix} 0.6726 \\ 1.7582 \end{bmatrix}$$

Finally, after substituting for the last iteration,

$$[x^{(4)}] = \begin{bmatrix} 0.6726 \\ 1.7582 \end{bmatrix} - \begin{bmatrix} 6.6198 & 2.0178 \\ 1.7582 & -6.3602 \end{bmatrix}^{-1} \begin{bmatrix} -0.0000 \\ 0.0000 \end{bmatrix}$$

$$= \begin{bmatrix} 0.6726 \\ 1.7582 \end{bmatrix}$$

Therefore, it is obvious that the iterations have rapidly converged toward the results of $x_1^{(4)} = 0.6726$ and $x_2^{(4)} = 1.7582$.

8.11 APPLICATION OF NEWTON–RAPHSON METHOD

The Newton–Raphson method is very reliable and extremely fast in convergence (especially with the introduction of the efficient sparsity programming). It is not sensitive to factors that cause poor or no convergence with other load flow methods (e.g., choice of slack bus, series capacitors, or negative resistances).[*] The rate of convergence is relatively independent of system size. Rectangular or polar coordinates can be used for the bus voltages [18]. In this method, the bus admittance matrix is used.

8.11.1 Application of Newton–Raphson Method to Load Flow Equations in Rectangular Coordinates[†]

As before, the slack bus, at which the magnitude and phase angle of the voltage is known, is not included in the iteration process. Therefore, the power at bus i in an n-bus system can be expressed as

$$\mathbf{S}_i = P_i - jQ_i = \mathbf{V}_i^* \mathbf{I}_i \tag{8.53}$$

or

$$\mathbf{S}_i = P_i - jQ_i = \mathbf{V}_i^* \sum_{j=1}^{n} \mathbf{Y}_{ij} \mathbf{V}_j \tag{8.54}$$

Let

$$\mathbf{V}_i \triangleq e_i + jf_i \tag{8.55}$$

---

[*] The representation of multiwound transformers and regulating transformers may involve negative resistances.
[†] For an in-depth explanation of the method and for excellent numerical examples, see Stagg and El-Abiad [73, Chapter 8].

$$\mathbf{V}_{ij} \triangleq G_{ij} - jB_{ij} \tag{8.56}$$

$$\mathbf{I}_i = \sum_{j=1}^{n} \mathbf{Y}_{ij}\mathbf{V}_{ij} \triangleq c_i + jd_i \tag{8.57}$$

where $\mathbf{I}_i$ is defined as the current flowing into bus i. Therefore, after the appropriate substitutions, Equation 8.54 can be reexpressed as

$$P_i - jQ_i = (e_i - jf_i)\sum_{j=1}^{n}(G_{ij} - jB_{ij})(e_j + jf_j) \tag{8.58}$$

or

$$P_i - jQ_i = (e_i - jf_i)\sum_{j=1}^{n}[(G_{ij}e_j + jB_{ij}f_j) + j(G_{ij}f_j - B_{ij}e_j)] \tag{8.59}$$

Therefore, it can be shown that

$$P_i = \sum_{j=1}^{n}[(e_iG_{ij}e_j + jB_{ij}f_j) + f_i(G_{ij}f_j - B_{ij}e_j)] \tag{8.60}$$

and

$$Q_i = \sum_{j=1}^{n}[(f_iG_{ij}e_j + jB_{ij}f_j) - e_i(G_{ij}f_j - B_{ij}e_j)] \tag{8.61}$$

Note that, P_i and Q_i are functions of e_i, e_j, f_i, and f_j. For each PQ load bus, P_i and Q_i can be calculated from Equations 8.60 and 8.61, respectively, for some estimated values of e and f. After each iteration, the calculated values of $P_{i,\text{calc}}$ and $Q_{i,\text{calc}}$ are compared against the known (or specified) values of $P_{j,\text{spec}}$ and $Q_{j,\text{spec}}$. Similarly, for each PV generator bus, the magnitude of the bus voltage can be calculated from the estimated values of e and f as

$$|\mathbf{V}_i|^2 = e_i^2 + f_i^2 \tag{8.62}$$

Then, the calculated voltage magnitude is compared with its specified value. Thus, the corrected values for the kth iteration can be expressed as

$$\Delta P_i^{(k)} = P_{i,\text{spec}} - P_{i,\text{calc}}^{(k)} \tag{8.63}$$

$$\Delta Q_i^{(k)} = Q_{i,\text{spec}} - Q_{i,\text{calc}}^{(k)} \tag{8.64}$$

Power Flow Analysis

$$\Delta|\mathbf{V}_i|^2 = |\mathbf{V}_{i,\text{spec}}|^2 - |\mathbf{V}_{i,\text{calc}}|^2 \tag{8.65}$$

The resultant values of $\Delta P_i^{(k)}$, $\Delta Q_i^{(k)}$, and $\Delta|\mathbf{V}_i|^2$ can be used to determine the changes in the real and imaginary components of the bus voltages from Equation 8.66. Since the changes in P, Q, and V^2 are related to the changes in e and f by Equations 8.60 through 8.62, it is possible to express them in a general form as

$$\begin{bmatrix} \Delta P_2^{(k)} \\ \vdots \\ \Delta P_n^{(k)} \\ \hline \Delta Q_2^{(k)} \\ \vdots \\ \Delta Q_n^{(k)} \\ \hline \Delta|\mathbf{V}_n^{(k)}|^2 \end{bmatrix} = \begin{bmatrix} \dfrac{\partial P_2}{\partial e_2} & \cdots & \dfrac{\partial P_2}{\partial e_2} & \dfrac{\partial P_2}{\partial f_2} & \cdots & \dfrac{\partial P_2}{\partial e_n} \\ \vdots & & \vdots & \vdots & & \vdots \\ \dfrac{\partial P_n}{\partial e_2} & \cdots & \dfrac{\partial P_n}{\partial e_n} & \dfrac{\partial P_2}{\partial f_2} & \cdots & \dfrac{\partial P_2}{\partial f_n} \\ \dfrac{\partial Q_2}{\partial e_2} & \cdots & \dfrac{\partial Q_2}{\partial e_n} & \dfrac{\partial Q_2}{\partial f_2} & \cdots & \dfrac{\partial Q_2}{\partial f_n} \\ \vdots & & \vdots & \vdots & & \vdots \\ \dfrac{\partial Q_{n-1}}{\partial e_2} & \cdots & \dfrac{\partial Q_{n-1}}{\partial e_n} & \dfrac{\partial Q_{n-1}}{\partial f_2} & \cdots & \dfrac{\partial Q_{n-1}}{\partial f_n} \\ \dfrac{\partial|\mathbf{V}_n|^2}{\partial e_2} & \cdots & \dfrac{\partial|\mathbf{V}_n|^2}{\partial e_n} & \dfrac{\partial|\mathbf{V}_n|^2}{\partial f_2} & \cdots & \dfrac{\partial|\mathbf{V}_n|^2}{\partial f_n} \end{bmatrix} \begin{bmatrix} \Delta e_2^{(k)} \\ \vdots \\ \Delta e_n^{(k)} \\ \hline \Delta f_2^{(k)} \\ \vdots \\ \Delta f_{n-1}^{(k)} \\ \Delta f_n^{(k)} \end{bmatrix} \tag{8.66}$$

In Equation 8.66, the coefficient matrix of partial derivatives is called the Jacobian. Thus, in matrix form, Equation 8.66 can be expressed as

$$\begin{bmatrix} \Delta P \\ \hline \Delta Q \\ \hline \Delta|\mathbf{V}|^2 \end{bmatrix} = \begin{bmatrix} J_1 & J_2 \\ \hline J_3 & J_4 \\ \hline J_5 & J_6 \end{bmatrix} \times \begin{bmatrix} \Delta e \\ \hline \Delta f \end{bmatrix} \tag{8.67}$$

Note that, it is assumed that bus 1 is the slack bus and that bus n is the *PV* generator (or voltage-controlled) bus. In Equation 8.67, the unknowns are the changes (or corrections) in the real and imaginary components of the voltages (i.e., Δe's and Δf's). It can be observed from Equation 8.66 that two equations are needed for each bus (excluding the slack bus and reference bus) in order to include both real and imaginary terms. All the elements of the Jacobian are functions of e's and f's. Thus, they can be calculated by substituting the initial assumed values for the first iteration, or computed in the last iteration, into the partial derivative equations. Thus, the unknown values of Δe's and Δf's can be determined from Equation 8.66 after inverting the Jacobian matrix. The resultant values of Δe's and Δf's can be used in the following equation to determine the new estimates for bus voltages for the next iteration:

$$e_i^{(k+1)} = e_i^{(k)} + \Delta e_i^{(k)} \tag{8.68}$$

$$f_i^{(k+1)} = f_i^{(k)} + \Delta f_i^{(k)} \tag{8.69}$$

The process is repeated until the values of ΔP, ΔQ, and $\Delta |V|^2$ determined from Equations 8.63 through 8.65 are less than the specified precision indexes.

It is possible to develop a convenient expression for bus current given by Equation 8.57 as

$$\mathbf{I}_i = \sum_{j=1}^{n} \mathbf{Y}_{ij} \mathbf{V}_j = \sum_{j=1}^{n} (G_{ij} - jB_{ij})(e_j + jf_j) \tag{8.70}$$

or

$$\mathbf{I}_i = \sum_{j=1}^{n} (e_j G_{ij} + jf_j B_{ij}) + j \sum_{j=1}^{n} (f_j G_{ij} - e_j B_{ij}) \tag{8.71}$$

or

$$\mathbf{I}_i = c_i + jd_i \tag{8.72}$$

where

$$c_i = \sum_{j=1}^{n} (e_j G_{ij} + f_j B_{ij}) \tag{8.73}$$

or

$$d_i = \sum_{j=1}^{n} (f_j G_{ij} - e_j B_{ij}) \tag{8.74}$$

or

$$c_i = e_i G_{ii} + f_i B_{ii} + \sum_{\substack{j=1 \\ j \neq i}}^{n} (e_j G_{ij} + f_j B_{ij}) \tag{8.75}$$

$$d_i = f_i G_{ii} - e_i B_{ii} + \sum_{\substack{j=1 \\ j \neq i}}^{n} (f_j G_{ij} - e_j B_{ij}) \tag{8.76}$$

From Equation 8.60, the real power of bus i can be expressed as

$$P_i = e_i \sum_{j=1}^{n} (e_j G_{ij} + f_j B_{ij}) + f_i \sum_{j=1}^{n} (f_j G_{ij} - e_j B_{ij}) \tag{8.77}$$

or substituting Equation 8.73 and 8.74 into Equation 8.77,

$$P_i = e_i c_i + f_i d_i \tag{8.78}$$

Similarly, from Equation 8.61, the reactive power of bus i can be expressed as

$$Q_i = f_i \sum_{j=1}^{n} (e_j G_{ij} + f_j B_{ij}) - e_i \sum_{j=1}^{n} (f_j G_{ij} - e_j B_{ij}) \tag{8.79}$$

or

$$Q = f_i c_i - e_i d_i \tag{8.80}$$

Thus, the elements of submatrices of the Jacobian matrix, given in Equation 8.67, can be evaluated for the values of P, Q, and $\mathbf{V}^2$ at each iteration as follows. For submatrix J_1, from Equation 8.77, the off-diagonal and diagonal elements, respectively, can be found as

$$\frac{\partial P_i}{\partial e_j} = e_i G_{ij} - f_i B_{ij} \quad i \neq j \tag{8.81}$$

and

$$\frac{\partial P_i}{\partial e_i} = (e_i G_{ii} - f_i B_{ii}) + c_i \quad i = j \tag{8.82}$$

Similarly, for the matrix J_2,

$$\frac{\partial P_i}{\partial f_j} = e_i B_{ij} + f_i G_{ij} \quad i \neq j \tag{8.83}$$

$$\frac{\partial P_i}{\partial f_i} = (e_i B_{ii} + f_i G_{ii}) + d_i \quad i = j \tag{8.84}$$

For submatrix J_3, from Equation 8.79,

$$\frac{\partial Q_i}{\partial e_j} = e_i B_{ij} + f_i G_{ij} \quad i \neq j \tag{8.85}$$

$$\frac{\partial Q_i}{\partial e_i} = (e_i B_{ii} + f_i G_{ii}) \quad i = j \tag{8.86}$$

Similarly, for submatrix J_4

$$\frac{\partial Q_i}{\partial f_j} = -(e_i G_{ij} - f_i B_{ij}) \quad i \neq j \tag{8.87}$$

$$\frac{\partial Q_i}{\partial f_j} = -(e_i G_{ii} - f_i B_{ii}) + c_i \quad i = j \tag{8.88}$$

Note the similarity between the elements of submatrices J_1 and J_4 and also between submatrices J_2 and J_3. It can be convenient to define the following expressions for diagonal and off-diagonal elements as

$$T_{ij} = e_i G_{ij} - f_i B_{ij} \quad \text{for all} \quad i, j \tag{8.89}$$

and

$$U_{ij} = e_i B_{ij} + f_i G_{ij} \quad \text{for all} \quad i, j \tag{8.90}$$

Therefore, the Jacobian matrix can be expressed as

$$\left[\begin{array}{c|c} J_1 & J_2 \\ \hline J_3 & J_4 \end{array} \right] = \left[\begin{array}{c|c} T & U \\ \hline U & -T \end{array} \right] \left[\begin{array}{c|c} c_i & d_i \\ \hline -d_i & c_i \end{array} \right] \tag{8.91}$$

Note that, the off-diagonal elements in the c_i and d_i submatrices are all zero.

The elements of the submatrices for the voltage-controlled bus i can be found from Equation 8.62 similarly. For submatrix J_5, the off-diagonal and diagonal elements, respectively, can be found as

$$\frac{\partial |\mathbf{V}_i|^2}{\partial e_j} = 0 \quad i \neq j \tag{8.92}$$

and

$$\frac{\partial |\mathbf{V}_i|^2}{\partial e_i} = 2e_i \quad i = j \tag{8.93}$$

Similarly, for submatrix J_6,

$$\frac{\partial |\mathbf{V}_i|^2}{\partial f_j} = 0 \quad i \neq j \tag{8.94}$$

and

$$\frac{\partial |\mathbf{V}_i|^2}{\partial f_i} = 2f_i \quad i = j \tag{8.95}$$

Thus, by substituting Equations 8.78 through 8.80 and 8.62 into Equations 8.63 through 8.65, respectively, it is possible to express them as

$$\Delta P_i^{(k)} = P_{i,\text{spec}} - (e_i c_i + f_i d_i) \tag{8.96}$$

Power Flow Analysis

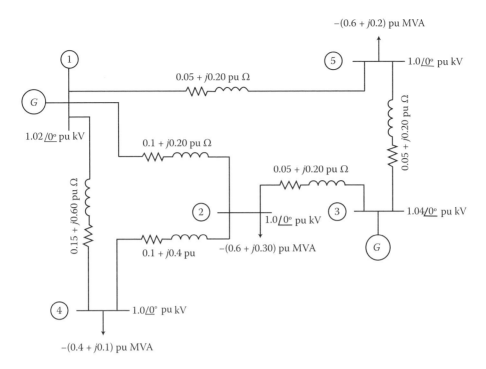

FIGURE 8.5 Five-bus system for Example 8.3.

$$\Delta Q_i^{(k)} = Q_{i,\text{spec}} - (f_i c_i - e_i d_i) \tag{8.97}$$

$$\Delta |\mathbf{V}_i|^2 = |\mathbf{V}_{p,\text{spec}}|^2 - (e_i^2 + f_i^2) \tag{8.98}$$

EXAMPLE 8.3

Consider the one-line diagram of a five-bus* power system as shown in Figure 8.5. Assume that generators are connected to buses 1 and 3 and that bus 1 is used as the slack bus. Therefore, the voltage magnitude and angle of bus 1 and the voltage magnitude of bus 3 will be kept constant. Assume that inductive loads are connected to buses 2, 4, and 5, as indicated. Note that, since the real and reactive powers are associated with the inductive load buses, they are negative, as shown in the figure. The minimum and maximum reactive power limits of the generator bus (i.e., bus 3) are 0.0 and 10.0 pu, respectively. Its real power is 1.0 pu based on 1.0 MVA as the megavolt-ampere base. Use the Newton–Raphson method in rectangular coordinates to obtain a power flow solution with tolerances of 0.0001 pu for the changes in the real and reactive bus powers and for the changes in bus voltages, respectively. Note that, the buses in Figure 8.5 are not necessary for the solution of the example. However, the proper ordering of buses can be important for large power systems in terms of faster convergence of the solutions. Tables 8.2a and 8.2b provide impedances for the six lines that are identified by the buses that are connected. Table 8.3 gives the values of real and reactive powers. The positive power indicates the power input to the network at each bus. Hence, the negative values indicate the inductive loads. The voltage values shown in Figure 8.2 represent the original estimates. Note that, the magnitude of the voltage value and the

---

* The input data for the example is adopted from Stevenson [74]. Included with permission from McGraw-Hill Book Company.

TABLE 8.2a
Line Impedances

Line (Bus to Bus)	R (Per Unit)	X (Per Unit)
1–2	0.10	0.40
1–4	0.15	0.60
1–5	0.05	0.20
2–3	0.05	0.20
3–5	0.05	0.20

TABLE 8.2b
Power and Voltage Data

Bus	P (Per Unit)	Q (Per Unit)	V (Per Unit)	Remarks
1	–	–	$1.02\angle 0°$	Swing bus
2	–0.6	–0.3	$1.00\angle 0°$	Load bus (inductive)
3	1.0	–	$1.00\angle 0°$	Voltage bus (inductive)
4	–0.4	–0.1	$1.00\angle 0°$	Load bus (inductive)
5	–0.6	–0.2	$1.00\angle 0°$	Load bus (inductive)

TABLE 8.3
Line Admittances and Self- and Mutual Admittances

Line (Bus to Bus)	G (Per Unit)	B (Per Unit)
1–2	0.588235	–2.352941
1–4	0.392157	–1.568627
1–5	1.176471	–4.705882
2–3	1.176471	–4.705882
2–4	0.588235	–2.352941
3–5	1.176471	–4.705882

angle are to remain constant at bus 3. Let the iterative computations begin at bus 2 and determine the value of $\mathbf{V}_2$ for the first iteration.

(a) Determine the ΔP and ΔQ changes for the four buses (2, 3, 4, and 5 after the first iteration).
(b) Determine the Δe and Δf voltage changes for the four buses after the first iteration.
(c) Determine the new voltages for the four buses after the first iteration.
(d) Repeat part (a) for the second iteration.
(e) Repeat part (b) for the second iteration.
(f) Repeat part (c) for the second iteration.
(g) Repeat part (a) for the third iteration.
(h) Repeat part (b) for the third iteration.
(i) Repeat part (c) for the third iteration.
(j) Determine the real and reactive power flows in all transmission lines after the third iteration.
(k) Determine the real and reactive powers summed at each of the five buses after the third iteration.
(l) Determine the total real and reactive power losses in the system after the third iteration.
(m) Determine the bus currents.

Power Flow Analysis

Solution

(a) The computer program output shows that after the first iteration,

$$\Delta \mathbf{S}_2^{(1)} = \Delta P_2^{(1)} + j\Delta Q_2^{(1)} = -0.5411748 + j0.06470662 \text{ pu}$$

$$\Delta \mathbf{S}_3^{(1)} = \Delta P_3^{(1)} + j\Delta Q_3^{(1)} = 0.9021182 + j0.0000000 \text{ pu}$$

$$\Delta \mathbf{S}_4^{(1)} = \Delta P_4^{(1)} + j\Delta Q_4^{(1)} = -0.3921567 - j0.0686270 \text{ pu}$$

$$\Delta \mathbf{S}_5^{(1)} = \Delta P_5^{(1)} + j\Delta Q_5^{(1)} = -0.5294129 + j0.8235341 \text{ pu}$$

(b) The computer program output shows that after the first iteration,

$$\Delta \mathbf{V}_2^{(1)} = \Delta e_2^{(1)} + j\Delta f_2^{(1)} = -0.03651552 - j0.06227919 \text{ pu}$$

$$\Delta \mathbf{V}_3^{(1)} = \Delta e_3^{(1)} + j\Delta f_3^{(1)} = 0.1093983 \times 10^{-6} + j0.004050539 \text{ pu}$$

$$\Delta \mathbf{V}_4^{(1)} = \Delta e_4^{(1)} + j\Delta f_4^{(1)} = 0.06240950 - j0.1263565 \text{ pu}$$

$$\Delta \mathbf{V}_5^{(1)} = \Delta e_5^{(1)} + j\Delta f_5^{(1)} = -0.005154848 - j0.3373529 \text{ pu}$$

(c) The computer program shows that the new bus voltages after the first iteration are

$$\mathbf{V}_2^{(1)} = e_2^{(1)} + jf_2^{(1)} = 0.963485 - j0.06227919 \text{ pu}$$

$$\mathbf{V}_3^{(1)} = e_3^{(1)} + jf_3^{(1)} = 1.040000 + j0.04050539 \text{ pu}$$

$$\mathbf{V}_4^{(1)} = e_4^{(1)} + jf_4^{(1)} = 0.9375905 - j0.1263565 \text{ pu}$$

$$\mathbf{V}_5^{(1)} = e_5^{(1)} + jf_5^{(1)} = 0.9948452 - j0.03373529 \text{ pu}$$

(d) After the second iteration,

$$\Delta \mathbf{S}_2^{(2)} = \Delta P_2^{(2)} + j\Delta Q_2^{(2)} = -0.01451135 - j0.03573412 \text{ pu}$$

$$\Delta \mathbf{S}_3^{(2)} = \Delta P_3^{(2)} + j\Delta Q_3^{(2)} = -0.4950496 \times 10^{-3} - j0.164032 \times 10^{-2} \text{ pu}$$

$$\Delta \mathbf{S}_4^{(2)} = \Delta P_4^{(2)} + j\Delta Q_4^{(2)} = -0.02258435 - j0.002303781 \text{ pu}$$

$$\Delta \mathbf{S}_5^{(2)} = \Delta P_5^{(2)} + j\Delta Q_5^{(2)} = -0.005330801 - j0.01714361 \text{ pu}$$

(e) After the second iteration,

$$\Delta \mathbf{V}_2^{(2)} = \Delta e_2^{(2)} + j\Delta f_2^{(2)} = -0.01077833 - j0.00310017 \text{ pu}$$

$$\Delta \mathbf{V}_3^{(2)} = \Delta e_3^{(2)} + j\Delta f_3^{(2)} = -0.6285994 \times 10^{-3} - j0.4108835 \times 10^{-2} \text{ pu}$$

$$\Delta \mathbf{V}_4^{(2)} = \Delta e_4^{(2)} + j\Delta f_4^{(2)} = -0.02258435 - j0.002303781 \text{ pu}$$

$$\Delta \mathbf{V}_5^{(2)} = \Delta e_5^{(2)} + j\Delta f_5^{(2)} = -0.002374709 - j0.002142731 \text{ pu}$$

(f) After the second iteration,

$$\mathbf{V}_2^{(2)} = e_2^{(2)} + jf_2^{(2)} = 0.9527062 - j0.06558919 \text{ pu}$$

$$\mathbf{V}_3^{(2)} = e_3^{(2)} + jf_3^{(2)} = 1.039371 + j0.03639655 \text{ pu}$$

$$\mathbf{V}_4^{(2)} = e_4^{(2)} + jf_4^{(2)} = 0.9150062 - j0.1286603 \text{ pu}$$

$$\mathbf{V}_5^{(2)} = e_5^{(2)} + jf_5^{(2)} = 0.9924704 - j0.03587802 \text{ pu}$$

(g) After the third iteration,

$$\Delta \mathbf{S}_2^{(3)} = \Delta P_2^{(3)} + j\Delta Q_2^{(3)} = -0.4726648 \times 10^{-4} - j0.5348325 \times 10^{-3} \text{ pu}$$

$$\Delta \mathbf{S}_3^{(3)} = \Delta P_3^{(3)} + j\Delta Q_3^{(3)} = -0.2393723 \times 10^{-3} - j0.1621246 \times 10^{-4} \text{ pu}$$

$$\Delta \mathbf{S}_4^{(3)} = \Delta P_4^{(3)} + j\Delta Q_4^{(3)} = -0.2399683 \times 10^{-3} - j0.1458764 \times 10^{-2} \text{ pu}$$

$$\Delta \mathbf{S}_5^{(3)} = \Delta P_5^{(3)} + j\Delta Q_5^{(3)} = 0.2741814 \times 10^{-4} - j0.6151199 \times 10^{-4} \text{ pu}$$

(h) After the third iteration,

$$\Delta \mathbf{V}_2^{(3)} = \Delta e_2^{(3)} + j\Delta f_2^{(3)} = -0.2121947 \times 10^{-3} - j0.3475162 \times 10^{-4} \text{ pu}$$

$$\Delta \mathbf{V}_3^{(3)} = \Delta e_3^{(3)} + j\Delta f_3^{(3)} = -0.4772861 \times 10^{-5} - j0.8642096 \times 10^{-4} \text{ pu}$$

$$\Delta \mathbf{V}_4^{(2)} = \Delta e_4^{(2)} + j\Delta f_4^{(2)} = -0.5581642 \times 10^{-3} + j0.1678166 \times 10^{-4} \text{ pu}$$

$$\Delta \mathbf{V}_5^{(3)} = \Delta e_5^{(3)} + j\Delta f_5^{(3)} = -0.1013366 \times 10^{-4} - j0.377548 \times 10^{-4} \text{ pu}$$

(i) After the third iteration,

$$\mathbf{V}_2^{(3)} = e_2^{(3)} + jf_2^{(3)} = 0.952494 - j0.06562394 \text{ pu}$$

Power Flow Analysis

$$\mathbf{V}_3^{(3)} = e_3^{(3)} + jf_3^{(3)} = 1.039366 + j0.03631013 \text{ pu}$$

$$\mathbf{V}_4^{(3)} = e_4^{(3)} + jf_4^{(3)} = 0.914448 - j0.1286435 \text{ pu}$$

$$\mathbf{V}_5^{(3)} = e_5^{(3)} + jf_5^{(3)} = 0.9924603 - j0.03591578 \text{ pu}$$

(j) As can be observed from parts (g), (h), and (i), the power flow problem solution is converged. Therefore, the computer program output gives the real and reactive power flows in all transmission lines after the third iteration as

$$\mathbf{S}_{12}^{(3)} = 0.19800 + j0.12264 \text{ MVA}$$

$$\mathbf{S}_{14}^{(3)} = 0.24805 + j0.11743 \text{ MVA}$$

$$\mathbf{S}_{15}^{(3)} = 0.20544 + j0.08910 \text{ MVA}$$

$$\mathbf{S}_{21}^{(3)} = -0.19279 - j0.10179 \text{ MVA}$$

$$\mathbf{S}_{23}^{(3)} = -0.57321 - j0.23698 \text{ MVA}$$

$$\mathbf{S}_{24}^{(3)} = 0.16600 + j0.03876 \text{ MVA}$$

$$\mathbf{S}_{32}^{(3)} = 0.59431 + j0.32138 \text{ MVA}$$

$$\mathbf{S}_{35}^{(3)} = 0.40569 + j0.15545 \text{ MVA}$$

$$\mathbf{S}_{41}^{(3)} = -0.23719 - j0.07399 \text{ MVA}$$

$$\mathbf{S}_{42}^{(3)} = -0.16281 - j0.20601 \text{ MVA}$$

$$\mathbf{S}_{51}^{(3)} = -0.20303 - j0.07945 \text{ MVA}$$

$$\mathbf{S}_{53}^{(3)} = -0.39697 - j0.12054 \text{ MVA}$$

(k) The real and reactive powers summed at each of the five buses are given by the computer output as

$$\mathbf{S}_1^{(3)} = 0.65149 + j0.32917 \text{ MVA}$$

$$\mathbf{S}_2^{(3)} = -0.60000 - j0.30000 \text{ MVA}$$

$$\mathbf{S}_3^{(3)} = 1.00000 + j0.47683 \text{ MVA}$$

$$\mathbf{S}_4^{(3)} = -0.40000 - j0.10000 \text{ MVA}$$

$$\mathbf{S}_5^{(3)} = -0.60000 - j0.19999 \text{ MVA}$$

(l) The total real and reactive power losses in the system are given as

$$\mathbf{S}_{\text{loss}}^{(3)} = 0.05150 + j0.20600 \text{ MVA}$$

(m) The bus currents are given as

$$\mathbf{I}_2^{(3)} = -0.6052267 + j0.355998 \text{ pu}$$

$$\mathbf{I}_3^{(3)} = 0.9771826 - j0.4235277 \text{ pu}$$

$$\mathbf{I}_4^{(3)} = -0.4135732 + j0.1658478 \text{ pu}$$

$$\mathbf{I}_5^{(3)} = -0.5965176 + j0.2230196 \text{ pu}$$

8.11.2 Application of Newton–Raphson Method to Load Flow Equations in Polar Coordinates

Assume that the basic variables are given in polar coordinates, that is, in terms of magnitude and angles. Let

$$\mathbf{V}_i \triangleq |\mathbf{V}_i| \angle \delta_i \tag{8.99}$$

$$\mathbf{V}_i \triangleq |\mathbf{V}_i| \angle -\theta_{ij} \tag{8.100}$$

Hence,

$$\mathbf{I}_i = \sum_{j=1}^n \mathbf{Y}_{ij} \mathbf{V}_j = \sum_{j=1}^n |\mathbf{Y}_{ij}||\mathbf{V}_j| \angle -\theta_{ij} + \delta_j \tag{8.101}$$

Thus,

$$P_i - jQ_i = \mathbf{V}_i^* \mathbf{I}_i = \sum_{j=1}^n |\mathbf{V}_i||\mathbf{Y}_{ij}||\mathbf{V}_j| \angle -(\theta_{ij} + \delta_i - \delta_j) \tag{8.102}$$

where

$$e^{-j(\theta_{ij} + \delta_i - \delta_j)} \triangleq \cos(\theta_{ij} + \delta_i - \delta_j) - j\sin(\theta_{ij} + \delta_i - \delta_j) \tag{8.103}$$

Power Flow Analysis

Hence, the real and reactive powers can be expressed as

$$P_i = \sum_{j=1}^{n} |\mathbf{V}_i||\mathbf{Y}_{ij}||\mathbf{V}_j|\cos(\theta_{ij} + \delta_i - \delta_j) \tag{8.104}$$

and

$$Q_i = \sum_{j=1}^{n} |\mathbf{V}_i||\mathbf{Y}_{ij}||\mathbf{V}_j|\sin(\theta_{ij} + \delta_i - \delta_j) \tag{8.105}$$

It can be shown that the changes in power are related to the changes in voltage magnitudes and phase angles.

8.11.2.1 Method 1. First Type of Formulation of Jacobian Matrix

$$\begin{bmatrix} \Delta P \\ \Delta Q \end{bmatrix} = \begin{bmatrix} J_1 & J_2 \\ J_3 & J_4 \end{bmatrix} \begin{bmatrix} \Delta \delta \\ \Delta |\mathbf{V}| \end{bmatrix} \tag{8.106}$$

The elements of submatrices of this Jacobian matrix can be found as follows [16–18]. For submatrix J_1, from Equation 8.104, the off-diagonal and diagonal elements, respectively, can be determined as

$$\frac{\partial P_i}{\partial \delta_j} = |\mathbf{V}_i||\mathbf{Y}_{ij}||\mathbf{V}_j|\sin(\theta_{ij} + \delta_i - \delta_j) \quad i \neq j \tag{8.107a}$$

$$\frac{\partial P_i}{\partial \delta_j} = \sum_{\substack{j=1 \\ i \neq j}}^{n} |\mathbf{V}_i||\mathbf{Y}_{ij}||\mathbf{V}_j|\sin(\theta_{ij} + \delta_i - \delta_j) \tag{8.107b}$$

$$= |\mathbf{V}_i|^2|\mathbf{Y}_{ii}|\sin\theta_{ii} - Q_i \quad i = j \tag{8.108}$$

Similarly, for submatrix J_2,

$$\frac{\partial P_i}{\partial |\mathbf{V}_j|} = |\mathbf{V}_i||\mathbf{Y}_{ij}||\mathbf{V}_j|\sin(\theta_{ij} + \delta_i - \delta_j) \quad i \neq j \tag{8.109a}$$

$$\frac{\partial P_i}{\partial |\mathbf{V}_i|} = \sum_{j=1}^{n} |\mathbf{V}_j||\mathbf{Y}_{ij}|\cos(\theta_{ij} + \delta_i - \delta_j) + |\mathbf{V}_i||\mathbf{Y}_{ii}|\cos\theta_{ii} \tag{8.109b}$$

$$= \frac{P_i}{|\mathbf{V}_i|} + |\mathbf{V}_i||\mathbf{Y}_{ii}|\cos\theta_{ii} \quad i = j \tag{8.110}$$

For submatrix J_3, from Equation 8.105,

$$\frac{\partial Q_i}{\partial \delta_j} = |\mathbf{V}_i||\mathbf{Y}_{ij}||\mathbf{V}_j|\sin(\theta_{ij} + \delta_i - \delta_j) \quad i \neq j \tag{8.111a}$$

$$\frac{\partial Q_i}{\partial \delta_i} = \sum_{j=1}^{n}|\mathbf{V}_j||\mathbf{Y}_{ij}|\cos(\theta_{ij} + \delta_i - \delta_j) + |\mathbf{V}_i||\mathbf{Y}_{ii}|\cos\theta_{ii} \tag{8.111b}$$

$$= -|\mathbf{V}_i|^2|\mathbf{Y}_{ii}|\sin\theta_{ii} + P_i \quad i = j \tag{8.112}$$

Similarly, for submatrix J_4,

$$\frac{\partial Q_i}{\partial |\mathbf{V}_j|} = |\mathbf{V}_i||\mathbf{Y}_{ij}|\sin(\theta_{ij} + \delta_i - \delta_j) \quad i \neq j \tag{8.113a}$$

$$\frac{\partial Q_i}{\partial |\mathbf{V}_i|} = |\mathbf{V}_i||\mathbf{Y}_{ii}|\cos\theta_{ii} + \sum_{j=1}^{n}|\mathbf{V}_j||\mathbf{Y}_{ij}|\cos(\theta_{ij} + \delta_i - \delta_j) \tag{8.113b}$$

$$= |\mathbf{V}_i||\mathbf{Y}_{ii}|\sin\theta_{ii} + \frac{Q_i}{|\mathbf{V}_i|} \quad i = j \tag{8.114}$$

In the event that a voltage-controlled bus is present, the general Equation 8.106 has to be modified as

$$\begin{bmatrix} \Delta P \\ \hline \Delta Q \\ \hline \Delta|\mathbf{V}| \end{bmatrix} = \begin{bmatrix} J_1 & J_2 \\ \hline J_3 & J_4 \\ J_5 & J_6 \end{bmatrix} \begin{bmatrix} \Delta\delta \\ \hline \Delta|\mathbf{V}| \end{bmatrix} \tag{8.115}$$

The elements of the submatrices for the voltage-controlled bus i can be found from the equation

$$|\mathbf{V}_i| = |\mathbf{V}_i| \tag{8.116}$$

Thus, submatrix J_5,

$$\frac{\partial |\mathbf{V}_i|}{\partial \delta_j} = 0 \quad \text{for all} \quad i,j \tag{8.117}$$

that is, the J_5 row matrix has only zeros as its elements. Similarly, for submatrix J_6,

$$\frac{\partial |\mathbf{V}_i|}{\partial |\mathbf{V}_j|} = 0 \quad i \neq j \tag{8.118}$$

and

$$\frac{\partial |\mathbf{V}_i|}{\partial |\mathbf{V}_i|} = 1 \quad i = j \tag{8.119}$$

that is, the J_5 row matrix also has zeros except one element. Note that, the angle $\Delta\delta$ in Equations 8.106 and 8.115 must be in radians.

It is interesting to observe the similarities between the elements of the submatrices of the Jacobian matrix. Therefore, let

$$\Phi_{ij} \triangleq \theta_{ij} + \delta_i - \delta_j \tag{8.120}$$

$$K_{ij} \triangleq |\mathbf{V}_i||\mathbf{Y}_{ij}|\cos\Phi_{ij} = |\mathbf{V}_i|G_{ij} \tag{8.121}$$

$$L_{ij} \triangleq |\mathbf{V}_i||\mathbf{Y}_{ij}|\sin\Phi_{ij} = |\mathbf{V}_i|B_{ij} \tag{8.122}$$

$$K'_{ij} \triangleq |\mathbf{V}_j| K_{ij} \tag{8.123}$$

$$L'_{ij} \triangleq |\mathbf{V}_j| L_{ij} \tag{8.124}$$

Therefore, the elements of each Jacobian submatrix can be redefined. For submatrix J_1, from Equations 8.107 and 8.108, the off-diagonal and diagonal elements, respectively, can be expressed as

$$\frac{\partial P_i}{\partial \delta_j} = L'_{ij} \quad i \neq j \tag{8.125}$$

$$\frac{\partial P_i}{\partial \delta_i} = L'_{ii} - Q_i \quad i = j \tag{8.126}$$

Thus, submatrix J_1 can be expressed as,*

$$[J_1] = [L'] - \begin{bmatrix} & & 0 \\ & Q_i & \\ 0 & & \end{bmatrix} \tag{8.127}$$

For submatrix J_2,

$$\frac{\partial P_i}{\partial |\mathbf{V}_j|} = K_{ij} \quad i \neq j \tag{8.128}$$

---

* Note that the off-diagonal elements of the Q_i matrix in Equation 8.127 are all equal to zero.

$$\frac{\partial P_i}{\partial |\mathbf{V}_i|} = K_{ii} + \frac{P_i}{|\mathbf{V}_i|} \quad i = j \tag{8.129}$$

Therefore,

$$[J_2] = [K] + \begin{bmatrix} & & 0 \\ & \frac{P_i}{|\mathbf{V}_i|} & \\ 0 & & \end{bmatrix} \tag{8.130}$$

For submatrix J_3,

$$\frac{\partial Q_i}{\partial \delta_j} = -K'_{ij} \quad i \neq j \tag{8.131}$$

$$\frac{\partial P_i}{\partial \delta_i} = K'_{ii} + P_i \quad i = j \tag{8.132}$$

Thus,

$$[J_3] = -[K'] + \begin{bmatrix} & & 0 \\ & P_i & \\ 0 & & \end{bmatrix} \tag{8.133}$$

For submatrix J_4,

$$\frac{\partial Q_i}{\partial |\mathbf{V}_j|} = L_{ij} \quad i \neq j \tag{8.134}$$

$$\frac{\partial Q_i}{\partial |\mathbf{V}_i|} = L_{ii} + \frac{Q_i}{|\mathbf{V}_i|} \quad i = j \tag{8.135}$$

Hence,

$$[J_4] = [L] + \begin{bmatrix} & & 0 \\ & \frac{Q_i}{|\mathbf{V}_i|} & \\ 0 & & \end{bmatrix} \tag{8.136}$$

Power Flow Analysis

Thus, the Jacobian matrix can be expressed as

$$\left[\begin{array}{c|c} J_1 & J_2 \\ \hline J_3 & J_4 \end{array}\right] = \left[\begin{array}{c|c} L' & K \\ \hline -K' & L \end{array}\right] + \left[\begin{array}{c|c} -Q_i & \dfrac{P_i}{|\mathbf{V}_i|} \\ \hline P_i & \dfrac{Q_i}{|\mathbf{V}_i|} \end{array}\right] \quad (8.137)$$

Note that, the off-diagonal elements in the submatrices of the second matrix are all zero. The method of solution in the polar coordinates is very similar to that in the rectangular coordinates. Therefore, it will not be repeated here.

8.11.2.2 Method 2. Second Type of Formulation of Jacobian Matrix

It has been proved that it is more efficient if Equation 8.106 is modified as follows [18]:

$$\left[\dfrac{\Delta P}{\Delta Q}\right] = \left[\begin{array}{c|c} J_1 & J_2 \\ \hline J_3 & J_4 \end{array}\right] \left[\dfrac{\Delta \delta}{\dfrac{\Delta |\mathbf{V}|}{|\mathbf{V}|}}\right] \quad (8.138)$$

where submatrices J_1 and J_3 have remained unchanged, as given in Equation 8.137. However, submatrices J_2 and J_4 are to be modified. For example, in the matrix multiplication, the ijth term in submatrix J_2 becomes

$$\left(\dfrac{\partial P_i}{\partial |\mathbf{V}_j|}\right) \Delta |\mathbf{V}_j| = |\mathbf{V}_i||\mathbf{Y}_{ij}|\cos(\theta_{ij} + \delta_i - \delta_j)\Delta|\mathbf{V}_j| \quad (8.139)$$

By multiplying the right side of Equation 8.138 by $|\mathbf{V}_j|/|\mathbf{V}_j|$,

$$\left(\dfrac{\partial P_i}{\partial |\mathbf{V}_j|}\right) \Delta |\mathbf{V}_j| = |\mathbf{V}_i||\mathbf{Y}_{ij}||\mathbf{V}_j|\cos(\theta_{ij} + \delta_i - \delta_j)\dfrac{\Delta|\mathbf{V}_j|}{|\mathbf{V}_j|} \quad (8.140)$$

or

$$\left(\dfrac{\partial P_i}{\partial |\mathbf{V}_j|}\right) \Delta |\mathbf{V}_j| = K'_{ij} \dfrac{\Delta|\mathbf{V}_j|}{|\mathbf{V}_j|} \quad (8.141)$$

Since

$$K'_{ij} = |\mathbf{V}_i||\mathbf{Y}_{ij}||\mathbf{V}_j|\cos(\theta_{ij} + \delta_i - \delta_j)$$

Therefore, submatrix J_2 becomes

$$[J_2] = [K'] + \begin{bmatrix} & & 0 \\ & P_i & \\ 0 & & \end{bmatrix} \qquad (8.142)$$

Similarly, submatrix J_4 takes the form

$$[J_4] = [L'] + \begin{bmatrix} & & 0 \\ & Q_i & \\ 0 & & \end{bmatrix} \qquad (8.143)$$

The voltage-controlled buses can be included as before. Submatrix J_5 remains the same (i.e., its elements are all zero). In submatrix J_6, all off-diagonal elements are zero and the diagonal element is $|V_i|$. Therefore, the modified Jacobian matrix can be expressed as

$$\begin{bmatrix} J_1 & J_2 \\ J_3 & J_4 \\ J_5 & J_6 \end{bmatrix} = \begin{bmatrix} L' & K' \\ -K' & L' \\ 0 & |V_i| \end{bmatrix} + \begin{bmatrix} -Q_i & P_i \\ P_i & Q_i \\ 0 & 0 \end{bmatrix} \qquad (8.144)$$

8.12 DECOUPLED POWER FLOW METHOD*

Consider the power flow problem defined in polar coordinating Equation 8.138,

$$\begin{bmatrix} \Delta P \\ \Delta Q \end{bmatrix} = \begin{bmatrix} J_1 & J_2 \\ J_3 & J_4 \end{bmatrix} \begin{bmatrix} \Delta \delta \\ \dfrac{\Delta |V|}{|V|} \end{bmatrix} \qquad (8.138)$$

which is known in the literature [16–18] as

$$\begin{bmatrix} \Delta P \\ \Delta Q \end{bmatrix} = \begin{bmatrix} H & N \\ J & L \end{bmatrix} \begin{bmatrix} \Delta \delta \\ \dfrac{\Delta |V|}{|V|} \end{bmatrix} \qquad (8.145)$$

In general, in a given power system, the real power flow is considered to be much less sensitive to the change in voltage magnitude than to the change in voltage angle. Therefore, the elements of submatrix N (or J_2) can be considered to be zero as an approximation. Likewise, the reactive power

---

* It is also known as the *decoupled Newton–Raphson method*.

Power Flow Analysis

is much less sensitive to change in voltage angle than voltage magnitude. Thus, the elements of submatrix J (or J_3) can be considered to be zero. Therefore, Equation 8.145 becomes

$$\left[\frac{\Delta P}{\Delta Q}\right] = \left[\begin{array}{c|c} H & 0 \\ \hline 0 & L \end{array}\right] \left[\begin{array}{c} \Delta \delta \\ \hline \frac{\Delta |V|}{|V|} \end{array}\right] \quad (8.146)$$

from which

$$[\Delta P] = [H][\Delta \delta] \quad (8.147)$$

$$[\Delta Q] = [L]\left[\frac{\Delta |V|}{|V|}\right] \quad (8.148)$$

which are known as the two *decoupled power flow equations* in the literature [28]. They are solved separately, so that

$$[\Delta \delta] = [H]^{-1}[\Delta P] \quad (8.149)$$

$$\left[\frac{\Delta |V|}{|V|}\right] = [L]^{-1}[\Delta Q] \quad (8.150)$$

Note that the solution involves the inversion of H and L matrices whose dimensions are approximately one-fourth the full-size Jacobian matrix. It is apparent that this approach significantly reduces not only the computation time for each iteration but also the computer memory requirement.

8.13 FAST DECOUPLED POWER FLOW METHOD

Consider the decoupled load flow of Equations 8.47 and 8.48 where the elements of submatrices H and L can be expressed as

$$[\Delta P] = [H][\Delta \delta] \quad (8.147)$$

$$[\Delta Q] = [L]\left[\frac{\Delta |V|}{|V|}\right] \quad (8.148)$$

$$H_{ij} = V_i V_j (G_{ij} \sin \delta_{ij} - B_{ij} \cos \delta_{ij}) \quad i \neq j \quad (8.151)$$

$$H_{ii} = -B_{ii} V_i^2 - Q_i \quad i = j \quad (8.152)$$

$$L_{ij} = V_i V_j (G_{ij} \sin \delta_{ij} - B_{ij} \cos \delta_{ij}) = H_{ij} \quad i \neq j \quad (8.153)$$

$$L_{ii} = -B_{ii} V_i^2 - Q_i \quad i = j \quad (8.154)$$

One can observe that even though the decoupled load flow method reduces the memory storage requirements considerably, it still requires significant amounts of computational effort. Thus, Stott and Alsac [30] have developed the *fast decoupled power flow method*. The basic assumptions used are the following:

1. The power systems have high X/R ratios. Hence,

$$G_{ij} \sin \delta_{ij} \ll B_{ij} \tag{8.155}$$

2. The difference between adjacent bus voltage angle is very small. Thus,

$$\sin \delta_{ij} = \sin(\delta_i - \delta_j) \cong \delta_i - \delta_j = \delta_{ij} \tag{8.156}$$

$$\cos \delta_{ij} = \cos(\delta_i - \delta_j) \cong 1.0 \tag{8.157}$$

3. Also,

$$Q_i \ll B_{ii} V_i^2 \tag{8.158}$$

Therefore, Equations 8.147 and 8.148 can be further approximated as

$$[\Delta P] = [V \times B' \times V][\Delta \delta] \tag{8.159}$$

$$[\Delta Q] = [V \times B'' \times V] \left[\frac{\Delta V}{V} \right] \tag{8.160}$$

where the elements of the matrices $[B']$ and $[B'']$ are the elements of the matrix $[-B]$. Thus,

$$B'_{ij} = -\frac{1}{X_{ij}} \quad i \neq j \tag{8.161}$$

$$B'_{ii} = \sum_{j=1}^{n} \frac{1}{X_{ij}} \quad i = j \tag{8.162}$$

$$B''_{ij} = -B_{ij} \tag{8.163}$$

The decoupling process in the fast decoupled power flow can be concluded [42] after additional modifications based on the simplifying assumptions as

$$\left[\frac{\Delta P}{V} \right] = [B'][\Delta \delta] \tag{8.164}$$

$$\left[\frac{\Delta Q}{V} \right] = [B''][\Delta V] \tag{8.165}$$

Note that, both matrices B' and B'' are real, sparse, and symmetrical owing to the fact that shunt susceptance, transformer off-nominal taps, and phase shifts are omitted. Therefore, both matrices B' and B'' are always symmetrical. Because of this, their constant sparse upper triangular factors can be computed and stored only once at the start of the solution. The method is very fast and reliable.

8.14 THE DC POWER FLOW METHOD

In certain power system studies (e.g., *reliability studies*), a very large number of power flow runs may be needed. Therefore, a very fast (and not necessarily accurate, due to the linear approximation involved) method can be used for such studies. The method of calculating the real flows by solving first for the bus angles is known as the *dc power flow method*, in contrast with the exact nonlinear solution, which is known as the *ac solution*.

Assume that bus i is connected to bus j over an impedance of Z_{ij}. Thus, the active power flow can be expressed as

$$P_{ij} = \frac{V_i V_j}{Z_{ij}} \sin(\delta_i - \delta_j) \qquad (8.166)$$

where

$$\mathbf{V}_i = |\mathbf{V}_i| \angle \delta_i \quad \mathbf{V}_j = |\mathbf{V}_j| \angle \delta_j$$

The following simplifying approximations are made:

$$X_{ij} \cong \mathbf{Z}_{ij} \text{ since } X_{ij} \gg R_{ij}$$

$$|\mathbf{V}_i| \cong 1.0 \text{ pu}$$

$$|\mathbf{V}_j| \cong 1.0 \text{ pu}$$

$$\sin(\delta_i - \delta_j) \cong \delta_i - \delta_j$$

Hence, the active power flow Equation 8.166 can be expressed as

$$P_{ij} \cong \frac{\delta_i - \delta i}{X_{ij}} \cong B_{ij}(\delta_i - \delta_j) \qquad (8.167)$$

Thus, in matrix form,

$$[P] = [B][\delta] \qquad (8.168)$$

from which

$$[\delta] = [B]^{-1}[P] \qquad (8.169)$$

or

$$[\delta] = [X][P] \qquad (8.170)$$

where the [B] matrix is an $(n-1) \times (n-1)$ matrix dimensionally for an n-bus system. The diagonal and off-diagonal elements of the [B] matrix can be found by adding the series susceptances of the branches connected to bus i and by setting them equal to the negated series susceptance of branch ij, respectively. The linear Equation 8.170 can be solved for δ by using matrix techniques. It is possible with the dc power flow method to carry out the thousands of power flow runs that are required for comprehensive contingency analysis on large-scale systems. Of course, the adequacy assessment provided by this representation is restricted to overload-related system problems.

In summary, the choice of a power flow method is a matter of choice between speed and accuracy. For a given degree of accuracy, the speed depends on the size, complexity, and configuration of the power system and on the numerical approach chosen.

EXAMPLE 8.4

Consider the IEEE five-bus system given in Example 8.1 and resolve by using the MATLAB® code.

Solution

(a) Here is the input data in a summary format:

```
basemva = 1; accuracy = 0.0001; maxiter = 50;

%          Bus  Bus Voltage  Angle   ---Load--    -Generator-----  Injected
%          No   code  Mag.   Degree  MW    Mvar   MW  Mvar Qmin Qmax Inj.Mvar
busdata = [1    1     1.02   0.0     0.0   0      0.0 0.0  0    0    0
           2    0     1.00   0.0     0.7   0.2    0.0 0.0  0    0    0
           3    0     1.00   0.0     0.4   0.1    0.0 0.0  0    0    0
           4    2     1.05   0.0     0     0      1.0 0.0  0    0    2
           5    0     1.00   0.0     0.6   0.3    0.0 0.0  0    0    0];

%                                    Line code
%           Bus bus    R        X      1/2 B      = 1 for lines
%           nl  nr     p.u.     p.u.   p.u.       > 1 or < 1 tr. tap at bus nl
Linedata = [1   2      0.200    0.400  0.0000     1
            1   5      0.100    0.200  0.0000     1
            2   3      0.000    0.200  0.0000     0.9091
            2   4      0.100    0.200  0.0000     1
            4   5      0.200    0.400  0.0000     1];

lfybus          % Form the bus admittance matrix

Lfgauss         % Load flow solution by Gauss-Seidel method
```

Power Flow Analysis

```
busout          % Prints the power flow solution on the screen

lineflow        % Computes and displays the line flow and losses
```

(b) Here is the MATLAB output:

```
= = = = = = = = = = = = = = = = = = = = = = = = =
Main MATLAB code:
= = = = = = = = = = = = = = = = = = = = = = = = =

Ybus =
3.0000 - 6.0000i      -1.0000 + 2.0000i      0           0         -2.0000 + 4.0000i
-1.0000 + 2.0000i    3.0000 -12.0499i    0 + 5.4999i   -2.0000 + 4.0000i      0
0                    0 + 5.4999i         0 - 5.0000i        0                 0
0         -2.0000 + 4.0000i         0           3.0000 - 6.0000i    -1.0000 + 2.0000i
-2.0000 + 4.0000i           0              0          -1.0000 + 2.0000i     3.0000 - 6.0000i

Iteration # =

  1

 0.950 -2.137
Iteration # =

  2

 0.938 -3.949
Iteration # =

  3

 0.927 -4.907
Iteration # =

  4

 0.919 -5.664
Iteration # =

  5

 0.914 -6.279
Iteration # =

  6

 0.910 -6.772
Iteration # =

  7

 0.908 -7.166
Iteration # =

  8
```

```
 0.906  -7.478
Iteration # =

    9

 0.905  -7.725
Iteration # =

   10

 0.903  -7.922
Iteration # =

   11

 0.903  -8.077
Iteration # =

   12

 0.902  -8.201
Iteration # =

   13

 0.901  -8.299
Iteration # =

   14

 0.901  -8.376
Iteration # =

   15

 0.901  -8.438
Iteration # =

   16

 0.900  -8.487
Iteration # =

   17

 0.900  -8.526
Iteration # =

   18

 0.900  -8.557
Iteration # =

   19
```

Power Flow Analysis

```
 0.900 -8.582
Iteration # =

  20

 0.900 -8.601
Iteration # =

  21

 0.900 -8.617
Iteration # =

  22

 0.900 -8.629
Iteration # =

  23

 0.899 -8.639
Iteration # =

  24

 0.899 -8.647
Iteration # =

  25

 0.899 -8.653
Iteration # =

  26

 0.899 -8.658
Iteration # =

  27

 0.899 -8.662
Iteration # =

  28

 0.899 -8.665
Iteration # =

  29

 0.899 -8.667
Iteration # =

  30
```

```
 0.899  -8.669
Iteration # =

  31

 0.899  -8.671
Iteration # =

  32

 0.899  -8.672
Iteration # =

  33

 0.899  -8.673
Iteration # =

  34

 0.899  -8.674
Iteration # =

  35

 0.899  -8.675
Iteration # =

  36

 0.899  -8.675       Power Flow Solution by Gauss-Seidel Method
              Maximum Power Mismatch = 9.22778e-005
                    No. of Iterations = 36

Bus    Voltage   Angle   ------Load------   ---Generation---   Injected
No.    Mag.      Degree   MW       Mvar      MW       Mvar      Mvar

1      1.020     0.000    0.000    0.000     0.845    0.316     0.000
2      0.899    -8.675    0.700    0.200     0.000    0.000     0.000
3      0.965   -13.483    0.400    0.100     0.000    0.000     0.000
4      1.050    -1.548    0.000    0.000     1.000   -1.389     2.000
5      0.943    -4.067    0.600    0.300     0.000    0.000     0.000

Total                     1.700    0.600     1.845   -1.073     2.000
```

Line Flow and Losses

| --Line-- | | Power at bus & line flow | | | --Line loss-- | | Transformer |
from	to	MW	Mvar	MVA	MW	Mvar	tap
1		0.845	0.316	0.902			
	2	0.410	0.129	0.430	0.036	0.071	
	5	0.435	0.187	0.473	0.022	0.043	
2		-0.700	-0.200	0.728			
	1	-0.375	-0.058	0.379	0.036	0.071	
	3	0.400	0.137	0.423	-0.000	0.037	0.909
	4	-0.725	-0.279	0.777	0.075	0.149	
3		-0.400	-0.100	0.412			
	2	-0.400	-0.100	0.412	-0.000	0.037	
4		1.000	0.611	1.172			
	2	0.800	0.428	0.907	0.075	0.149	
	5	0.200	0.183	0.271	0.013	0.027	
5		-0.600	-0.300	0.671			
	1	-0.413	-0.144	0.437	0.022	0.043	
	4	-0.187	-0.156	0.244	0.013	0.027	
Total loss					0.145	0.327	

```
>>
```

EXAMPLE 8.5

Consider the IEEE five-bus system given in Example 8.3 and resolved by using the MATLAB:

(a) By using the Newton–Raphson method
(b) By using the decoupled method
(c) By using the Gauss–Seidel method

Solution

(a) By using the Newton–Raphson method

Main MATLAB code:

```
= = = = = = = = = = = = = = = = = = = = = = = =
%IEEE 6-Bus System: Power Flow

basemva = 1.0; accuracy = 0.0001; maxiter = 2;
```

```
%           Bus   Bus   Volt.   Angle   ---Load---    ---Generator---
%           No    code  Mag.    Degree  MW    Mvar   MW    Mvar   Qmin   Qmax   Inj.Mvar

busdata = [ 1     1     1.02    0.0     0.0   0      0.0   0.0    0      0      0
            2     0     1.00    0.0     0.6   0.3    0.0   0.0    0      0      0
            3     2     1.04    0.0     0.0   0      1.0   0.0    0      10     0
            4     0     1.00    0.0     0.4   0.1    0.0   0.0    0      0      0
            5     0     1.00    0.0     0.6   0.2    0.0   0.0    0      0      0];

%                                       Line code
%           Bus    bus    R       X      1/2 B     = 1 for lines

%           nl     nr     p.u.    p.u.   p.u.      > 1 or < 1 tr. tap at bus nl

linedata = [ 1  2  0.100   0.400  0.0000    1
             1  4  0.150   0.600  0.0000    1
             1  5  0.050   0.200  0.0000    1
             2  3  0.050   0.200  0.0000    1
             2  4  0.100   0.400  0.0000    1
             3  5  0.050   0.200  0.0000    1];

lfybus         % form the bus admittance matrix

lfnewton       % Load flow solution by Newton-Raphson method

% decouple     % Load flow solution by Fast Decoupled method

% Lfgauss      % Load flow solution by Gauss-Seidel method

busout         % Prints the power flow solution on the screen

lineflow       % Computes and displays the line flow and losses
```

= =
Load flow solution by the Newton-Raphson method (3 Iterations, accuracy of 0.0001)
= =

Ybus =

```
 2.1569 - 8.6275i   -0.5882 + 2.3529i    0                   -0.3922 + 1.5686i   -1.1765+ 4.7059i
-0.5882 + 2.3529i    2.3529 - 9.4118i   -1.1765 + 4.7059i    -0.5882 + 2.3529i    0
 0                  -1.1765 + 4.7059i    2.3529 - 9.4118i     0                  -1.1765 + 4.7059i
-0.3922 + 1.5686i   -0.5882 + 2.3529i    0                    0.9804 - 3.9216i    0
-1.1765 + 4.7059i    0                  -1.1765 + 4.7059i     0                   2.3529 - 9.4118i
```

WARNING: Iterative solution did not converged after 2 iterations.

Press Enter to terminate the iterations and print the results

```
                 ITERATIVE SOLUTION DID NOT CONVERGE
                 Maximum Power Mismatch = 0.0457402
                       No. of Iterations = 3
```

Bus No.	Voltage Mag.	Angle Degree	---Load--- MW	Mvar	---Generation--- MW	Mvar	Injected Mvar
1	1.020	0.000	0.000	0.000	0.651	0.328	0.000
2	0.955	-3.882	0.600	0.300	0.000	0.000	0.000
3	1.050	2.101	0.000	0.000	1.000	0.476	0.000
4	0.923	-7.973	0.400	0.100	0.000	0.000	0.000
5	0.998	-1.986	0.600	0.200	0.000	0.000	0.000
Total			1.600	0.600	1.651	0.805	0.000

Power Flow Analysis

```
                    Line Flow and Losses
---Line---      Power at bus & line flow      ---Line loss--- Transformer
from    to       MW       Mvar      MVA        MW      Mvar       tap

  1              0.651    0.328     0.729
         2       0.196    0.123     0.231     0.005    0.021
         4       0.247    0.118     0.274     0.011    0.043
         5       0.193    0.065     0.204     0.002    0.008

  2             -0.600   -0.300     0.671
         1      -0.190   -0.103     0.216     0.005    0.021
         3      -0.592   -0.279     0.655     0.024    0.094
         4       0.167    0.039     0.171     0.003    0.013

         3       1.000    0.476     1.108
         2       0.616    0.374     0.720     0.024    0.094
         5       0.419    0.180     0.456     0.009    0.038

  4             -0.400   -0.100     0.412
         1      -0.236   -0.074     0.248     0.011    0.043
         2      -0.164   -0.026     0.166     0.003    0.013

  5             -0.600   -0.200     0.632
         1      -0.191   -0.057     0.199     0.002    0.008
         3      -0.409   -0.142     0.433     0.009    0.038

Total loss                                    0.054    0.216
```

Main MATLAB code:

= =

```
%IEEE 6-Bus System: Power Flow

basemva = 1.0; accuracy = 0.0001; maxiter = 2;

%         Bus  Bus  Voltage  Angle   ---Load---    -----Generator-----   Injected
%         No   code Mag.     Degree  MW    Mvar    MW   Mvar  Qmin Qmax  Mvar
busdata = [1    1   1.02     0.0     0.0   0       0.0  0.0   0    0     0
           2    0   1.00     0.0     0.6   0.3     0.0  0.0   0    0     0
           3    2   1.04     0.0     0.0   0       1.0  0.0   0    10    0
           4    0   1.00     0.0     0.4   0.1     0.0  0.0   0    0     0
           5    0   1.00     0.0     0.6   0.2     0.0  0.0   0    0     0];

%                            Line code
%         Bus  bus   R       X       1/2 B    = 1 for lines
%         nl   nr    p.u.    p.u.    p.u.     > 1 or < 1 tr. tap at bus nl
linedata = [1   2    0.100   0.400   0.0000   1
            1   4    0.150   0.600   0.0000   1
            1   5    0.050   0.200   0.0000   1
            2   3    0.050   0.200   0.0000   1
            2   4    0.100   0.400   0.0000   1
            3   5    0.050   0.200   0.0000   1];

lfybus         % Form the bus admittance matrix

lfnewton       % Load flow solution by Newton-Raphson method

% decouple     % Load flow solution by Fast Decoupled method

% Lfgauss      % Load flow solution by Gauss-Seidel method
```

```
busout       % Prints the power flow solution on the screen

lineflow     % Computes and displays the line flow and losses
```

Load flow solution by the Newton–Raphson method (3 Iterations, accuracy of 0.0001)

= =

Ybus =

```
 2.1569 - 8.6275i  -0.5882 + 2.3529i   0                 -0.3922 + 1.5686i  -1.1765 + 4.7059i
-0.5882 + 2.3529i   2.3529 - 9.4118i  -1.1765 + 4.7059i  -0.5882 + 2.3529i   0
 0                 -1.1765 + 4.7059i   2.3529 - 9.4118i   0                 -1.1765 + 4.7059i
-0.3922 + 1.5686i  -0.5882 + 2.3529i   0                  0.9804 - 3.9216i   0
-1.1765 + 4.7059i   0                 -1.1765 + 4.7059i   0                  2.3529 - 9.4118i
```

WARNING: Iterative solution did not converged after 2 iterations.

Press Enter to terminate the iterations and print the results

```
                 ITERATIVE SOLUTION DID NOT CONVERGE
                 Maximum Power Mismatch = 0.0457402
                       No. of Iterations = 3
```

Bus No.	Voltage Mag.	Angle Degree	---Load--- MW	Mvar	---Generation--- MW	Mvar	Injected Mvar
1	1.020	0.000	0.000	0.000	0.651	0.328	0.000
2	0.955	-3.882	0.600	0.300	0.000	0.000	0.000
3	1.050	2.101	0.000	0.000	1.000	0.476	0.000
4	0.923	-7.973	0.400	0.100	0.000	0.000	0.000
5	0.998	-1.986	0.600	0.200	0.000	0.000	0.000
Total			1.600	0.600	1.651	0.805	0.000

Line Flow and Losses

---Line--- from	to	Power at bus & line flow MW	Mvar	MVA	---Line loss--- MW	Mvar	Transformer tap
1		0.651	0.328	0.729			
	2	0.196	0.123	0.231	0.005	0.021	
	4	0.247	0.118	0.274	0.011	0.043	
	5	0.193	0.065	0.204	0.002	0.008	
2		-0.600	-0.300	0.671			
	1	-0.190	-0.103	0.216	0.005	0.021	
	3	-0.592	-0.279	0.655	0.024	0.094	
	4	0.167	0.039	0.171	0.003	0.013	
3		1.000	0.476	1.108			
	2	0.616	0.374	0.720	0.024	0.094	
	5	0.419	0.180	0.456	0.009	0.038	
4		-0.400	-0.100	0.412			
	1	-0.236	-0.074	0.248	0.011	0.043	
	2	-0.164	-0.026	0.166	0.003	0.013	
5		-0.600	-0.200	0.632			
	1	-0.191	-0.057	0.199	0.002	0.008	
	3	-0.409	-0.142	0.433	0.009	0.038	
Total loss					0.054	0.216	

Power Flow Analysis

= =

(b) Load flow solution by the fast decoupled method (3 Iterations, accuracy of 0.0001)

= =

Ybus =

```
 2.1569 - 8.6275i  -0.5882 + 2.3529i   0                 -0.3922 + 1.5686i  -1.1765 + 4.7059i
-0.5882 + 2.3529i   2.3529 - 9.4118i  -1.1765 + 4.7059i  -0.5882 + 2.3529i   0
 0                 -1.1765 + 4.7059i   2.3529 - 9.4118i   0                 -1.1765 + 4.7059i
-0.3922 + 1.5686i  -0.5882 + 2.3529i   0                  0.9804 - 3.9216i   0
-1.1765 + 4.7059i   0                 -1.1765 + 4.7059i   0                  2.3529 - 9.4118i
```

WARNING: Iterative solution did not converged after 2 iterations.

Press Enter to terminate the iterations and print the results
          ITERATIVE SOLUTION DID NOT CONVERGE
          Maximum Power Mismatch = 0.0364484
                  No. of Iterations = 3

Bus No.	Voltage Mag.	Angle Degree	Load MW	Load Mvar	Generation MW	Generation Mvar	Injected Mvar
1	1.020	0.000	0.000	0.000	0.682	0.329	0.000
2	0.953	-3.928	0.600	0.300	0.000	0.000	0.000
3	1.040	2.010	0.000	0.000	1.000	0.478	0.000
4	0.922	-8.009	0.400	0.100	0.000	0.000	0.000
5	0.992	-2.061	0.600	0.200	0.000	0.000	0.000
Total			1.600	0.600	1.682	0.807	0.000

Line Flow and Losses

Line from	to	Power at bus & line flow MW	Mvar	MVA	Line loss MW	Mvar	Transformer tap
1		0.682	0.329	0.757			
	2	0.198	0.127	0.235	0.005	0.021	
	4	0.248	0.120	0.276	0.011	0.044	
	5	0.205	0.094	0.226	0.002	0.010	
2		-0.600	-0.300	0.671			
	1	-0.193	-0.105	0.220	0.005	0.021	
	3	-0.574	-0.244	0.623	0.021	0.086	
	4	0.166	0.038	0.170	0.003	0.013	
3		1.000	0.478	1.108			
	2	0.595	0.329	0.680	0.021	0.086	
	5	0.406	0.160	0.437	0.009	0.035	
4		-0.400	-0.100	0.412			
	1	-0.237	-0.076	0.249	0.011	0.044	
	2	-0.163	-0.025	0.165	0.003	0.013	
5		-0.600	-0.200	0.632			
	1	-0.203	-0.084	0.220	0.002	0.010	
	3	-0.397	-0.125	0.417	0.009	0.035	
Total loss					0.052	0.208	

```
>>
```

= =

(c) Load flow solution by the Gauss–Seidel method (3 Iterations, accuracy of 0.0001)

= =

Ybus =

```
 2.1569 - 8.6275i  -0.5882 + 2.3529i   0                 -0.3922 + 1.5686i  -1.1765 + 4.7059i
-0.5882 + 2.3529i   2.3529 - 9.4118i  -1.1765 + 4.7059i  -0.5882 + 2.3529i   0
 0                 -1.1765 + 4.7059i   2.3529 - 9.4118i   0                 -1.1765 + 4.7059i
-0.3922 + 1.5686i  -0.5882 + 2.3529i   0                  0.9804 - 3.9216i   0
-1.1765 + 4.7059i   0                 -1.1765 + 4.7059i   0                  2.3529 - 9.4118i
```

iter =

   1

   0.981 -3.066
WARNING: Iterative solution did not converged after 2 iterations.

Press Enter to terminate the iterations and print the results

iter =

   2

   0.964 -3.063
iter =

   3

   0.957 -3.332 ITERATIVE SOLUTION DID NOT CONVERGE
        Maximum Power Mismatch = 0.0555341
              No. of Iterations = 3

Bus No.	Voltage Mag.	Angle Degree	---Load--- MW	Mvar	---Generation--- MW	Mvar	Injected Mvar
1	1.020	0.000	0.000	0.000	0.556	0.310	0.000
2	0.957	-3.332	0.600	0.300	0.000	0.000	0.000
3	1.040	2.553	0.000	0.000	1.000	0.454	0.000
4	0.926	-7.614	0.400	0.100	0.000	0.000	0.000
5	0.993	-1.799	0.600	0.200	0.000	0.000	0.000
Total			1.600	0.600	1.556	0.765	0.000

Line Flow and Losses

---Line--- from	to	Power at bus & line flow MW	Mvar	MVA	---Line loss--- MW	Mvar	Transformer tap
1		0.556	0.310	0.636			
	2	0.172	0.121	0.210	0.004	0.017	
	4	0.237	0.114	0.263	0.010	0.040	
	5	0.182	0.094	0.205	0.002	0.008	
2		-0.600	-0.300	0.671			
	1	-0.168	-0.104	0.197	0.004	0.017	
	3	-0.567	-0.227	0.611	0.020	0.082	
	4	0.175	0.037	0.179	0.003	0.014	

```
3                  1.000      0.454      1.098
          2        0.588      0.309      0.664      0.020      0.082
          5        0.430      0.151      0.455      0.010      0.038

4                 -0.400     -0.100      0.412
          1       -0.227     -0.074      0.239      0.010      0.040
          2       -0.171     -0.023      0.173      0.003      0.014

5                 -0.600     -0.200      0.632
          1       -0.180     -0.086      0.200      0.002      0.008
          3       -0.420     -0.113      0.435      0.010      0.038

Total loss                                          0.050      0.199

>>
```

REFERENCES

1. Brown, H. E., *Solution of Large Networks by Matrix Methods*. Wiley, New York, 1975.
2. Shipley, R. B., *Introduction to Matrices and Power Systems*. Wiley, New York, 1976.
3. Dunstan, L. E., Digital load flow studies. *Trans. Am. Inst. Electr. Eng.*, Part 3A 73, 825–831 (1954).
4. Ward, J. B., and Hale, H. W., Digital computer solution of power flow problems. *Trans. Am. Inst. Electr. Eng., Part 3* 75, 398–404 (1956).
5. Glimn, A. F., and Stagg, G. W., Automatic calculation of load flows. *Trans. Am. Inst. Electr. Eng., Part 3* 76, 817–828 (1957).
6. Trevino, C., Cases of difficult convergence in load flow problems. *IEEE Power Eng. Soc. Winter Power Meet.,* 1971, Pap. No. 71 CP 62-PWR (1971).
7. Hubert, F. J., and Hayes, D. R., A rapid digital computer solution for power system network load-flow. *IEEE Trans. Power Appar. Syst.* PAS-90, 934–940 (1971).
8. Maslin, W. W. et al., A power system planning computer program package emphasizing flexibility and compatibility. *IEEE Power Eng. Soc. Summer Power Meet., 1970*, Pap. No. 70 CP 684-PWR (1970).
9. Podmore, R., and Undrill, J. M., Modified nodal iterative load flow algorithm to handle series capacitive branches. *IEEE Trans. Power Appar. Syst.* PAS-92, 1379–1387 (1973).
10. Treece, J. A., Bootstrap Gauss–Seidel load flow. *Proc. IEEE* 116, 866–870 (1969).
11. Gupta, P. P., and Humphrey Davies, M. W., Digital computers in power system analysis. *Proc. Inst. Electr. Eng., Part A* 109, 383–404 (1961).
12. Brameller, A., and Denmead, J. K., Some improved methods of digital network analysis. *Proc. Inst. Electr. Eng. Part A* 109, 109–116 (1962).
13. Brown, H. E., Cater, O. K., Happ, H. H., and Person, C. E., Power flow solution by impedance matrix iterative method. *IEEE Trans. Power Appar. Syst.* PAS-82, 1–10 (1963).
14. Freris, L. L., and Sasson, A. M., Investigations on the load-flow problem. *Proc. Inst. Electr. Eng.* 114, 1960 (1967).
15. Brown, H. E., Cater, G. K., Happ, H. H., and Person, C. E., Z-matrix algorithms in load-flow programs. *IEEE Trans. Power Appar. Syst.* PAS·S7, 807–814 (1968).
16. Van Ness, J. E., Iteration methods for digital load flow studies. *Trans. Am. Inst. Electr. Eng., Part 3* 78, 583–588 (1959).
17. Van Ness, J. E., Convergence of iterative load flow studies. *Trans. Am. Inst. Electr. Eng.* 78, 1590–1597 (1960).
18. Van Ness, J. E., and Griffin, J. H., Elimination methods for load-flow studies. *Trans. Am. Inst. Electr. Eng.* 80, 299–304 (1961).
19. Tinney, W. F., and Hart, C. E., Power flow solution by Newton's method. *IEEE Trans. Power Appar. Syst.* PAS-86, 1449–1456 (1967).
20. Britton, J. P., Improved area interchange control for Newton's method load flows. *IEEE Trans. Power Appar. Syst.* PAS-88, 1577–1581 (1969).
21. Dommel, H. W., Tinney, W. F., and Powell, W. L., Further developments in Newton's method for power system applications. *IEEE Power Eng. Soc. Winter Power Meet., 1970*, Pap. No. 70 CP 161-PWR (1970).
22. Stott, B., Effective starting process for Newton–Raphson load flows. *Proc. Inst. Electr. Eng.* 11S, 983–987 (1971).

23. Britton, J. P., Improved load flow performance through a more general equation form. *IEEE Trans. Power Appar. Syst.* PAS-90, 109–116 (1971).
24. Peterson, N. M., and Meyer, W. S., Automatic adjustment of transformer and phase-shifter taps in the Newton power flow. *IEEE Trans. Power Appar. Syst.* PAS-90, 103–108 (1971).
24a. Sato, N., and Tinney, W. F., Techniques for exploring the sparsity of the network admittance matrix. *Trans. Am. Inst. Electr. Eng., Part 3* 82, 944–949 (1983).
24b. Tinney, W. P., and Walker, J. W., Direct solutions of sparse network equations by optimally ordered triangular factorization. *Proc. IEEE* 55, 1801–1809 (1967).
25. Stagg, G. W., and Phadke, A. G, Real-time evaluation of power-system contingencies detection of steady state overloads. *IEEE Power Eng. Soc. Summer Power Meet., 1970*, Pap. No. 70 CP 692-PWR (1970).
26. Despotovic, S. T., Babic, B. S., and Mastilovic, V. P., A rapid and reliable method for solving load flow problems. *IEEE Trans. Power Appar. Syst.* PAS-90, 123–130 (1971).
27. Uemura, K., Power flow solution by a Z-matrix type method and its application to contingency evaluation. *Proc. IEEE Power Ind. Comput. Appl. Conf.* 1971, 151–159 (1971).
28. Stott, B., Decoupled Newton load flow. *IEEE Trans. Power Appar. Syst.* PAS-91, 1955–1959 (1972).
29. Peterson, N. M., Tinney, W. F., and Bree, D. W., Jr., Iterative linear ac power flow solution for fast approximate outage studies. *IEEE Trans. Power Appar. Syst.* PAS-91, 2048–2056 (1972).
30. Stott, B., and Alsac, O., Fast decoupled load flow. *IEEE Trans. Power Appar. Syst.* PAS-93, 859–869 (1974).
31. Wallach, Y., Gradient methods for load flow problems. *IEEE Trans. Power Appar. Syst.* PAS-87, 1314–1318 (1968).
32. Sasson, A. M., Nonlinear programming solutions for the load flow, minimum-loss, and economic dispatching problems. *IEEE Trans. Power Appar. Syst.* PAS-SS, 399–409 (1969).
33. Galloway, R. H., Hogg, W. D., and Scott, M., New approach to power-system load-flow analysis in a digital computer. *Proc. Inst. Electr. Eng.* 117, 165–169 (1970).
34. Dusonchet, Y. P., Talukdar, S. N., Sinnott, H. E., and El-Abiad, A. H., Load flow using a combination of point Jacobi and Newton's methods. *IEEE Trans. Power Appar. Syst.* PAS-90, 941–949 (1971).
35. Sasson, A. M., Trevino, C., and Aboytes, F., Improved Newton's load flow through a minimization technique. *Proc. IEEE Power Ind. Comput. Appl. Conf.* 1971, 160–169 (1971).
36. Dommel, H. W., and Tinney, W. F., Optimal power flow solutions. *IEEE Trans. Power Appar. Syst.* PAS-87, 1866–1876 (1968).
37. Brown, H. E., Contingencies evaluated by a Z-matrix method. *IEEE Trans. Power Appar. Syst.* PAS-88, 409–412 (1969).
38. Dommel, H. W., and Sato, N., Fast transient stability solutions. *IEEE Trans. Power Appar. Syst.* PAS-91, 1643–1650 (1972).
39. Fox, B., and Revington, A. M., Network calculations for on-line control of a power system. *Proc. IEEE Conf. Comput. Power Syst. Oper. Control, 1972*, 261–275 (1972).
40. Sachdev, M. S., and Ibrahim, S. A., A fast approximate technique for outage studies in power system planning and operation. *IEEE Power Eng. Soc. Summer Power Meet., 1973*, Pap. No. T 73 469-4 (1973).
41. Alsac, O., and Stott, B., Optimal load flow with steady-state security. *IEEE Power Eng. Soc. Summer Power Meet.*, 1973, Pap. No. T 73 484-3 (1973).
42. Stott, B., Review of load-flow calculation methods. *Proc. IEEE* 62 (7), 916–929 (1974).
43. Takahashi, K., Fagan, J., and Chen, M., Formulation of a sparse bus impedance matrix and its application to short circuit study. *Proc. IEEE Power Ind. Comput. Appl. Conf., 1973*, 41–50 (1973).
44. Sasson, A. M., and Merrill, H. M., Some applications of optimization techniques to power systems problems. *Proc. IEEE* 62 (7), 959–972 (1974).
45. DyLiacco, T. E., Real-time computer control of power systems. *Proc. IEEE* 62 (7), 884–891 (1974).
46. Undrill, J. M. et al., Interactive computation in power system analysis. *Proc. IEEE* 62 (7), 1009–1018 (1974).
47. Korsak, A. J., On the question of uniqueness of stable load-flow solutions. *IEEE Trans. Power Appar. Syst.* PAS-9t, 1093–1100 (1972).
48. Bosarge, W. E., Jordan, J. A., and Murray, W. A., A non-linear block SOR-Newton load flow algorithm. *IEEE Power Ind. Eng. Soc., Summer Meet., 1973*, Pap. No. C 73 644-5 (1973).
49. Van Slyck, L. S., and Dopazo, J. F., Conventional load flow not suited for real-time power system monitoring. *IEEE Power Ind. Comput. Appl. Conf., 1973*, 12–21 (1973).
50. Borkowska, B., Probabilistic load flow. *IEEE Power Eng. Soc. Summer Meet., 1973*, Pap. No. T 73 485-10 (1973).

51. Dopazo, J. F., Klitin, O. A., and Sasson, A. M., Stochastic load flows. *IEEE Power Eng. Soc. Summer Meet., 1974*, Pap. No. T 74 308-3 (1974).
52. Happ, H. H., Diakoptics—The solution of system problems by tearing *Proc. IEEE* 62 (7), 930–940 (1974).
53. Sachdev, M. S., and Medicherla, T. K. P., A second order load flow technique. *IEEE Trans. Power Appar. Syst.* PAS·96 (1), 189–197 (1977).
54. Wu, F. F., Theoretical study of the convergence of the fast decoupled load flow. *IEEE Trans. Power Appar. Syst.* PAS-96 (1), 268–275 (1977).
55. Lewis, A. H. et al., *Large Scale System Effectiveness Analysis*, Rep. U.S. Dept. of Energy, Systems Control, Inc., Palo Alto, CA, 1978.
56. Brown, R. J., and Tinney, W. F., Digital solutions for large power networks. *Trans. Am. Inst. Electr. Eng., Part 3* 76, 347–355 (1957).
57. Jordan, R. H., Rapidly converging digital load flows. *Trans. Am. Inst. Electr. Eng., Part 3* 76, 1433–1438 (1957).
58. Conner, U. A., Representative bibliography on load-flow analysis and related topics. *IEEE Power Eng. Soc. Winter Meet., 1973*, Pap. No. C 73 104-7 (1973).
59. Bennett, J. M., Digital computers and the load flow problem. *Proc. Inst. Electr. Eng., Part B* 103, Suppl. 1, 16 (1955).
60. Sasson, A. M. and Jaimes, F. J., Digital methods applied to power flow studies. *IEEE Trans. Power Appar. Syst.* PAS·86, 860–867 (1967).
61. Laughton, M. A., and Humphrey Davis, M. W. Numerical techniques in the solution of power system load flow problems. *Proc. Inst. Electr. Eng.* 111, 1575–1588.
62. Ogbuobiri, E. C. et al., Sparsity-directed decomposition for Gaussian elimination on matrices. *IEEE Power Ind. Comput. Appl. Conf., 1969*, 131–140 (1969).
63. Carpentier, J., Application de la methode de Newton au calcul des reseaux. *Proc. 1st Power Syst. Comput. Conf. (PSCC)*. Queen Mary College, London, 1963.
64. Tamura, Y., Mori, H., and Iwamato, S., Relationship between voltage instability and multiple load flow solutions in electric power systems. *IEEE Trans. Power Appar. Syst.* PAS-102, 1115–1125 (1983).
65. Kron, G., *Diakoptics: The Piecewise Solution of Large Scale Systems*. MacDonald & Co., London, 1963.
66. M. Happ, H. H., *Diakoptics and Network*. Academic Press, New York, 1971.
67. Happ, H. H., and Undrill, J. M., Multicomputer configurations and diakoptics: Real power flow in power pools. *IEEE Trans. Power Appar. Syst.* PAS-88, 789–796 (1969).
68. Happ, H. H., Diakoptics and piecewise methods. *IEEE Trans. Power Appar. Syst.* PAS-89, 1373–796 (1969).
69. Happ, H. H., and Young, C. C., Tearing algorithms for large scale network programs. *IEEE Trans. Power Appar. Syst.* PAS-90, 2639–2650 (1971).
70. Happ, H. H., *Piecewise Methods and Applications to Power Systems*. Wiley, New York, 1980.
71. Heydt, G. T., Stochastic power How calculations. *IEEE Power Eng. Soc. Summer Meet., 1975*, Pap. No. A 75 530-6 (1975).
72. Allan, R. N., Borkowska, B., and Grizz, C. H., Probabilistic analysis of power flows. *Proc. IEEE* 121 (12), 1551–1556 (1974).
73. Stagg, G. W., and El-Abiad, A. H., *Computer Methods in Power System Analysis*. McGraw-Hill, New York, 1968.
74. Stevenson, W. D., *Elements of Power System Analysis*, 3rd ed. McGraw-Hill, New York, 1975.

GENERAL REFERENCES

Anderson, P. M., *Analysis of Faulted Power Systems*. Iowa State Univ. Press, Ames, IA, 1973.
Eigerd, O., *Electric Energy Systems Theory*. McGraw-Hill, New York, 1971.
Gross, C. A., *Power System Analysis*. Wiley, New York, 1979.
Guile, A. E., and Paterson, W., *Electrical Power Systems*, 2nd ed., Vol. 2. Pergamon Press, New York, 1977.
Knight, U. G., *Power Systems Engineering and Mathematics*. Pergamon Press, New York, 1972.
Mortlock, J. R., and Humphrey Davis, M. W., *Power System Analysis*. Chapman & Hall, London, 1952.
Neuenswander, J. R., *Modern Power Systems*. International Textbook Co., Scranton, Pennsylvania, 1971.
Sterling, M. J. H., *Power System Control*. Peter Peregrinus Ltd., Stevenage, UK, 1978.
Sullivan, R. L., *Power Systems Planning*. McGraw-Hill, New York, 1977.
Weedy, B. M., Electric Power Systems, 2nd ed. Wiley, New York, 1972.

PROBLEMS

1. Use the results of Example 8.1 and the Gauss–Seidel method and determine the following:
   (a) Value of voltage V_3 for the first iteration
   (b) Value of voltage V_4 for the first iteration
2. Consider the following the nonlinear equations

$$f_1(x_1,x_2) = x_1^2 + 2x_2 - 10 = 0$$

$$f_2(x_1, x_2) = x_1 x_2 - 3x_2^2 + 5 = 0$$

   Determine the values of x_1 and x_2 by using the Newton–Raphson method. Use 3 and 2 as the initial approximation for x_1 and x_2, respectively.
3. Use the rectangular form for voltages and the bus admittance quantities as follows and develop an expression for the real power loss in line i to j:

$$Y_{ii} = G_{ij} - jB_{ij}$$

$$V_i = e_i + jf_i$$

$$V_j = e_j + jf_j$$

4. Use the resultant real power loss expression from the solution of Problem 3 and determine the real power loss in line 1–2. Assume that $Z_{12} = 0.1 + j0.3$ pu, $V_1 = 0.98 - j0.06$ pu, and $V_2 = 1.04 - j1.04$ pu.
5. Assume that the branch admittance of a line connecting buses 1 and 2 is given as $1 - j4$ pu and that the bus voltages are given as $0.98 + j0.06$ pu and $1.04 + 0.00$ pu for buses 1 and 2, respectively. Neglect the line capacitance and determine the following:
   (a) Real and reactive powers in line at the bus 1 end
   (b) Real and reactive power losses in the line
6. Assume that line 1–2 given in Problem 3 has a significant amount of line capacitance. Half of the total line-charging admittance, $1/2$ y', is given as $j0.05$ pu and is placed at each end of the line in the π representation. Determine the following:
   (a) Real and reactive powers in line at the bus 2 end
   (b) Real and reactive power losses in the line
7. Assume that the data in Table P8.1 have been given for Example 8.3 and determine the following:
   (a) All individual branch admittances
   (b) All elements of bus admittance matrix
8. Use the results of Problem 7 in the application of the Newton–Raphson method in rectangular coordinates to Example 8.3 to determine the following after the first iteration:
   (a) Jacobian submatrix J_1
   (b) Jacobian submatrix J_2
   (c) Jacobian submatrices J_3 and J_5
   (d) Jacobian submatrices J_4 and J_6
   (e) Bus currents in terms of $c_i + jd_i$
   (f) The Δe and Δf voltage changes for four buses
   (g) The ΔP, ΔQ, and $\Delta |V|^2$ changes

Power Flow Analysis

TABLE P8.1
Line Impedans Data for Problem 7

Line (Bus to Bus)	R (pu)	X (pu)
1–2	0.20	0.40
1–4	0.20	0.60
1–5	0.10	0.20
2–3	0.10	0.20
2–4	0.20	0.40
3–5	0.10	0.20

9. Modify Example 8.3 using the data in Tables P8.2a and P8.2b and apply the Newton–Raphson method in polar coordinates. Using the second type of formulation of the Jacobian matrix, determine the following:
   (a) Jacobian submatrix J_3 for the second iteration
   (b) Jacobian submatrix J_5 for the second iteration

10. Consider a four-bus power system in which bus 1 is the slack bus with constant-voltage magnitude and angle. A generator is connected to bus 3 where the voltage magnitude is maintained at 1.02 pu. Buses 2 and 4 are designated as the load buses. Apply the Newton–Raphson method in rectangular coordinates. Using the following data, given in per units, determine the value of the terms indicated in black (i.e., the elements in ΔP_2, $J_{1(2,2)}$ and $J_{3(2,3)}$) for the next iteration (Table P8.3).

$$[\mathbf{Y}_{bus}] = \begin{array}{c} \\ 1 \\ 2 \\ 3 \\ 4 \end{array} \begin{bmatrix} 3-j9 & -2+j5 & 0 & -1+4 \\ -2+j5 & 4-j10 & -2+j5 & 0 \\ 0 & -2+j5 & 3-j9 & -1+j4 \\ -1+j4 & 0 & -1+j4 & 2-j8 \end{bmatrix}$$

$$\begin{bmatrix} \boxed{\Delta P_2} \\ \Delta P_3 \\ \Delta P_4 \\ \Delta Q_2 \\ \Delta Q_3 \\ \Delta |V_3|^2 \end{bmatrix} = \begin{bmatrix} \boxed{} & & \\ J_1 & & J_2 \\ & & \\ \boxed{} & & \\ J_3 & & J_4 \\ J_5 & & J_6 \end{bmatrix} \times \begin{bmatrix} \Delta e_2 \\ \Delta e_3 \\ \Delta e_4 \\ \Delta f_2 \\ \Delta f_3 \\ \Delta f_4 \end{bmatrix}$$

11. Consider the four-bus power system given in Problem 10. Apply the modified version of the Newton–Raphson method in polar coordinates. Assume that the specified voltages are given in polar coordinates as $1.03\angle 0°$, $0.9055\angle -6.3402°$, $1.02\angle 0°$, and $1.0012\angle -2.8624°$ for $\mathbf{V}_1$, $\mathbf{V}_2$, $\mathbf{V}_3$, and $\mathbf{V}_4$, respectively. Determine the values of the terms indicated by bold boxes (i.e., the elements in ΔP_3, $J_{1(2,3)}$, $J_{2(3,2)}$, and $J_{4(2,4)}$).

$$\begin{bmatrix} \Delta P_2 \\ \boxed{\Delta P_3} \\ \Delta P_4 \\ \Delta Q_2 \\ \Delta Q_4 \end{bmatrix} \begin{matrix} 2 \\ 3 \\ 4 \\ 2 \\ 4 \end{matrix} = \begin{bmatrix} \overset{2}{\Box} & \overset{3}{} & \overset{4}{} & \overset{2}{} & \overset{4}{} \\ & J_1 & & \Box & J_2 \\ & & & & \\ & J_3 & & & \boxed{J_4} \\ & & & & \end{bmatrix} \times \begin{bmatrix} \Delta \delta_2 \\ \Delta \delta_3 \\ \Delta \delta_4 \\ \dfrac{\Delta |V_2|}{|V_2|} \\ \dfrac{\Delta |V_4|}{|V_4|} \end{bmatrix}$$

12. Consider a three-bus power system as shown in Figure P8.1. Assume that bus 1 is the slack bus with a voltage of $1.02 + j0.000$ pu. Bus 2 is the regulated generator bus with a voltage magnitude of 1.02 pu. Use $1.02 + j0.000$ pu and 0.5 pu as its initial voltage and real power values, respectively. Bus 3 is the load bus with an initial voltage of $0.0.95 - j0.05$ pu. Its load is made up of 1.0 pu resistive load and 0.4 pu inductive load. Assume that the current components calculated

TABLE P8.2a
Conductance, Susceptance, and Admittance Data of the Lines for Problem 9

i	j	G_{ij}	B_{ij}	Y_{ij}	Φ_{ij}
1	2	−1	−2	2.236	−116.565°
2	4	−1	−2	2.236	−116.565°
1	4	−0.5	−1.5	1.581	−108.435°
1	5	−2	−4	4.472	−116.565°
2	3	−2	−4	4.472	−116.565°
3	5	−2	−4	4.472	−116.565°
2	2	4	8	8.944	63.435°
3	3	4	8	8.944	63.435°
5	5	4	8	8.944	63.435°
1	1	3.5	7.5	8.2765	63.435°
4	4	1.5	7.5	3.808	66.801°

TABLE P8.2b
Specified Values, Initial Values, and Calculated Values of P, Q, and $|V|$ at Each Bus for Problem 9

	Specified Values			Initial Values		Calculated Values (After First Iteration)	
Bus	P_i	Q_i	$|V_i|$	$|V_i|$	δ_i	P_i	Q_i
1	–	–	1.02	1.02	0°	–	–
2	−0.6	−0.3	–	1.0	0°	−0.1	−0.02
3	1.0	–	1.04	1.04	0°	0.1664	–[a]
4	−0.4	−0.1	–	1.0	0°	−0.1	−0.03
5	−0.6	−0.2	–	1.0	0°	−0.12	−0.24

[a] Calculated $|V_i| = 1.04$ after the first iteration.

Power Flow Analysis

TABLE P8.3
For Problem 10

Bus	Specified Values			Calculated Values (After kth Iteration)	
	V_i	P_i	Q_i	c_i	d_i
1	$1.03 + j0.0$	–	–	–	–
2	$0.9 + j0.1$	–1.2	–0.4	–1.50	–0.85
3	$1.02 + j0.0$	1.0	–	0.96	–0.43
4	$1.0 - j0.05$	–0.5	–0.1	–0.45	0.10

at each bus are d_2, c_3, and d_3 with values of –0.200, –1.375, and 1.000 pu, respectively. Use the Newton–Raphson method in rectangular coordinates and determine the following:
   (a) Current component c_3 for bus 2
   (b) Real and reactive power changes, ΔP_3 and ΔQ_3, for bus 3
   (c) Submatrices J_1, J_3, and J_5

13. Use the appropriate data given in Problem 12 and the Gauss–Seidel method and determine the following:
    (a) Value of voltage $\mathbf{V}_3$ for the first iteration
    (b) Value of voltage $\mathbf{V}_2$ for the first iteration

14. Assume that the LTC transformer shown in Figure P8.2 has an off-nominal turns ratio as indicated. (Note that, in the event that the ratio is equal to the system nominal voltage ratio, it is represented as a 1:1 ratio in the per-unit system.) Use the two-port network theory and verify the equation

$$\begin{bmatrix} \mathbf{I}_1 \\ \mathbf{I}_2 \end{bmatrix} = \begin{bmatrix} \dfrac{Y}{a^2} & -\dfrac{Y}{a} \\ -\dfrac{Y}{a} & Y \end{bmatrix} \begin{bmatrix} \mathbf{V} \\ \mathbf{V}_2 \end{bmatrix}$$

15. Use the results of Problem 14 and verify the equivalent π circuit representation of the transformer shown in Figure 8.3b.

532 Modern Power System Analysis

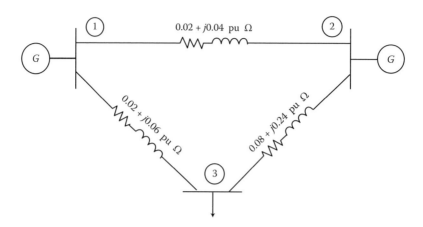

FIGURE P8.1 Three-bus system for Problem 12.

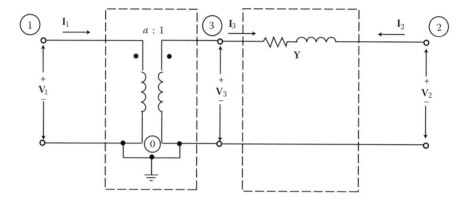

FIGURE P8.2 LTC transformer connection for Problem 14.

Appendix A: Impedance Tables for Overhead Lines, Transformers, and Underground Cables

TABLE A.1
Characteristics of Copper Conductors, Hard Drawn, 97.3% Conductivity [1,2]

Size of Conductor		No. of Strands	Diameter of Individual Strands (in.)	Outside Diameter (in.)	Breaking Strength (lb)	Weight (lb/mi)	Approx. Current-Carrying Capacity (A)	Geometric Mean Radius at 60 Cycles (ft)
Circular Mils	AWG or B&S							
1,000,000	...	37	0.1644	1.151	43,830	16,300	1,300	0.0368
900,000	...	37	0.1560	1.092	39,510	14,670	1,220	0.0349
800,000	...	37	0.1470	1.029	35,120	13,040	1,130	0.0329
750,000	...	37	0.1424	0.997	33,400	12,230	1,090	0.0319
700,000	...	37	0.1375	0.963	31,170	11,410	1,040	0.0308
600,000	...	37	0.1273	0.891	27,020	9,781	940	0.0285
500,000	...	37	0.1162	0.814	22,510	8,151	840	0.0260
500,000	...	19	0.1622	0.811	21,590	8,151	840	0.0256
450,000	...	19	0.1539	0.770	19,750	7,336	780	0.0243
400,000	...	19	0.1451	0.726	17,560	6,521	730	0.0229
350,000	...	19	0.1357	0.679	15,590	5,706	670	0.0214
350,000	...	12	0.1708	0.710	15,140	5,706	670	0.0225
300,000	...	19	0.1257	0.629	13,510	4,891	610	0.01987
300,000	...	12	0.1581	0.657	13,170	4,891	610	0.0208
250,000	...	19	0.1147	0.574	11,360	4,076	540	0.01813
250,000	...	12	0.1443	0.600	11,130	4,076	540	0.01902
211,600	4/0	19	0.1055	0.528	9,617	3,450	480	0.01668
211,600	4/0	12	0.1328	0.552	9,483	3,450	490	0.01750
211,600	4/0	7	0.1739	0.522	9,154	3,450	480	0.01579
167,800	3/0	12	0.1183	0.492	7,556	2,736	420	0.01559

167,800	3/0	7	0.1548	0.464	7,366	2,736	420	0.01404
133,100	2/0	7	0.1379	0.414	5,926	2,170	360	0.01252
105,500	1/0	7	0.1228	0.368	4,752	1,720	310	0.01113
83,690	1	7	0.1093	0.328	3,804	1,364	270	0.00992
66,370	2	7	0.0974	0.292	3,045	1,082	230	0.00883
66,370	2	7	0.1487	0.320	2,913	1,071	240	0.00903
66,370	2	3	.....	0.258	3,003	1,061	220	0.00836
83,690	1	3	0.1670	0.360	3,620	1,351	270	0.01016
66,370	2	7	0.0974	0.292	3,045	1,082	230	0.00883
66,370	2	3	0.1487	0.320	2,913	1,071	240	0.00903
66,370	2	1	.....	0.258	3,003	1,061	220	0.00836
52,630	3	7	0.0867	0.260	2,433	858	200	0.0078
52,630	3	3	0.1326	0.285	2,359	850	200	0.00805
52,630	3	1	.....	0.229	2,439	841	190	0.00745
41,740	4	3	0.1180	0.254	1,879	674	180	0.00717
41,740	4	1	.....	0.204	1,970	667	170	0.00663
33,100	5	3	0.1050	0.226	1,505	534	150	0.00638
33,100	5	1	.....	0.1819	1,591	529	140	0.00590
26,250	5	3	0.0935	0.201	1,205	424	130	0.00568
36,250	6	1	.....	0.1620	1,280	420	120	0.00526
20,820	7	1	.....	0.1443	1,030	333	110	0.00468
16,510	8	1	.....	0.1285	825	264	90	0.00417

(continued)

TABLE A.1 (Continued)
Characteristics of Copper Conductors, Hard Drawn, 97.3% Conductivity [1,2]

	T_a Resistance (Ω/Conductor/mi)								x_a Inductive Reactance (Ω/ Conductor/mi) at 1-ft Spacing				x_a' Shunt Capacitive Reactance (MΩ/Conductor/mi) at 1-ft Spacing			
	25°C (77°F)				50°C (122°F)											
dc	25 Cycles	50 Cycles	60 Cycles	dc	25 Cycles	50 Cycles	60 Cycles		25 Cycles	50 Cycles	60 Cycles		25 Cycles	50 Cycles	60 Cycles
0.0585	0.0594	0.0620	0.0634	0.0640	0.0648	0.572	0.0685		0.1666	0.333	0.400		0.216	0.1081	0.0901
0.0650	0.0658	0.0682	0.0695	0.0711	0.0718	0.0740	0.0752		0.1693	0.339	0.406		0.6220	0.1100	0.0916
0.0731	0.0739	0.0760	0.0772	0.0800	0.0806	0.0826	0.0837		0.1722	0.0344	0.413		0.224	0.1121	0.0934
0.0780	0.0787	0.0807	0.0818	0.0853	0.0859	0.0878	0.0888		0.1739	0.348	0.417		0.225	0.1132	0.0963
0.0836	0.0842	0.0861	0.0871	0.0914	0.0920	0.0937	0.0947		0.759	0.352	0.422		0.229	0.1145	0.0954
0.0975	0.0981	0.0997	0.1006	0.1066	0.1071	0.1086	0.1095		0.1799	0.360	0.432		0.235	0.1173	0.0977
0.1170	0.1175	0.1188	0.1196	0.1280	0.1283	0.1296	0.1303		0.1845	0.369	0.443		0.241	0.1205	0.1004
0.1170	0.1175	0.1188	0.1196	0.1280	0.1283	0.1296	0.1303		0.1853	0.371	0.445		.241	0.1206	0.1006
0.1300	0.1304	0.1316	0.1323	0.1422	0.1426	0.1437	0.1443		0.1879	0.376	0.451		0.245	0.1224	0.1020
0.1462	0.1466	0.1477	0.1484	0.1600	0.1603	0.1613	0.1619		0.1909	0.382	0.458		0.249	0.1245	0.1038
0.1671	0.1675	0.1684	0.1690	0.1828	0.1831	0.1840	0.1845		0.1943	0.389	0.466		0.254	0.1269	0.1058
0.1671	0.1675	0.1684	0.1690	0.1828	0.1831	0.1840	0.1845		0.1918	0.384	0.460		0.251	0.1253	0.1044
0.1950	0.1953	0.1961	0.1966	0.213	0.214	0.214	0.215		0.1982	0.396	0.76		0.259	0.1296	0.1080
0.1950	0.1953	0.196	0.1966	0.213	0.214	0.214	0.215		0.1957	0.392	0.470		0.256	0.1281	0.1088
0.234	0.234	0.235	0.235	0.256	0.256	0.257	0.257		0.203	0.406	0.487		0.266	0.1329	0.1108
0.234	0.234	0.235	0.235	0.256	0.256	0.257	0.0257		0.200	0.401	0.481		0.263	0.1313	0.1094
0.276	0.277	0.277	0.278	0.302	0.302	0.303	0.303		0.207	0.414	0.497		0.272	0.1359	0.1132
0.276	0.277	0.277	0.278	0.302	0.303	0.303	0.303		0.205	0.409	0.491		0.269	0.1343	0.1119
0.276	0.277	0.277	0.277	0.278	0.302	0.303	0.303		0.210	0.420	0.503		0.273	0.1363	0.1136
0.349	0.349	0.349	0.350	0.381	0.381	0.382	0.382		0.210	0.421	0.505		0.277	0.1384	0.1153
0.349	0.349	0.349	0.350	0.381	0.381	0.382	0.382		0.216	0.3431	0.518		0.281	0.1405	0.1171

Appendix A

0.440	0.440	0.440	0.440	0.481	0.481	0.481	0.481	0.222	0.443	0.532	0.289	0.1445	0.1206
0.555	0.555	0.555	0.555	0.606	0.607	0.607	0.607	0.227	0.455	0.546	0.298	0.1488	0.1240
0.399	0.699	0.699	0.699	0.765				0.233	0.467	0.560	0.306	0.1528	0.1274
0.881	0.882	0.882	0.882	0.964				0.232	0.464	0.557	0.299	0.1495	0.1246
0.873				0.955				0.238	0.476	0.571	0.307	0.1537	0.1281
0.864				0.945				0.242	0.484	0.581	0.323	0.1614	0.1345
0.692	0.692	0.692		0.757				0.232	0.464	0.557	0.299	0.1495	0.1246
0.881	0.882	0.882		0.964				0.239	0.478	0.574	0.314	0.1570	0.1308
0.873				0.955				0.238	0.476	0.571	0.307	0.1537	0.1281
0.864				0.945				0.242	0.484	0.581	0.323	0.1614	0.1345
1.112				1.216				0.245	0.490	0.588	0.322	0.1611	0.1343
1.101				1.204				0.248	0.488	0.585	0.318	0.1578	0.1315
1.090	Same as dc			1.192	Same as dc			0.248	0.496	0.595	0.331	0.1656	0.1380
1.388				1.518				0.250	0.499	0.599	0.324	0.1619	0.1349
1.374				1.503				0.254	0.507	0.609	0.339	0.1697	0.1415
1.750				1.914				0.256	0.511	0.613	0.332	0.1661	0.1384
1.733				1.895				0.260	0.519	0.623	0.348	0.1738	0.1449
2.21				2.41				0.262	0.523	0.628	0.341	0.1703	0.1419
2.18				2.39				0.265	0.531	0.637	0.356	0.1779	0.1483
2.75				3.01				0.271	0.542	0.651	0.364	0.1821	0.1517
3.47				3.80				0.277	0.554	0.665	0.372	0.1882	0.1552

[a] For conductor at 75°C, air at 25°C, wind 1.4 mi/h (2 ft/s), frequency = 60 cycles.

TABLE A.2
Characteristics of Aluminum Conductors, Hard Drawn, 61% Conductivity [1,2] (Aluminum Company of America)

Size of Conductor Circular Mils or AWG	No. of Strands	Diameter of Individual Strands (in.)	Outside Diameter (in.)	Ultimate Strength (/mi)	Weight (lb/mi)	Geometric Mean Radius at 60 Cycles (ft)	Approx. Current-Carrying Capacity[a] (A)
6	7	0.0612	0.184	528	130	0.00556	100
4	7	0.0772	0.232	826	207	0.00700	134
3	7	0.0867	0.260	1,022	261	0.00787	155
2	7	0.974	0.292	1,266	329	0.00883	180
1	7	0.1094	0.328	1,537	414	0.00992	209
1/0	7	0.1228	0.368	1,865	523	0.01113	242
1/0	19	0.0745	0.373	2,090	523	0.01177	244
2/0	7	0.1379	0.414	2,350	659	0.01251	282
2/0	19	0.0837	0.419	2,586	659	0.1321	283
3/0	7	0.1548	0.464	2,845	832	0.1404	327
3/0	19	0.0940	0.470	3,200	832	0.01483	328
4/0	7	0.1739	0.522	3,590	1,049	0.01577	380
4/0	19	0.1055	0.528	3,890	1,049	0.01666	381
250,000	37	0.0822	0.575	4,860	1,239	0.01841	425
266,800	7	0.1953	0.586	4,525	1,322	0.01771	441
266,800	37	0.0849	0.594	5,180	1,322	0.01902	443
300,000	19	0.1257	0.629	5,300	1,487	0.01983	478
300,000	37	0.0900	0.630	5,830	1,487	0.02017	478
336,400	19	0.1331	0.666	5,940	667	0.02100	514
336,400	37	0.0954	0.668	6,400	1,667	0.02135	514
350,000	37	0.0973	0.681	6,680	1,735	0.02178	528
397,500	19	0.1447	0.724	6,880	1,967	0.02283	575
477,000	19	0.1585	0.793	8,090	2,364	0.02501	646

500,000	19	0.1623	0.812	8,475	2,478	0.02560	664
500,000	37	0.1162	0.813	9,010	2,478	0.02603	664
556,500	19	0.1711	0.856	9,440	2,758	0.02701	710
636,000	37	0.1311	0.918	11,240	3,152	0.02936	776
715,500	37	0.1391	0.974	12,640	3,546	0.03114	817
750,000	37	0.1424	0.997	12,980	377	0.03188	864
795,000	37	0.1466	1.026	13,770	3,940	0.03283	897
874,500	37	0.1538	1.077	14,830	4,334	0.03443	949
954,000	37	0.1606	1.024	16,180	4,728	0.03596	1,000
1,000,000	61	0.01280	1.152	17,670	4,956	0.03707	1,030
1,000,000	91	0.1048	1.153	18,380	4,956	0.03720	1,030
1,033,500	37	0.1672	1.170	18,260	5,122	0.03743	1,050
1,113,000	61	0.1351	1.216	19,660	5,517	0.03910	1,110
1,192,500	61	0.1398	1.258	21,000	5,908	0.04048	1,160
1,192,500	91	0.1145	1.259	21,400	5,908	0.04062	1,160
1,272,000	61	0.1444	1.300	22,000	6,299	0.04180	1,210
1,351,500	61	0.1489	1.340	23,400	6,700	0.04309	1,250
1,431,000	61	0.1532	1.379	24,300	7,091	0.04434	1,300
1,510,500	61	0.1574	1.417	25,600	7,487	0.04556	1,320
1,590,000	61	0.1615	1.454	27,000	7,883	0.04674	1,380
1,590,000	91	0.1322	1.454	28,100	7,883	0.04691	1,380

(continued)

TABLE A.2 (Continued)
Characteristics of Aluminum Conductors, Hard Drawn, 61% Conductivity [1,2] (Aluminum Company of America)

	T_a Resistance (Ω/Conductor/mi)									x_a Inductive Reactance (Ω/ Conductor/mi) at 1-ft spacing			x_a' Shunt Capacitive Reactance (MΩ·Conductor/mi) at 1-ft Spacing		
	25°C (77°F)				50°C (122°F)										
dc	25 Cycles	50 Cycles	60 Cycles	dc	25 Cycles	50 Cycles	60 Cycles		25 Cycles	50 Cycles	60 Cycles		25 Cycles	50 Cycles	60 Cycles
3.56	3.56	3.56	3.56	3.91	3.91	3.91	3.91		0.2626	0.5251	0.6301		0.3468	0.1734	0.1445
2.24	2.24	2.24	2.24	2.46	2.46	2.46	2.46		0.2509	0.5017	0.6201		0.3302	0.1651	0.1376
1.77	1.77	1.77	1.77	1.95	1.95	1.95	1.95		0.2450	0.4899	0.5879		0.3221	0.1610	0.1342
1.41	1.41	1.41	1.41	1.55	1.55	1.55	1.55		0.2391	0.4782	0.5739		0.3139	0.1570	0.1308
1.12	1.12	1.12	1.12	1.23	1.23	1.23	1.23		0.2333	0.4665	0.5598		0.3055	0.1528	0.1273
0.885	0.8851	0.8853	0.885	0.973	0.9731	0.9732	0.973		0.2264	0.4528	0.5434		0.2976	0.1488	0.1240
0.885	0.8851	0.8853	0.885	0.973	0.9731	0.9732	0.973		0.2264	0.4492	0.5391		0.2964	0.1482	0.1235
0.702	0.7021	0.7024	0.702	0.771	0.7711	0.7713	0.771		0.2216	0.4431	0.5317		0.2890	0.1445	0.1204
0.702	0.7021	0.7024	0.702	0.771	0.7711	0.7713	0.771		0.2188	0.4376	0.5251		0.2882	0.1441	0.1201
0.557	0.5571	0.5574	0.558	0.612	0.6121	0.6124	0.613		0.2157	0.4314	0.5177		0.2810	0.1405	0.1171
0.557	0.5571	0.5574	0.558	0.612	0.6121	0.6124	0.613		0.2129	0.4258	0.5110		0.2801	0.1400	0.1167
0.441	0.4411	0.4415	0.442	0.485	0.4851	0.4855	0.486		0.2099	0.4196	0.5036		0.2726	0.1363	0.1136
0.441	0.4411	0.4415	0.442	0.485	0.4851	0.4855	0.486		0.2071	0.4141	0.4969		0.2717	0.1358	0.1132
0.374	0.3741	0.3746	0.375	0.411	0.4111	0.4115	0.412		0.2020	0.4040	0.4848		0.2657	0.13258	0.1107
0.350	0.3502	0.3506	0.351	0.385	0.3852	0.3855	0.386		0.2040	0.4079	0.4895		0.2642	0.1321	0.1101
0.350	0.3502	0.3506	0.351	0.385	0.3852	0.3855	0.386		0.2004	0.4007	0.4809		0.2633	0.1316	0.1097
0.311	0.3112	0.3117	0.312	0.342	0.3422	0.3426	0.343		0.1983	0.3965	0.4758		0.2592	0.1296	0.1080
0.311	0.3112	0.3117	0.312	0.342	0.3422	0.3426	0.343		0.1974	0.3947	0.4737		0.2592	0.1296	0.1080
0.278	0.2782	0.2788	0.279	0.306	0.3062	0.3067	0.307		0.1953	0.3907	0.4688		0.2551	0.1276	0.1063
0.278	0.2782	0.2788	0.279	0.306	0.3062	0.3067	0.307		0.1945	0.3890	0.4668		0.2549	0.1274	0.1062

Appendix A

0.267	0.2672	0.2678	0.268	0.294	0.2942	0.2947	0.295	0.1935	0.3870	0.4644	0.2537	0.1268	0.1057
0.235	0.2352	0.2359	0.236	0.258	0.2582	0.2589	0.259	0.1911	0.3822	0.4587	0.2491	0.1246	0.1038
0.196	0.1963	0.1971	0.198	0.215	0.2153	0.2160	0.216	0.1865	0.3730	0.4476	0.2429	0.1214	0.1012
0.187	0.1873	0.1882	0.189	0.206	0.2062	0.2070	0.208	0.1853	0.3730	0.4476	0.2429	0.1206	0.1005
0.187	0.1873	0.1882	0.189	0.206	0.2062	0.2070	0.208	0.1845	0.3689	0.4427	0.2410	0.1205	0.1004
0.168	0.1683	0.1693	0.170	0.185	0.1853	0.1862	0.187	0.1826	0.3652	0.4383	0.2374	0.1187	0.0989
0.147	0.1474	0.1484	0.149	0.162	0.1623	0.1633	0.164	0.1785	0.3569	0.4283	0.2323	0.1162	0.0968
0.167	0.1314	0.1326	0.133	0.144	0.1444	0.1455	0.146	0.1754	0.3508	0.4210	0.2282	0.1141	0.0951
0.125	0.1254	0.1267	0.127	0.137	0.1374	0.1385	0.139	0.1743	0.3485	0.4182	0.2266	0.1133	0.0944
0.117	0.1175	0.1188	0.120	0.129	0.1294	0.1306	0.131	0.1728	0.3455	0.4146	0.2244	0.1122	0.0935
0.107	0.1075	0.1089	0.110	0.118	0.1185	0.1198	0.121	0.1703	0.3407	0.4088	0.2210	0.1105	0.0921
0.0979	0.0985	0.1002	0.100	0.108	0.1085	0.1100	0.111	0.01682	0.3363	0.4036	0.2179	0.1090	0.0908
0.0934	0.0940	0.0956	0.0966	0.103	0.1035	0.1050	0.106	0.1666	0.3332	0.3998	0.2162	0.1081	0.0901
0.0934	0.0940	0.0956	0.0966	0.103	0.1035	0.1050	0.106	0.1664	0.3328	0.3994	0.2160	0.1080	0.0900
0.0904	0.0910	0.0927	0.0936	0.0994	0.0999	0.1015	0.102	0.1661	0.3322	0.3987	0.2150	0.1075	0.0896
0.0839	0.0845	0.0864	0.0874	0.0922	0.0928	0.0945	0.0954	0.1639	0.3278	0.3934	0.2124	0.1062	0.0885
0.0783	0.0790	0.0810	0.0821	0.08860	0.0866	0.0884	0.0895	0.1622	0.3243	0.3892	0.2100	0.1050	0.0875
0.0783	0.0790	0.0810	0.0821	0.0860	0.0866	0.0884	0.0895	0.1620	0.3240	0.3888	0.2098	0.1049	0.0874
0.0734	0.0741	0.0762	0.0774	0.0806	0.0813	0.0832	0.0843	0.1606	0.211	0.3853	0.2076	0.1038	0.0865
0.0691	0.0699	0.0721	0.0733	0.0760	0.0767	0.0787	0.0798	0.1590	0.3180	0.3816	0.2054	0.1027	0.0856
0.0653	0.0661	0.0685	0.0697	0.0718	0.0725	0.0747	0.0759	0.1576	0.3152	0.3782	0.2033	0.1016	0.0847
0.0618	0.0627	0.0651	0.0665	0.0679	0.0687	0.0710	0.0722	0.1562	0.3123	0.3748	0.2014	0.1007	0.0839
0.0537	0.0596	0.0622	0.0636	0.0645	0.0653	0.0677	0.0690	0.1549	0.3098	0.3718	0.1997	0.0998	0.0832
0.0587	0.0596	0.0622	0.0636	0.0645	0.0653	0.0677	0.0690	0.1547	0.03094	0.3713	0.0997	0.0998	0.0832

[a] For conductor at 75°C, wind 1.4 mi/h (2 ft/s), frequency = 60 cycles.

TABLE A.3
Characteristics of Aluminum Cable, Steel Reinforced [1] (Aluminum Company of America)

Circular Mils or AWG Aluminum	Aluminum			Steel		Outside Dia. (in.)	Copper Equivalent (cmil or AWG)	Ultimate Strength (lb)	Wt. (lb/mi)	Geometric Mean Radius at 60 Cycles (ft)	Approx. Current-Carrying Capacity[b] (A)
	Strands	Layers	Strand Dia. (in.)	Strands	Strand Dia. (in.)						
1,590,000	54	3	0.1716	19	0.1030	1.545	1,000,000	56,000	10,777	0.0520	1380
1,510,500	54	3	0.1673	19	0.1004	1.506	950,000	53,200	10,237	0.0507	1340
1,431,000	54	3	0.1628	19	0.0977	1.465	900,000	50,400	9,699	0.0493	1300
1,351,000	54	3	0.1628	19	0.0949	1424	850,000	47,600	9,160	0.0479	1250
1,272,000	54	3	0.1535	19	0.0921	13.82	800,000	44,800	8,621	0.0465	1200
1,192,500	54	3	0.1486	19	0.0892	1.338	750,000	43,100	8,082	0.0450	1160
1,113,000	54	3	0.1438	19	0.0862	1.293	700,000	40,200	7,544	0.0435	1110
1,033,500	54	3	0.1384	7	0.1384	1.246	650,000	37,100	7,019	0.0420	1060
954,000	54	3	0.1329	7	0.1329	1.196	600,000	34,200	6,479	0.0403	1010
900,000	54	3	0.1291	7	0.1291	1.162	566,000	32,300	6,112	0.0391	970
874,500	54	3	0.1273	7	0.1273	1.146	550,000	31,400	5,940	0.0388	950
795,000	54	3	0.1214	7	0.1214	1.093	500,000	28,500	5,399	0.368	900
795,000	26	2	0.1749	7	0.1360	1.108	500,000	31,200	5,770	0.075	900
795,000	30	2	0.1628	19	0.0977	1.140	500,000	38,400	6,517	0.0393	910
715,500	54	3	0.1151	7	0.1151	1.036	450,000	26,300	4,859	0.0349	830
715,500	26	2	0.1659	7	0.1290	1.051	450,000	28,100	5,193	0.0355	840
715,500	30	2	0.1544	19	0.0926	1.081	450,000	34,500	5,865	0.0372	840
666,600	54	3	0.1111	7	0.1111	1.000	419,000	24,500	4,527	0.0337	800
636,000	54	3	0.1085	7	0.1085	0.977	400,000	23,600	4,319	0.0329	770
636,000	26	2	0.1564	7	0.1216	0.990	400,000	25,000	4,616	0.0335	780
636,000	30	2	0.1456	19	0.0874	1.019	400,000	31,500	5,213	0.0351	780
605,000	54	3	0.1059	7	0.1059	0.953	380,500	22,500	4,109	0.0321	750
605,000	26	2	0.1525	7	0.1186	0.966	380,500	24,100	4,391	0.0327	760
556,500	26	2	0.1463	7	0.1138	0.927	350,000	22,400	4,039	0.0313	730
556,500	30	2	0.1362	7	0.1362	0.953	350,000	27,200	4,588	0.0328	730
500,000	30	2	0.1291	7	0.1291	0.904	314,500	24,400	4,122	0.0311	690

477,000	26	2	0.1355	7	0.1054	0.858	300,000	19,430	3,462	0.0290	670
477,000	30	2	0.1261	7	0.1261	0.883	300,000	23,300	3,933	0.0304	670
397,500	26	2	0.236	7	0.0961	0.783	250,000	16,190	2,885	0.0265	590
397,500	30	2	0.1151	7	0.1551	0.806	250,000	19,980	3,277	0.0278	600
336,400	26	2	0.1138	7	0.0885	0.721	4/0	14,050	2,442	0.0244	530
336,400	30	2	0.1059	7	0.1059	0.741	4/0	17,040	2,774	0.0255	530
300,000	26	2	0.1074	7	0.0835	0.680	188,700	12,650	2,178	0.0230	490
300,000	30	2	0.1000	7	0.1000	0.700	18,870	15,430	2,473	0.0241	500
266,800	26	2	0.1013	7	0.0788	0.642	3/0	11,250	1,936	0.0217	460

										For Current Approx. 75% Capacity[c]	
266,800	6	1	0.2109	7	0.0703	0.633	3/0	9,645	1,802	0.00684	460
4/0	6	1	0.1878	1	0.1878	0.583	2/0	8,420	1,542	0.0081	340
3/0	6	1	0.1672	1	0.1672	0.502	1/0	6,675	1,223	0.00600	300
2/0	6	1	0.1490	1	0.1490	0.447	1	5,345	970	0.00510	270
1/0	6	1	0.327	1	0.1327	0.398	2	4,280	769	0.00446	230
1	6	1	0.1182	1	0.1182	0.355	3	3,480	610	0.00418	200
2	6	1	0.1052	1	0.1052	0.316	4	2,790	484	0.00418	180
2	7	1	0.0974	1	0.1299	0.325	4	3,525	566	0.00504	180
3	6	1	0.0937	1	0.0937	0.281	5	2,250	384	0.00430	166
4	6	1	0.0834	1	0.834	0.250	6	1,830	304	0.00437	40
4	7	1	0.0772	1	0.1029	0.257	6	2,288	356	0.00452	140
5	6	1	0.0743	1	0.0743	0.223	7	1,460	241	0.00416	120
6	6	1	0.0661	1	0.0661	0.198	8	1,170	191	0.00394	100

(continued)

TABLE A.3 (Continued)
Characteristics of Aluminum Cable, Steel Reinforced [1] (Aluminum Company of America)

T_a Resistance (Ω/Conductor/mi)									x_a Inductive Reactance (Ω/Conductor/mi) at 1-ft Spacing, All Currents				x'_a Shunt Capacitive Reactance (MΩ/Conductor/mi) at 1-ft Spacing		
	25°C (77°F) Small Currents				50°C (122°F) Current Approx. 75% Capacity[c]										
dc	25 Cycles	50 Cycles	60 Cycles	dc	25 Cycles	50 Cycles	60 Cycles		25 Cycles	50 Cycles	60 Cycles		25 Cycles	50 Cycles	60 Cycles
0.0587	0.0588	0.0590	0.0591	0.0646	0.0656	0.0675	0.0684		0.1495	0.299	0.359		0.1953	0.0977	0.0814
0.0618	0.0619	0.0621	0.0622	0.0680	0.0690	0.0710	0.0720		0.1508	0.302	0.362		0.1971	0.0986	1.5105
0.0652	0.0653	0.0655	0.0656	0.0718	0.0729	0.0749	0.0760		0.1522	0.304	0.365		0.1991	0.0996	0.0830
0.0691	0.0692	0.0694	0.0695	0.0761	0.0771	0.0792	0.0803		0.1536	0.307	0.369		0.201	0.1006	0.0838
0.0734	0.0735	0.0737	0.0738	0.0808	0.0819	0.8040	0.0851		0.1551	0.310	0.372		0.203	0.1016	0.0847
0.0783	0.0784	0.0786	0.0788	0.0862	0.0872	0.0894	0.0906		0.1568	0.314	0.376		0.206	0.1028	0.0857
0.0839	0.0840	0.0842	0.0844	0.0924	0.0935	0.0957	0.0969		0.1585	0.317	0.380		0.208	0.1040	0.0867
0.0903	0.0905	0.0907	0.0909	0.0994	0.1005	0.1025	0.1035		0.1603	0.321	0.385		0.211	0.1053	0.0878
0.0979	0.0980	0.0981	0.0982	0.1078	0.1088	0.1118	0.1128		0.1624	0.325	0.390		0.214	0.1068	0.0890
0.104	0.104	0.104	0.104	0.1145	0.1155	0.1175	0.1185		0.1639	0.328	0.393		0.216	0.1078	0.0895
0.107	0.107	0.107	0.108	0.1178	0.1188	0.1218	0.1228		0.1646	0.329	0.395		0.217	0.1083	0.0903
0.117	0.118	0.118	0.119	0.1288	0.1308	0.1358	0.1378		0.1670	0.334	0.401		0.220	0.1100	0.0917
0.117	0.117	0.117	0.117	0.1288	0.1288	0.1288	0.1288		0.1660	0.332	0.339		0.219	0.1095	0.0912
0.117	0.117	0.117	0.17	0.1288	0.1288	0.1288	0.1288		0.1637	0.327	0.393		0.217	0.1085	0.0904
0.131	0.131	0.132	0.1442	0.1452	0.1472	0.1472	0.1482		0.1697	0.339	0.407		0.224	0.1119	0.0932
0.131	0.131	0.131	0.131	0.1442	0.1442	0.1442	0.1442		0.1687	0.337	0.405		0.223	0.1114	0.0928
0.131	0.131	0.131	0.131	0.1442	0.1442	0.1442	0.1442		0.1664	0.333	0.399		0.221	0.1104	0.0920
0.140	0.140	0.141	0.141	0.1541	0.1571	0.1591	0.1601		0.1715	0.343	0.412		0.226	0.1132	0.0943
0.147	0.147	0.148	0.148	0.1618	0.1638	0.1678	0.1688		0.1728	0.345	0.414		0.228	0.1140	0.0950
0.147	0.147	0.147	0.147	0.1618	0.1618	0.1618	0.1618		0.718	0.344	0.412		0.227	0.1135	0.0945
0.147	0.147	0.147	0.147	0.1618	0.1618	0.1618	0.1618		0.1693	0.339	0.406		0.225	0.1125	0.0937
0.154	0.155	0.155	0.155	0.1695	0.1715	0.1755	0.1775		0.1739	0.348	0.417		0.230	0.1149	0.0957
0.154	0.154	0.154	0.154	0.1700	0.1720	0.1720	0.1720		0.1730	0.346	0.415		0.229	0.1144	0.0953
0.168	0.168	0.168	0.168	0.1849	0.1859	0.1859	0.1859		0.1751	0.350	0.420		0.232	0.1159	0.0965
0.168	0.168	0.168	0.168	0.1849	0.1859	0.159	0.1859		0.1728	0.346	0.415		0.230	0.1149	0.0957

Appendix A

0.187	0.187	0.187	0.187	0.206					0.1754		0.42	0.234	0.1167	0.0973
0.3196	0.196	0.196	0.196	0.216					0.1790		0.351	0.430	0.237	0.1186	0.0988
0.196	0.196	0.196	0.196	0.216					0.1766		0.358	0.424	0.235	0.1176	0.0980
0.235				0.259					0.1836		0.353	0.441	0.244	0.1219	0.1015
0.235	Same as dc			0.259	Same as dc				0.1812		0.367	0.435	0.242	0.1208	0.1006
0.278				0.306					0.1872		0.362	0.451	0.250	0.1248	0.1039
0.278				0.306					0.1855		0.376	0.445	0.248	0.1238	0.1032
0.311				0.342					0.1908		0.371	0.458	0.254	0.1269	0.1057
0.311				0.342					0.1883		0.382	0.452	0.252	0.1258	0.1049
0.350				0.385					0.1936		0.377	0.465	0.258	0.1289	0.1074
											0.387				

Single-Layer Conductors

							Small Currents			Current Approx. 75% Capacity[c]		
							25 Cycles	50 Cycles	60 Cycles	25 Cycles	50 Cycles	60 Cycles
0.351	0.351	0.351	0.352	0.386	0.430	0.510	0.552	0.194	0.388	0.466	0.252	0.504	0.605	0.259	0.1294	0.1079
0.441	0.442	0.444	0.445	0.485	0.514	0.567	0.592	0.218	0.437	0.524	0.242	0.484	0.581	0.267	0.1336	0.1113
0.556	0.557	0.559	0.560	0.612	0.642	0.697	0.723	0.225	0.450	0.540	0.259	0.517	0.621	0.275	0.1377	0.1147
0.702	0.702	0.704	0.706	0.773	0.806	0.866	0.895	0.231	0.462	0.554	0.267	0.534	0.641	0.284	0.1418	0.1182
0.885	0.885	0.887	0.888	0.974	0.01	1.08	1.12	0.237	0.473	0.568	0.273	0.547	0.656	0.292	0.1460	0.1216
1.12	1.12	1.12	1.12	1.23	1.27	1.34	1.38	0.242	0.483	0.580	0.277	0.554	0.665	0.300	0.1500	0.1250
1.41	1.41	1.41	1.41	1.55	1.59	156	1.69	0.247	0.493	0.592	0.277	0.554	0.665	0.308	0.1542	0.1285
1.41	1.141	1.41	1.41	1.55	1.59	1.62	1.65	0.247	0.493	0.592	0.267	0.535	0.642	0.306	0.1532	0.1276
1.78	1.78	1.78	1.78	1.95	1.98	2.04	2.07	0.252	0.503	0.604	0.275	0.551	0.661	0.317	0.1583	0.1320
2.24	2.24	2.24	2.24	2.47	2.50	2.54	2.57	0.257	0.514	0.611	0.274	0.549	0.659	0.325	0.1627	0.1355
2.24	2.24	2.24	2.24	2.47	2.50	2.54	2.57	0.257	0.515	0.618	0.273	0.545	0.655	0.323	0.615	0.1345
2.82	2.82	2.82	2.82	3.10	3.12	3.16	3.18	0.262	0.525	0.630	0.279	0.557	0.665	0.333	0.1666	0.1388
3.55	3.56	3.56	3.56	3.92	3.94	3.97	3.98	0.268	0.536	0.643	0.281	0.561	0.673	0.342	0.1708	0.1423

[a] Based on copper 97% aluminum 61% conductivity.
[b] For conductor at 75°C, air at 25°C, wind 1.4 mi/h (2 ft/s), frequency = 60 cycles.
[c] "Current Approx. 75% Capacity" is 75% of the "Approx. Current-Carrying Capacity (A)," and is approximately the current that will produce a 50°C conductor temperature (25°C rise).

TABLE A.4
Characteristics of "Expanded" Aluminum Cable, Steel Reinforced [1] (Aluminum Company America)

Circular Mils or AWG Aluminum	Aluminum		Steel				Filler Section				Outside Diameter (in.)	Copper Equivalent (cmil or AWG)	Ultimate Strength (lb)	Weight (lb/mi)
							Aluminum		Paper					
	Strands	Layers	Strand Diameter (in.)	Strands	Strand Diameter (in.)	Strands	Strand Diameter (in.)	Strands	Layers				
850,000	54	2	0.1255	19	0.834	4	0.1182	23	2	1.38	534,000	35,371	7,200
1,150,000	54	2	0.1409	19	0.0921	4	0.1353	24	2	1.55	724,000	41,900	9,070
1,338,000	66	2	0.1350	19	0.100	4	0.184	18	2	1.75	840,000	49,278	11,340

Geometric Mean Radius at 60 Cycles (ft)	Approx. Current-Carrying Capacity (A)	T_a Resistance (Ω/Conductor/mi)								x_a Inductive Reactance (Ω/Conductor/mi) at 1-ft Spacing, All Currents				x_a Shunt Capacitive Reactance (MΩ/Conductor/mi) at 1-ft Spacing		
		25°C (77°F) Small Currents				50°C (122°F) Current Approx. 75% Capacity				25 Cycles	50 Cycles	60 Cycles		25 Cycles	50 Cycles	60 Cycles
		dc	25 Cycles	50 Cycles	60 Cycles	dc	25 Cycle	50 Cycles	60 Cycles							
a	a	a	a	a	a	a	a	a	a	a	a	a		a	a	a

a Electrical characteristics not available until laboratory measurements are completed.

TABLE A.5
Characteristics of Copperweld–Copper Conductors [1,2] (Copperweld Steel Company)

Nominal Designation	Size of Conductor		Outside Diameter (in.)	Copper Equivalent (cmil or AWG)	Rated Breaking Load (lb)	Weight (lb/mi)	Geometric Mean Radius at 60 Cycles (ft)	Approx. Current-Carrying Capacity at 60 Cycles (A)[a]	r_a Resistance (Ω/Conductor/mi) at 25°C (77°F), Small Currents				r_a Resistance (Ω/Conductor/mi) at 50°C (122°F) Current, Approx. 75% Capacity				x_a Inductive Reactance (Ω/Conductor/mi) at 1-ft Spacing, Average Currents			x_a' Capacitive Reactance (MΩ/Conductor/mi) at 1-ft Spacing			
	Number and Diameter of Wires								dc	25 Cycles	50 Cycles	60 Cycles	dc	25 Cycles	50 Cycles	60 Cycles	25 Cycles	50 Cycles	60 Cycles	25 Cycles	50 Cycles	60 Cycles
	Copperweld	Copper																					
350 E	7×.1576″	12×.1576″	0.788	350,000	32,420	7,409	0.220	660	0.1658	0.1728	0.1780	0.812	0.1812	0.1915	0.201	0.204	0.1929	0.386	0.463	0.243	0.1216	0.1014
350 EK	4×.1470″	15×.1470″	0.735	350,000	23,850	6,536	0.0245	680	0.1658	0.1682	0.1700	0.1705	0.1812	0.1845	0.1873	0.1882	0.1875	0.1375	0.450	0.248	0.1241	0.1034
350 V	3×.1751″	9×.1893″	0.754	350,000	23,480	6,578	0.0226	650	0.1655	0.1725	0.1800	0.1828	0.1809	0.1910	0.202	0.206	0.1915	0.383	0.460	0.246	0.1232	0.1027
300 E	7×.1459″	12×.1459″	0.729	300,000	27,770	6,351	0.0204	600	0.1934	0.200	0.207	0.209	0.211	0.222	0.0232	0.235	0.1989	0.394	0.473	0.249	0.1244	0.1037
300 EK	4×.1361″	15×.1361″	0.680	300,000	20,960	5,602	0.0227	610	0.1934	0.1958	0.1976	0.198	0.211	0.215	0.218	0.219	0.1914	0.383	0.460	0.254	0.1269	0.1057
300 V	3×.1621″	9×.1752″	0.698	300,000	20,730	5,639	0.0209	590	0.1930	0.200	0.208	0.210	0.211	0.222	0.233	0.237	0.1954	0.391	0.469	0.252	0.1259	0.1050
250 E	7×.1332″	12×.1332″	0.666	250,000	23,920	5,292	0.01859	540	0.232	0.239	0.25	0.248	0.254	0.265	0.275	0.279	0.202	0.403	0.484	0.255	0.1276	0.1604
250 EK	4×.1242″	15×.1242″	0.621	250,000	17,840	4,669	0.0207	540	0.232	0.235	0.236	0.237	0.254	0.258	0.261	0.261	0.1960	0.392	0.471	0.260	0.1301	0.1084
250 V	3×.1480″	9×.1600″	0.637	250,000	17,420	4,699	0.01911	530	0.232	0.239	0.246	0.249	0.253	0.264	0.276	0.281	0.200	0.400	0.480	0.258	0.1292	0.1077
4/0 E	7×.1332″	12×.1332″	0.613	4/0	20,730	4,479	0.01711	480	0.274	0.281	0.287	0.290	0.300	0.312	0.323	0.326	0.206	0.411	0.493	0.261	0.1306	0.1088
4/0 G	2×.1944″	5×.1944″	0.583	4/0	15,640	4,168	0.01409	460	0.273	0.284	0.294	0.298	0.299	0.318	0.336	0.342	0.215	0.431	0.517	0.265	0.1324	0.1103
4/0 EK	4×.1143″	15×.1143″	0.571	4/0	15,370	3,951	0.01903	490	0.274	0.277	0.278	0.279	0.300	0.304	0.307	0.308	0.200	0.401	0.481	0.266	0.1331	0.1109
4/0 V	3×.1361″	9×.1472″	0.586	4/0	15,000	3,977	0.01758	470	0.274	0.281	0.288	0.291	0.299	0.311	0.323	0.328	0.204	0.409	0.490	0.264	0.1322	0.1101
4/0 F	1×.1833″	6×.1833″	0.550	4/0	12,290	3,750	0.01558	470	0.273	0.280	0.285	0.287	0.299	0.309	0.318	0.322	0.210	0.421	0.505	0.269	0.1344	0.1220
3/0 E	7×.1091″	12×.1091″	0.545	3/0	16,800	3,552	0.01521	420	0.346	0.353	0.359	0.361	0.378	0.391	0.402	0.407	0.212	0.423	0.508	0.270	0.1348	0.1123
3/0 J	3×.1851″	4×.1851″	0.555	3/0	16,170	3,732	0.01156	410	0.344	0.356	0.367	0.372	0.377	0.398	0.419	0.428	0.225	0.451	0.541	0.268	0.1341	0.1118
3/0 G	2×.1731″	5×.1731″	0.519	3/0	12,860	3,305	0.01254	400	0.344	0.355	0.365	0.369	0.377	0.397	0.416	0.423	0.221	0.443	0.531	0.273	0.136	50.1137
3/0 EK	4×.1018″	4×.1018″	0.509	3/0	12,370	3,134	0.01697	420	0.346	0.348	0.350	0.351	0.378	0.382	0.386	0.386	0.206	0.412	0.495	0.274	0.1372	0.1143
3/0 V	3×.1311″	9×.1311″	0.522	3/0	12,220	3,154	0.01566	410	0.345	0.352	0.360	0.362	0.377	0.390	0.403	0.408	0.210	0.420	0.504	0.273	0.1363	0.1136
3/0 F	1×.1632″	6×.1632″	0.490	3/0	9,980	2,974	0.01388	410	0.344	0.351	0.355	0.358	0.377	0.388	0.397	0.401	0.216	0.432	0.519	0.277	0.1385	0.1156
2/0 K	4×.1780″	3×.1780″	0.534	2/0	17,600	3,411	0.00912	360	0.434	0.447	0.459	0.466	0.475	0.499	0.524	0.535	0.237	0.475	0.570	0.271	0.1355	0.1129
2/0 J	3×.1648″	4×.1648″	0.494	2/0	13,430	2,960	0.01029	350	0.434	0.446	0.457	0.462	0.475	0.498	0.520	0.530	0.231	0.463	0.555	0.277	0.1383	0.1152
2/0 G	2×.1542″	5×.1542″	0.463	2/0	10,510	2,622	0.1119	350	0.434	0.445	0.456	0.459	0.475	0.497	0.518	0.525	0.227	0.454	0.545	0.281	0.1406	0.1171
2/0 Y	3×.1080″	9×.1167″	0.465	2/0	9,846	2,502	0.01395	360	0.435	0.445	0.450	0.452	0.476	0.489	0.504	0.509	0.216	0.432	0.518	0.281	0.1404	0.1170
2/0 F	1×.1454″	6×.1454″	0.436	2/0	8,094	2,359	0.01235	350	0.434	0.441	0.446	0.448	0.475	0.487	0.497	0.501	0.222	0.444	0.533	0.285	0.1427	0.1189
1/0 K	4×.1585″	3×.1585″	0.475	1/0	14,490	2,703	0.00812	310	0.548	0.560	0.573	0.579	0.599	0.625	0.652	0.664	0.243	0.487	0.584	0.279	0.1397	0.1164

(continued)

TABLE A.5 (Continued)
Characteristics of Copperweld–Copper Conductors [1,2] (Copperweld Steel Company)

Nominal Designation	Size of Conductor — Number and Diameter of Wires: Copperweld	Copper	Outside Diameter (in.)	Copper Equivalent (cmil or AWG)	Rated Breaking Load (lb)	Weight (lb/mi)	Geometric Mean Radius at 60 Cycles (ft)	Approx. Current-Carrying Capacity at 60 Cycles (A)[a]	r_a Resistance (Ω/Conductor/mi) at 25°C (77°F), Small Currents — dc	25 Cycles	50 Cycles	60 Cycles	r_a Resistance (Ω/Conductor/mi) at 50°C (122°F) Current, Approx. 75% Capacity — dc	25 Cycles	50 Cycles	60 Cycles	x_a Inductive Reactance (Ω/Conductor/mi) at 1-ft Spacing, Average Currents — 25 Cycles	50 Cycles	60 Cycles	x_a' Capacitive Reactance (MΩ/Conductor/mi) at 1-ft Spacing — 25 Cycles	50 Cycles	60 Cycles
1/0 J	3×.1467″	4×.1467″	0.440	1/0	10,970	2,346	0.00917	310	0.548	0.559	0.570	0.576	0.599	0.624	0.648	0.659	0.237	0.474	0.569	0.285	0.1423	0.1188
1/0 G	2×.1373″	5×.1373″	0.412	1/0	8,563	2,078	0.00996	310	0.548	0.559	0.568	0.573	0.599	0.623	0.645	0.654	0.233	0.466	0.559	0.289	0.1447	0.1206
1/0 F	1×.1294″	6×.1294″	0.388	1/0	6,536	1,870	0.01099	310	0.548	0.554	0.559	0.562	0.599	0.612	0.622	0.627	0.228	0.456	0.547	0.294	0.1469	0.1224
1 N	5×.1546″	2×.1546″	0.404	1	15,410	2,541	0.00638	280	0.691	0.705	0.719	0.725	0.755	0.787	0.818	0.832	0.256	0.512	0.614	0.281	0.1405	0.1171
1 K	4×.1412″	3×.1412″	0.423	1	11,900	2,144	0.00723	270	0.691	0.704	0.716	0.722	0.755	0.784	0.813	0.825	0.249	0.498	0.598	0.288	0.1438	0.1198
1 J	3×.1307″	4×.1307″	0.392	1	9,000	1,861	0.00817	270	0.691	0.703	0.714	0.719	0.755	0.783	0.808	0.820	0.243	0.486	0.583	0.293	0.1465	0.1221
1 G	2×.1222″	5×.1222″	0.367	1	6,956	1,649	0.00887	270	0.691	0.702	0.712	0.716	0.755	0.781	0.805	0.815	0.239	0.478	0.573	0.298	0.1488	0.1240
1 F	1×.1153″	6×.1153″	0.346	1	5,266	1,483	0.00980	270	0.691	0.698	0.704	0.705	0.755	0.769	0.781	0.786	0.234	0.468	0.561	0.302	0.1509	0.1258
2 P	6×.1540″	1×.1540″	0.462	2	16,870	2,487	0.00501	250	0.871	0.886	0.901	0.909	0.952	0.988	1.024	1.040	0.268	0.536	0.643	0.281	0.1406	0.1172
2 N	5×.1377″	2×.1377″	0.413	2	12,680	2,015	0.00568	250	0.871	0.885	0.899	0.906	0.952	0.986	1.020	1.035	0.261	0.523	0.627	0.289	0.1446	0.1205
2 K	4×.1257″	3×.1257″	0.377	2	9,730	1,701	0.00644	240	0.871	0.885	0.899	0.906	0.952	0.983	1.014	1.028	0.255	0.510	0.612	0.296	0.1479	0.1232
2 J	3×.1164″	4×.1164″	0.349	2	7,322	1,476	0.00727	240	0.871	0.883	0.894	0.899	0.952	0.982	1.010	1.022	0.249	0.498	0.598	0.301	0.1506	0.1255
2 A	1×.1699″	2×.1699″	0.366	2	5,876	1,356	0.00763	240	0.869	0.875	0.88	0.882	0.950	0.962	0.973	0.979	0.247	0.493	0.592	0.298	0.1489	0.1241
2 G	2×.1089″	5×.1089″	0.327	2	5,626	1,307	0.00790	230	0.871	0.882	0.892	0.896	0.952	0.986	1.006	1.016	0.245	0.489	0.587	0.306	0.1529	0.1275
2 F	1×.1026″	6×.1026″	0.308	2	4,233	1,176	0.00873	230	0.871	0.878	0.884	0.885	0.952	0.967	0.979	0.985	0.230	0.479	0.575	0.310	0.1551	0.1292
3 P	6×.1371″	1×.1371″	0.411	3	13,910	1,973	0.00445	220	1.098	1.113	1.127	1.136	1.200	1.239	1.273	1.296	0.274	0.547	0.657	0.290	0.1448	0.1207
3 N	5×.1226″	2×.1226″	0.368	3	10,390	1,598	0.00506	210	1.098	1.112	1.126	1.133	1.200	1.237	1.273	1.289	0.267	0.534	0.641	0.298	0.1487	0.1239

Appendix A

3 K	4×.1120"	3×.1120"	0.336	3	7,910	1,349	0.00574	210	1.098	1.111	1.123	1.129	1.200	1.233	1.267	1.281	0.261	0.522	0.626	0.304	0.1520	0.1254
3 J	3×.1036"	4×.1036"	0.311	3	5,955	1,171	0.00648	200	10.98	1.110	1.121	1.126	1.200	1.232	1.262	1.275	0.255	0.509	0.611	0.309	0.1547	0.1289
3 A	1×.1513"	2×.1513"	0.325	3	4,810	1,075	0.00679	210	1.096	1.102	1.107	1.109	1.198	1.211	1.225	1.229	0.252	0.505	0.606	0.306	0.1531	0.1275
4 P	6×.1221"	1×.1221"	0.366	4	11,420	1,564	0.00397	190	1.385	1.400	1.414	1.423	1.514	1.555	1.598	1.616	0.280	0.559	0.671	0.298	0.1489	0.1241
4 N	5×.1092"	2×.1092"	0.328	4	8,460	1,267	0.00451	180	1.385	1.399	1.413	1.420	1.514	1.554	1.593	1.610	0.273	0.546	0.655	0.306	0.1528	0.1274
4 D	2×.1615"	1×.1615"	0.348	4	7,340	1,191	0.00566	190	1.382	1.389	1.396	1.399	1.511	1.529	1.544	1.542	0.262	0.523	0.628	0.301	0.1507	0.1256
4 D	1×.1347"	2×.1347"	0.290	4	3,938	853	0.00604	180	1.382	1.388	1.393	1.395	1.511	1.525	1.540	1.545	0.258	0.517	0.620	0.314	0.1572	0.1310
4 A	6×.1087"	1×.1087"	0.326	4	9,311	1,240	0.00353	160	1.747	1.762	1.776	1.785	1.909	1.954	2.00	2.02	0.285	0.571	0.685	0.306	0.1531	0.1275
5 P	2×.1438"	1×.1438"	0.310	5	6,035	944	0.00504	160	1.742	1.749	1.756	1.759	1.905	1.924	1.941	1.939	0.268	0.535	0.642	0.310	0.1548	0.1290
5 D	1×.1200"	2×.1200"	0.258	5	3,193	676	0.00538	160	1.742	1.748	1.753	1.755	1.906	1.920	1.936	1.941	0.264	0.528	0.634	0.323	0.1614	0.1345
5 A	2×.1281"	1×.1281"	0.276	6	4,942	749	0.00449	140	2.20	2.21	2.21	2.22	2.40	2.42	2.44	2.44	0.273	0.547	0.656	0.318	0.1590	0.1326
6 D	1×.1068"	2×.1068"	0.230	6	2,585	536	0.00479	140	2.20	2.20	2.21	2.21	2.40	2.42	2.44	2.44	0.270	0.540	0.648	0.331	0.1666	0.1379
6 A	1×.1046"	2×.1046"	0.225	6	2,143	514	0.00469	130	2.20	2.20	2.21	2.21	2.40	2.42	2.44	2.44	0.271	0.542	0.651	0.333	0.1663	0.1384
6 C	2×.1141"	1×.1141"	0.246	6	4,022	594	0.00400	120	2.77	2.78	2.79	2.79	3.03	3.05	3.07	3.07	0.279	0.558	0.670	0.326	0.1631	0.1359
7 D	1×.1266"	2×.0895"	0.223	7	2,754	495	0.00441	120	2.77	2.78	2.78	2.78	3.06	3.06	3.07	3.07	0.274	0.548	0.658	0.333	0.1666	0.1388
7 A	2×.1016"	1×.1016"	0.219	8	3,255	471	0.00356	110	3.49	3.50	3.51	3.51	3.82	3.84	3.86	3.86	0.286	0.570	0.684	0.334	0.1672	0.1393
8 D	1×.1127"	2×.0797"	0.199	8	2,233	392	0.00394	100	3.49	3.50	3.51	3.51	3.82	3.84	3.86	3.87	0.280	0.560	0.672	0.341	0.1706	0.1422
8 A	1×.0808"	2×.0834"	0.179	8	1,362	320	0.00373	100	3.49	3.50	3.51	3.51	3.82	3.84	3.86	3.86	0.283	0.565	0.679	0.349	0.1744	0.1453
8 C	2×.0808"	1×.0808"	0.174	9½	1,743	298	0.00283	85	4.19	4.92	4.92	4.93	5.37	5.39	5.42	5.42	0.297	0.593	0.712	0.351	0.1754	0.1462
9½ D	1×.0808"																					

[a] Based on a conductor temperature of 75°C and an ambient temperature of 25°C, wind 1.4 mi/h (2 ft/s) frequency = 60 cycles, average tarnished surface.
[b] Resistances at 50°C total temperature, based on an ambient temperature of 25°C plus 25°C rise due to the hasting effect of current rule. The approximate magnitude of the current necessary to produce the 25°C rise is 75% of the "Approximate Current-Carrying Capacity at 60 Cycles."

TABLE A.6
Characteristics of Copperweld Conductors [1,2] (Copperweld Steel Company)

Nominal Conductor Sing	Number and Size of Wires	Outside Diameter (in.)	Area of Conductor (cmil)	Rated Breaking Load (lb) High Strength	Rated Breaking Load (lb) Extra Strength	Weight (lb/mi)	Geometric Mean Radius at 60 Cycles and Average Currents (ft)	Approx. Carrying Capacity (A) at 60 Cycles	r_a Resistance (Ω/Conductor/mi) at 25°C (77°F) Small Currents dc	25 Cycles	50 Cycles	60 Cycles	dc	r_a Resistance (Ω/Conductor/mi) at 75°C (167°C) Current Approx. 75% of Capacity[b] 25 Cycles	50 Cycles	60 Cycles	x_a Inductive Reactance (Ω/Conductor/mi) at 1-ft spacing, Average Currents 25 Cycles	50 Cycles	60 Cycles	x_a' Capacitive Reactance (MΩ/Conductor/mi) at 1-ft Spacing 25 Cycles	50 Cycles	60 Cycles
									30% Conductivity													
7/8"	19 No. 5	0.910	628,900	55,570	66,910	9,344	0.00758	620	0.306	0.316	0.326	0.331	0.363	0.419	0.476	0.499	0.261	0.493	0.592	0.233	0.1165	0.0971
13/16"	19 No. 6	0.810	496,800	45,830	55,530	7,410	0.00675	540	0.386	0.396	0.405	0.411	0.458	0.518	0.580	0.605	0.267	0.505	0.606	0.241	0.1206	0.1006
23/32"	19 No.7	0.721	395,500	37,740	45,850	5,877	0.00601	470	0.486	0.495	0.506	0.511	0.577	0.643	0.710	0.737	0.273	0.517	0.621	0.250	0.1244	0.1040
21/32"	19 No.8	0.642	313,700	313,700	347,690	4,660	0.00535	410	0.613	0.623	0.633	0.638	0.728	0.799	0.872	0.902	0.279	0.529	0.635	0.258	0.1289	0.1074
9/16"	19 No. 9	0.572	248,800	25,500	30,610	3,696	0.00477	360	0.773	0.783	0.793	0.798	0.917	0.995	1.075	1.106	0.285	0.541	0.649	0.266	0.1330	0.1109
5/8"	7 No. 4	0.613	292,200	24,780	29,430	4,324	0.00511	410	0.656	0.664	0.672	0.676	0.778	0.824	0.870	0.887	0.281	0.533	0.640	0.261	0.1306	0.1088
9/16"	7 No.5	0.546	231,700	20,470	24,650	3,429	0.00455	360	0.827	0.835	0.843	0.847	0.981	1.030	1.080	1.099	0.287	0.545	0.654	0.269	0.1347	0.1122
1/2"	7 No.6	0.486	183,800	16,890	20,460	2,719	0.00405	310	1.042	1.050	1.058	1.062	1.237	1.290	1.343	1.364	0.293	0.557	0.668	0.278	0.1388	0.1157
7/16"	7 No. 7	0.433	145,700	13,910	16,890	2,157	0.00361	270	1.315	1.323	1.331	1.335	1.560	1.617	1.675	1.697	0.299	0.569	0.683	0.286	0.1429	0.1191
3/8"	7 No. 8	0.385	115,600	11,440	13,890	1,710	0.00321	230	1.658	1.666	1.674	1.678	1.967	2.03	2.09	2.12	0.305	0.581	0.697	0.294	0.1471	0.1226
11/32"	7 No. 9	0.343	91,650	9,393	11,280	1,356	0.00286	200	2.09	2.10	2.1	2.11	2.48	2.55	2.61	2.64	0.311	0.592	0.711	0.303	0.1512	0.1260
5/10"	7 No. 10	0.306	72,680	7,758	91,96	1,076	0.00255	170	2.64	2.64	2.65	2.66	3313	3.20	3.27	3.30	0.316	0.604	0.725	0.31	0.1553	0.1294
3 No. 8	3 No. 5	0.392	99,310	9,262	11,860	1,467	0.00457	220	1.926	1.931	1.936	1.938	2.29	2.31	2.34	2.35	0.289	0.545	0.654	0.2936	0.1465	0.1221
3 No. 6	3 No. 6	0.349	78,750	7,639	9,754	1,163	0.00407	190	2.43	2.43	2.44	2.44	2.88	2.91	2.94	2.95	0.295	0.556	0.668	0.301	0.1506	0.1255
3 No. 7	3 No. 7	0.311	62,450	6,291	7,922	922.4	0.00363	160	3.06	3.07	3.07	3.07	3.63	3.66	3.70	3.71	0.301	0.568	0.682	0.310	0.1547	0.1289
3 No. 8	3 No. 8	0.277	49,530	5,174	6,282	731.5	0.00323	140	3.86	3.87	3.87	3.87	4.58	4.61	4.65	4.66	0.307	0.580	0.696	0.318	0.1589	0.1324
3 No. 9	3 No. 9	0.247	39,280	4,250	5,129	58.1	0.00288	120	4.87	4.87	4.88	4.88	5.78	5.81	5.85	5.86	0.313	0.591	0.710	0.26	0.1629	0.1358
3 No. 10	3 No. 10	0.220	31,150	3,509	4,160	460.0	0.00257	110	6.14	6.14	6.15	6.15	7.28	7.32	7.36	7.38	0.319	0.603	0.724	0.344	0.1671	0.1392

Appendix A

								40% Conductivity											
7/8"	19 No. 5	0.910	628,900	50,240	....	9,344	0.01175	690	0.229	0.239	0.249	0.254	0.272	0.321	0.371	0.391	0.236	0.449	0.539	0.233	0.1165	0.0971
13/16"	19 No. 6	0.810	498,800	41,600	....	7,410	0.01046	610	0.289	0.299	0.309	0.314	0.343	0.396	0.450	0.472	0.241	0.461	0.553	0.241	0.1206	0.1005
23/32"	19 No. 7	0.721	395,500	34,390	....	5,877	0.00931	530	0.365	0.375	0.385	0.390	0.433	0.490	0.549	0.573	0.247	0.473	0.567	0.250	0.1248	0.1040
21/32"	19 No.8	0.642	313,700	28,380	....	4,660	0.00829	470	0.460	0.470	0.480	0.485	0.546	0.608	0.672	0.698	0.253	0.485	0.582	0.258	0.1289	0.1074
9/18"	19 No. 9	0.572	248,800	23,390	....	3,696	0.00739	410	0.580	0.590	0.600	0.605	0.688	0.756	0.826	0.753	0.259	0.496	0.595	0.266	0.1330	0.1109
5/8"	7 No. 4	0.613	292,200	22,310	....	4,324	0.00792	470	0.492	0.500	0.508	0.512	0.584	0.624	0.664	0.680	0.255	0.489	0.587	0.261	0.1306	0.1088
9/16"	7 No. 5	0.546	231,700	18,510	....	3,429	0.00705	410	0.520	0.628	0.536	0.640	0.736	0.780	0.843	0.840	0.261	0.501	0.601	0.269	0.1347	0.1122
1/2"	7 No. 6	0.486	183,800	15,330	....	2,719	0.00628	350	0.782	0.790	0.798	0.802	.98	0.975	1.021	1.040	0.267	0.513	0.615	0.278	0.1388	0.1157
7/6"	7 No. 7	0.435	145,700	12,670	....	2,157	0.00559	310	0.986	0.994	1.002	1.006	1.170	1.220	1.271	1.291	0.273	0.524	0.629	0.286	0.1429	0.1191
3/8"	7 No. 8	0.385	115,600	10,460	....	1,710	0.00497	270	1.244	1.252	1.260	1.264	1.476	1.530	1.584	1.606	0.279	0.536	0.644	0.294	0.1471	0.1226
11/32"	7 No.9	0.343	91,650	8,616	....	1,356	0.00443	230	1.568	1.576	1.584	1.588	1.861	1.919	1.978	2.00	0.285	0.548	0.658	0.303	0.1512	0.1260
5/16"	7 No.10	0.306	72,680	7,121	....	1,076	0.00395	200	1.978	1.986	1.944	1.998	2.35	2.41	2.47	2.50	0.291	0.559	0.671	0.311	0.1553	0.1294
3 No. 5	3 No. 5	0.392	99,310	8,373	....	1,467	0.00621	250	1.445	1.450	1.455	1.457	1.714	1.738	1.762	1.772	0.269	0.514	0.617	0.293	0.1465	0.1221
3 No. 6	3 No. 6	0.349	78,750	6,934	....	1,163	0.00553	220	1.821	1.826	1.831	1.833	2.16	2.19	2.21	2.22	0.275	0.526	0.631	0.301	0.1506	0.1255
3 No. 7	3 No. 7	0.31	52,450	5,732	....	922.4	0.00492	190	2.30	2.30	2.31	2.31	2.73	2.75	2.78	2.79	0.281	0.537	0.645	0.310	0.1547	0.1289
3 No. 8	3 No. 8	0.277	49,530	4,730	....	731.5	0.00439	160	2.90	2.90	2.91	2.91	3.44	3.47	3.50	3.51	0.286	0.549	0659	0.318	0.1589	0.1324
3 No. 9	3 No. 9	0.247	39,280	3,896	....	580.1	0.00391	140	3.65	3.66	3.66	3.66	4.32	4.37	4.40	4.41	0.292	0.561	0.673	0.326	0.1629	0.1358
3 No. 10	3 No. 10	1.220	31,150	3,221	....	460.0	0.00348	120	4.61	4.61	4.62	4.62	5.46	5.50	5.53	5.55	0.297	0.572	0.687	0.334	0.1671	0.1392
3 No. 12	3 No. 12	0.174	19,590	2,236	....	289.3	0.00276	90	7.32	7.33	7.33	7.34	8.69	8.73	8.77	8.78	0.310	0.596	0.715	0.351	0.1754	0.1462

[a] Based on a conductor temperature of 75°C and an ambient temperature of 25°C.

[b] Resistances at 50°C total temperature, based on an ambient temperature of 25°C plus 25°C rise due to the hasting effect of current. The approximate magnitude of the current necessary to produce the 25°C rise is 75% of the "Approximate Current-Carrying Capacity at 60 Cycles."

TABLE A.7
Electrical Characteristics of Overhead Ground Wires [3]

Strand (AWG)	Resistance (Ω/mi)				60-Hz Reactance for 1-ft Radius		60-Hz Geometric Mean Radius (ft)
	Small Currents		75% of Capacity		Inductive (Ω/mi)	Capacitive (MΩ-mi)	
	25°C oc	25°C 60 Hz	75°C oc	75°C 60 Hz			
7 No. 5	1.217	1.240	1.432	1.669	0.707	0.1122	0.002958
7 No. 6	1.507	1.536	1.773	2.010	0.721	0.1157	0.002633
7 No. 7	1.900	1.937	2.240	2.470	0.735	0.1191	0.002345
7 No. 8	2.400	2.440	2.820	3.060	0.749	0.1226	0.002085
7 No. 9	3.020	3.080	3.560	3.800	0.763	0.1260	0.001858
7 No. 10	3.810	3.880	4.480	4.730	0.777	0.1294	0.001658
3 No. 5	2.780	2.780	3.270	3.560	0.707	0.1221	0.002940
3 No. 6	3.510	3.510	4.130	4.410	0.721	0.1255	0.002618
3 No. 7	4.420	4.420	5.210	5.470	0.735	0.1289	0.002333
3 No. 8	5.580	5.580	6.570	6.820	0.749	0.1324	0.002078
3 No. 9	7.040	7.040	8.280	8.520	0.763	0.1358	0.001853
3 No. 10	8.870	8.870	10.440	10.670	0.777	0.1392	0.001650

Part B: Single-Layer ACSR

Code	Resistance (Ω/mi)				60-Hz Reactance for 1-ft Radius			Capacitive (MΩ-mi)
	25°C dc	60 Hz, 75°C			Inductive (Ω/mi) at 72°C			
		$I = 0$ A	$I = 100$ A	$I = 200$ A	$I = 0$ A	$I = 100$ A	$I = 200$ A	
Brahma	0.394	0.470	0.510	0.565	0.500	0.520	0.545	0.1043
Cochin	0.400	0.480	0.520	0.590	0.505	0.515	0.550	0.1065
Dorking	0.443	0.535	0.575	0.650	0.515	0.530	0.565	0.1079
Dotterel	0.479	0.565	0.620	0.705	0.515	0.530	0.575	0.1091
Guinea	0.531	0.630	0.685	0.780	0.520	0.545	0.590	0.1106
Leghorn	0.630	0.760	0.810	0.930	0.530	0.550	0.605	0.1131
Minorca	0.765	0.915	0.980	1.130	0.540	0.570	0.640	0.1460
Petrel	0.830	1.000	1.065	1.220	0.550	0.580	0.655	0.1172
Grouse	1.080	1.295	1.420	1.520	0.570	0.640	0.675	0.1240

(continued)

TABLE A.7 (Continued)
Electrical Characteristics of Overhead Ground Wires [3]

Part C: Steel Conductors

| | | Resistance (Ω/mi) at 60 Hz | | | 60-Hz Reactance for 1-ft Radius | | | |
| | | | | | Inductive (Ω/mi) | | | Capacitive |
		$I = 0$ A	$I = 30$ A	$I = 60$ A	$I = 0$ A	$I = 30$ A	$I = 60$ A	(MΩ-mi)
Ordinary	1/4	9.5	11.4	11.3	1.3970	3.7431	3.4379	0.1354
Ordinary	9/32	7.1	9.2	9.0	1.2027	3.0734	2.5146	0.1319
Ordinary	5/16	5.4	7.5	7.8	0.8382	2.5146	2.0409	0.1288
Ordinary	3/8	4.3	6.5	6.6	0.8382	2.2352	1.9687	0.1234
Ordinary	1/2	2.3	4.3	5.0	0.7049	1.6893	1.4236	0.1148
E.B.	1/4	8.0	12.0	10.1	1.2027	4.4704	3.1565	0.1354
E.B.	9/32	6.0	10.0	8.7	1.1305	3.7783	2.6255	0.1319
E.B.	5/16	4.9	8.0	7.0	0.9843	2.9401	2.5146	0.1288
E.B.	3/8	3.7	7.0	6.3	0.8382	2.5997	2.4303	0.1234
E.B.	1/2	2.1	4.9	5.0	0.7049	1.8715	1.7616	0.1148
E.B.B.	1/4	7.0	12.8	10.9	1.6764	5.1401	3.9482	0.1354
E.B.B.	9/32	5.4	10.9	8.7	1.1305	4.4833	3.7783	0.1319
E.B.B.	5/16	4.0	9.0	6.8	0.9843	3.6322	3.0734	0.1288
E.B.B.	3/8	3.5	7.9	6.0	0.8382	3.1168	2.7940	0.1234
E.B.B.	1/2	2.0	5.7	4.7	0.7049	2.3461	2.2352	0.1148

TABLE A.8
Inductive Reactance Spacing Factor X_d (Ω/mi/Conductor) at 60 Hz [1]

Ft	0.0	0.1	0.2	0.3	0.4	0.5	0.6	0.7	0.8	0.9
0		−0.2794	−0.1953	−0.1461	−0.1112	−0.0841	−0.0620	−0.0433	−0.0271	−0.0128
1	0.0	0.0116	0.0221	0.0318	0.0408	0.0492	0.0570	0.0644	0.0713	0.0779
2	0.0841	0.0900	0.0957	0.1011	0.1062	0.01112	0.0119	0.1205	0.1249	0.1292
3	0.1333	0.1373	0.1411	0.1449	0.1485	0.1520	0.1554	0.1588	0.1620	0.1651
4	0.1682	0.1712	0.1741	0.1770	0.1798	0.1825	0.1852	0.1878	0.1903	0.1928
5	0.1953	0.1977	0.2001	0.2024	0.2046	0.2069	0.2090	0.2112	0.2133	0.2154
6	0.2174	0.2194	0.2214	0.2233	0.2252	0.2271	0.2290	0.2308	0.2326	0.2344
7	0.2361	0.2378	0.2395	0.2412	0.2429	0.2445	0.2461	0.2477	0.2493	0.2508
8	0.2523	0.2538	0.2553	0.2568	0.2582	0.2597	0.2611	0.2625	0.2639	0.2653
9	0.2666	0.2680	0.2693	0.2706	0.2719	0.2732	0.2744	0.2757	0.2769	0.2782
10	0.2794	0.2806	0.2818	0.2830	0.2842	0.2853	0.2865	0.2876	0.2887	0.2899
11	0.2910	0.2921	0.2932	0.2942	0.2953	0.2964	0.2974	0.2985	0.2995	0.3005
12	0.3015	0.3025	0.3035	0.3045	0.3055	0.3065	0.3074	0.3084	0.3094	0.3103

(*continued*)

TABLE A.8 (Continued)
Inductive Reactance Spacing Factor X_d (Ω/mi/Conductor) at 60 Hz [1]

Ft	0.0	0.1	0.2	0.3	0.4	0.5	0.6	0.7	0.8	0.9
13	0.3112	0.3122	0.3131	0.3140	0.3149	0.3158	0.3167	0.3176	0.3185	0.3194
14	0.3202	0.3211	0.3219	0.3228	0.3236	0.3245	0.3253	0.3261	0.3270	0.3278
15	0.3286	0.3294	0.3302	0.3310	0.3318	0.3326	0.3334	0.3261	0.3270	0.3278
16	0.3364	0.3372	0.3379	0.3387	0.3394	0.3402	0.3409	0.3416	0.3424	0.3431
17	0.3438	0.3445	0.3452	0.3459	0.3466	0.3473	0.3480	0.3487	0.3494	0.3500
18	0.3507	0.3514	0.3521	0.3527	0.3534	0.3540	0.3547	0.3554	0.3560	0.3566
19	0.3537	0.3579	0.3586	0.3592	0.3598	0.3604	0.3611	0.3617	0.3623	0.3629
20	0.3635	0.3641	0.3647	0.3563	0.3659	0.3665	0.3671	0.3677	0.3683	0.3688
21	0.3694	0.3700	0.3706	0.3711	0.3717	0.3723	0.3728	0.3734	0.3740	0.3745
22	0.3751	0.3756	0.3762	0.3767	0.3773	0.3778	0.3783	0.3789	0.3794	0.3799
23	0.3805	0.3810	0.3815	0.3820	0.3826	0.3831	0.3836	0.3841	0.3846	0.3851
24	0.3856	0.3861	0.3866	0.3871	0.3876	0.3881	0.3886	0.3891	0.3896	0.3901
25	0.3906	0.3911	0.3916	0.3920	0.3925	0.3930	0.3935	0.3939	0.3944	0.3949
26	0.3953	0.3958	0.3963	0.3967	0.3972	0.3977	0.3981	0.3986	0.3990	0.3995
27	0.3999	0.4004	0.4008	0.4013	0.4017	0.4021	0.4026	0.4030	0.4035	0.4039
28	0.4043	0.4048	0.4052	0.4056	0.4061	0.4065	0.4069	0.4073	0.4078	0.4082
29	0.4086	0.4090	0.4094	0.4098	0.4103	0.4107	0.4111	0.4115	0.4119	0.4123
30	0.4127	0.4131	0.4135	0.4139	0.4143	0.4147	0.4151	0.4155	0.4159	0.4163
31	0.4167	0.4171	0.4175	0.4179	0.4182	0.4186	0.4190	0.4194	0.4198	0.4202
32	0.4205	0.4209	0.4213	0.4217	0.4220	0.4224	0.4228	0.4232	0.4235	0.4239
33	0.4243	0.4246	0.4250	0.4254	0.4257	0.4261	0.4265	0.4268	0.4272	0.4275
34	0.4279	0.4283	0.4286	0.4290	0.4293	0.4297	0.4300	0.4304	0.4307	0.4311
35	0.4314	0.4318	0.4321	0.4324	0.4328	0.4331	0.4335	0.4338	0.4342	0.4345
36	0.4348	0.4352	0.4355	0.4358	0.4362	0.4365	0.4368	0.4372	0.4375	0.4378
37	0.4382	0.4385	0.4388	0.4391	0.4395	0.4398	0.4401	0.4404	0.4408	0.4411
38	0.4414	0.4417	0.4420	0.4423	0.4427	0.4430	0.4433	0.4436	0.4439	0.4442
39	0.4445	0.4449	0.4452	0.4455	0.4458	0.4461	0.4464	0.4467	0.4470	0.4473
40	0.4476	0.4479	0.4492	0.4485	0.4488	0.4491	0.4494	0.4497	0.4500	0.4503
41	0.4506	0.4509	0.4512	0.4515	0.4518	0.4521	0.4524	0.4527	0.4530	0.4532
42	0.4535	0.4538	0.4541	0.4544	0.4547	0.4550	0.4553	0.4555	0.4558	0.4561
43	0.4564	0.4567	0.4570	0.4572	0.4575	0.4578	0.4581	0.4584	0.4586	0.4589
44	0.4592	0.4595	0.4597	0.4600	0.4603	0.4606	0.4608	0.4611	0.4614	0.4616
45	0.4619	0.4622	0.4624	0.4627	0.4630	0.4632	0.4635	0.4638	0.4640	0.4643
46	0.4646	0.4648	0.4651	0.4654	0.4656	0.4659	0.4661	0.4664	0.4667	0.4669
47	0.4672	0.4674	0.4677	0.4680	0.4682	0.4685	0.4687	0.4690	0.4692	0.4695
48	0.4697	0.4700	0.4702	0.4705	0.4707	0.4710	0.4712	0.4715	0.4717	0.4720
49	0.4722	0.4725	0.4727	0.4730	0.4732	0.4735	0.4737	0.4740	0.4742	0.4744
50	0.4747	0.4749	0.4752	0.4754	0.4757	0.4759	0.4761	0.4764	0.4766	0.4769
51	0.4771	0.4773	0.4776	0.4778	0.4780	0.4783	0.4785	0.4787	0.4790	0.4792
52	0.4795	0.4797	0.4799	0.4801	0.4804	0.4806	0.4808	0.4811	0.4813	0.4815
53	0.4818	0.4820	0.4822	0.4824	0.4827	0.4829	0.4831	0.4834	0.4836	0.4838
54	0.4840	0.4843	0.4845	0.4847	0.4849	0.4851	0.4854	0.4856	0.4858	0.4860
55	0.4863	0.4865	0.4867	0.4869	0.4871	0.4874	0.4876	0.4878	0.4880	0.4882
56	0.4884	0.4887	0.4889	0.4891	0.4893	0.4895	0.4897	0.4900	0.4902	0.4904
57	0.4906	0.908	0.4910	0.912	0.4914	0.4917	0.4919	0.4921	0.4923	0.4925
58	0.4927	0.4929	0.4931	0.4933	0.4935	0.4937	0.4940	0.4942	0.4944	0.4946
59	0.4948	0.4950	0.4952	0.4954	0.4956	0.4958	0.4960	0.4962	0.4964	0.4966

(continued)

TABLE A.8 (Continued)
Inductive Reactance Spacing Factor X_d (Ω/mi/Conductor) at 60 Hz [1]

Ft	0.0	0.1	0.2	0.3	0.4	0.5	0.6	0.7	0.8	0.9
60	0.4968	0.4970	0.4972	0.4974	0.4976	0.4978	0.4980	0.4982	0.4984	0.4986
61	0.4988	0.4990	0.4992	0.4994	0.4996	0.4998	0.5000	0.5002	0.5004	0.5006
62	0.5008	0.510	0.5012	0.5014	0.5016	0.5018	0.5020	0.5022	0.5023	0.5025
63	0.5027	0.5029	0.5031	0.5033	0.5035	0.5037	0.5039	0.0541	0.5043	0.5045
64	0.5046	0.5048	0.5050	0.5052	0.5054	0.5056	0.5058	0.5060	0.5062	0.5063
65	0.5065	0.5067	0.5069	0.5071	0.5073	0.5075	0.5076	0.5078	0.5080	0.5082
66	0.5084	0.5086	0.5087	0.5089	0.5091	0.5093	0.5095	0.5097	0.5098	0.5100
67	0.5102	0.5104	0.5106	0.5107	0.5109	0.5111	0.5113	0.5115	0.5116	0.5118
68	0.5120	0.5122	0.5124	0.5125	0.5127	0.5129	0.5131	0.5132	0.5134	0.5136
69	0.5138	0.5139	0.5141	0.5143	0.5145	0.5147	0.5148	0.5150	0.5152	0.5153
70	0.5155	0.5157	0.5159	0.5160	0.5162	0.5164	0.5166	0.5167	0.5169	0.5171
71	0.5172	0.5174	0.5176	0.5178	0.5179	0.5181	0.5183	0.5184	0.5186	0.5188
72	0.5189	0.5191	0.5193	0.5194	0.5196	0.5198	0.5199	0.5201	0.5203	0.5204
73	0.5206	0.5208	0.5209	0.5211	0.5213	0.5214	0.5216	0.5218	0.5219	0.5221
74	0.5223	0.5224	0.5226	0.5228	0.5229	0.5231	0.5232	0.5234	0.5236	0.5237
75	0.5239	0.5241	0.5242	0.5244	0.5245	0.5247	0.5249	0.5250	0.5252	0.5253
76	0.5255	0.5257	0.5258	0.5260	0.5261	0.5263	0.5265	0.5266	0.5268	0.5269
77	0.5271	0.5272	0.5274	0.5276	0.5277	0.5279	0.5280	0.5282	0.5283	0.5285
78	0.5287	0.5288	0.5290	0.5291	0.5293	0.5294	0.5296	0.5297	0.5299	0.5300
79	0.5302	0.5304	0.5305	0.5307	0.5308	0.5310	0.5311	0.5313	0.5314	0.5316
80	0.5317	0.5319	0.5320	0.5322	0.5323	0.5325	0.5326	0.5328	0.5329	0.5331
81	0.5332	0.5334	0.5335	0.5337	0.5338	0.5340	0.5341	0.5343	0.5344	0.5346
82	0.5347	0.5349	0.5350	0.5352	0.5353	0.5355	0.5356	0.5358	0.5359	0.5360
83	0.5362	0.5363	0.5365	0.5366	0.5368	0.5369	0.5371	0.5372	0.5374	0.5375
84	0.5376	0.5378	0.5379	0.53841	0.5382	0.5384	0.5385	0.5387	0.5388	0.5389
85	0.5391	0.5392	0.5394	0.5395	0.5396	0.5398	0.5399	0.5401	0.5402	0.5404
86	0.5405	0.5406	0.5408	0.5409	0.5411	0.5412	0.5413	0.5415	0.5416	0.5418
87	0.5419	0.5420	0.5422	0.5423	0.5425	0.5426	0.5427	0.5429	0.5430	0.5432
88	0.5433	0.5434	0.5436	0.5437	0.5438	0.5440	0.5441	0.5442	0.5444	0.5445
89	0.5447	0.5448	0.5449	0.5451	0.5452	0.5453	0.5455	0.5456	0.5457	0.5459
90	0.5460	0.5461	0.5463	0.5464	0.5466	0.5467	0.5468	0.5470	0.5471	0.5472
91	0.5474	0.5475	0.5476	0.5478	0.5479	0.5480	0.5482	0.5483	0.5484	0.5486
92	0.5487	0.5488	0.5489	0.5491	0.5492	0.5493	0.5495	0.5496	0.5497	0.5944
93	0.5500	0.5501	0.5503	0.5504	0.5505	0.5506	0.5508	0.5509	0.5510	0.5512
94	0.5513	0.5514	0.5515	0.5517	0.5518	0.5519	0.5521	0.5522	0.5523	0.5524
95	0.5526	0.5527	0.5528	0.5530	0.5531	0.5532	0.5533	0.5535	0.5536	0.5537
96	0.5538	0.5540	0.5541	0.5542	0.5544	0.5545	0.5546	0.5547	0.5549	0.5550
97	0.5551	0.5552	0.5554	0.5555	0.5556	0.5557	0.5559	0.5560	0.5561	0.5562
98	0.5563	0.5565	0.5566	0.5567	0.5568	0.5570	0.5571	0.5572	0.5573	0.5575
99	0.5576	0.5577	0.5578	0.5579	0.5581	0.5582	0.5583	0.5584	0.5586	0.5587
100	0.5588	0.5589	0.5590	0.5592	0.5593	0.5594	0.5595	0.5596	0.5598	0.5599

(*continued*)

TABLE A.8 (Continued)
Inductive Reactance Spacing Factor X_d (Ω/mi/Conductor) at 60 Hz [1]

	ρ (Ω·m)	r_e, x_e (f = 60 Hz)[a]
r_e	All	0.2860
	1	2.050
	5	2.343
	10	2.469
x_e	50	2.762
	100[b]	2.888[b]
	500	3.181
	1000	3.307
	5000	3.600
	10,000	3.726

[a] From the following formulas:

$$r_e = 0.004764 f$$

$$x_e = 0.006985 f \log_{10} 4{,}665{,}660 \frac{\rho}{f}$$

where

f = frequency

ρ = resistivity, Ω·m

[b] This is an average value that may be used in the absence of definite information. Fundamental equations:

$$z_1 = z_2 = r_e + j(x_a + x_d)$$
$$z_0 = r_a + r_e + j(x_a + x_e - 2x_d)$$

where

$x_d = \omega k \ln d$

d = separation, ft

TABLE A.9
Shunt Capacitive Reactance Spacing Factor X'_d (MΩ/mi/Conductor) at 60 Hz [1]

Ft	0.0	0.1	0.2	0.3	0.4	0.5	0.6	0.7	0.8	0.9
0		−0.0683	−0.0477	−0.0357	−0.0272	−0.0206	−0.0152	−0.0106	−0.0066	−0.0031
1	0.0000	0.0028	0.0054	0.0078	0.0100	0.0120	0.0139	0.0157	0.0174	0.0190
2	0.0206	0.0220	0.0234	0.0247	0.0260	0.0272	0.0283	0.0295	0.0305	0.0316
3	0.0326	0.0336	0.0345	0.0354	0.0363	0.0372	0.0380	0.0388	0.0396	0.0404
4	0.0411	0.0419	0.0426	0.0433	0.0440	0.0446	0.0453	0.0459	0.0465	0.0471
5	0.0477	0.0483	0.0489	0.0495	0.0500	0.0506	0.0511	0.0516	0.0521	0.0527
6	0.0532	0.0536	0.0541	0.0546	0.0551	0.0555	0.0560	0.0564	0.0569	0.0573
7	0.0577	0.0581	0.0586	0.0590	0.0594	0.0598	0.0602	0.0606	0.0609	0.0613
8	0.0617	0.0621	0.0624	0.0628	0.0631	0.0635	0.0638	0.0642	0.0645	0.0649
9	0.0652	0.0655	0.0658	0.0662	0.0665	0.0668	0.0671	0.0674	0.0677	0.0680
10	0.0683	0.0686	0.0689	0.0692	0.0695	0.0698	0.0700	0.0703	0.0706	0.0709
11	0.0711	0.0714	0.0717	0.0719	0.0722	0.0725	0.0727	0.0730	0.0732	0.0735
12	0.0737	0.0740	0.0742	0.0745	0.0747	0.0479	0.0752	0.0754	0.0756	0.0759
13	0.0761	0.0763	0.0765	0.0768	0.0770	0.0772	0.0774	0.0776	0.0779	0.0781
14	0.0783	0.0785	0.0787	0.0789	0.0791	0.0793	0.0795	0.0797	0.0799	0.0801
15	0.0803	0.0805	0.0807	0.0809	0.0811	0.0813	0.0815	0.0817	0.0819	0.0821
16	0.0823	0.0824	0.0826	0.0828	0.0830	0.0832	0.0833	0.0835	0.0837	0.0839
17	0.0841	0.0842	0.0844	0.0846	0.0847	0.0849	0.0851	0.0852	0.0854	0.0856
18	0.0857	0.0859	0.0861	0.0862	0.0864	0.0866	0.0867	0.0869	0.0870	0.0872
19	0.0874	0.0875	0.0877	0.0878	0.080	0.0881	0.0883	0.0884	0.0886	0.0887
20	0.0889	0.0890	0.0892	0.0893	0.0895	0.0896	0.0898	0.0899	0.0900	0.0902
21	0.0903	0.0905	0.0906	0.0907	0.0909	0.0910	0.0912	0.0913	0.0914	0.0916
22	0.0917	0.0918	0.0920	0.0921	0.0922	0.0924	0.0925	0.0926	0.0928	0.0929
23	0.0930	0.0931	0.0933	0.0934	0.0935	0.0937	0.0938	0.0939	0.0940	0.0942
24	0.0943	0.0944	0.0945	0.0947	0.0948	0.0949	0.0950	0.0951	0.0953	0.0954
25	0.0955	0.0956	0.0957	0.0958	0.0960	0.0961	0.0962	0.0963	0.0964	0.9565
26	0.0697	0.0968	0.0969	0.0970	0.0971	0.0972	0.0973	0.0974	0.0976	0.0977
27	0.0978	0.0979	0.0980	0.0981	0.0982	0.0983	0.0984	0.0985	0.0986	0.0987
28	0.0989	0.0990	0.00991	0.0992	0.0993	0.0994	0.0995	0.0996	0.0997	0.0998
29	0.0999	0.01000	0.01001	0.1002	0.1003	0.1004	0.1005	0.1006	0.01007	0.1008
30	0.1009	0.1010	0.1011	0.1012	0.1013	0.1014	0.1015	0.1016	0.01017	0.1018
31	0.1019	0.1020	0.0121	0.1022	0.1023	0.1023	0.1024	0.1025	0.1026	0.1027
32	0.1028	0.1029	0.1030	0.1031	0.1032	0.1033	0.1034	0.1035	0.1035	0.1036
33	0.1037	0.1038	0.1039	0.1040	0.1041	0.1042	0.1043	0.1044	0.1044	0.1045
34	0.1046	0.1047	0.1048	0.1049	0.1050	0.1050	0.1051	0.1052	0.1053	0.1054
35	0.1055	0.1056	0.1056	0.1057	01058	0.1059	0.1060	0.1061	0.1061	0.1062
36	0.1063	0.1064	0.1065	0.1066	0.1066	0.1067	0.1068	0.1069	0.1070	0.1070
37	0.1071	0.1072	0.1073	0.1074	0.1074	0.1075	0.1076	0.1077	0.1078	0.1078
38	0.1079	0.1080	0.1081	0.1081	0.1082	0.1083	0.1084	0.1085	0.1085	0.1086
39	0.1087	0.1088	0.1088	0.1089	0.1090	0.1091	0.1091	0.1092	0.1093	0.1094
40	0.1094	0.1095	0.1096	0.1097	0.1097	0.1098	0.1099	0.1100	0.1100	0.1101
41	0.1102	0.1102	0.1103	0.1104	0.1105	0.1105	0.1106	0.1107	0.1107	0.1108
42	0.1109	0.1110	0.1110	0.1111	0.1112	0.1112	0.1113	0.1114	0.1114	0.1115
43	0.1116	0.1117	0.1117	0.1118	0.1119	0.1119	0.1120	0.1121	0.1121	0.1122
44	0.1123	0.1123	0.1124	0.1125	0.1125	0.1126	0.1127	0.1127	0.1128	0.1129
45	0.1129	0.1130	0.1131	0.1131	0.1132	0.1133	0.1133	0.1134	0.1335	0.1135
46	0.1136	0.1136	0.1137	0.1138	0.1138	0.1139	0.1140	0.1140	0.1141	0.1142

(*continued*)

TABLE A.9 (Continued)
Shunt Capacitive Reactance Spacing Factor X'_d (MΩ/mi/Conductor) at 60 Hz [1]

Ft	0.0	0.1	0.2	0.3	0.4	0.5	0.6	0.7	0.8	0.9
47	0.1142	0.1143	0.1143	0.1144	0.1145	0.1145	0.1146	0.1147	0.1147	0.1148
48	0.1148	0.1149	0.1150	0.1150	0.1151	0.1152	0.1152	0.1153	0.1153	0.1154
49	0.1155	0.1155	0.1156	0.1156	0.1157	0.1158	0.1158	0.1159	0.1159	0.1160
50	0.1161	0.1161	0.1162	0.1162	0.1163	0.1164	0.1164	0.1165	0.1165	0.1166
51	0.1166	0.1167	0.1168	0.1168	0.1169	0.1169	0.1170	0.1170	0.1171	0.1172
52	0.1172	0.1173	0.1173	0.1174	0.1174	0.1175	0.1176	0.1176	0.1177	0.1177
53	0.1178	0.1178	0.1179	0.1180	0.1180	0.1181	0.1181	0.1182	0.1182	0.1183
54	0.1183	0.1184	0.1184	0.1185	0.1186	0.1186	0.1187	0.1187	0.1188	0.1188
55	0.1189	0.1189	0.1190	0.1190	0.1191	0.1192	0.1192	0.1193	0.1193	0.1194
56	0.1194	0.1195	0.1195	0.1196	0.1196	0.1197	0.1197	0.1198	0.1198	0.1199
57	0.1199	0.1200	0.1200	0.1201	0.1202	0.1202	0.1203	0.1203	0.1204	0.1204
58	0.1205	0.1205	0.1206	0.1206	0.1207	0.1207	0.1208	0.1208	0.1209	0.1209
59	0.1210	0.1210	0.1211	0.1211	0.1212	0.1212	0.1213	0.1213	0.1214	0.1214
60	0.1215	0.1215	0.1216	0.1216	0.1217	0.1217	0.1218	0.1218	0.1219	0.1219
61	0.1220	0.1220	0.1221	0.1221	0.1221	0.1222	0.1222	0.1223	0.1223	0.1224
62	0.1224	0.1225	0.1225	0.1226	0.1226	0.1227	0.1227	0.1228	0.1228	0.1229
63	0.1229	0.1230	0.1230	0.1231	0.1231	0.1231	0.1232	0.1232	0.1233	0.1233
64	0.1234	0.1234	0.1235	0.1235	0.1236	0.1236	0.1237	0.1237	0.1237	0.1238
65	0.1238	0.1239	0.1239	0.124	0.1240	0.1241	0.1241	0.1242	0.1242	0.1242
66	0.1243	0.1243	0.1244	0.1244	0.1245	0.1245	0.1246	0.1246	0.1247	0.1247
67	0.1247	0.1248	0.1248	0.1249	0.1249	0.1250	0.1250	0.150	0.1251	0.1251
68	0.1252	0.1252	0.1253	0.1253	0.1254	0.1254	0.1254	0.1255	0.1255	0.1256
69	0.1256	0.1257	0.1257	0.1257	0.1258	0.1258	0.1259	0.1259	0.1260	0.1260
70	0.1260	0.1261	0.1261	0.1262	0.1262	0.1262	0.1263	0.1263	0.1264	0.1264
71	0.1265	0.1265	0.1265	0.1266	0.1266	0.1267	0.1267	0.1268	0.1268	0.1268
72	0.1269	0.1269	0.1270	0.1270	0.1274	0.10271	0.1271	0.1272	0.1272	0.1272
73	0.1273	0.1273	0.1274	0.1274	0.1274	0.1275	0.1275	0.1276	0.1276	0.1276
74	0.1277	0.1277	0.1278	0.1278	0.1278	0.1279	0.1279	0.1280	0.1280	0.1280
75	0.1281	0.1281	0.1282	0.1282	0.1282	0.1283	0.1283	0.1284	0.1284	0.1284
76	0.1285	0.1285	0.286	0.1286	0.1286	0.1287	0.1287	0.1288	0.1288	0.1288
77	0.1289	0.1289	0.1289	0.1290	0.1290	0.1291	0.1291	0.1291	0.1292	0.1292
78	0.1292	0.1293	0.1293	0.1294	0.1294	0.1294	0.1295	0.1295	0.1296	0.1296
79	0.1296	0.1297	0.1297	0.1297	0.1298	0.1298	0.1299	0.1299	0.1299	0.1300
80	0.1300	0.1300	0.1301	0.1301	0.1301	0.1302	0.1302	0.1303	0.1303	0.1303
81	0.1304	0.1304	0.1304	0.1305	0.1305	0.1306	0.1306	0.1306	0.1307	0.1307
82	0.1307	0.1308	0.1308	0.1308	0.1309	0.1309	0.1309	0.1310	0.1310	0.1311
83	0.1311	0.1311	0.1312	0.1312	0.1312	0.1313	0.1313	0.1313	0.1314	0.1314
84	0.1314	0.1315	0.1315	0.1316	0.1316	0.1316	0.1317	0.1317	0.1317	0.1318
85	0.1318	0.1318	0.1319	0.1319	0.1319	0.1320	0.1320	0.1320	0.1321	0.1321
86	0.1321	0.1322	0.1322	0.1322	0.1323	0.1323	0.1324	0.1324	0.1324	0.1325
87	0.1325	0.1325	0.1326	0.1326	0.1326	0.1327	0.1327	0.1327	0.1328	0.1328
89	0.1332	0.1332	0.1332	0.1333	0.1333	0.1333	0.1334	0.1334	0.1334	0.1335
90	0.1335	0.1335	0.1336	0.1336	0.1336	0.1337	0.1337	0.1377	0.1338	0.1338
91	0.1338	0.1339	0.1339	0.1339	0.1340	0.1340	0.1340	0.1340	0.1341	0.1341
92	0.1341	0.1342	0.1342	0.1342	0.1343	0.1343	0.1343	0.1344	0.1344	0.1344
93	0.1345	0.1345	0.1345	0.1346	0.1346	0.1346	0.1347	0.1347	0.1347	0.1348
94	0.1348	0.1348	0.1348	0.1349	0.1349	0.1349	0.1350	0.1350	0.1350	0.1351

(continued)

TABLE A.9 (Continued)
Shunt Capacitive Reactance Spacing Factor X'_d (MΩ/mi/Conductor) at 60 Hz [1]

Ft	0.0	0.1	0.2	0.3	0.4	0.5	0.6	0.7	0.8	0.9
95	0.1351	0.1351	0.1352	0.1352	0.1352	0.1353	0.1353	0.1353	0.1353	0.1354
96	0.1354	0.1354	0.1355	0.1355	0.1355	0.1356	0.1356	0.1356	0.1357	0.1357
97	0.1357	0.1357	0.1358	0.1358	0.1358	0.1359	0.1359	0.1359	0.1360	0.1360
98	0.1360	0.1361	0.1361	0.1361	0.1361	0.1362	0.1362	0.1362	0.1363	0.1363
99	0.1363	0.1364	0.1364	0.1364	0.1364	0.1365	0.1365	0.1365	0.1366	0.1366
100	0.1366	0.1366	0.1367	0.1367	0.1367	0.1368	0.1368	0.1368	0.1369	0.1369

Conductor Height Above Ground (ft)	$x'_0 (f = 60$ Hz)
10	0.267
15	0.303
20	0.328
25	0.318
30	0.364
40	0.390
50	0.410
60	0.426
70	0.440
80	0.452
90	0.462
100	0.472

Notes:

$$x'_0 = \frac{12.30}{f} \log_{10} 2h$$

where
  h = height above ground
  f = frequency
Fundamental equations:

$$x'_0 = x'_2 = x'_a = x'_d$$
$$x'_0 = x'_a + x'_c = -2x'_d$$

where
  $x'_d = \omega k \ln d$
  d = separation, ft

TABLE A.10
Standard Impedances for Power Transformers 10,000 kVA and Below [4]

Highest-Voltage Winding (BIL kV)	Low-Voltage Winding (BIL kV) (For Intermediate BIL, Use Value for Next Higher BIL Listed)	At kVA Base Equal to 55°C Rating of Largest Capacity Winding Self-Cooled (OA), Self-Cooled Rating Self-Cooled/Forced-Air Cooled (OA/FA) Standard Impedance (%)	
		Ungrounded Neutral Operation	Grounded Neutral Operation
110 and below	45	5.75	
	60, 75, 95, 110	5.5	
150	45	5.75	
	60, 75, 95, 110	5.5	
200	45	6.25	
	60, 75, 95, 110	6.0	
	150	6.5	
250	45	6.75	
	60, 150	6.5	
	200	7.0	
350	200	7.0	
	250	7.5	
450	200	7.5	7.00
	250	8.0	7.50
	350	8.5	8.00
550	200	8.0	7.50
	350	9.0	8.25
	450	10.0	9.25
650	200	8.5	8.00
	350	9.5	8.50
	550	10.5	9.50
750	250	9.0	8.50
	450	10.0	9.50
	650	11.0	10.25

TABLE A.11
Standard Impedance Limits for Power Transformers above 10,000 kVA [4]

Highest-Voltage Winding (BIL kV)	Low-Voltage Winding (BIL kV) (For Intermediate BIL, Use Value for Next Higher BIL Listed)	At kVA Base Equal to 55°C Rating of Largest Capacity Winding							
		Self-Cooled (OA), Self-Cooled Rating of Self-Cooled/Forced-Air Cooled (OA/FA), Self-Cooled Rating of Self-Cooled/Forced-Air Forced-Oil Cooled (OA/FOA) Standard Impedance (%)				Forced-Oil Cooled (FOA and FOW) Standard Impedance (%)			
		Ungrounded Neutral Operation		Grounded Neutral Operation		Ungrounded Neutral Operation		Grounded Neutral Operation	
		Min.	Max.	Min.	Max.	Min.	Max.	Min.	Max.
110 and below	110 and below	5.0	6.25			8.25	10.5		
150	110	5.0	6.25			8.25	10.5		
200	110	5.5	7.0			9.0	12.0		
	150	5.75	7.5			9.75	12.75		
250	150	5.75	7.5			9.5	12.75		
	200	6.25	8.5			10.5	14.25		
350	200	6.25	8.5			10.25	14.25		
	250	6.75	9.5			11.25	15.75		
450	200	6.75	9.5	6.0	8.75	11.25	15.75	10.5	14.5
	250	7.25	10.75	6.75	9.5	12.0	17.25	11.25	16.0
	350	7.75	11.75	7.0	10.25	12.75	18.0	12.0	17.25
550	200	7.25	10.75	605	9.75	12.0	18.0	10.75	16.5
	350	8.25	13.0	7.25	10.75	13.25	21.0	12.0	18.0
	450	8.5	13.5	7.75	11.75	14.0	22.5	12.75	19.5
650	200	7.75	11.75	7.0	10.75	12.75	19.5	11.75	18.0
	350	8.5	13.5	7.75	12.0	14.0	22.5	12.75	19.5
	450	9.25	14.0	8.5	13.5	15.25	24.5	14.0	22.5
750	250	8.0	12.75	7.5	11.5	13.5	21.25	12.5	19.25
	450	9.0	13.75	8.25	13.0	15.0	24.0	13.75	21.5
	650	10.25	15.0	9.25	14.0	16.5	25.0	15.0	24.0

(*continued*)

TABLE A.11 (Continued)
Standard Impedance Limits for Power Transformers above 10,000 kVA [4]

Highest-Voltage Winding (BIL kV)	Low-Voltage Winding (BIL kV) (For Intermediate BIL, Use Value for Next Higher BIL Listed)	At kVA Base Equal to 55°C Rating of Largest Capacity Winding							
		Self-Cooled (OA), Self-cooled Rating of Self-Cooled/Forced-Air Cooled (OA/FA), Self-Cooled Rating of Self-Cooled/Forced-Air Forced-Oil Cooled (OA/FOA) Standard Impedance (%)				Forced-Oil Cooled (FOA and FOW) Standard Impedance (%)			
		Ungrounded Neutral Operation		Grounded Neutral Operation		Ungrounded Neutral Operation		Grounded Neutral Operation	
		Min.	Max.	Min.	Max.	Min.	Max.	Min.	Max.
825	250	8.5	13.5	7.75	12.0	14.25	22.5	13.0	20.0
	450	9.5	14.25	8.75	13.5	15.75	24.0	14.5	22.25
	650	10.75	15.75	9.75	15.0	17.25	26.25	15.75	24.0
900	250			8.25	12.5			13.75	21.0
	450			9.25	14.0			15.25	23.5
	750			10.25	15.0			16.5	25.5
1050	250			8.75	13.5			14.75	22.0
	550			10.0	15.0			16.75	25.0
	825			11.0	16.5			18.25	27.5
1175	250			9.25	14.0			15.5	23.0
	550			10.5	15.75			17.5	25.5
	900			12.0	17.5			19.5	29.0
1300	250			9.75	14.5			16.25	24.0
	550			11.25	17.0			18.75	27.0
	1050			12.5	18.25			20.75	30.5

TABLE A.12
60-Hz Characteristics of Three-Conductor Belted Paper-Insulated Cable [1]

Voltage Class	Insulation Thickness (mil) Conductor	Insulation Thickness (mil) Belt	Circular Mils or AWG (B&S)	Type of Conductor[e]	Weight per 1,000 ft	Diameter or Sector Depth (in.)	Resistance (Ω/Conductor/mi)[a]	GMR of One Conductor (in.)[a]	Positive and Negative Sequence Series Reactance (Ω/Conductor/mi)	Positive and Negative Sequence Shunt Capacitive Reactance (Ω/mi)[a]	GMR—Three Conductors	Zero Sequence Series Resistance (Ω/mi)[d]	Zero Sequence Series Reactance (Ω/mi)[d]	Zero Sequence Shunt Capacitive Reactance (Ω/mi)[a]	Sheath Thickness (mil)	Sheath Resistance (Ω/mi) at 50°C
1 kV	60	35	6	SR	1,500	0.184	2.50	0.067	0.185	6,300	0.184	10.66	0.315	11,600	85	2.69
	60	35	4	SR	1,910	0.232	1.58	0.084	0.175	5,400	0.218	8.39	0.293	10,200	90	2.27
	60	35	2	SR	2,390	0.292	0.987	0.106	0.165	4,700	0.262	6.99	0.273	9,000	90	2.00
	60	35	1	SR	2,820	0.332	0.786	0.126	0.155	4,300	0.295	6.07	0.256	84,700	95	1.76
	60	35	0	SR	3,210	0.373	0.622	0.142	0.152	4,000	0.326	5.54	0.246	7,900	95	1.64
	60	35	00	CS	3,160	0.323	0.495	0.151	0.138	2,800	0.290	5.96	0.250	5,400	95	1.82
	60	35	000	CS	3,650	0.364	0.392	0.171	0.134	2,300	0.320	4.56	0.241	4,500	95	1.69
	60	35	0000	CS	4,390	0.417	0.310	0.391	0.131	2,000	0.355	4.72	0.237	4,000	100	1.47
	60	35	250,000	CS	4,900	0.455	0.263	0.210	0.129	1,800	0.387	4.46	0.224	3,600	100	1.40
	60	35	300,000	CS	5,660	0.497	0.220	0.230	0.128	1,700	0.415	3.97	0.221	3,400	105	1.25
	60	35	350,000	CS	6,310	0.539	0.190	0.249	0.126	1,700	0.415	3.97	0.221	3,400	105	1.18
	60	35	400,000	CS	7,080	0.572	0.166	0.265	0.124	1,500	0.467	3.41	0.214	2,900	110	1.08
	60	35	500,000	CS	8,310	0.642	0.134	0.297	0.123	1,300	0.517	3.11	0.208	2,600	110	0.993
	65	40	600,000	CS	9,800	0.700	0.113	0.327	0.122	1,200	0.567	2.74	0.197	2,400	115	0.877
	65	40	750,000	CS	11,800	0.780	0.091	0.366	0.121	1,100	0.623	2.40	0.194	2,100	120	0.771
3 kV	70	40	6	SR	1,630	0.184	2.50	0.067	0.192	6,700	0.192	9.67	0.322	12,500	90	2.39
	70	40	4	SR	2,030	0.232	1.58	0.084	0.181	5,800	0.227	8.06	0.298	11,200	90	2.16
	70	40	2	SR	2,600	0.292	0.987	0.106	0.171	5,100	0.271	6.39	0.278	9,800	95	1.80
	70	40	1	SR	2,930	0.323	0.786	0.126	0.161	4,700	0.304	5.83	0.263	9,200	95	1.68
	70	40	0	SR	3,440	0.373	0.622	0.142	0.156	44,000	0.335	5.06	0.256	8,600	100	1.48
	70	40	00	CS	3,300	0.323	0.495	0.151	0.142	3,500	0.297	5.69	0.259	6,700	95	1.73
	70	40	000	CS	3,890	0.364	0.392	0.171	0.138	2,700	0.329	5.28	0.246	5,100	95	1.63
	70	40	0000	CS	4,530	0.417	0.310	0.191	0.135	2,400	0.367	4.57	0.237	4,600	100	1.42

(continued)

TABLE A.12 (Continued)
60-Hz Characteristics of Three-Conductor Belted Paper-Insulated Cable [1]

Voltage Class	Insulation Thickness (mil) Conductor	Insulation Thickness (mil) Belt	Circular Mils or AWG (B&S)	Type of Conductor[e]	Weight per 1,000 ft	Diameter or Sector Depth (in.)	Resistance (Ω/Conductor/mi)[a]	GMR of One Conductor (in.)[a]	Positive and Negative Sequence Series Reactance (Ω/Conductor/mi)	Positive and Negative Sequence Shunt Capacitive Reactance (Ω/mi)[a]	GMR—Three Conductors	Zero Sequence Series Resistance (Ω/mi)[d]	Zero Sequence Series Reactance (Ω/mi)[d]	Zero Sequence Shunt Capacitive Reactance (Ω/mi)[a]	Sheath Thickness (mil)	Sheath Resistance (Ω/mi) at 50°C
3 kV	70	40	250,000	CS	5,160	0.455	0.263	0.210	0.132	2,100	0.396	4.07	0.231	4,200	105	1.27
	70	40	300,000	CS	5,810	0.497	0.220	0.230	0.130	1,900	0.424	3.61	0.219	3,700	105	1.14
	70	40	350,000	CS	6,470	0.539	0.190	0.249	0.129	1,800	0.455	3.61	0.219	3,700	105	1.14
	70	40	400,000	CS	7,240	0.572	0.166	0.265	0.128	1,700	0.478	3.32	0.218	3,400	110	1.05
	70	40	500,000	CS	8,660	0.642	0.134	0.297	0.126	1,500	0.527	2.89	0.214	3,000	155	0.918
	75	40	600,000	CS	9,910	0.400	0.113	0.327	0.125	1,400	0.577	2.37	0.204	2,500	120	0.758
	75	40	750,000	CS	11,920	0.780	0.091	0.366	0.123	1,300	0.633	2.37	0.204	2,500	120	0.758
5 kV	105	55	6	SR	2,150	0.184	2.50	0.067	0.215	8,500	0.218	8.14	0.342	15,000	95	1.88
	100	55	4	SR	2,470	0.232	1.58	0.084	0.199	7,600	0.250	6.86	0.317	13,600	95	1.76
	95	50	2	SR	2,900	0.292	0.987	0.106	0.184	6,100	0.291	5.88	0.290	11,300	95	1.63
	90	45	1	SR	3,280	0.322	0.786	0.126	0.171	5,400	0.321	5.23	0.270	10,200	100	1.39
	90	45	0	SR	3,660	0.373	0.622	0.142	0.165	5,000	0.352	4.79	0.259	9,600	100	1.39
	85	45	00	CS	3,480	0.323	0.495	0.151	0.148	3,600	0.312	5.42	0.263	9,300	95	1.64
	85	45	000	CS	4,080	0.364	0.392	0.171	0.143	3,200	0.343	4.74	0.254	6,700	100	1.45
	85	45	0000	CS	4,720	0.417	0.310	0.191	0.141	2,800	0.380	4.33	0.245	8,300	100	1.34
	85	45	250,000	CS	5,370	0.455	0.263	0.210	0.138	2,600	0.410	3.89	0.237	7,800	105	1.21
	85	45	300,000	CS	6,050	0.497	0.220	0.230	0.135	2,400	0.438	3.67	0.231	7,400	105	1.15
	85	45	350,000	CS	6,830	0.539	0.190	0.249	0.133	2,200	0.470	3.31	0.225	7,000	110	1.04
	85	45	400,000	CS	7,480	0.572	0.166	0.265	0.131	2,000	0.493	3.17	0.221	6,700	110	1.00
	85	45	500,000	CS	8,890	0.642	0.134	0.297	0.129	1,800	0.542	2.79	0.216	6,200	115	0.885
	85	45	600,000	CS	10,300	0.700	0.113	0.327	0.128	1,600	0.587	2.51	0.210	5,800	120	0.798
	85	45	750,000	CS	12,340	0.780	0.091	0.366	0.125	1,500	0.643	2.21	0.206	5,400	125	0.707

Appendix A

8 kV	130	65	6	SR	2,450	0.184	2.50	0.067	0.230	9,600	0.236	7.57	0.353	16,300	95	1.69
	125	65	4	SR	2,900	0.232	1.58	0.084	0.212	8,300	0.269	6.08	0.329	14,500	100	1.50
	115	60	2	SR	3,280	0.292	0.987	0.106	0.193	6,800	0.307	5.25	0.302	12,500	100	1.42
	110	55	1	SR	3,560	0.332	0.786	0.126	0.179	6,100	0.338	4.90	0.280	11,400	100	1.37
	110	55	0	SR	4,090	0.373	0.622	0.142	0.174	5,700	0.368	4.31	0.272	10,700	105	1.23
	105	55	00	CS	3,870	0.323	0.495	0.151	0.156	4,300	0.330	4.79	0.273	8,300	100	1.43
	105	55	000	CS	4,390	0.364	0.392	0.171	0.151	3,800	0.362	4.41	0.263	7,400	100	1.34
	105	55	0000	CS	5,150	0.417	0.310	0.191	0.147	3,500	0.399	3.88	0.254	6,600	105	1.19
	105	55	250,000	CS	5,830	0.455	0.263	0.210	0.144	3,200	0.428	3.50	0.246	6,200	110	1.08
	105	55	300,000	CS	6,500	0.497	0.220	0.230	0.141	2,900	0.458	3.31	0.239	5,600	110	1.03
	105	55	350,000	CS	7,160	0.539	0.190	0.249	0.139	2,700	0.489	3.12	0.233	5,200	110	0.978
	105	55	400,000	CS	7,980	0.572	0.166	0.265	0.137	2,500	0.513	2.86	0.230	4,900	115	0.899
	105	55	500,000	CS	9,430	0.642	0.134	0.297	0.135	2,200	0.563	2.53	0.224	4,300	120	0.800
	105	55	600,000	CS	10,680	0.700	0.113	0.327	0.132	2,000	0.606	2.39	0.218	3,900	120	0.758
	105	55	750,000	CS	12,740	0.780	0.091	0.366	0.129	1,800	0.663	2.11	0.211	3,500	125	0.673
15 kV	170	85	2	SR	4,350	0.292	0.987	0.106	0.217	8,600	0.349	4.20	0.323	15,000	110	1.07
	165	80	1	SR	4,640	0.332	0.786	0.126	0.202	7,800	0.381	3.88	0.305	13,800	110	1.03
	160	75	0	SR	4,990	0.373	0.622	0.142	0.193	7,100	0.109	3.62	0.288	12,800	110	1.00
	155	75	00	SR	5,600	0.419	0.495	0.159	0.185	6,500	0.439	3.25	0.280	12,000	115	0.918
	155	75	000	SR	6,230	0.470	0.392	0.178	0.180	6,000	0.476	2.99	0.272	11,300	115	0.867
	155	75	0000	SR	7,180	0.528	0.310	0.200	0.174	5,600	0.520	2.64	0.263	10,600	120	0.778
	155	75	250,000	SR	7,840	0.575	0.263	0.218	0.168	5,300	0.555	2.50	0.256	10,200	120	0.744
	155	75	300,000	CS	7,480	0.497	0.220	0.230	0.155	5,400	0.507	2.79	0.254	7,900	115	0.855
	155	75	350,000	CS	7,340	0.539	0.190	0.249	0.152	5,100	0.536	2.54	0.250	7,200	120	0.784
	155	75	400,000	CS	9,030	0.572	0.166	0.265	0.149	4,900	0.561	2.44	0.245	6,900	120	0.758
	155	75	500,000	CS	10,550	0.642	0.134	0.297	0.145	4,600	0.611	2.26	0.239	6,200	125	0.680
	155	75	600,000	CS	12,030	0.700	0.133	0.327	0.142	4,300	0.656	1.97	0.231	5,700	130	0.620
	155	75	750,000	CS	14,790	0.780	0.091	0.366	0.139	4,000	0.712	1.77	0.266	5,100	135	0.558

a Ac resistance based on 100% conductivity at 65°C including 2% allowance for standing.
b The GMR of sector-shaped conductors is an approximate figure close enough for most practical applications.
c For dielectric constant = 3.7.
d Based on all return current in the sheath; none in ground.
e The following symbols are used to designate the cable types: SR—stranded round; CS—compact sector.

TABLE A.13
60-Hz Characteristics of Three-Conductor Shielded Paper-Insulated Cables [1]

Voltage Class	Insulation Thickness (mil)	Circular Mils or AWG (B&S)	Type of Conductor[e]	Weight per 1,000 ft	Diameter or Sector Depth (in.)	Resistance (Ω/mi)[a]	GMR of One Conductor (in.)[b]	Positive and Negative Sequence — Series Reactance (Ω/mi)	Positive and Negative Sequence — Shunt-Capacitive Reactance (Ω/mi)[c]	GMR—Three Conductors	Zero Sequence — Series Resistance (Ω/mi)[d]	Zero Sequence — Series Reactance (Ω/mi)[d]	Zero Sequence — Shunt-Capacitive Reactance (Ω/mi)[c]	Sheath — Thickness (mil)	Sheath — Resistance (Ω/mi) at 50°C
15 kV	205	4	SR	3,860	0.232	1.58	0.084	0.248	8,200	0.328	5.15	0.325	8,200	105	1.19
	190	2	SR	4,260	0.292	0.987	0.106	0.226	6,700	0.365	4.44	0.298	6,700	105	1.15
	185	1	SR	4,740	0.332	0.786	0.126	0.210	6,000	0.398	3.91	0.285	6,000	110	1.04
	180	0	SR	5,090	0.373	0.622	0.141	0.201	5,400	0.425	3.695	0.275	5,400	110	1.01
	175	00	CS	4,790	0.323	0.495	0.151	0.178	5,200	0.397	3.95	0.268	5,200	105	1.15
	175	000	CS	5,510	0.364	0.392	0.171	0.170	480	0.432	3.48	0.256	4,800	110	1.03
	175	0000	CS	6,180	0.417	0.310	0.191	0.166	4,400	0.468	3.24	0.249	4,400	110	0.975
	175	250,000	CS	6,910	0.455	0.263	0.210	0.158	4,100	0.498	2.95	0.243	4,100	115	0.897
	175	300,000	CS	7,610	0.497	0.220	0.230	0.156	3,800	0.530	2.80	0.237	3,800	115	0.860
	175	350,000	CS	8,480	0.539	0.190	0.249	0.153	3,600	0.561	2.53	0.233	3,600	120	0.783
	175	400,000	CS	9,170	0.572	0.166	0.265	0.151	3,400	0.585	2.45	0.228	3,400	120	0.761
	175	500,000	CS	10,710	0.642	0.134	0.297	0.146	3,100	0.636	2.19	0.222	3,100	125	0.684
	175	600,000	CS	12,230	0.700	0.113	0.327	0.143	2,900	0.681	1.98	0.215	2,900	130	0.623
	175	750,000	CS	14,380	0.780	0.091	0.366	0.139	2,600	0.737	1.78	0.211	2,600	135	0.562

Appendix A

kV		Size	Type												
23 kV	265	2	SR	5,590	0.292	0.987	0.106	0.250	8,300	0.418	3.60	0.317	8,300	115	0.0870
	250	1	SR	5,860	0.332	0.786	0.126	0.232	7,500	0.450	3.26	0.298	7,500	115	0.851
	250	0	SR	6,440	0.373	0.622	0.141	0.222	6,800	0.477	2.99	0.290	6,800	120	0.788
	240	00	CS	6,060	0.323	0.495	0.151	0.196	6,600	0.446	3.16	0.285	6,600	115	0.890
	240	000	CS	6,620	0.364	0.392	0.171	0.188	6,000	0.480	2.95	0.285	6,000	115	0.851
	240	0000	CS	7,480	0.410	0.310	0.191	0.181	5,600	0.515	2.64	0.268	5,600	120	0.775
	240	250,000	CS	8,070	0.447	0.263	0.210	0.177	5,200	0.545	2.50	0.261	5,200	120	0.747
	240	300,000	CS	8,990	0.490	0.220	0.230	0.171	4,900	0.579	2.29	0.252	4,900	125	0.690
	240	350,000	CS	9,720	0.532	0.190	0.249	0.167	4,600	0.610	2.10	0.249	4,600	125	0.665
	240	400,000	CS	10,650	0.566	0.166	0.265	0.165	4,400	0.633	2.03	0.246	4,400	130	0.620
	240	500,000	CS	12,280	0.635	0.134	0.297	0.159	3,900	0.687	1.82	0.237	3,900	135	0.562
	240	600,000	CS	13,610	0.690	0.113	0.327	0.154	3,700	0.730	1.73	0.230	3,700	135	0.540
	240	750,000	CS	15,830	0.767	0.091	0.366	0.151	3,400	0.787	1.56	0.225	3,400	140	0.488
35 kV	355	0	SR	8,520	0.288	0.622	0.141	0.239	9,900	0.523	2.40	0.330	9,900	130	0.594
	345	00	SR	9,180	0.323	0.495	0.159	0.226	9,100	0.548	2.17	0.322	9,100	135	0.559
	345	000	SR	9,900	0.364	0.392	0.178	0.217	8,500	0.585	2.01	0.312	8,500	135	0.538
	345	0000	CS	9,830	0.410	0.310	0.191	0.204	7,200	0.594	2.00	0.290	7,200	135	0.563
	345	250,000	CS	10,470	0.447	0.263	0.210	0.197	6,800	0.628	1.90	0.280	6,800	135	0.545
	345	300,000	CS	11,290	0.490	0.220	0.230	0.191	6,400	0.663	1.80	0.273	6,400	135	0.527
	345	350,000	CS	12,280	0.532	0.190	0.249	0.187	6,000	0.693	1.66	0.270	6,000	140	0.491
	345	400,000	CS	13,030	0.566	0.166	0.265	0.183	5,700	0.721	1.61	0.265	5,700	140	0.480
	345	500,000	CS	14,760	0.635	0.134	0.297	0.177	5,200	0.773	1.46	0.257	5,200	145	0.441
	345	600,000	CS	16,420	0.690	0.113	0.327	0.171	4,900	0.819	1.35	0.248	4,900	150	0.412
	345	750,000	CS	18,860	0.767	0.091	0.366	0.165	4,500	0.879	1.22	0.243	4,500	155	0.377

[a] ac resistance based on 100% conductivity at 65°C including 2% allowance for stranding.
[b] The GMR of sector-shaped conductors is an approximate figure close enough for most practical applications.
[c] For dielectric constant = 3.7.
[d] Based on all return current in the sheath; none in the ground.
[e] The following symbols are used to designate conductor types: SR—stranded round; CS—compact sector.

TABLE A.14
60-Hz Characteristics of Three-Conductor Oil-Filled Paper-Insulated Cables [1]

Voltage Class	Insulation Thickness (mil)	Circular Mils or AWG (B&S)	Type of Conductor[f]	Weight per 1,000 ft	Diameter Sector Depth (in.)[e]	Resistance (Ω/mi)[a]	GMR of One Conductor (in.)[b]	Positive and Negative Seq. Series Reactance (Ω/mi)	Positive and Negative Seq. Shunt Capacitive Reactance (Ω/mi)[b]	GMR—Three Conductors	Zero Sequence Series Resistance (Ω/mi)[d]	Zero Sequence Series Reactance (Ω/mi)[d]	Zero Sequence Shunt Capacitive Reactance (Ω/mi)[c]	Sheath Thickness (mil)	Sheath Resistance (Ω/mi) at 50°C
35 kV	190	00	CS	5,590	0.323	0.495	0.151	0.185	6,030	0.406	3.56	0.265	6,030	115	1.02
		000	CS	6,150	0.364	0.392	0.171	0.178	5,480	0.439	3.30	0.256	5,480	115	0.970
		0000	CS	6,860	0.417	0.310	0.191	0.172	4,840	0.478	3.06	0.243	4,840	115	0.918
		250,000	CS	7,680	0.455	0.263	0.210	0.168	4,570	0.508	2.72	0.238	4,570	125	0.820
		300,000	CS	9,090	0.497	0.220	0.230	0.164	4,200	0.539	2.58	0.232	4,200	125	0.788
		350,000	CS	9,180	0.539	0.190	0.249	0.160	3,900	0.570	2.44	0.227	3,900	125	0.752
		400,000	CS	9,900	0.572	0.166	0.265	0.157	3,690	0.595	2.35	0.223	3,090	125	0.723
		500,000	CS	11,550	0.3642	0.134	0.297	0.153	3,400	0.646	2.04	0.217	3,400	135	0.636
		600,000	CS	12,900	0.700	0.113	0.327	0.150	3,200	0.391	1.94	0.210	3,200	135	0.608
		750,000	CS	15,660	0.780	0.091	0.366	0.148	3,070	0.763	1.73	0.202	3,070	140	0.584

Appendix A

46 kV	225	00	CS	6,360	0.323	0.495	0.151	0.195	6,700	0.436	3.28	0.272	6,700	115	0.928
		000	CS	6,940	0.364	0.392	0.171	0.188	6,100	0.468	2.87	0.265	6,100	125	0.826
		0000	CS	7,660	0.410	0.310	0.191	0.180	5,520	0.503	2.67	0.258	5,520	125	0.788
		250,000	CS	8,280	0.447	0.263	0.210	0.177	5,180	0.533	2.55	0.247	5,180	125	0.761
		300,000	CS	9,690	0.490	0.220	0.230	0.172	4,820	0.566	2.41	0.241	4,820	125	0.729
		350,000	CS	10,100	0.532	0.190	0.249	0.168	4,490	0.596	2.16	0.237	4,490	135	0.658
		400,000	CS	10,820	0.566	0.166	0.265	0.165	4,220	0.623	2.08	0.232	4,220	135	0.639
		500,000	CS	12,220	0.635	0.134	0.297	0.160	3,870	0.672	1.94	0.226	3,870	135	0.639
		600,000	CS	13,930	0.690	0.113	0.327	0.156	3,670	0.178	1.74	0.219	3,670	140	0.542
		750,000	CS	16,040	0.767	0.091	0.366	0.151	3,350	0.773	1.62	0.213	3,350	140	0.510
		1,000,000													
65 kV	315	00	CR	8,240	0.376	0.495	0.147	0.234	8,330	0.532	2.41	0.290	8,330	135	0.639
		000	CS	8,830	0.364	0.392	0.171	0.208	7,560	0.538	2.32	0.284	7,580	135	0.642
		0000	CS	9,660	0.410	0.310	0.191	0.200	6,840	0.375	2.16	0.274	6,840	135	0.618
		250,000	CS	10,330	0.447	0.263	0.210	0.195	6,500	0.607	2.06	0.266	6,500	135	0.597
		300,000	CS	11,540	0.490	0.220	0.230	0.190	6,030	0.640	1.85	0.260	6,030	140	0.543
		350,000	CS	12,230	0.532	0.190	0.249	0.185	5,700	0.672	1.77	0.254	5,700	140	0.527
		400,000	CS	13,040	0.566	0.166	0.265	0.181	5,430	0.700	1.55	0.248	5,430	140	0.513
		500,000	CS	14,880	0.635	0.134	0.297	0.176	5,050	0.750	1.51	0.242	5,050	150	0.460
		600,000	CS	16,320	0.690	0.113	0.327	0.171	4,740	0.797	1.44	0.235	4,740	150	0.442
		750,000	CS	18,980	0.767	0.091	0.366	0.165	4,360	0.854	1.29	0.230	4,360	155	0.399
		1,000,000													

[a] ac resistance based on 100% conductivity at 65°C including 2% allowance for stranding.
[b] GMR of sector-shaped conductors is an approximate figure close enough for most practical applications.
[c] For dielectric constant = 3.5.
[d] Based on all return current in sheath, none in ground.
[e] See Figure 7.
[f] The following symbols are used to designate the cable types: CR—compact round; CS—compact sector.

TABLE A.15
60-Hz Characteristics of Single-Conductor Concentric-Strand Paper-Insulated Cables [1]

Voltage Class	Insulation Thickness (mil)	Circular Mils or AWG (B&S)	Weight per 1,000 ft	Diameter of Conductor (in.)	GMR of One Conductor (in.)	x_a Reactance at 12 in. (Ω/Phase/mi)	x_b Reactance of Sheath (Ω/mi)	r_c Resistance of One Conductor (Ω/Phase/mi)[b]	r_s Resistance of Sheath (Ω/Phase/mi) at 50°C	Shunt Capacitive Reactance (Ω/Phase/mi)[c]	Lead Sheath Thickness (mil)
1 kV	60	6	560	0.184	0.067	0.628	0.489	2.50	6.20	4,040	75
	60	4	670	0.232	0.084	0.602	0.475	1.58	5.56	3,360	75
	60	2	880	0.292	0.106	0.573	0.458	0.987	4.55	2,760	80
	60	1	990	0.332	0.126	0.552	0.450	0.786	4.25	2,490	80
	60	0	1,110	0.373	0.141	0.539	0.442	0.622	3.61	2,250	80
	60	00	1,270	0.418	0.159	0.524	0.434	0.495	3.34	2,040	80
	60	000	1,510	0.470	0.178	0.512	0.425	0.392	3.23	1,840	85
	60	0000	1,740	0.528	0.200	0.496	0.414	0.310	2.98	1,650	85
	60	250,000	1,930	0.575	0.221	0.484	0.408	0.263	2.81	1,530	85
	60	350,000	2,490	0.581	0.262	0.464	0.392	0.190	2.31	1,300	90
	60	500,000	3,180	0.814	0.313	0.442	0.378	0.134	2.06	1,090	90
	60	750,000	4,380	0.998	0.385	0.417	0.358	0.091	1.65	885	95
	60	1,000,000	5,560	1.152	0.445	0.400	0.344	0.070	1.40	800	100
	60	1,500,000	8,000	1.412	0.543	0.374	0.319	0.050	1.05	645	110
	60	2,000,000	10,190	1.632	0.633	0.356	0.305	0.041	0.894	555	115
3 kV	75	6	600	0.184	0.067	0.628	0.481	2.50	5.80	4,810	75
	75	4	720	0.232	0.084	0.602	0.467	1.58	5.23	4,020	75
	75	2	930	0.292	0.106	0.573	0.453	0.987	4.31	3,300	80
	75	1	1,040	0.332	0.126	0.552	0.445	0.786	4.03	2,990	80
	75	0	1,170	0.373	0.141	0.539	0.436	0.622	3.79	2,670	80
	75	00	1,320	0.418	0.159	0.524	0.428	0.495	3.52	2,450	80
	75	000	1,570	0.470	0.178	0.512	0.420	0.392	3.10	2,210	85
	75	0000	1,800	0.528	0.200	0.496	0.412	0.310	2.87	2,010	85
	75	250,000	1,990	0.575	0.221	0.484	0.403	0.263	2.70	1,860	85
	75	350,000	2,550	0.681	0.262	0.464	0.389	0.190	2.27	1,610	85
	75	500,000	3,340	0.814	0.313	0.442	0.375	0.134	1.89	1,340	90

Appendix A 571

5 kV	75	750,000	4,570	0.998	0.385	0.417	0.352	0.091	1.53	1,060	95
	75	1,000,000	5,640	1.152	0.445	0.400	0.341	0.070	1.37	980	100
	75	1,500,000	8,090	1.412	0.543	0.374	0.316	0.050	1.02	805	110
	75	2,000,000	10,300	1.632	0.633	0.356	0.302	0.041	0.877	685	115
	120	6	740	0.184	0.067	0.628	0.456	2.50	4.47	6,700	80
	115	4	890	0.232	0.084	0.573	0.447	1.58	4.17	5,540	80
	110	2	1,040	0.292	0.106	0.573	0.439	0.987	3.85	4,520	80
	110	1	1,160	0.332	0.126	0.552	0.431	0.786	3.62	4,100	80
	105	0	1,270	0.373	0.141	0.539	0.425	0.622	3.47	3,600	80
	100	00	1,520	0.418	0.159	0.524	0.420	0.495	3.09	3,140	85
	100	000	1,710	0.470	0.178	0.512	0.412	0.392	2.91	2,860	85
	95	0000	1,870	0.525	0.200	0.496	0.406	0.310	2.74	2,480	85
	90	250,000	2,080	0.575	0.221	0.484	0.400	0.263	2.62	2,180	85
	90	350,000	2,620	0.681	0.262	0.464	0.386	0.190	2.20	1,890	90
	90	500,000	3,410	0.814	0.313	0.442	0.369	0.134	1.85	1,610	95
	90	750,000	4,650	0.998	0.385	0.417	0.350	0.091	1.49	1,350	100
	90	1,000,000	5,850	1.152	0.445	0.400	0.339	0.070	1.27	1,140	105
	90	1,500,000	8,160	1.412	0.543	0.374	0.316	0.050	1.02	950	110
	90	2,000,000	10,370	1.632	0.663	0.356	0.302	0.041	0.870	820	115
8 kV	150	6	890	0.184	0.067	0.628	0.431	2.50	3.62	7,780	80
	150	4	1,010	0.232	0.084	0.602	0.425	1.58	3.62	6,660	85
	140	2	1,150	0.292	0.106	0.573	0.417	0.987	3.06	5,400	85
	140	1	1,330	0.332	0.126	0.552	0.411	0.786	2.91	4,920	85
	135	0	1,450	0.373	0.141	0.539	0.408	0.622	2.83	4,390	85
	130	00	1,590	0.418	0.159	0.524	0.403	0.495	2.70	3,890	85
	125	000	1,760	0.470	0.178	0.512	0.397	0.392	2.59	3,440	85
	120	0000	1,980	0.528	0.200	0.496	0.389	0.310	2.29	3,020	90
	120	250,000	2,250	0.575	0.221	0.484	0.383	0.263	2.18	2,790	90
	115	350,000	2,730	0.681	0.262	0.464	0.375	0.190	1.90	2,350	95
	115	500,000	3,530	0.814	0.313	0.442	0.361	0.134	1.69	2,010	95
	115	750,000	4,790	0.998	0.385	0.417	0.341	0.091	1.39	1,670	100
	115	1,000,000	6,000	1.152	0.415	0.400	0.330	0.070	1.25	1,470	105
	115	1,500,000	8,250	1.412	0.543	0.374	0.310	0.050	0.975	1,210	110
	115	2,000,000	10,480	1.632	0.663	0.356	0.297	0.041	0.797	1,055	120

(continued)

TABLE A.15 (Continued)
60-Hz Characteristics of Single-Conductor Concentric-Strand Paper-Insulated Cables [1]

Voltage Class	Insulation Thickness (mil)	Circular Mils or AWG (B&S)	Weight per 1,000 ft	Diameter of Conductor (in.)	GMR of One Conductor (in.)	x_a Reactance at 12 in. (Ω/Phase/mi)	x_b Reactance of Sheath (Ω/mi)	r_c Resistance of One Conductor (Ω/Phase/mi)[b]	r_s Resistance of Sheath (Ω/Phase/mi) at 50°C	Shunt Capacitive Reactance (Ω/Phase/mi)[c]	Lead Sheath Thickness (mil)
15 kV	220	4	1,340	0.232	0.084	0.602	0.412	1.58	2.91	8,560	85
	215	2	1,500	0.292	0.106	0.573	0.408	0.987	2.74	7,270	85
	210	1	1,610	0.332	0.126	0.552	0.400	0.786	2.64	6,580	85
	200	0	1,710	0.373	0.141	0.539	0.397	0.622	2.59	5,880	85
	195	00	1,940	0.418	0.159	0.524	0.391	0.495	2.32	5,290	90
	185	000	2,100	0.470	0.178	0.512	0.386	0.392	2.24	4,680	90
	180	0000	2,300	0.528	0.200	0.496	0.380	0.310	2.14	4,200	90
	175	250,000	2,500	0.575	0.221	0.484	0.377	0.263	2.06	3,820	90
	175	350,000	3,110	0.681	0.262	0.464	0.366	0.190	1.98	3,340	95
	175	500,000	3,940	0.814	0.313	0.442	0.352	0.134	1.51	2,870	100
	175	750,000	5,240	0.998	0.385	0.417	0.336	0.091	1.26	2,420	105
	175	1,000,000	6,350	1.152	0.445	0.400	0.325	0.070	1.15	2,130	105
	175	1,500,000	8,810	1.412	0.546	0.374	0.305	0.050	0.90	1,790	115
	175	2,000,000	11,080	1.632	0.633	0.356	0.294	0.041	0.772	1,570	120
23 kV	295	2	1,920	0.292	0.106	0.573	0.383	0.987	2.16	8,890	90
	285	1	2,010	0.332	0.126	0.552	0.380	0.786	2.12	8,050	90
	275	0	2,120	0.373	0.141	0.539	0.377	0.622	2.08	7,300	90
	265	00	2,250	0.418	0.159	0.524	0.375	0.495	2.02	6,580	90
	260	000	2,530	0.470	0.178	0.512	0.370	0.392	1.85	6,000	95
	250	0000	2,740	0.528	0.200	0.496	0.366	0.310	1.78	5,350	95
	245	250,000	2,930	0.575	0.221	0.484	0.361	0.263	1.72	4,950	95
	240	350,000	3,550	0.681	0.262	0.464	0.352	0.190	1.51	4,310	100
	240	500,000	4,300	0.814	0.313	0.442	0.341	0.134	1.38	3,720	100
	240	750,000	5,630	0.998	0.385	0.417	0.325	0.091	1.15	3,170	105
	240	1,000,000	6,910	1.152	0.445	0.400	0.313	0.070	1.01	2,800	110
	240	1,500,000	9,460	1.412	0.546	0.374	0.296	0.050	0.806	2,350	120
	240	2,000,000	11,790	1.632	0.633	0.356	0.285	0.041	0.697	2,070	125

Appendix A

35 kV	395	0	2,900	0.373	0.141	0.539	0.352	0.622	1.51	9,150	100
	385	00	3,040	0.418	0.159	0.524	0.350	0.495	1.48	8,420	100
	370	000	3,190	0.470	0.178	0.512	0.347	0.392	1.46	7,620	100
	355	0000	3,380	0.528	0.200	0.496	0.344	0.310	1.43	6,870	100
	350	250,000	3,590	0.575	0.221	0.484	0.342	0.263	1.39	6,410	100
	345	350,000	4,230	0.681	0.262	0.464	0.366	0.190	1.24	5,640	105
	345	500,000	5,040	0.814	0.313	0.442	0.352	0.134	1.15	4,940	105
	345	750,000	6,430	0.998	0.385	0.417	0.311	0.091	0.975	4,250	110
	345	1,000,000	7,780	1.152	0.445	0.400	0.302	0.070	0.866	3,780	115
	345	1,500,000	10,420	1.412	0.546	0.374	0.285	0.050	0.700	3,210	125
	345	2,000,000	12,830	1.632	0.633	0.356	0.274	0.041	0.611	2,830	130
46 kV	475	000	3,910	0.470	0.178	0.512	0.331	0.392	1.20	8,890	105
	460	0000	4,080	0.528	0.200	0.496	0.329	0.310	1.19	8,100	105
	450	250,000	4,290	0.575	0.221	0.484	0.326	0.263	1.16	7,570	105
	445	350,000	4,990	0.681	0.262	0.464	0.319	0.190	1.05	6,720	110
	4456	500,000	5,820	0.814	0.313	0.442	0.310	0.134	0.930	5,950	115
	445	750,000	7,450	0.998	0.385	0.417	0.298	0.091	0.807	5,130	120
	445	1,000,000	8,680	1.152	0.445	0.400	0.290	0.070	0.752	4,610	120
	445	1,500,000	11,420	1.412	0.546	0.374	0.275	0.050	0.615	3,930	130
	445	2,000,000	13,910	1.632	0.633	0.356	0.264	0.041	0.543	3,520	135
69 kV	650	350,000	6,720	0.681	0.262	0.464	0.292	0.190	0.773	8,590	120
	650	500,000	7,810	0.814	0.313	0.442	0.284	0.134	0.695	7,680	125
	650	750,000	9,420	0.998	0.385	0.417	0.275	0.091	0.615	6,780	130
	650	1,000,000	10,940	1.152	0.445	0.400	0.267	0.070	0.557	6,060	135
	650	1,500,000	13,680	1.412	0.546	0.374	0.256	0.050	0.488	5,250	140
	650	2,000,000	16,320	1.632	0.633	0.356	0.246	0.041	0.437	4,710	145

TABLE A.16
60-Hz Characteristics of Single-Conductor Oil-Filled (Hollow-Core) Paper-Insulated Cables [1]

Inside Diameter of Spring Core: 0.5 in.

Voltage Class	Insulation Thickness (mil)	Circular Mils or AWG (B&S)	Weight per 1,000 ft	Diameter of conductor (in.)	GMR of One Conductor (in.)	x_a Reactance at 12 in. (Ω/Phase/mi)	x_b Reactance Sheath (Ω/Phase/mi)	r_c Resistance of One Conductor (Ω/Phase/mi)[a]	r_s Resistance of Sheath (Ω/Phase/mi) at 50°C	Shunt Capacitive Reactance (Ω/Phase/mi)[b]	Load Sheath Thickness (mil)
69 kV	315	00	3,980	0.736	0.345	0.431	0.333	0.495	1.182	5,240	110
		000	4,090	0.768	0.356	0.427	0.331	0.392	1.157	5,070	110
		0000	4,320	0.807	0.373	0.421	0.328	0.310	1.130	4,900	110
		250,000	4,650	0.837	0.381	0.418	0.325	0.263	1.057	4,790	115
		350,000	5,180	0.918	0.408	0.410	0.320	0.188	1.009	4,470	115
		500,000	6,100	1.028	0.448	0.399	0.312	0.133	0.905	4,070	120
		750,000	7,310	1.180	0.05	0.381	0.302	0.089	0.838	3,620	120
		1,000,000	8,630	1.310	0.550	0.374	0.294	0.068	0.752	3,380	125
		1,500,000	11,050	1.547	0.639	0.356	0.281	0.048	0.649	2,920	130
		200,000	13,750	1.760	0.716	0.342	0.270	0.039	0.550	2,570	140
115 kV	480	0000	5,720	0.807	0.373	0.421	0.303	0.310	0.805	6,650	120
		250,000	5,930	0.837	0.381	0.418	0.303	0.263	0.793	6,500	120
		350,000	6,390	0.918	0.448	0.410	0.298	0.188	0.730	6,090	125
		500,000	7,480	1.028	0.448	0.399	0.291	0.133	0.692	5,600	125
		750,000	8,950	1.180	0.505	0.384	0.283	0.089	0.625	5,040	130
		1,000,000	10,350	1.310	0.550	0.374	0.276	0.068	0.568	4,700	135
		1,500,000	12,869	1.547	0.639	0.356	0.265	0.048	0.500	4,110	140
		2,000,000	15,530	1.760	0.716	0.342	0.255	0.039	0.447	3,710	145

138 kV	560	0000	6,480	0.807	0.373	0.421	0.295	0.310	0.758	7,410	125
		250,000	6,700	0.837	0.381	0.418	0.293	0.263	0.746	7,240	125
		350,000	7,460	0.918	0.408	0.410	0.288	0.188	0.690	6,820	130
		500,000	8,310	1.028	0.448	0.399	0.282	0.133	0.658	6,260	130
		750,000	9,800	1.180	0.505	0.384	0.274	0.089	0.592	5,680	135
		1,000,000	11,270	1.310	0.550	0.374	0.268	0.068	0.541	5,240	140
		1,500,000	13,720	1.547	0.639	0.356	0.257	0.048	0.477	4,670	145
		2,000,000	16,080	1.760	0.716	0.342	0.248	0.039	0.427	4,170	150
161 kV	650	250,000	7,600	0.837	0.381	0.418	0.283	0.263	0.660	7,980	130
		350,000	8,390	0.918	0.408	0.410	0.279	0.188	0.611	7,520	135
		500,000	9,270	1.028	0.448	0.399	0.273	0.133	0.585	6,980	135
		750,000	10,840	1.180	0.505	0.384	0.266	0.089	0.532	6,320	140
		1,000,000	12,340	1.310	0.550	0.374	0.259	0.068	0.483	5,880	145
		1,500,000	15,090	1.547	0.639	0.356	0.246	0.048	0.433	5,190	150
		2,000,000	18,000	1.760	0.716	0.342	0.241	0.039	0.391	4,710	155

(continued)

TABLE A.16 (Continued)
60-Hz Characteristics of Single-Conductor Oil-Filled (Hollow-Core) Paper-Insulated Cables [1]

Inside Diameter of Spring Core: 0.69 in.

Voltage Class	Insulation Thickness (mil)	Circular Mils or AWG (B&S)	Weight per 1,000 ft	Diameter of Conductor (in.)	GMR of One Conductor (in.)[b]	x_a Reactance at 12 in. (Ω/Phase/mi)	x_b Reactance Sheath (Ω/Phase/mi)	r_c Resistance of One Conductor (Ω/Phase/mi)[a]	r_s Resistance of Sheath (Ω/Phase/mi) at 50°C	Shunt Capacitive Reactance (Ω/Phase/mi)[b]	Load Sheath Thickness (mil)
69 kV	315	000	4,860	0.924	0.439	0.399	0.320	0.392	1.007	4,450	115
		0000	5,090	0.956	0.450	0.398	0.317	0.310	0.985	4,350	115
		250,000	5,290	0.983	0.460	0.396	0.315	0.263	0.975	4,230	115
		350,000	5,950	1.050	0.483	0.390	0.310	0.188	0.897	40,000	120
		500,000	6,700	1.145	0.516	0.382	0.304	0.188	0.897	4,000	120
		750,000	8,080	1.286	0.550	0.374	0.295	0.089	0.759	3,410	125
		1,000,000	9,440	1.416	0.612	0.360	0.288	0.057	0.688	3,140	130
		1,500,000	11,970	1.635	0.692	0.346	0.278	0.047	0.601	2,750	135
		2,000,000	14,450	1.835	0.763	0.334	0.266	0.038	0.533	2,510	140
138 kV	560	0000	6,590	0.956	0.450	0.398	0.295	0.310	0.760	5,950	125
		250,000	6,800	0.983	0.460	0.398	0.294	0.263	0.752	5,790	125
		350,000	7,340	1.050	0.483	0.390	0.290	0.188	0.729	5,540	125
		500,000	8,320	1.145	0.516	0.382	0.284	0.132	0.669	5,150	130
		750,000	9,790	1.286	0.550	0.374	0.277	0.089	0.607	4,770	135
		1,000,000	11,080	1.416	0.612	0.360	0.270	0.067	0.573	4,430	135
		1,500,000	11,3900	1.635	0.692	0.346	0.260	0.047	0.490	3,920	145
		2,000,000	16,610	1.835	0.763	0.334	0.251	0.038	0.440	3,580	150

Appendix A

138 kV	560	0000	7,390	0.956	0.450	0.398	0.786	0.310	0.678	6,500	130
		250,000	7,610	0.983	0.450	0.396	0.285	0.263	0.669	6,480	130
		350,000	8,170	1.050	0.483	0.390	0.281	0.188	0.649	6,180	130
		500,000	9,180	1.145	0.516	0.382	0.276	0.132	0.601	5,790	135
		750,000	10,660	1.286	0.550	0.374	0.269	0.089	0.545	5,320	140
		1,000,000	12,010	1.416	0.612	0.369	9.263	0.067	0.519	4,940	140
		1,500,000	14,450	1.635	0.692	0.346	0.253	0.047	0.462	4,460	145
		2,000,000	16,820	1.835	0.763	0.334	0.245	0.038	0.404	4,060	155
161 kV	650	250,000	8,560	0.983	0.460	0.369	0.276	0.263	0.596	7,210	135
		350,000	9,140	1.050	0.483	0.390	0.272	0.188	0.580	6,860	135
		5,000,000	10,280	1.145	0.156	0.382	0.267	0.132	0.537	6,430	140
		750,000	11,770	1.286	0.550	0.374	0.261	0.089	0.492	5,980	145
		1,000,000	13,110	1.416	0.612	0.360	0.255	0.067	0.469	5,540	145
		1,500,000	15,840	1.635	0.692	0.346	0.246	0.047	0.421	4,980	150
		2,000,000	18,840	1.835	0.763	0.334	0.238	0.038	0.369	4,600	160
230 kV	925	750,000	15,360	1.286	0.550	0.374	0.238	0.069	0.369	7,610	180
		1,000,000	16,790	1.416	0.612	0.360	0.233	0.067	0.355	7,140	160
		2,000,000	22,990	1.835	0.763	0.334	0.219	0.038	0.315	5,960	170

[a] ac resistance based on 100% conductivity at 65°C including 2% allowance for stranding. Above values were calculated from *A Set of Curves for Skin Effect in Isolated Tubular Conductors*, by A. W. Ewan, G. E. Review, Vol. 33, April 1930.
[b] For dielectric constant = 3.5.
[c] Calculate for circular tube as given in *Symmetrical Components* by Wagner & Evans, Ch. VII, p. 138.

TABLE A.17
Current-Carrying Capacity of Three-Conductor Belted Paper-Insulated Cables [1]

No. of Equally Loaded Cables in Duct Bank

4500 V — Copper Temperature 85°C

Amperes per Conductor[a]

Conductor Size AWG of 1000 CM	Conductor Type[a]	One				Three				Six				Nine				Twelve			
		30	50	75	100	30	50	75	100	30	50	75	100	30	50	75	100	30	50	75	100
6	S	82	80	78	75	81	78	73	68	79	74	68	63	78	72	65	58	79	69	61	54
4	SR	109	106	103	98	108	102	96	89	104	97	89	81	102	94	84	74	100	90	79	69
2	SR	143	139	134	128	139	133	124	115	136	127	115	104	133	121	108	95	130	117	101	89
1	SR	164	161	153	146	159	152	141	130	156	145	130	118	152	138	122	108	148	133	115	100
0	CS	189	184	177	168	184	175	162	149	180	166	149	134	175	159	140	122	170	152	130	114
00	CS	218	211	203	192	211	201	185	170	208	190	170	152	201	181	158	138	195	173	148	128
000	CS	250	242	232	219	242	229	211	193	237	217	193	172	229	206	179	156	223	197	167	145
0000	CS	286	276	264	249	276	260	240	218	270	246	218	194	261	234	202	176	254	223	189	163
250	CS	316	305	291	273	305	288	263	239	297	271	239	212	288	258	221	192	279	244	206	177
300	CS	354	340	324	304	340	321	292	264	332	301	264	234	321	285	245	211	310	271	227	195
350	CS	392	376	357	334	375	353	320	288	366	330	288	255	351	311	266	229	341	296	248	211
400	CS	424	406	385	359	406	380	344	309	395	355	309	272	380	334	285	244	367	317	264	224
500	CS	487	465	439	408	465	433	390	348	451	403	348	305	433	378	320	273	417	357	296	251
600	CS	544	517	487	450	517	480	430	383	501	444	383	334	480	416	350	298	462	393	323	273
750	CS	618	581	550	505	585	541	482	427	566	500	427	371	541	466	390	331	519	439	359	302

(1.07 at 10°C, 0.92 at 30°C, 0.83 at 40°C, 0.73 at 50°C)[S]

Appendix A

		7500 V											Copper Temperature 83°C			
6	S	81	80	77	74	79	76	67	72	78	74	67	62	77	71	64	57	69	75	60	53
4	SR	107	105	101	97	104	100	87	94	103	96	87	79	100	92	82	73	89	98	77	68
2	SR	140	137	132	126	136	131	113	122	134	125	113	102	130	119	105	93	114	127	99	87
1	SR	161	156	150	143	156	149	128	138	153	142	128	115	149	136	120	105	130	145	112	98
0	CS	186	180	171	165	180	172	146	156	177	163	148	131	172	155	136	120	149	167	128	111
00	CS	214	206	198	188	206	196	166	181	202	186	166	148	196	177	155	135	169	191	145	125
000	CS	243	236	226	214	236	224	188	206	230	211	188	168	223	200	174	152	192	217	163	141
0000	CS	280	270	258	243	270	255	214	235	264	241	213	190	255	229	198	172	218	247	184	159
250	CS	311	300	287	269	300	283	235	259	293	266	235	208	282	262	217	188	240	273	202	174
300	CS	349	336	320	300	335	316	260	288	326	296	259	230	315	279	240	207	265	304	223	190
350	CS	385	369	351	328	369	346	283	315	359	323	282	249	345	305	261	224	289	333	242	206
400	CS	417	399	378	353	398	373	303	338	388	348	303	267	371	317	279	239	309	360	257	220
500	CS	476	454	429	399	454	423	341	381	440	392	340	298	422	369	312	267	348	406	288	245
600	CS	534	508	479	443	507	471	376	422	491	436	375	327	469	408	343	291	384	451	315	267
750	CS	607	576	540	497	575	532	418	473	555	489	418	363	529	455	381	323	428	507	350	295
		(1.08 at 10°C, 0.92 at 30°C, 0.83 at 40°C, 0.72 at 50°C)[S]				(1.08 at 10°C, 0.92 at 30°C, 0.83 at 40°C, 0.72 at 50°C)[S]				(1.08 at 10°C, 0.92 at 30°C, 0.83 at 40°C, 0.72 at 50°C)[S]				(1.08 at 10°C, 0.92 at 30°C, 0.83 at 40°C, 0.72 at 50°C)[S]							
		1500 V												Copper Temperature 75°C			
6	S	78	77	74	71	76	74	64	69	75	70	64	59	73	68	61	54	65	72	57	50
4	SR	102	99	96	92	98	95	83	89	97	91	83	75	95	87	78	69	85	93	73	64
2	SR	132	129	125	119	129	123	106	115	126	117	106	96	123	112	99	88	108	120	93	82
1	SR	151	147	142	135	146	140	120	131	144	133	120	109	140	128	112	99	122	136	107	92
0	CS	175	170	163	155	169	161	138	150	166	153	137	123	161	146	128	112	139	156	120	104
00	CS	200	194	187	177	194	184	156	170	189	175	156	139	183	166	145	127	158	178	135	117
000	CS	230	223	214	202	222	211	178	195	217	199	177	158	210	189	165	143	180	203	153	132
0000	CS	266	257	245	232	253	242	202	222	249	228	201	179	240	215	187	158	205	233	173	149

(continued)

TABLE A.17 (Continued)
Current-Carrying Capacity of Three-Conductor Belted Paper-Insulated Cables [1]

No. of Equally Loaded Cables in Duct Bank

Per Cent Load Factor

Amperes per Conductor[a]

Conductor Size AWG of 1000 CM	Conductor Type[a]	One				Three				Six				Nine				Twelve			
		30	50	75	100	30	50	75	100	30	50	75	100	30	50	75	100	30	50	75	100
250	CS	295	284	271	255	281	268	245	221	276	251	220	196	266	239	204	177	257	225	189	163
300	CS	330	317	301	283	316	297	271	245	307	278	244	215	295	264	225	194	285	248	208	178
350	CS	365	349	332	310	348	327	297	267	339	305	266	235	324	289	245	211	313	271	227	193
400	CS	394	377	357	333	375	352	319	286	365	327	285	251	349	307	262	224	336	290	241	206
500	CS	449	429	406	377	428	399	359	321	414	396	319	280	396	346	293	250	379	326	269	229
600	CS	502	479	450	417	476	443	396	352	459	409	351	306	438	380	319	273	420	358	294	249
750	CS	572	543	510	468	540	499	444	393	520	458	391	341	494	425	356	302	471	399	326	275

(1.09 at 10°C, 0.92 at 30°C, 0.83 at 40°C, 0.67 at 50°C)[5] (1.09 at 10°C, 0.92 at 30°C, 0.83 at 40°C, 0.67 at 50°C)[5] (1.09 at 10°C, 0.92 at 30°C, 0.83 at 40°C, 0.66 at 50°C)[5] (1.09 at 10°C, 0.92 at 30°C, 0.83 at 40°C, 0.66 at 50°C)[5] (1.09 at 10°C, 0.92 at 30°C, 0.83 at 40°C, 0.66 at 50°C)[5]

[a] The following symbols are used here to designate conductor types: S—solid copper; SR—standard round concentric-stranded; CS—compact-sector stranded.

[b] Current ratings are based on the following conditions:
1. Ambient earth temperature: 20°C
2. Sixty-cycle alternating current
3. Ratings include dielectric loss and induced ac losses
4. One cable per duct, all cables equally loaded and in outside ducts only

[c] Multiply tabulated currents by these factors when the earth temperature is other than 20°C.

TABLE A.18
Current-Carrying Capacity of Three-Conductor Shielded Paper-Insulated Cables [1]

No. of Equally Loaded Cables in Duct Bank

Conductor Size AWG or 1000 CM	Conductor Type[a]	One				Three				Six				Nine				Twelve			
		30	50	75	100	30	50	75	100	30	50	75	100	30	50	75	100	30	50	75	100
								Percent Load Factor													
										Amperes per Conductor[a]											
		15,000 V																Copper Temperature 81°C			
6	S	94	91	88	83	91	87	81	75	89	83	74	66	87	78	69	60	84	75	64	56
4	SR	123	120	115	107	119	114	104	95	116	108	95	85	113	102	89	77	109	96	83	72
2	SR	159	154	146	137	153	144	139	121	149	136	12	107	144	129	112	97	139	123	104	90
1	SR	179	174	166	156	172	163	149	136	168	153	136	121	162	145	125	109	158	138	117	100
0	CS	203	195	182	176	196	185	169	154	190	173	154	137	183	164	141	122	178	156	131	112
00	CS	234	224	215	202	225	212	193	175	218	198	174	156	211	187	162	139	203	177	148	127
000	CS	270	258	245	230	258	242	220	198	249	225	198	174	241	212	182	157	232	202	108	144
0000	CS	308	295	281	261	295	276	250	223	285	257	224	196	275	241	205	176	265	227	189	162
250	CS	341	327	310	290	325	305	276	246	315	283	245	215	303	265	224	193	291	250	207	177
300	CS	383	365	344	320	364	339	305	272	351	313	271	236	337	293	246	211	322	276	227	194
350	CS	417	397	375	346	397	369	330	293	383	340	293	255	366	318	267	227	350	301	245	208
400	CS	153	428	403	373	429	396	354	314	413	366	313	273	394	340	285	242	376	320	262	222
500	CS	513	487	450	418	483	446	399	350	467	410	350	303	444	381	318	269	419	358	292	247
600	CS	567	537	501	460	534	491	437	385	513	450	384	330	488	416	346	293	465	390	317	269
750	CS	643	606	562	514	602	551	485	426	576	502	423	365	545	465	383	323	519	432	348	293
		(1.08 at 10°C, 0.9 at 30°C, 0.82 at 40°C 0.71 at 50°C)[S]				(1.08 at 10°C, 0.91 at 30°C, 0.82 at 40°C, 0.71 at 50°C)[S]				(1.08 at 10°C, 0.91 at 30°C, 0.82 at 40°C, 0.71 at 50°C)[S]				(1.08 at 10°C, 0.91 at 30°C, 0.82 at 40°C, 0.71 at 50°C)[S]				(1.08 at 10°C, 0.91 at 30°C, 0.81 at 40°C, 0.70 at 50°C)[S]			
		23,000 V												Copper Temperature 77°C							
2	SR	156	150	143	134	149	141	130	117	145	132	117	105	140	125	107	84	134	119	100	88
1	SR	177	170	162	152	170	160	145	133	164	149	132	117	159	140	121	105	154	133	112	97

(*continued*)

TABLE A.18 (Continued)
Current-Carrying Capacity of Three-Conductor Shielded Paper-Insulated Cables [1]

No. of Equally Loaded Cables in Duct Bank

Conductor Size AWG or 1000 CM	Conductor Type[a]	One				Three				Six				Nine				Twelve			
		30	50	75	100	30	50	75	100	30	50	75	100	30	50	75	100	30	50	75	100
		Percent Load Factor																			
		Amperes per Conductor[a]																			
		23,000 V												Copper Temperature 77°C							
0	CS	200	192	183	172	192	182	166	149	186	169	147	132	178	158	136	118	173	148	125	109
00	CS	227	220	210	197	221	208	189	170	212	193	168	149	202	181	156	134	196	172	144	123
000	CS	262	251	238	223	254	238	216	193	242	220	191	169	230	206	175	150	222	195	162	139
0000	CS	301	289	271	251	291	273	246	219	278	250	215	190	264	233	197	169	255	221	182	157
250	CS	334	315	298	277	321	299	270	239	308	275	236	207	290	258	216	184	279	242	199	170
300	CS	373	349	328	306	354	329	297	263	341	302	259	227	320	283	232	202	309	266	217	186
350	CS	405	379	358	331	384	356	3185	283	369	327	280	243	347	305	255	217	335	285	233	199
400	CS	434	409	386	356	412	379	340	302	396	348	298	260	374	325	273	232	359	303	247	211
500	CS	492	465	436	401	461	427	379	335	443	391	333	288	424	363	302	257	400	336	275	230
600	CS	543	516	484	440	512	470	414	366	489	428	365	313	464	396	329	279	441	367	299	248
750	CS	616	583	541	495	577	528	465	407	550	479	402	347	520	439	364	306	490	408	329	276
		(1.09 at 10°C, 0.90 at 30°C, 0.80 at 40°C, 0.67 at 50°C)[c]				(1.09 at 10°C, 0.90 at 30°C, 0.80 at 40°C, 0.67 at 50°C)[c]				(1.09 at 10°C, 0.90 at 30°C, 0.79 at 40°C, 0.67 at 50°C)[c]				(1.09 at 10°C, 0.90 at 30°C, 0.79 at 40°C, 0.66 at 50°C)[c]				(1.09 at 10°C, 0.90 at 30°C, 0.79 at 40°C, 0.65 at 50°C)[c]			
		34,500 V												Copper Temperature 70°C							
0	CS	193	185	176	165	184	174	158	141	178	161	140	124	171	149	129	111	164	142	119	103
00	CS	219	209	199	187	208	197	178	160	202	182	158	140	194	170	145	126	185	161	134	115
000	CS	250	238	225	211	238	222	202	182	229	206	179	158	220	193	165	141	209	182	152	128

0000	CS	288	275	260	241	273	256	229	205	263	234	203	179	251	219	186	160	238	205	170	144
250	CS	316	302	285	266	301	280	253	224	289	258	222	196	276	240	202	174	262	222	189	157
300	CS	352	335	315	293	334	310	278	246	320	284	244	213	304	264	221	190	288	244	203	171
350	CS	384	364	342	318	363	336	301	267	346	308	264	229	329	285	238	204	311	263	217	184
400	CS	413	392	367	3341	384	360	321	284	372	329	281	244	352	303	254	216	334	282	232	195
500	CS	468	442	414	381	436	402	358	317	418	367	312	271	393	337	281	238	372	313	256	215
600	CS	514	487	455	416	481	440	391	344	459	401	340	294	430	367	304	259	406	340	277	232
750	CS	584	548	510	466	541	496	435	383	515	447	378	324	481	409	3367	284	452	377	304	255

(1.10 at 10°C, 0.89 at 30°C, 0.76 at 40°C, 0.61 at 50°C)[c] (1.10 at 10°C, 0.89 at 30°C, 0.76 at 40°C, 0.60 at 50°C)[c] (1.10 at 10°C, 0.89 at 30°C, 0.76 at 40°C, 0.60 at 50°C)[c] (1.10 at 10°C, 0.88 at 30°C, 0.75 at 40°C, 0.58 at 50°C)[c] (1.10 at 10°C, 0.88 at 30°C, 0.74 at 40°C, 0.56 at 50°C)[c]

[a] The following symbols are used here to designate conductor types: S—solid copper; SR—standard round concentric–standard; CS—compact-sector standard.

[b] Current ratings are based on the following conditions:
1. Ambient earth temperature: 20°C
2. Sixty-cycle alternating current
3. Ratings include dielectric loss, and all include ac losses
4. One cable per duct, all cables equally loaded and in outside ducts only

[c] Multiply tabulated currents by these factors when earth temperature is other than 20°C.

TABLE A.19
Current-Carrying Capacity of Single-Conductor Solid Paper-Insulated Cables

No. of Equally Loaded Cables in Duct Bank

Amperes per Conductor[a]

Copper Temperature 85°C

7500 V

Conductor Size AWG or MGM	Three				Six					Nine					Twelve			
	30	50	75	100	30	50	75	100		30	50	75	100		30	50	75	100
	\multicolumn{4}{c}{Percent Load Factor}														

Conductor Size AWG or MGM	30	50	75	100	30	50	75	100	30	50	75	100	30	50	75	100
6	116	113	109	103	115	110	103	96	113	107	98	90	111	104	94	85
4	154	149	142	135	152	144	134	125	149	140	128	116	147	1336	122	110
2	202	196	186	175	199	189	175	162	196	183	167	151	192	178	159	142
1	234	226	214	201	230	218	201	185	226	210	190	172	222	204	181	162
0	270	262	245	232	266	251	231	212	261	242	219	196	256	234	208	184
00	311	300	283	262	309	290	270	241	303	278	250	224	295	268	236	208
000	356	344	324	300	356	333	303	275	348	319	285	255	340	308	270	236
0000	412	395	371	345	408	380	347	314	398	364	325	290	390	352	307	269
250	456	438	409	379	449	418	379	344	437	400	356	316	427	386	336	294
300	512	491	459	423	499	464	420	380	486	442	394	349	474	428	371	325
400	607	580	540	496	593	548	493	445	576	522	461	407	560	502	434	378
500	692	660	611	561	676	626	560	504	659	597	524	459	641	571	490	427
600	772	735	679	621	757	696	621	557	733	663	579	506	714	632	542	470
700	846	804	741	677	827	758	674	604	802	721	629	548	779	688	587	508
750	881	837	771	702	860	789	700	627	835	750	651	568	810	714	609	526
800	914	866	797	725	892	817	725	648	865	776	674	588	840	740	630	544
1,000	1,037	980	898	816	1,012	922	815	725	980	874	756	657	950	832	705	606
1,250	1,176	1,108	1,012	914	1,145	1,039	914	809	1,104	981	845	730	1,068	941	784	673
1,500	1,300	1,224	1,110	1,000	1,268	1,146	1,000	884	1,220	1,078	922	794	1,178	1,032	855	731

Appendix A

Size																
1,750	1,420	1,332	1,204	1,080	1,382	1,240	1,078	949	1,342	1,166	992	851	1,280	1,103	919	783
2,000	1,546	1,442	1,300	1,162	1,500	1,343	1,162	1,019	1,442	1,260	1,068	914	1,385	1,190	986	839
	(1.07 at 10°C, 0.92 at 30°C, 0.83 at 40°C, 0.73 at 50°C)[c]				(1.07 at 10°C, 0.92 at 30°C, 0.83 at 40°C, 0.73 at 50°C)[c]				(1.07 at 10°C, 0.92 at 30°C, 0.83 at 40°C, 0.73 at 50°C)[c]				(1.07 at 10°C, 0.92 at 30°C, 0.83 at 40°C, 0.73 at 50°C)[c]			

15,000 V   **Copper Temperature 81°C**

Size																
6	113	110	105	100	112	107	100	93	110	104	96	87	108	101	92	83
4	149	145	138	131	147	140	131	117	144	136	125	114	142	132	119	107
2	195	190	180	170	193	183	170	157	189	177	161	146	186	172	154	137
1	226	218	208	195	222	211	195	179	218	204	185	167	214	197	175	157
0	256	248	234	220	252	239	220	203	247	230	209	188	242	223	198	177
00	297	287	271	254	295	278	253	232	287	265	239	214	283	257	226	202
000	344	330	312	290	341	320	293	267	333	306	274	245	327	296	260	230
0000	399	384	361	335	392	367	335	305	383	352	315	280	374	340	298	263
250	440	423	396	367	432	404	367	334	422	387	345	306	412	372	325	286
300	490	470	439	406	481	449	406	369	470	429	382	338	457	413	359	316
350	539	516	481	444	527	491	443	401	514	468	416	367	501	450	391	342
400	586	561	522	480	575	530	478	432	556	506	447	395	542	485	419	366
500	669	639	592	543	655	605	542	488	636	577	507	445	618	551	474	412
600	746	710	656	601	727	668	598	537	705	637	557	488	685	608	521	452
700	810	772	712	652	790	726	647	581	766	691	604	528	744	656	564	488
750	840	797	736	674	821	753	672	602	795	716	625	547	772	684	584	505
800	869	825	762	696	850	780	695	622	823	741	646	565	800	707	604	522
1,000	991	939	864	785	968	882	782	697	933	832	724	631	903	794	675	581
1,250	1,130	1,067	975	864	1,102	1,000	883	784	1,063	941	816	706	1,026	898	759	650
1,500	1,250	1,176	1,072	966	1,220	1,105	972	856	1,175	1,037	892	772	1,133	987	828	707
1,750	1,368	1,282	1,162	1,044	1,330	1,198	1,042	919	1,278	1,124	958	824	1,230	1,063	886	755
2,000	1,464	1,368	1,233	1,106	1,422	1,274	1,105	970	1,360	1,192	1,013	869	1,308	1,125	935	795
	(1.08 at 10°C, 0.92 at 30°C, 0.82 at 40°C, 0.71 at 50°C)[c]				(1.08 at 10°C, 0.92 at 30°C, 0.82 at 40°C, 0.71 at 50°C)[c]				(1.08 at 10°C, 0.92 at 30°C, 0.82 at 40°C, 0.71 at 50°C)[c]				(1.08 at 10°C, 0.92 at 30°C, 0.82 at 40°C, 0.71 at 50°C)[c]			

(continued)

TABLE A.19 (Continued)
Current-Carrying Capacity of Single-Conductor Solid Paper-Insulated Cables

No. of Equally Loaded Cables in Duct Bank

23,000 V

Amperes per Conductor[a]

Copper Temperature 77°C

Conductor Size AWG or MGM	Three						Six						Nine						Twelve				
	30	50	75	100	30	50	75	100	30	50	75	100	30	50	75	100	30	50	75	100
	Percent Load Factor																							
2	186	181	172	162	184	175	162	150	180	169	154	140	178	164	147	132
1	214	207	197	186	211	200	185	171	206	193	176	159	203	187	167	150
0	247	239	227	213	244	230	213	196	239	222	197	182	234	216	192	171
00	283	273	258	242	278	263	243	221	275	253	225	205	267	245	217	193
000	326	314	296	277	320	302	276	252	315	290	259	233	307	280	247	220
0000	376	362	340	317	367	345	315	288	360	332	297	265	351	320	281	250
250	412	396	373	346	405	380	346	316	396	365	326	290	386	351	307	272
300	463	444	416	386	450	422	382	349	438	404	360	319	428	389	340	301
350	508	488	466	422	493	461	418	380	481	442	393	347	468	424	369	326
400	548	525	491	454	536	498	451	409	521	478	423	373	507	458	398	349
500	627	600	559	514	615	570	514	464	597	546	480	423	580	521	450	392
600	695	663	316	566	684	632	568	511	663	603	529	466	645	577	496	431
700	765	729	675	620	744	689	617	554	725	656	574	503	703	627	538	467
750	797	759	702	643	779	717	641	574	754	681	596	527	732	650	538	467
800	826	786	726	665	808	743	663	595	782	706	617	540	759	674	576	500
1,000	946	898	827	752	921	842	747	667	889	797	692	603	860	759	646	580
1,250	1,080	1,020	935	848	1,052	957	845	751	1,014	904	781	676	980	858	725	630
1,500	1,192	1,122	1,025	925	1,162	1,053	926	818	1,118	993	855	736	1,081	940	791	682

Appendix A

Size																
1,750	1,296	1,215	1,106	994	1,258	1,130	991	875	1,206	1,067	911	785	1,162	1,007	843	720
2,000	1,390	1,302	1,180	1,058	1,352	1,213	1,053	928	1,293	1,137	967	831	1,240	1,073	893	760

(1.09 at 10°C, 0.90 at 30°C, 0.80 at 40°C, 0.68 at 50°C)^c (1.09 at 10°C, 0.90 at 30°C, 0.80 at 40°C, 0.68 at 50°C)^c (1.09 at 10°C, 0.90 at 30°C, 0.80 at 40°C, 0.62 at 50°C)^c

34,500 V

Amperes per Conductor — **Copper Temperature 70°C**

Size																
0	227	221	200	209	197	225	213	197	220	205	187	169	215	199	177	158
00	260	251	230	239	224	255	242	224	249	234	211	190	245	226	200	179
000	299	280	273	256	256	295	278	256	288	268	242	217	282	259	230	204
0000	341	330	312	291	291	336	317	291	328	304	274	246	321	293	259	230
250	380	367	345	322	322	374	352	321	364	337	303	270	356	324	286	253
300	422	408	382	355	355	416	390	356	405	374	334	298	395	359	315	278
350	464	446	419	389	389	455	426	388	443	408	364	324	432	392	343	302
400	502	484	451	419	419	491	460	417	478	440	390	347	466	421	368	323
500	575	551	514	476	476	562	524	747	547	500	442	392	532	479	416	364
600	644	616	573	528	528	629	584	526	610	556	491	433	593	532	459	401
700	710	675	626	577	577	690	639	574	669	608	535	470	649	580	500	435
750	736	702	651	598	598	718	664	595	696	631	554	486	675	602	518	450
800	765	730	676	620	620	747	690	617	723	654	574	503	700	624	535	465
1,000	875	832	766	701	701	852	783	698	823	741	646	564	796	706	601	520
1,250	994	941	864	786	786	967	882	782	930	833	722	628	898	790	670	577
1,500	1,095	1,036	949	859	859	1,068	972	856	1,025	914	788	682	988	865	730	626
1,750	1,192	1,123	1,023	925	925	1,156	1,048	919	1,109	984	845	730	1,065	929	780	668
2,000	1,275	1,197	1,088	981	981	1,234	1,115	975	1,182	1,045	893	770	1,135	985	824	704
2,500	1,418	1,324	1,196	1,072	1,072	1,367	1,225	1,064	1,305	1,144	973	834	1,248	1,075	893	760

(1.10 at 10°C, 0.89 at 30°C, 0.76 at 40°C, 0.61 at 50°C)^b (1.10 at 10°C, 0.89 at 30°C, 0.76 at 40°C, 0.61 at 50°C)^b (1.10 at 10°C, 0.89 at 30°C, 0.76 at 40°C, 0.61 at 50°C)^b (1.10 at 10°C, 0.89 at 30°C, 0.76 at 40°C, 0.61 at 50°C)^b

(*continued*)

TABLE A.19 (Continued)
Current-Carrying Capacity of Single-Conductor Solid Paper-Insulated Cables

No. of Equally Loaded Cables in Duct Bank

Percent Load Factor

Amperes per Conductor[a]

Copper Temperature 65°C

Conductor Size AWG or MGM	Three				Six				Nine				Twelve			
	30	50	75	100	30	50	75	100	30	50	75	100	30	50	75	100
	46,000 V															
000	279	270	258	240	274	259	239	221	249	226	204	191	262	241	214
0000	322	312	294	276	317	299	274	251	287	259	232	217	302	276	244
250	352	340	321	300	346	326	299	274	313	282	252	236	329	301	266
300	394	380	358	334	385	364	332	304	349	313	280	260	367	335	295
350	433	417	392	365	425	398	364	331	382	341	304	283	403	366	321
400	469	451	423	393	459	430	391	356	411	367	326	307	433	394	344
500	534	512	482	444	522	487	441	400	464	412	365	339	492	444	386
600	602	577	538	496	589	546	494	447	520	460	406	377	553	497	430
700	663	633	589	542	645	598	538	486	569	502	441	408	605	542	468
750	689	658	611	561	672	622	559	504	590	520	457	422	629	562	485
800	717	683	638	583	698	645	578	522	612	538	472	436	652	582	501
1,000	816	776	718	657	794	731	653	585	691	604	528	487	740	657	562
1,250	927	879	810	738	900	825	732	654	777	675	589	541	834	736	626
1,500	1,020	968	887	805	992	904	799	703	850	735	638	585	914	802	679
1,750	1,110	1,047	959	867	1,074	976	859	762	915	788	682	623	987	862	726
2,000	1,184	1,115	1,016	918	1,144	1,035	909	805	970	833	718	656	1,048	913	766
2,500	1,314	1,232	1,115	1,002	1,265	1,138	994	875	1,062	905	778	708	1,151	996	830

Appendix A

Size	69,000 V								Copper Temperature 60°C							
350	395	382	360	336	387	364	333	305	375	348	312	279	365	332	293	259
400	428	413	389	362	418	393	358	328	405	375	335	300	394	358	315	278
500	489	470	441	409	477	446	406	370	461	425	379	337	447	405	354	312
600	545	524	490	454	532	496	450	409	513	471	419	371	497	448	391	343
700	599	573	536	495	582	543	490	444	561	514	455	403	542	489	425	372
750	623	597	556	514	605	562	508	460	583	533	472	417	653	506	439	384
800	644	617	575	531	626	582	525	475	603	554	487	430	582	523	453	396
1,000	736	702	652	599	713	660	592	533	685	622	547	481	660	589	508	442
1,250	832	792	734	672	806	742	664	595	772	698	610	535	741	659	564	489
1,500	918	872	804	733	886	814	724	647	848	763	664	580	812	718	612	529
1,750	994	942	865	788	957	876	776	692	913	818	711	618	873	770	653	563
2,000	1,066	1,008	924	840	1,020	931	822	732	972	868	750	651	927	814	688	592
2,500	1,163	1,096	1,001	903	1,115	1,013	892	791	1,060	942	811	700	1,007	880	741	635

(1.13 at 10°C, 0.85 at 30°C, 0.67 at 40°C, 0.42 at 50°C)[c] (1.13 at 10°C, 0.85 at 30°C 0.66 at 40°C, 0.40 at 50°C)[c] (1.13 at 10°C, 0.84 at 30°C 0.65 at 40°C, 0.36 at 50°C)[c] (1.14 at 10°C, 0.84 at 30°C 0.64 at 40°C, 0.32 at 50°C)[c]

[a] Current ratings are based on the following conditions:
1. Ambient earth temperature: 20°C
2. Sixty-cycle alternating current
3. Sheaths bonded and grounded at one point only (open circuited sheaths)
4. Standard concentric standard conductors
5. Ratings include dielectric loss and skin effect
6. One cable per duct, all cables equally loaded and in outside ducts only

[b] Multiply tabulated values by these factors when earth temperature is other than 20°C.

TABLE A.20
60-Hz Characteristics of Self-Supporting Rubber-Insulated, Neoprene-Jacketed Aerial Cable [1]

Voltage Class	Conductor Size	Stranding	Insulation Thickness	Shielding	Jacket Thickness	Diameter	Messenger Used with Copper Conductor	Wt. per 1,000 ft Messenger and Copper	Messenger Used with Aluminum Conductor
3-kV Ungrounded Neutral, 5-kV Grounded Neutral	6	7	10/64	No	3/64	0.59	[$] 30% CCS	1,020	[$] 30% CCS
	4	7	10/64	No	3/64	0.67	[$] 30% CCS	1,230	[$] 30% CCS
	2	7	10/64	No	3/64	0.73	[$] 30% CCS	1,530	[$] 30% CCS
	1	19	10/64	No	3/64	0.77	[$] 30% CCS	1,780	[$] 30% CCS
	1/0	19	10/64	No	3/64	0.81	[$] 30% CCS	2,070	[$] 30% CCS
	2/0	19	10/64	No	3/64	0.85	[$] 30% CCS	2,510	[$] 30% CCS
	3/0	19	10/64	No	3/64	0.91	[$] 30% CCS	2,890	[$] 30% CCS
	4/0	19	10/64	No	4/64	0.99	1/2" 30% CCS	3,570	[$] 30% CCS
	250	37	11/64	No	4/64	1.08	1/2" 30% CCS	4,080	[$] 30% CCS
	300	37	11/64	No	4/64	1.13	1/2" 30% CCS	4,620	[$] 30% CCS
	350	37	11/64	No	4/64	1.18	1/2" 30% CCS	5,290	[$] 30% CCS
	400	37	11/64	No	4/64	1.23	1/2" 30% CCS	5,800	[$] 30% CCS
	500	37	11/64	No	4/64	1.32	1/2" 30% CCS	6,860	[$] 30% CCS

5-kV Ungrounded Neutral	6	7	10/64			4/64	0.74	[S] 30% CCS	1,310	[S] 30% CCS
	4	7	10/64			4/64	0.79	[S] 30% CCS	1,540	[S] 30% CCS
	2	7	10/64			5/64	0.88	[S] 30% CCS	1,950	[S] 30% CCS
	1	19	10/64	Yes		5/64	0.92	[S] 30% CCS	2,180	[S] 30% CCS
	1/0	19	10/64	Yes		5/64	0.96	[S] 30% CCS	2,450	[S] 30% CCS
	2/0	19	10/64	Yes		5/64	1.00	[S] 30% CCS	2,910	[S] 30% CCS
	3/0	19	10/64	Yes		5/64	1.06	[S] 30% CCS	3,320	[S] 30% CCS
	4/0	19	10/64	Yes		5/64	1.11	[S] 30% CCS	4,030	[S] 30% CCS
	250	37	11/64	Yes		5/64	1.20	1/2″ 30% CCS	4,570	[S] 30% CCS
	300	37	11/64	Yes		6/64	1.29	1/2″ 30% CCS	5,260	[S] 30% CCS
	350	37	11/64	Yes		6/64	1.34	1/2″ 30% CCS	5,840	[S] 30% CCS
	400	37	11/64	Yes		6/64	1.39	1/2″ 30% CCS	6,380	[S] 30% CCS
	500	37	11/64	Yes		6/64	1.47	1/2″ 30% CCS	7,470	[S] 30% CCS
15-kV Grounded Neutral	6	19	10/64	Yes		6/64	1.05	[S] 30% CCS	2,090	[S] 30% CCS
	4	19	10/64	Yes		6/64	1.10	[S] 30% CCS	2,350	[S] 30% CCS
	2	19	10/64	Yes		6/64	1.16	[S] 30% CCS	2,860	[S] 30% CCS
	1	19	10/64	Yes		6/64	1.20	[S] 30% CCS	3,120	[S] 30% CCS
	1/0	19	10/64	Yes		6/64	1.27	[S] 30% CCS	3,560	[S] 30% CCS
	2/0	19	10/64	Yes		6/64	1.32	1/2″ 30% CCS	4,120	[S] 30% CCS
	3/0	19	10/64	Yes		6/64	1.37	1/2″ 30% CCS	4,580	[S] 30% CCS
	4/0	19	10/64	Yes		6/64	1.43	1/2″ 30% CCS	5,150	[S] 30% CCS
	250	37	10/64	Yes		6/64	1.47	1/2″ 30% CCS	5,590	[S] 30% CCS
	300	37	10/64	Yes		6/64	1.53	1/2″ 30% CCS	6,260	[S] 30% CCS
	350	37	10/64	Yes		6/64	1.59	1/2″ 30% CCS	6,870	[S] 30% CCS
	400	37	10/64	Yes		6/64	1.63	1/2″ 30% CCS	7,450	[S] 30% CCS
	500	37	10/64	Yes		7/64	1.75	1/2″ 30% CCS	8,970	[S] 30% CCS

(*continued*)

TABLE A.20 (Continued)
60-Hz Characteristics of Self-Supporting Rubber-Insulated, Neoprene-Jacketed Aerial Cable [1]

Wt. per 1,000 ft Messenger and Aluminum	Positive Sequence 60~ ac (Ω/mi)				Zero Sequence[c] 60~ ac (Ω/mi)				
	Resistance[a]		Reactance		Resistance[a]		Reactance		Shunt Capacitive[b]
			Series Inductive	Shunt Capacitive[b]			Series Inductive		
	Copper	Aluminum			Copper	Aluminum	Copper	Aluminum	
854	2.52	4.13	0.258	......	3.592	5.082	3.712	3.712	......
956	1.58	2.58	0.246	......	2.632	3.572	3.662	3.662	......
1,100	1.00	1.64	0.229	......	2.025	2.605	3.615	3.615	......
1,250	0.791	1.29	0.211	......	1.815	2.275	3.582	3.582	......
1,390	0.635	1.03	0.207	......	1.644	2.015	3.555	3.555	......
1,530	0.501	0.816	0.200	......	1.622	1.803	3.162	3.526	......
1,690	0.402	0.644	0.194	......	1.517	1.637	3.135	3.499	......
1,900	0.318	0.518	0.191	......	1.401	1.508	2.665	3.459	......
2,160	0.269	0.437	0.189	......	1.351	1.430	2.635	3.429	......
2,500	0.228	0.366	0.184	......	1.308	1.465	2.612	3.042	......
2,780	0.197	0.316	0.180	......	1.277	1.415	2.591	3.021	......
3,040	0.172	0.276	0.176	......	1.252	1.377	2.576	3.006	......
3,650	0.141	0.223	0.172	......	1.219	1.290	2.543	2.543	......
1,140	2.52	4.13	0.292	4,970	......	......	......	......	......
1,270	1.58	2.58	0.272	4,320	......	......	......	......	......
1,520	1.00	1.64	0.257	3,630	......	......	......	......	......
1,640	0.791	1.29	0.241	3,330	......	......	......	......	......

1,770	0.655	1.03	0.233	3,080	.....	.....	.....	.....	.....
1,930	0.501	0.816	0.223	2,830	.....	.....	.....	.....	.....
2,120	0.402	0.644	0.215	2,580	.....	.....	.....	.....	.....
2,350	0.318	0.518	0.207	2,380	.....	.....	.....	.....	.....
2,770	0.269	0.437	0.206	2,380	.....	.....	.....	.....	.....
3,140	0.228	0.366	0.203	2,280	.....	.....	.....	.....	.....
3,380	0.197	0.316	0.199	2,090	.....	.....	.....	.....	.....
3,610	0.172	0.276	0.194	1,890	.....	.....	.....	.....	.....
4,240	0.141	0.223	0.187	1,740	.....	.....	.....	.....	.....
1,920	2.52	4.13	0.326	7,150	3.846	5.346	3.396	3.396	7,150
2,080	1.58	2.58	0.302	6,260	2.901	3.831	3.364	3.364	6,260
2,430	1.00	1.64	0.279	5,460	2.459	3.039	3.851	3.851	5,460
2,580	0.791	1.29	0.268	5,110	2.238	2.701	2.837	2.837	5,110
2,880	0.655	1.03	0.260	4,720	2.052	2.426	2.825	2.825	4,720
3,070	0.501	0.816	0.249	4,370	1.896	2.214	2.251	2.801	4,370
3,510	0.402	0.644	0.241	4,120	1.782	2.008	2.240	2.240	4,120
3,790	0.318	0.518	0.231	3,770	1.681	1.864	2.235	2.235	3,770
3,980	0.269	0.437	0.223	3,570	1.630	1.782	2.227	2.227	3,570
4,330	0.228	0.366	0.217	3,330	1.577	1.701	2.226	2.226	3,330
4,600	0.197	0.316	0.212	3,130	1.536	1.640	2.226	2.226	3,130
4,880	0.172	0.276	0.208	2,980	1.500	1.592	2.216	2.216	2,980
5,560	0.141	0.223	0.204	2,830	1.454	1.524	2.198	2.198	2,830

[a] ac resistance based on 65°C with allowance for stranding, skin effect, and proximity effect.
[b] Dielectric constant assumed 6.0.
[c] Zero-sequence impedance based on return current both in the messenger and in 100 m·Ω earth.

TABLE A.21
Inductive Reactance of ACSR Bundled Conductors at 60 Hz [3]

| | | Strands | | Dia. | GMR | Single | 60-Hz Inductive Reactance x_a in Ω/mi for 1-ft Radius | | | | | | | | | |
| | | | | | | | Two-Conductor Spacing (in.) | | | | | Two-Conductor Spacing (in.) | | | | |
Code	Area (cmil)	Al	St	(in.)	(ft)	Cond.	6	9	12	15	18	6	9	12	15	18
Expanded	3,108,000	62/8	19	2.500	0.0900	0.2922	0.1881	0.1635	0.1461	0.1326	0.1215	0.1535	0.1207	0.0974	0.0793	0.0646
Expanded	2,294,000	66/6	19	2.320	0.0858	0.2980	0.1910	0.1664	0.1490	0.1355	0.1244	0.1554	0.1226	0.0993	0.0813	0.0665
Expanded	1,414,000	58/4	19	1.750	0.0640	0.3336	0.2088	0.1842	0.1668	0.1532	0.1422	0.1673	0.1345	0.1112	0.0931	0.0784
Expanded	1,275,000	50/4	19	1.600	0.0578	0.3459	0.2150	0.1904	0.1730	0.1594	0.1484	0.1714	0.1386	0.1153	0.0973	0.0825
Kiwi	2,167,000	72	7	1.737	0.0571	0.3474	0.2158	0.1912	0.1737	0.1602	0.1491	0.1719	0.1391	0.1158	0.0977	0.0830
Bluebird	2,156,000	84	19	1.762	0.0588	0.3438	0.2140	0.1894	0.1719	0.1584	0.1473	0.1707	0.1379	0.1146	0.0966	0.0818
Chukar	1,780,000	84	19	1.602	0.0536	0.3551	0.2196	0.1950	0.1775	0.1640	0.1529	0.1744	0.1416	0.1184	0.1003	0.0856
Falcon	1,590,000	54	19	1.545	0.0523	0.3580	0.2211	0.1965	0.1790	0.1655	0.1544	0.1754	0.1426	0.1193	0.1013	0.0866
Lapwing	1,590,000	45	7	1.502	0.498	0.3640	0.2241	0.1995	0.1820	0.1685	0.1574	0.1774	0.1446	0.1213	0.1033	0.0885
Parrot	1,510,500	54	19	1.506	0.0506	0.3621	0.2231	0.1985	0.1810	0.1675	0.1564	0.1768	0.1440	0.1207	0.1026	0.0879
Nuthatch	1,510,500	45	7	1.466	0.486	0.3670	0.2255	0.2009	0.1835	0.1699	0.1589	0.1784	0.1456	0.1223	0.1043	0.0895
Plover	1,431,000	54	19	1.465	0.0494	0.3650	0.2245	0.1999	0.1825	0.1689	0.1579	0.1777	0.1449	0.1217	0.1036	0.0889
Bobolink	1,431,000	45	7	1.427	0.0470	0.3710	0.2276	0.2030	0.1855	0.1720	0.1609	0.1797	0.1469	0.1237	0.1056	0.0909
Martin	1,351,500	54	19	1.424	0.0482	0.3680	0.2260	0.2014	0.1840	0.1704	0.1594	0.1787	0.1459	0.1227	0.1046	0.0899
Dipper	1,351,500	45	7	1.385	0.0459	0.3739	0.2290	0.2044	0.1869	0.1734	0.1623	0.1807	0.1479	0.1246	0.1066	0.0918
Pheasant	1,272,000	54	19	1.382	0.0466	0.3721	0.2281	0.2035	0.1860	0.1725	0.1614	0.1801	0.1473	0.1240	0.1060	0.0912
Bittern	1,272,000	45	7	1.345	0.0444	0.3779	0.2310	0.2064	0.1890	0.1754	0.1644	0.1820	0.1492	0.1260	0.1079	0.0932
Grackle	1,192,500	54	19	1.333	0.0451	0.3760	0.2301	0.2055	0.1880	0.1745	0.1634	0.1814	0.1486	0.1253	0.1073	0.0925
Bunting	1,195,000	45	7	1.302	0.0429	0.3821	0.2331	0.2085	0.1910	0.1775	0.1664	0.1834	0.1506	0.1274	0.1093	0.0946
Finch	1,113,000	54	19	1.293	0.0436	0.3801	0.2321	0.2075	0.1901	0.1765	0.1655	0.1828	0.1500	0.1267	0.1087	0.0939
Bluejay	1,113,000	45	7	1.259	0.0415	0.3861	0.2351	0.2105	0.1931	0.1795	0.1685	0.1848	0.1520	0.1287	0.1107	0.0959
Curlew	1,033,500	54	7	1.246	0.0420	0.3847	0.2344	0.2098	0.1923	0.1788	0.1677	0.1843	0.1515	0.1282	0.1102	0.0954
Ortolan	1,033,500	45	7	1.213	0.0402	0.3900	0.2370	0.2124	0.1950	0.1815	0.1704	0.1861	0.1533	0.1300	0.1119	0.0972
Tanager	1,033,500	36	1	1.186	0.0384	0.3955	0.2398	0.2152	0.1978	0.1842	0.1732	0.1879	0.1551	0.1318	0.1138	0.0990
Cardinal	954,000	54	7	1.196	0.0402	0.3900	0.2370	0.2124	0.1950	0.1815	0.1704	0.1861	0.1533	0.1300	0.1119	0.0972
Rail	954,000	45	7	1.165	0.0386	0.3949	0.2395	0.2149	0.1975	0.1839	0.1729	0.1877	0.1549	0.1316	0.1136	0.0988
Catbird	954,000	36	1	1.140	0.0370	0.4000	0.2421	0.2175	0.2000	0.1865	0.1754	0.1894	0.1566	0.1333	0.1153	0.1005

Canary	900,000	54	7	1.162	0.0392	0.3930	0.2386	0.2140	0.1965	0.1830	0.1719	0.1871	0.1543	0.1310	0.1130	0.0982
Ruddy	900,000	45	7	1.131	0.0374	0.3987	0.2414	0.2168	0.1994	0.1858	0.1748	0.1890	0.1562	0.1329	0.1149	0.1001
Mallard	795,000	30	19	1.140	0.0392	0.3930	0.2386	0.2140	0.1965	0.1830	0.1719	0.1871	0.1543	0.1310	0.1130	0.0982
Drake	795,000	26	7	1.108	0.0373	0.3930	0.2416	0.2170	0.1995	0.1860	0.1749	0.1891	0.1563	0.1330	0.1150	0.1002
Condor	795,000	54	7	1.093	0.0370	0.3991	0.2421	0.2175	0.2000	0.1865	0.1754	0.1894	0.1566	0.1333	0.1153	0.1005
Cuckoo	795,000	24	7	1.092	0.0366	0.4000	0.2427	0.2181	0.2007	0.1871	0.1761	0.1899	0.1571	0.1338	0.1157	0.1010
Tern	795,000	45	7	1.063	0.0352	0.4014	0.2451	0.2205	0.2030	0.1895	0.1784	0.1914	0.1586	0.1354	0.1173	0.1026
Coot	795,000	36	1	1.040	0.0377	0.3978	0.2409	0.2163	0.1989	0.1853	0.1743	0.1887	0.1559	0.1326	0.1145	0.0998
Redwing	715,500	30	19	1.081	0.0373	0.3991	0.2416	0.2170	0.1995	0.1860	0.1749	0.1891	0.1563	0.1330	0.1150	0.1002
Starling	715,500	26	7	1.051	0.0355	0.4051	0.2446	0.2200	0.2025	0.1890	0.1779	0.1911	0.1583	0.1350	0.1170	0.1022
Stilt	715,500	24	7	1.036	0.0347	0.4078	0.2460	0.2214	0.2039	0.1904	0.1793	0.1920	0.1592	0.1359	0.1179	0.1031
Gannet	666,600	26	7	1.014	0.0343	0.4092	0.2467	0.2221	0.2046	0.1911	0.1800	0.1920	0.1597	0.1364	0.1184	0.1036
Flamingo	666,600	24	7	1.000	0.0355	0.4121	0.2481	0.2235	0.2061	0.1925	0.1815	0.1934	0.1606	0.1374	0.1193	0.1046
— —	653,900	18	3	0.953	0.0308	0.4223	0.2532	0.2286	0.2111	0.1976	0.1865	0.1968	0.1640	0.1408	0.1227	0.1080
Egret	636,000	30	19	1.019	0.0352	0.4061	0.2451	0.2205	0.2030	0.1895	0.1784	0.1914	0.1586	0.1354	0.1173	0.1026
Grosbeak	636,000	26	7	0.990	0.0335	0.4121	0.2481	0.2235	0.2061	0.1925	0.1815	0.1934	0.1606	0.1374	0.1193	0.1046
Rook	636,000	24	7	0.997	0.0327	0.4150	0.2496	0.2250	0.2075	0.1940	0.1829	0.1944	0.1616	0.1383	0.1203	0.1055
Kingbird	636,000	18	1	0.940	0.0304	0.4239	0.2540	0.2294	0.2119	0.1984	0.1873	0.1974	0.1646	0.1413	0.1232	0.1085
Swift	636,000	36	1	0.930	0.0301	0.4251	0.2546	0.2300	0.2125	0.1990	0.1879	0.1978	0.1650	0.1417	0.1236	0.1089
Teal	605,000	30	19	0.994	0.0341	0.4099	0.2470	0.2224	0.2050	0.1914	0.1804	0.1927	0.1599	0.1366	0.1186	0.1038
Squab	605,000	26	7	0.966	0.0327	0.4150	0.2496	0.2250	0.2075	0.1940	0.1829	0.1944	0.1616	0.1383	0.1203	0.1055
Peacock	605,000	24	7	0.953	0.0319	0.4180	0.2511	0.2265	0.2090	0.1955	0.1844	0.1954	0.1626	0.1393	0.1213	0.1065
Eagle	556,500	30	7	0.953	0.0327	0.4150	0.2496	0.2250	0.2075	0.1940	0.1829	0.1944	0.1616	0.1383	0.1203	0.1055
Dove	556,500	26	7	0.927	0.0314	0.4200	0.2520	0.2274	0.2100	0.1964	0.1854	0.1961	0.1633	0.1400	0.1219	0.1072
Parakeet	556,500	24	7	0.914	0.0306	0.4231	0.2536	0.2290	0.2115	0.1980	0.1869	0.1971	0.1643	0.1410	0.1230	0.1082
Osprey	556,500	18	1	0.879	0.0284	0.4321	0.2581	0.2335	0.2161	0.2025	0.1915	0.2001	0.1673	0.1440	0.1260	0.1112
Hen	477,000	30	7	0.883	0.0304	0.4239	0.2540	0.2294	0.2119	0.1984	0.1873	0.1974	0.1646	0.1413	0.1232	0.1085
Hawk	477,000	26	7	0.858	0.0289	0.4300	0.2571	0.2325	0.2150	0.2015	0.1904	0.1994	0.1666	0.1433	0.1253	0.1105
Flicker	477,000	24	7	0.846	0.0284	0.4321	0.2581	0.2335	0.2161	0.2025	0.1915	0.2001	0.1673	0.1440	0.1260	0.1112
Pelican	477,000	18	1	0.814	0.0264	0.4410	0.2626	0.2380	0.2205	0.2070	0.1959	0.031	0.1703	0.1470	0.1289	0.1142
Lark	397,500	30	7	0.806	0.0277	0.4352	0.2596	0.2350	0.2176	0.2040	0.1930	0.2011	0.1683	0.1451	0.1270	0.1123
Ibis	397,500	26	7	0.783	0.0264	0.4410	0.2626	0.2380	0.2205	0.2070	0.1959	0.2031	0.1703	0.1470	0.1289	0.1142
Brant	397,500	24	7	0.772	0.0258	0.4438	0.2639	0.2394	0.2219	0.2084	0.1973	0.2040	0.1712	0.1479	0.1299	0.1151

(continued)

TABLE A.21 (Continued)
Inductive Reactance of ACSR Bundled Conductors at 60 Hz [3]

						60-Hz Inductive Reactance[a] x_a in Ω/mi for 1-ft Radius									
	Area	Strands		Dia.	GMR	Single	Two-Conductor Spacing (in.)					Two-Conductor Spacing (in.)			
Code	(cmil)	Al	St	(in.)	(ft)	Cond.	6	9	12	15	18	6	9	12	15	18
Chickadee	397,500	18	1	0.743	0.0241	0.4521	0.2681	0.2435	0.2260	0.2125	0.2014	0.2068	0.1740	0.1507	0.1326	0.1179
Oriole	336,400	30	7	0.741	0.0255	0.4452	0.2647	0.2401	0.2226	0.2091	0.1980	0.2045	0.1717	0.1484	0.1304	0.1156
Linnet	336,400	26	7	0.721	0.0243	0.4511	0.2676	0.2430	0.2255	0.2120	0.2009	0.2064	0.1736	0.1504	0.1323	0.1176
Merlin	336,400	18	1	0.684	0.0222	0.4620	0.2731	0.2485	0.2310	0.2175	0.2064	0.2101	0.1773	0.1540	0.1360	0.1212
Ostrich	300,000	26	7	0.680	0.0229	0.4583	0.2712	0.2466	0.2291	0.2156	0.2045	0.2088	0.1760	0.1528	0.1347	0.1200

[a] x_a is the component of inductive reactance due to the magnetic flux within a 1-ft radius. The remaining component of inductive reactance, $x_{d'}$ is that due to other phases. The total inductive reactance per phase is the sum of x_a and x_d. The following formula can be used to calculate additional values of x_a. x_d is obtained from the formula below.

$$x_a = 0.2794 \log_{10} \left[\frac{1}{\left[n(\text{GMR})(a)^{n-1} \right]^{\frac{1}{n}}} \right] \ \Omega/\text{mi}$$

$x_d = 0.2794 \log_{10} (\text{GMD}) \ \Omega/\text{mi}$
where GMD = geometric mean distance between phases in feet
where GMR = geometric mean radius in feet
n = number of conductors per phase
$a = s/(2 \sin (\pi/n); n > 1)$
$a = 0; 0° \equiv 1: n = 1$
s = bundle spacing in feet

TABLE A.22
Inductive Reactance of ACSR Bundled Conductors at 60 Hz [3]

		Strands		Dia.	GMR	60-Hz Inductive Reactance^a x_a in Ω/mi for 1-ft Radius										
						Four-Conductor Spacing (in.)						Six-Conductor Spacing (in.)				
Code	Area (cmil)	Al	St	(in.)	(ft)	6	9	12	15	18	6	9	12	15	18
Expanded	3,108,000	62/8	19	2.500	0.0900	0.1256	0.0887	0.0625	0.0422	0.0256	0.0826	0.0416	0.0125	−0.0101	−0.0285
Expanded	2,294,000	66/6	19	2.320	0.0858	0.1271	0.0902	0.0640	0.0437	0.0271	0.0835	0.0425	0.0134	−0.0091	−0.0276
Expanded	1,414,000	58/4	19	1.750	0.0640	0.1360	0.0991	0.0729	0.0526	0.0350	0.0894	0.0484	0.0194	−0.0032	−0.0216
Expanded	1,275,000	50/4	19	1.600	0.0578	0.1390	0.1021	0.0760	0.0557	0.0391	0.0915	0.0505	0.0214	−0.0011	−0.0196
Kiwi	2,167,000	72	7	1.737	0.0571	0.1394	0.1025	0.0763	0.0560	0.0394	0.0918	0.0508	0.0217	−0.0009	−0.0193
Bluebird	2,156,000	84	19	1.762	0.0588	0.1385	0.1016	0.0754	0.0551	0.0385	0.0912	0.0502	0.0211	−0.0015	−0.0199
Chukar	1,780,000	84	19	1.602	0.0536	0.1413	0.1044	0.0783	0.0579	0.0414	0.0930	0.0520	0.0229	0.0004	−0.0181
Falcon	1,590,000	54	19	1.545	0.0523	0.1421	0.1052	0.0790	0.0587	0.0421	0.0935	0.0525	0.0234	0.0019	−0.0176
Lapwing	1,590,000	45	7	1.502	0.0498	0.1436	0.1067	0.0805	0.0602	0.0436	0.0945	0.0535	0.0244	0.0015	−0.0166
Parrot	1,590,000	54	19	1.506	0.0506	0.1443	0.1062	0.0800	0.0597	0.0431	0.0942	0.0532	0.0241	0.0024	−0.0169
Nuthatch	1,590,000	45	7	1.466	0.0486	0.1443	0.1074	0.0812	0.0609	0.0443	0.0950	0.0540	0.0249	0.0025	−0.0161
Plover	1,431,000	54	19	1.465	0.0486	0.1438	0.1069	0.0807	0.0604	0.0438	0.0947	0.0537	0.0246	0.0030	−0.0164
Bobolink	1,431,000	45	7	1.427	0.0470	0.1453	0.1084	0.0822	0.0619	0.0453	0.0957	0.0547	0.0256	0.0030	−0.0154
Martin	1,351,000	54	19	1.424	0.0482	0.1446	0.1077	0.0815	0.0612	0.0446	0.0952	0.0542	0.0251	0.0025	−0.0159
Dipper	1,351,000	45	7	1.385	0.0459	0.1460	0.1091	0.0830	0.0627	0.0461	0.0962	0.0552	0.0261	0.0035	−0.0149
Pheasant	1,272,000	54	19	1.382	0.0466	0.1456	0.1087	0.0825	0.0622	0.0456	0.0959	0.0549	0.0258	0.0032	−0.0152
Bittern	1,272,000	45	7	1.345	0.044	0.1470	0.1101	0.0840	0.0637	0.0471	0.0968	0.0558	0.0268	0.0042	−0.0142
Grackle	1,192,500	54	19	1.333	0.0451	0.1466	0.1097	0.0835	0.0632	0.0466	0.0965	0.0555	0.0264	0.0039	−0.0146
Bunting	1,192,500	45	7	1.302	0.0429	0.1481	0.1097	0.0850	0.0647	0.0481	0.0975	0.0565	0.0274	0.0049	−0.0136
Finch	1,113,000	54	19	1.293	0.0436	0.1476	0.1112	0.0845	0.0642	0.0476	0.0972	0.0562	0.0271	0.0046	−0.0139
Bluejay	1,113,000	45	7	1.259	0.0415	0.1491	0.1107	0.0860	0.0657	0.0491	0.0982	0.0572	0.0281	0.0056	−0.0129
Curlew	1,033,500	54	7	1.246	0.0420	0.1487	0.1118	0.0857	0.0653	0.0488	0.0980	0.0570	0.0279	0.0053	−0.0131
Ortolan	1,033,500	45	7	1.213	0.0402	0.1501	0.1132	0.0870	0.0667	0.0501	0.0989	0.0579	0.0288	0.0062	−0.0122
Tanager	1,033,500	36	1	1.186	0.0384	0.1515	0.1146	0.0884	0.0681	0.0515	0.0998	0.0588	0.0297	0.0071	−0.0113
Cardinal	954,000	54	7	1.196	0.0402	0.1501	0.1132	0.0870	0.0667	0.0501	0.0989	0.0579	0.0288	0.0062	−0.0122
Rail	954,000	45	7	1.165	0.0384	0.1513	0.1144	0.0882	0.0679	0.0513	0.0997	0.0587	0.0296	0.0070	−0.0114

(*continued*)

TABLE A.22 (Continued)
Inductive Reactance of ACSR Bundled Conductors at 60 Hz [3]

Code	Area (cmil)	Strands Al	Strands St	Dia. (in.)	GMR (ft)	60-Hz Inductive Reactance^a x_a in Ω/mi for 1-ft Radius												
						Four-Conductor Spacing (in.)						Six-Conductor Spacing (in.)					
						6	9	12	15	18		6	9	12	15	18
Catbird	954,000	36	1	1.140	0.0402	0.1526	0.1157	0.0895	0.0692	0.0526		0.1005	0.0595	0.0304	0.0679	−0.0106
Canary	900,000	54	7	1.162	0.0386	0.1508	0.1139	0.0877	0.0674	0.0508		0.0994	0.0584	0.0293	0.0067	−0.0108
Ruddy	900,000	45	7	1.131	0.0370	0.1523	0.1154	0.0892	0.0689	0.0523		0.1003	0.0593	0.0302	0.0077	−0.0117
Mallard	795,000	30	19	1.140	0.0392	0.1508	0.1139	0.0877	0.0674	0.0508		0.0994	0.0584	0.0293	0.0067	−0.0107
Drake	795,000	26	7	1.108	0.0373	0.1523	0.1154	0.0893	0.0689	0.0524		0.1004	0.0594	0.0303	0.0077	−0.0106
Condor	795,000	54	7	1.093	0.0370	0.1526	0.1157	0.0895	0.0692	0.0526		0.1005	0.0595	0.0304	0.0079	−0.0103
Cuckoo	795,000	24	7	1.092	0.0366	0.1529	0.1160	0.0898	0.0695	0.0529		0.1007	0.0597	0.0307	0.0081	−0.0096
Tern	795,000	45	7	1.063	0.0352	0.1541	0.1172	0.0910	0.0707	0.0541		0.1015	0.0605	0.0314	0.0089	−0.0109
Coot	795,000	36	1	1.040	0.0377	0.1520	0.1151	0.0889	0.0686	0.0520		0.1001	0.0605	0.0301	0.0075	−0.0109
Redwing	715,500	30	19	1.081	0.0373	0.1523	0.1154	0.0893	0.0689	0.0524		0.1004	0.0594	0.0303	0.0077	−0.0107
Starling	715,500	26	7	1.051	0.0355	0.1538	0.1169	0.0908	0.0704	0.0539		0.1014	0.0604	0.0313	0.0087	−0.0097
Stilt	715,500	24	7	1.036	0.0347	0.1545	0.1176	0.0914	0.0711	0.0545		0.1018	0.0608	0.0317	0.0092	−0.0093
Gannet	666,600	26	7	1.014	0.0343	0.1549	0.1180	0.0918	0.0715	0.0549		0.1021	0.0611	0.0317	0.0094	−0.0090
Flamingo	666,600	24	7	1.000	0.0355	0.1556	0.1187	0.0925	0.0722	0.0556		0.1025	0.0615	0.0320	0.0099	−0.0086
— — —	653,900	18	3	0.953	0.0308	0.1581	0.1212	0.0951	0.0748	0.0582		0.1042	0.0632	0.0341	0.0116	−0.0069
Egret	636,000	30	19	1.019	0.0352	0.1541	0.1172	0.0910	0.0707	0.0541		0.1015	0.0605	0.0314	0.0089	−0.0096
Grosbeak	636,000	26	7	0.990	0.0335	0.1556	0.1187	0.0925	0.0722	0.0556		0.1025	0.0615	0.0324	0.0099	−0.0086
Rook	636,000	24	7	0.977	0.0327	0.1563	0.1194	0.0932	0.0729	0.0563		0.1030	0.0620	0.0329	0.0104	−0.0081
Kingbird	636,000	18	1	0.940	0.0304	0.1585	0.1216	0.0955	0.0752	0.0563		0.1045	0.0635	0.0344	0.0118	−0.0066
Swift	636,000	36	1	0.930	0.0301	0.1588	0.1219	0.0958	0.0755	0.0589		0.1047	0.0637	0.0346	0.0120	−0.0064
Teal	605,000	30	19	0.994	0.0341	0.1551	0.1182	0.0920	0.0717	0.0551		0.1022	0.0612	0.0321	0.0095	−0.0089
Squab	605,000	26	7	0.966	0.0327	0.1563	0.1194	0.0932	0.0729	0.0563		0.1030	0.0620	0.0329	0.0104	−0.0081
Peacock	605,000	24	7	0.953	0.0319	0.1571	0.1202	0.0940	0.0737	0.0571		0.1035	0.0625	0.0334	0.0109	−0.0076
Eagle	556,500	30	7	0.953	0.0327	0.1563	0.1194	0.0932	0.0729	0.0563		0.1030	0.0620	0.0329	0.0104	−0.0081

Appendix A

Dove	556,500	26	7	0.927	0.0314	0.1576	0.1207	0.0945	0.0742	0.0576	0.1038	0.0628	0.0338	0.0112	−0.0072
Parakeet	556,500	24	7	0.914	0.0306	0.1583	0.1214	0.0953	0.0750	0.0584	0.1044	0.0634	0.0343	0.0117	−0.0067
Osprey	556,500	18	1	0.879	0.0284	0.1506	0.1237	0.0975	0.0772	0.0606	0.1059	0.0649	0.0358	0.0132	−0.00521
Hen	477,000	30	7	0.883	0.0304	0.1585	0.1216	0.0955	0.0752	0.0586	0.1045	0.0635	0.0344	0.0118	−0.0066
Hawk	477,000	26	7	0.858	0.0289	0.1601	0.1232	0.0970	0.0767	0.0601	0.1055	0.0645	0.0354	0.0129	−0.0056
Flicker	477,000	24	7	0.846	0.0284	0.1628	0.1237	0.0975	0.0772	0.0606	0.1059	0.0649	0.0358	0.0132	−0.0052
Pelican	477,000	18	1	0.814	0.0264	0.1614	0.1259	0.0997	0.0794	0.0628	0.1074	0.0664	0.0373	0.0147	−0.0037
Lark	397,500	30	7	0.806	0.0277	0.1628	0.1245	0.0983	0.0780	0.0614	0.1064	0.0654	0.0363	0.0137	−0.0047
Ibis	397,500	26	7	0.783	0.0264	0.1614	0.1259	0.0997	0.794	0.0628	0.1074	0.0664	0.073	0.0147	−0.0037
Brant	397,500	24	7	0.772	0.0258	0.1635	0.1266	0.1004	0.0801	0.0635	0.1078	0.0668	0.0377	0.0152	−0.0033
Chickadee	397,500	18	1	0.743	0.0241	0.1656	0.1287	0.1025	0.0822	0.0656	0.1092	0.0682	0.0391	0.0165	−0.0019
Oriole	336,400	30	7	0.741	0.0255	0.1639	0.1270	0.1008	0.0805	0.0639	0.1081	0.0671	0.0380	0.0154	−0.0030
Linnet	336,400	26	7	0.721	0.0243	0.1653	0.1284	0.1023	0.0819	0.0654	0.1090	0.0680	0.0389	0.0164	−0.0021
Merlin	336,400	18	1	0.684	0.0222	0.1681	0.1312	0.1050	0.0847	0.0681	0.1109	0.0699	0.0408	0.0182	−0.0002
Ostrich	300,000	26	7	0.680	0.0229	0.1671	0.1302	0.1041	0.0837	0.0672	0.1102	0.0692	0.0401	0.0176	−0.0009

[a] x_a is the component of inductive reactance due to the magnetic flux within a 1-ft radius. The remaining component of inductive reactance, x_d, is that due to other phases. The total inductive reactance per phase is the sum of x_a and x_d. The following formula can be used to calculate additional values of x_a. x_d is obtained from the formula below.

$$x_a = 0.2794 \log_{10} \left[\frac{1}{\left[n(\mathrm{GMR})(a)^{n-1} \right]^{\frac{1}{n}}} \right] \; \Omega/\mathrm{mi}$$

$x_d = 0.2794 \log_{10} (\mathrm{GMD}) \; \Omega/\mathrm{mi}$

where GMD = geometric mean distance between phases in feet
where GMR = geometric mean radius in feet
n = number of conductors per phase
$a = s/(2 \sin (\pi/n); n > 1)$
$a = 0; 0° \equiv 1: n = 1$
s = bundle spacing in feet

TABLE A.23
Inductive Reactance of ACAR Bundled Conductors at 60 Hz [3]

Area 62% Eq. EC-Al (cmil)	Strands EC/6201		Dia. (in.)	GMR (ft)	Single Cond.	60-H_z Inductive Reactance^a x_a in Ω/mi for 1-ft Radius									
						Two-Conductor Spacing (in.)				Three-Conductor Spacing (in.)					
						6	9	12	15	18	6	9	12	15	18
2,413,000	72	19	1.821	0.0596	0.3422	0.2132	0.1886	0.1711	0.1576	0.1465	0.1701	0.1373	0.1141	0.0960	0.0813
2,375,000	63	28	1.821	0.0596	0.3422	0.2132	0.1886	0.1711	0.1576	0.1465	0.1701	0.1373	0.1141	0.0960	0.0813
2,338,000	54	37	1.821	0.0599	0.3416	0.2128	0.1882	0.1708	0.1573	0.1462	0.1699	0.1371	0.1139	0.0958	0.0811
2,297,000	54	7	1.762	0.0571	0.3474	0.2158	0.1912	0.1737	0.1602	0.1491	0.1719	0.1391	0.1158	0.0977	0.0830
2,262,000	48	13	1.762	0.0578	0.3459	0.2150	0.1904	0.1730	0.1594	0.1484	0.1714	0.1386	0.1153	0.0973	0.0825
2,226,000	42	19	1.762	0.0578	0.3459	0.2150	0.1904	0.1730	0.1594	0.1484	0.1714	0.1386	0.1153	0.0973	0.0825
2,227,000	54	7	1.735	0.0561	0.3495	0.2168	0.1922	0.1748	0.1612	0.1502	0.1726	0.1398	0.1165	0.0985	0.0837
2,193,000	48	13	1.735	0.0530	0.3564	0.2203	0.1957	0.1782	0.1647	0.1536	0.1749	0.1421	0.1188	0.1008	0.0860
2,159,000	42	19	1.735	0.0568	0.3480	0.2161	0.1915	0.1740	0.1605	0.1494	0.1721	0.1393	0.1160	0.0980	0.0832
1,899,000	54	7	1.602	0.0519	0.3590	0.2215	0.1969	0.1795	0.1660	0.1549	0.1757	0.1429	0.1197	0.1016	0.0869
1,870,000	48	13	1.602	0.0490	0.3660	0.2250	0.2004	0.1830	0.1694	0.1584	0.1781	0.1453	0.1220	0.1039	0.0892
1,841,000	42	19	1.602	0.0526	0.3574	0.2207	0.1961	0.1787	0.1651	0.1541	0.1752	0.1424	0.1191	0.1011	0.0863
1,673,000	54	7	1.504	0.0486	0.3670	0.2255	0.2009	0.1835	0.1699	0.1589	0.1784	0.1456	0.1223	0.1043	0.0895
1,647,000	48	13	1.504	0.0461	0.3734	0.2287	0.2041	0.1867	0.1731	0.1621	0.1805	0.1477	0.1245	0.1064	0.0917
1,622,000	42	19	1.504	0.0495	0.3647	0.2244	0.1998	0.1824	0.1688	0.1578	0.1776	0.1448	0.1216	0.1035	0.0888
1,337,000	54	7	1.345	0.0436	0.3801	0.2321	0.2075	0.1901	0.1765	0.1655	0.1828	0.1500	0.1267	0.1087	0.0939
1,296,000	42	19	1.345	0.0440	0.3790	0.2316	0.2070	0.1895	0.1760	0.1649	0.1824	0.1496	0.1263	0.1083	0.0935
1,243,000	30	7	1.302	0.0421	0.3844	0.2342	0.2096	0.1922	0.1786	0.1676	0.1842	0.1514	0.1281	0.1101	0.0953
1,211,000	24	13	1.302	0.0417	0.3855	0.2348	0.2102	0.1928	0.1792	0.168	0.1846	0.1518	0.1285	0.1105	0.0957
1,179,000	18	19	1.302	0.0426	0.389	0.2335	0.2089	0.1915	0.1779	0.1669	0.1837	0.1509	0.1276	0.1096	0.0948
1,163,000	30	7	1.259	0.0407	0.3885	0.2363	0.2117	0.1942	0.1807	0.1696	0.1856	0.1528	0.1295	0.1114	0.0967
1,133,000	24	13	1.259	0.0408	0.3882	0.2361	0.2115	0.1941	0.1806	0.1695	0.1855	0.1527	0.1294	0.1113	0.0966
1,104,000	18	19	1.259	0.0412	0.3870	0.2356	0.2110	0.1935	0.1800	0.1689	0.1851	0.1523	0.1290	0.1109	0.0962
1,153,000	33	4	1.246	0.0401	0.3903	0.2372	0.2126	0.1951	0.1816	0.1705	0.1862	0.1534	0.1301	0.1120	0.0973

1,138,000	30	7	1.246	0.0403	0.3897	0.2369	0.2123	0.1948	0.1813	0.1702	0.1860	0.1532	0.1299	0.1118	0.0971
1,109,000	24	13	1.246	0.0405	0.3891	0.2366	0.2120	0.1945	0.1810	0.1699	0.1858	0.1530	0.1297	0.1116	0.0969
1,080,000	18	19	1.246	0.0407	0.3885	0.2363	0.2117	0.1942	0.1807	0.1696	0.1856	0.1528	0.1295	0.1114	0.0967
1,077,000	30	7	1.212	0.0393	0.3927	0.2384	0.2138	0.1964	0.1828	0.1718	0.1870	0.1542	0.1309	0.1129	0.0981
1,049,000	24	13	1.212	0.0389	0.3940	0.2390	0.2144	0.1970	0.1834	0.1724	0.1874	0.1546	0.1313	0.1133	0.0985
1,022,000	18	19	1.212	0.0396	0.3918	0.2380	0.2134	0.1959	0.1824	0.1713	0.1867	0.1539	0.1306	0.1125	0.0978
1,050,000	30	7	1.196	0.0388	0.3943	0.2392	0.2146	0.1971	0.1836	0.1725	0.1875	0.1547	0.1314	0.1134	0.0986
1,023,000	24	13	1.196	0.0384	0.3955	0.2398	0.2152	0.1978	0.1842	0.1732	0.1879	0.1551	0.1318	0.1138	0.0990
996,000	18	19	1.196	0.0391	0.3933	0.2387	0.2141	0.1967	0.1831	0.1721	0.1872	0.1544	0.1311	0.1131	0.0983
994,800	30	7	1.165	0.0376	0.3981	0.2411	0.2165	0.1990	0.1855	0.1744	0.1888	0.1560	0.1327	0.1146	0.0999
954,600	30	7	1.141	0.0369	0.4004	0.2422	0.2176	0.2002	0.1866	0.1756	0.1895	0.1567	0.1335	0.1154	0.1007
969,300	24	13	1.165	0.0374	0.3987	0.2414	0.2168	0.1994	0.1858	0.1748	0.1890	0.156	0.1329	0.1149	0.1001
958,000	24	13	1.158	0.0371	0.3997	0.2419	0.2173	0.1999	0.1863	0.1753	0.1893	0.1565	0.1329	0.1142	0.1004
943,900	18	19	1.165	0.0381	0.3965	0.2403	0.2173	0.1982	0.1847	0.1736	0.1882	0.1554	0.1322	0.1141	0.0994
900,300	30	7	1.108	0.0358	0.4040	0.2441	0.2195	0.2020	0.1885	0.1774	0.1908	0.1580	0.1347	0.1166	0.1019
795,000	30	7	1.042	0.0334	0.4125	0.2483	0.2237	0.2062	0.1927	0.1816	0.1936	0.1608	0.1375	0.1194	0.1047
877,300	24	13	1.108	0.0355	0.4051	0.2446	0.2200	0.2025	0.1890	0.1779	0.1583	0.1583	0.1350	0.1170	0.1022
795,000	24	13	1.055	0.0339	0.4107	0.2474	0.2228	0.2053	0.1918	0.1807	0.1602	0.1602	0.1369	0.1188	0.1041
854,200	18	19	1.108	0.0361	0.4030	0.2436	0.2190	0.2015	0.1880	0.1769	0.1576	0.1576	0.1343	0.1163	0.1015
795,000	18	19	1.069	0.0349	0.4071	0.2456	0.2210	0.2036	0.1900	0.1790	0.1590	0.1590	0.1357	0.1177	0.1029
829,000	30	7	1.063	0.0343	0.4092	0.2467	0.2221	0.2046	0.1911	0.1800	0.1925	0.1597	0.1364	0.1184	0.1036
807,700	24	13	1.063	0.0342	0.4096	0.2468	0.2222	0.2048	0.1913	0.1802	0.1926	0.1598	0.1365	0.1185	0.1037
786,500	18	19	1.063	0.0348	0.4075	0.2458	0.2212	0.2037	0.1902	0.1791	0.1919	0.1591	0.1358	0.1178	0.1030
727,500	33	4	0.990	0.0319	0.4180	0.2511	0.2265	0.2090	0.1955	0.1844	0.1954	0.1626	0.1393	0.1213	0.1065
718,300	30	7	0.990	0.030	0.4177	0.2509	0.2288	0.2088	0.1953	0.1842	0.1953	0.1625	0.1392	0.1212	0.1064
700,000	24	13	0.990	0.0317	0.4188	0.2515	0.2269	0.2094	0.1959	0.1848	0.1957	0.1629	0.1396	0.1215	0.1068
681,600	18	19	0.990	0.0324	0.4162	0.2501	0.2255	0.2081	0.1945	0.1835	0.1948	0.1620	0.1387	0.1207	0.1059
632,000	15	4	0.927	0.0296	0.4271	0.2556	0.2310	0.2136	0.2000	0.1890	0.1984	0.1656	0.1424	0.1243	0.1096

(continued)

TABLE A.23 (Continued)
Inductive Reactance of ACAR Bundled Conductors at 60 Hz [3]

Area 62% Eq. EC-Al (cmil)	Strands EC/6201	Dia. (in.)	GMR (ft)	Single Cond.	60-Hz Inductive Reactance^a x_a in Ω/mi for 1-ft Radius									
					Two-Conductor Spacing (in.)					Three-Conductor Spacing (in.)				
					6	9	12	15	18	6	9	12	15	18
616,200	12 7	0.927	0.0291	0.4292	0.2566	0.2320	0.2146	0.2011	0.1900	0.1991	0.1663	0.1431	0.1250	0.1103
487,400	15 4	0.814	0.0260	0.4429	0.2635	0.2389	0.2214	0.2079	0.1968	0.2037	0.1709	0.1476	0.1296	0.1148
475,200	12 7	0.814	0.0261	0.4424	0.2632	0.2386	0.2212	0.2077	0.1966	0.2035	0.1707	0.1475	0.1294	0.1147
343,600	15 4	0.684	0.0221	0.4626	0.2733	0.2487	0.2313	0.2177	0.2067	0.2103	0.1775	0.1542	0.1361	0.1214
335,000	12 7	0.684	0.0219	0.4637	0.2739	0.2493	0.2318	0.2183	0.2072	0.2106	0.1778	0.1546	0.1365	0.1218

^{a} x_a is the component of inductive reactance due to the magnetic flux within a 1-ft radius. The remaining component of inductive reactance, x_d, is that due to other phases. The total inductive reactance per phase is the sum of x_a and x_d. The following formula can be used to calculate additional values of x_a. x_d is obtained from the formula below:

$$x_a = 0.2794 \log_{10} \left[\frac{1}{\left[n(\text{GMR})(a)^{n-1}\right]^{\frac{1}{n}}} \right] \; \Omega/\text{mi}$$

$x_d = 0.2794 \log_{10} (\text{GMD}) \; \Omega/\text{mi}$

where GMD = geometric mean distance between phases in feet
where GMR = geometric mean radius in feet
n = number of conductors per phase
$a = s/(2 \sin (\pi/n)); n > 1$
$a = 0; 0° \equiv 1 : n = 1$
s = bundle spacing in feet

Appendix A

TABLE A.24
Inductive Reactance of ACAR Bundled Conductors at 60 Hz [3]

Area 62% Eq. EC-Al (cmil)	Strands EC/6201		Dia. (in.)	GMR (ft)	60-Hz Inductive Reactance^a x_a in Ω/mi for 1-ft Radius										
					Four-Conductor Spacing (in.)				Six-Conductor Spacing (in.)						
					6	9	12	15	18	6	9	12	15	18
2,413,000	72	19	1.821	0.0596	0.1381	0.1012	0.0750	0.0547	0.0381	0.909	0.499	0.0208	−0.0018	−0.0202
2,375,000	63	28	1.821	0.0596	0.1381	0.1012	0.0750	0.0547	0.0381	0.0909	0.0499	0.0208	−0.0018	−0.0202
2,338,000	54	37	1.821	0.0599	0.1380	0.1011	0.0749	0.0546	0.0380	0.0908	0.0498	0.0207	−0.0019	−0.0203
2,297,000	54	7	1.762	0.0571	0.1394	0.1025	0.0763	0.0560	0.0394	0.0918	0.0508	0.0217	−0.0009	−0.0196
2,262,000	48	13	1.762	0.0578	0.1390	0.1021	0.0760	0.0557	0.0391	0.0915	0.0505	0.0214	−0.0011	−0.0196
2,226,000	42	19	1.762	0.0578	0.1390	0.1021	0.0760	0.0557	0.0391	0.0915	0.0505	0.0214	−0.0011	−0.0196
2,227,000	54	7	1.735	0.0561	0.1400	0.1031	0.0769	0.0566	0.0400	0.0921	0.0511	0.0220	−0.0005	−0.0190
2,193,000	48	13	1.735	0.0530	0.1417	0.1048	0.0786	0.0583	0.0417	0.0933	0.0523	0.0232	0.0006	−0.0178
2,159,000	42	19	1.735	0.0568	0.1396	0.1027	0.0765	0.0562	0.0396	0.0919	0.0509	0.0218	−0.0008	−0.0192
1,899,000	54	7	1.602	0.0519	0.1423	0.1054	0.0792	0.0589	0.0423	0.0937	0.0527	0.0236	0.0010	−0.0174
1,870,000	48	13	1.602	0.0490	0.1441	0.1072	0.810	0.0607	0.0441	0.048	0.0538	0.0248	0.0022	−0.0162
1,841,000	42	19	1.602	0.0526	0.1419	0.1050	0.0788	0.0585	0.0419	0.0934	0.0524	0.0233	0.0008	−0.0177
1,673,000	54	7	1.504	0.0486	0.1443	0.1074	0.0812	0.0609	0.0443	0.0950	0.0540	0.0249	0.0024	−0.0161
1,647,000	48	13	1.504	0.0461	0.1459	0.1090	0.0828	0.0625	0.0459	0.0961	0.0551	0.0260	0.0034	−0.0150
1,622,000	42	19	1.504	0.0495	0.1437	0.1068	0.0807	0.0604	0.0438	0.0946	0.0536	0.0246	0.0020	−0.0164
1,337,000	54	7	1.345	0.0436	0.1476	0.1107	0.0845	0.0642	0.0476	0.0972	0.0562	0.0271	0.0046	−0.0139
1,296,000	42	19	1.345	0.0440	0.1473	0.1104	0.0842	0.0639	0.0473	0.0970	0.0560	0.0269	0.0044	−0.0141
1,243,000	30	7	1.302	0.0421	0.1487	0.1118	0.0856	0.0653	0.0487	0.0979	0.0569	0.0278	0.0053	−0.0132
1,211,000	24	13	1.302	0.0417	0.1490	0.1121	0.0859	0.0656	0.0490	0.0981	0.0571	0.0280	0.0055	−0.0130
1,179,000	18	19	1.302	0.0426	0.1483	0.1114	0.0852	0.0649	0.0483	0.0977	0.0567	0.0276	0.0050	−0.0134
1,163,000	30	7	1.259	0.0407	0.1497	0.1128	0.0866	0.0663	0.0497	0.0986	0.0576	0.0285	0.0059	−0.0125
1,133,000	24	13	1.259	0.0408	0.1496	0.1127	0.0865	0.0662	0.0496	0.0986	0.0576	0.0285	0.0059	−0.0125
1,104,000	18	19	1.259	0.0412	0.1493	0.1124	0.0862	0.0659	0.0493	0.0984	0.0574	0.0283	0.0057	−0.0127
1,153,000	33	4	1.246	0.0401	0.1501	0.1132	0.0871	0.0667	0.0502	0.0989	0.0579	0.0288	0.0062	−0.0122
1,138,000	30	7	1.246	0.0403	0.1500	0.1131	0.0869	0.0666	0.0500	0.0988	0.0578	0.0287	0.0061	−0.0123
1,109,000	24	13	1.246	0.0405	0.1498	0.1129	0.0868	0.0664	0.0499	0.0987	0.0577	0.0286	0.0060	−0.0124

(continued)

TABLE A.24 (Continued)
Inductive Reactance of ACAR Bundled Conductors at 60 Hz [3]

Area 62% Eq. EC-Al (cmil)	Strands EC/6201		Dia. (in.)	GMR (ft)	60-H$_z$ Inductive Reactance^a x_a in Ω/mi for 1-ft Radius								
					Four-Conductor Spacing (in.)				Six-Conductor Spacing (in.)				
					6	9	12	15	18	6	9	12	15	18
1,080,000	18	19	1.246	0.0407	0.1497	0.1128	0.0866	0.0663	0.0497	0.0986	0.0576	0.0285	0.0059	−0.0125
1,077,000	30	7	1.212	0.0393	0.1507	0.1138	0.0877	0.0674	0.0508	0.0993	0.0583	0.0292	0.0067	−0.0118
1,049,000	24	13	1.212	0.0389	0.1511	0.1142	0.0880	0.0677	0.0511	0.0995	0.0585	0.0294	0.0069	−0.0116
1,022,000	18	19	1.212	0.0396	0.1505	0.1136	0.0874	0.0671	0.0505	0.0992	0.0582	0.0291	0.0065	−0.0119
1,050,000	30	7	1.196	0.0388	0.1511	0.1142	0.0881	0.0677	0.0512	0.0996	0.0586	0.0295	0.0069	−0.0115
1,023,000	24	13	1.196	0.0384	0.0151	0.1146	0.0884	0.0681	0.0515	0.0998	0.0588	0.0297	0.0071	−0.0113
996,000	18	19	1.196	0.0391	0.1509	0.1140	0.0878	0.0675	0.0509	0.0994	0.0584	0.0293	0.0068	−0.0117
994,800	30	7	1.165	0.0376	0.1521	0.1152	0.0890	0.0687	0.0521	0.1002	0.0592	0.0301	0.0075	−0.0109
954,600	30	7	1.141	0.0369	0.1527	0.1158	0.0896	0.0693	0.0527	0.1006	0.0596	0.0305	0.0079	−0.0105
969,300	24	13	1.165	0.0374	0.1523	0.1154	0.0892	0.0689	0.0523	0.1003	0.0593	0.0302	0.0077	−0.0108
958,000	24	13	1.158	0.0371	0.1525	0.1156	0.0894	0.0691	0.0525	0.1005	0.0595	0.0304	0.0078	−0.0106
943,900	18	19	1.165	0.0381	0.1517	0.1148	0.0886	0.0683	0.0517	0.0999	0.0589	0.0298	0.0073	−0.0112
900,300	30	7	1.108	0.0358	0.1536	0.1167	0.0905	0.0702	0.0536	0.1012	0.0602	0.0311	0.0085	−0.0099
795,000	30	7	1.042	0.334	0.1557	0.1188	0.0926	0.0723	0.0557	0.1026	0.0616	0.0325	0.0099	−0.0085
877,300	24	13	1.108	0.0355	0.1538	0.1169	0.0908	0.0704	0.0539	0.1014	0.0604	0.0313	0.0087	−0.0097
795,000	24	13	1.055	0.0339	0.1552	0.1183	0.0922	0.0718	0.0553	0.1023	0.0613	0.0322	0.0096	−0.0088
854,200	18	19	1.108	0.0361	0.1533	0.1164	0.0902	0.0699	0.0533	0.1010	0.0600	0.0309	0.0084	−0.0101
795,000	18	19	1.069	0.0349	0.1544	0.1175	0.0913	0.0710	0.0544	0.1017	0.0607	0.0316	0.0091	−0.0094
829,000	30	7	1.063	0.343	0.1549	0.1180	0.0918	0.0715	0.0549	0.1021	0.0611	0.0320	0.0094	−0.0090
807,700	24	13	1.063	0.0342	0.1550	0.1181	0.0919	0.0716	0.0550	0.1021	0.0611	0.0320	0.0095	−0.0090

Appendix A

786,500	18	19	1.063	0.0348	0.1544	0.1175	0.0914	0.0710	0.0545	0.1018	0.0608	0.0317	0.0091	-0.0093
727,500	33	4	0.990	0.319	0.1571	0.1202	0.0940	0.0737	0.0571	0.1035	0.0625	0.0334	0.0109	-0.0076
718,300	30	7	0.990	0.0320	0.1570	0.1201	0.0939	0.0736	0.0570	0.1035	0.0625	0.0334	0.0108	-0.0076
700,000	24	13	0.990	0.0317	0.1573	0.1204	0.0942	0.0739	0.0573	0.1037	0.0627	0.0336	0.0110	-0.0074
681,600	18	19	0.990	0.0324	0.1566	0.1197	0.0935	0.0732	0.0566	0.1032	0.0622	0.0331	0.0106	-0.0079
632,000	15	4	0.927	0.0296	0.1593	0.1224	0.0963	0.0760	0.0594	0.1050	0.0640	0.0350	0.0124	-0.0060
616,200	12	7	0.927	0.0291	0.1599	0.1230	0.0968	0.0765	0.0599	0.1054	0.0644	0.0353	0.0127	-0.0057
487,400	15	4	0.814	0.0260	0.1633	0.1264	0.1002	0.0799	0.0633	0.1077	0.0667	0.0376	0.0150	-0.0034
475,200	12	7	0.814	0.0261	0.1632	0.1263	0.1001	0.0798	0.0632	0.1076	0.0666	0.0375	0.0149	-0.0035
343,600	15	4	0.684	0.0221	0.1682	0.1313	0.1051	0.0848	0.0682	0.1109	0.0699	0.0409	0.0183	-0.0001
335,000	12	7	0.684	0.0219	0.1685	0.1316	0.1054	0.0851	0.0685	0.1111	0.0701	0.0410	0.0185	0.0000

[a] x_a is the component of inductive reactance due to the magnetic flux within a 1-ft radius. The remaining component of inductive reactance, x_d, is that due to other phases. The total inductive reactance per phase is the sum of x_a and x_d. The following formula can be used to calculate additional values of x_a. x_d is obtained from the formula below:

$$x_a = 0.2794 \log_{10} \left[\frac{1}{\left[n(\text{GMR})(a)^{n-1} \right]^{\frac{1}{n}}} \right] \, \Omega/\text{mi}$$

$x_d = 0.2794 \log_{10} (\text{GMD}) \; \Omega/\text{mi}$

where GMD = geometric mean distance between phases in feet

where GMR = geometric mean radius in feet

n = number of conductors per phase

$a = s/(2 \sin (\pi/n)); n > 1$

$a = 0; 0° \equiv 1: n = 1$

s = bundle spacing in feet

TABLE A.25
Inductive Reactance of ACAR Bundled Conductors at 60 Hz [3]

60-Hz Capacitive Reactance[a] x'_a in MΩ-mi for 1-ft Radius

Code	Area (cmil)	Strands Al	Strands St	Dia. (in.)	Single Cond.	Two-Conductor Spacing (in.) 6	9	12	15	18	Three-Conductor Spacing (in.) 6	9	12	15	18
Expanded	3,108,000	62/8	19	2.500	0.0671	0.0438	0.0378	0.0336	0.0302	0.0275	0.0361	0.0281	0.0224	0.0180	0.0143
Expanded	2,294,000	66/6	19	2.320	0.0693	0.0449	0.0389	0.0347	0.0313	0.0285	0.0362	0.0288	0.0231	0.0187	0.0151
Expanded	1,414,000	58/4	19	1.750	0.0777	0.0491	0.0431	0.0388	0.0355	0.0328	0.0396	0.0316	0.0259	0.0215	0.0179
Expanded	1,275,000	50/4	19	1.600	0.0803	0.0505	0.0444	0.0402	0.0369	0.0342	0.0405	0.0325	0.0268	0.0224	0.0188
Kiwi	2,167,000	72	7	1.737	0.0779	0.0492	0.0432	0.0390	0.0356	0.0329	0.0397	0.0317	0.0280	0.0216	0.0179
Bluebird	2,156,000	84	19	1.762	0.0775	0.0490	0.0430	0.0387	0.0354	0.0327	0.0395	0.0315	0.0258	0.0214	0.0178
Chukar	1,780,000	84	19	1.602	0.0803	0.0504	0.0444	0.0402	0.0368	0.0341	0.0405	0.0325	0.0268	0.0224	0.0187
Falcon	1,590,000	54	19	1.545	0.0814	0.0510	0.0450	0.0407	0.0374	0.0347	0.0408	0.0328	0.0271	0.0227	0.0191
Lapwing	1,590,000	45	7	1.502	0.0822	0.0514	0.0454	0.0411	0.0378	0.0351	0.0411	0.0331	0.0274	0.0230	0.0194
Parrot	1,510,500	54	19	1.506	0.0821	0.0514	0.0453	0.0411	0.0378	0.0351	0.0411	0.0331	0.0274	0.0230	0.0194
Nuthatch	1,510,500	45	7	1.466	0.0829	0.0518	0.0457	0.0415	0.0382	0.0355	0.0414	0.0333	0.0276	0.0232	0.0196
Plover	1,431,000	54	19	1.465	0.0830	0.0518	0.0457	0.0415	0.0382	0.0355	0.0414	0.0333	0.0277	0.0232	0.0196
Bobolink	1,431,000	45	7	1.427	0.0837	0.0522	0.0461	0.0419	0.0386	0.0359	0.0416	0.0336	0.0279	0.0235	0.0199
Martin	1,351,500	54	19	1.424	0.0838	0.0522	0.0462	0.0419	0.0386	0.0359	0.0416	0.0336	0.0279	0.0235	0.0199
Dipper	1,351,500	45	7	1.385	0.0846	0.0526	0.0466	0.0423	0.0390	0.0363	0.0419	0.0339	0.0282	0.0238	0.0202
Pheasant	1,272,000	54	19	1.382	0.0847	0.0526	0.0466	0.0423	0.0390	0.0363	0.0149	0.0339	0.0282	0.0238	0.0202
Bittern	1,272,000	45	7	1.345	0.0855	0.0530	0.0470	0.0427	0.0394	0.0367	0.0422	0.0342	0.0285	0.0241	0.0205
Grackle	1,192,500	54	19	1.333	0.0858	0.0532	0.0471	0.0429	0.0396	0.0369	0.0423	0.0343	0.0286	0.0242	0.0206
Bunting	1,192,500	45	7	1.302	0.0865	0.0532	0.0475	0.0432	0.0399	0.0372	0.0425	0.0345	0.0288	0.0244	0.0208
Finch	1,113,000	54	19	1.293	0.0867	0.0536	0.0476	0.0433	0.0400	0.0373	0.0426	0.0346	0.0289	0.0245	0.0209
Bluejay	1,113,000	45	7	1.259	0.0875	0.0540	0.0480	0.0437	0.0404	0.0377	0.0429	0.0348	0.0292	0.0247	0.0211
Curlew	1,033,500	54	7	1.246	0.0878	0.0542	0.0481	0.0739	0.0406	0.0379	0.0430	0.0349	0.0293	0.0248	0.0212
Ortolan	1,033,500	45	7	1.213	0.0886	0.0546	0.0485	0.0443	0.0410	0.0383	0.0432	0.0352	0.0295	0.0251	0.0215
Tanager	1,033,500	36	1	1.186	0.0892	0.0549	0.0489	0.0446	0.0413	0.0386	0.0435	0.0354	0.0297	0.0253	0.0217
Cardinal	954,000	54	7	1.196	0.0890	0.0548	0.0488	0.0445	0.0412	0.0385	0.0434	0.0353	0.0297	0.0252	0.0216
Rail	954,000	45	7	1.165	0.0898	0.0552	0.0491	0.0449	0.0416	0.0389	0.0436	0.0356	0.0299	0.0255	0.0219
Catbird	954,000	36	1	1.140	0.0904	0.0555	0.0495	0.0452	0.0419	0.0392	0.0438	0.0358	0.0301	0.0257	0.0221

Canary	900,000	54	7	1.162	0.0898	0.0552	0.0492	0.0449	0.0416	0.0389	0.0437	0.0356	0.0299	0.0255	0.0219
Ruddy	900,000	45	7	1.131	0.0906	0.0556	0.0496	0.0453	0.0420	0.0393	0.0439	0.0359	0.0302	0.0258	0.0222
Mallard	795,000	30	19	1.140	0.0904	0.0555	0.0495	0.0452	0.0419	0.0392	0.0438	0.0358	0.0301	0.0257	0.0221
Drake	795,000	26	7	1.108	0.0912	0.0559	0.0499	0.0456	0.0423	0.0396	0.0441	0.0361	0.0304	0.0260	0.0224
Condor	795,000	54	7	1.093	0.0916	0.0561	0.0501	0.0458	0.0425	0.0398	0.0443	0.0362	0.0305	0.0267	0.0225
Cuckoo	795,000	24	7	1.092	0.0917	0.0561	0.0501	0.0458	0.0425	0.0398	0.0443	0.0362	0.0306	0.0261	0.0225
Tern	795,000	45	7	1.063	0.0925	0.0565	0.0505	0.0462	0.0429	0.0402	0.0445	0.0365	0.0308	0.0264	0.0228
Coot	795,000	36	1	1.040	0.0931	0.0568	0.0508	0.0466	0.0433	0.0405	0.0448	0.0367	0.0310	0.0266	0.0330
Redwing	715,500	30	19	1.081	0.0920	0.0563	0.0503	0.0460	0.0427	0.0400	0.0444	0.0363	0.0307	0.0262	0.0226
Starling	715,500	26	7	1.051	0.0928	0.0567	0.0507	0.0464	0.0431	0.0404	0.0446	0.0366	0.0309	0.0265	0.0229
Stilt	715,500	24	7	1.036	0.0932	0.0569	0.0508	0.0466	0.0433	0.0406	0.0448	0.0368	0.0311	0.0267	0.0231
Gannet	666,600	26	7	1.014	0.0939	0.0572	0.0512	0.0469	0.0436	0.0409	0.0450	0.0370	0.0313	0.0269	0.0233
Flamingo	666,600	24	7	1.000	0.0943	0.0574	0.0514	0.0471	0.0438	0.0411	0.0451	0.0371	0.0314	0.0270	0.0234
———	653,900	18	3	0.953	0.0957	0.0581	0.0521	0.0479	0.0445	0.0418	0.0456	0.0376	0.0319	0.0275	0.0239
Egret	636,000	30	19	1.019	0.0937	0.0571	0.0511	0.0469	0.0436	0.0408	0.0450	0.0369	0.0319	0.0268	0.0232
Grosbeak	636,000	26	7	0.990	0.0946	0.0576	0.0516	0.0473	0.0440	0.0413	0.0452	0.0372	0.0312	0.0271	0.0235
Rook	636,000	24	7	0.977	0.0950	0.0578	0.0518	0.0475	0.0442	0.0415	0.0454	0.0373	0.0315	0.0272	0.0236
Kingbird	636,000	18	1	0.940	0.0961	0.0583	0.0523	0.0481	0.0448	0.0420	0.0458	0.0377	0.0317	0.0276	0.0240
Swift	636,000	36	1	0.930	0.0964	0.0578	0.0525	0.0482	0.0449	0.0422	0.0459	0.0378	0.0321	0.0277	0.0241
Teal	605,000	30	19	0.994	0.0945	0.0575	0.0515	0.0472	0.0439	0.0412	0.0452	0.0372	0.0315	0.0271	0.0235
Squab	655,000	26	7	0.966	0.0953	0.0579	0.0519	0.0477	0.0443	0.0416	0.0455	0.0375	0.0318	0.0274	0.0238
Peacock	605,000	24	7	0.953	0.0957	0.0581	0.0521	0.0479	0.0445	0.0418	0.0456	0.0376	0.0319	0.0275	0.0239
Eagle	556,500	30	7	0.953	0.0957	0.0581	0.0521	0.0479	0.0445	0.0418	0.0456	0.0376	0.0319	0.0275	0.0239
Dove	556,500	26	7	0.927	0.0965	0.0586	0.0525	0.0483	0.0450	0.0423	0.0459	0.0379	0.0322	0.0278	0.0242
Parakeet	556,500	24	7	0.914	0.0970	0.0588	0.0527	0.0485	0.0452	0.0425	0.0460	0.0380	0.0323	0.0279	0.0243
Osprey	556,500	18	1	0.879	0.0981	0.0593	0.0533	0.0491	0.0457	0.0430	0.0464	0.0384	0.0327	0.0283	0.0247
Hen	477,000	30	7	0.883	0.0980	0.0593	0.0533	0.0490	0.0457	0.0430	0.0464	0.0383	0.0327	0.0282	0.0246
Hawk	477,000	26	7	0.858	0.0988	0.0597	0.0537	0.0494	0.0461	0.0434	0.0467	0.0386	0.0329	0.0285	0.0249
Flicker	477,000	24	7	0.846	0.0992	0.0599	0.0539	0.0496	0.0463	0.0436	0.0468	0.0388	0.0331	0.0287	0.0251
Pelican	477,000	18	1	0.814	0.1004	0.0606	0.0545	0.0502	0.0469	0.0442	0.0472	0.0392	0.0335	0.0291	0.0254
Lark	397,500	30	7	0.806	0.1007	0.0606	0.0546	0.0503	0.0470	0.0443	0.0473	0.0393	0.0336	0.0291	0.0255
Ibis	397,500	26	7	0.783	0.1015	0.0611	0.0550	0.0508	0.0475	0.0448	0.0476	0.0395	0.0338	0.0294	0.0258
Brant	397,500	24	7	0.772	0.1020	0.0613	0.0552	0.0510	0.0477	0.0450	0.0477	0.0397	0.0340	0.0296	0.0260

(continued)

TABLE A.25 (Continued)
Inductive Reactance of ACAR Bundled Conductors at 60 Hz [3]

					60-Hz Capacitive Reactance[a] x'_a in MΩ-mi for 1-ft Radius									
	Area	Strands		Dia.	Single	Two-Conductor Spacing (in.)					Three-Conductor Spacing (in.)			
Code	(cmil)	Al	St	(in.)	Cond.	6	9	12	15	18	6	9	12	15	18
Chickadee	397,500	18	1	0.743	0.1031	0.0618	0.0558	0.0516	0.0482	0.0455	0.0481	0.0401	0.0344	0.0300	0.0263
Oriole	336,400	30	7	0.741	0.1032	0.0619	0.0559	0.0516	0.0483	0.0456	0.0481	0.0401	0.0344	0.0300	0.0264
Linnet	336,400	26	7	0.721	0.1040	0.0623	0.0563	0.0520	0.0487	0.0460	0.0484	0.0404	0.0347	0.0303	0.0266
Merlin	336,400	18	1	0.684	0.1056	0.0631	0.0570	0.0528	0.0495	0.0468	0.0489	0.0409	0.0352	0.0308	0.0272
Ostrich	300,000	26	7	0.680	0.1057	0.0631	0.0571	0.0529	0.0496	0.0468	0.0490	0.0409	0.0352	0.0308	0.0272

[a] x'_a is the component of capacitive reactance due to the electrostatic flux within a 1-ft radius. The remaining component of capacitive reactance, x'_d, accounts for the flux between the 1-ft radius and the other phases. The total capacitive reactance per phase is the sum of x'_a and x'_d. The following formula can be used to calculate additional values of x'_a. x'_d is obtained from the formula below.

$$x'_a = 0.0683 \log_{10}\left[\frac{1}{[n(r)(a)^{n-1}]^{\frac{1}{N}}}\right] \text{ M}\Omega\text{-mi}$$

where r = conductor radius in feet
n = number of conductors per phase
$a = s/(2 \sin(\pi/n))$: $n > 1$
$a = 0$: $0° \equiv 1$: $n = 1$
s = bundle spacing in feet
$x'_d = 0.0683 \log_{10}$ (GMD) MΩ-mi
where GMD = geometric mean distance between phases in feet

TABLE A.26
Capacitive Reactance of ACSR Bundled Conductors at 60 Hz [3]

60-Hz Capacitive Reactance[a] x'_a in MΩ-mi for 1-ft Radius

Code	Area (cmil)	Strands Al	Strands St	Dia. (in.)	Four-Conductor Spacing (in.) 6	9	12	15	18	Six-Conductor Spacing (in.) 6	9	12	15	18
Expanded	3,108,000	62/8	19	2.500	0.0296	0.0206	0.0142	0.0092	0.0052	0.0196	0.0094	0.0023	−0.0032	−0.0077
Expanded	2,294,000	66/6	19	2.320	0.0302	0.0212	0.0148	0.0098	0.0057	0.0198	0.0098	0.0027	−0.0028	−0.0073
Expanded	1,414,000	58/4	19	1.750	0.0323	0.0233	0.0169	0.0119	0.0078	0.0212	0.0112	0.0041	−0.0014	−0.0059
Expanded	1,275,000	50/4	19	1.600	0.0329	0.0239	0.0175	0.0126	0.0085	0.0217	0.0116	0.0045	−0.0010	−0.0055
Kiwi	2,167,000	72	7	1.737	0.0323	0.0233	0.0169	0.0119	0.0079	0.0213	0.0112	0.0041	−0.0014	−0.0059
Bluebird	2,156,000	84	19	1.762	0.0322	0.0232	0.0168	0.0118	0.0078	0.0212	0.0112	0.0041	−0.0014	−0.0060
Chukar	1,780,000	84	19	1.602	0.0329	0.0239	0.0175	0.0125	0.0085	0.0217	0.0116	0.0045	−0.0010	−0.0055
Falcon	1,590,000	54	19	1.545	0.0332	0.0242	0.0178	0.0128	0.0088	0.0218	0.0118	0.0047	−0.0008	−0.0053
Lapwing	1,590,000	45	7	1.502	0.0334	0.0244	0.0180	0.0130	0.0090	0.0220	0.0120	0.0048	−0.0007	−0.0052
Parrot	1,510,500	54	19	1.506	0.0334	0.0244	0.0180	0.0130	0.0089	0.0220	0.0119	0.0048	−0.0007	−0.0052
Nuthatch	1,510,500	45	7	1.466	0.0336	0.0246	0.0182	0.0132	0.0091	0.0221	0.0121	0.0050	−0.0006	−0.0051
Plover	1,431,000	54	19	1.465	0.0336	0.0246	0.0182	0.0132	0.0091	0.0221	0.0121	0.0050	−0.0006	−0.0051
Bobolink	1,431,000	45	7	1.427	0.0338	0.0248	0.0184	0.0134	0.0093	0.0222	0.0122	0.0051	−0.0004	−0.0049
Martin	1,351,500	54	19	1.424	0.0338	0.0248	0.0184	0.0134	0.0094	0.0222	0.0122	0.0051	−0.0004	−0.0049
Dipper	1,351,500	45	7	1.385	0.0340	0.0250	0.0186	0.0136	0.0096	0.0224	0.0124	0.0052	−0.0003	−0.0048
Pheasant	1,272,000	54	19	1.382	0.0340	0.0250	0.0186	0.0136	0.0096	0.0224	0.0124	0.0053	−0.0003	−0.0048
Bittern	1,272,000	45	7	1.345	0.0342	0.0252	0.0188	0.0138	0.0098	0.0225	0.0125	0.0054	−0.0001	−0.0046
Grackle	1,192,500	54	19	1.333	0.0343	0.0253	0.0189	0.0139	0.0098	0.0226	0.0125	0.0054	−0.0001	−0.0046
Bunting	1,192,500	45	7	1.302	0.0345	0.0254	0.0190	0.0141	0.0100	0.0227	0.0127	0.0055	0.0000	−0.0045
Finch	1,113,000	54	19	1.293	0.0345	0.0255	0.0191	0.0141	0.0101	0.0227	0.0127	0.0056	0.0001	−0.0044
Bluejay	1,113,000	45	7	1.259	0.0347	0.0257	0.0193	0.0143	0.0103	0.0229	0.0128	0.0057	0.0002	−0.0043
Curlew	1,033,500	54	7	1.246	0.0348	0.0258	0.0194	0.0144	0.0103	0.0229	0.0129	0.0058	0.0003	−0.0043
Ortolan	1,033,500	45	7	1.213	0.0350	0.0260	0.0196	0.0146	0.0105	0.0230	0.0130	0.0059	0.0004	−0.0041
Tanager	1,033,500	36	1	1.186	0.0352	0.0261	0.0197	0.0148	0.0107	0.0231	0.0131	0.0060	0.0005	−0.0040
Cardinal	954,000	54	7	1.196	0.0351	0.0261	0.0197	0.0147	0.0107	0.0231	0.0131	0.0060	0.0005	−0.0041
Rail	954,000	45	7	1.165	0.0353	0.0263	0.0199	0.0149	0.0108	0.0232	0.0132	0.0061	0.0006	−0.0039

(continued)

TABLE A.26 (Continued)
Capacitive Reactance of ACSR Bundled Conductors at 60 Hz [3]

	Area	Strands		Dia. (in.)	60-Hz Capacitive Reactance^a x'_a in MΩ-mi for 1-ft Radius											
					Four-Conductor Spacing (in.)					Six-Conductor Spacing (in.)						
Code	(cmil)	Al	St		6	9	12	15	18		6	9	12	15	18	
Catbird	954,000	36	1	1.140	0.0355	0.0264	0.0200	0.0151	0.0110		0.0233	0.0133	0.0062	0.0007	−0.0038	
Canary	900,000	54	7	1.162	0.0353	0.0263	0.0199	0.0149	0.0109		0.0232	0.0132	0.0061	0.0006	−0.0039	
Ruddy	900,000	45	7	1.131	0.0355	0.0265	0.0201	0.0151	0.0111		0.0234	0.0134	0.0062	0.0007	−0.0038	
Mallard	795,000	30	19	1.140	0.0355	0.0264	0.0200	0.0151	0.0110		0.0233	0.0133	0.0062	0.0007	−0.0038	
Drake	795,000	26	7	1.108	0.0357	0.0266	0.0202	0.0153	0.0112		0.0235	0.0135	0.0063	0.0008	−0.0037	
Condor	795,000	54	7	1.093	0.0358	0.0267	0.0203	0.0154	0.0113		0.0236	0.0135	0.0064	0.0009	−0.0036	
Cuckoo	795,000	24	7	1.092	0.0358	0.0267	0.0203	0.0154	0.0113		0.0236	0.0135	0.0064	0.0009	−0.0036	
Tern	795,000	45	7	1.063	0.0360	0.0269	0.0205	0.0156	0.0115		0.0237	0.0137	0.0066	0.0010	−0.0035	
Coot	795,000	36	1	1.040	0.0361	0.0271	0.0207	0.0157	0.0117		0.0238	0.0138	0.0067	0.0011	−0.0034	
Redwing	715,500	30	19	1.081	0.0358	0.0268	0.0204	0.0155	0.0114		0.0236	0.0136	0.0065	0.0010	−0.0036	
Starling	715,500	26	7	1.051	0.0361	0.0270	0.0206	0.0157	0.0116		0.0237	0.0137	0.0066	0.0011	−0.0034	
Stilt	715,500	24	7	1.036	0.0362	0.0271	0.0207	0.0158	0.0117		0.0238	0.0138	0.0067	0.0012	−0.0033	
Gannet	666,600	26	7	1.014	0.0363	0.0273	0.0209	0.0159	0.0119		0.0239	0.0139	0.0068	0.0013	−0.0032	
Flamingo	666,600	24	7	1.000	0.0364	0.0274	0.0210	0.0160	0.0120		0.0240	0.0140	0.0069	0.0013	−0.0032	
—	653,900	18	3	0.953	0.0368	0.0278	0.0214	0.0164	0.0123		0.0242	0.0142	0.0071	0.0016	−0.0029	
Egret	636,000	30	19	1.019	0.0363	0.0273	0.0209	0.0159	0.0118		0.0239	0.0139	0.0068	0.0012	−0.0033	
Grosbeak	636,000	26	7	0.990	0.0365	0.0275	0.0211	0.0161	0.0121		0.0240	0.0140	0.0069	0.0014	−0.0031	
Rook	636,000	24	7	0.977	0.0366	0.0276	0.0212	0.0162	0.0122		0.0241	0.0141	0.0070	0.0015	−0.0031	
Kingbird	636,000	18	1	0.940	0.0369	0.0279	0.0215	0.0165	0.0124		0.0243	0.0143	0.0072	0.0016	−0.0029	
Swift	636,000	36	1	0.930	0.0370	0.0279	0.0215	0.0166	0.0125		0.0244	0.0143	0.0072	0.0017	−0.0028	
Teal	605,000	30	19	0.994	0.0365	0.0274	0.0210	0.0161	0.0120		0.0240	0.0140	0.0069	0.0014	−0.0031	
Squab	655,000	26	7	0.966	0.0367	0.0277	0.0213	0.0163	0.0122		0.0242	0.0141	0.0070	0.0015	−0.0030	
Peacock	605,000	24	7	0.953	0.0368	0.0278	0.0214	0.0164	0.0123		0.0242	0.0142	0.0071	0.0016	−0.0029	
Eagle	556,500	30	7	0.953	0.0368	0.0278	0.0214	0.0164	0.0123		0.0242	0.0142	0.0071	0.0016	−0.0029	
Dove	556,500	26	7	0.927	0.0370	0.0280	0.0216	0.0166	0.0125		0.0244	0.0143	0.0072	0.0017	−0.0028	
Parakeet	556,500	24	7	0.914	0.0371	0.0281	0.0217	0.0167	0.0126		0.0244	0.0144	0.0073	0.0018	−0.0027	
Osprey	556,500	18	1	0.879	0.0374	0.0284	0.0220	0.0170	0.0129		0.0246	0.0146	0.0075	0.0020	−0.0025	

Hen	477,000	30	7	0.883	0.0373	0.0283	0.0219	0.0170	0.0129	0.0246	0.0146	0.0075	0.0020	-0.0026
Hawk	477,000	26	7	0.858	0.0376	0.0285	0.0221	0.0172	0.0131	0.0247	0.0147	0.0076	0.0021	-0.0024
Flicker	477,000	24	7	0.846	0.0377	0.0286	0.0222	0.0173	0.0132	0.0248	0.0148	0.0077	0.0022	-0.0023
Pelican	477,000	18	1	0.814	0.0380	0.0289	0.0225	0.0176	0.0135	0.0250	0.0150	0.0079	0.0024	-0.0022
Lark	397,500	30	7	0.806	0.0380	0.0290	0.0226	0.0176	0.0136	0.0251	0.0150	0.0079	0.0024	-0.0021
Ibis	397,500	26	7	0.783	0.0382	0.0292	0.0228	0.0179	0.0138	0.0252	0.0152	0.0081	0.0025	-0.0020
Brant	397,500	24	7	0.772	0.0383	0.0293	0.0229	0.0180	0.0139	0.0253	0.0152	0.0081	0.0026	-0.0019
Chickadee	397,500	18	1	0.743	0.0386	0.0296	0.0232	0.0182	0.0142	0.0255	0.0154	0.0083	0.0028	-0.0017
Oriole	336,400	30	7	0.741	0.0386	0.0296	0.0232	0.0183	0.0142	0.0255	0.0154	0.0089	0.0028	-0.0017
Linnet	336,400	26	7	0.721	0.0389	0.0298	0.0234	0.0185	0.0144	0.0256	0.0156	0.0085	0.0030	-0.0016
Merlin	336,400	18	1	0.684	0.0392	0.0302	0.0238	0.0189	0.0148	0.0253	0.0158	0.0087	0.0032	-0.0013
Ostrich	300,000	26	7	0.680	0.0393	0.0303	0.0239	0.0189	0.0148	0.0253	0.0159	0.0088	0.0032	-0.0013

[a] x'_a is the component of capacitive reactance due to the electrostatic flux within a 1-ft radius. The remaining component of capacitive reactance, x_d, accounts for the flux between the 1-ft radius and the other phases. The total capacitive reactance per phase is the sum of x'_a and x'_d. The following formula can be used to calculate additional values of x'_a. x'_a is obtained from the formula below.

$$x'_a = 0.0683 \log_{10} \left[\frac{1}{\left[n(r)(a)^{n-1}\right]^{\frac{1}{n}}} \right] \text{ M}\Omega\text{-mi}$$

where r = conductor radius in feet
n = number of conductors per phase
$a = s/(2 \sin(\pi/n))$: $n > 1$
$a = 0$: $0° \equiv 1$: $n = 1$
s = bundle spacing in feet
$x'_d = 0.0683 \log_{10} (\text{GMD})$ MΩ-mi
where GMD = geometric mean distance between phases in feet

TABLE A.27
Capacitive Reactance of ACAR Bundled Conductors at 60 Hz [3]

Area 62% Eq. EC-Al (cmil)	Strands EC/6201		Dia. (in.)	Single Cond.	60-Hz Capacitive Reactance[a] x_a' in MΩ-mi for 1-ft Radius									
					Two-Conductor Spacing (in.)					Three-Conductor Spacing (in.)				
					6	9	12	15	18	6	9	12	15	18
2,413,000	72	19	1.821	0.0765	0.0485	0.0425	0.0383	0.0349	0.0322	0.0392	0.0312	0.0255	0.0211	0.0175
2,375,000	63	28	1.821											
2,338,000	54	37	1.821											
2,297,000	54	7	1.762	0.0775	0.0490	0.0430	0.0387	0.0354	0.0327	0.0395	0.0315	0.0258	0.0214	0.0178
2,262,000	48	13	1.762											
2,226,000	42	19	1.762											
2,227,000	54	7	1.735	0.0779	0.0493	0.0432	0.0390	0.0357	0.0330	0.0397	0.0317	0.0260	0.0216	0.0180
2,193,000	48	13	1.735											
2,159,000	42	19	1.735											
1,899,000	54	7	1.602	0.0803	0.0504	0.0444	0.0402	0.0368	0.0341	0.0405	0.0325	0.0268	0.0224	0.0187
1,870,000	48	13	1.602											
1,841,000	42	19	1.602											
1,673,030	54	7	1.504	0.0822	0.0514	0.0454	0.0411	0.0378	0.0351	0.0411	0.0331	0.0274	0.0230	0.0194
1,647,030	48	13	1.504											
1,622,000	42	19	1.504											
1,337,000	54	7	1.345	0.0855	0.0530	0.0470	0.0427	0.0394	0.0367	0.0422	0.0342	0.0285	0.0241	0.0205
1,296,000	42	19	1.345											
1,243,000	30	7	1.302	0.0865	0.0535	0.0475	0.0432	0.0399	0.0372	0.0425	0.0345	0.0288	0.0244	0.0208
1,211,000	24	13	1.302											
1,179,000	18	19	1.302											

Appendix A

1,163,000	30	7	1.259							0.0248	0.0212
1,133,000	24	13	1.259								
1,104,000	18	19	1.259								
1,135,000	33	4	1.246	0.0875	0.0542	0.0481	0.0439	0.0406	0.0379	0.0430	0.0349	0.0293	0.0248	0.0212
1,138,000	30	7	1.246								
1,109,000	24	13	1.246								
1,080,000	18	19	1.246								
1,077,000	30	7	1.212	0.0878							
1,049,000	24	13	1.212								
1,022,000	18	19	1.212								
1,050,000	30	7	1.196	0.0886	0.0546	0.0486	0.0443	0.0410	0.0383	0.0432	0.0352	0.0295	0.0251	0.0215
1,023,000	24	13	1.196								
996,000	18	19	1.196	0.0890	0.0548	0.0488	0.0445	0.0412	0.0385	0.0434	0.0353	0.0297	0.0252	0.0216
994,800	30	7	1.165	0.0898	0.0552	0.0491	0.0449	0.0416	0.0389	0.0436	0.0356	0.0299	0.0255	0.0219
954,600	30	7	1.141	0.0904	0.0555	0.0495	0.0452	0.0419	0.0392	0.0438	0.0358	0.0301	0.0257	0.0221
969,300	24	13	1.165	0.0898	0.0552	0.0491	0.0449	0.0416	0.0389	0.0436	0.0356	0.0299	0.0255	0.0219
958,000	24	13	1.158	0.0899	0.0552	0.0492	0.0450	0.0417	0.0390	0.0437	0.0357	0.0300	0.0256	0.0220
943,900	18	19	1.165	0.0898	0.0552	0.0491	0.0449	0.0416	0.0389	0.0436	0.0356	0.0299	0.0255	0.0219
900,300	30	7	1.108	0.0912	0.0559	0.0499	0.0456	0.0423	0.0396	0.0441	0.0361	0.0304	0.0260	0.0224
795,000	30	7	1.042	0.0931	0.0568	0.0508	0.0465	0.0432	0.0405	0.0447	0.0367	0.0310	0.0266	0.0230
877,300	24	13	1.108	0.0912	0.0559	0.0499	0.0456	0.0423	0.0396	0.0441	0.0361	0.0304	0.0260	0.0224
795,000	24	13	1.055	0.0927	0.0566	0.0506	0.0463	0.0430	0.0403	0.0446	0.0366	0.0309	0.0265	0.0229
854,200	18	19	1.108	0.0912	0.0559	0.0499	0.0456	0.0423	0.0396	0.0441	0.0361	0.0304	0.0260	0.0224
795,000	18	19	1.069	0.0923	0.0564	0.0504	0.0462	0.0428	0.0401	0.0445	0.0365	0.0308	0.0264	0.0227
829,000	30	7	1.063	0.0925	0.0565	0.0505	0.0462	0.0429	0.0402	0.0445	0.0365	0.0308	0.0264	0.0228
807,700	24	13	1.063								
786,500	18	19	1.063								

(*continued*)

TABLE A.27 (Continued)
Capacitive Reactance of ACAR Bundled Conductors at 60 Hz [3]

Area 62% EC-Al (cmil)	Strands EC/6201		Dia. (in.)	60-Hz Capacitive Reactance[a] x'_a in MΩ-mi for 1-ft Radius											
				Single Cond.	Two-Conductor Spacing (in.)					Three-Conductor Spacing (in.)					
					6	9	12	15	18	6	9	12	15	18
727,500	33	4	0.990	0.0946	0.0576	0.0516	0.0473	0.0440	0.0413	0.0452	0.0372	0.0315	0.0271	0.0235
718,300	30	7	0.990											
700,000	24	13	0.990											
681,600	18	19	0.990											
632,000	15	4	0.927	0.0965	0.0586	0.0545	0.0483	0.0450	0.0423	0.0459	0.0379	0.0322	0.0278	0.0242
616,200	12	7	0.927											
487,400	15	4	0.814	0.1004	0.0605	0.0570	0.0502	0.0469	0.0442	0.0472	0.0392	0.0335	0.0291	0.0254
475,200	12	7	0.814											
343,600	15	4	0.684	0.1056	0.0631	0.0570	0.0528	0.0495	0.0468	0.0489	0.0409	0.0352	0.0308	0.0272
335,000	12	7	0.684											

[a] x'_a is the component of capacitive reactance due to the electrostatic flux within a 1-ft radius. The remaining component of capacitive reactance, x'_d, accounts for the flux between the 1-ft radius and the other phases. The total capacitive reactance per phase is the sum of x'_a and x'_d. The following formula can be used to calculate additional values of x'_a. x'_d is obtained from the formula below.

$$x'_a = 0.0683 \log_{10} \left[\frac{1}{\left[n(r)(a)^{n-1} \right]^{\frac{1}{N}}} \right] \text{M}\Omega\text{-mi}$$

where r = conductor radius in feet
n = number of conductors per phase
$a = s/(2 \sin (\pi/n))$: $n > 1$
$a = 0$: $0° \equiv 1$: $n = 1$
s = bundle spacing in feet
$x'_d = 0.0683 \log_{10}$ (GMD) MΩ-mi
where GMD = geometric mean distance between phases in feet

TABLE A.28
Capacitive Reactance of ACAR Bundled Conductors at 60 Hz [3]

Area 62% Eq. EC-Al (cmil)	Strands EC/6201		Dia. (in.)	60-Hz Capacitive Reactance^a x_a' in MΩ-mi for 1-ft Radius									
				Four-Conductor Spacing (in.)				Six-Conductor Spacing (in.)					
				6	9	12	15	18	6	9	12	15	18
2,413,000	72	19	1.821	0.0320	0.0230	0.0166	0.0116	0.0075	0.0210	0.0110	0.0039	-0.0016	-0.0061
2,375,000	63	28	1.821										
2,338,000	54	37	1.821										
2,297,000	54	7	1.762	0.0322	0.0232	0.0168	0.0118	0.0078	0.0212	0.0112	0.0041	-0.0015	-0.0060
2,262,000	48	13	1.762										
2,226,000	42	19	1.762										
2,227,000	54	7	1.735	0.0333	0.0233	0.0169	0.0119	0.0079	0.0213	0.0112	0.0041	-0.0014	-0.0059
2,193,000	48	13	1.735										
2,159,000	42	19	1.735										
1,899,000	54	7	1.602	0.0329	.0239	0.0175	0.0125	0.0085	0.0217	0.0116	0.0045	-0.0010	-0.0055
1,870,000	48	13	1.602										
1,841,000	42	19	1.602										
1,673,000	54	7	1.504	0.0342	0.0252	0.0188	0.0138	0.0098	0.0225	0.0125	0.0054	-0.0001	-0.0046
1,647,000	48	13	1.504										
1,622,000	42	19	1.504										
1,337,000	54	7	1.345	0.0342	0.0252	0.0188	0.0138	0.0098	0.0225	0.0125	0.0054	-0.0001	-0.0046
1,296,000	42	19	1.345										
1,243,000	30	7	1.302	0.0345	0.0254	0.0190	0.0141	0.0100	0.0227	0.0127	0.0055	0.0000	-0.0045
1,211,000	24	13	1.302										
1,179,000	18	19	1.302										
1,163,000	30	7	1.259	0.0347	0.0257	0.0193	0.0143	0.0103	0.0229	0.0128	0.0057	0.0002	-0.0043
1,133,000	24	13	1.259										
1,104,000	18	19	1.259										
1,153,000	33	4	1.246	0.0348	0.0258	0.0194	0.0144	0.0103	0.0229	0.0129	0.0058	0.0003	-0.0043
1,138,000	30	7	1.246										
1,109,000	24	13	1.246										

(*continued*)

TABLE A.28 (Continued)
Capacitive Reactance of ACAR Bundled Conductors at 60 Hz [3]

60-Hz Capacitive Reactance[a] x'_a in MΩ-mi for 1-ft Radius

Area 62% Eq. EC-Al (cmil)	Strands EC/6201		Dia. (in.)	Four-Conductor Spacing (in.)					Six-Conductor Spacing (in.)				
				6	9	12	15	18	6	9	12	15	18
1,080,000	18	19	1.246										
1,077,000	30	7	1.212	0.0350	0.0260	0.0196	0.0146	0.0106	0.0230	0.0130	0.0059	0.0004	−0.0041
1,049,000	24	13	1.212										
1,022,000	18	19	1.212										
1,050,000	30	7	1.196	0.0351	0.0261	0.0197	0.0147	0.0107	0.0231	0.0131	0.0060	0.0005	−0.0041
1,023,000	24	13	1.196										
996,000	18	19	1.196										
994,800	30	7	1.165	0.0353	0.0263	0.0199	0.0149	0.0108	0.0232	0.0132	0.0061	0.0006	−0.0039
954,600	30	7	1.141	0.0354	0.0264	0.0200	0.0151	0.0110	0.0233	0.0133	0.0062	0.0007	−0.0038
969,300	24	13	1.165	0.0353	0.0263	0.0199	0.0149	0.0108	0.0232	0.0132	0.0061	0.0006	−0.0039
958,000	24	13	1.158	0.0353	0.0263	0.0199	0.0149	0.0109	0.0233	0.0132	0.0061	0.0006	−0.0039
943,900	18	19	1.165	0.0353	0.0263	0.0199	0.0149	0.0108	0.0232	0.0132	0.0061	0.0006	−0.0039
900,300	30	7	1.108	0.0357	0.0266	0.0202	0.0153	0.0112	0.0235	0.0135	0.0063	0.0008	−0.0037
795,000	30	7	1.042	0.0361	0.0271	0.0207	0.0157	0.0117	0.0238	0.0138	0.0067	0.0011	−0.0034
877,300	24	13	1.108	0.0357	0.0266	0.0202	0.0153	0.0112	0.0235	0.0135	0.0063	0.0008	−0.0037
795,000	24	13	1.055	0.0360	0.0270	0.0206	0.0156	0.0116	0.0237	0.0137	0.0066	0.0011	−0.0037
854,200	18	19	1.108	0.0357	0.0266	0.0202	0.0153	0.0112	0.0235	0.0135	0.0063	0.0008	−0.0037
795,000	18	19	1.069	0.0359	0.0269	0.0205	0.0155	0.0115	0.0237	0.0136	0.0065	0.0010	−0.0035
829,000	30	7	1.063	0.0360	0.0269	0.0205	0.0158	0.0105	0.0237	0.0137	0.0066	0.0010	−0.0035
807,700	24	13	1.063										
786,500	18	19	1.063										
727,500	33	4	0.990	0.0365	0.0275	0.0211	0.0161	0.0121	0.0240	0.0140	0.0069	0.0014	−0.0031

718,300	30	7	0.990									
700,000	24	13	0.990									
681,600	18	19	0.990									
632,000	15	4	0.814	0.0380	0.0289	0.0225	0.0176	0.0135	0.0250	0.0150	0.0079	0.0024	-0.0022
616,200	12	7	0.814									
343,600	15	4	0.684	0.0392	0.0302	0.0238	0.0189	0.0148	0.0259	0.0158	0.0087	0.0032	-0.0013
335,000	12	7	0.684									

[a] x'_a is the component of capacitive reactance due to the electrostatic flux within a 1-ft radius. The remaining component of capacitive reactance, x'_d, accounts for the flux between the 1-ft radius and the other phases. The total capacitive reactance per phase is the sum of x'_a and x'_d. x'_d is obtained from the formula below.

$$x'_a = 0.0683 \log_{10} \left[\frac{1}{\left[n(r)(a)^{n-1} \right]^{\frac{1}{n}}} \right] \text{ M}\Omega\text{-mi}$$

where

r = conductor radius in feet
n = number of conductors per phase
$a = s/(2 \sin (\pi/n)); n > 1$
$a = 0: 0° \equiv 1; n = 1$
s = bundle spacing in feet
$x'_d = 0.0683 \log_{10} (\text{GMD})$ MΩ-mi
where
GMD = geometric mean distance between phases in feet

TABLE A.29
Resistance of ACSR Conductors (Ω/mi)

Expanded	3,108,000	62/8	19	2.500	0.0294	0.0333	0.0362	0.0389	0.0418
Expanded	2,294,000	66/6	19	2.320	0.0399	0.0412	0.0453	0.0493	0.0533
Expanded	1,414,000	58/4	19	1.750	0.0644	0.0663	0.0728	0.0793	0.0859
Expanded	1,275,000	50/4	19	1.600	0.0716	0.0736	0.0808	0.0881	0.0953
Kiwi	2,167,000	72	7	1.737	0.0421	0.0473	0.0515	0.0552	0.0586
Bluebird	2,156,000	84	19	1.762	0.0420	0.0464	0.0507	0.0545	0.0586
Chukar	1,780,000	84	19	1.602	0.0510	0.0548	0.0599	0.0647	0.0696
Falcon	1,590,000	54	19	1.545	0.0567	0.0594	0.0653	0.0707	0.0763
Lapwing	1,590,000	45	7	1.502	0.0571	0.0608	0.0686	0.0719	0.0774
Parrot	1,510,500	54	19	1.506	0.0597	0.0625	0.0686	0.0744	0.0802
Nuthatch	1,510,500	45	7	1.466	0.0602	0.0636	0.0697	0.0755	0.0813
Plover	1,431,000	54	19	1.465	0.0630	0.0657	0.0721	0.0782	0.0843
Bobolink	1,431,000	45	7	1.427	0.0636	0.0668	0.0733	0.0794	0.0856
Martin	1,351,000	54	19	1.424	0.0667	0.0692	0.0760	0.0825	0.0890
Dipper	1,351,000	45	7	1.385	0.0672	0.0705	0.0771	0.0836	0.0901
Pheasant	1,272,000	54	19	1.382	0.0709	0.0732	0.0805	0.0874	0.0944
Bittern	1,272,000	45	7	1.345	0.0715	0.0746	0.0817	0.0886	0.0956
Grackle	1,192,500	54	19	1.333	0.0756	0.0778	0.0855	0.0929	0.1000
Bunting	1,192,500	45	7	1.302	0.0762	0.0792	0.0867	0.0942	0.1002
Finch	1,113,000	54	19	1.293	0.0810	0.0832	0.0914	0.0993	0.1080
Bluejay	1,113,000	45	7	1.259	0.0818	0.0844	0.0926	0.1010	0.1090
Curlew	1,033,500	54	7	1.246	0.0871	0.0893	0.0979	0.1070	0.1150
Ortolan	1,033,500	45	7	1.213	0.0881	0.0905	0.0994	0.1080	0.1170
Tanager	1,033,500	36	1	1.186	0.0885	0.0905	0.0994	0.1080	0.1170
Cardinal	954,000	54	7	1.196	0.0944	0.0963	0.1060	0.1150	0.1250
Rail	954,000	45	7	1.165	0.0954	0.0978	0.1080	0.1170	0.1260
Catbird	954,000	36	1	1.140	0.0959	0.0987	0.1090	0.1180	0.1270
Canary	900,000	54	7	1.162	0.1000	0.1020	0.1120	0.1220	0.1320
Ruddy	900,000	45	7	1.131	0.1010	0.1030	0.1130	0.1230	0.1340
Mallard	795,000	30	19	1.140	0.111	0.114	0.125	0.137	0.147
Drake	795,000	26	7	1.108	0.112	0.114	0.125	0.137	0.147
Condor	795,000	54	7	1.093	0.113	0.115	0.127	0.138	0.149
Cuckoo	795,000	24	7	1.092	0.113	0.114	0.127	0.137	0.148
Tern	795,000	45	7	1.063	0.114	0.116	0.128	0.139	0.150
Coot	795,000	36	1	1.040	0.115	0.117	0.129	0.141	0.152
Redwing	715,500	30	19	1.081	0.124	0.126	0.139	0.151	0.164
Starling	715,500	26	7	0.051	0.125	0.126	0.139	0.151	0.164
Stilt	715,500	24	7	1.036	0.126	0.127	0.141	0.153	0.165
Gannet	666,600	25	7	1.014	0.134	0.135	0.149	0.162	0.176
Flamingo	666,600	24	7	1.000	0.135	0.137	0.151	0.164	0.177
———	653,900	18	3	0.953	0.140	0.142	0.156	0.171	0.184
Egret	636,000	30	19	1.019	1.139	0.143	0.157	0.172	0.186
Grosbeak	636,000	26	7	0.990	0.140	0.142	0.156	0.170	0.184
Rook	636,000	24	7	0.977	0.142	0.143	0.157	0.172	0.186
Kingbird	636,000	18	1	0.940	0.143	0.145	0.160	0.174	0.188
Swift	636,000	36	1	0.930	0.144	0.146	0.161	0.175	0.189
Teal	605,000	30	19	0.994	0.146	0.150	0.165	0.180	0.195
Squab	605,000	26	7	0.966	0.147	0.149	0.164	0.179	0.193

(continued)

TABLE A.29 (Continued)
Resistance of ACSR Conductors (Ω/mi)

Peacock	605,000	24	7	0.953	0.149	0.150	0.165	0.180	0.195
Eagle	556,500	30	7	0.953	0.158	0.163	0.179	0.196	0.212
Dove	556,500	26	7	0.927	0.160	0.162	0.178	0.194	0.211
Parakeet	556,500	24	7	0.914	0.162	0.163	0.179	0.196	0.212
Osprey	556,500	18	1	0.879	0.163	0.166	0.183	0.199	0.215
Hen	477,000	30	7	0.883	0.185	0.190	0.209	0.228	0.247
Hawk	477,000	26	7	0.858	0.187	0.188	0.207	0.226	0.247
Flicker	477,000	24	7	0.846	0.189	0.190	0.209	0.228	0.247
Pelican	477,000	18	1	0.824	0.191	0.193	0.212	0.232	0.250
Lark	397,500	30	7	0.806	0.222	0.227	0.250	0.273	0.295
Ibis	397,500	26	7	0.783	0.224	0.226	0.249	0.271	0.294
Brant	397,500	24	7	0.772	0.226	0.227	0.250	0.273	0.295
Chickadee	397,500	18	1	0.743	0.229	0.231	0.254	0.277	0.300
Oriole	336,400	30	7	0.741	0.262	0.268	0.295	0.322	0.349
Linnet	336,400	26	7	0.721	0.265	0.267	0.294	0.321	0.347
Merlin	336,400	18	1	0.684	0.270	0.273	0.300	0.328	0.355
Ostrich	300,000	26	7	0.680	0.297	0.299	0.329	0.359	0.3829

TABLE A.30
Resistance of ACAR Conductors (Ω/mi)

Area 62% Eq. EC-Al (cmil)	Strands EC/6201		Dia. (in.)	dc 20°C	ac-60 Hz 30°C	50°C	75°C	100°C
2,413,000	72	19	1.821	0.0373	0.0456	0.0483	0.0516	0.0565
2,375,000	63	28	1.821	0.0379	0.0462	0.0488	0.0521	0.0555
2,338,000	54	37	1.821	0.0385	0.0467	0.0493	0.0527	0.0561
2,297,000	54	7	1.762	0.0392	0.0474	0.0502	0.0538	0.0573
2,262,000	48	13	1.762	0.0399	0.0479	0.0507	0.0543	0.0580
2,226,000	42	19	1.762	0.0405	0.0485	0.0513	0.0549	0.0585
2,227,000	54	7	1.735	0.0405	0.0485	0.0514	0.0551	0.089
2,159,000	42	19	1.735	0.0417	0.0497	0.0526	0.0562	0.0601
1,899,000	54	7	1.602	0.0474	0.0550	0.0585	0.0629	0.0674
1,870,000	48	13	1.602	0.0482	0.0557	0.0592	0.0636	0.0682
1,841,000	42	19	1.602	0.0489	0.0564	0.0599	0.0644	0.0689
1,673,000	54	7	1.504	0.0539	0.0611	0.0651	0.0702	0.0754
1,647,000	48	13	1.504	0.0546	0.0619	0.0659	0.0711	0.0762
1,622,000	42	19	1.504	0.0555	0.0627	0.0667	0.0719	0.0771
1,337,000	54	7	1.345	0.0674	0.0742	0.0794	0.0860	0.0925
1,296,000	42	19	1.345	0.0695	0.0763	0.0815	0.0881	0.0947
1,243,000	30	7	1.302	0.0725	0.0793	0.0849	0.0919	0.0989
1,211,000	24	13	1.302	0.0744	0.0812	0.0868	0.0937	0.1008
1,179,000	18	19	1.302	0.0764	0.0831	0.0887	0.0957	0.1028
1,163,000	30	7	1.259	0.0775	0.0842	0.0902	0.0977	0.1052
1,133,000	24	13	1.259	0.0795	0.0862	0.0922	0.0997	0.1073
1,104,000	18	19	1.259	0.0816	0.0882	0.0942	0.1018	0.1095

(continued)

TABLE A.30 (Continued)
Resistance of ACAR Conductors (Ω/mi)

Area 62% Eq. EC-Al (cmil)	Strands EC/6201		Dia. (in.)	dc 20°C	ac-60 Hz 30°C	50°C	75°C	100°C
1,153,000	33	4	1.246	0.0781	0.0850	0.0910	0.0987	0.1064
1,138,000	30	7	1.246	0.0791	0.0859	0.0920	0.0997	0.1074
1,109,000	24	13	1.246	0.0812	0.0880	0.0941	0.1017	0.1095
1,080,000	18	19	1.246	0.0834	0.0900	0.0962	0.1039	0.1117
1,077,000	30	7	1.212	0.0836	0.0904	0.0969	0.1050	0.1132
1,049,000	24	13	1.212	0.0859	0.0926	0.0991	0.1072	0.1154
1,022,000	18	19	1.212	0.0882	0.0948	0.1013	0.1096	0.1177
1,050,000	30	7	1.196	0.0859	0.0926	0.0993	0.1076	0.1160
1,023,000	24	13	1.196	0.0881	0.0948	0.1015	0.1099	0.1160
996,000	18	19	1.196	0.0881	0.0971	0.1038	0.1123	0.1207
994,800	30	7	1.165	0.0906	0.0974	0.1044	0.1133	0.1221
954,600	30	7	1.141	0.0944	0.1072	0.1086	0.1178	0.1271
969,300	24	13	1.165	0.0929	0.0997	0.1068	0.1156	0.1246
958,000	24	13	1.158	0.0941	0.1008	0.1080	0.1170	0.1260
943,900	18	19	1.165	0.0955	0.1021	0.1092	0.1182	0.1271
900,300	30	7	1.108	0.1001	0.1070	0.1148	0.1246	0.1345
795,000	30	7	1.042	0.1133	0.1204	0.1293	0.1404	0.1516
877,300	24	13	0.108	0.1027	0.1096	0.1174	0.1272	0.1372
795,000	24	13	1.055	0.1133	0.1204	0.1290	0.1400	0.1509
854,200	18	19	1.108	0.1054	0.1123	0.1201	0.1300	0.1400
795,000	18	19	1.069	0.1134	0.1203	0.1288	0.1394	0.1501
807,700	24	13	1.063	0.1115	0.1185	0.1271	0.1378	0.1486
786,500	18	19	1.063	0.1145	0.1216	0.1302	0.1410	0.1518
727,500	33	4	0.990	0.1238	0.1312	0.1411	0.1534	0.1658
718,300	30	7	0.990	0.1254	0.1327	0.1427	0.1550	0.1675
700,000	24	13	0.990	0.1287	0.1360	0.1459	0.1584	0.1709
681,600	18	19	0.990	0.1322	0.1394	0.1494	0.1619	0.1743
632,000	15	4	0.927	0.1425	0.1503	0.1615	0.1756	0.1897
616,200	12	7	0.927	0.1462	0.1539	0.1652	0.1794	0.1935
487,400	15	4	0.814	0.1849	0.1938	0.2085	0.2268	0.2453
475,200	12	7	0.814	0.1896	0.1986	0.2133	0.2317	0.2501
343,600	15	4	0.684	0.2623	0.2739	0.2948	0.3209	0.3470
335,000	12	7	0.684	0.2690	0.2806	0.3015	0.3278	0.3540

Appendix B: Standard Device Numbers Used in Protection Systems

Some of the frequently used device numbers are listed below. A complete list and definitions are given in the American National Standards Institute/Institute of Electrical and Electronics Engineers (ANSI/IEEE) Standard C37.2-1079.

1. Master element: normally used for hand-operated devices
2. Time-delay starting or closing relay
3. Checking or interlocking relay
4. Master contactor
5. Stopping device
6. Starting circuit breaker
7. Anode circuit breaker
8. Control power-disconnecting device
9. Reversing device
10. Unit sequence switch
12. Synchronous-speed device
14. Underspeed device
15. Speed- or frequency-matching device
17. Shunting or discharge switch
18. Accelerating or decelerating device
20. Electrically operated valve
21. Distance relay
23. Temperature control device
25. Synchronizing or synchronism-check device
26. Apparatus thermal device
27. Undervoltage relay
29. Isolating contactor
30. Annunciator relay
32. Directional power relay
37. Undercurrent or underpower relay
46. Reverse-phase or phase-balance relay
47. Phase-sequence voltage relay
48. Incomplete-sequence relay
49. Machine or transformer thermal relay
50. Instantaneous overcurrent or rate-of-rise relay
51. Ac time overcurrent relay
52. Ac circuit breaker: the mechanism-operated contacts are
    (a) 52a, 52aa: open when breaker, closed when breaker contacts closed
    (b) 52b, 52bb: operates just as the mechanism motion start; known as high-speed contacts
55. Power factor relay
57. Short-circuiting or grounding device

59. Overvoltage relay
60. Voltage or current balance relay
62. Time-delay stopping or opening relay
64. Ground detector relay
67. Ac directional overcurrent relay
68. Blocking relay
69. Permissive control device
72. Ac circuit breaker
74. Alarm relay
76. Dc overcurrent relay
78. Phase-angle measuring or out-of-step protective relay
79. Ac reclosing relay
80. Flow switch
81. Frequency relay
82. Dc reclosing relay
83. Automatic selective control or transfer relay
84. Operating mechanism
85. Carrier or pilot-wire receiver relay
86. Lockout relay
87. Differential protective relay
89. Line switch
90. Regulating device
91. Voltage directional relay
92. Voltage and power directional relay
93. Field-changing contactor
94. Tripping or trip-free relay

Appendix C: Unit Conversions from English System to SI System

The following are useful when converting from the English system to the SI system:

Length:	1 in. = 2.54 cm = 0.0245 m
	1 ft = 30.5 cm = 0.305 m
	1 mile = 1609 m
Area:	1 square mile = 2.59×10^6 m^2
	1 in.^{2} = 0.000645 m^2
	1 in.^{2} = 6.45 cm^2
Volume:	1 ft^3 = 0.0283 m^3
Linear speed:	1 ft/s = 0.305 m/s = 30.3 cm/s
	1 mph = 0.447 m/s
	1 in./s = 0.0254 m/s = 2.54 cm/s
Rotational speed:	1 rev/min = 0.105 rad/s = 6 deg/s
Force:	1 lb = 4.45 N
Power:	1 hp = 746 W = 0.746 kW
Torque:	1 ft-lb = 1.356 N-m
Magnetic flux:	1 line = 1 maxwell = 10^{-8} Wb
	1 kiloline = 1000 maxwells = 10^{-5} Wb
Magnetic flux density:	1 line/in^2 = 15.5×10^{-6} T
	100 kilolines/in^2 = 1.55 T = 1.55 Wb/m^2
Magnetomotive force:	1 ampere-turn = 1 A
Magnetic field intensity:	1 A-turn/in. = 39.37 A/m

Appendix D: Unit Conversions from SI System to English System

The following are useful when converting from the SI system to the English system:

Length:	1 m = 100 cm = 39.7 in.
	1 m = 3.28 ft
	1 m = 6.22×10^{-4} mile
Area:	1 m^2 = 0.386×10^{-6} mile^2
	1 m^2 = 1550 in.^{2}
	1 cm^2 = 0.155 in.^{2}
Volume:	1 m^3 = 35.3 ft^3
Linear speed:	1 m/s = 100 cm/s = 3.28 ft/s
	1 m/s = 2.237 mph
	1 m/s = 39.37 in./s
Rotational speed:	1 rad/s = 9.55 rev/min = 57.3 deg/s
Force:	1 N = 0.225 lb
Power:	1 kW = 1000 W = 1.34 hp
Torque:	1 N-m = 0.737 lb
Magnetic flux:	1 Wb = 10^8 lines = 10^8 maxwells
	1 Wb = 105 kilolines
Magnetic flux density:	1 T = 6.45×10^4/line/in^2
	1 T = 64.5 kilolines/in^2
	1 T = 1 Wb/m^2
Magnetomotive force:	1 A = 1 A-turn
Magnetic field intensity:	1 A/m = 0.0254 A-turn/in.

Appendix E: Prefixes

E.1 PREFIXES

The prefixes indicating decimal multiples or submultiples of units and their symbols are given in Table E.1.

TABLE E.1
Recommended Prefixes

Multiple	Prefix	Symbol
10^{12}	tera	T
10^{9}	giga	G
10^{6}	mega	M
10^{3}	kilo	k
10^{2}	hecto	h
10	deca	da
10^{-1}	deci	d
10^{-2}	centi	c
10^{-3}	mili	m
10^{-6}	micro	A
10^{-9}	nano	n
10^{-12}	pico	p
10^{-15}	femto	f
10^{-18}	atto	a

Appendix F: Greek Alphabet Used for Symbols

Table F.1 presents capital and lowercase Greek alphabet symbols.

TABLE F.1
Greek Alphabet Symbols

Greek Letter	Greek Name	English Equivalent	Greek Letter	Greek Name	English Equivalent
Αα	Alpha	a	Νν	Nu	n
Ββ	Beta	b	Ξξ	Xi	x
Γγ	Gamma	g	Οο	Omicron	ŏ
Δδ	Delta	d	Ππ	Pi	p
Εε	Epsilon	ĕ	Ρρ	Rho	r
Ζζ	Zeta	z	Σσς	Sigma	s
Ηη	Eta	ē	Ττ	Tau	t
Θθ	Theta	th	Υυ	Upsilon	u
Ιι	Iota	i	Φφ	Phi	ph
Κκ	Kappa	k	Χχ	Chi	ch
Λλ	Lambda	l	Ψψ	Psi	ps
Μμ	Mu	m	Ωω	Omega	ō

Appendix G: Additional Solved Examples of Shunt Faults

Example G.1

Consider the system shown in Figure G.1 and the following data:

Generator G_1: 15 kV, 50 MVA, $X_1 = X_2 = 0.10$ pu and $X_0 = 0.05$ pu based on its own ratings
Synchronous motor M: 15 kV, 20 MVA, $X_1 = X_2 = 0.20$ pu and $X_0 = 0.07$ pu based on its own ratings
Transformer T_1: 15/115 kV, 30 MVA, $X_1 = X_2 = X_0 = 0.06$ pu based on its own ratings
Transformer T_2: 115/15 kV, 25 MVA, $X_1 = X_2 = X_0 = 0.07$ pu based on its own ratings
Transmission line TL_{23}: $X_1 = X_2 = 0.03$ pu and $X_0 = 0.10$ pu based on its own ratings

Assume a single line-to-ground (SLG) fault at bus 4 and determine the fault current in per units and amperes. Use 50 MVA as the megavolt-ampere base and assume that Z_f is $j0.1$ pu based on 50 MVA.

Solution

Assuming an SLG fault at bus 4 with a $Z_f = j0.1$ pu on a 50 MVA base, the given reactance has to be adjusted on the basis of the new S_B. Hence, using

$$Z_{adjusted} = X_{pu(old)} \times \left(\frac{S_{B(new)}}{S_{B(old)}} \right)$$

where in this example $S_{B(new)} = 50$ MVA. Therefore, for generator G_1:

$$Z_1 = Z_2 = j0.10 \times \left(\frac{50 \text{ MVA}}{50 \text{ MVA}} \right) = j0.10 \text{ pu}$$

$$Z_0 = j0.05 \times \left(\frac{50 \text{ MVA}}{50 \text{ MVA}} \right) = j0.00 \text{ pu}$$

For transformer T_1:

$$Z_1 = Z_2 = Z_0 = j0.06 \times \left(\frac{50 \text{ MVA}}{50 \text{ MVA}} \right) = j0.01 \text{ pu}$$

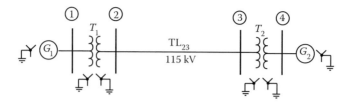

FIGURE G.1 System for Problem G.1.

For transmission line TL_{23}:

$$\mathbf{Z}_1 = \mathbf{Z}_2 = j0.03 \times \left(\frac{50 \text{ MVA}}{50 \text{ MVA}} \right) = j0.03 \text{ pu}$$

and

$$\mathbf{Z}_0 = j0.10 \times \left(\frac{50 \text{ MVA}}{50 \text{ MVA}} \right) = j0.10 \text{ pu}$$

For transformer T_2:

$$\mathbf{Z}_1 = \mathbf{Z}_2 = \mathbf{Z}_0 = j0.07 \times \left(\frac{50 \text{ MVA}}{25 \text{ MVA}} \right) = j0.14 \text{ pu}$$

For synchronous motor M:

$$\mathbf{Z}_1 = \mathbf{Z}_2 = j0.20 \times \left(\frac{50 \text{ MVA}}{20 \text{ MVA}} \right) = j0.5 \text{ pu}$$

and

$$\mathbf{Z}_0 = j0.07 \times \left(\frac{50 \text{ MVA}}{20 \text{ MVA}} \right) = j0.175 \text{ pu}$$

Thus, the Thévenin impedance at the faulted bus 4 is

$$\mathbf{Z}_{1,\text{th}} = \frac{(\mathbf{Z}_{1,G_1} + \mathbf{Z}_{1,T_1} + \mathbf{Z}_{1,TL} + \mathbf{Z}_{1,T_2})(\mathbf{Z}_{1,M})}{\mathbf{Z}_{1,G_1} + \mathbf{Z}_{1,T_1} + \mathbf{Z}_{1,TL} + \mathbf{Z}_{1,T_2} + \mathbf{Z}_{1,M}}$$

$$= j \frac{(0.10 + 0.10 + 0.03 + 0.14)(0.5)}{(0.10 + 0.10 + 0.03 + 0.14 + 0.5)}$$

$$= j0.213 \text{ pu}$$

and since

$$\mathbf{Z}_{2,\text{th}} = \mathbf{Z}_{1,\text{th}} = j0.213 \text{ pu}$$

and

$$Z_{0,th} = \frac{(Z_{0,G} + Z_{0,T_1} + Z_{0,TL} + Z_{0,M})(Z_{0,M})}{Z_{0,G} + Z_{0,T_1} + Z_{0,TL} + Z_{0,T_2} + Z_{0,M}}$$

$$= j\frac{(0.05 + 0.10 + 0.03 + 0.14)(0.175)}{(0.05 + 0.10 + 0.10 + 0.14 + 0.175)}$$

$$= j0.1208 \text{ pu}$$

Since the voltage at the faulted bus 4 before the fault took place is $1.0\angle 0°$ pu V, then

$$\mathbf{I}_{af} = 3\mathbf{I}_{a1} = 3\left(\frac{\mathbf{V}_F}{(\mathbf{Z}_{0,th} + \mathbf{Z}_{1,th} + \mathbf{Z}_{2,th}) + 3\mathbf{Z}f}\right)$$

$$= \frac{3.0\angle 0°}{j0.1208 + j0.2126 + j0.2126 + 3(j0.1)}$$

$$\cong 3.555\angle -90° \text{ pu}$$

Since the current base at bus 4 is

$$I_B = \frac{S_{B(3\phi)}}{\sqrt{3} \times V_{L-L}}$$

$$= \frac{50 \times 10^6 \text{ VA}}{\sqrt{3} \times (15 \times 10^3 \text{ V})}$$

$$= 192.45 \text{ A}$$

Thus, the phase fault current in amperes is

$$I_f = |\mathbf{I}_{af}| \times I_B$$

$$= (5.555 \text{ pu})(192.45 \text{ A})$$

$$\cong 684.16 \text{ A}$$

Example G.2

Consider the system given in Example G.1 and assume that there is a line-to-line fault at bus 3 involving phases b and c. Determine the fault currents for both phases in per units and amperes.
Consider the system shown in Figure G.1 and the following data:

Generator G_1: 15 kV, 50 MVA, $X_1 = X_2 = 0.10$ pu and $X_0 = 0.05$ pu based on its own ratings
Synchronous motor M: 15 kV, 20 MVA, $X_1 = X_2 = 0.20$ pu and $X_0 = 0.07$ pu based on its own ratings
Transformer T_1: 15/115 kV, 30 MVA, $X_1 = X_2 = X_0 = 0.06$ pu based on its own ratings
Transformer T_2: 115/15 kV, 25 MVA, $X_1 = X_2 = X_0 = 0.07$ pu based on its own ratings
Transmission line TL_{23}: $X_1 = X_2 = 0.03$ pu and $X_0 = 0.10$ pu based on its own ratings

Assume an SLG fault at bus 3 and determine the fault current in per units and amperes. Use 50 MVA as the megavolt-ampere base and assume that $\mathbf{Z}_f$ is $j0.1$ pu based on 50 MVA.

Solution

Assuming a line-to-line fault at bus 3 with a $Z_f = j0.1$ pu on a 50 MVA base, the given reactance has already been adjusted on the basis of the new $S_{B(\text{new})} = 50$ MVA. Hence,

For generator G_1:

$$Z_1 = Z_2 = j0.10 \text{ pu}$$

and

$$Z_0 = j0.05 \text{ pu}$$

For transformer T_1:

$$Z_1 = Z_2 = Z_0 = j0.01 \text{ pu}$$

For transmission line TL_{23}:

$$Z_1 = Z_2 = j0.03 \text{ pu}$$

and

$$Z_0 = j0.10 \text{ pu}$$

For transformer T_2:

$$Z_1 = Z_2 = Z_0 = j0.14 \text{ pu}$$

For synchronous motor M:

$$Z_1 = Z_2 = j0.5 \text{ pu}$$

and

$$Z_0 = j0.175 \text{ pu}$$

Thus, the Thévenin impedance at the faulted bus 3 is

$$Z_{1,th} = \frac{(Z_{1,G_1} + Z_{1,T_1} + Z_{1,TL_{23}})(Z_{1,M} + Z_{1,T_2})}{Z_{1,G_1} + Z_{1,T_1} + Z_{1,TL_{23}} + Z_{1,T_2} + Z_{1,M}}$$

$$= j\frac{(0.10 + 0.10 + 0.03)(0.14 + 0.5)}{(0.10 + 0.10 + 0.03 + 0.14 + 0.5)}$$

$$= j0.1692 \text{ pu}$$

and since

$$Z_{2,th} = Z_{1,th} = j0.1692 \text{ pu}$$

Since the voltage at the faulted bus 3 before the fault took place is $1.0\angle 0°$ pu V, then

$$I_{a1} = \left(\frac{V_F}{Z_{1,th} + Z_{2,th}}\right)$$

$$= \frac{1.0\angle 0°}{j0.1692 + j0.1692}$$

$$= 2.9552\angle -90° \text{ pu}$$

Appendix G

Hence, the faulted phase currents for phases b and c are

$$\mathbf{I}_{bf} = \sqrt{3}\mathbf{I}_{a1}\angle -90°$$
$$= \sqrt{3}(-j2.9552\,\text{pu})\angle -90°$$
$$= -5.1186\,\text{pu}$$

where

$$\angle -90° = -j$$

Since current base 3 is

$$I_B = \frac{S_{B(3\phi)}}{\sqrt{3}\times V_{L-L}}$$
$$= \frac{50\times 10^6\,\text{VA}}{\sqrt{3}\times (115\times 10^3\,\text{V})}$$
$$= 251.02\,\text{A}$$

Thus, phase fault current in amperes is

$$I_{bf} = |\mathbf{I}_{bf}|\times I_B$$
$$= (5.1186\,\text{pu})(251.02\,\text{A})$$
$$= -1284.9\,\text{A}$$

and

$$\mathbf{I}_{cf} = -\mathbf{I}_{bf} = -(-5.1186) - 5.1186\,\text{pu}$$
$$\mathbf{I}_{cf} = -\mathbf{I}_{bf} = -(-1284.9\,\text{A}) = 1284.9\,\text{A}$$

Example G.3

Consider the system given in Example G.1 and assume that there is a DLG fault at bus 2, involving phases b and c. Assume that $\mathbf{Z}_f$ is $j0.1$ pu and $\mathbf{Z}_g$ is $j0.2$ pu (where $\mathbf{Z}_g$ is the neutral-to-ground impedance) both based on 50 VA. Consider the system shown in Figure G.1 and the following data:

Generator G_1: 15 kV, 50 MVA, $X_1 = X_2 = 0.10$ pu and $X_0 = 0.05$ pu based on its own ratings
Synchronous motor: 15 kV, 20 MVA, $X_1 = X_2 = 0.20$ pu and $X_0 = 0.07$ pu based on its own ratings
Transformer T_1: 15/115 kV, 30 MVA, $X_1 = X_2 = X_0 = 0.06$ pu based on its own ratings
Transformer T_2: 115/15 kV, 25 MVA, $X_1 = X_2 = X_0 = 0.07$ pu based on its own ratings
Transmission line TL_{23}: $X_1 = X_2 = 0.03$ pu and $X_0 = 0.10$ pu based on its own ratings

Assume a DLG fault at bus 2, involving phases b and c, and determine the fault current in per units and amperes. Use 50 MVA as the megavolt-ampere base and assume that $\mathbf{Z}_f$ is $j0.1$ pu and $\mathbf{Z}_g$ is $j0.2$ pu (where $\mathbf{Z}_g$ is the neutral-to-ground impedance) both based on 50 MVA.

Solution

The Thévenin impedance at the faulted bus 2 is

$$\mathbf{Z}_{1,th} = \frac{(\mathbf{Z}_{1,G} + \mathbf{Z}_{1,T_1})(\mathbf{Z}_{1,TL_{23}} + \mathbf{Z}_{1,T_2} + \mathbf{Z}_{1,M})}{\mathbf{Z}_{1,G} + \mathbf{Z}_{1,T_1} + \mathbf{Z}_{1,TL_{23}} + \mathbf{Z}_{1,T_2} + \mathbf{Z}_{1,M}}$$

$$= j\frac{(0.10 + 0.10)(0.03 + 0.14 + 0.05)}{(0.10 + 0.10)(0.03 + 0.14 + 0.05)}$$

$$= j0.15402 \text{ pu}$$

and since

$$\mathbf{Z}_{2,th} = \mathbf{Z}_{1,th} = j0.15402 \text{ pu}$$

and

$$\mathbf{Z}_{0,th} = \frac{(\mathbf{Z}_{0,G} + \mathbf{Z}_{0,T_1})(\mathbf{Z}_{0,TL_{23}} + \mathbf{Z}_{0,T_2} + \mathbf{Z}_{0,M})}{\mathbf{Z}_{0,G} + \mathbf{Z}_{0,T_1} + \mathbf{Z}_{0,TL_{23}} + \mathbf{Z}_{0,T_2} + \mathbf{Z}_{0,M}}$$

$$= j\frac{(0.05 + 0.10)(0.10 + 0.14 + 0.175)}{(0.05 + 0.10)(0.10 + 0.14 + 0.175)}$$

$$= j0.11018 \text{ pu}$$

thus

$$\mathbf{I}_{a1} = \frac{\mathbf{V}_F}{\mathbf{Z}_f + \mathbf{Z}_{1,th} + \frac{(\mathbf{Z}_{2,th} + \mathbf{Z}_f)(\mathbf{Z}_{0,TL_{23}} + \mathbf{Z}_f + 3\mathbf{Z}_g)}{\mathbf{Z}_{2,th} + \mathbf{Z}_{0,th} + 2\mathbf{Z}_f + 3\mathbf{Z}_g}}$$

$$= -j\frac{1.0\angle 0°}{0.1 + 0.154 + \frac{(0.154 + 0.1)(0.110 + 0.1 + 0.6)}{0.154 + 0.110 + 0.2 + 0.6}}$$

$$= -j2.2351 \text{ pu}$$

Here, by applying current division,

$$\mathbf{I}_{a0} = -\left(\frac{\mathbf{Z}_{2,th} + \mathbf{Z}_f}{\mathbf{Z}_{2,th} + \mathbf{Z}_{0,th} + 2\mathbf{Z}_f + 3\mathbf{Z}_g}\right)\mathbf{I}_{a1}$$

$$= -\left(\frac{(j0.254)(-j2.2354)}{j0.154 + j0.110 + j0.2 + j0.6}\right)(-j2.2351)$$

$$= j0.53351 \text{ pu}$$

and similarly

$$\mathbf{I}_{a2} = -\left(\frac{\mathbf{Z}_{0,th} + \mathbf{Z}_f + 3\mathbf{Z}_g}{\mathbf{Z}_{2,th} + \mathbf{Z}_{0,th} + 2\mathbf{Z}_f + 3\mathbf{Z}_g}\right)\mathbf{I}_{a1}$$

$$= -\left(\frac{(j0.254)(-j2.2352)}{j0.154 + j0.110 + j0.2 + j0.6}\right)(-j2.2351)$$

$$= j1.70161\,\text{pu}$$

or

$$\mathbf{I}_{a2} = -(\mathbf{I}_{a1} + \mathbf{I}_{a0})$$

$$= -(-j2.2351 + j0.53351)$$

$$\cong j1.70161\,\text{pu}$$

Hence, the ground current is

$$\mathbf{I}_G = 3\mathbf{I}_{a0}$$

$$= 3(-0.53351\angle -90°)$$

$$\cong 1.6005\angle -90°\,\text{pu}$$

or since

$$I_B = \frac{50,000\,\text{kVA}}{\sqrt{3}(115\,\text{kV})}$$

$$= 251.02\,\text{A}$$

or

$$I_G = I_{G,\text{pu}} \times I_B$$

$$= |(1.6005\angle -90°)|(251.02)$$

$$= 401.77\,\text{A}$$

Thus, the faulted phase current is

$$\mathbf{I}_{bf} = \mathbf{I}_{a0} + a^2\mathbf{I}_{a1} + a\mathbf{I}_{a2}$$

$$= (0.53351\angle -90°) + (1\angle 240°)(2.2351\angle 90°) + (1\angle 120°)(1.70161\angle 90°)$$

$$= 3.5017\angle 166.79°\,\text{pu}$$

or

$$I_{bf} = |I_{bf,pu}| \times I_B$$
$$= 3.5017 \times 251.02$$
$$= 879.62 \text{ A}$$

and

$$\mathbf{I}_{cf} = \mathbf{I}_{a0} + a\mathbf{I}_{a1} + a^2\mathbf{I}_{a2}$$
$$= (0.5335\angle -90°) + (1\angle 120°)(2.2351\angle -90°) + (1\angle 240°)(1.70161\angle -90°)$$
$$= 3.5017\angle 13.21° \text{ pu}$$

or

$$I_{cf} = |I_{cf,pu}| \times I_B$$
$$= 3.5017 \times 251.02$$
$$= 879.62 \text{ A}$$

Example G.4

Consider the system shown in Figure G.2 and assume that the generator is loaded and running at the rated voltage with the circuit breaker open at bus 3. Assume that the reactance values of the generator are given as $X''_d = X_1 = X_2 = 0.14$ pu and $X_0 = 0.08$ pu based on its ratings. The transformer impedances are $Z_1 = Z_2 = Z_0 = j0.05$ pu based on its ratings. The transmission line TL$_{23}$ has $\mathbf{Z}_1 = \mathbf{Z}_2 = j0.04$ pu and $\mathbf{Z}_0 = j0.10$ pu. Assume that the fault point is located on bus 1. Select 25 MVA as the megavolt-ampere base, and 8.5 and 138 kV as the low-voltage and high-voltage bases, respectively, and determine the following:

(a) Subtransient fault current for a three-phase fault in per units and amperes
(b) Line-to-ground fault [Also find the ratio of this line-to-ground fault current to the three-phase fault current found in part (a)]

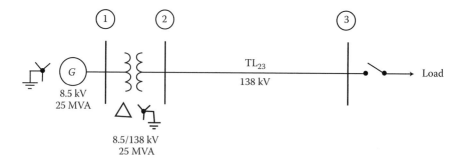

FIGURE G.2 System for Problem G.4.

Appendix G

(c) Line-to-line fault (Also find the ratio of this line-to-line fault current to previously calculated three-phase fault current)
(d) Double line-to-ground (DLG) fault

Solution

(a) The subtransient fault current is

$$\mathbf{I}''_{f,3\phi} = \frac{1.0\angle 0°}{\mathbf{Z}_1}$$

$$= \frac{1.0\angle 0°}{j0.14}$$

$$= 0.7143\angle -90° \text{ pu}$$

Since $\mathbf{I}_{f(L-L)}$ is about 86.6% of $\mathbf{I}_{f(3\phi)}$ so that

$$\left|\mathbf{I}_{f(L-L)}\right| = \frac{\sqrt{3}}{2}\left|\mathbf{I}_{f(3\phi)}\right|$$

$$= \frac{\sqrt{3}}{2}|8.3333|$$

$$= 7.2169 \text{ A}$$

(b) The SLG fault current is

$$\mathbf{I}_{af} = \mathbf{I}_{a(L-G)} = 3\mathbf{I}_{a0} = \frac{3(1.0\angle 0°)}{\mathbf{Z}_0 + \mathbf{Z}_1 + \mathbf{Z}_2}$$

$$= \frac{3.0\angle 0°}{j(0.08+0.14+0.14)}$$

$$= -j8.3333 \text{ pu}$$

and

$$\frac{I_{f(L-G)}}{I_{f(3\phi)}} = \frac{8.3333}{7.2169} = 1.1547$$

(c) Since $\mathbf{I}_{a0} = 0$ and $\mathbf{I}_{af} = 0$,

$$\mathbf{I}_{a1} = -\mathbf{I}_{a2} = \frac{1.0\angle 0°}{\mathbf{Z}_1 + \mathbf{Z}_2}$$

$$= \frac{1.0\angle 0°}{j0.14 + j0.14}$$

$$= -j3.571 \text{ pu}$$

Therefore,

$$\mathbf{I}_{cf} = -\sqrt{3}\mathbf{I}_{a1}\angle -90°$$
$$= \sqrt{3}(3.571)\angle -90°$$
$$= -6.186 \text{ pu}$$

and

$$\mathbf{I}_{cf} = -\mathbf{I}_{bf}$$
$$= (6.186 \text{ pu})(1698.089 \text{ A})$$
$$= 10,504.2 \text{ A}$$

and the ratio is

$$\frac{I_{f(L-L)}}{I_{f(3\phi)}} = \frac{6.186}{7.143} = 0.866$$

thus

$$I_{f(L-L)} = 86.6\% \text{ of } I_{f(3\phi)}$$

(d) To calculate the DLG fault current,

$$\mathbf{I}_{a1} = \frac{1.0\angle 0°}{\mathbf{Z}_1 + \dfrac{\mathbf{Z}_2 \times \mathbf{Z}_0}{\mathbf{Z}_2 + \mathbf{Z}_0}}$$

$$= \frac{1.0\angle 0°}{j0.14 + j\dfrac{0.14 \times 0.08}{0.14 + 0.08}}$$

$$= -j5.2381 \text{ pu}$$

where

$$\mathbf{V}_{a1} = 1.0\angle 0° - \mathbf{I}_{a1}\mathbf{Z}_{1,G}$$
$$= 1 - (-j5.2381)(j0.14)$$
$$= 0.26667 \text{ pu}$$

so that

$$\mathbf{I}_{a2} = -\frac{\mathbf{V}_{a1}}{\mathbf{Z}_2}$$
$$= -\frac{0.26667}{j0.14}$$
$$= j1.9048 \text{ pu}$$

Appendix G

and

$$\mathbf{I}_{a0} = -\frac{\mathbf{V}_{a1}}{\mathbf{Z}_0}$$

$$= -\frac{2.6667}{j0.08}$$

$$= j3.3333\,\text{pu}$$

also,

$$\mathbf{I}_{nf} = \text{neutral current at fault}$$
$$= 3\mathbf{I}_{a0}$$
$$= 3(j3.3333)$$
$$= j10.013\,\text{pu}$$

The faulted phase currents

$$\mathbf{I}_{bf} = \mathbf{I}_{a0} + a^2\mathbf{I}_{a1} + a\mathbf{I}_{a2}$$
$$= j3.3333 + (1.0\angle 240°)(-j5.238) + (1.0\angle 120°)(j1.905)$$
$$\cong -6.1857 + j5.0001$$
$$\cong 7.954\angle 218.95°\,\text{pu}$$

and

$$\mathbf{I}_{cf} = \mathbf{I}_{a0} + a\mathbf{I}_{a1} + a^2\mathbf{I}_{a2}$$
$$= j3.3333 + (1.0\angle 120°)(-j5.238) + (1.0\angle 240°)(j1.905)$$
$$\cong 6.1857 + j5.0001$$
$$\cong 7.954\angle 38.95°\,\text{pu}$$

hence

$$I_{bf} = I_{cf} = \left|\mathbf{I}_{bf,\text{pu}}\right| I_B$$
$$= 7.954 \times 1698.089$$
$$= 13.501\,\text{A}$$

Example G.5

Repeat Example G.4 assuming that the fault is located on bus 2.

Solution

(a) The subtransient fault current is

$$I''_{f,3\phi} = \frac{1.0\angle 0°}{Z_{1,G} + Z_{1,T_1}}$$

$$= \frac{1.0\angle 0°}{j0.14 + j0.05}$$

$$= \frac{1.0\angle 0°}{j0.19}$$

$$= 5.2632\angle -90° \text{ pu}$$

Since $I_{f(L-L)}$ is about 86.6% of $I_{f(3\phi)}$ so that

$$\left|I_{f(L-L)}\right| = \frac{\sqrt{3}}{2}\left|I_{f(3\phi)}\right|$$

$$= \frac{\sqrt{3}}{2}|5.2632|$$

$$= 4.5580 \text{ A}$$

(c) The SLG fault current is

$$I_{af} = I_{a(L-G)} = 3I_{a0} = \frac{3(1.0\angle 0°)}{Z_0 + Z_1 + Z_2}$$

$$= \frac{3.0\angle 0°}{j(0.05 + 0.19 + 0.19)}$$

$$= -j6.9767 \text{ pu}$$

where

$$Z_1 = Z_{1,G} + Z_{1,T_1} = j0.14 + j0.05 = j0.19 \text{ pu}$$

$$Z_2 = Z_{2,G} + Z_{2,T_1} = j0.14 + j0.05 = j0.19 \text{ pu}$$

$$Z_0 = Z_{0,T_1} = j0.05 = j0.05 \text{ pu}$$

and

$$I_B = \frac{25 \times 10^6}{\sqrt{3}(138 \times 10^3)} 104.59 \text{ A}$$

Appendix G

Also,

$$\mathbf{I}_{af} = |\mathbf{I}_{af}| \times I_B$$
$$= |-j6.9767| \times (104.59\text{ A})$$
$$= 729.71\text{ A}$$

and

$$\frac{I_{f(L-G)}}{I_{f(3\phi)}} = \frac{6.9767}{5.2632} = 1.3255$$

(c) Since $\mathbf{I}_{a0} = 0$ and $\mathbf{I}_{af} = 0$,

$$\mathbf{I}_{a1} = -\mathbf{I}_{a2} = \frac{1.0\angle 0°}{\mathbf{Z}_1 + \mathbf{Z}_2}$$
$$= \frac{1.0\angle 0°}{j0.19 + j0.19}$$
$$= -j2.6316\text{ pu}$$

Therefore,

$$\mathbf{I}_{cf} = -\sqrt{3}\mathbf{I}_{a1}\angle -90°$$
$$= \sqrt{3}(2.6316)\angle -90°$$
$$= -4.558\text{ pu}$$

and

$$\mathbf{I}_{cf} = -\mathbf{I}_{bf}$$
$$= (4.558\text{ pu})(104.592\text{ A})$$
$$= 476.73\text{ A}$$

and the ratio is

$$\frac{I_{f(L-L)}}{I_{f(3\phi)}} = \frac{4.558}{5.2632} = 0.866$$

thus

$$I_{f(L-L)} = 86.6\% \text{ of } I_{f(3\phi)}$$

(d) To calculate the DLG fault current,

$$\mathbf{I}_{a1} = \frac{1.0\angle 0°}{\mathbf{Z}_1 + \dfrac{\mathbf{Z}_2 \times \mathbf{Z}_0}{\mathbf{Z}_2 + \mathbf{Z}_0}}$$

$$= \frac{1.0\angle 0°}{j0.19 + j\dfrac{0.19 \times 0.08}{0.19 + 0.08}}$$

$$= -j4.3557 \text{ pu}$$

where

$$\mathbf{V}_{a1} = 1.0\angle 0° - \mathbf{I}_{a1}\mathbf{Z}_{1,G}$$

$$= 1 - (-j4.3557)(j0.19)$$

$$= 0.17241 \text{ pu}$$

so that

$$\mathbf{I}_{a2} = -\frac{\mathbf{V}_{a1}}{\mathbf{Z}_2}$$

$$= -\frac{0.17241}{j0.19}$$

$$= j1.90744 \text{ pu}$$

and

$$\mathbf{I}_{a0} = -\frac{\mathbf{V}_{a1}}{\mathbf{Z}_0}$$

$$= -\frac{0.17241}{j0.05}$$

$$= j3.4483 \text{ pu}$$

also,

$$\mathbf{I}_{nf} = \text{neutral current at fault}$$

$$= 3\mathbf{I}_{a0}$$

$$= 3(j3.4483)$$

$$= j10.345 \text{ pu}$$

Appendix G

The faulted phase currents

$$\mathbf{I}_{bf} = \mathbf{I}_{a0} + a^2 \mathbf{I}_{a1} + a \mathbf{I}_{a2}$$
$$= j3.4483 + (1.0\angle 240°)(-j4.3557) + (1.0\angle 120°)(j0.90744)$$
$$\cong -4.5579 + j5.1722$$
$$\cong 6.8939\angle 228.6° \text{ pu}$$

and

$$\mathbf{I}_{cf} = \mathbf{I}_{a0} + a \mathbf{I}_{a1} + a^2 \mathbf{I}_{a2}$$
$$= j3.483 + (1.0\angle 120°)(-j4.3557) + (1.0\angle 240°)(j0.90744)$$
$$\cong 4.186 + j5.1722$$
$$\cong 6.8939\angle 48.6° \text{ pu}$$

hence

$$I_{bf} = I_{cf} = |\mathbf{I}_{bf,\text{pu}}| I_B$$
$$= 6.8939 \times 104.59$$
$$= 721.03 \text{ A}$$

Example G.6

Repeat Example G.4 assuming that the fault is located on bus 3.

Solution

(b) The subtransient fault current is

$$\mathbf{I}''_{f,3\phi} = \frac{1.0\angle 0°}{\mathbf{Z}_{1,G} + \mathbf{Z}_{1,T_1} + \mathbf{Z}_{TL_{23}}}$$
$$= \frac{1.0\angle 0°}{j0.14 + j0.05 + j0.05}$$
$$= \frac{1.0\angle 0°}{j0.23}$$
$$= 4.918\angle -90° \text{ pu}$$

Since $\mathbf{I}_{f(L-L)}$ is about 86.6% of $\mathbf{I}_{f(3\phi)}$ so that

$$|\mathbf{I}_{f(L-L)}| = \frac{\sqrt{3}}{2}|\mathbf{I}_{f(3\phi)}|$$
$$= \frac{\sqrt{3}}{2}|4.918|$$
$$= 4.259 \text{ A}$$

(d) The SLG fault current is

$$\mathbf{I}_{af} = \mathbf{I}_{a(L-G)} = 3\mathbf{I}_{a0} = \frac{3(1.0\angle 0°)}{\mathbf{Z}_0 + \mathbf{Z}_1 + \mathbf{Z}_2}$$

$$= \frac{3.0\angle 0°}{j(0.15 + 0.23 + 0.23)}$$

$$= -j4.918\,\text{pu}$$

where

$$\mathbf{Z}_1 = \mathbf{Z}_{1,G} + \mathbf{Z}_{1,TL_{23}} = j0.14 + j0.05 + j0.04 = j0.23\,\text{pu}$$

$$\mathbf{Z}_2 = \mathbf{Z}_{2,G} + \mathbf{Z}_{2,T_1} + \mathbf{Z}_{2,TL_{23}} = j0.14 + j0.05 + j0.04 = j0.23\,\text{pu}$$

$$\mathbf{Z}_0 = \mathbf{Z}_{0,T_1} + \mathbf{Z}_{0,TL_{23}} = j0.05 + j0.10 = j0.15\,\text{pu}$$

and

$$I_B = \frac{25 \times 10^6}{\sqrt{3}(138 \times 10^3)} 104.59\,\text{A}$$

Also,

$$\mathbf{I}_{af} = \left|\mathbf{I}_{af}\right| \times I_B$$

$$= \left|-j4.9181\right| \times (104.59\,\text{A})$$

$$= 514.39\,\text{A}$$

and

$$\frac{I_{f(L-G)}}{I_{f(3\phi)}} = \frac{4.918}{4.3478} \cong 1.1311$$

(c) Since $\mathbf{I}_{a0} = 0$ and $\mathbf{I}_{af} = 0$ for a line-to-line fault,

$$\mathbf{I}_{a1} = -\mathbf{I}_{a2} = \frac{1.0\angle 0°}{\mathbf{Z}_1 + \mathbf{Z}_2}$$

$$= \frac{1.0\angle 0°}{j0.23 + j0.23}$$

$$= -j2.1739\,\text{pu}$$

Appendix G

Therefore,

$$\mathbf{I}_{bf} = -\sqrt{3}\mathbf{I}_{a1}\angle -90°$$
$$= \sqrt{3}(2.1739)\angle -90°$$
$$= -3.7653\,\text{pu}$$

or

$$\mathbf{I}_{bf} = |\mathbf{I}_{bf,pu}| \times I_B$$
$$= \sqrt{2}(2.1739\angle -90°)$$
$$= -3.7653\,\text{pu}$$

and

$$\mathbf{I}_{cf} = -\mathbf{I}_{bf}$$
$$= (3.7653\,\text{pu})(104.592\,\text{A})$$
$$= 393.82\,\text{A}$$

and the ratio is

$$\frac{I_{f(L-L)}}{I_{f(3\phi)}} = \frac{3.7653}{4.3478} = 0.866$$

thus

$$I_{f(L-L)} = 86.6\% \text{ of } I_{f(3\phi)}$$

(d) To calculate the DLG fault current,

$$\mathbf{I}_{a1} = \frac{1.0\angle 0°}{\mathbf{Z}_1 + \dfrac{\mathbf{Z}_2 \times \mathbf{Z}_0}{\mathbf{Z}_2 + \mathbf{Z}_0}}$$

$$= \frac{1.0\angle 0°}{j0.23 + j\dfrac{0.23 \times 0.15}{0.23 + 0.15}}$$

$$= -j3.1173\,\text{pu}$$

where

$$\mathbf{V}_{a1} = 1.0\angle 0° - \mathbf{I}_{a1}\mathbf{Z}_{1,G}$$
$$= 1 - (-j3.1173)(j0.23)$$
$$= 0.28302\,\text{pu}$$

so that

$$\mathbf{I}_{a2} = -\frac{\mathbf{V}_{a1}}{\mathbf{Z}_2}$$

$$= -\frac{0.28302}{j0.23}$$

$$= j1.2305\,\text{pu}$$

and

$$\mathbf{I}_{a0} = -\frac{\mathbf{V}_{a1}}{\mathbf{Z}_0}$$

$$= -\frac{0.28302}{j0.15}$$

$$= j1.8868\,\text{pu}$$

also,

$$\mathbf{I}_{nf} = \text{neutral current at fault}$$

$$= 3\mathbf{I}_{a0}$$

$$= 3(j1.8868)$$

$$= j5.6604\,\text{pu}$$

The faulted phase currents

$$\mathbf{I}_{bf} = \mathbf{I}_{a0} + a^2\mathbf{I}_{a1} + a\mathbf{I}_{a2}$$

$$= j1.8868 + (1.0\angle 240°)(-j3.1173) + (1.0\angle 120°)(j1.2305)$$

$$\cong -3.7652 + j2.2801$$

$$\cong 4.4018\angle 211.2°\,\text{pu}$$

and

$$\mathbf{I}_{cf} = \mathbf{I}_{a0} + a\mathbf{I}_{a1} + a^2\mathbf{I}_{a2}$$

$$= j1.8868 + (1.0\angle 120°)(-j3.1173) + (1.0\angle 240°)(j1.2305)$$

$$\cong 3.7652 + j2.8302$$

$$\cong 4.7103\angle 36.93°\,\text{pu}$$

hence

$$I_{bf} = I_{cf} = |\mathbf{I}_{bf,\text{pu}}|I_B$$

$$= 4.7103 \times 104.59$$

$$= 492.66\,\text{A}$$

Appendix G

Example G.7

Consider the system shown in Figure G.3. Assume that loads, line capacitance, and transformer-magnetizing currents are neglected and that the following data are given based on 20 MVA and the line-to-line voltages as shown in Figure G.3. Do not neglect the resistance of the transmission line TL_{23}. The prefault positive-sequence voltage at bus 3 is $V_{an} = 1.0\angle 0°$ pu, as shown in Figure G.3.

Generator G_1: $X_1 = 0.20$ pu, $X_2 = 0.10$ pu, $X_0 = 0.05$ pu
Transformer T_1: $X_1 = X_2 = 0.05$ pu, $X_0 = X_1$ (looking into the high-voltage side)
Transformer T_2: $X_1 = X_2 = 0.05$ pu, $X_0 = \infty$ (looking into the high-voltage side)
Transmission line TL_{23}: $Z_1 = Z_2 = 0.2 + j0.2$ pu, $Z_0 = 0.6 + j0.6$ pu

Assume that there is a bolted (i.e., with zero fault impedance) line-to-line fault on phases b and c at bus 3 and determine the following:

(a) Fault current I_{bf} in per units and amperes
(b) Phase voltages V_a, V_b, and V_c at bus 2 in per units and kilovolts
(c) Line-to-line voltages V_{ab}, V_{bc}, and V_{ca} at bus 2 in kilovolts
(d) Generator line currents I_a, I_b, and I_c

Given: Per-unit positive-sequence currents on the low-voltage side, of the delta–wye-connected transformer bank, lag positive-sequence currents on the high-voltage side by 30°, and similarly, for negative-sequence currents, on the low-voltage side of the transformer bank, lead positive-sequence currents on the high-voltage side by 30°.

Solution

When fault is located on bus 3,

$$I_B = \frac{20 \times 10^6}{\sqrt{3}(20 \times 10^3)}$$

$$= 577.35 \text{ A}$$

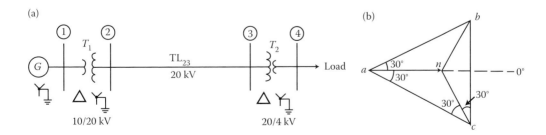

FIGURE G.3 System for Example G.7.

(a) Since there is a bolted line-to-line fault, $I_{a0} = 0$, $I_{af} = 0$, and $Z_f = 0$, then

$$I_{a1} = -I_{a2} = \frac{1.0\angle 0°}{Z_1 + Z_2}$$

$$= \frac{1.0\angle 0°}{(0.2 + j0.45) + (0.2 + j0.35)}$$

$$= \frac{1.0\angle 0°}{(0.4 + j)0.80}$$

$$= 1.118\angle -63.43° \text{ pu}$$

where

$$Z_1 = Z_{1,G} + Z_{1,T_1} + Z_{1,TL_{23}}$$

$$= j0.2 + j0.05 + (0.2 + j0.2)$$

$$= 0.2 + j0.45 \text{ pu}$$

and

$$Z_2 = Z_{2,G} + Z_{2,T_1} + Z_{2,TL_{23}}$$

$$= j0.1 + j0.05 + (0.2 + j0.2)$$

$$= 0.2 + j0.35 \text{ pu}$$

Therefore,

$$I_{bf} = \sqrt{3} I_{a1} \angle -90°$$

$$= \sqrt{3}(3.571)\angle -90°$$

$$= -6.186 \text{ pu}$$

and

$$I_{bf} = I_{a0} + a^2 I_{a1} + a I_{a2}$$

$$= 0 + a^2 I_{a1} + a I_{a2}$$

$$= (a^2 - a) I_{a1}$$

$$= -j\sqrt{3}(1.118\angle -63.4°)$$

$$\cong 1.94\angle 26.6° \text{ pu}$$

$$\cong (1.94\angle 26.6° \text{ pu})(577.35 \text{ A})$$

$$= 1120 \text{ A}$$

Appendix G

(b) The positive-sequence voltage at bus 2 is

$$\mathbf{V}_{a1}^{(2)} = 1.0\angle 0° - \mathbf{I}_{a1}\mathbf{Z}_1^{(2)}$$

$$= 1 - \frac{j0.25}{0.895\angle 63.4°}$$

$$= 0.75 - j0.121 \text{ pu}$$

$$= 0.76\angle -9.2° \text{ pu}$$

so that

$$\mathbf{V}_{a2}^{(2)} = -\mathbf{Z}_2^{(2)} \times \mathbf{I}_{a2}$$

$$= \frac{j0.15}{0.895\angle 63.4°}$$

$$= 0.15 - j0.084$$

$$= 0.168\angle 26.6° \text{ pu}$$

and the faulted phase voltages are

$$\mathbf{V}_a^{(2)} = \mathbf{V}_{a0}^{(2)} + \mathbf{V}_{a1}^{(2)} + \mathbf{V}_{a2}^{(2)}$$

$$= 0 + 0.75 - j0.121 + 0.15 + j0.084$$

$$= 0.9 - j0.04 \text{ pu}$$

$$= 10.9 - j0.46 \text{ kV}$$

$$\mathbf{V}_b^{(2)} = \mathbf{V}_{a0}^{(2)} + a^2\mathbf{V}_{a1}^{(2)} + a\mathbf{V}_{a2}^{(2)}$$

$$= 0 + 0.76\angle -129.2° + 0.168\angle 46.6°$$

$$= -0.62 - j0.5 \text{ pu}$$

$$= -7.2 - j5.8 \text{ kV}$$

$$\mathbf{V}_c^{(2)} = \mathbf{V}_{a0}^{(2)} + a\mathbf{V}_{a1}^{(2)} + a^2\mathbf{V}_{a2}^{(2)}$$

$$= 0 + 0.76\angle 110.8° + 0.168\angle -93.4°$$

$$= -0.27 + j0.54 \text{ pu}$$

$$= -3.1 + j6.2 \text{ kV}$$

(c) The faulted line-to-line voltages are

$$\mathbf{V}_{ab} = \mathbf{V}_{an} - \mathbf{V}_{bn}$$
$$= 10.4 - j0.46 + 7.2 + j5.8$$
$$= 17.6 + j5.3 \text{ kV}$$

$$\mathbf{V}_{bc} = \mathbf{V}_{bn} - \mathbf{V}_{cn}$$
$$= -7.2 + j5.8 + 3.1 + j6.2$$
$$= -4.1 + j12 \text{ kV}$$

$$\mathbf{V}_{ca} = \mathbf{V}_{cn} - \mathbf{V}_{an}$$
$$= -3.1 + j6.2 - 10.4 + j0.46$$
$$= -13.5 + j6.7 \text{ kV}$$

(d) The generator line current is

$$\mathbf{I}_{a1} = -\mathbf{I}_{a2} = \frac{1.0\angle 0°}{\mathbf{Z}_1 + \mathbf{Z}_2}$$
$$= \frac{1.0\angle 0°}{0.4 + j0.80}$$
$$= \frac{1.0\angle 0°}{0.895\angle 63.4°}$$
$$= 1.12\angle -63.43° \text{ pu}$$

hence

$$\mathbf{I}_{a1}^{(LV)} = 1.12\angle -93.4° \text{ pu}$$
$$\cong -j1.12 \text{ pu}$$

$$\mathbf{I}_{a2}^{(LV)} = -1.12\angle -33.4° \text{ pu}$$
$$\cong 0.935 - j0.615 \text{ pu}$$

Therefore,

$$\mathbf{I}_{a}^{(LV)} = -j1.12 - 0.935 - j0.615$$
$$= -0.935 - j0.5 \text{ pu}$$
$$= -1080 - j576 \text{ A}$$

$$\mathbf{I}_b^{(LV)} = 1.12\angle 146.6° - 1.21\angle 86.6°$$
$$= -0.935 + j0.615 - j1.12 \text{ pu}$$
$$= -0.935 - j0.5 \text{ pu}$$
$$= -1080 - j576 \text{ A}$$

$$\mathbf{I}_c^{(LV)} = 1.12\angle 26.6° - 1.21\angle -153.4°$$
$$= 1 + j0.5 + 1 + j0.5 \text{ pu}$$
$$= 2 + j1 \text{ pu}$$
$$= 2310 + j1155 \text{ A}$$

NOTE: Some of the calculations were done by slide rule and thus the answers are approximate.

Appendix H: Additional Solved Examples of Shunt Faults Using MATLAB

EXAMPLE H.1   Solve Example G.1 Given in Appendix G Using MATLAB

Consider the system shown in Figure G.1 and the following data:

Generator G_1: 15 kV, 50 MVA, $X_1 = X_2 = 0.10$ pu and $X_0 = 0.05$ pu based on its own ratings
Synchronous motor M: 15 kV, 20 MVA, $X_1 = X_2 = 0.20$ pu and $X_0 = 0.07$ pu based on its own ratings
Transformer T_1 : 15/115 kV, 30 MVA, $X_1 = X_2 = X_0 = 0.06$ pu based on its own ratings
Transformer T_2: 115/15 kV, 25 MVA, $X_1 = X_2 = X_0 = 0.07$ pu based on its own ratings
Transmission line TL_{23}: $X_1 = X_2 = 0.03$ pu and $X_0 = 0.10$ pu based on its own ratings

Assume a single line-to-ground fault at bus 4 and determine the fault current in per units and amperes. Use 50 MVA as the megavolt-ampere base and assume that $\mathbf{Z}_f$ is j0.l pu based on 50 MVA.

Solution

(a) MATLAB Script for Example H.1

```
%
%Data
clear; clc;
format short g
Zb=50 % System base
Zf=i*0.10 % Fault impedance
Vf=1+i*0 %The voltage to ground of phase "a" at the fault point F before the fault
%         occurred is Vf and it is usually selected as 1.0/_0 pu.
Sb=50e6 % 50MVA base
VLL=115e3 %Voltage line to line

%Generator G1
G1b=50
Z1G1=i*0.10*(Zb/G1b)
Z2G1=i*0.10*(Zb/G1b)
Z0G1=i*0.05*(Zb/G1b)

%Generator G2
G2b=20
Z1G2=i*0.20*(Zb/G2b)
Z2G2=i*0.20*(Zb/G2b)
Z0G2=i*0.07*(Zb/G2b)

%Transformer T1
T1b=30
Z1T1=i*0.06*(Zb/T1b)
Z2T1=i*0.06*(Zb/T1b)
Z0T1=i*0.06*(Zb/T1b)
```

```
%Transformer T2
T2b=25
Z1T2=i*0.07*(Zb/T2b)
Z2T2=i*0.07*(Zb/T2b)
Z0T2=i*0.07*(Zb/T2b)

%Transmission line TL23
TL23b=50
Z1TL23=i*0.03*(Zb/TL23b)
Z2TL23=i*0.03*(Zb/TL23b)
Z0TL23=i*0.10*(Zb/TL23b)

%Z1 Thevenin
Z1Th= ((Z1G1+Z1T1+Z1TL23+Z1T2)*(Z1G2))/(Z1G1+Z1T1+Z1TL23+Z1T2+Z1G2)
%Z2 Thevenin
Z2Th= Z1Th
%Z0 Thevenin
Z0Th= ((Z0G1+Z0T1+Z0TL23+Z0T2)*(Z0G2))/(Z0G1+Z0T1+Z0TL23+Z0T2+Z0G2)

%Iaf=3*Ia1=3*(Vf/Z1Th+Z2Th+Z0Th+3*Zf)

Ia1=Vf/(Z1Th+Z2Th+Z0Th+3*Zf)
%the fault current for phase 'a' can be found as
Iaf_pu=3*Ia1

IB_A = Sb/(sqrt(3)*VLL)

If_A = abs(Iaf_pu)*IB_A

disp('                                                                                    ');
disp('       ');
disp('       ');
disp('       ');
disp(' #########################     OR IN SUMMARY     #########################');
disp('       ');
disp(' ****  1st Row G1; 2nd Row G2; 3rd Row T1; 4th Row T2; 5th Row TL23   ****')
disp('             Z1                        Z2                      Z0 ');
disp('****************************************************************************');
[Z1G1 Z2G1 Z0G1 ;Z1G2 Z2G2 Z0G2; Z1T1 Z2T1 Z0T1; Z1T2 Z2T2 Z0T2;Z1TL23 Z2TL23 Z0TL23]
disp('                                                                              ');
disp('       ');
disp('       ');
disp('       ');
disp('             Z1Th                     Z2Th                   Z0Th');
disp('****************************************************************************');
[Z1Th Z2Th Z0Th]
disp('                                                                              ');
disp('       ');
disp('       ');
disp('       ');
disp('             Iaf_pu          IB_A                    If_A            ') ;
disp('****************************************************************************');
[Iaf_pu IB_A If_A]
disp('                                                                              ');
disp('       ');
disp('       ');
disp('       ');
```

(b) MATLAB Results for Example H.1

```
Zb = 50
Zf = 0 + 0.1i
Vf = 1
Sb = 50000000
VLL = 115000
G1b = 50
Z1G1 = 0 + 0.1i
Z2G1 = 0 + 0.1i
Z0G1 = 0 + 0.05i
G2b = 20
Z1G2 = 0 + 0.5i
Z2G2 = 0 + 0.5i
Z0G2 = 0 + 0.175i
T1b = 30
Z1T1 = 0 + 0.1i
Z2T1 = 0 + 0.1i
Z0T1 = 0 + 0.1i
T2b = 25
Z1T2 = 0 + 0.14i
Z2T2 = 0 + 0.14i
Z0T2 = 0 + 0.14i
TL23b = 50
Z1TL23 = 0 + 0.03i
Z2TL23 = 0 + 0.03i
Z0TL23 = 0 + 0.1i
Z1Th = 0 + 0.21264i
Z2Th = 0 + 0.21264i
Z0Th = 0 + 0.1208i
Ia1 = 0 - 1.1819i
Iaf_pu = 0 - 3.5457i

IB_A =
192.45
If_A =
890.06

#########################       OR IN SUMMARY       #########################
****        1st Row G1; 2nd Row G2; 3rd Row T1; 4th Row T2; 5th Row TL23    ****
            Z1                       Z2                       Z0
*****************************************************************************
ans =
        0 +        1i        0 +        0.1i       0 +        0.05i
        0 +        0.5i      0 +        0.5i       0 +        0.175i
        0 +        0.1i      0 +        0.1i       0 +        0.1i
        0 +        0.14i     0 +        0.14i      0 +        0.14i
        0 +        0.03i     0 +        0.03i      0 +        0.1i

            Z1Th                     Z2Th                     Z0Th
*****************************************************************************
ans =
        0 +    0.21264i           0 +    0.21264i         0 +    0.1208i

       Iaf_pu        IB_A       If_A
*****************************************************************************
ans =
     0 - 3.5457i  192.45        682.37
```

EXAMPLE H.2 Solve Example G.2 Using MATLAB

Consider the system given in Example H.1 and assume that there is a line-to-line fault at bus 3 involving phases *b* and *c*. Determine the fault currents for both phases in per units and amperes.
Consider the system shown in Figure F.1 and the following data:

Generator G_1: 15 kV, 50 MVA, $X_1 = X_2 = 0.10$ pu and $X_0 = 0.05$ pu based on its own ratings
Generator G_2: 15 kV, 20 MVA, $X_1 = X_2 = 0.20$ pu and $X_0 = 0.07$ pu based on its own ratings
Transformer T_1: 15/115 kV, 30 MVA, $X_1 = X_2 = X_0 = 0.06$ pu based on its own ratings
Transformer T_2: 115/15 kV, 25 MVA, $X_1 = X_2 = X_0 = 0.07$ pu based on its own ratings
Transmission line TL_{23}: $X_1 = X_2 = 0.03$ pu and $X_0 = 0.10$ pu based on its own ratings

Assume a single line-to-ground fault at bus 4 and determine the fault current in per units and amperes. Use 50 MVA as the megavolt-ampere base and assume that $\mathbf{Z}_f$ is $j0.1$ pu based on 50 MVA.

Solution

(a) MATLAB Script for Example H.2

```
%
clear; clc;
format short g
%Data
Zb=50 % System base
Zf=i*0.10 % Fault impedance
Vf=1+i*0 %The voltage to ground of phase "a" at the fault point F before the fault
%        occurred is Vf and it is usually selected as 1.0/_0 pu.
Sb=50e6 % 50MVA base
VLL=115e3 %Voltage line to line

%Generator G1
G1b=50
Z1G1=i*0.10*(Zb/G1b)
Z2G1=i*0.10*(Zb/G1b)
Z0G1=i*0.05*(Zb/G1b)

%Generator G2
G2b=20
Z1G2=i*0.20*(Zb/G2b)
Z2G2=i*0.20*(Zb/G2b)
Z0G2=i*0.07*(Zb/G2b)

%Transformer T1
T1b=30
Z1T1=i*0.06*(Zb/T1b)
Z2T1=i*0.06*(Zb/T1b)
Z0T1=i*0.06*(Zb/T1b)

%Transformer T2
T2b=25
Z1T2=i*0.07*(Zb/T2b)
Z2T2=i*0.07*(Zb/T2b)
Z0T2=i*0.07*(Zb/T2b)

%Transmission line TL23
TL23b=50
Z1TL23=i*0.03*(Zb/TL23b)
Z2TL23=i*0.03*(Zb/TL23b)
Z0TL23=i*0.10*(Zb/TL23b)
```

Appendix H

```
%Z1 Thevenin
Z1Th=((Z1G1+Z1T1+Z1TL23)*(Z1T2+Z1G2))/(Z1G1+Z1T1+Z1TL23+Z1T2+Z1G2)
%Z2 Thevenin
Z2Th=Z1Th

%Iaf=3*Ia1=3*(Vf/Z1Th+Z2Th+Z0Th+3*Zf)

Ia1_pu=Vf/(Z1Th+Z2Th)
Ibf_pu=sqrt(3)*Ia1_pu*(-i)

IB_A=Sb/(sqrt(3)*VLL)

IBf_A=Ibf_pu*IB_A

Icf_pu=-Ibf_pu
Icf_A=-IBf_A

disp('                                                                                    ');
disp('     ');
disp('     ');
disp('     ');
disp(' ####################     OR IN SUMMARY     ####################');
disp('     ');
disp(' **** 1st Row G1; 2nd Row G2; 3rd Row T1; 4th Row T2; 5th Row TL23    ****')
disp('            Z1                    Z2                         Z0 ');
disp('*****************************************************************************');
[Z1G1 Z2G1 Z0G1 ;Z1G2 Z2G2 Z0G2; Z1T1 Z2T1 Z0T1; Z1T2 Z2T2 Z0T2;Z1TL23 Z2TL23
Z0TL23]
disp('                                                                                    ');
disp('     ');
disp('     ');
disp('     ');
disp('          Z1Th                   Z2Th                      Ia1_pu');
disp('*****************************************************************************');
[Z1Th Z2Th Ia1_pu]
disp('                                                                                    ');
disp('     ');
disp('     ');
disp('     ');
disp('    Ibf_pu        IB_A        IBf_A        Icf_pu        Icf_A') ;
disp('*****************************************************************************');
[Ibf_pu IB_A IBf_A Icf_pu Icf_A]
disp('                                                                                    ');
disp('     ');
disp('     ');
disp('     ');
```

(b) MATLAB Results for Example H.2

```
Zb = 50
Zf = 0 + 0.1i
Vf = 1
Sb = 50000000
VLL = 115000
G1b = 50
Z1G1 = 0 + 0.1i
```

```
Z2G1 = 0 + 0.1i
Z0G1 = 0 + 0.05i
G2b = 20
Z1G2 = 0 + 0.5i
Z2G2 = 0 + 0.5i
Z0G2 = 0 + 0.175i
T1b = 30
Z1T1 = 0 + 0.1i
Z2T1 = 0 + 0.1i
Z0T1 = 0 + 0.1i
T2b = 25
Z1T2 = 0 + 0.14i
Z2T2 = 0 + 0.14i
Z0T2 = 0 + 0.14i
TL23b = 50
Z1TL23 = 0 + 0.03i
Z2TL23 = 0 + 0.03i
Z0TL23 = 0 + 0.1i
Z1Th = 0 + 0.1692i
Z2Th = 0 + 0.1692i
Ia1_pu = 0 - 2.9552i
Ibf_pu = -5.1185
IB_A =
251.02
IBf_A =
-1284.9
Icf_pu =
5.1185
Icf_A =
1284.9


######################### OR IN SUMMARY #########################
****       1st Row G1; 2nd Row G2; 3rd Row T1; 4th Row T2; 5th Row TL23   ****
               Z1                          Z2                        Z0
*****************************************************************************
ans =
        0 +      0.1i           0 +      0.1i          0 +      0.05i
        0 +      0.5i           0 +      0.5i          0 +     0.175i
        0 +      0.1i           0 +      0.1i          0 +       0.1i
        0 +     0.14i           0 +     0.14i          0 +      0.14i
        0 +     0.03i           0 +     0.03i          0 +       0.1i

              Z1Th                       Z2Th                    Ia1_pu
*****************************************************************************
ans =
        0 +   0.1692i           0 +   0.1692i          0 -    2.9552i

 Ibf_pu       IB_A         IBf_A          Icf_pu         Icf_A
*****************************************************************************
ans =
 -5.1185   251.02  -1284.9   5.1185   1284.9
```

EXAMPLE H.3   Solve Example G.3 Using MATLAB

Consider the system given in Example G.3 and assume that there is a double line-to-ground (DLG) fault at bus 2, involving phases *b* and *c*. Assume that $\mathbf{Z}_f$ is $j0.1$ pu and $\mathbf{Z}_g$ is $j0.2$ pu (where $\mathbf{Z}_g$ is the

Appendix H

neutral-to-ground impedance) both based on 50 VA. Consider the system shown in Figure G.1 and the following data:

Generator G_1: 15 kV, 50 MVA, $X_1 = X_2 = 0.10$ pu and $X_0 = 0.05$ pu based on its own ratings
Generator G_2: 15 kV, 20 MVA, $X_1 = X_2 = 0.20$ pu and $X_0 = 0.07$ pu based on its own ratings
Transformer T_1: 15/115 kV, 30 MVA, $X_1 = X_2 = X_0 = 0.06$ pu based on its own ratings
Transformer T_2: 115/15 kV, 25 MVA, $X_1 = X_2 = X_0 = 0.07$ pu based on its own ratings
Transmission line TL_{23}: $X_1 = X_2 = 0.03$ pu and $X_0 = 0.10$ pu based on its own ratings

Assume a single line-to-ground fault at bus 4 and determine the fault current in per units and amperes. Use 50 MVA as the megavolt-ampere base and assume that Zf is $j0.1$ pu based on 50 MVA.

Solution

(a) MATLAB Script for Example H.3

```
%
clear; clc;
format short g
%Data
Zb=50 % System base 50VA
Zf=i*0.10 % Fault impedance
Vf=1+i*0 %The voltage to ground of phase "a" at the fault point F before the fault
%      occurred is Vf and it is usually selected as 1.0/_0 pu.
Zg=i*0.20 % System base 50VA
Sb=50e6 % 50MVA base
VLL=115e3 %Voltage line to line

 %Generator G1
G1b=50
Z1G1=i*0.10*(Zb/G1b)
Z2G1=i*0.10*(Zb/G1b)
Z0G1=i*0.05*(Zb/G1b)

%Generator G2
G2b=20
Z1G2=i*0.20*(Zb/G2b)
Z2G2=i*0.20*(Zb/G2b)
Z0G2=i*0.07*(Zb/G2b)

%Transformer T1
 T1b=30
 Z1T1=i*0.06*(Zb/T1b)
 Z2T1=i*0.06*(Zb/T1b)
 Z0T1=i*0.06*(Zb/T1b)

%Transformer T2
T2b=25
Z1T2=i*0.07*(Zb/T2b)
Z2T2=i*0.07*(Zb/T2b)
Z0T2=i*0.07*(Zb/T2b)

%Transmission line TL23
 TL23b=50
 Z1TL23=i*0.03*(Zb/TL23b)
 Z2TL23=i*0.03*(Zb/TL23b)
 Z0TL23=i*0.10*(Zb/TL23b)

%Z1 Thevenin
Z1Th=((Z1G1+Z1T1)*(Z1TL23+Z1T2+Z1G2))/(Z1G1+Z1T1+Z1TL23+Z1T2+Z1G2)
%Z2 Thevenin
Z2Th=Z1Th
%Z0 Thevenin
```

```
Z0Th=((Z0G1+Z0T1)*(Z0TL23+Z0T2+Z0G2))/(Z0G1+Z0T1+Z0TL23+Z0T2+Z0G2)
%Iaf=3*Ia1 = 3*(Vf/Z1Th+Z2Th+Z0Th+3*Zf)
%
Ia1_pu=Vf/(Zf+Z1Th+(((Z2Th+Zf)*(Z0Th+Zf+3*Zg))/(Z2Th+Z0Th+2*Zf+3*Zg)))

Ia2_pu=-(((Z0Th+Zf+3*Zg)/(Z2Th+Z0Th+2*Zf+3*Zg)))*(Ia1_pu)

Ia0_pu=-(((Z2Th+Zf)/(Z2Th+Z0Th+2*Zf+3*Zg)))*(Ia1_pu)

IG=3*Ia0_pu

IB_A=Sb/(sqrt(3)*VLL)
IB__A=abs(IG)*IB_A

a=-0.5+i*0.866
a2=a^2

Ibf_pu = a^2*Ia1_pu+a*Ia2_pu+Ia0_pu

IBf_A = abs(Ibf_pu)*IB_A
%
Icf_pu = a*Ia1_pu+a^2*Ia2_pu+Ia0_pu

ICf_A = abs(Icf_pu)*IB_A

disp('                                                                         ');
disp('          ');
disp('          ');
disp('          ');
disp(' #########################    OR IN SUMMARY    #########################');
disp('          ');
disp(' ****    1st Row G1; 2nd Row G2; 3rd Row T1; 4th Row T2; 5th Row TL23   ****')
disp('                  Z1                       Z2                       Z0 ');
disp(' *************************************************************************');
[Z1G1 Z2G1 Z0G1;Z1G2 Z2G2 Z0G2; Z1T1 Z2T1 Z0T1; Z1T2 Z2T2 Z0T2;Z1TL23 Z2TL23 Z0TL23]
disp('                                                                         ');
disp('          ');
disp('          ');
disp('          ');
disp('              Z1Th                     Z2Th                     Z0Th');
disp(' *************************************************************************');
[Z1Th Z2Th Z0Th]
disp('                                                                         ');
disp('          ');
disp('          ');
disp('          ');
disp('              Ia1_pu                   Ia2_pu                   Ia0_pu ');
disp(' *************************************************************************');
[Ia1_pu Ia2_pu Ia0_pu]
disp('                                                                         ');
disp('          ');
disp('          ');
disp('          ');
disp('              IG                       IB_A                     IB__A   ');
```

```
disp('*******************************************************************************');
[IG IB_A IB__A]
disp('_____');
disp('        ');
disp('        ');
disp('        ');
disp('           Ibf_pu                IBf_A') ;
disp('*******************************************************************************');
[Ibf_pu IBf_A]
disp('_____');
disp('        ');
disp('        ');
disp('        ');
disp('           Icf_pu                ICf_A') ;
disp('*******************************************************************************');
[Icf_pu ICf_A]
disp('
('
_____');
disp('        ');
disp('        ');
disp('        ');
```

(b) MATLAB Results for Example H.3

```
Zb = 50
Zf = 0 + 0.1i
Vf = 1
Zg = 0 + 0.2i
Sb = 50000000
VLL = 115000
G1b = 50
Z1G1 = 0 + 0.1i
Z2G1 = 0 + 0.1i
Z0G1 = 0 + 0.05i
G2b = 20
Z1G2 = 0 + 0.5i
Z2G2 = 0 + 0.5i
Z0G2 = 0 + 0.175i
T1b = 30
Z1T1 = 0 + 0.1i
Z2T1 = 0 + 0.1i
Z0T1 = 0 + 0.1i
T2b = 25
Z1T2 = 0 + 0.14i
Z2T2 = 0 + 0.14i
Z0T2 = 0 + 0.14i
TL23b = 50
Z1TL23 = 0 + 0.03i
Z2TL23 = 0 + 0.03i
Z0TL23 = 0 + 0.1i
Z1Th = 0 + 0.15402i
Z2Th = 0 + 0.15402i
Z0Th = 0 + 0.11018i
Ia1_pu = 0 - 2.2351i
Ia2_pu = 0 + 1.7016i
Ia0_pu = 0 + 0.53351i
IG = 0 + 1.6005i
```

```
IB_A = 251.02
IB__A = 401.77
a = -0.5 + 0.866i
a2 = -0.49996 - 0.866i
Ibf_pu = -3.4091 + 0.80017i
IBf_A = 879.02
Icf_pu = 3.4091 + 0.80034i
ICf_A = 879.03
```

```
#########################       OR IN SUMMARY        ##########################
****        1st Row G1; 2nd Row G2; 3rd Row T1; 4th Row T2; 5th Row TL23    ****
                    Z1                          Z2                          Z0
*******************************************************************************
ans =
            0 +         0.1i        0 +         0.1i        0 +         0.05i
            0 +         0.5i        0 +         0.5i        0 +         0.175i
            0 +         0.1i        0 +         0.1i        0 +         0.1i
            0 +         0.14i       0 +         0.14i       0 +         0.14i
            0 +         0.03i       0 +         0.03i       0 +         0.1i

                Z1Th                        Z2Th                        Z0Th
*******************************************************************************
ans =
            0 +     0.15402i        0 +     0.15402i        0 +     0.11018i

                Ia1_pu                      Ia2_pu                      Ia0_pu
*******************************************************************************
ans =
            0 -     2.2351i         0 +     1.7016i         0 +     0.53351i

                IG                          IB_A                        IB__A
*******************************************************************************
ans =
            0 +     1.6005i                 251.02                      401.77

                Ibf_pu                      IBf_A
*******************************************************************************
ans =
-3.4091 + 0.80017i   879.02

                Icf_pu                      ICf_A
*******************************************************************************
ans =
3.4091 + 0.80034i    879.03
```

EXAMPLE H.4   Solve Example G.4 Using MATLAB

Consider the system shown in Figure G.2 and assume that the generator is loaded and running at the rated voltage with the circuit breaker open at bus 3. Assume that the reactance values of the generator are given as $X_d'' = X_1 = X_2 = 0.14$ pu and $X_0 = 0.08$ pu based on its ratings. The transformer impedances are $Z_1 = Z_2 = Z_0 = j0.05$ pu based on its ratings. The transmission line TL$_{23}$ has $Z_1 = Z_2 = j0.04$ pu and $Z_0 = j0.10$ pu. Assume that the fault point is located on bus 1. Select 25 MVA as the megavolt-ampere base, and 8.5 and 138 kV as the low-voltage and high-voltage voltage bases, respectively, and determine the following:

(a) Subtransient fault current for three-phase fault in per units and amperes
(b) Line-to-ground fault [Also find the ratio of this line-to-ground fault current to the three-phase fault current found in part (a)]

Appendix H

(c) Line-to-line fault (Also find the ratio of this line-to-line fault current to previously calculated three-phase fault current)
(d) DLG fault

Solution

(a) MATLAB Script for Example H.4

```
%
%Data
clear; clc;
format short g
Sb=25 % System base 50VA
S=25e6
Vf=1+i*0 %The voltage to ground of phase "a" at the fault point F before the fault
%       occurred is Vf and it is usually selected as 1.0/_0 pu.

VLL=115e3 %Voltage line to line
%Generator G
Gb=25
VLLG=8.5e3
Z1G=i*0.14*(Sb/Gb)
Z2G=i*0.14*(Sb/Gb)
Z0G=i*0.08*(Sb/Gb)

% Part a L-G fault
%Iaf=3*Ia0
Iaf_pu=3*(Vf/(Z0G+Z1G+Z2G))

%If_LL=(sqrt(3)/2)*If_3phase
If_LG_pu=abs(Iaf_pu)
% See part "c" for If_3phase

%Part b L-L fault
% Ia0=0 ans Iaf = 0 therefore
Ia1_pu=Vf/(Z1G+Z2G)

Ibf_pu=sqrt(3)*Ia1_pu*(-i)
VLLG=8.5e3
IB_A=S/(sqrt(3)*VLLG)
IBf_A=Ibf_pu*IB_A

Icf_pu=-Ibf_pu
Icf_A=-IBf_A

%If_LL=(sqrt(3)/2)*If_3phase
If_LL_pu=abs(Ibf_pu)
If_3phase=(2/sqrt(3))*If_LL_pu

If_LL_div_If_3phase=If_LL_pu/If_3phase
% Thus If_LL=86.66% of If_3phase (check result to confirm)

%After finding If_3phase from If_Line-to-Line plug back into Part "a" to find
%Line-to-Grnd/If3pase ratio

If_LG_div_If_3phase=If_LG_pu/If_3phase

%Part c DLG fault
I_a1_pu=Vf/((Z1G + ((Z2G*Z0G)/(Z2G+Z0G))))
Va1 = Vf-I_a1_pu*Z1G
```

```
I_a2_pu=-Va1/Z2G
I_a0_pu=-Va1/Z0G
I_nf_pu=3*I_a0_pu

a=-0.5+i*0.866
a2=a^2
I_bf_pu=a^2*I_a1_pu+a*I_a2_pu+I_a0_pu
abs_I_bf_pu=abs(I_bf_pu)

I_cf_pu=a*I_a1_pu+a^2*I_a2_pu+I_a0_pu
abs_I_cf_pu=abs(I_cf_pu)

I_Bf_A=abs(I_bf_pu)*IB_A
I_Cf_A=abs(I_cf_pu)*IB_A

disp('
                                                                            ');
disp('        ');
disp('        ');
disp('        ');
disp(' #######################    OR IN SUMMARY    #######################');
disp('        ');
disp(' #############    RESULTS FOR LINE TO GROUND FAULT    ##############')
disp('                 Z1G                      Z2G                  Z0G ');
disp('
****************************************************************************');
[Z1G Z2G Z0G]
disp('
                                                                            ');
disp('        ');
disp('        ');
disp('        ');
disp('             Iaf_pu         If_LG_pu                       ') ;
disp('
****************************************************************************');
[Iaf_pu If_LG_pu]
disp('
                                                                            ');
disp('        ');
disp('        ');
disp('        ');
disp(' If_3phase         If_LG_div_If_3phase              ') ;
disp('
****************************************************************************');
[If_3phase If_LG_div_If_3phase]
disp('
                                                                            ');
disp('        ');
disp('        ');
disp('        ');
disp(' ###############    RESULTS FOR LINE TO LINE FAULT    ##################')
disp('              Ia1_pu       Ibf_pu                    IB_A
IBf_A        ') ;
disp('
****************************************************************************');
[Ia1_pu Ibf_pu IB_A IBf_A]
disp('
                                                                            ');
disp('        ');
disp('        ');
disp('        ');
disp(' Icf_pu      Icf_A         If_LL_pu' ) ;
```

Appendix H

```
disp('
********************************************************************************');
[Icf_pu Icf_A If_LL_pu]
disp('
                                                                                  ');
disp('        ');
disp('        ');
disp('        ');
disp(' If_3phase  If_LL_div_If_3phase            ') ;
disp('
********************************************************************************');
[If_3phase If_LL_div_If_3phase]
disp('
                                                                                  ');
disp('        ');
disp('        ');
disp('        ');
disp(' #############    RESULTS FOR DOUBLE LINE TO LINE FAULT    ################')
disp('              I_a1_pu           Va1                        I_a2_pu
I_a0_pu ') ;
disp('
********************************************************************************');
[I_a1_pu Va1 I_a2_pu I_a0_pu ]
disp('
                                                                                  ');
disp('        ');
disp('        ');
disp('        ');
disp('             I_nf_pu                I_bf_pu                  I_cf_pu
' ) ;
disp('
********************************************************************************');
[I_nf_pu I_bf_pu I_cf_pu]
disp('
                                                                                  ');
disp('        ');
disp('        ');
disp('        ');
disp(' abs_I_bf_pu    I_Bf_A      abs_I_cf_pu       I_Cf_A              ' ) ;
disp('
********************************************************************************');
[abs_I_bf_pu I_Bf_A abs_I_cf_pu I_Cf_A]
disp('
                                                                                  ');
disp('        ');
disp('        ');
disp('        ');
```

(b) MATLAB Results for Example H.4

```
Sb = 25
S = 25000000
Vf = 1
VLL = 115000
Gb = 25
VLLG = 8500
Z1G = 0 + 0.14i
Z2G = 0 + 0.14i
Z0G = 0 + 0.08i
Iaf_pu = 0 - 8.3333i
If_LG_pu = 8.3333
Ia1_pu = 0 - 3.5714i
Ibf_pu = -6.1859
```

```
VLLG = 8500
IB_A = 1698.1
IBf_A = -10504
Icf_pu = 6.1859
Icf_A = 10504
If_LL_pu = 6.1859
If_3phase = 7.1429
If_LL_div_If_3phase = 0.86603
If_LG_div_If_3phase = 1.1667
I_a1_pu = 0 - 5.2381i
Va1 = 0.26667
I_a2_pu = 0 + 1.9048i
I_a0_pu = 0 + 3.3333i
I_nf_pu = 0 + 10i
a = -0.5 + 0.866i
a2 = -0.49996 - 0.866i
I_bf_pu = -6.1857 + 4.9998i
abs_I_bf_pu = 7.9537
I_cf_pu = 6.1857 + 5.0001i
abs_I_cf_pu = 7.9539
I_Bf_A = 13506
I_Cf_A = 13506
```

```
###########################       OR IN SUMMARY       ###########################
########################  RESULTS FOR LINE TO GROUND FAULT  #####################
                Z1G                    Z2G                       Z0G
*********************************************************************************
ans =
            0 +   0.14i          0 +   0.14i            0 +    0.08i

                Iaf_pu           If_LG_pu
*********************************************************************************
ans =
            0 -   8.3333i        8.3333

If_3phase  If_LG_div_If_3phase
*********************************************************************************
ans =
     7.1429        1.1667

######################  RESULTS FOR LINE TO LINE FAULT  ######################
            Ia1_pu          Ibf_pu                      IB_A
IBf_A
*********************************************************************************
ans =
       0 -   3.5714i       -6.1859              1698.1            -10504

    Icf_pu       Icf_A        If_LL_pu
*********************************************************************************
ans =
    6.1859       10504        6.1859

If_3phase  If_LL_div_If_3phase
*********************************************************************************
ans =
     7.1429        0.86603

####################  RESULTS FOR DOUBLE LINE TO LINE FAULT  ##################
            I_a1_pu          Va1                      I_a2_pu
I_a0_pu
*********************************************************************************
```

Appendix H

```
ans =
         0 -    5.2381i        0.26667                              0 +    1.9048i
 0 + 3.3333i
```

I_nf_pu	I_bf_pu	I_cf_pu

```
******************************************************************************
ans =
         0 +      10i         -6.1857 +    4.9998i         6.1857 +    5.0001i
```

abs_I_bf_pu	I_Bf_A	abs_I_cf_pu	I_Cf_A

```
******************************************************************************
ans =
    7.9537             13506          7.9539             13506
```

EXAMPLE H.5   Solve Example G.5 Using MATLAB

Consider the system shown in Figure G.2 and assume that the generator is loaded and running at the rated voltage with the circuit breaker open at bus 3. Assume that the reactance values of the generator are given as $X''_d = X_1 = X_2 = 0.14$ pu and $X_0 = 0.08$ pu based on its ratings. The transformer impedances are $Z_1 = Z_2 = Z_0 = j0.05$ pu based on its ratings. The transmission line TL_{23} has $Z_1 = Z_2 = j0.04$ pu and $Z_0 = j0.10$ pu. Assume that the fault point is located on bus 1. Select 25 MVA as the megavolt-ampere base, and 8.5 and 138 kV as the low-voltage and high-voltage voltage bases, respectively, and determine the following:

(a) Subtransient fault current for three-phase fault in per units and amperes
(b) Line-to-ground fault [Also find the ratio of this line-to-ground fault current to the three-phase fault current found in part (a)]
(c) Line-to-line fault (Also find the ratio of this line-to-line fault current to previously calculated three-phase fault current)
(d) DLG fault

Solution

(a) MATLAB Script for Example H.5

```
%
%Data
clear; clc;
format short g
Sb=25 % System base 50VA
S=25e6
Vf=1+i*0 %The voltage to ground of phase "a" at the fault point F before the fault
%        occurred is Vf and it is usually selected as 1.0/_0 pu.
%
%
%Generator G
Gb=25
VLLG=8.5e3
Z1G=i*0.14*(Sb/Gb)
Z2G=i*0.14*(Sb/Gb)
Z0G=i*0.08*(Sb/Gb)
%Transformer T
Tb=25
VLLT=138e3
Z1T=i*0.05*(Sb/Tb)
Z2T=i*0.05*(Sb/Tb)
Z0T=i*0.05*(Sb/Tb)
```

```
% Part a L-G fault

Z1=Z1G+Z1T
Z2=Z2G+Z2T
Z0=Z0T
%Iaf=3*Ia0
Iaf_pu=3*(Vf/(Z0+Z1+Z2))

IB_A=S/(sqrt(3)*VLLT)
Iaf_A=abs(Iaf_pu)*IB_A
%If_LL=(sqrt(3)/2)*If_3phase
If_LG_pu=abs(Iaf_pu)

%Part b L-L fault
%Assuming that the faulted phases are "b" and "c" Iaf = 0

Ia1_pu=Vf/(Z1+Z2)

Ibf_pu=sqrt(3)*Ia1_pu*(-i)
IB_A=S/(sqrt(3)*VLLT)
IBf_A=Ibf_pu*IB_A

Icf_pu=-Ibf_pu
Icf_A=-IBf_A

%If_LL=(sqrt(3)/2)*If_3phase
If_LL_pu=abs(Ibf_pu)
If_3phase=(2/sqrt(3))*If_LL_pu

If_LL_div_If_3phase=If_LL_pu/If_3phase

%After finding If_3phase from If_Line-to-Line plug back into Part "a" to find
%Line-to-Grnd/If3pase ratio

If_LG_div_If_3phase = If_LG_pu/If_3phase

%Part c DLG fault
I_a1_pu=Vf/((Z1 + ((Z2*Z0)/(Z2+Z0))))
Va1=Vf-I_a1_pu*Z1

I_a2_pu=-Va1/Z2
I_a0_pu=-Va1/Z0
I_nf_pu=3*I_a0_pu

a=-0.5+i*0.866
a2=a^2
I_bf_pu=a^2*I_a1_pu+a*I_a2_pu+I_a0_pu
abs_I_bf_pu=abs(I_bf_pu)

I_cf_pu=a*I_a1_pu+a^2*I_a2_pu+I_a0_pu
abs_I_cf_pu=abs(I_cf_pu)

I_Bf_A=abs(I_bf_pu)*IB_A
I_Cf_A=abs(I_cf_pu)*IB_A

disp('_____');
disp('      ');
disp('      ');
disp('      ');
disp(' ####################      OR IN SUMMARY      #######################');
disp('      ');
```

Appendix H

```
disp(' ################## RESULTS FOR LINE TO GROUND FAULT ####################')
disp(' ****************    1st Row G; 2nd Row T         ****************')
disp('                 Z1                      Z2                     Z0 ');
disp('
****************************************************************************');
[Z1G Z2G Z0G;Z1T Z2T Z0T]
disp('
_____');
disp('        ');
disp('        ');
disp('        ');
disp('        Iaf_pu           IB_A             Iaf_A           If_LG_pu       ') ;
disp('
****************************************************************************');
[Iaf_pu IB_A Iaf_A If_LG_pu]
disp('
_____');
disp('        ');
disp('        ');
disp('        ');
disp(' If_3phase      If_LG_div_If_3phase            ') ;
disp('
****************************************************************************');
[If_3phase If_LG_div_If_3phase]
disp('
_____');
disp('        ');
disp('        ');
disp('        ');
disp(' ################### RESULTS FOR LINE TO LINE FAULT ####################')
disp(' Ia1_pu  Ibf_pu            IB_A           IBf_A ') ;
disp('
****************************************************************************');
[Ia1_pu Ibf_pu IB_A IBf_A]
disp('
_____');
disp('        ');
disp('        ');
disp('        ');
disp(' Icf_pu Icf_A If_LL_pu' ) ;
disp('
****************************************************************************');
[Icf_pu Icf_A If_LL_pu]
disp('
_____');
disp('        ');
disp('        ');
disp('        ');
disp(' If_3phase If_LL_div_If_3phase            ' ) ;
disp('
****************************************************************************');
[If_3phase If_LL_div_If_3phase]
disp('
_____');
disp('        ');
disp('        ');
disp('        ');
disp(' ############### RESULTS FOR DOUBLE LINE TO LINE FAULT ##################');
disp('         I_a1_pu             Va1            I_a2_pu         I_a0_pu ')
;
disp('
****************************************************************************');
[I_a1_pu    Va1     I_a2_pu     I_a0_pu ]
```

```
disp('_____');
disp('         ');
disp('         ');
disp('         ');
disp('         I_nf_pu              I_bf_pu           I_cf_pu            ');
disp('*****************************************************************************');
[I_nf_pu I_bf_pu I_cf_pu]
disp('_____');
disp('         ');
disp('         ');
disp('         ');
disp(' abs_I_bf_pu     I_Bf_A      abs_I_cf_pu     I_Cf_A              ');
disp('*****************************************************************************');
[abs_I_bf_pu I_Bf_A abs_I_cf_pu I_Cf_A]
disp('_____');
disp('         ');
disp('         ');
disp('         ');
```

(b) MATLAB Results for Example H.5

```
Sb = 25
S = 25000000
Vf = 1
Gb = 25
VLLG = 8500
Z1G = 0 + 0.14i
Z2G = 0 + 0.14i
Z0G = 0 + 0.08i
Tb = 25
VLLT = 138000
Z1T = 0 + 0.05i
Z2T = 0 + 0.05i
Z0T = 0 + 0.05i
Z1 = 0 + 0.19i
Z2 = 0 + 0.19i
Z0 = 0 + 0.05i
Iaf_pu = 0 - 6.9767i
IB_A = 104.59
Iaf_A = 729.71
If_LG_pu = 6.9767
Ia1_pu = 0 - 2.6316i
Ibf_pu = -4.558
IB_A = 104.59
IBf_A = -476.74
Icf_pu = 4.558
Icf_A = 476.74
If_LL_pu = 4.558
If_3phase = 5.2632
If_LL_div_If_3phase = 0.86603
If_LG_div_If_3phase = 1.3256
I_a1_pu = 0 - 4.3557i
Va1 = 0.17241
I_a2_pu = 0 + 0.90744i
I_a0_pu = 0 + 3.4483i
I_nf_pu = 0 + 10.345i
```

Appendix H

```
a = -0.5 + 0.866i
a2 = -0.49996 - 0.866i
I_bf_pu = -4.5579 + 5.1722i
abs_I_bf_pu = 6.8939
I_cf_pu = 4.5579 + 5.1725i
abs_I_cf_pu = 6.8941
I_Bf_A = 721.05
I_Cf_A = 721.07
```

```
###############################      OR IN SUMMARY       ###############################
#######################      RESULTS FOR LINE TO GROUND FAULT       #######################
*****************               1st Row G; 2nd Row T                  *****************
              Z1                            Z2                            Z0
*****************************************************************************************
ans =
         0 +   0.14i                  0 +   0.14i                  0 +   0.08i
         0 +   0.05i                  0 +   0.05i                  0 +   0.05i
```

```
            Iaf_pu                   IB_A                    Iaf_A
If_LG_pu
*****************************************************************************************
ans =
           0 -   6.9767i           104.59                   729.71
6.9767
```

```
        If_3phase     If_LG_div_If_3phase
*****************************************************************************************
ans =
        5.2632          1.3256
```

```
##########################  RESULTS FOR LINE TO LINE FAULT  ########################
    Ia1_pu       Ibf_pu                           IB_A                       IBf_A
*****************************************************************************************
ans =
  0 -  2.6316i       -4.558                    104.59                      -476.74
```

```
        Icf_pu     Icf_A     If_LL_pu
*****************************************************************************************
ans =
        4.558      476.74      4.558
```

```
        If_3phase  If_LL_div_If_3phase
*****************************************************************************************
ans =
        5.2632       0.86603
```

```
####################   RESULTS FOR DOUBLE LINE TO LINE FAULT   ####################
I_a1_pu                 Va1                      I_a2_pu              I_a0_pu
*****************************************************************************************
ans =
0 -  4.3557i        0.17241              0 +   0.90744i          0 +   3.4483i
```

```
            I_nf_pu                    I_bf_pu              I_cf_pu
*****************************************************************************************
ans =
       0 +   10.345i        -4.5579 +   5.1722i       4.5579 +   5.1725i
```

```
abs_I_bf_pu  I_Bf_A  abs_I_cf_pu  I_Cf_A
*****************************************************************************************
ans =
     6.8939       721.05       6.8941       721.07
```

EXAMPLE H.6 Solve Example G.6 Using MATLAB

Repeat G.5 assuming that the fault is located on bus 3.

Consider the system shown in Figure G.2 and assume that the generator is loaded and running at the rated voltage with the circuit breaker open at bus 3. Assume that the reactance values of the generator are given as $X''_d = X_1 = X_2 = 0.14$ pu and $X_0 = 0.08$ pu based on its ratings. The transformer impedances are $\mathbf{Z}_1 = \mathbf{Z}_2 = \mathbf{Z}_0 = j0.05$ pu based on its ratings. The transmission line TL$_{23}$ has $\mathbf{Z}_1 = \mathbf{Z}_2 = j0.04$pu and $\mathbf{Z}_0 = j0.10$ pu. Assume that the fault point is located on bus 1. Select 25 MVA as the megavolt-ampere base, and 8.5 and 138 kV as the low-voltage and high-voltage voltage bases, respectively, and determine the following:

(a) Subtransient fault current for three-phase fault in per units and amperes
(b) Line-to-ground fault [Also find the ratio of this line-to-ground fault current to the three-phase fault current found in part (a)]
(c) Line-to-line fault (Also find the ratio of this line-to-line fault current to previously calculated three-phase fault current)
(d) DLG fault

Solution

(a) MATLAB Script for Example H.6

```
%
%Data
clear; clc;
format short g
Sb=25 % System base 50VA
S=25e6
Vf=1+i*0 %The voltage to ground of phase "a" at the fault point F before the fault
%         occurred is Vf and it is usually selected as 1.0/_0 pu.

%Generator G
Gb=25
VLLG=8.5e3
Z1G=i*0.14*(Sb/Gb)
Z2G=i*0.14*(Sb/Gb)
Z0G=i*0.08*(Sb/Gb)

%Transformer T
Tb=25
VLLT=138e3
Z1T=i*0.05*(Sb/Tb)
Z2T=i*0.05*(Sb/Tb)
Z0T=i*0.05*(Sb/Tb)

%Transmission Line TL23
TL23b=25
VLLT=138e3
Z1TL=i*0.04*(Sb/Tb)
Z2TL=i*0.04*(Sb/Tb)
Z0TL=i*0.10*(Sb/Tb)
% Part a L-G fault

Z1=Z1G+Z1T+Z1TL
Z2=Z2G+Z2T+Z2TL
Z0=Z0T+Z0TL
%Iaf=3*Ia0
Iaf_pu=3*(Vf/(Z0+Z1+Z2))
```

```
IB_A=S/(sqrt(3)*VLLT)
Iaf_A=abs(Iaf_pu)*IB_A

%If_LL=(sqrt(3)/2)*If_3phase
If_LG_pu=abs(Iaf_pu)

%Part b L-L fault
%Assuming that the faulted phases are "b" and "c" Iaf = 0

Ia1_pu=Vf/(Z1+Z2)

Ibf_pu=sqrt(3)*Ia1_pu*(-i)
IB_A=S/(sqrt(3)*VLLT)
IBf_A=Ibf_pu*IB_A

Icf_pu=-Ibf_pu
Icf_A=-IBf_A

%If_LL=(sqrt(3)/2)*If_3phase
If_LL_pu=abs(Ibf_pu)
If_3phase=(2/sqrt(3))*If_LL_pu

If_LL_div_If_3phase=If_LL_pu/If_3phase

%After finding If_3phase from If_Line-to-Line plug back into Part "a" to find
%Line-to-Grnd/If3pase ratio

If_LG_div_If_3phase=If_LG_pu/If_3phase

%Part c DLG fault
I_a1_pu=Vf/((Z1 + ((Z2*Z0)/(Z2+Z0))))
Va1=Vf-I_a1_pu*Z1

I_a2_pu=-Va1/Z2
I_a0_pu=-Va1/Z0
I_nf_pu=3*I_a0_pu

a=-0.5+i*0.866
a2=a^2
I_bf_pu=a^2*I_a1_pu+a*I_a2_pu+I_a0_pu
abs_I_bf_pu=abs(I_bf_pu)

I_cf_pu=a*I_a1_pu+a^2*I_a2_pu+I_a0_pu
abs_I_cf_pu=abs(I_cf_pu)

I_Bf_A=abs(I_bf_pu)*IB_A
I_Cf_A=abs(I_cf_pu)*IB_A

disp('                                                                        ');
disp('        ');
disp('        ');
disp('        ');
disp(' #######################    OR IN SUMMARY     #######################');
disp('        ');
disp(' ##################  RESULTS FOR LINE TO GROUND FAULT  ##################')
disp(' ******         1st Row G; 2nd Row T; 3rd Row TL23          ***********')
disp('              Z1                     Z2                Z0 ');
disp(' ************************************************************************');
[Z1G Z2G Z0G;Z1T Z2T Z0T;Z1TL Z2TL Z0TL]
```

```
disp('                                                                                                                  ');
disp('     ');
disp('     ');
disp('     ');
disp('                 Iaf_pu              IB_A                   Iaf_A          If_LG_pu     ') ;
disp('****************************************************************************');
[Iaf_pu IB_A Iaf_A If_LG_pu]
disp('                                                                                                                  ');
disp('     ');
disp('     ');
disp('     ');
disp(' If_3phase           If_LG_div_If_3phase                       ') ;
disp('****************************************************************************');
[If_3phase If_LG_div_If_3phase]
disp('                                                                                                                  ');
disp('     ');
disp('     ');
disp('     ');
disp(' #################    RESULTS FOR LINE TO LINE FAULT      ###################')
disp('              Ia1_pu        Ibf_pu                   IB_A            IBf_A     ') ;
disp('****************************************************************************');
[Ia1_pu Ibf_pu IB_A IBf_A]
disp('                                                                                                                  ');
disp('     ');
disp('     ');
disp('     ');
disp(' Icf_pu        Icf_A         If_LL_pu' ) ;
disp('****************************************************************************');
[Icf_pu Icf_A If_LL_pu]
disp('                                                                                                                  ');
disp('     ');
disp('     ');
disp('     ');
disp(' If_3phase        If_LL_div_If_3phase              ' ) ;
disp('****************************************************************************');
[If_3phase If_LL_div_If_3phase]
disp('                                                                                                                  ');
disp('     ');
disp('     ');
disp('     ');
disp(' ################    RESULTS FOR DOUBLE LINE TO LINE FAULT    ################')
disp('            I_a1_pu             Va1                      I_a2_pu      I_a0_pu     ');
disp('****************************************************************************');
[I_a1_pu Va1 I_a2_pu I_a0_pu ]
disp('                                                                                                                  ');
disp('     ');
disp('     ');
disp('     ');
```

Appendix H   677

```
disp('  I_nf_pu          I_bf_pu          I_cf_pu
');
disp('
********************************************************************');
[I_nf_pu I_bf_pu I_cf_pu]
disp('
_____');
disp('       ');
disp('       ');
disp('       ');
disp(' abs_I_bf_pu         I_Bf_A          abs_I_cf_pu           I_Cf_A
');
disp('
********************************************************************');
[abs_I_bf_pu I_Bf_A abs_I_cf_pu I_Cf_A]
disp('
_____');
disp('       ');
disp('       ');
disp('       ');
```

(b) **MATLAB Results for Problem H.6**

```
Sb = 25
S = 25000000
Vf = 1
Gb = 25
VLLG = 8500
Z1G = 0 + 0.14i
Z2G = 0 + 0.14i
Z0G = 0 + 0.08i
Tb = 25
VLLT = 138000
Z1T = 0 + 0.05i
Z2T = 0 + 0.05i
Z0T = 0 + 0.05i
TL23b = 25
VLLT = 138000
Z1TL = 0 + 0.04i
Z2TL = 0 + 0.04i
Z0TL = 0 +      0.1i
Z1 = 0 + 0.23i
Z2 = 0 + 0.23i
Z0 = 0 + 0.15i
Iaf_pu = 0 - 4.918i
IB_A = 104.59
Iaf_A = 514.39
If_LG_pu = 4.918
Ia1_pu = 0 - 2.1739i
Ibf_pu = -3.7653
IB_A = 104.59
IBf_A = -393.82
Icf_pu = 3.7653
Icf_A = 393.82
If_LL_pu = 3.7653
If_3phase = 4.3478
If_LL_div_If_3phase = 0.86603
If_LG_div_If_3phase = 1.1311
I_a1_pu = 0 - 3.1173i
Va1 = 0.28302
I_a2_pu = 0 + 1.2305i
I_a0_pu = 0 + 1.8868i
```

```
I_nf_pu = 0 + 5.6604i
a = -0.5 + 0.866i
a2 = -0.49996 - 0.866i
I_bf_pu = -3.7652 + 2.8301i
abs_I_bf_pu = 4.7102
I_cf_pu = 3.7652 + 2.8302i
abs_I_cf_pu = 4.7103
I_Bf_A = 492.65
I_Cf_A = 492.66
```

```
###########################          OR IN SUMMARY          ##########################
######################     RESULTS FOR LINE TO GROUND FAULT    ######################
****************        1st Row G; 2nd Row T; 3rd Row TL23        ****************
                  Z1                           Z2                      Z0
*************************************************************************************
ans =
       0 +  0.14i              0 +  0.14i              0 +  0.08i
       0 +  0.05i              0 +  0.05i              0 +  0.05i
       0 +  0.04i              0 +  0.04i              0 +  0.1i

       Iaf_pu              IB_A                 Iaf_A                If_LG_pu
*************************************************************************************
ans =
      0 -   4.918i          104.59              514.39                 4.918

      If_3phase      If_LG_div_If_3phase
*************************************************************************************
ans =
       4.3478            1.1311

########################    RESULTS FOR LINE TO LINE FAULT    ######################
     Ia1_pu            Ibf_pu                  IB_A                   IBf_A
*************************************************************************************
ans =
     0 -   2.1739i       -3.7653              104.59                 -393.82

          Icf_pu        Icf_A         If_LL_pu
*************************************************************************************
ans =
       3.7653         393.82         3.7653

If_3phase       If_LL_div_If_3phase
*************************************************************************************
ans =
       4.3478           0.86603

###################### RESULTS FOR DOUBLE LINE TO LINE FAULT #####################
     I_a1_pu              Va1                 I_a2_pu               I_a0_pu
*************************************************************************************
ans =
     0 -   3.1173i       0.28302            0 +   1.2305i          0 + 1.8868i

              I_nf_pu              I_bf_pu                 I_cf_pu
*************************************************************************************
ans =
       0 +   5.6604i       -3.7652 +     2.8301i        3.7652 +     2.8302i

         abs_I_bf_pu          I_Bf_A         abs_I_cf_pu          I_Cf_A
*************************************************************************************
ans =
       4.7102           492.65            4.7103            492.66
```

Appendix H

EXAMPLE H.7 Solve Example G.7 Using MATLAB

Consider the system shown in Figure G.3. Assume that loads, line capacitance, and transformer-magnetizing currents are neglected and that the following data are given based on 20 MVA and the line-to-line voltages as shown in Figure G.3. Do not neglect the resistance of the transmission line TL_{23}. The prefault positive-sequence voltage at bus 3 is Van = 1.0/_0 pu, as shown in Figure G.3.

Generator G_1: $X_1 = 0.20$ pu, $X_2 = 0.10$ pu, $X_0 = 0.05$ pu
Transformer T_1: $X_1 = X_2 = 0.05$ pu, $X_0 = X_1$ (looking into the high-voltage side)
Transformer T_2: $X_1 = X_2 = 0.05$ pu, $X_0 = $ inf. (looking into the high-voltage side)
Transmission line TL_{23}: $\mathbf{Z}_1 = \mathbf{Z}_2 = 0.2 + j0.2$ pu, $\mathbf{Z}_0 = 0.6 + j0.6$ pu

Assume that there is a bolted (i.e., with zero fault impedance) line-to-line fault on phases b and c at bus 3 and determine the following:

(a) Fault current $\mathbf{I}_{bf}$ in per units and amperes
(b) Phase voltages $\mathbf{V}_a$, $\mathbf{V}_b$, and $\mathbf{V}_c$ at bus 2 in per units and kilovolts

Solution

(a) MATLAB Script for Example H.7

```
%
%Data
clear;
clc;
format short g
%data
S=20 % System base 20 MVA
Sb=20e6
VLL=20e3
Vf=1+i*0 %The voltage to ground of phase "a" at the fault point F before the fault
%        occurred is Vf and it is usually selected as 1.0/_0 pu.

%Generator G
Gb=20
VLLG=10e3
Z1G=i*0.2*(S/Gb); Z2G = i*0.1*(S/Gb)
Z0G=i*0.05*(S/Gb)

%Transformer T1
T1b=20
VLLT1=20e3
Z1T1=i*0.05*(S/T1b)
Z2T1=i*0.05*(S/T1b)
Z0T1=i*0.05*(S/T1b)

%Transformer T2
T2b=20
VLLT2=4e3
Z1T2=i*0.05*(S/T2b)
Z2T2=i*0.05*(S/T2b)
Z0T2=i*Inf*(S/T2b)

%Transmission Line TL23
TLb=20
VLLT=20e3
Z1TL=0.2+i*0.2
Z2TL=0.2+i*0.2
Z0TL=0.6+i*0.6
```

```matlab
% Part (a) Fault current Ibf in per units and amperes.

Z1=Z1G+Z1T1+Z1TL
Z2=Z2G+Z2T1+Z2TL

Ia1_pu=Vf/(Z1+Z2)
abs_Ia1_pu=abs(Ia1_pu)
angle_Ia1_rad=angle(Ia1_pu)
angle_Ia1_deg=angle(Ia1_pu)*(180/pi)

%'a' in rectangular form
        a=(-0.5+i*0.866)        % polar: 1/_120
        a2=(-0.5-i*0.866)       % polar: 1/_-120

a2=a^2
Ibf_pu=(a^2-a)*Ia1_pu
abs_Ibf_pu=abs(Ibf_pu)
angle_Ibf_rad=angle(Ibf_pu)
angle_Ibf_deg=angle(Ibf_pu)*(180/pi) %Convert rad into degrees

IB_A=Sb/(sqrt(3)*VLL)
Ibf_A=abs(Ibf_pu)*IB_A
angle_Ibf_deg=angle_Ibf_deg+360 %Convert negative to positive angle (+360)

% Part (b) Phase voltages Va, Vb, and Vc at bus 2 in per units and kilovolts.
% The positive-sequence voltage at bus 2 is
Z1_2=Z1G+Z1T1
Va1_2_pu=Vf-Z1_2*Ia1_pu
abs_Va1_2_pu=abs(Va1_2_pu)
angle_Va1_2_pu=angle(Va1_2_pu)
angle_Va1_2_deg_pu=angle_Va1_2_pu*(180/pi)

% The negative-sequence voltage at bus 2 is
Z2_2=Z2G+Z2T1
Va2_2_pu=-Z2_2*Ia1_pu
abs_Va2_2_pu=abs(Va2_2_pu)
angle_Va2_2_rad_pu=angle(Va2_2_pu)
angle_Va2_2_deg_pu=angle(Va2_2_pu)*(180/pi)
angle_Va2_2_deg_pu=angle_Va2_2_deg_pu + 360

disp('_____');
disp('      ');
disp('      ');
disp('      ');
disp('########################      OR IN SUMMARY         ###################');
disp('      ');
disp('############## RESULTS FOR FAULT CURRENT Ibf IN pu AND AMPER ###############')
disp('************** 1st Row G; 2nd Row T1; 3rd Row T2; 4th Row TL23 *************')
disp('              Z1                      Z2                    Z0 ');
disp(' ***********************************************************************');
[Z1G Z2G Z0G;Z1T1 Z2T1 Z0T1;Z1T1 Z2T1 Z0T1; Z1TL Z2TL Z0TL]
disp('_____');
disp('      ');
disp('      ');
disp('      ');
disp('              Z1                      Z2                       ') ;
disp('***********************************************************************');
[Z1 Z2]
disp('_____');
```

```
disp('      ');
disp('      ');
disp('      ');
disp('              Ia1_pu            abs_Ia1_pu          angle_Ia1_deg
') ;
disp('
********************************************************************************');
[Ia1_pu abs_Ia1_pu angle_Ia1_deg]
disp('
_____');
disp('    ');
disp('    ');
disp('    ');
disp('              Ibf_pu            abs_Ibf_pu          angle_Ibf_deg
') ;
disp('
********************************************************************************');
[Ibf_pu abs_Ibf_pu angle_Ibf_deg]
disp('
_____');
disp('    ');
disp('    ');
disp('    ');
disp(' IB_A        Ibf_A        angle_Ibf_deg            ' ) ;
disp('
********************************************************************************');
[IB_A   Ibf_A    angle_Ibf_deg]
disp('
_____');
disp('    ');
disp('    ');
disp('    ');
```

(b) MATLAB Results for Example H.7

```
S = 20
Sb = 20000000
VLL = 20000
Vf = 1
Gb = 20
VLLG = 10000
Z2G = 0 + 0.1i
Z0G = 0 + 0.05i
T1b = 20
VLLT1 = 20000
Z1T1 = 0 + 0.05i
Z2T1 = 0 + 0.05i
Z0T1 = 0 + 0.05i
T2b = 20
VLLT2 = 4000
Z1T2 = 0 + 0.05i
Z2T2 = 0 + 0.05i
Z0T2 = 0 + Infi
TLb = 20
VLLT = 20000
Z1TL = 0.2 + 0.2i
Z2TL = 0.2 + 0.2i
Z0TL = 0.6 + 0.6i
Z1 = 0.2 + 0.45i
Z2 = 0.2 + 0.35i
Ia1_pu = 0.5 - 1i
abs_Ia1_pu = 1.118
angle_Ia1_rad = -1.1071
```

```
angle_Ia1_deg = -63.435
a = -0.5 + 0.866i
a2 = -0.5 - 0.866i
a2 = -0.49996 - 0.866i
Ibf_pu = -1.732 - 0.86604i
abs_Ibf_pu = 1.9364
angle_Ibf_rad = -2.6779
angle_Ibf_deg = -153.43
IB_A = 577.35
Ibf_A = 1118
angle_Ibf_deg = 206.57
Z1_2 = 0 + 0.25i
Va1_2_pu = 0.75 - 0.125i
abs_Va1_2_pu = 0.76035
angle_Va1_2_pu = -0.16515
angle_Va1_2_deg_pu = -9.4623
Z2_2 = 0 + 0.15i
Va2_2_pu = -0.15 - 0.075i
abs_Va2_2_pu = 0.16771
angle_Va2_2_rad_pu = -2.6779
angle_Va2_2_deg_pu = -153.43
angle_Va2_2_deg_pu = 206.57
```

```
#################          OR IN SUMMARY            #####################
###############    RESULTS FOR FAULT CURRENT Ibf IN pu AND AMPERS  ##############
***************   1st Row G; 2nd Row T1; 3rd Row T2; 4th Row TL23  ***************
                Z1                       Z2                      Z0
*********************************************************************************
ans =
         0 +    0.2i            0 +    0.1i            0 +   0.05i
         0 +   0.05i            0 +   0.05i            0 +   0.05i
         0 +   0.05i            0 +   0.05i            0 +   0.05i
       0.2 +    0.2i          0.2 +    0.2i          0.6 +    0.6i

                Z1                       Z2
*********************************************************************************
ans =
       0.2 +   0.45i          0.2 +   0.35i

             Ia1_pu                  abs_Ia1_pu             angle_Ia1_deg
*********************************************************************************
ans =
       0.5 -     1i            1.118              -63.435

              Ibf_pu                 abs_Ibf_pu             angle_Ibf_deg
*********************************************************************************
ans =
      -1.732 -   0.86604i       1.9364              206.57

               IB_A                     Ibf_A               angle_Ibf_deg
*********************************************************************************
ans =
        577.35             1118   206.57
```

Appendix I: Glossary for Modern Power System Analysis Terminology

Some of the most commonly used terms, both in this book and in general usage, are defined on the following pages. Most of the definitions given in this glossary are based on References 1 through 8.

AA: Abbreviation for all-aluminum conductors.
AAAC: Abbreviation for all-aluminum-alloy conductors. Aluminum-alloy conductors have higher strength than that of the ordinary electric-conductor grade of aluminum.
AC circuit breaker: A circuit breaker whose principal function is usually to interrupt short circuit or fault currents.
ACAR: Abbreviation for aluminum conductor alloy-reinforced. It has a central core of higher-strength aluminum surrounded by layers of electric-conductor-grade aluminum.
Accuracy classification: The accuracy of an instrument transformer at specified burdens. The number used to indicate accuracy is the maximum allowable error of the transformer for specified burdens. For example, a 0.2 accuracy class means the maximum error will not exceed 0.2% at rated burdens.
ACSR: Abbreviation for aluminum conductor, steel-reinforced. It consists of a central core of steel strands surrounded by layers of aluminum strands.
Admittance: The ratio of the phasor equivalent of the steady-state sine-wave current to the phasor equivalent of the corresponding voltage.
Adverse weather: Weather conditions that cause an abnormally high rate of forced outages for exposed components during the periods such conditions persist, but that do not qualify as major storm disasters. Adverse weather conditions can be defined for a particular system by selecting the proper values and combinations of conditions reported by the Weather Bureau: thunderstorms, tornadoes, wind velocities, precipitation, temperature, etc.
Air-blast transformer: A transformer cooled by forced circulation of air through its core and coils.
Air circuit breaker: A circuit breaker in which the interruption occurs in air.
Air switch: A switch in which the interruptions of the circuit occur in air.
Al: Symbol for aluminum.
Ampacity: Current rating in amperes, as of a conductor.
ANSI: Abbreviation for American National Standards Institute.
Apparent sag (at any point): The departure of the wire at the particular point in the span from the straight line between the two points of the span, at 60°F, with no wind loading.
Arcback: A malfunctioning phenomenon in which a valve conducts in the reverse direction.
Arcing time of fuse: The time elapsing from the severance of the fuse link to the final interruption of the circuit under specified conditions.
Arc-over of insulator: A discharge of power current in the form of an arc following a surface discharge over an insulator.
Armored cable: A cable provided with a wrapping of metal, usually steel wires, primarily for the purpose of mechanical protection.
Askarel: A generic term for a group of nonflammable synthetic chlorinated hydrocarbons used as electrical insulating media. Askarels of various compositional types are used. Under arcing conditions, the gases produced, while consisting predominantly of noncombustible hydrogen chloride, can include varying amounts of combustible gases depending on the askarel type. Because of environmental concerns, it is not used in new installations anymore.

Automatic reclosing: An intervention that is not manual. It probably requires specific interlocking such as a full or check synchronizing, voltage or switching device checks, or other safety or operating constrains. It can be high speed or delayed.

Automatic substations: Those in which switching operations are so controlled by relays that transformers or converting equipment are brought into or taken out of service as variations in load may require, and feeder circuit breakers are closed and reclosed after being opened by overload relays.

Autotransformer: A transformer in which at least two windings have a common section.

Auxiliary relay: A relay that operates in response to the opening or closing of its operating circuit to assist another relay in the performance of its function.

AWG: Abbreviation for American Wire Gauge. It is also sometimes called the Brown and Sharpe Wire Gauge.

Base load: The minimum load over a given time.

Basic impulse insulation level: See **BIL**.

BIL: Abbreviation for basic impulse insulation levels, which are reference levels expressed in impulse-crest voltage with a standard wave not longer than 1.5×50 μs. The impulse waves are defined by a combination of two numbers. The first number is the time from the start of the wave to the instant crest value; the second number is the time from the start to the instant of half-crest value on the tail of the wave.

Blocking: Preventing the relay from tripping either due to its own characteristic or to an additional relay.

Breakdown: Also termed puncture, denoting a disruptive discharge through insulation.

Breaker, primary-feeder: A breaker located at the supply end of a primary feeder that opens on a primary-feeder fault if the fault current is of sufficient magnitude.

Breaker-and-a-half scheme: A scheme that provides the facilities of a double main bus at a reduction in equipment cost by using three circuit breakers for each two circuits.

Bundled conductor: In 345-kV and above voltages, instead of having one large conductor per phase, two or more conductors of approximately the same cross section are suspended per phase in close proximity compared with the spacing between phase; the voltage gradient at the conductor surface is significantly reduced. The use of bundled conductors provide (1) reduced voltage gradient; (2) reduced line inductive reactance; and (3) increased corona critical voltage, hence less corona power loss, radio noise, and audible noise. If the subconductors of a bundle are transposed, the current will be divided exactly between the conductors of the bundle. The disadvantages of using bundled conductors are (1) increased cost, (2) increased clearance requirements at structures, (3) increased wind and ice loading, (4) increased tendency of conductor galloping (or dancing), (5) requirement for more complex suspension, (6) increased requirement for duplex or quadruple insulator strings, and (7) increased charging kilovolt-amperes per conductor. The bundles used at the extra-high-voltage level usually have two, three, and four subconductors. On the other hand, the bundle conductors used at the ultrahigh-voltage level may have 8, 12, 16, and even higher number of conductors. Bundled conductors are also called duplex, triplex, and so on, conductors, referring to grouped or multiple conductors.

Burden: The loading imposed by the circuits of the relay on the energizing input power source or sources, that is, the relay burden is the power required to operate the relay.

Bus: A conductor or group of conductors that serves as a common connection for two or more circuits in a switchgear assembly.

Bus, auxiliary: See **Transfer bus**.

Bus (or busbar): An electrical connection of zero impedance joining several items such as lines, loads, etc. "Bus" in a one-line diagram is essentially the same as a "node" in a circuit diagram. It is the term used for a main bar or conductor carrying an electric current to which many connections may be made. Buses are simply convenient means of connecting switches and other equipment into various arrangements. They can be in a variety of sizes and shapes. They can be made of rectangular bars, round solid bars, square tubes, open pairs, or even stranded cables. In substations, they are built above the head and supported by insulated metal structures. Bus materials, in general use, are aluminum and copper, with hard-drawn aluminum, especially in the tubular shape, being the most widely used in high-voltage and extra-high-voltage open-type outdoor stations. Copper or aluminum tubing as well as special shapes are sometimes used for low-voltage-distribution substation buses.

Bus–tie circuit breaker: A circuit breaker that serves to connect buses or bus sections together.

Bus, transfer: A bus to which one circuit at a time can be transferred from the main bus.

Bushing: An insulating structure including a through conductor, or providing a passageway for such a conductor, with provision for mounting on a barrier, conductor or otherwise, for the purpose of insulating the conductor from the barrier and conducting from one side of the barrier to the other.

BVR: Abbreviation for bus voltage regulator or regulation.

BW: Abbreviation for bandwidth.

BX cable: A cable with galvanized interlocked steel spiral armor. It is known as an alternating current (ac) cable and used in a damp or wet location in buildings at low voltage.

Cable: Either a standard conductor (single-conductor cable) or a combination of conductors insulated from one another (multiple-conductor cable).

Cable fault: A partial or total load failure in the insulation or continuity of the conductor.

Capability: The maximum load-carrying ability expressed in kilovolt-amperes or kilowatts of generating equipment or other electric apparatus under specified conditions for a given time interval.

Capability, net: The maximum generation expressed in kilowatt-hours per hour that a generating unit, station, power source, or system can be expected to supply under optimum operating conditions.

Capacitor bank: An assembly at one location of capacitors and all necessary accessories (switching equipment, protective equipment, controls, etc.) required for a complete operating installation.

Capacity: The rated load-carrying ability expressed in kilovolt-amperes or kilowatts of generating equipment or other electric apparatus.

Capacity factor: The ratio of the average load on a machine or equipment for the period considered to the capacity of the machine or equipment.

Characteristic quantity: The quantity or the value that characterizes the operation of the relay.

Characteristics (of a relay in steady state): The locus of the pickup or reset when draw on a graph.

Charge: The amount paid for a service rendered or facilities used or made available for use.

Choppe-wave insulation level: It is determined by test using waves of the same shape to determine the BIL, with exception that the wave is chopped after about 3 μs.

CIGRÉ: The international conference of large high-voltage electric systems. It is recognized as a permanent nongovernmental and nonprofit international association based in France. It focuses on issues related to the planning and operation of power systems, as well as the design, construction, maintenance, and disposal of high-voltage equipment and plants.

Circuit, earth (ground) return: An electric circuit in which the earth serves to complete a path for current.

Circuit breaker: A device that interrupts a circuit without injury to itself so that it can be reset and reused over again.

Circuit-breaker mounting: Supporting structure for a circuit breaker.

Circular mil: A unit of area equal to i/4 of a square mil (= 0.7854 square mil). The cross-sectional area of a circle in circular mils is therefore equal to the square of its diameter in mils. A circular inch is equal to 1 million circular mils. A mil is one-thousandth of an inch. There are 1974 circular mils in a square millimeter. Abbreviated cmil.

CL: Abbreviation for current-limiting (fuse).

cmil: Abbreviation for circular mil.

Component: A piece of equipment, a line, a section of a line, or a group of items that is viewed as an entity.

Computer usage:
    Offline usage: It includes research, routine calculations of the system performance, and data assimilations and retrieval.
    Online usage: It includes data logging and the monitoring of the system state, including switching, safe interlocking, plant loading, postfault control, and load shedding.

Condenser: Also termed capacitor; a device whose primary purpose is to introduce capacitance into an electric circuit. The term condenser is deprecated.

Conductor: A substance that has free electrons or other charge carriers that permit charge flow when an electromotive force is applied across the substance.

Conductor tension, final unloaded: The longitudinal tension in a conductor after the conductor has been stretched by the application for an appreciable period, with subsequent release, of the loadings of ice and wind, at the temperature decrease assumed for the loading district in which the conductor is strung (or equivalent loading).

Congestion cost: The difference between the actual price of electricity at the point of usage and the lowest price on the grid.

Contactor: An electric power switch, not operated manually and designed for frequent operation.

Counterpoise: A grounding system that is made of buried metal (usually galvanized steel wire) strips, wires, or cables that is buried into ground under the transmission line.

Conventional RTU: Designated primarily for hardwired input/output (I/O) and has little or no capability to talk to downstream intelligent electronic devices.

Converter: A machine, device, or system for changing ac power to dc power or vice versa.

Cress factor: A value that is displayed on many power quality monitoring instruments representing the ratio of the crest value of the measured waveform to the root mean square (rms) value of the waveform. For example, the cress factor of a sinusoidal wave is 1.414.

Critical flashover voltage (CFO): The peak voltage for a 50% probability of flashover or disruptive discharge.

CT: Abbreviation for current transformers.

Cu: Symbol for copper.

Current transformers: They are usually rated on the basis of 5-A secondary current and used to reduce primary current to usable levels for transformer-rated meters and to insulate and isolate meters from high-voltage circuits.

Current transformer burdens: CT burdens are normally expressed in ohms impedance such as B-0.1, B-0.2, B-0.5, B-0.9, or B-1.8. Corresponding volt-ampere values are 2.5, 5.0, 12.5, 22.5, and 45.

Current transformer ratio: CT ratio is the ratio of primary to secondary current. For current transformer rated 200:5, the ratio is 200:5, or 40:1.

Demand: The load at the receiving terminals averaged over a specified interval of time.

Demand factor: The ratio of the maximum coincident demand of a system, or part of a system, to the total connected load of the system, or part of the system, under consideration.

Demand, instantaneous: The load at any instant.

Demand, integrated: The demand integrated over a specified period.

Demand interval: The period of time during which the electric energy flow is integrated in determining demand.

Dependability (in protection): The certainty that a relay will respond correctly for all faults for which it is designed and applied to operate.

Dependability (in relays): The ability of a relay or relay system to provide correct operation when required.

Appendix I

Dependent time-delay relay: A time-delay relay in which the time delay varies with the value of the energizing quantity.
Depreciation: The component that represents an approximation of the value of the portion of plant consumed or "used up" in a given period by a utility.
Differential current relay: A fault-detecting relay that functions on a differential current of a given percentage or amount.
Directional (or directional overcurrent) relay: A relay that functions on a desired value of power flow in a given direction on a desired value of overcurrent with ac power flow in a given direction.
Disconnecting or isolating switch: A mechanical switching device used for changing the connections in a circuit or for isolating a circuit or equipment from the source of power.
Disconnector: A switch that is intended to open a circuit only after the load has been thrown off by other means. Manual switches designed for opening loaded circuits are usually installed in a circuit with disconnectors to provide a safe means for opening the circuit under load.
Displacement factor (DPF): The ratio of active power (watts) to apparent power (volt-amperes).
Distance relay: A relay that responds to input quantities as a function of the electrical circuit distance between the relay location and the point of faults.
Distribution center: A point of installation for automatic overload protective devices connected to buses where an electric supply is subdivided into feeders and/or branch circuits.
Distribution switchboard: A power switchboard used for the distribution of electric energy at the voltages common for such distribution within a building.
Distribution system: That portion of an electric system that delivers electric energy from transformation points in the transmission, or bulk power system, to the consumers.
Distribution transformer: A transformer for transferring electric energy from a primary distribution circuit to a secondary distribution circuit or consumer's service circuit; it is usually rated in the order of 5–500 kVA.
Diversity factor: The ratio of the sum of the individual maximum demands of the various subdivisions of a system to the maximum demand of the whole system.
Double line-to-ground (DLG) fault: Fault that exists between two lines and the ground.
Dropout or reset: A relay drops out when it moves from the energized position to the unenergized position.
Effectively grounded: Grounded by means of a ground connection of sufficiently low impedance that fault grounds that may occur cannot build up voltages dangerous to connected equipment.
EHV: Abbreviation for extra high voltage.
Electric system loss: Total electric energy loss in the electric system. It consists of transmission, transformation, and distribution losses between sources of supply and points of delivery.
Electrical fields: They exist whenever voltage exists on a conductor. They are not dependent on the current.
Electrical reserve: The capability in excess of that required to carry the system load.
Element: See **Unit**.
Emergency rating: Capability of installed equipment for a short time interval.
Energizing quantity: The electrical quantity, that is, current or voltage either alone or in combination with other electrical quantities required for the function of the relay.
Energy: That which does work or is capable of doing work. As used by electric utilities, it is generally a reference to electric energy and is measured in kilowatt-hours.
Energy loss: The difference between energy input and output as a result of transfer of energy between two points.
Energy management system (EMS): A computer system that monitors, controls, and optimizes the transmission and generation facilities with advanced applications. A SCADA system is a subject of an EMS.

Equivalent commutating resistance (R_c): The ratio of drop of direct voltage to direct current. However, it does not consume any power.

Express feeder: A feeder that serves the most distant networks and that must traverse the systems closest to the bulk power source.

Extra high voltage: A term applied to voltage levels higher than 230 kV. Abbreviated EHV.

Facilities charge: The amount paid by the customer as a lump sum, or periodically, as reimbursement for facilities furnished. The charge may include operation and maintenance as well as fixed costs.

Fault: A malfunctioning of the network, usually due to the short circuiting of two or more conductors or live conductors connecting to the earth.

Feeder: A set of conductors originating at a main distribution center and supplying one or more secondary distribution centers, one or more branch-circuit distribution centers, or any combination of these two types of load.

Feeder, multiple: Two or more feeders connected in parallel.

Feeder, tie: A feeder that connects two or more independent sources of power and has no tapped load between the terminals. The source of power may be a generating system, substation, or feeding point.

Fiber optics cable: It is made up of varying numbers of either single or multimode fibers, with a strength member in the center of the cable and additional outer layers to provide support and protection against physical damage to the cable. Large amounts of data as high as gigabytes per second can be transmitted over the fiber. They have inherent immunity from electromagnetic interference and have high bandwidth. Two types of them used by utilities: (1) optical power grid wire (OPGW) type or (2) all dielectric self-supporting (ADSS) type.

First-contingency outage: The outage of one primary feeder.

Fixed-capacitor bank: A capacitor bank with fixed, not switchable, capacitors.

Flash: A term encompassing the entire electrical discharge from cloud to the stricken object.

Flashover: An electrical discharge completed from an energized conductor to a grounded support. It may clear itself and trip a circuit breaker.

Flexible ac transmission systems (FACTS): They are the converter stations for ac transmission. It is an application of power electronics for control of the ac system to improve the power flow, operation, and control of the ac system.

Flicker: Impression of unsteadiness of visual sensation induced by a light stimulus whose luminance or spectral distribution fluctuates with time.

Forced interruption: An interruption caused by a forced outage.

Forced outage: An outage that results from emergency conditions directly associated with a component, requiring that component to be taken out of service immediately, either automatically or as soon as switching operations can be performed, or an outage caused by improper operation of equipment or human error.

Frequency deviation: An increase or decrease in the power frequency. Its duration varies from a few cycles to several hours.

Fuse: An overcurrent protective device with a circuit-opening fusible part that is heated and severed by the passage of overcurrent through it.

Fuse cutout: An assembly consisting of a fuse support and holder; it may also include a fuse link.

Gas-insulated transmission line (GIL): A system for the transmission of electricity at high power ratings over long distances. The GIL consists of three single-phase encapsulated aluminum tubes that can be directly buried into the ground, laid in a tunnel, or installed on steel structures at heights of 1–5 m above the ground.

Gauss–Seidel iterative method: It is based on the Gauss iterative method. The only difference is that a more efficient substitution technique is used in this method.

Grip, conductor: A device designed to permit the pulling of conductor without splicing on fittings.

Ground: Also termed earth; a conductor connected between a circuit and the soil; an accidental ground occurs due to cable insulation faults, an insulator defect, etc.

Ground protective relay: Relay that functions on the failure of insulation of a machine, transformer, or other apparatus to ground.

Ground wire: A conductor, having grounding connections at intervals, that is suspended usually above but not necessarily over the line conductor to provide a degree of protection against lightning discharges.

Grounding: The connection of a conductor or frame of a device to the main body of the earth. Thus, it must be done in a way to keep the resistance between the item and the earth under the limits. It is often that the burial of large assemblies of conducting rods in the earth, and the use of connectors in large cross diameters are needed.

Grounding resistance: The resistance of a buried electrode that has functions of (1) the resistance of the electrode itself and connections to it, (2) contact resistance between the electrode and the surrounding soil, and (3) resistance of the surrounding soil, from the electrode surface outward.

GTOs: Abbreviation for gate turn-off thyristors.

Harmonic distortion: Periodic distortion of the sine wave.

Harmonic resonance: A condition in which the power system is resonating near one of the major harmonics being produced by nonlinear elements in the system, hence increasing the harmonic distortion.

Harmonics: Sinusoidal voltages or currents having frequencies that are integer multiples of the fundamental frequency at which the supply system is designed to operate.

Hazardous open circulating (in CTs): The operation of the CTs with the secondary winding open can result in a high voltage across the secondary terminals, which may be dangerous to personnel or equipment. Therefore, the secondary terminals should always be short-circuited before a meter is removed from service.

High-speed relay: A relay that operates in less than a specified time. The specified time in present practice is 50 ms (i.e., three cycles on a 60-Hz system).

HV: Abbreviation for high voltage.

IED: Any device incorporating one or more processors with the capability to receive or send data or control from or to an external source (e.g., electronic multifunction meters, digital relays, or controllers).

IED integration: Integration of protection, control, and data acquisition functions into a minimal number of platforms to reduce capital and operating costs, reduce panel and control room space, and eliminate redundant equipment and database.

Impedance: The ratio of the phasor equivalent of a steady-state sine-wave voltage to the phasor equivalent of a steady-state sine-wave current.

Impedance relay: A relay that operates for all impedance values that are less than its setting, that is, for all points within the cross-hatched circles.

Impulse ratio (flashover or puncture of insulation): The ratio of impulse peak voltage to the peak values of the 60-Hz voltage to cause flashover or puncture.

Impulsive transient: A sudden (nonpower) frequency change in the steady-state condition of the voltage or current that is unidirectional in polarity.

Incremental energy costs: The additional cost of producing or transmitting electric energy above some base cost.

Independent time-delay relay: A time-delay relay in which the time delay is independent of the energizing quantity.

Index of reliability: A ratio of cumulative customer minutes that service was available during a year to total customer minutes demanded; can be used by the utility for feeder reliability comparisons.

Infinite bus: A bus that represents a very large external system. It is considered that at such bus, voltage and frequency are constant. Typically, a large power system is considered as an infinite bus.

Installed reserve: The reserve capability installed on a system.

Instantaneous relay: A relay that operates and resets with no intentional time delay. Such relay operates as soon as a secure decision is made. No intentional time delay is introduced to slow down the relay response.

Instrument transformer: A transformer that is used to produce safety for the operator and equipment from high voltage and to permit proper insulation levels and current-carrying capability in relays, meters, and other measurements.

Insulation coordination: The process of determining the proper insulation levels of various components in a power system and their arrangements. That is, it is the selection of an insulation structure that will withstand the voltage stresses to which the system or equipment will be subjected together with the proper surge arrester.

Insulator: A material that prevents the flow of an electric current and can be used to support electrical conductors.

Insulator flashover: If an overhead line is built along the seashore, especially in California, it will be subjected to winds blowing in from the ocean, which carry a fine salt vapor that deposits salt crystals on the windward side of the insulator. However, if the line is built in areas where rain is seasonal, the insulator surface leakage resistance may become so low during the dry seasons that insulators flash over without warning. Also, if the line is going to be built near gravel pits, cement mills, and refineries, its insulators may become so contaminated that extra insulation is required. Contamination flashovers on transmission systems are initiated by airborne particles deposited on the insulators. These particles may be of natural origins or they may be generated by pollution that is mostly a result of industrial, agricultural, or construction activities, Hence, when line insulators are contaminated, many insulator flashovers occur during light fogs unless arcing rings protect the insulators or special fog-type insulators are used.

Insulator testing: The insulators used on overhead lines are subjected to tests that can be classified as (1) design tests, (2) performance tests, and (3) routine tests.

Integrated services digital network: A switched, end-to-end wide-area network designed to combine digital telephony and data transport services.

Intelligent electronic devices (IED): Any device incorporating one or more processors with the capability to receive or send data and control from or to an external source (e.g., electronic multifunction meters, digital relays, and controllers).

Interconnections: See Tie lines.

International Electrotechnical Commission (IEC): An international organization whose mission is to prepare and publish standards for all electrical, electronic, and related technologies.

Interruptible load: A load that can be interrupted as defined by contract.

Interruption: The loss of service to one or more consumers or other facilities and is the result of one or more component outages, depending on system configuration.

Interruption duration: The period from the initiation of an interruption to a consumer until service has been restored to that consumer.

Inverse time-delay relay: A dependent time-delay relay having an operating time that is an inverse function of the electrical characteristic quantity.

Inverse time-delay relay with definite minimum: A relay in which the time delay varies inversely with the characteristic quantity up to a certain value, after which the time delay becomes substantially independent.

Inverter: A converter for changing direct current to alternating current.

Investment-related charges: Those certain charges incurred by a utility that are directly related to the capital investment of the utility.

ISO: Independent system operator.

Isokeraunic level: The average number of thunder-days per year at that locality (i.e., the average number of thunder that will be heard during a 24-h period).

Isokeraunic map: A map showing mean annual days of thunderstorm activity within the continental United States.

Isolated ground: It originates at an isolated ground-type receptacle or equipment input terminal block and terminates at the point where neutral and ground are bonded at the power source. Its conductor is insulated from the metallic raceway and all ground points throughout its length.

K-factor: A factor used to quantify the load impact of electric arc furnaces on the power system.

kcmil: Abbreviation for a thousand circular mils.

Keraunic level: See **Isokeraunic level**.

Knee-point emf: That sinusoidal electromotive force (emf) applied to the secondary terminals of a current transformer, which, when increased by 10%, causes the exciting current to increase by 50%.

L–L: Abbreviation for line to line.

L–L fault: Fault that exists between two phases.

L–N: Abbreviation for line to neutral.

Lag: Denotes that a given sine wave passes through its peak at a later time than a reference time wave.

Lambda: The incremental operating cost at the load center, commonly expressed in mils per kilowatt-hour.

Let-go current: The maximum current level at which a human holding an energized conductor can control his muscles enough to release it.

Lightning arrestor: A device that reduces the voltage of a surge applied to its terminals and restores itself to its original operating condition.

Line: A component part of a system extending between adjacent stations or from a station to an adjacent interconnection point. A line may consist of one or more circuits.

Line loss: Energy loss on a transmission or distribution line.

Line, pilot: A lightweight line, normally synthetic fiber rope, or wire rope, used to pull heavier pulling lines that in turn are used to pull the conductor.

Line, pulling: A high-strength line, normally synthetic fiber rope, used to pull the conductor.

Load: It may be used in a number of ways to indicate a device or collection of devices that consume electricity, or to indicate the power required from a given supply circuit, or the power or current being passed through a line or machine.

Load, interruptible: A load that can be interrupted as defined by contract.

Load center: A point at which the load of a given area is assumed to be concentrated.

Load diversity: The difference between the sum of the maxima of two or more individual loads and the coincident or combined maximum load, usually measured in kilowatts over a specified period.

Load duration curve: A curve of loads, plotted in descending order of magnitude, against time intervals for a specified period.

Load factor: The ratio of the average load over a designated period to the peak load occurring in that period.

Load-interrupter switch: An interrupter switch designed to interrupt currents not in excess of the continuous-current rating of the switch.

Load losses, transformer: Those losses that are incident to the carrying of a specified load. They include I^2R loss in the winding due to load and eddy currents, stray loss due to leakage fluxes in the windings, etc., and the loss due to circulating currents in parallel windings.

Load management: (also called **demand-side management**): It extends remote supervision and control to subtransmission and distribution circuits, including control of residential, commercial, and industrial loads.

Load tap changer: A selector switch device applied to power transformers to maintain a constant low-side or secondary voltage with a variable primary voltage supply, or to hold a constant voltage out along the feeders on the low-voltage side for varying load conditions on the low-voltage side. Abbreviated LTC.

Load-tap-changing transformer: A transformer used to vary the voltage, or phase angle, or both, of a regulated circuit in steps by means of a device that connects different taps of tapped winding(s) without interrupting the load. The voltage magnitude at a given bus is controlled by changing its taps, and hence, it can be kept constant or within certain limits.

Local backup: Those relays that do not suffer from the same difficulties as remote backup, but they are installed in the same substation and use some of the same elements as the primary protection.

Loss factor: The ratio of the average power loss to the peak-load power loss during a specified period.

Low-side surges: The current surge that appears to be injected into the transformer secondary terminals upon a lighting strike to grounded conductors in the vicinity.

LTC: Abbreviation for load tap changer.

LV: Abbreviation for low voltage.

Magnetic fields: Such fields exist whenever current flows in a conductor. They are not voltage dependent.

Main bus: A bus that is normally used. It has a more elaborate system of instruments, relays, and so on, associated with it.

Main distribution center: A distribution center supplied directly by mains.

Maintenance expenses: The expense required to keep the system or plant in proper operating repair.

Maximum demand: The largest of a particular type of demand occurring within a specified period.

Messenger cable: A galvanized steel or copperweld cable used in construction to support a suspended current-carrying cable.

Metal-clad switchgear, outdoor: A switchgear that can be mounted in suitable weatherproof enclosures for outdoor installations. The base units are the same for both indoor and outdoor applications. The weatherproof housing is constructed integrally with the basic structure and is not merely a steel enclosure. The basic structure, including the mounting details and withdrawal mechanisms for the circuit breakers, bus compartments, transformer compartments, etc., is the same as that of indoor metal-clad switchgear. (Used in distribution systems.)

Minimum demand: The smallest of a particular type of demand occurring within a specified period.

Momentary interruption: An interruption of duration limited to the period required to restore service by automatic or supervisory-controlled switching operations or by manual switching at locations where an operator is immediately available. Such switching operations are typically completed in a few minutes.

Monthly peak duration curve: A curve showing the total number of days within the month during which the net 60-min clock-hour integrated peak demand equals or exceeds the percent of monthly peak values shown.

MOV: The metal oxide varistor that is built from zinc oxide disks connected in series and parallel arrangements to achieve the required protective level and energy requirement. It is similar to a high-voltage surge arrester.

N.C.: Abbreviation for normally closed.

NESC: Abbreviation for National Electrical Safety Code.

Net system energy: Energy requirements of a system, including losses, defined as (1) net generation of the system, plus (2) energy received from others, less (3) energy delivered to other systems.

Appendix I   693

Network configurator: An application that determines the configuration of the power system based on telemetered breaker and switch statuses.

Network transmission system: A transmission system that has more than one simultaneous path of power flow to the load.

Newton–Raphson method: This method of power-flow solution technique is very reliable and extremely fast in convergence. It is not sensitive to factors that cause poor or no convergence with other power-flow techniques. Rate of convergence is relatively independent of system size. Rectangular or polar coordinates can be used for the bus voltages.

N.O.: Abbreviation for normally open.

Noise: An unwanted electrical signal with a less than 200 kHz superimposed on the power system voltage or current in phase conductors, or found on neutral conductors or signal lines. It is not a harmonic distortion or transient. It disturbs microcomputers and programmable controllers.

No-load current: The current demand of a transformer primary when no current demand is made on the secondary.

No-load loss: Energy losses in an electric facility when energized at rated voltage and frequency but not carrying load.

Nonlinear load: An electrical load that draws current discontinuously or whose impedances vary throughout the cycle of the input ac voltage waveform.

Normal rating: Capacity of installed equipment.

Normal weather: All weather not designated as adverse or major storm disaster.

Normally closed: Denotes the automatic closure of contacts in a relay when deenergized. Abbreviated N.C.

Normally open: Denotes the automatic opening of contacts in a relay when deenergized. Abbreviated N.O.

NSW: Abbreviation for nonswitched.

NX: Abbreviation for nonexpulsion (fuse).

Off-peak energy: Energy supplied during designated periods of relatively low system demands.

OH: Abbreviation for overhead.

On-peak energy: Energy supplied during designated periods of relatively high system demands.

One line open (OPO): A series fault having one of the phases open.

Open systems: A computer system that embodies supplier-independent standards so that software can be applied on many different platforms and can interoperate with other applications on local and remote systems.

Operating expenses: The labor and material costs for operating the plant involved.

Operational data: Also called SCADA data, and are instantaneous values of power system analog and status points (e.g., volts, amperes, MW, Mvar, circuit breaker status, switch positions).

Oscillatory transient: A sudden and nonpower frequency change in the steady-state condition of voltage or current that includes both positive and negative polarity values; usually somewhere between 20° and 25° at full load.

Outage: It describes the state of a component when it is not available to perform its intended function due to some event directly associated with that component. An outage may or may not cause an interruption of service to consumers depending on system configuration.

Outage duration: The period from the initiation of an outage until the affected component or its replacement once again becomes available to perform its intended function.

Outage rate: For a particular classification of outage and type of component, the mean number of outages per unit exposure time per component.

Overhead expenses: The costs that, in addition to direct labor and materials, are incurred by all utilities.

Overload: Loading in excess of the normal rating of equipment.

Overload protection: Interruption or reduction of current under conditions of excessive demand, provided by a protective device.

Overshoot time: The time during which stored operating energy dissipated after the characteristic quantity has been suddenly restored from a specified value to the value it had at the initial position of the relay.

Overvoltage: A voltage that has a value at least 10% above the nominal voltage for a period greater than 1 min.

Passive filter: A combination of inductors, capacitors, and resistors designed to eliminate one or more harmonics. The most common variety is simply an inductor in series with a shunt capacitor, which short circuits the major distorting harmonic component from the system.

PE: An abbreviation used for polyethylene (cable insulation).

Peak current: The maximum value (crest value) of an alternating current.

Peak voltage: The maximum value (crest value) of an alternating voltage.

Peaking station: A generating station that is normally operated to provide power only during maximum load periods.

Peak-to-peak value: The value of an ac waveform from its positive peak to its negative peak. In the case of a sine wave, the peak-to-peak value is double the peak value.

Pedestal: A bottom support or base of a pillar, statue, etc.

Percent regulation: See **Percent voltage drop**.

Percent voltage drop: The ratio of voltage drop in a circuit to voltage delivered by the circuit, multiplied by 100 to convert to percent.

Permanent forced outage: An outage whose cause is not immediately self-clearing but must be corrected by eliminating the hazard or by repairing or replacing the component before it can be returned to service. An example of a permanent forced outage is a lightning flashover that shatters an insulator, thereby disabling the component until repair or replacement can be made.

Permanent forced outage duration: The period from the initiation of the outage until the component is replaced or repaired.

Persistent forced outage: A component outage whose cause is not immediately self-clearing but must be corrected by eliminating the hazard or by repairing or replacing the affected component before it can be returned to service. An example of a persistent forced outage is a lightning flashover that shatters an insulator, thereby disabling the component until repair or replacement can be made.

Phase: The time of occurrence of the peak value of an ac waveform with respect to the time of occurrence of the peak value of a reference waveform.

Phase angle: An angular expression of phase difference.

Phase-angle measuring relay: A relay that functions at a predetermined phase angle between voltage and current.

Phase shift: The displacement in time of one voltage waveform relative to other voltage waveform(s).

Pick up: A relay is said to pick up when it moves from the unenergized position to the energized position (by closing its contacts).

Pilot channel: A means of interconnection between relaying points for the purpose of protection.

Planning, conceptual: A long range of guidelines for decision.

Planning, preliminary: A state of project decisions.

Polarity: The relative polarity of the primary and secondary windings of a CT is indicated by polarity marks, associated with one end of each winding. When a current enters at the polarity end of the primary winding, a current in phase with it leaves the polarity end of the secondary winding.

Pole: A column of wood or steel, or some other material, supporting overhead conductors, usually by means of arms or brackets.

Pole fixture: A structure installed in lieu of a single pole to increase the strength of a pole line or to provide better support for attachments than would be provided by a single pole. Examples are A fixtures, H fixtures.

Port: A communication pathway into or out of a computer, or networked device such as a server. Well-known applications have standard port numbers.

Power: The rate (in kilowatts) of generating, transferring, or using energy.

Power factor: The ratio of active power to apparent power.

Power flow (load flow): The solution for normal balanced three-phase steady-state operating conditions of an electric power system. The data obtained from load flow studies are used for the studies of normal operating mode, contingency analysis, outage security assessment, and optimal dispatching and stability. They are performed for power system planning and operational planning, and in connection with system operation and control. Power flow techniques are based on the fact that in any power transmission network operating in the steady state, the coupling (i.e., interdependence) between $P-\theta$ (i.e., active powers and bus voltage angles) and $Q-V$ (i.e., reactive powers and bus voltage magnitudes) is relatively weak, contrary to the strong coupling between P and θ, and Q and V. Hence, these methods solve the power flow problem by "decoupling" (i.e., solving separately) the $P-\theta$ and $Q-V$ problems. A given power flow algorithm must provide a fast convergence.

Power line carrier (PLC): Systems operating on narrow channels between 30 and 50 kHz are frequently used for high-voltage line-protective relaying applications.

Power pool: A group of power systems operating as an interconnected system and pooling their resources.

Power system stability: The ability of an electric power system, for a given initial operating condition, to regain a state of operating equilibrium after being subjected to a physical disturbance.

Power transformer: A transformer that transfers electric energy in any part of the circuit between the generator and the distribution primary circuits.

Power, active: The product of the rms value of the voltage and the rms value of the in-phase component of the current.

Power, apparent: The product of the rms value of the voltage and the rms value of the current.

Power, instantaneous: The product of the instantaneous voltage multiplied by the instantaneous current.

Power, reactive: The product of the rms value of the voltage and the rms value of the quadrature component of the current.

Primary disconnecting devices: Self-coupling separable contacts provided to connect and disconnect the main circuits between the removable element and the housing.

Primary distribution feeder: A feeder operating at primary voltage supplying a distribution circuit.

Primary distribution mains: The conductors that feed from the center of distribution to direct primary loads or to transformers that feed secondary circuits.

Primary distribution network: A network consisting of primary distribution mains.

Primary distribution system: A system of ac distribution for supplying the primaries of distribution transformers from the generating station or substation distribution buses.

Primary distribution trunk line: A line acting as a main source of supply to a distribution system.

Primary feeder: That portion of the primary conductors between the substation or point of supply and the center of distribution.

Primary lateral: That portion of a primary distribution feeder that is supplied by a main feeder or other laterals and extends through the load area with connections to distribution transformers or primary loads.

Primary main feeder: The higher-capacity portion of a primary distribution feeder that acts as a main source of supply to primary laterals or directly connected distribution transformers and primary loads.

Primary network: A network supplying the primaries of transformers whose secondaries may be independent or connected to a secondary network.

Primary open-loop service: A service that consists of a single distribution transformer with dual primary switching, supplied from a single primary circuit that is arranged in an open-loop configuration.

Primary selective service: A service that consists of a single distribution transformer with primary throw-over switching, supplied by two independent primary circuits.

Primary transmission feeder: A feeder connected to a primary transmission circuit.

Primary unit substation: A unit substation in which the low-voltage section is rated above 1000 V.

Protective gear: The apparatus, including protective relays, transformers, and auxiliary equipment, for use in a protective system.

Protective relay: An electrical device whose function is to detect defective lines or apparatus or other power system conditions of an abnormal or dangerous nature and to initiate isolation of a part of an electrical system, or to operate an alarm signal in the case of a fault or other abnormal condition.

Protective scheme: The coordinated arrangements for the protection of a power system.

Protective system: A combination of protective gears designed to secure, under predetermined conditions, usually abnormal, the disconnection of an element of a power system, or to give an alarm signal, or both.

Protective system usage: In protection systems, it is used to compare relevant quantities and to replace slower, more conventional devices, based on the high-speed measurement of system parameters.

PT: Abbreviation for potential transformers.

pu: Abbreviation for per unit.

Puller, reel: A device designed to pull a conductor during stringing operations.

Pulse number (p): The number of pulsations (i.e., cycles of ripple) of the direct voltage per cycle of alternating voltage (e.g., pulse numbers for three-phase one-way and three-phase two-way rectifier bridges are 3 and 6, respectively).

Radial distribution system: A distribution system that has a single simultaneous path of power flow to the load.

Radial service: A service that consists of a single distribution transformer supplied by a single primary circuit.

Radial system, complete: A radial system that consists of a radial subtransmission circuit, a single substation, and a radial primary feeder with several distribution transformers each supplying radial secondaries; has the lowest degrees of service continuity.

Ratchet demand: The maximum past or present demands that are taken into account to establish billings for previous or subsequent periods.

Rate base: The net plant investment or valuation base specified by a regulatory authority upon which a utility is permitted to earn a specified rate of return.

Rated burden: The load that may be imposed on the transformer secondaries by associated meter coils, leads, and other connected devices without causing an error greater than the stated accuracy classification.

Rated continuous current: The maximum 60-Hz rms current that the breaker can carry continuously while it is in the closed position without overheating.

Rated impulse withstand voltage: The maximum crest voltage of a voltage pulse with standard rise and delay times that the breaker insulation can withstand.

Rated insulation class: Denotes the nominal (line-to-line) voltage of a circuit on which it should be used.

Rated interrupting MVA: For a three-phase circuit breaker, it is $\sqrt{3}$ times the rated maximum voltage in kV times the rated short-circuit current in kA. It is more common to work with current and voltage ratings than with MVA rating.

Rated interrupting time: The time in cycles on a 60-Hz basis from the instant the trip coil is energized to the instant the fault current is cleared.

Rated low-frequency withstanding voltage: The maximum 60-Hz rms line-to-line voltage that the circuit breaker can withstand without insulation damage.

Rated maximum voltage: Designated the maximum rms line-to-line operating voltage. The breaker should be used in systems with an operating voltage less than or equal to this rating.

Rated momentary current: The maximum rms asymmetrical current that the breaker can withstand while in the closed position without damage. Rated momentary current for standard breakers is 1.6 times the symmetrical interrupting capacity.

Rated short-circuit current: The maximum rms symmetrical current that the breaker can safely interrupt at rated maximum voltage.

Rated voltage range factor K: The range of voltage for which the symmetrical interrupting capability times the operating voltage is constant.

Ratio correction factor: The factor by which the marked ratio of a CT must be multiplied to obtain the true ratio.

Reach: A distance relay operates whenever the impedance seen by the relay is less than a prescribed value. This impedance or the corresponding distance is known as the reach of the relay.

Reactive power compensation: Shunt reactors, shunt capacitors, static var systems, and synchronous condensers are used to control voltage. Series capacitors are used to reduce line impedance.

Reactor: An inductive reactor between the dc output of the converter and the load. It is used to smooth the ripple in the direct current adequately, to reduce harmonic voltages and currents in the dc line, and to limit the magnitude of fault current. It is also called a **smoothing reactor**.

Recloser: A dual-timing device that can be set to operate quickly to prevent downline fuses from blowing.

Reclosing device: A control device that initiates the reclosing of a circuit after it has been opened by a protective relay.

Reclosing fuse: A combination of two or more fuse holders, fuse units, or fuse links mounted on a fuse support(s), mechanically or electrically interlocked, so that one fuse can be connected into the circuit at a time and the functioning of that fuse automatically connects the next fuse into the circuit, thereby permitting one or more service restorations without replacement of fuse links, refill units, or fuse units.

Reclosing relay: A programming relay whose function is to initiate the automatic reclosing of a circuit breaker.

Reclosure: The automatic closing of a circuit-interrupting device following automatic tripping. Reclosing may be programmed for any combination of instantaneous, time-delay, single-shot, multiple-shot, synchronism-check, dead line–live bus, or dead bus–live line operation.

Recovery voltage: The voltage that occurs across the terminals of a pole of a circuit-interrupting device upon interruption of the current.

Rectifier: A converter for changing alternating current to direct current.

Relays: A low-powered electrical device used to activate a high-powered electrical device. (In T & D systems, it is the job of relays to give the tripping commands to the right circuit breakers.)

Remote access: Access to a control system or IED by a user whose operations terminal is not directly connected to the control systems or IED.

Remote backup: Those relays that are located in a separate location and are completely independent of the relays, transducers, batteries, and circuit breakers that they are backing up.

Remote terminal unit (RTU): A hardware that telemeters system-wide data from various field locations (i.e., substations, generating plants to a central location). It includes the entire complement of devices, functional modules, and assemblies that are electrically interconnected to affect the remote station supervisory functions.

Required reserve: The system planned reserve capability needed to ensure a specified standard of service.

Resetting value: The maximum value of the energizing quantity that is insufficient to hold the relay contacts closed after operating.

Resistance: The real part of impedance.

Return on capital: The requirement that is necessary to pay for the cost of investment funds used by the utility.

Ripple: The ac component from dc power supply arising from sources within the power supply. It is expressed in peak, peak-to-peak, rms volts, or as percent rms. Since high-voltage dc converters have large dc smoothing reactors, approximately 1 H, the resultant direct current is constant (i.e., free from ripple). However, the direct voltage on the valve side of the smoothing reactor has ripple.

Ripple amplitude: The maximum value of the instantaneous difference between the average and instantaneous value of a pulsating unidirectional wave.

Risk: The probability that a particular threat will exploit a particular vulnerability of an equipment, plant, or system.

Risk management: Decisions to accept exposure or to reduce vulnerabilities by either mitigating the risks or applying cost-effective controls.

SA: Deployment of substation and feeder operating functions and applications ranging from supervisory control and data acquisition (SCADA) and alarm processing to integrated volt/var control in order to optimize the management of capital assets and enhance operational and maintenance efficiencies with minimal human intervention.

Sag: The distance measured vertically from a conductor to the straight line joining its two points of support. Unless otherwise stated, the sag referred to is the sag at the midpoint of the span.

Sag: A decrease to between 0.1 and 0.9 pu in rms voltage and current at the power frequency for a duration of 0.5 cycles to 1 min.

SAG of a conductor (at any point in a span): The distance measured vertically from the particular point in the conductor to a straight line between its two points of support.

Sag, final unloaded: The sag of a conductor after it has been subjected for an appreciable period to the loading prescribed for the loading district in which it is situated, or equivalent loading, and the loading removed. Final unloaded sag includes the effect of inelastic deformation.

Sag, initial unloaded: The sag of a conductor before the application of any external load.

Sag section: The section of line between snub structures. More than one sag section may be required to properly sag the actual length of conductor that has been strung.

Sag span: A span selected within a sag section and used as a control to determine the proper sag of the conductor, thus establishing the proper conductor level and tension. A minimum of two, but normally three, sag spans are required within a sag section to sag properly. In mountainous terrain or where span lengths vary radically, more than three sag spans could be required within a sag section.

SCADA: Abbreviation for supervisory control and data acquisition.

SCADA communication line: The communication link between the utility's control center and the RTU at the substation.

Scheduled interruption: An interruption caused by a scheduled outage.

Scheduled outage: An outage that results when a component is deliberately taken out of service at a selected time, usually for purposes of construction, preventive maintenance, or repair. The key test to determine if an outage should be classified as forced or scheduled is as follows. If it is possible to defer the outage when such deferment is desirable, the outage is a scheduled outage; otherwise, the outage is a forced outage. Deferring an outage may be desirable, for example, to prevent overload of facilities or an interruption of service to consumers.

Scheduled outage duration: The period from the initiation of the outage until construction, preventive maintenance, or repair work is completed.

Scheduled maintenance (generation): Capability that has been scheduled to be out of service for maintenance.

SCV: Abbreviation for steam-cured (cable insulation).

Seasonal diversity: Load diversity between two (or more) electric systems that occurs when their peak loads are in different seasons of the year.

Secondary current rating: The secondary current existing when the transformer is delivering rated kilovolt-amperes at a rated secondary voltage.

Secondary disconnecting devices: Self-coupling separable contacts provided to connect and disconnect the auxiliary and control circuits between the removable element and the housing.

Secondary distributed network: A service consisting of a number of network transformer units at a number of locations in an urban load area connected to an extensive secondary cable grid system.

Secondary distribution feeder: A feeder operating at secondary voltage supplying a distribution circuit.

Secondary distribution mains: The conductors connected to the secondaries of distribution transformers from which consumers' services are supplied.

Secondary distribution network: A network consisting of secondary distribution mains.

Secondary fuse: A fuse used on the secondary-side circuits, restricted for use on a low-voltage secondary distribution system that connects the secondaries of distribution transformers to consumers' services.

Secondary mains: Those that operate at utilization voltage and serve as the local distribution main. In radial systems, secondary mains that supply general lighting and small power are usually separate from mains that supply three-phase power because of the dip in voltage caused by starting motors. This dip in voltage, if sufficiently large, causes an objectionable lamp flicker.

Secondary network: It consists of two or more network transformer units connected to a common secondary system and operating continuously in parallel.

Secondary network service: A service that consists of two or more network transformer units connected to a common secondary system and operating continuously in parallel.

Secondary system, banked: A system that consists of several transformers supplied from a single primary feeder, with the low-voltage terminals connected together through the secondary mains.

Secondary unit substation: A unit substation whose low-voltage section is rated 1000 V and below.

Secondary voltage regulation: A voltage drop caused by the secondary system; it includes the drop in the transformer and in the secondary and service cables.

Second-contingency outage: The outage of a secondary primary feeder in addition to the first one.

Sectionalizer: A device that resembles an oil circuit recloser but lacks the interrupting capability.

Security: The measure that a relay will not operate incorrectly for any faults.

Security (in protection): The measure that a relay will not operate incorrectly for any fault.

Security (in relays): The ability of a relay or relaying system never to operate falsely.

Selector: See **Transfer switches**.

Sequence filters: They are used in three-phase systems to measure (and therefore to indicate the presence of) symmetrical components of current and voltage.

Service area: Territory in which a utility system is required or has the right to supply or make available electric service to ultimate consumers.

Service availability index: See **Index of reliability**.

Service drop: The overhead conductors, through which electric service is supplied, between the last utility company pole and the point of their connection to the service facilities located at the building or other support used for the purpose.

Service entrance: All components between the point of termination of the overhead service drop or underground service lateral and the building main disconnecting device, with the exception of the utility company's metering equipment.

Service entrance conductors: The conductors between the point of termination of the overhead service drop or underground service lateral and the main disconnecting device in the building.

Service entrance equipment: Equipment located at the service entrance of a given building that provides overcurrent protection to the feeder and service conductors, provides a means of disconnecting the feeders from energized service conductors, and provides a means of measuring the energy used by the use of metering equipment.

Service lateral: The underground conductors, through which electric service is supplied, between the utility company's distribution facilities and the first point of their connection to the building or area service facilities located at the building or other support used for the purpose.

Setting: The actual value of the energizing or characteristic quantity of which the relay designed to operate under given conditions.

SF_6: Formula for sulfur hexafluoride (gas).

Shielding, effective: A shielding that has zero unprotective width.

Short-circuit selective relay: A relay that functions instantaneously on an excessive value of current.

Shunt capacitor bank: A large number of capacitor units connected in series and parallel arrangement to make up the required voltage and current ratings, and connected between line and neutral, or between line and line.

Skin effect: The phenomenon by which alternative current tends to flow in the outer layer of a conductor. It is a function of conductor size, frequency, and the relative resistance of the conductor material.

Simultaneous fault: The same fault point having two separate faults.

St: Abbreviation for steel.

Stability: The quality whereby a protective system remains in operation under all conditions other than those for which it is specifically designed to operate.

STATCOM: A static compensator. It provides variable lagging or leading reactive powers without using inductors or capacitors for var generation.

Static var system: A static var compensator that can also control mechanical switching of shunt capacitor banks or reactors.

Strand: One of the wires, or groups of wires, of any stranded conductor.

Stranded conductor: A conductor composed of a group of wires, or of any combination of groups of wires. Usually, the wires are twisted together.

Strike distance: The distance that is jumped by an approaching flash to make contact.

Stroke: The high-current components in a flash. A single flash may contain several strokes.

Submarine cable: A cable designed for service under water. It is usually a lead-covered cable with a steel armor applied between layers of jute.

Submersible transformer: A transformer so constructed as to be successfully operable when submerged in water under predetermined conditions of pressure and time.

Substation: An assemblage of equipment for purposes other than generation or utilization, through which electric energy in bulk is passed for the purpose of switching or modifying its characteristics. The term substation includes all stations classified as switching, collector bus, distribution, transmission, or bulk-power substations.

Substation grounding: A grounding system that is buried into the ground throughout the substation area. It is connected to every individual equipment, structure, and installation so that it can provide the means by which grounding currents are connected to remote areas, and therefore a proper grounding is accomplished to provide safety. It is crucial for the substation ground to have a low ground resistance, adequate current-carrying capacity, and safety features for personnel. Thus, the substation ground resistance has to be kept very low so that the total rise of the grounding system potential will not reach values that are unsafe for human contact. Hence, the substation grounding system is normally made up of buried horizontal conductors and the ground rods driven into ground interconnected by clamping, welding, and brazing to form a continuous grid (also called **mat**) network.

Substation LAN: A communications network, typically high speed, within the substation and extending into the switchyard.

Substation local area network (LAN): A technology that is used in a substation environment and facilitate interfacing to process-level equipment (IEDs and PLCs) while providing immunity and isolation to substation noise.

Substation voltage regulation: The regulation of the substation voltage by means of the voltage regulation equipment which can be LTC (load-tap-changing) mechanisms in the substation transformer, a separate regulator between the transformer and low-voltage bus, switched capacitors at the low-voltage bus, or separate regulators located in each individual feeder in the substation.

Subsynchronous: Electrical and mechanical quantities associated with frequencies below the synchronous frequency of a power system.

Subsynchronous oscillation: The exchange of energy between the electric network and the mechanical spring-mass system of the turbine generator at subsynchronous frequencies.

Subsynchronous resonance: An electric power system condition where the electric power network exchanges energy with a turbine generator at one or more of the natural frequencies of the combined system below the synchronous frequency of the system.

Subtransmission: That part of the distribution system between bulk power source(s) (generating stations or power substations) and the distribution substation.

Supersynchronous: Electrical or mechanical quantities associated with frequencies above the synchronous frequency of a power system.

Supervisory control and data acquisition (SCADA): A computer system that performs data acquisition and remote control of a power system.

Supply security: Provision must be made to ensure continuity of supply to consumers even with certain items of plant out of action. Usually two circuits in parallel are used and a system is said to be secure when continuity is assured. It is the prerequisite in design and operation.

Susceptance: The imaginary part of admittance.

Sustained interruption: The complete loss of voltage (<0.1 pu) on one or more phase conductors for a time greater than 1 min.

SVC: Static var compensator.

Swell: An increase to between 1.1 and 1.8 pu in rms voltage or current at the power frequency for durations from 0.5 cycle to 1 min.

Switch: A device for opening and closing or for changing connections in a circuit.

Switching: Connecting or disconnecting parts of the system from each other. It is accomplished using breakers and/or switches.

Switch, isolating: An auxiliary switch for isolating an electric circuit from its source of power; it is operated only after the circuit has been opened by other means.

Switchboard: A large single panel, frame, or assembly of panels on which switches, fuses, buses, and usually instruments are mounted (on the face, or back, or both).

Switched-capacitor bank: A capacitor bank with switchable capacitors.

Switchgear: A general term covering switching or interrupting devices and their combination with associated control, instrumentation, metering, protective, and regulating devices; also assemblies of these devices with associated interconnections, accessories, and supporting structures.

Switching time: The period from the time a switching operation is required due to a forced outage until that switching operation is performed.

System: A group of components connected together in some fashion to provide flow of power from one point or points to another point or points.

System interruption duration index: The ratio of the sum of all customer interruption durations per year to the number of customers served. It gives the number of minutes out per customer per year.

Systems: It is used to describe the complete electrical network, generators, loads, and prime movers.

TCR: Abbreviation for thyristor-controlled reactor.

TCSC: Abbreviation for thyristor-controlled series compensation. It provides fast control and variation of the impedance of the series capacitor bank. It is part of the flexible system (FACTS).

Three-phase fault: The fault that exists between three phases.

Thyristor (SCR): A thyristor (silicon-controlled rectifier) is a semiconductor device with an anode, a cathode terminal, and a gate for the control of the firing.

Tie lines: The transmission lines between the electrical power systems of separate utility companies.

Time delay: An intentional time delay is inserted between the relay decision time and the initiation of the trip action.

Time delay relay: A relay having an intentional delaying device.

Total demand distortion (TDD): The ratio of the root mean square of the harmonic current to the rms value of the rated or maximum demand fundamental current, expressed as a percent.

Total harmonic distortion (THD): The ratio of the root mean square of the harmonic content to the rms value of the fundamental quantity, expressed as a percent of the fundamental.

Transfer bus: A bus used for the purpose of transferring a load.

Transfer switches: The switches that permit feeders or equipment to be connected to a bus.

Transformer ratio (TR): The total ratio of current and voltage transformers. For 200:5 CT and 480:120 VT, TR = 40 × 4 = 160.

Transient forced outage: A component outage whose cause is immediately self-clearing so that the affected component can be restored to service either automatically or as soon as a switch or a circuit breaker can be reclosed or a fuse replaced. An example of a transient forced outage is a lightning flashover that does not permanently disable the flashed component.

Transposition: When the conductors of a three-phase line are not equally spaced, the problem of computing capacitance (and/or its capacitive reactance) becomes more difficult. In a transposed line, each conductor occupies the same position as every other phase conductor over equal distance along the transposition cycle. In an untransposed line, the average capacitances of each phase to neutral of any phase to neutral are unequal. Thus, in a transposed line, the average capacitance to neutral of (and/or its capacitive reactance) any other phase, for a given complete transposition cycle, is the same for all the phases. A complete transposition is achieved by transposing the line three times at equal distances.

Traveler: A sheave complete with suspension arm or frame used separately or in groups and suspended from structures to permit the stringing of conductors.

Triplen harmonics: A term frequently used to refer to the odd multiples of the third harmonic, which deserve special attention because of their natural tendency to be zero sequence.

Tripout: A flashover of a line that does not clear itself. It must be cleared by operation of a circuit breaker.

True power factor (TPF): The ratio of the active power of the fundamental wave, in watts, to the apparent power of the fundamental wave, in rms volt-amperes (including the harmonic components).

TSC: Abbreviation for thyristor switched capacitor.

Two-lines open (TLO) fault: A series fault having two phases open.

Ultra high-speed: A term that is not included in the relay standards but is commonly considered to be in operation in 4 ms or less.

Underground distribution system: That portion of a primary or secondary distribution system that is constructed below the earth's surface. Transformers and equipment enclosures for such a system may be located either above or below the surface as long as the served and serving conductors are located underground.

Undervoltage: A voltage that has a value at least 10% below the nominal voltage for a period greater than 1 min.

Undervoltage relay: A relay that functions on a given value of single-phase ac under voltage.

Unit: A self-contained relay unit that, in conjunction with one or more other relay units, performs a complex relay function.

Unit substation: A substation consisting primarily of one or more transformers that are mechanically and electrically connected to and coordinated in design with one or more switchgear or motor control assemblies or combinations thereof.

Unreach: The tendency of the relay to restrain at impedances larger than its setting. That is, it is due to error in relay measurement resulting in wrong operation.

URD: Abbreviation for underground residential distribution.

Utilization factor: The ratio of the maximum demand of a system to the rated capacity of the system.

VD: Abbreviation for voltage drop.

VDIP: Abbreviation for voltage dip.

Voltage collapse: The process by which voltage instability leads to a very low voltage profile in a significant part of the system.

Voltage dip: A voltage change resulting from a motor starting.

Voltage drop: The difference between the voltage at the transmitting and receiving ends of a feeder, main or service.

Voltage fluctuation: A series of voltage changes or a cyclical variation of the voltage envelope.

Voltage imbalance (or unbalance): The maximum deviation from the average of the three-phase voltages or currents, divided by the average of the three-phase voltages or currents, expressed in percent.

Voltage interruption: Disappearance of the supply voltage on one or more phases. It can be momentary, temporary, or sustained.

Voltage magnification: The magnification of capacitor switching oscillatory transient voltage on the primary side by capacitors on the secondary side of a transformer.

Voltage regulation: The percent voltage drop of a line with reference to the receiving-end voltage.

$$\% \text{ regulation} = \frac{|\bar{E}_s| - |\bar{E}_r|}{|\bar{E}_r|} \times 100$$

where $|\bar{E}_s|$ is the magnitude of the sending-end voltage and $|\bar{E}_r|$ is the magnitude of the receiving-end voltage.

Voltage regulator: An induction device having one or more windings in shunt with, and excited from, the primary circuit, and having one or more windings in series between the primary circuit and the regulated circuit, all suitably adapted and arranged for the control of the voltage, or of the phase angle, or of both, of the regulated circuit.

Voltage spread: The difference between the maximum and minimum voltages.

Voltage stability: The ability of a power system to maintain steady voltages at all buses in the system after being subjected to a disturbance from a given initial operational condition. It can be either fast (short term, with voltage collapse in the order of fractions of a few seconds), or slow (long term, with voltage collapse in minutes or hours).

Voltage stability problems: It is manifested by low system voltage profiles, heavy reactive line flows, inadequate reactive support, and heavy-loaded power systems.

Voltage transformation: It is done by substation power transformers by raising or lowering the voltage.

Voltage transformer: The transformer that is connected across the points at which the voltage is to be measured.

Voltage transformer burdens: The voltage transformer burdens are normally expressed as volt-amperes at a designated power factor. It may be a W, X, M, Y, or Z where W is 12.5 VA at 0.10 power factor, X is 25 VA at 0.70 power factor, M is 35 VA at 0.20 power factor, Y is 75 VA at 0.85 power factor, and Z is 200 VA at 0.85 power factor. The complete expression for a current transformer accuracy classification might be 0.3 at B-0.1, B-0.2, and B-0.5 while the potential transformer might be 0.3 at W, X, M, and Y.

Voltage transformer ratio: Also called "VT ratio." It is the ratio of primary to secondary voltage. For a voltage transformer rated 480:120, the ratio is 4:1 and for a voltage transformer rated 7200:120, it is 60:1.

Voltage, base: A reference value that is a common denominator to the nominal voltage ratings of transmission and distribution lines, transmission and distribution equipment, and utilization equipment.

Voltage, maximum: The greatest 5-min average or mean voltage.

Voltage, minimum: The least 5-min average or mean voltage.

Voltage, nominal: A nominal value assigned to a circuit or system of a given voltage class for the purpose of convenient designation.

Voltage, rated: The voltage at which operating and performance characteristics of equipment are referred.

Voltage, service: Voltage measured at the terminals of the service entrance equipment.

Voltage, utilization: Voltage measured at the terminals of the machine or device.

VRR: Abbreviation for voltage-regulating relay.

Waveform distortion: A steady-state deviation from an ideal sine wave of power frequency principally characterized by the special content of the deviation.

Weatherability: The ability to operate in all weather conditions. For example, transformers are rated as indoor or outdoor, depending on their construction (including hardware).

Withstand voltage: The BIL that can be repeatedly applied to an equipment without any flashover, disruptive charge, puncture, or other electrical failure, under specified test conditions.

XLPE: Abbreviation for cross-linked polyethylene (cable insulation).

REFERENCES

1. IEEE Committee Report, Proposed definitions of terms for reporting and analyzing outages of electrical transmission and distribution facilities and interruptions, *IEEE Trans. Power Appar. Syst.* PAS-87 (5) 1318–1323 (1968).
2. IEEE Committee Report, Guidelines for use in developing a specific underground distribution system design standard, *IEEE Trans. Power Appar. Syst.* PAS-97 (3) 810–827 (1978).
3. *IEEE Standard Definitions in Power Operations Terminology.* IEEE Standard 346-1973, Nov. 2, 1973.
4. *Proposed Standard Definitions of General Electrical and Electronics Terms.* IEEE Standard 270, 1966.
5. Pender, H., and Del Mar, W. A., *Electrical Engineers' Handbook—Electrical Power*, 4th ed. Wiley, New York, 1962.
6. *National Electrical Safety Code*, 1977 ed., ANSI C2, IEEE, New York, November, 1977.
7. Fink, D. G., and Carroll, J. M. (eds.), *Standard Handbook for Electrical Engineers*, 10th ed., McGraw-Hill, New York, 1969.
8. *IEEE Standard Dictionary of Electrical and Electronics Terms.* IEEE, New York, 1972.

Index

Page numbers followed by f, t and n denotes figures, tables and notes, respectively.

A

ACAR (aluminum conductor alloy reinforced) bundled conductors
    capacitive reactance of, 612–617
Acceleration factors, application of, 482, 482f
AC relay connection, 383
AC solution, 513
ACSR (aluminum conductor steel reinforced) bundled conductors
    capacitive reactance of, 609–611
Active power flow, 513
Admittance, 500t
    diagram, 410
    matrix, 335, 336, 358, 483
    in megohms, 28
Admittance (mho) relay, 429–431, 430f, 431f
Air capacitances, 189, 191
Air-cored-type reactors, 181
Alpha diagram, 411
Aluminum cable
    characteristics of, 542–545
        expanded, 546
Aluminum conductors, 58. *See also* ACAR; ACSR
    characteristics of, 538–541
American National Standards Institute (ANSI) standards, 169
American Wire Gauge (AWG) standard, 52
Amplitude
    comparator, 409, 410f, 413–414, 414f, 416t, 417t
    modulation, 459
Analysis equations, 249, 250
Angle impedance relay, 428, 433
Apparatus insulation, 159
Apparent inrush current, 454
Arcing ground, 346
Arcing horns, 186
Arcing ring, 186
Arc resistance, 430, 430f
Arc suppression coil, 349
Arc voltage, 163
Area power interchange control, 484–488, 485f
Armature, 386
Asymmetrical fault currents, 220
Asymmetrical π and T networks, 117–118, 117f, 118f
Asymmetrical T network, 118f
Atabekov's fault diagram, 323
Automatic adjustment features, 482
Automatic circuit recloser, three-phase, 376f
Automatic tap-changing feature, 483
Autotransformers, 41–43, 41f
    advantages, 43
    disadvantages, 43
Auxiliary capacitor, 403
Auxiliary relay, 377

B

Backup protection, 382–385, 383, 385, 425, 432f, 459
Balanced-beam-type relay, 388, 389f
Balanced faults, 164, 164f, 166f
Balanced three-phase faults
    at full load, 175–181, 175f
    at no load, 164–168, 164f, 165f, 166f
Balanced transmission line, complex power in, 13–16
Balance point of relay, 420
Bare conductor, 16
Base current, 19
Beta diagram, 411
Biddle Megger ground resistance tester, 208
Bimetallic strip, 388
Blinders, 433
Block, 390
    diagram, 412f
Blocking, 378, 411
Brown and Sharpe Wire Gauge (B&S), 52
Buchholz relay, 454. *See also* Relay(s)
Bulk power substation, 375f
Bundled conductors, 147–151, 147f
Burden, 377, 397
Buried conductors, 207
Bus
    admittance matrix, 473, 486, 487
    capacitive generator, 475
    capacitive load, 475
    classification, 474t
    current, 496
    defined, 13, 15
    differential protection of, 455f
    faults, 432
    generator, 474
    impedance matrix, 488
    inductive generator, 475
    inductive load, 475
    power, 473
    voltage-controlled, 498, 506, 510
    voltages, 473
Bushing-type current transformer, 397

C

Cable(s), 52
    aluminum, 542–546
    critical length of, 134
    345-kV pipe-type, 136t–137t
    paper-insulated, 563–569
    self-supporting rubber-insulated, neoprene-jacketed aerial, 590–593
    single-conductor concentric-strand paper-insulated, 570–573

705

single-conductor oil-filled (hollow-core) paper-insulated, 574–577
single-conductor solid paper-insulated, 584–589
three-conductor belted paper-insulated, 578–581
three-conductor shielded paper-insulated, 582–583
Cable system, 138f
Cable transmission
extra-high-voltage (EHV) underground, 134–142, 135f, 136t–137t, 138f, 141f, 142f
Callaway nuclear power plant switchyard, 374f
Capacitance, 16, 151, 189, 191, 193, 271. *See also* Transmission lines, steady-state performance of
Capacitance-bushing voltage divider, 403
Capacitive generator bus, 475
Capacitive load bus, 475
Capacitive reactance, 66, 134, 151. *See* Transmission lines, steady-state performance of
of ACAR bundled conductors, 612–617
of ACSR bundled conductors, 609–611
single-phase overhead lines, 61–64, 61f, 63f
three-phase overhead lines, 64–65, 64f
Capacitor voltage transformers (CVT), 402, 402f, 403f
Cap-and-pin insulator unit, 189, 190f
Carrier current phase comparison, 461f
Cascading, 377
Characteristic angle, 378
Characteristic quantity, 378
Characteristics of relay in steady state, 378
Charging current per phase, 65
Circuit breaker (CB), 169, 169n, 170t, 171t, 172, 376, 382
Circuit constants, 113t–114t
determination of, 111–112, 113t–114t, 115t
parameters measurement by test, 112, 115–116
of transformer, 116–117, 116f, 117f
Clapper relays, 387
Coefficient of grounding, defined, 347
Coil reactance, 181
Common winding, 41
Comparators
amplitude, 409, 410f, 413–414, 414f, 416t, 417t
general equation of, 412, 412f
phase, 414–418, 416t, 417t
phase and amplitude, 409, 410f
Complete transposition, 357
Complex planes, 410–412, 411f
Complex power in balanced transmission line, 13–16
Composite sequence current filters, 395f
Computer applications in protective relaying
computer relaying, 462–464, 463f
in relay settings and coordination, 462
Computer relaying, 462–464, 463f, 464
Conductance, 16, 68, 84, 107, 530t
Conductor
bundled, 147–151, 147f
buried, 207
resistance, 58
size, 51–58, 52f, 53f, 54f, 55f, 56f, 57t, 226
Constant impedance representation of loads, 38–39
Contaminants, types of, 195
Contamination
flashover, 194
on insulator, 192
insulator flashover for, 196
Continuous counterpoise, 233, 234f

Coordination delay time (CDT), 446, 447
Coordination time delay, 385
Copper conductors
characteristics of, 534–537
Copperweld–copper conductors
characteristics of, 547–551
Co-type curve shapes, 440, 441, 441f
Counterpoise, 16, 207, 233
Coupling-capacitor voltage divider, 403
Critical length of cable, 134
Crowfoot type counterpoise, 233
Current base determination, 20
Current differential relaying, 450
Current grading method, 444
Current-limiting reactors, 181–185, 183f, 184f
Current-tap settings (CTS), 441
Current transformers (CT), 397–400, 453
delta connections of, 397f
equivalent circuit of, 399f
error, 399
formula method, 400
saturation curve method, 401–402, 401f
Cyclic transposition, 357
Cylinder-type magnetic induction relay, 387
Cylinder units, 387, 388f
Cylindrical-rotor machine, 167
Cylindrical-rotor synchronous machine, 275

D

D'Arsonval unit, 388, 388f
DC power flow method, 513–525
DC structures, 56f
Dead-end insulator, 186
Decoupled Newton–Raphson method. *See* Decoupled power flow method
Decoupled power flow method, 510–511
Decrement factor, 218
Delta-connected load, 280f, 281
Delta connection, 425
Delta currents, 425, 427
Delta–delta bank, 281
Delta–delta connection, 29f
Delta voltages, 425
Delta–wye connection, 404f
Delta–wye transformation, 43, 44f, 85
Dependent time-delay relay, 379
Desensitization, 454
Design criteria for protective system
economics, 382
reliability, 381
selectivity, 380
simplicity, 382
speed, 380
Design tests, 187
Device numbers, 621–622
Diakoptics, 472
Differential protection, 450–459, 450f, 452f, 455f, 456f, 457f
Differential relays, 456f, 457f, 459. *See also* Relay(s)
use of, 453
Digital computers, 462
Direct axis, 167
reactances, 167
synchronous reactance, 275

Index

Direct current (dc) resistance, 58
Directional time-overcurrent relay, 439f
Discrimination, 380
Distance relays. *See also under* Relays; System protection
    application of, 434f
    connections, 426f
Distributed constant circuit, 68, 68f
Distribution substation, 375f
Distribution system
    loads, 9
    planning, 9
        block diagram, 10f
Double-circuit transmission lines, 358–361
Double contingency, 471
Double line-to-ground (DLG) faults, 307–312, 307f, 311f
Dropout, 378
Dry bands, 195
Dry-process porcelain, 194
Dry-type reactors, 181
Duplex conductors, 147

E

Earth resistivity, 258
Economics, 153, 380, 382, 472
Electrical characteristics of overhead ground wires, 552–553
Electrical load, 2
Electrical power system, structure of, 2f
Electric current (mA) on men/women, 200t
Electric shock
    and effects on humans, 197–204, 198t, 199f, 200t, 201f, 202f
    hazard situations, 201f
Electric utility plants
    growth in, 4, 4f
    investment growth in, 5f
Electromagnetic unbalance factors, 355–357
Electromechanical relays, 386
Electrostatic induction coefficients, 269
Element, defined, 378
Energizing quantity, 378
Energy
    conversion, 1, 2, 3
    defined, 1
    sink, 2
    source, 1, 2
    utilization, 2
Environmental planning, 9
Equipment grounding. *See* Grounding
Equivalent circuit, 68, 175f, 192f, 272f, 277f
    of long transmission lines, 100–103, 100f, 102f, 103f
    of short transmission line, 68f
Equivalent impedance, 16, 17f
    short circuit MVA and. *See* Short-circuit MVA
Equivalent positive-/negative-sequence network, 334–339, 335f, 337f, 339f
Equivalent spacing, 60
Equivalent transfer matrix, 117
Equivalent zero-sequence networks, 333–334
Error margin, 446
Examples, 15–16, 30–37, 35f, 37f, 47–49, 262–267, 263f
    disturbance in normal operation, 172–174, 173f, 178–181, 182–185, 183f, 184f
    power in symmetrical components, 254–255

steady-state performance of transmission lines, 66–67, 74–78, 79–80, 86–89, 95–100, 105–107, 109–110, 120, 122, 125–127, 130–134, 130f, 134–142, 143–147, 149
    on system protection, 400, 401, 407, 417–418, 433–439, 447–449, 456–459
    three-phase unbalanced system, 251–252
    on unbalanced faults, analysis of, 297–302, 305–306, 310–312, 315–317, 335–339, 339–343, 349–352
Excitation current, 399
External faults, 451, 459, 461
Extra-high-voltage (EHV) transmission systems, 383
Extra-high-voltage (EHV) underground cable transmission, 134–142, 135f, 136t–137t, 138f, 141f, 142f

F

Fast decoupled power flow method, 511–513
Fault. *See also* Unbalanced faults
    analysis/types, 161–164, 163f
    arc resistance, 162, 163
    classification of, 160
    clearance, 435f
    defined, 159
    impedance, 163, 293, 293n, 307, 314
    interruption, 168–174, 170t, 171t, 173f
    megavolt-ampere, 169, 178
Fault clearance time, function of, 382f
Fault current, 168, 177
    determination, 323
    division factor, 216
Fault currents/voltages, 294t–295t
Fault diagram, generalized
    for series fault, 339–343, 340f, 340t, 341f, 342f
    for shunt faults, 323–328, 323f, 324f, 324t, 326f, 328f
Fault impedance, 293n
Financial planning, 9
Fine-to-line voltages, 298
Five-bus system, 485f, 499f, 519
Flashover characteristics of suspension insulator, 188t
Flashovers, contaminated, 195t, 196f
Flashover voltage, 187
Flat line, 105
Floating bus, 474n
Foot-to-foot currents, 203
Four-conductor bundle, 147f
Four-terminal transmission network, 119, 121
Fuel supply system, 3f

G

Gas-insulated switchgear (GIS), 143
Gas-insulated transmission line (GIL), 142–147
Gauss iterative method, 476–477
Gauss-Seidel iterative method, 477–478
    application of, 478–481, 481f, 488–489
General circuit constants. *See under* Transmission lines, steady-state performance of
General circuit parameter matrix, 83
Generalized fault diagram
    for series fault, 339–343, 340f, 340t, 341f, 342f
    for shunt faults, 323–328, 323f, 324f, 324t, 326f, 328f
General network, 113t
Generator bus, 474

Geometric mean distance (GMD), 59, 270
Geometric mean radius (GMR), 59, 65, 147
Glass pin insulator, 185
Grading shield, 186, 193
Greek alphabet symbols, 629
Grid current, 228
    maximum, 217
    symmetrical, 218
Grid resistance, 215, 228
Ground
    defined, 207
    on line capacitance, 151, 152f
    resistance, 207–208, 208t, 209f, 210t–211t, 212f
    resistor, 17
Ground conductor sizing factors, 218–220, 219t
Ground distance relays, 391
Grounded system, 343, 344f
    effectively, defined, 347n
    system characteristics, 348t
Ground fault neutralizer, 349
Ground faults, types of
    line-to-line-to-ground fault, 223–224
    single-line-to-ground fault, 224
Grounding
    electric shock, 197–204, 198t, 199f, 200t, 201f, 202f
    electric shock on humans, 197–204, 198t, 199f, 200t, 201f, 202f
    ground potential rise (GPR), 206–207, 207t
    ground resistance, 207–208, 208t, 209f, 210t–211t, 212f
    reduction of factor C_s, 204–206, 205f, 206t
    soil resistivity measurements, 209, 213–214
    substation, 214–218, 217f
    transformers, 235f
Grounding material, material constants of, 219t
Grounding (grid) system design, 236f, 237f
Ground potential rise (GPR), 206–207, 207t, 224–233, 226f, 227f, 228t, 229, 229t
Ground relay, 427. *See also* Relay(s)
    connections, 426f, 445f
Ground resistivity distribution (in US), 212f
Guard frequency, 459

H

Hand-to-foot currents, 203
Harmonic restrained-percentage differential relay, 454
High-speed reclosing, 160
High-voltage dc (HVDC) lines, 196
High-voltage transmission lines, 233
Hinged armature type construction, 387

I

Impedance, 16, 38, 38f, 40, 44, 45, 162, 163, 177
    algorithm, 464
    base determination, 20
    diagram, 403, 408f, 410
    matrix, 357, 358, 483
Impedance relay, 422–427, 423f, 424f, 426f. *See also* Relay(s)
    disadvantages of, 425
Impulse flashover voltage, 188
Impulse ratio, 188
Impulse voltages, 402
Incident voltage, 103–107

Incremental charging current, 90
Independent time-delay relay, 379
Inductance
    average, 148
    single-phase overhead line, 59–60, 59f
    three-phase overhead lines, 60–61, 61f
Induction cup-type unit, 387
Induction disk, 387, 388f
Induction-type overcurrent relays, 445f
Inductive generator bus, 475
Inductive load bus, 475
Inductive reactance, 59–61, 59f, 61f, 65, 182
    of ACAR bundled conductors, 600–608
    of ACSR bundled conductors, 594–599
Inductive reactance spacing factor, 553–556
Infinite line, 105
Initial reactance, 167
Initial symmetrical rms current, 168
Input impedance, 124, 125
Installed recloser, three-phase, 377f
Instantaneous overcurrent relay, 440
Instantaneous relay, 379
Instrument transformers, 396–397
    current, 397–402, 397f, 399f, 401f
    voltage, 402–403, 402f, 403f
Insulated conductors, 52
Insulating materials, 194
Insulation (line/apparatus), 159
Insulators
    cleaning, cost for, 196
    contamination, 159
    testing of, 187–189, 188t
    types of, 185–187, 186f, 187f
    voltage distribution, 189–193, 190f, 191f, 192f, 193f
Insulator flashover
    for contamination, 194–196, 195t, 196f
    on overhead high-voltage DC lines, 196–197
Interconnection, 9, 235
Inverse time-delay relay, 379
    with definite minimum, 379
Inverse-time overcurrent relays, 446
Iron-cored reactors, 182
Isolated system, 344f, 345

J

Jacobian matrix, 491, 492
    formulation of, 505–510

K

Kelvin's law, 52, 58
Kirchhoff's current law, 200
Kirchhoff's voltage law, 277
Knee-point electromotive force, 378
Kron, G., 472
Kron reduction, 267, 272, 358
Kron's method, 323

L

Leakage
    current, 194
    impedance measurement, 40

Index

reactance, 167, 350
resistance, 68
resistance current, 193
Let-go currents, 198, 199f
Lightning generator, 188
Line
  admittances, 500t
  impedance, 27
  insulation, 159
  resistance, 27
Linear couplers, 455
Line constants, tables of, 65–67
Line impedances, 330, 421, 500t
Line post insulators, 185
Line protection by pilot relaying, 460f
Line shunt capacitive reactance, 27
Line-to-line (L-L) capacitance, 63f
Line-to-line (L-L) fault, 302–306, 303f, 305f
Line-to-line (L-L)-ground fault, 223–224
Line-to-line (L-L) voltages, 100, 304, 306, 312, 315, 317
Line-to-neutral capacitance, 16, 63, 63f, 64, 151
Line-to-neutral susceptance, 63, 64
Line-to-neutral voltages, 71
Load, 2, 378
  bus, 474
  currents, 177, 394, 439
  flow. *See* Power flow analysis
  forecasting, 7
  impedance, 125, 437
  representations, 38–39, 38f
Load flow equations
  in polar coordinates, 504–505
    Jacobian matrix, formulation of, 505–510
  in rectangular coordinates, 493–504, 499f, 500t
Load-tap-changing (LTC) transformers, 483, 483f
Logic circuits, 391
Longitudinal faults. *See* Series faults
Long-range planning, 5
Long transmission lines. *See under* Transmission lines, steady-state performance of
Lossless line, 107, 108

M

Magnetic attraction relays, 386, 386f
Magnetic field
  effects, 153
  of single-phase line, 59f
Magnetic induction relays, 387, 388f
Magnetizing inrush current, 454
Magnitude comparators, 391
Mat, 215
MATLAB, 655–682
Matrix of potential coefficients, 269
Maxwell's coefficients, 269, 272
Medium-length transmission lines, 80–89, 80f, 81f
Merz–Price protection, 451
Mesh coefficient determination, 221
Mesh voltage, 202, 215n, 216
  calculations, 230
  design calculations, 221–223
Metal-to-metal touch voltage, 202
Mho relay, 430f, 431f. *See also* Relay(s)

Microwave channel, 459
Minicomputers, 464
Modular transmission line protection, 392f, 393f
Multiwound transformers, 493n
Mutual admittances, 500t
Mutual impedance, 256
  of short lines, 79–80, 79f, 80f

N

Natural loading. *See* Surge impedance loading (SIL)
Negative-sequence currents, 308, 394, 395f
Negative-sequence network, 339f
Negative-sequence system, 245, 246f
Negative-sequence unbalance, 273
Neglecting conductance, 84
Net-circulating unbalances, 359, 360
Net-through unbalances, 359, 360
Network conversion formulas, 115t
Networks connected in parallel, 121–123, 121f, 122f
Networks connected in series, 119–120, 119f, 120f
Neutral grounding, advantages of, 346–347
Newton-Raphson method, 472, 489–493
  application to load flow equations
    in polar coordinates, 504–510
    in rectangular coordinates, 493–504, 499f, 500t
  decoupled, 510–511
Nodal voltages, 473
Noise measurements, 152
Nonsalient-pole machine, 167
Normalization process, 21
N-1 secure, 471

O

Offset inrush current, 454
Offset mho (modified impedance) relay, 431–432, 432f. *See also* Relay(s)
Ohm relay, 433–439, 433f, 434f, 435f, 436f. *See also* Relay(s)
Oil circuit breaker, 377f
Oil-immersed reactors, 181
One-line diagram, 16–19
  of power line carrier, 460f
  power system representations, 17f
  of radial system, 447
  symbols in, 18t
One line open (OLO) faults, 322f, 329f, 330, 340t
Open circuit, measurement of, 112, 115–116
Open-circuit impedance parameter determination, 332
Open-circuit test, 112
Open-circuit voltage, 395
Open–delta connection, 404
Operator a, 247–248, 247f, 248t
  powers and functions of, 248t
Oscillations, 109n
Outdoor circuit breaker ratings, 171
Overcurrent relays, 391f, 439–449, 439f, 440f, 441f, 442f, 443f, 445f, 446f, 447f
Overhead ground wires, 358
  electrical characteristics of, 552–553
  three-phase transmission line and, 268–275, 272f
Overhead high-voltage DC lines, insulator flashover on, 196–197

Overhead transmission lines, 68, 152
   conductors in, 58
   environmental effects of, 152–153
   single-phase, 59–60, 59f, 61–64, 61f, 63f
   three-phase, 60–61, 61f, 64–65, 64f
Overinsulation, 197
Overlapping protection, 383, 384f
Overload, 160
Overreaching zone, 425
Overshoot time, 378

P

Paper-insulated cable
   characteristics of three-conductor belted, 563–569
Parallel type counterpoise, 233, 234f
Π circuit, 81f, 83, 85
Peak momentary current, 169, 172
Percent voltage regulation, 73, 86
Performance tests, 187
Permanent faults, 160
Per-unit admittance, 28
Per-unit system
   advantages, 19–20
   change of base, 24–25
   defined, 19
   physical values, converting to, 24
   single-phase system, 20–24
   three-phase systems, 25–37, 29f, 37t
Peterson coil (PC), 349–352, 349f, 350f, 351, 352
Phase and amplitude comparators, 409, 410f
Phase arrangements, 358
Phase comparator, 414–418, 416t, 417t
Phase currents, 298, 301
Phase impedance matrix, 355
Phase relay(s), 425
   connections, 426f
Phase selectors, 391, 393
Phase shift, 104, 302, 353
Phase shifters, 323, 324t, 340t
Phase-shifting transformers, 483
Phase voltages, 249, 309, 312
Phasor(s), 245, 246f
   three-phase unbalanced system of, 248–252
Phasor diagram, 348f, 412
   of powers and functions (operator **a**), 247f
   of short transmission line, 69f
Physical quantity, 19
Pickup, 378
Pilot channel, 378
Pilot relaying, 459–462, 460f, 461f
Pilot wires, 459
Pin-type insulators, 185, 186f
Pipe-type cable, characteristics of, 136t–137t
Plunger relays, 386, 387
Polar-type relays, 387
Porcelain insulator disk, 187f
Porcelain pin insulator, 185
Porosity test, 189
Positive-sequence currents, 308
Positive-sequence network, 339f
Positive-sequence system, 245, 246f
Positive-sequence voltage, 394, 395f

Post insulator, 185, 186f
Power, 1
   devices, 2
   in symmetrical components, 252–255
Power flow analysis
   acceleration factors, application of, 482, 482f
   area power interchange control, 484–488, 485f
   automatic adjustment features, 482
   DC power flow method, 513–525
   decoupled power flow method, 510–511
   fast decoupled power flow method, 511–513
   Gauss iterative method, 476–477
   Gauss-Seidel iterative method, 477–478
      application of, 478–481, 481f, 488–489
   load-tap-changing (LTC) transformers, 483, 483f
   Newton-Raphson method, 489–493
      application of, 493–510, 499f, 500t
   overview of, 471–472
   phase-shifting transformers, 483
   power flow problem, defined, 473–475, 474t
   sign of power (real/reactive), 475
Power line carrier, 459
   one-line diagram, 460f
Power relations (using A, B, C, D line constants), 127–134, 130f
Power system arrangement, 420f
Power system planning, 5–10
   activities, 8f
   factors, 6f
   load forecasting, 7
   long-range, 5
   objective of, 6
   organizational chart, 7f
   short-range, 5
   transmission planning, 8–9
Power transformers, 452, 456f, 457f
   standard impedances for, 560–562
Power transmission limits, 135f
PQ bus (load bus), 474
Prefault bus voltages, 162
Prefault load current, 176
Prefault voltage, 178
Prefixes, 627
Primary circuit, 28
Primary distribution system loads, 9
Primary protection, 382–385, 383, 383f, 384f
Primary shock currents, 197, 198
Primary winding, 41
Problems
   disturbance of normal operating conditions, 240–243, 240f, 241f, 242f, 243f
   on power flow analysis, 528–532
   on steady-state performance of transmission lines, 154–157, 155f, 156f
   symmetrical components/sequence impedances, 289–292, 292f
   on system protection, 465–470
   on unbalanced faults, 362–371
Protected zone, defined, 383
Protection, 382
   zones in power system, 384f
Protective device coordination program, 462, 463f
Protective gear, 378

Index

Protective relay, 377
Protective relaying, computer applications in
    computer relaying, 462–464, 463f
    in relay settings and coordination, 462
Protective scheme, 378
Protective system, 378
    selection, factors affecting, 380, 381f
Puncture, 189

Q

Quadrature axis, 167

R

Radial system, one-line diagram of, 447f
Radial-type counterpoise, 233, 234f
Reach of relay, 378
Reactance, 15, 30, 162
    grounded system, 347
    relay, 427–429, 429f
    of three-phase synchronous machines, 276t
Reactive power, 71, 129
    absorbed, 86
Reactors, current-limiting, 181–185, 183f, 184f
Real/reactive power, sign of, 475
Receiving-end current limits, 135f
Receiving-end impedance, 407
Receiving-end power, 99
Receiving-end voltage, 100
Reciprocal transposition, 357
Reference bus, 474n
Reference rods, 213
Reflected voltages, 103–107
Regulating generator, 484
Relay(s), 385–394, 386f, 388f, 389f, 390f, 391f, 392f, 393f
    admittance (mho), 429–431, 430f, 431f
    burden, 377
    characteristics of, in steady state, 378
    clapper, 387
    as comparators, 409
    coordination, 444
    differential, 453, 456f, 457f
    distance. *See under* System protection
    general equation of, 418–419, 419f
    impedance, 422–427, 423f, 424f, 426f
    offset mho (modified impedance), 431–432, 432f
    ohm, 433–439, 433f, 434f, 435f, 436f
    overcurrent, 439–449, 439f, 440f, 441f, 442f, 443f, 445f, 446f, 447f
    phase, 425, 426f
    protective, 377
    reach of, 378
    reactance, 427–429, 429f
Reliability, 381, 513
Remote backup relaying, 385
Resetting value, 378
Resistance, 58–59, 109n
    of earth, 209f
    to ground, 210t–211t
Resistance-grounded system, 347
Resistance of ACAR conductors, 619–620
Resistance of ACSR conductors, 618–619

Resistivity of soils, 206t, 258t
Resistor–capacitor (RC) delay circuit, 391
Resonant grounding, 349
Root mean square (rms), 168
R–X diagram, 403–409, 404f, 405f, 406f, 406t, 408f

S

Salient-pole generators, 276t
Salient-pole machine, 167
Secondary circuit, 28
Secondary shock currents, 197, 198
Selectivity, 380, 444
Self-impedances, 79, 256, 257
Self-supporting rubber-insulated, neoprene-jacketed aerial cable, 590–593
Sending-end charging current, 99
Sending-end current, 135f
Sending-end impedance, 407
Sending-end power, 98, 99
Sending-end voltage, 98, 104
Sequence admittance matrix, 268, 359
Sequence capacitances, 273
    three-phase transmission line
        with overhead ground wire, 271–275, 272f
        without overhead ground wire, 268–271
Sequence currents, 298, 303, 310
Sequence filters, 394–396, 395f
Sequence impedances
    matrix, 268, 355
    of synchronous machines, 275, 276t, 277–279, 277f, 279f
    of transformers, 281–288, 282f, 283f, 284f, 284t, 285f, 286f, 287f
Sequence impedances of transmission lines
    electromagnetic unbalances, 260–267, 263f
    of transposed lines, 257–260, 258t, 259f
    of untransposed lines, 255–257, 255f
        with overhead ground wire, 267–268
Sequence network equivalents, determination of. *See under* Unbalanced faults
Sequence networks. *See also* Sequence filters
    of synchronous machine, 279f
    of transmission line, 259f
Sequence voltages, 249, 309, 312
Series
    capacitance, 189
    impedance, 90, 95, 113t
    reactance, 16
    winding, 41
Series-common winding, 41
Series faults, 164, 329–331, 329f, 331f
    generalized fault diagram for, 339–343, 340f, 340t, 341f, 342f
    one line open, 330
    sequence network equivalents for, 332–339, 332f, 335f, 337f, 339f
    two lines open, 330–331, 331f
Series faults, determination of sequence network equivalents for
    equivalent positive- and negative-sequence networks, 334–339
    equivalent zero-sequence networks, 333–334
    two-port network, 332–333

Setting, defined, 378
Shock currents, 197
Short circuit, 159
  measurement of, 112, 115–116
  test, 112
Short-circuit admittance parameters, 333
Short-circuit MVA, 44
  single-phase-to-ground, 46–49, 49f, 49t
  three-phase, 45–46
Short lines, mutual impedance of, 79–80, 79f, 80f
Short-range planning, 5
Short transmission line, 68–71, 68f, 69f
  per-phase representation of, 14f
Shunt
  admittance, 95
    matrix, 271
  capacitance, 68
  capacitive compensation apparatus, 475
  capacitive reactance, 16
  capacitive reactance spacing factor, 557–559
  impedance, 113t
  susceptance, 27
Shunt faults, 164
  double line-to-ground (DLG) faults, 307–312, 307f, 311f
  generalized fault diagrams for, 323–328, 323f, 324f, 324t, 326f, 328f
  line-to-line (L-L) fault, 302–306, 303f, 305
  single line-to-ground (SLG) faults, 293–302, 294t–295t, 295f, 297f, 300f
  solved examples, 631–653
  symmetrical three-phase faults, 312–317, 313f, 316f
  unsymmetrical three-phase fault, 317, 318f, 319f, 320f, 321f, 322f
Shunt faults using MATLAB
  solved examples of, 655–682
Sign of power (real/reactive), 475
Simplicity, 382
Simultaneous faults, 164
Single-circuit transmission line, 376f
Single-conductor concentric-strand paper-insulated cables, 570–573
Single-conductor oil-filled (hollow-core) paper-insulated cables, 574–577
Single-conductor solid paper-insulated cables, 584–589
Single line-to-ground (SLG) fault, 224, 293–302, 294t–295t, 295f, 297f, 300f
  current elimination, 349–352, 349f, 350f
Single-phase overhead line, 59–60, 59f, 61–64, 61f, 63f
Single-phase system, 20–24
Single-phase-to-ground short-circuit MVA, 46–49, 49f, 49t
Six-phase systems. *See under* Unbalanced faults
Skin effect, 59
Slack bus, 474, 488
Soil resistivity, 206t, 215, 258t
  measurements
    driven-ground rod method, 213–214, 214f
    three-pin method, 213, 214f
    Wenner four-pin method, 209, 213, 213f
  moisture content on, 207t
  temperature on, 208t
Solar energy, 3
Solidly grounded system, 348f
Solid-state relay, 386, 390, 390f, 417
Speed, 380

Split factor, 216, 217
Stability, 378
Standard device, 377, 379t
Standard impedances for power transformers, 560–562
Star configuration, 43n
Starters, 391
Starting relay, 429
Static relays, 390, 390f
Static wire, 233
Steady-state internal voltage, 176
Steady-state performance. *See* Transmission lines, steady-state performance of
Steady-state power limit, 71–73, 129
Steo coefficient, 223
Step voltage, 202
  calculations, 231–232
  design calculations, 223
Strain insulators, 185, 186f
String efficiency, 192, 193
Substation grounding, 214–218, 217f
  design, 226
  system, 214, 215
Subsynchronous resonance, 109n
Subtransient reactance, 162, 166, 167, 175, 176
Subtransmission planning, 9
Subtransmission system, 349, 349f
Subtransmission voltage, 40
Sudden pressure relay, 454
Surface layer derating factor (C_s), 205f
  reduction of, 204–206, 205f, 206t
Surge impedance loading (SIL), 107–110, 108f, 109f
Susceptance, 63, 64
Suspension insulator, 160f, 185
  flashover characteristics, 188t
Swing bus, 474n, 488, 489
Switchgear, 373, 454n
Switching surges, 159
Symmetrical components, 353, 354f
  negative-sequence system, 245, 246f
  positive-sequence system, 245, 246f
  power in, 252–255
  zero-sequence system, 245, 246f
Symmetrical conductor current, 218
Symmetrical current-interrupting capability, 169, 170
Symmetrical grid current, 218
Symmetrical three-phase faults, 312–317, 313f, 316f
Synchronous condensers, 276t
Synchronous generator, 164, 164f
Synchronous machines, sequence impedances of, 275, 276t, 277–279, 277f, 279f
Synchronous motor, 176
Synchronous reactance, 167
Synthesis equations, 249
System grounding, 343–348, 344f, 345f, 346f, 348f, 348t, 421
System neutral ground, 343
System protection
  amplitude comparator, 413–414, 414f
  backup, 382–385
  comparators, general equation of, 412, 412f
  complex planes, 410–412, 411f
  definitions, 377–379
  design criteria
    economics, 382
    reliability, 381

Index

selectivity, 380
simplicity, 382
speed, 380
differential protection, 450–459, 450f, 452f, 455f, 456f, 457f
distance relays, 419–422, 420f, 422f
    admittance (MHO) relay, 429–431, 430f, 431f
    impedance relay, 422–427, 423f, 424f, 426f
    offset mho (modified impedance) relay, 431–432, 432f
    ohm relay, 433–439, 433f, 434f, 435f, 436f
    reactance relay, 427–429, 429f
factors affecting system design, 380, 381f
instrument transformers, 396–397
    current, 397–402, 397f, 399f, 401f
    voltage, 402–403, 402f, 403f
overcurrent relays, 439–449, 439f, 440f, 441f, 442f, 443f, 445f, 446f, 447f
overview of, 373–377, 374f, 375f, 376f, 377f
phase and amplitude comparators, 409, 410f
phase comparator, 414–418, 416t, 417t
pilot relaying, 459–462, 460f, 461f
primary protection, 382–385, 383f, 384f
protective relaying, computer applications in
    computer relaying, 462–464, 463f, 464
    in relay settings and coordination, 462
relays, 385–394, 386f, 388f, 389f, 390f, 391f, 392f, 393f
    as comparators, 409
    general equation of, 418–419, 419f
R–X diagram, 403–409, 404f, 405f, 406f, 406t, 408f
sequence filters, 394–396, 395f
standard device, 377, 379t

T

Taylor's series, 491
TCAP, 218, 219, 219t
    defined, 220
T circuit, 80, 80f, 83, 101, 103f
Temporary faults, 160
Terminal voltage, 175
Terminated transmission line, 123–127, 124f, 125f
Thermal unit, 388
Thévenin equivalent impedance, 224
Thévenin equivalent system reactance, 178
Thevenin impedance, 632, 634, 636
Thevenin's theorem, 332, 334
Thévenin voltage and impedance, 177
Three-conductor belted paper-insulated cables, 578–581
Three-conductor bundle, 147f
Three-conductor shielded paper-insulated cables, 582–583
Three-phase automatic circuit recloser, 376f
Three-phase faults
    balanced. *See* Balanced three-phase faults
    symmetrical, 312–317, 313f, 316f
    unsymmetrical, 317, 318f, 319f, 320f, 321f, 322f
Three-phase installed recloser, 377f
Three-phase line conductors, 152f
    capacitance of, 151, 152f
Three-phase overhead lines, 60–61, 61f, 64–65, 64f
Three-phase short-circuit MVA, 45–46
Three-phase systems, 25–37, 29f, 37t

Three-phase transmission line
    with overhead ground wire, 271–275, 272f
    without overhead ground wire, 268–271
Three-phase unbalanced system of phasors, 248–252
Three-winding transformers, 40–41, 40f
Three-zone mho relay, 431f
Tie lines, 8
Time–current curves, 442f, 443f
Time–current grading method, 444
Time-delay relay, 379. *See also* Relay(s)
Time-delay-type overcurrent relays, 440, 443
Time dial setting (TDS), 441
Time grading method, 444
Time-overcurrent relays, 439f, 440f
Torque, 387, 389, 418
Touch and step voltage criteria, 226
Touch voltage, 202, 203
Transfer matrix, 83, 116, 118
Transferred voltage, 202
Transformation(s), 353, 354–355
Transformation matrix, 359
Transformer(s), 113t, 116, 116f, 117f
    current, 397–402, 397f, 399f, 401f
    grounding, 235f
    instrument. *See under* System protection
    load-tap-changing (LTC), 483, 483f
    phase-shifting, 483
    sequence impedances of, 281–288, 282f, 283f, 284f, 284t, 285f, 286f, 287f
    step-up, 454
    three-winding, 40–41, 40f
    two-winding, 41, 41f
    voltage, 402–403, 402f, 403f
Transformer bank, 281
Transient currents, 197
Transient internal voltage, 176
Transient reactance, 167
Transmission
    bulk-power station, 374f
    efficiency, 71
    planning, 8
    system, 2, 436f
Transmission line
    circuit diagrams, 255f
    constants, 58
    double-circuit, 357–359
    efficiency, 99
    grounds, 233–237, 234f, 235f, 236f, 237f
Transmission lines, steady-state performance of
    bundled conductors, 147–151, 147f
    capacitance/capacitive reactance
        single-phase overhead lines, 61–64, 61f, 63f
        three-phase overhead lines, 64–65, 64f
    compact configurations, 52f
    conductor size, 51–58, 52f, 53f, 54f, 55f, 56f, 57t
    environmental effects, 152–153
    equivalent circuits, 68
    extra-high-voltage (EHV) underground cable transmission, 134–142, 135f, 136t–137t, 138f, 141f, 142f
    gas-insulated transmission line (GIL), 142–147
    general circuit constants, 110–111, 110f
        asymmetrical π and T networks, 117–118, 117f, 118f

determination of, 111–112, 113t–114t, 115t
measurement of open/short circuit, 112, 115–116
networks connected in parallel, 121–123, 121f, 122f
networks connected in series, 119–120, 119f, 120f
power relations, 127–134, 130f
terminated transmission line, 123–127, 124f, 125f
of transformer, 116–117, 116f, 117f
ground on line capacitance, 151, 152f
inductance/inductive reactance
single-phase overhead line, 59–60, 59f
three-phase overhead lines, 60–61, 61f
line constants, tables of, 65–67
long transmission lines, 90–100, 90f
equivalent circuit of, 100–103, 100f, 102f, 103f
incident and reflected voltages of, 103–107
surge impedance loading of, 107–110, 108f, 109f
medium-length transmission lines, 80–89, 80f, 81f
mutual impedance of short lines, representation of, 79–80, 79f, 80f
percent voltage regulation, 73–78
resistance, 58–59
short transmission lines, 68–71, 68f, 69f
steady-state power limit, 71–73
transmission line constants, 58
Transmission networks
in parallel, 121f, 122f
in series, 120f
Transmission systems
H-frame structures, 54f
pole/lattice structures of, 53f
Transposed lines
sequence impedances of, 257–260, 258t, 259f
Transposition cycle of three-phase line, 61f
Transposition on six-phase lines, 357–358
Triplex conductors, 147
Tripping, 383, 411, 433, 433f
Turbine generators, 276t
Two-conductor bundle, 147f, 149
Two lines open (TLO) fault, 330–331, 331f, 339, 340t, 341f
Two-port theory, 332–333, 332f
Two-winding transformers, 41, 41f

U

Unbalanced faults
Peterson coil, use of, 349–352, 349f, 350f
sequence network equivalents, determination of
equivalent positive-/negative-sequence network, 334–339, 335f, 337f, 339f
equivalent zero-sequence networks, 333–334
two-port theory, 332–333, 332f
series faults, 329–331, 329f, 331f
generalized fault diagram for, 339–343, 340f, 340t, 341f, 342f
one line open, 330
sequence network equivalents for, 332–339, 332f, 335f, 337f, 339f
two lines open, 330–331, 331f
shunt faults
double line-to-ground (DLG) faults, 307–312, 307f, 311f
generalized fault diagrams for, 323–328, 323f, 324f, 324t, 326f, 328f
line-to-line (L-L) fault, 302–306, 303f, 305f
single line-to-ground (SLG) faults, 293–302, 294t–295t, 295f, 297f, 300f
symmetrical three-phase faults, 312–317, 313f, 316f
unsymmetrical three-phase fault, 317, 318f, 319f, 320f, 321f, 322f
six-phase systems, 352
double-circuit transmission lines, 358–361
electromagnetic unbalance factors, 355–357
overhead ground wires, 358
phase arrangements, 358
symmetrical components, 353, 354f
transformations, 353, 354–355
transposition on, 357–358
system grounding, 343–348, 344f, 345f, 346f, 348f, 348t
Unbalanced voltage phasors, 246f
Uncoupled sequence networks, 339f
Underground cable circuit, 141f
Underreach, 379
Underreaching zone, 425
Undervoltages, 160
Ungrounded system, 344f, 345, 345f
voltage diagrams of, 346f
Unit/element, defined, 378
Unit generator–transformer protection, 454, 455f
Unit phasor, 247
Unsymmetrical three-phase fault, 317, 318f, 319f, 320f, 321f, 322f
Untransposed lines
electromagnetic unbalances, 260–266, 263f
with overhead ground wire, 267–268
sequence impedances, 255–257, 255f
with overhead ground wire, 267–268

V

Variable autotransformer, 41
Ventricular fibrillation, 197, 198
Voltage
distribution, 189, 190f, 191f, 193f
equivalent circuit for, 192f
drop, 59, 90
gradient, 147, 149
regulation, 73–74, 86
Voltage transformers (VT), 402–403, 402f, 403f

W

Wavelength, defined, 104
Westinghouse Electric Corporation, 462, 463f
Wet-process porcelain, 194
Winding, relay-operating, 451
Wire, 52
Wire circuits, 459
Wound-type current transformer, 397
Wye-connected load, 280f, 281
Wye connection, 427
Wye–delta connections, 301
Wye–delta transformations, 43–44, 44f
Wye winding, 281, 347
Wye–wye connection, 29f, 404f

Z

Zero-sequence capacitance, 350
Zero-sequence currents, 308, 453
Zero-sequence displacement, 273
Zero-sequence filters, 394, 395f
Zero-sequence network, 280–281, 280f, 283, 283f
Zero-sequence neutral displacement, defined, 273
Zero-sequence reactance, 350
Zero-sequence self-impedance, 256
Zero-sequence system, 245, 246f
Z-matrix methods, 471
Zones of protection, 423
Z-plane, 403